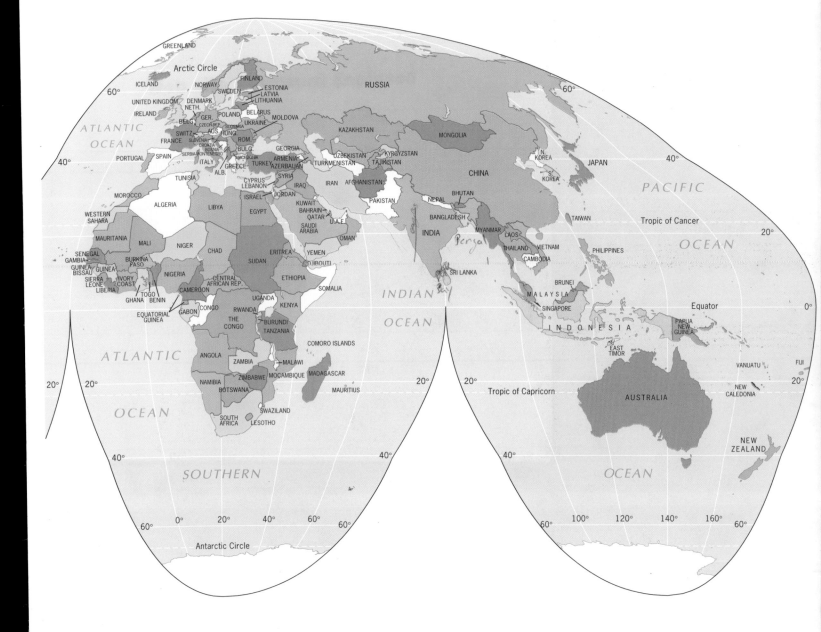

GREENLAND

Arctic Circle

ICELAND

NORWAY

FINLAND

RUSSIA

60°

60°

UNITED KINGDOM

SWEDEN

ESTONIA
LATVIA
LITHUANIA

DENMARK
NETH.

IRELAND

POLAND

BELARUS

KAZAKHSTAN

MONGOLIA

ATLANTIC

BELG.

GER.

CZECH REP.

SLOVAKIA

UKRAINE

MOLDOVA

N.
KOREA

JAPAN

OCEAN

SWITZ.

AUS.

HUNG.

ROM.

GEORGIA

UZBEKISTAN

KYRGYZSTAN

40°

FRANCE

SLOVENIA

CROATIA

BOSNIA

MACEDONIA

BULG.

TURKMENISTAN

TAJIKISTAN

CHINA

S.
KOREA

40°

PORTUGAL

SPAIN

SERBIA-MONTENEGRO

ITALY

GREECE

TURKEY

ARMENIA
AZERBAIJAN

PACIFIC

ALB.

PAKISTAN

TAIWAN

TUNISIA

CYPRUS
LEBANON

SYRIA

IRAQ

IRAN

AFGHANISTAN

Tropic of Cancer

MOROCCO

ISRAEL

JORDAN

NEPAL

BHUTAN

OCEAN

20°

WESTERN
SAHARA

ALGERIA

LIBYA

EGYPT

KUWAIT
BAHRAIN
QATAR

BANGLADESH

MAURITANIA

MALI

NIGER

CHAD

SAUDI
ARABIA

U.A.E.

OMAN

INDIA

MYANMAR

LAOS

SENEGAL

ERITREA

YEMEN

THAILAND

VIETNAM

PHILIPPINES

GAMBIA
GUINEA
BISSAU

GUINEA

BURKINA
FASO

SUDAN

DJIBOUTI

CAMBODIA

SIERRA
LEONE

NIGERIA

SRI LANKA

LIBERIA

IVORY
COAST

CENTRAL
AFRICAN REP.

ETHIOPIA

BRUNEI

TOGO

CAMEROON

SOMALIA

INDIAN

MALAYSIA

GHANA

BENIN

EQUATORIAL
GUINEA

GABON

CONGO

RWANDA

UGANDA

KENYA

OCEAN

SINGAPORE

INDONESIA

PAPUA
NEW
GUINEA

Equator

0°

THE
CONGO

BURUNDI

TANZANIA

EAST
TIMOR

COMORO ISLANDS

ATLANTIC

ANGOLA

ZAMBIA

MALAWI

VANUATU

FIJI

NAMIBIA

ZIMBABWE

MOZAMBIQUE

MADAGASCAR

NEW
CALEDONIA

20°

20°

BOTSWANA

MAURITIUS

20°

Tropic of Capricorn

AUSTRALIA

20°

OCEAN

SWAZILAND

SOUTH
AFRICA

LESOTHO

NEW
ZEALAND

40°

40°

40°

OCEAN

SOUTHERN

60°

0°

20°

40°

60°

60°

100°

120°

140°

160°

60°

Antarctic Circle

GEOGRAPHY

Realms, Regions, and Concepts

Eleventh Edition

GEOGRAPHY
Realms, Regions, and Concepts
Eleventh Edition

H. J. de Blij

Michigan State University

Peter O. Muller

University of Miami

The first map figure in each chapter that appears in this text
comes from *Goode's World Atlas*, Rand McNally,
R.L. 93-S-115, and is used with permission.

WILEY

John Wiley & Sons, Inc.

SENIOR ACQUISITIONS EDITOR:	Ryan Flahive
MARKETING MANAGER:	Clay Stone
ASSISTANT EDITOR:	Denise Powell
EDITORIAL ASSISTANT:	Aliyah Vinikoor
ASSOCIATE PRODUCTION MANAGER:	Kelly Tavares
SENIOR DESIGNER:	Karin Kincheloe
PHOTO EDITOR:	Jennifer MacMillan
PHOTO RESEARCHER:	Teresa Romito
ILLUSTRATION EDITOR:	Sigmund Malinowski
MEDIA EDITOR:	Tom Kulesa
TEXT DESIGN:	Lee Goldstein
COVER DESIGN:	David Levy
ELECTRONIC ILLUSTRATIONS:	Mapping Specialists
COVER PHOTOS:	H. J. de Blij

This book was set in Times Roman by UG / GGS Information Services, Inc. and printed and bound by Von Hoffman Press.

Copyright © 2004 by H. J. de Blij & Peter O. Muller for
Copyright © 2004 by John Wiley & Sons, Inc.

ISBN 0-471-15224-2

Printed in the United States of America

10 9 8 7 6 5 4 3 2 1

In this 100th anniversary year of the Association of American Geographers (AAG) we dedicate this edition of *Regions* to the memory of the distinguished faculty of the Department of Geography of Harvard University at the time of the AAG's creation, when Harvard was a leader in the discipline and Harvard geography professor William Morris Davis was the AAG's founding president.

Preface

OR MORE THAN THREE DECADES, *Geography: Realms, Regions, and Concepts* has reported (and sometimes anticipated) trends in the discipline of Geography and developments in the world at large. In ten preceding editions, *Regions*, as the book is generally called, has explained the modern world's great geographic realms and their physical and human contents, and has introduced geography itself, the discipline that links the study of human societies and natural environments through a fascinating, spatial approach. From old ideas to new, from environmental determinism to expansion diffusion from decolonization to devolution, *Regions* has provided geographic perspective on our transforming world.

The book before you, therefore, is an information highway to geographic literacy. The first edition appeared in 1971, at a time when school geography in the United States (though not in Canada) was a subject in decline. It was a precursor of a dangerous isolationism in America, and geographers foresaw the looming cost of geographic illiteracy. Sure enough, the media during the 1980s began to report that polls, public surveys, tests, and other instruments were recording a lack of geographic knowledge at a time when our world was changing ever faster and becoming more competitive by the day. Various institutions, including the National Geographic Society, banks, airline companies, and a consortium of scholarly organizations mobilized to confront an educational dilemma that had resulted substantially from a neglect of the very topics this book is about.

Before we can usefully discuss such commonplace topics as our "shrinking world," our "global village," and our "distant linkages," we should know what the parts are, the components that do the shrinking and linking. This is not just an academic exercise. You will find that much of what you encounter in this book is of immediate, practical value to you—as a citizen, a consumer, a traveller, a voter, a jobseeker. North America is a geographic realm with intensifying global interests and involvements. Those interests and involvements require countless, often instanta-

neous decisions. Such decisions must be based on the best possible knowledge of the world beyond our continent. That knowledge can be gained by studying the layout of our world, its environments, societies, resources, policies, traditions, and other properties—in short, its regional geography.

REALMS AND CONCEPTS

This book is organized into thirteen chapters. The Introduction discusses the world as a whole, outlining the physical stage on which the human drama is being played out, providing environmental information, demographic data, political background, and economic geographical context. Each of the remaining twelve chapters focuses on one of the world's major geographic realms.

Geographic concepts and ideas are placed in their regional settings in all 13 chapters. Most of these approximately 150 concepts are primarily geographical, but others are ideas about which, we believe, students of geography should have some knowledge. Although such concepts are listed on the opening page of every chapter, we have not, of course, enumerated every geographic notion used in that chapter. Many colleagues, we suspect, will want to make their own realm-concept associations, and as readers will readily perceive, the book's organization is quite flexible. It is possible, for example, to focus almost exclusively on substantive regional material, or, alternatively, to concentrate mainly on conceptual issues.

THE ELEVENTH EDITION

This Eleventh Edition of *Regions*, now in its thirty-third year, continues a tradition of comprehensive revision that is a hallmark of this book's context and currency. As those who have long been familiar with it are aware, previous revisions have invariably reflected major developments in the world as well as the discipline, ranging from the collapse of the Soviet Union to the impact of globalization.

A major structural change marks this edition. We have repositioned chapters six and seven, placing Subsaharan Africa after South America and before the chapter on the Islamic World (North Africa/Southwest Asia). In turn, the regional sequence in South America has been revised to place Brazil, the realm's largest entity, last—and thus closest to Africa, with which Brazil has historic as well as current ties. Also facilitated is the transition from the Islamic realm to South Asia, where linkages are changing as a result of parallel developments in Afghanistan, Pakistan, and Kashmir.

This edition also introduces a new feature, "Regional Issue." Two voices state opposing viewpoints regarding an issue that roils the realm and has geographic context. These comments are based on the senior author's four decades of note-taking in the field and international correspondence, and they are intended to stimulate discussion and debate in the classroom.

The reality of global terrorism has required consideration of the way the "War on Terror" is affecting countries large and small, powerful and weak, and references to this "war" occur now throughout the book as the geographic dimensions of terrorism range from diffusion processes (contagious as well as relocational) and resource origins (diamonds to "charities") to power and geopolitics (Afghanistan and Iraq). The domino effect of the fall of New York's World Trade Center towers, presaged by President Bush's reference to an "axis of evil," continues.

Major text revision marks every chapter of this edition except North America (3) and East Asia (9), which was revised exhaustively in the previous edition and where most of the modifications affect Xinjiang and the two Koreas. The Introduction, retitled Global Perspectives, has been streamlined without loss of focus on environmental and developmental topics. The chapter on Russia (2) has new emphasis on Moscow's struggle with its Islamic minorities. The chapter on Europe (1) also includes new material, including a new map on Muslim minorities, but the major geographic development in Europe is the expansion of the European Union by an anticipated 10 countries, a momentous transformation of the map. The continuing problem of Cyprus receives narrative as well as cartographic attention. Middle America (4) introduces new text on NAFTA, Mexico's regional geography, and the Central American republics. South America (5) has not only been restructured; it also contains new material on Brazil and especially on economically troubled Argentina, including a new map of its provinces. Current text in the chapter on Africa (6) centers on the continuing instability of parts of West Africa and Equatorial Africa, and new material (including a new map) focuses on the "Horn" of Africa where terrorist risks and strategic responses converge.

Major changes also were required in the chapter on North Africa/Southwest Asia (7) including two new maps (Iraq and Afghanistan) and revisions in the sections on the "Middle East" and Turkestan. The political, social, and economic geography of Iraq was in the balance as the easy conquest was followed by faltering reconstruction and deteriorating security. As this book went to press, the so-called "Road Map" that was designed to guide the Palestinians and Israel toward a territorial accommodation was encountering major obstacles efforts to secure details suggested that this was more a "road" than a "map" when the latter will be the key to the former. The War on Terror also necessitated changes in the South Asia chapter (8), notably in reference to Pakistan and Kashmir but also in the discussion of India. Attention is given to the economic problems of India, resulting in part from its responses to globalization but also, in its high-tech sector, from difficulties experienced halfway around the world. In the chapter on Southeast Asia (10) we illustrate the problems of developing countries by focusing on East Timor and its energy resources lying under the sea, coveted by Australia. When a poor and weak country faces the demands of a rich and powerful neighbor, United Nations regulations may not be enough to protect its interests. Australia (11) is one of the economic success stories of the decade, but challenges involving immigration and social diversification arise from its relative location and attractiveness. In the chapter on the Pacific Realm (12) we refer anew to the maritime rights of island-states and include an improved map on Antarctica.

The eleventh Edition of *Regions* also includes new photography and accompanying fieldnotes, updated bibliographies, and, most importantly in this category, a completely up-to-date table of population indicators (Table G-1). We hope that you will find this edition as satisfying and productive as you have our earlier ones, and we thank you for your interest in *Regions 11*.

Data Sources

For all matters geographical, of course, we consult *The Annals of the Association of American Geographers, The Professional Geographer, The Geographical Review, The Journal of Geography*, and many other academic journals published regularly in North America—plus an array of similar periodicals published in English-speaking countries from Scotland to New Zealand.

As with every new edition of this book, all quantitative information was updated to the year of publication and checked rigorously. In addition to the major revisions described above, hundreds of other modifications were made, many in response to readers' and reviewers' comments. Some readers found our habit of reporting urban population data within the text disruptive, so we continue to tabulate these at the beginning of the "Regions of the Realm" section of each chapter. The stream of new spellings of geographic names continues, and we pride ourselves in being a reliable source for current and correct usage.

The statistical data that constitute Table G-1 (pp. 33–37) are derived from numerous sources. As users of such data are aware, considerable inconsistency marks the reportage by various agencies, and it is often necessary to make informed decisions on contradictory information. For example, some sources still do not reflect the rapidly declining rates of population increase or life expectancies in AIDS-stricken African countries. Others list demographic averages without accounting for differences between males and females in this regard.

In formulating Table G-1 we have used among our sources the United Nations, the Population Reference Bureau, the World Bank, the Encyclopaedia Britannica *Books of the Year*, the *Economist* Intelligence Unit, the *Statesman's Year-Book*, and the *The New York Times Almanac*.

The urban population figures—which also entail major problems of reliability and comparability—are mainly drawn from the most recent database published by the United Nations' Population Division. For cities of less than 750,000, we developed our own estimates from a variety of other sources. At any rate, the urban population figures used here are estimates for 2004 and they represent metropolitan-area totals unless otherwise specified.

Cartography

This newest version of *Regions* continues the innovation begun in the Seventh Edition, when atlas-style maps from the most recently available edition of *Goode's World Atlas* (currently the 20th, published in 2000) were first used as opening maps for each chapter. In the Eighth Edition, two maps were specifically drawn in the Rand McNally style to serve as matching openers: those of North Africa/Southwest Asia and the Pacific Realm. The South Asia map was substantially expanded from its *Goode's* base.

Readers of earlier editions of this book will note that our tradition of updating, enhancing, and improving our own thematic cartography continues. In this Eleventh Edition of *Regions*, the following map figures are new:

1–12	Europe: Muslims as a Percentage of National Population, 2004
4–9	Mexico's Amerindian Minorities
4–14	Haiti and Dominican Republic
5–12	Provinces of Argentina
6–6	Ethnic Areas of Subsaharan Africa
6–20	Horn of Africa
7–13	Iraq
7–21	Afghanistan
10–13	The Timor gap
12–4	Antarctica

The following maps were significantly redrawn:

G–11	States and Economies of the World
1–10	European Supranationalism
1–20	Mediterranean Europe
1–26	Serbia-Montenegro and Its Neighbors
5–11	The Southern Cone: Mid-Latitude South America
7–8	North Africa/Southwest Asia: Oil and Natural Gas
8–9	India: population Growth Rate by State
9–15	China's Western Flank
9–16	Xinjiang-Uyghur Autonomous Region

Users of this book should note that the spelling of some names on these thematic maps does not always match that on the *Goode's World Atlas* maps. This is not unusual; you will even find inconsistencies among various atlases. Almost invariably, we have followed the very latest standards set by the United States Board of Geographic Names.

Photography

The map undoubtedly is geography's closest ally, but there are times when photography is not far behind. Whether from space, from an aircraft, from the tallest building in town, or on the ground, a photograph can, indeed, be worth a thousand words. When geographers perform field research in some area of the world, they are likely to maintain a written record that correlates with the photographic one.

This Eleventh Edition revision is not confined to the text and maps. As readers of *Regions* will note, the illustrations program was thoroughly overhauled with numerous new photographs, many by the authors, new accompanying "Field Notes" where appropriate, and new, detailed captions.

Pedagogy

We continue to devise ways to help students learn important geographic concepts and ideas, and to make sense of our complex and rapidly changing world. Continuing special features include the following:

Atlas Maps. As in previous editions, a comprehensive map of the region opens each chapter. The maps are reproduced from the 20th revised edition (2000) of *Goode's World Atlas* (the maps for Chapters 6 and 12 have been created in the Atlas style). In the Eleventh Edition, each of these maps is assigned the first figure number in each chapter, which better facilitates the integration of this cartographic material into the text.

Concepts, Ideas, and Terms. Each chapter begins with a boxed sequential listing of the key geographic concepts, ideas, and terms that appear in the pages that follow. These are noted by numbers in the margins that correspond to the introduction of each item in the text.

Two-Part Chapter Organization. To help the reader to logically organize the material within chapters, we have broken the regional chapters into two distinct parts: first, "Defining the Realm" includes the general physiographic, historical, and human-geographic background common to the realm, and the second section, "Regions of the Realm," presents each of the distinctive regions within the realm (denoted by the symbol ▶).

List of Regions. Also on the chapter-opening page, a list of the regions within the particular realm provides a preview and helps to organize the chapter. For ease of identification, the semi-circular symbol (shown 4 lines above) that denotes the regions list here also appears beside each region heading in the chapter.

Major Geographic Qualities. Near the beginning of each realm chapter, we list, in boxed format, the major geographic qualities that best summarize that portion of the Earth's surface.

Sidebar Boxes. Special topical issues are highlighted in boxed sections. These boxes allow us to include interesting and current topics without interrupting the flow of material within chapters.

Regional Issues (Boxes). In each regional chapter, a controversial issue has been singled out for debate by two opposing voices. As indicated above, these commentaries are drawn from the senior author's field notes and correspondence, and are intended to stimulate classroom discussion. Individual readers may "vote" their opinions by e-mail, as indicated at the bottom of each box.

Among the Realm's Great Cities (Boxes). This feature reflects the growing process and influence of urbanization worldwide. More than thirty profiles of the world's leading cities are presented, each accompanied by specially drawn maps.

Major Cities of the Realm Population Tables. Near the beginning of every "Regions" section of each chapter, we have included a table reporting the most up-to-date urban population data (based on 2004 estimates drawn from the sources listed above). Readers should find this format less disruptive than citation of the population when the city is mentioned in the text.

From the Field Notes. In the Eighth Edition we introduced a new feature that has proven effective in some of our other textbooks. Many of the photographs in this book were taken by the senior author while doing fieldwork. The more extensive captions, *From the Field Notes*, provide valuable insights into how a geographer observes and interprets information in the field.

Appendices, References, and Glossary. At the end of the book, the reader will find five sections that enrich and/or supplement the main text: (1) *Appendix A*, a guide to Using the Maps; (2) *Appendix B*, an overview of Career Opportunities in Geography; (3) *Appendix C*, a Pronunciation Guide; (4) a detailed Bibliography (*References and Further Readings*) that introduces the wide-ranging literature of the discipline and World Regional Geography; and (5) an extensive *Glossary*. The general index follows. A geographical index or *gazetteer* of the place names contained in our maps now appears in the book's Web site.

Ancillaries

A broad spectrum of print and electronic ancillaries are available to accompany the Eleventh Edition of *Regions*. Additional information, including prices and ISBNs for ordering, can be obtained by contacting John Wiley & Sons.

THE TEACHING AND LEARNING PACKAGE

This Eleventh Edition of *Geography: Realms, Regions and Concepts* is supported by a comprehensive supplements package that includes an extensive selection of print, visual and electronic materials.

Supplements to Help Teachers Teach

The Regions 11e Instructors' Site. This comprehensive website includes numerous resources to help you enhance your current presentations, create new presentations and employ our pre-made PowerPoint presentations. These resources include:

- A complete collection of PowerPoint presentations. Three for each chapter.
- All of the line art from the text.
- Access to photographs.
- A comprehensive collection of animations.
- A comprehensive test bank.

Instructor's Manual and Test Bank. This manual includes a test bank, chapter overviews and outlines, and lecture suggestions. Prepared by long-term Test Bank author

Ira Sheskin, University of Miami, the *Test Bank* for the eleventh edition of *Regions* contains over 3,000 test items including multiple-choice, fill-in, matching, and essay questions. It is distributed via the CD-Rom or through the secure Instructor's Web Site as electronic files, which can be saved into all major word processing programs. The Instructors Manual is also available on the password protected Instructor's website.

Overhead Transparency Set. The book's maps and diagrams are available in their entirety for transparency projection in beautifully rendered, 4-color format. All full-color textbook illustrations are resized and edited for maximum effectiveness in large lecture halls.

Course Management. On-line course management assets are available to accompany the Eleventh Edition of *Regions*.

Supplements to Help Students Learn

GeoDiscoveries Website *www.wiley.com/college/deblij.* This easy to use and student-focused website helps reinforce and illustrate key concepts from the text. It also provides interactive media content that helps students prepare for tests and improve their grades. This web site provides additional resources that complement the textbook and enhance your students' understanding of Geography and improve their grade by using the following resources:

- **Map Quizzes** help students master the place names that are crucial to their success in this course. Three game-formatted place name activities are provided for each chapter.
- **GeoDiscoveries Modules** allow students to explore key concepts in greater depth using videos, animations and interactive exercises.
- **Chapter Review Quizzes** provide immediate feedback to true/false, multiple choice, and short answer questions.
- **Concepts, Ideas and Terms Interactive Flashcards** help students review and quiz themselves on the concepts, ideas and terms discussed in each chapter.
- **Interactive Drag-and-Drop Exercises** challenge students to correctly label important illustrations from the textbook.
- **Audio Pronunciation** is provided for over 2000 key words and place names from the text.
- **Annotated Web Links** put useful electronic resources into context.
- **Area and Demographic Data** is provided for every country and world realm.

Student Study Guide. Text co-author Peter O. Muller and his geographer daughter, Elizabeth Muller Hames, have written a popular Study Guide to accompany the book that is packed with useful study and review tools. For each chapter in the textbook, the Study Guide gives students and faculty access to chapter objectives, content questions-and-answers, outline maps of each realm, sample tests, and more.

Goode's Atlas **from Rand McNally.** With the Eleventh Edition of *Regions*, we are delighted to be able to continue offering the *Goode's Atlas* at a deeply-discounted price when shrink-wrapped with the text. Economies of scale allow us to provide this at a net price that is close to our cost. Our partnership with Rand McNally and the widely-popular *Goode's Atlas* is an arrangement that is exclusive to John Wiley & Sons.

Social Geographies of the Modern World. This FREE 60-page supplement includes chapters that discuss:

- Global Disparities in Nutrition and Health;
- Geographies of Inequality: Race and Ethnicity; and
- Gender Inequities in Geographic Perspective.

To arrange for your students to receive this Free supplement with their copy of Regions 11e, please contact your local Wiley Sales Representative.

Microsoft Encarta Interactive Atlas CD-ROM. This award-winning atlas CD-ROM will captivate the imaginations of your students and engage them in a spatial adventure, all the while exposing them to an abundance of resources appropriate for university-level geography. Our arrangement with Microsoft enables us to offer the Encarta Interactive Atlas at a cost that is less than one-third the suggested retail price when shrink-wrapped with our text.

ACKNOWLEDGMENTS

Over the third of a century since the publication of the First Edition of *Geography: Realms, Regions, and Concepts*, we have been fortunate to receive advice and assistance from literally hundreds of people. One of the rewards associated with the publication of a book of this kind is the steady stream of correspondence and other feedback it generates. Geographers, economists, political scientists, education specialists, and others have written us, often with fascinating enclosures. We make it a point to respond personally to every such letter, and our editors have communicated with many of our correspondents as well. Moreover, we have considered every suggestion made—and many who wrote or transmitted their reactions through other channels will see their recommendations in print in this edition.

STUDENT RESPONSE

A major part of the correspondence we receive comes from student readers. We would like to take this opportunity to extend our deep appreciation to the several million students around the world who have studied from the first ten editions of our text. In particular, we thank the students from more than 100 different colleges across the United States who took the time to send us their opinions.

Students told us they found the maps and graphics attractive and functional. We have not only enhanced the map program with exhaustive updating, but have added a number of new maps to this Eleventh Edition as well as making significant changes in many others.

Generally, students have told us that they found the pedagogical devices quite useful. We have kept the study aids the students cited as effective: a boxed list of each chapter's key concepts, ideas, and terms (now numbered for quick reference in both the box and text margins); a box summarizing each realm's major geographic qualities; a pronunciation guide in Appendix C; and an extensive Glossary.

FACULTY FEEDBACK

Faculty members from a large number of North American colleges and universities continue to supply us with vital feedback and much-appreciated advice. Our publishers arranged several feedback sessions, and we are most grateful to the following professors for showing us where the written text could be strengthened and made more precise:

SIAW AKWAWUA, University of Northern Colorado

KLAUS J. BAYR, Keene State College (New Hampshire)

RANDY BERTOLAS, Wayne State College (Nebraska)

R. DENISE BLANCHARD BOEHM, Southwest Texas State University

SCOTT DOBLER, Western Kentucky University

ELLEN J. FOSTER, Southwest Texas State University

RAYMOND GREENE, Western Illinois University

RON HAGELMAN, University of New Orleans

CHANGJOO KIM, Ohio State University

MAX LU, Kansas State University

CHARLES MANYARA, Radford University (Virginia)

PATRICK MAY, Plymouth State College (New Hampshire)

DALTON E. MILLER, Mississippi State University

ROGER MILLER, Black Hills State University (South Dakota)

JAMES MILLS, State University of New York—College at Oneonta

BRIAN PLASTER, Southwest Texas State University

GINGER SCHMID, Southwest Texas State University

ROBERT M. SIMPSON, University of Tennessee at Martin

DAVID STEA, Southwest Texas State University

LINDA WANG, University of South Carolina, Aiken

MARK WILJANEN, Eastern Kentucky University

FORREST WILKERSON, Southwest Texas State University

In addition, several faculty colleagues from around the world assisted us with earlier editions, and their contributions continue to grace the pages of this book. Among them are:

JAMES P. ALLEN, California State University, Northridge

STEPHEN S. BIRDSALL, University of North Carolina

J. DOUGLAS EYRE, University of North Carolina

FANG YONG-MING, Shanghai, China

EDWARD J. FERNALD, Florida State University

RAY HENKEL, Arizona State University

RICHARD C. JONES, University of Texas at San Antonio

GIL LATZ, Portland State University (Oregon)

IAN MACLACHLAN, University of Lethbridge (Alberta)

MELINDA S. MEADE, University of North Carolina

HENRY N. MICHAEL, Temple University (Pennsylvania)

CLIFTON W. PANNELL, University of Georgia

JOHN ROBERT VICTOR PRESCOTT, University of Melbourne (Victoria)

JOHN D. STEPHENS, University of Washington

CANUTE VANDER MEER, University of Vermont

We also received input from a much wider circle of academic geographers. The list that follows is merely representative of a group of colleagues across North America to whom we are grateful for taking the time to share their thoughts and opinions with us:

MEL AAMODT, California State University-Stanislaus

WILLIAM V. ACKERMAN, Ohio State University

W. FRANK AINSLEY, University of North Carolina, Wilmington

SIAW AKWAWUA, University of Northern Colorado

DONALD P. ALBERT, Sam Houston State University

TONI ALEXANDER, Louisiana State University

R. GABRYS ALEXSON, University of Wisconsin-Superior

NIGEL ALLAN, University of California-Davis

JAMES P. ALLEN, California State University, Northridge

JOHN L. ALLEN, University of Connecticut

JOHN ALTHAUSEN, Salem State College

KRISTIN J. ALVAREZ, Keene State College

DAVID L. ANDERSON, Louisiana State University, Shreveport

KEAN ANDERSON, Freed-Hardeman University

JEFF ARNOLD, Southwestern Illinois College

HENRY ARUFFO, University of South Florida

JERRY R. ASCHERMANN, Missouri Western State College

JOSEPH M. ASHLEY, Montana State University

PATRICK ASHWOOD, Hawkeye Community College

THEODORE P. AUFDEMBERGE, Concordia College (Michigan)

JAIME M. AVILA, Sacramento City College

EDWARD BABIN, University of South Carolina-Spartanburg

ROBERT BAERENT, Randolph-Macon College

MARVIN W. BAKER, University of Oklahoma

GOURI BANERJEE, Boston University (Massachusetts)

MICHELE BARNABY, Pittsburgh State University

J. HENRY BARTON, Thiel College (Pennsylvania)

STEVEN BASS, Paradise Valley Community College (Arizona)

THOMAS F. BAUCOM, Jacksonville State University (Alabama)

KLAUS J. BAYR, Keene State College

DENIS A. BEKAERT, Middle Tennessee State University

JAMES BELL, Linn Benton Community College (Oregon)

KEITH M BELL, Volunteer State Community College

WILLIAM H. BERENTSEN, University of Connecticut

DONALD J. BERG, South Dakota State University

ROYAL BERGLEE, Morehead State University

RIVA BERLEANT-SCHILLER, University of Connecticut

LEE LUCAS BERMAN, Southern Connecticut State University

RANDY BERTOLAS, Wayne State College

THOMAS BITNER, University of Wisconsin

DENISE BLANCHARD, Southwest Texas State University

WARREN BLAND, California State University-Northridge

DAVIS BLEVINS, Huntington College (Alabama)

HUBERTUS BLOEMER, Ohio University

S. BO JUNG, Bellevue College (Nebraska)

MARTHA BONTE, Clinton Community College (Idaho)

GEORGE R. BOTJER, University of Tampa (Florida)

R. LYNN BRADLEY, Belleville Area College (Illinois)

KEN BREHOB, Elmhurst, Illinois

JAMES A. BREY, University of Wisconsin-Fox Valley

ROBERT BRINSON, Santa Fe Community College (Florida)

REUBEN H. BROOKS, Tennessee State University

PHILLIP BROUGHTON, Arizona Western College

LARRY BROWN, Ohio State University

LAWRENCE A. BROWN, Troy State-Dothan (Alabama)

ROBERT N. BROWN, Delta State University (Mississippi)

STANLEY D. BRUNN, University of Kentucky

RANDALL L. BUCHMAN, Defiance College (Ohio)

MICHAELE ANN BUELL, NorthWest Arkansas Community College

DANIEL BUNYE, South Plains College

MICHAEL BUSBY, Murray State University

MARY CAMERON, Slippery Rock University

MICHAEL CAMILLE, Univ. of Louisiana at Monroe

MARY CARAVELIS, Barry University

DIANA CASEY, Muskegon Community College

DiANN CASTEEL, Tusculum College (Tennessee)

BILL CHAPPELL, Keystone College

MICHAEL SEAN CHENOWETH, University of Wisconsin, Milwaukee

STANLEY CLARK, California State University Bakersfield

JOHN E. COFFMAN, University of Houston (Texas)

DAWYNE COLE, Grand Rapids Baptist College (Michigan)

JERRY COLEMAN, Mississippi Gulf Coast Community College

JONATHAN C. COMER, Oklahoma State University

BARBARA CONNELLY, Westchester Community College (New York)

FRED CONNINGTON, Southern Wesleyan University

WILLIS M. CONOVER, University of Scranton (Pennsylvania)

OMAR CONRAD, Maple Woods Community College (Missouri)

ALLAN D. COOPER, Otterbein College

DEBORAH CORCORAN, Southwest Missouri State University

WILLIAM COUCH, University of Alabama, Huntsville

BARBARA CRAGG, Aquinas College (Michigan)

GEORGES G. CRAVINS, University of Wisconsin

ELLEN K. CROMLEY, University of Connecticut

JOHN A. CROSS, University of Wisconsin-Oshkosh

SHANNON CRUM, University of Texas at San Antonio

WILLIAM CURRAN, South Suburban (Illinois)

KEVIN M. CURTIN, University of Texas at Dallas

JOSE A. DA CRUZ, Ozarks Technical Community College

ARMANDO DA SILVA, Towson State University (Maryland)

MARK DAMICO, Green Mountain College

DAVID D. DANIELS, Central Missouri State University

RUDOLPH L. DANIELS, Morningside College (Iowa)

SATISH K. DAVGUN, Bemidji State University (Minnesota)

CHARLES DAVIS, Mississippi Gulf Coast Community College

JAMES DAVIS, Illinois College

JAMES L. DAVIS, Western Kentucky University

PEGGY E. DAVIS, Pikeville College

ANN DEAKIN, State University of New York, College at Fredonia

KEITH DEBBAGE, University of North Carolina-Greensboro

MOLLY DEBYSINGH, California State University, Long Beach

DENNIS K. DEDRICK, Georgetown College (Kentucky)

JOEL DEICHMANN, Bentley College

STANFORD DEMARS, Rhode Island College

TOM DESULIS, Spoon River College

THOMAS DIMICELLI, William Paterson College (New Jersey)

MARY DOBBS, Highland Community College, Wamego

SCOTT DOBLER, Western Kentucky University

D.F. DOEPPERS, University of Wisconsin-Madison

JAMES DOERNER, University of Northern Colorado

ANN DOOLEN, Lincoln College (Illinois)

STEVEN DRIEVER, University of Missouri-Kansas City

WILLIAM DRUEN, Robert Western Kentucky University

ALASDAIR DRYSDALE, University of New Hampshire

KEITH A. DUCOTE, Cabrillo Community College (California)

ELIZABETH DUDLEY-MURPHY, University of Utah, Salt Lake City

WALTER N. DUFFET, University of Arizona

MIKE DUNNING, University of Alaska Southeast

CHRISTINA DUNPHY, Champlain College (Vermont)

ANTHONY DZIK, Shawnee State University (Kansas)

DENNIS EDGELL, Firelands BGSU (Ohio)

RUTH M. EDIGER, Seattle Pacific University

JAMES H. EDMONSON, Union University (Tennessee)

M.H. EDNEY, State University of New York-Binghamton

HAROLD M. ELLIOTT, Weber State University (Utah)

JAMES ELSNES, Western State College

LINDA ESSENSON

ROBERT W. EVANS, Fresno City College

EVERSON DICKEN, California State University, San Bernardino

DINO FIABANE, Community College of Philadelphia (Pennsylvania)

G.A. FINCHUM, Milligan College (Tennessee)

IRA FOGEL, Foothill College (California)

RICHARD FOLEY, Cumberland College

ROBERT G. FOOTE, Wayne State College (Nebraska)

RONALD FORESTA, University of Tennessee

ELLEN J. FOSTER, Southwest Texas State University

ELLEN J. FOSTER, Southwest Texas State University

G.S. FREEDOM, McNeese State University (Louisiana)

EDWARD T. FREELS, Carson-Newman College

PHILIP FRIEND, Inver Hills Community College

JAMES FRYMAN, University of Northern Iowa

OWEN FURUSETH, University of North Carolina-Charlotte

RICHARD FUSCH, Ohio Wesleyan University

GARY GAILE, University of Colorado-Boulder

EVELYN GALLEGOS, Eastern Michigan University & Schoolcraft College

GAIL GARBRANDT, Mount Union College, University of Akron

RICHARD GARRETT, Marymount Manhattan College

JERRY GERLACH, Winona State University (Minnesota)

MARK GISMONDI, Northwest Nazarene University

LORNE E. GLAIM, Pacific Union College (California)

SHARLEEN GONZALEZ, Baker College (Michigan)

DANIEL B. GOOD, Georgia Southern University

GARY C. GOODWIN, Suffolk Community College (New York)

S. GOPAL, Boston University (Massachusetts)

MARVIN GORDON, Lake Forest Graduate School of Business

ROBERT GOULD, Morehead State University (Kentucky)

MARY GRAHAM, York College of Pennsylvania

GORDON GRANT, Texas A&M University

PAUL GRAY, Arkansas, Tech University

DONALD GREEN, Baylor University (Texas)

GARY M. GREEN, University of North Alabama

STANLEY C. GREEN, Laredo State University (Texas)

RAYMOND GREENE, Western Illinois University

MARK GREER, Laramie County Community College (Wyoming)

WALT GUTTINGER, Flagler College

JOHN E. GYGAX, Wilkes Community College

LEE ANN HAGAN, College of Southern Idaho

RON HAGELMAN, University of New Orleans

W. GREGORY HAGER, Northwestern Connecticut Community College

RUTH F. HALE, University of Wisconsin-River Falls

JOHN W. HALL, Louisiana State University-Shreveport

PETER L. HALVORSON, University of Connecticut

DAVID HANSEN, Pennsylvania State University, Harrisburg & Schuylkill

MERVIN HANSON, Willmar Community College (Minnesota)

ROBERT C. HARDING, Lynchburg College

MICHELLE D. HARRIS, Bob Jones University

SCOTT HARRIS, CGCS Drury University

ROBERT J. HARTIG, Fort Valley State College (Georgia)

SUZANNA HARTLEY, Shelton State Community College

CAROL HAZARD, Meredith College

HARLOW Z. HEAD, Barton College

DOUG HEFFINGTON, Middle Tennessee State University

JAMES G. HEIDT, University of Wisconsin Center-Sheboygan

CATHERINE HELGELAND, University of Wisconsin-Manitowoc

NORMA HENDRIX, East Arkansas Community College

JAMES E. HERRELL, Otero Junior College

JAMES HERTZLER, Goshen College (Indiana)

JOHN HICKEY, Inver Hills Community College (Minnesota)

THOMAS HIGGINS, San Jacinto College (Texas)

EUGENE HILL, Westminster College (Missouri)

LOUISE HILL, University of South Carolina-Spartanburg

MIRIAM HELEN HILL, Indiana University Southeast

SUZY HILL, University of South Carolina-Spartanburg

ROBERT HILT, Pittsburg State University (Kansas)

SOPHIA HINSHALWOOD, Montclair State University (New Jersey)

PRISCILLA HOLLAND, University of North Alabama

MARK R. HOOPER, Freed-Hardeman University

R. HOSTETLER, Fresno City College (California)

ERICK HOWENSTINE, Northeastern Illinois University

LLOYD E. HUDMAN, Brigham Young University (Utah)

JANIS W. HUMBLE, University of Kentucky

JUANA IBANEZ, University of New Orleans

TONY IJOMAH, Harrisburg Area Community College

WILLIAM IMPERATORE, Appalachian State University (North Carolina)

RICHARD JACKSON, Brigham Young University (Utah)

MARY JACOB, Mount Holyoke College (Massachusetts)

GREGORY JEANE, Samford University (Alabama)

SCOTT JEFFREY, Catonsville Community College (Maryland)

JERZY JEMIOLO, Ball State University (Indiana)

NILS I. JOHANSEN, University of Southern Indiana

DAVID JOHNSON, University of Southwestern Louisiana

INGRID JOHNSON, Towson University

RICHARD JOHNSON, Oklahoma City University

SHARON JOHNSON, Marymount College (New York)

JEFFREY JONES, University of Kentucky

KRIS JONES, Saddleback College

MARCUS E. JONES, Claflin College (South Carolina)

M. KAMIAR, Florida Community College, Jacksonville

MOHAMMAD S. KAMIAR, Florida Community College, Jacksonville

MATTI E. KAUPS, University of Minnesota-Duluth

JO ANNE W. KAY, Brigham Young University, Idaho

DAVID KEELING, Western Kentucky University

COLLEEN KEEN, Gustavus Adolphus College (Minnesota)

ARTIMUS KEIFFER, Wittenberg University

GORDON F. KELLS, Mott Community College

KAREN J. KELLY, Palm Beach Community College, Boca Raton Campus

RYAN KELLY, Lexington Community College

VIRGINIA KERKHEIDE, Cuyahoga Community College and Cleveland State University

TOM KESSINGER, Xavier University

MASOUD KHEIRABADI, Marylhurst University

SUSANNE KIBLER-HACKER, Unity College (Maine)

CHANGJOO KIM, Ohio State University

JAMES W. KING, University of Utah

JOHN C. KINWORTHY, Concordia College (Nebraska)

ALBERT KITCHEN, Paine College

TED KLIMASEWSKI, Jacksonville State University (Alabama)

ROBERT D. KLINGENSMITH, Ohio State University-Newark

LAWRENCE M. KNOPP, JR., University of Minnesota-Duluth

Lynn Koehnemann, Gulf Coast Community College

Terrill J. Kramer, University of Nevada

Joe Krause, Community College of Indiana at Lafayette

Brenda Krekeler, Northern Kentucky University, Miami University

Arthur J. Krim, Cambridge, Massachusetts

Michael A. Kukral, Rose-Hulman Institute of Technology

Elroy Lang, El Camino Community College (California)

Richard L. Langill, Saint Martin's College

Christopher Lant, Southern Illinois University-Carbondale

A.J. Larson, University of Illinois-Chicago

Paul R. Larson, Southern Utah University

Larry League, Dickinson State University (North Dakota)

David R. Lee, Florida Atlantic University

Wook Lee, Ohio State University

Joe Leeper, Humboldt State University (California)

Yechiel M. Lehavy, Atlantic Community College (New Jersey)

Scott Leith, Central Connecticut State University

James Leonard, Marshall University

Elizabeth J. Leppman, St. Cloud State University

John C. Lewis, Northeast Louisiana University

Dan Lewman Jr., Southwest Mississippi Community College

Caedmon S. Liburd, University of Alaska-Anchorage

T. Ligibel, Eastern Michigan University

Z.L. Lipchinsky, Berea College (Kentucky)

Allan L. Lippert, Manatee Community College (Florida)

John H. Litcher, Wake Forest University (North Carolina)

Lee Liu, Central Montana State University

Li Liu, Stephen F. Austin State University (Texas)

William R. Livingston, Baker College (Michigan)

Catherine M. Lockwood, Chadron State College

George E. Longenecker, Vermont Technical College

Cynthia Longstreet, Ohio State University

Jack Looney, University of Massachusetts, Boston

Tom Love, Linfield College (Oregon)

K.J. Lowrey, Miami University (Ohio)

James Lowry, Stephen F. Austin State University

Max Lu, Kansas State University

Robin R. Lyons, University of Hawai'i-Leeward Community College

Susan M. Macey, Southwest Texas State University

Michael Madsen, Brigham Young University, Idaho

Ronald Magden, Tacoma Community College

Christiane Mainzer, Oxnard College (California)

Laura Makey, California State University, San Marcos

Harley I. Manner, University of Guam

Anthony Paul Mannion, Kansas State University

Gary Manson, Michigan State University

Charles Manyara, Radford University

James T. Markley, Lord Fairfax Community College (Virginia)

Sister May Lenore Martin, Saint Mary College (Kansas)

Kent Mathewson, Louisiana State University

Patrick May, Plymouth State College

Dick Mayer, Maui Community College (Hawai'i)

Sara Mayfield, San Jacinto College, Central (California)

Dean R. Mayhew, Maine Maritime Academy

J.P. McFadden, Orange Coast College (California)

Bernard McGonigle, Community College of Philadelphia (Pennsylvania)

Paul D. Meartz, Mayville State University (North Dakota)

Dianne Meredith, California State University-Sacramento

Gary C. Meyer, University of Wisconsin, Stevens Point

JUDITH L. MEYER, Southwest Missouri State University

MARK MICOZZI, East Central University

JOHN MILBAUER, Northeastern State University

DALTON W. MILLER JR., W. Mississippi State University

RAOUL MILLER, University of Minnesota, Duluth

ROGER MILLER, Black Hills State University

JAMES MILLS, State University of New York, College at Oneonta

INES MIYARES, Hunter College, CUNY (New York)

BOB MONAHAN, Western Carolina University

KEITH MONTGOMERY, University of Wisconsin-Marathon

DAN MORGAN, University of South Carolina, Beaufort

DEBBIE MORIMOTO, Merced College

JOHN MORTON, Benedict College (South Carolina)

ANNE MOSHER, Syracuse University (New York)

BARRY MOWELL, Broward Community College (Florida)

TOM MUELLER, California University of Pennsylvania

DONALD MYERS, Central Connecticut State University

ROBERT R. MYERS, West Georgia College

GARY NACHTIGALL, Fresno Pacific University

YASER M. NAJJAR, Framingham State College (Massachusetts)

KATHERINE NASHLEANAS, Southeast Community College

JEFFREY W. NEFF, Western Carolina University

DAVID NEMETH, University of Toledo (Ohio)

ROBERT NEWCOMER, East Central University

WILLIAM NIETER, St. Johns University

WILLIAM N. NOLL, Highland Community College

RAYMOND O'BRIEN, Bucks County Community College (Pennsylvania)

PATRICK O'SULLIVAN, Florida State University

DIANE O'CONNELL, Schoolcraft College

JOHN ODLAND, Indiana University

DOUG OETTER, Georgia College & State University

ANNE O'HARA, White, Marygrove College

PATRICK OLSEN, University of Idaho

JOSEPH R. OPPONG, University of North Texas

LYNN ORLANDO, Holy Family University

MARK A. OUMETTE, Hardin-Simmons University

RICHARD OUTWATER, California State University, Long Beach

CISSIE OWEN, Lamar University

MARY ANN OWOC, Mercyhurst College

STEVE PALLADINO, Ventura College

BIMAL K. PAUL, Kansas State University

SELINA PEARSON, Northwest-Shoals Community College

MAURI PELTO, Nichols College

JAMES PENN, Southeastern Louisiana University

LINDA PETT-CONKLIN, University of St. Thomas

DIANE PHILEN, Lower Brule Community College

PAUL PHILLIPS, Fort Hays State University (Kansas)

MICHAEL PHOENIX, ESRI (California)

JERRY PITZL, Macalester College (Minnesota)

BRIAN PLASTER, Southwest Texas State University

ARMAND POLICICCHIO, Slippery Rock University

ROSANN POLTRONE, Arapahoe Community College

BILLIE E. POOL, Holmes Community College (Mississippi)

GREGORY POPE, Montclair State University

JEFF POPKE, East Carolina University

VINTON M. PRINCE, Wilmington College (North Carolina)

GEORGE PUHRMANN, Drury University

DONALD N. RALLIS, Mary Washington College

RHONDA REAGAN, Blinn College (Texas)

DANNY I. REAMS, Southeast Community College (Nebraska)

JIM RECK, Golden West College (California)

ROGER REEDE, Southwest State University (Minnesota)

JOHN RESSLER, Central Washington University

JOHN B. RICHARDS, Southern Oregon State College

DAVID C. RICHARDSON, Evangel College (Missouri)

GARY RINGLEY, Virginia Highlands Community College

SUSAN ROBERTS, University of Kentucky

CURT ROBINSON, California State University, Sacramento

WOLF RODER, University of Cincinnati

JAMES ROGERS, University of Central Oklahoma

PAUL A. ROLLINSON, AICP, Southwest Missouri State University

JAMES C. ROSE, Tompkins/Cortland Community College (New York)

THOMAS E. ROSS, Pembroke State University (North Carolina)

THOMAS A. RUMNEY, State University of New York-Plattsburgh

GEORGE H. RUSSELL, University of Connecticut

BILL RUTHERFORD, Martin Methodist College

RAJAGOPAL RYALI, Auburn University at Montgomery (Alabama)

PERRY RYAN, Mott Community College

JAMES SAKU, Frostburg State University

DAVID R. SALLEE, North Lake College

RICHARD A. SAMBROOK, Eastern Kentucky University

EDUARDO SANCHEZ, Grand Valley State University

JOHN SANTOSUOSSO, Florida Southern College

GINGER SCHMID, Southwest Texas State University

BRENDA THOMPSON SCHOOLFIELD, Bob Jones University

ADENA SCHUTZBERG, Middlesex Community College (Massachusetts)

ROGERM. SELYA, University of Cincinnati

RENEE SHAFFER, University of South Carolina

WENDY SHAW, Southern Illinois University, Edwardsville

SIDNEY R. SHERTER, Long Island University (New York)

HAROLD SHILK, Wharton County Jr. College

NANDA SHRESTHA, Florida A&M University

WILLIAM R. SIDDALL, Kansas State University

DAVID SILVA, Bee County College (Texas)

JOSE ANTONIO SIMENTAL, Marshall University

MORRIS SIMON, Stillman College (Alabama)

ROBERT MARK SIMPSON, University of Tennessee at Martin

KENN E. SINCLAIR, Holyoke Community College (Massachusetts)

ROBERT SINCLAIR, Wayne State University (Michigan)

JIM SKINNER, Southwest Missouri State University

BRUCE SMITH, Bowling Green State University

EVERETT G. SMITH JR., University of Oregon

PEGGY SMITH, California State University, Fullerton

RICHARD V. SMITH, Miami University (Ohio)

JAMES SNADEN, Charter Oak State College

DAVID SORENSON, Augustana College

SISTER CONSUELO SPARKS, Immaculata University

CAROLYN D. SPATTA, California State University-Hayward

M.R. SPONBERG, Laredo Junior College (Texas)

DONALD L. STAHL, Towson State University (Maryland)

DAVID STEA, Southwest Texas State University

ELAINE STEINBERG, Central Florida Community College

D.J. STEPHENSON, Ohio University Eastern

HERSCHEL STERN, Mira Costa College (California)

REED F. STEWART, Bridgewater State College (Massachusetts)

NOEL L. STIRRAT, College of Lake County (Illinois)

JOSEPH P. STOLTMAN, Western Michigan University

DEAN B. STONE, Scott Community College

WILLIAM M. STONE, Saint Xavier University, Chicago

GEORGE STOOPS, Mankato State University (Minnesota)

DEBRA STRAUSSFOGEL, University of New Hampshire

JAMIE STRICKLAND, University of North Carolina, Charlotte

WAYNE STRICKLAND, Roanoke College

PHILIP STURM, Ohio Valley College, Vienna

PHILIP SUCKLING, University of Northern Iowa

ALLEN SULLIVAN, Central Washington University

SELIMA SULTANA, Auburn University

RAY SUMNER, Long Beach City College

CHRISTOPHER SUTTON, Western Illinois University

T. L. TARLOS, Orange Coast College (California)

WESLEY TERAOKA, Leeward Community College

MICHAEL THEDE, North Iowa Area Community College

DERRICK J. THOM, Utah State University

CURTIS THOMSON, University of Idaho

BEN TILLMAN, Texas Christian University

CLIFF TODD, University of Nebraska, Omaha

S. TOOPS, Miami University (Ohio)

RICHARD J. TORZ, St. Joseph's College, New York

HARRY TRENDELL, Kennesaw State University

ROGER T. TRINDELL, Mansfield University of Pennsylvania

DAN TURBEVILLE, East Oregon State College

NORMAN TYLER, Eastern Michigan University

GEORGE VAN OTTEN, Northern Arizona University

C.S. VERMA, Weber State College (Utah)

KELLY ANN VICTOR, Eastern Michigan university

SEAN WAGNER, Tri-State University

GRAHAM T. WALKER, Metropolitan State College of Denver

MONTGOMERY WALKER, Yakima Valley Community College

DEBORAH WALLIN, Skagit Valley College (Washington)

MIKE WALTERS, Henderson Community College (Kentucky)

LINDA WANG, University of South Carolina, Aiken

J.L. WATKINS, Midwestern State University (Texas)

KIT W. WESLER, Murray State University

PETER W. WHALEY, Murray State University

MACEL WHEELER, Northern Kentucky University

P. GARY WHITE, Western Carolina University (North Carolina)

W.R. WHITE, Western Oregon University

GARY WHITTON, Fairbanks, Alaska

REBECCA WIECHEL, Wilmington College

MARK WILJANEN, Eastern Kentucky University

GENE C. WILKEN, Colorado State University

FORREST WILKERSON, Southwest Texas State University

P. WILLIAMS, Baldwin-Wallace College

STEPHEN A. WILLIAMS, Methodist College

DEBORAH WILSON, Sandhills Community College

MORTON D. WINSBERG, Florida State University

ROGER WINSOR, Appalachian State University (North Carolina)

WILLIAM A. WITHINGTON, University of Kentucky

A. WOLF, Appalachian State University, N.C.)

JOSEPH WOOD, University of Southern Maine

RICHARD WOOD, Seminole Junior College (Florida)

GEORGE I. WOODALL, Winthrop College (North Carolina)

STEPHEN E. WRIGHT, James Madison University (Virginia)

LEON YACHER, Southern Connecticut State University

KYONG YUP CHU, Bergen Community College, Paramus, NJ

FIROOZ E. ZADEH, Colorado Mountain College

DONALD J. ZEIGLER, Old Dominion University (Virginia)

ROBERT C. ZIEGENFUS, Kutztown University of Pennsylvania

In assembling the Eleventh Edition, we are indebted to the following people for advising us on a number of matters:

KHALED MD. ALI, Geologist, Dhaka, Bangladesh

THOMAS L. BELL, University of Tennessee

CATHY BARTSCH, Temple University (Pennsylvania)

KATHLEEN BRADEN, Seattle Pacific University

STUART E. CORBRIDGE, University of Miami

DEBORAH CORCORAN, Southwest Missouri State University

WILLIAM V. DAVIDSON, Louisiana State University

GARY A. FULLER, University of Hawai'i

RICHARD J. GRANT, University of Miami

MARGARET M. GRIPSHOVER, University of Tennessee

TRUMAN A. HARTSHORN, Georgia State University

PHILIP L. KEATING, Indiana University

RICHARD LISICHENKO, Fort Hays State University (Kansas)

DALTON E. MILLER, Mississippi State University

VALIANT C. NORMAN, Lexington Community College (Kentucky)

PAI YUNG-FENG, New York City

EUGENE J. PALKA, U.S. Military Academy (New York)

DAVID WELK, Reedley College (California)

MARK WHITE, National University of Ireland, Galway

For assistance with the map of North American indigenous peoples, we are greatly indebted to Jack Weatherford, Professor of Anthropology at Macalester College (Minnesota); Henry T. Wright, Professor and Curator of Anthropology at the University of Michigan; and George E. Stuart, President of the Center for Maya Research (North Carolina), who also assisted with the Altun Ha site in Belize. The map of Russia's federal regions could not have been compiled without the invaluable help of David B. Miller, Senior Edit Cartographer at the National Geographic Society, and Leo Dillon of the Russia Desk of the U.S. Department of State. And special thanks, too, go to Charles Pirtle, Professor of Geography at Georgetown University's School of Foreign Service for his exhaustive review of the entire book, and to Charles Fahrer of the Department of Geography at the University of South Carolina for his suggestions on Chapter 7.

We also record our appreciation to those geographers who ensured the quality of this book's ancillary products: Ira M. Sheskin (University of Miami) prepared the *Test Bank* and manipulated a large body of demographic data to derive the tabular display in Table G-1; Eugene J. Palka (U.S. Military Academy) prepared the *Instructor's Manual*; and Elizabeth Muller Hames (M.A. in Geography, University of Miami) co-authored the *Study Guide*, recompiled the Geographical Index found on the Web site, and handled the data preparation for Table G-1. At the University of Miami's Department of Geography and Regional Studies, Peter Muller is most grateful for the advice he received from faculty colleagues Tom Boswell, Stuart Corbridge, Douglas Fuller, Richard Grant, Janek Mandel, Jan Nijman, Rinku Roy Chowdhury, and Ira Sheskin as well as GIS Lab Manager Chris Hanson.

PERSONAL APPRECIATION

We are privileged to work with a team of professionals at John Wiley & Sons that is unsurpassed in the college textbook publishing industry. As authors we are acutely aware of these talents on a daily basis during the crucial production stage, especially the outstanding coordination and leadership skills of Production Manager Kelly Tavares, Illustration Editor Sigmund Malinowski, Photo Coordinator Jennifer MacMillan, and Production Assistants Rebecca Rothaug and Carmen Hernandez. Others who played a leading role in this process were Senior Designer Karin Kincheloe, Copy Editor Betty Pessagno, Photo Director Marge Graham, Photo Manager Hilary Newman, and Don Larson and Emily Van Cleave of Mapping Specialists, Ltd. in Madison, Wisconsin. Geography Editor Ryan Flahive was the prime mover in introducing a number of innovations for the Eleventh Edition, and together with his assistants, Denise Powell and Aliyah Vinikoor, smoothly propelled and guided this latest revision. Our College Marketing Manager, Clay Stone, gave us considerable attention throughout the revision process, and his advice and enthusiasm were always most welcome. Beyond this immediate circle, we acknowledge the support and encouragement we have received over the years from others at Wiley including Publisher Anne Smith, Vice-President for Production Ann Berlin, and Executive Editor Kaye Pace.

Finally, and most of all, we thank our wives, Bonnie and Nancy, for once again seeing us through the challenging schedule of our eighth collaboration on this volume in the past 20 years.

H. J. de Blij
Boca Grande, Florida

Peter O. Muller
Coral Gables, Florida

September 25, 2003

Brief Table of Contents

Contents

GEOGRAPHY
Realms, Regions, and Concepts
Eleventh Edition

CONCEPTS, IDEAS, AND TERMS

1 Geographic realm	14 Hinterland	27 Interglaciation	40 Urbanization
2 Spatial perspective	15 Functional region	28 Hydrologic cycle	41 State
3 Taxonomy	16 Scale	29 Climatic region	42 European state model
4 Transition zone	17 Natural landscape	30 Physiography	43 Development
5 Geographic change	18 Physical geography	31 Culture	44 Economic geography
6 New World Order	19 Continental drift	32 Regional character	45 Core area
7 Regional concept	20 Tectonic plate	33 Cultural landscape	46 Periphery
8 Regional boundaries	21 Subduction	34 Central business district (CBD)	47 Regional disparity
9 Location	22 Pacific Ring of Fire	35 Ethnicity	48 Advantage
10 Absolute location	23 Climate	36 Population distribution	49 Neocolonialism
11 Relative location	24 Desertification	37 Population density	50 Globalization
12 Formal region	25 Glaciation	38 Megalopolis	51 Regional geography
13 Spatial system	26 Ice age	39 Cartogram	52 Systematic geography

PHYSICAL

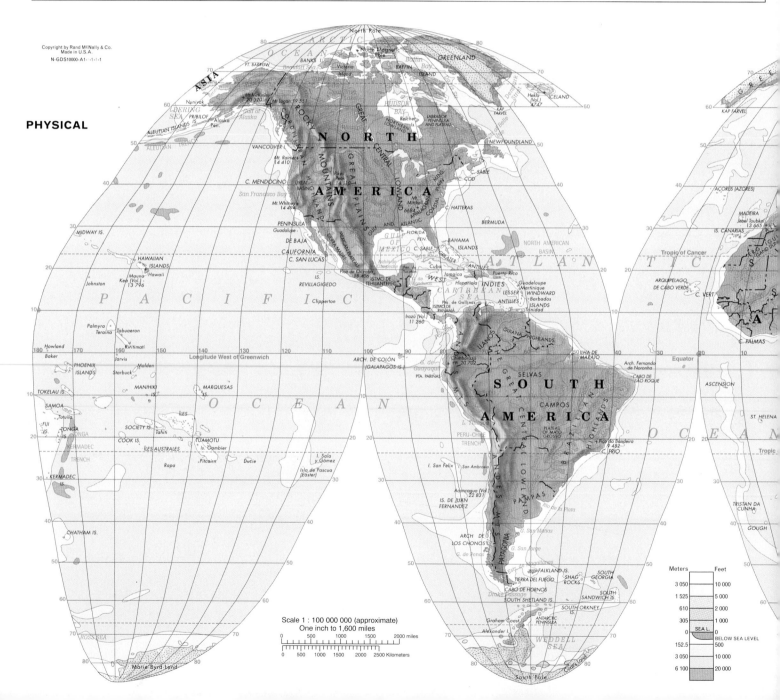

Scale 1 : 100 000 000 (approximate)
One inch to 1,600 miles

World Regional Geography: Global Perspectives

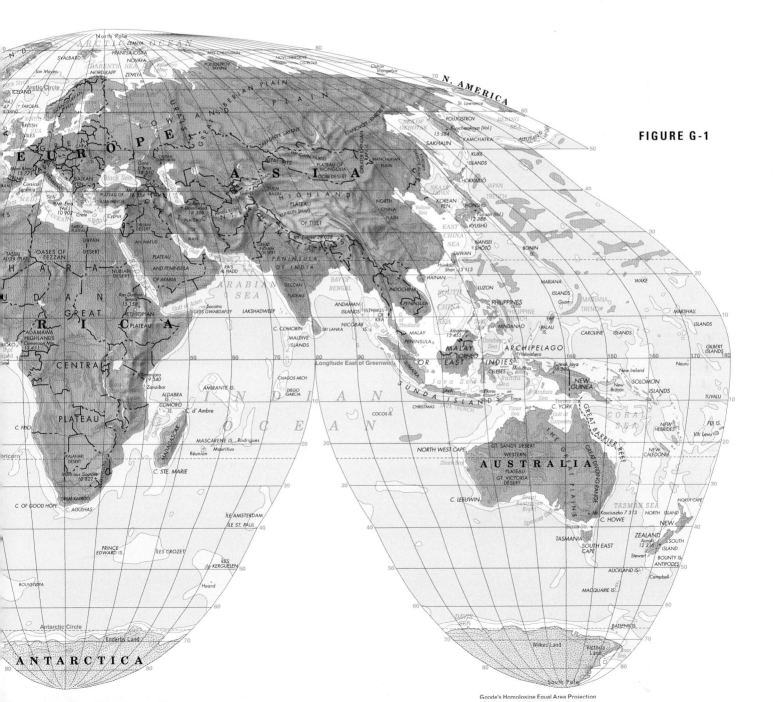

FIGURE G-1

Goode's Homolosine Equal Area Projection

*W*HAT A TIME this is to be studying geography! The world is undergoing a historic transformation, one of those momentous political, economic, and social upheavals about which we have read in history books. It is happening today, and as in the past, new alliances are forming, old unions are fracturing, novel ideas are spreading, older notions are fading. We hear from our political leaders talk about the emergence of a "New World Order," but such pronouncements are premature. A future world order is indeed in the offing, but it is too early to tell what it will be like. Of this we can be sure: the new world map will look quite different from the old. Our task is to understand the ongoing changes, to make sense of the new directions our world is taking. Geography is our most powerful ally in this mission.

GEOGRAPHIC PERSPECTIVES

In this book we take a penetrating look at the geographic framework of the contemporary world, the grand design that is the product of thousands of years of human achievement and failure, movement and stagnation, revolution and stability, interaction and isolation. Ours is an interconnected world of travel and trade, tourism and television, a global village—but the village still has neighborhoods. Their names are Europe, South America, Southeast Asia, and others familiar to all of us. We call such global neighborhoods **geographic realms**, and when we subject these realms to geographic scrutiny, we find that each has its own identity and distinctiveness.

Geographers study the location and distribution of features on the Earth's surface. These features may be the landmarks of human occupation, the properties of the natural environment, or both: one of the most interesting themes in geography is the relationship between natural environments and human societies. Geographers investigate the reasons for these distributions. Their approach is guided by a **spatial perspective**. Just as historians focus on chronology, geographers concentrate on space and place. The spatial structure of cities, the layout of farms and fields, the networks of transportation, the system of rivers, the pattern of climate—all these and more go into the study of a geographic realm. As you will discover, geography is full of spatial terms: area, distance, direction, clustering, proximity, accessibility, isolation, and many others which we will encounter in the pages that follow.

In this book we use the geographic perspective and geography's spatial terminology to investigate the world's great geographic realms. We will find that each of these realms possesses a special combination of cultural, organizational, and environmental properties. These characteristic qualities are imprinted on the landscape, giving each realm its own traditional attributes and social settings. As we come to understand the human and natural makeup of those geographic realms, we learn not only *where* they are located (always a crucial question in geography, and often the answer is not as simple as it may appear), but also *why they are located where they are*, how they are constituted, and what their future is likely to be in our changing world.

It is not enough, however, to study the world from a geographic viewpoint without learning something about the discipline of geography itself. Beginning in this introductory chapter, we introduce the fundamental ideas and concepts that make modern geography what it is. Not only will these geographic ideas enhance your awareness of the many dimensions of our complex, multicultural, interconnected world, but also many of them will remain useful to you long after you have closed this book. Welcome to geography . . . realms, regions, *and* concepts.

REALMS AND REGIONS

Geographers, like other scholars, seek to establish order from the countless data that confront them. Biologists have established a system of classification, or **taxonomy**, to categorize the many millions of plants and animals into a hierarchical system of seven ranks. In descending order, we humans belong to the animal *kingdom*, the *phylum* (division) named chordata, the *class* of mammals, the *order* of primates, the *family* of hominids, the *genus* designated *Homo*, and the *species* known as *Homo sapiens*. Geologists classify the Earth's rocks into three major (and many subsidiary) categories and then fit these categories into a complicated geologic time scale that spans hundreds of millions of years. Historians define eras, ages, and periods to conceptualize the sequence of the events they study.

Geography, too, employs systems of classification. When geographers deal with urban problems, for instance, they use a classification scheme based on the sizes and functions of the places involved. Some of the terms in this classification are part of our everyday language: megalopolis, metropolis, city, town, village, hamlet.

In regional geography, which is the focus of this book, our challenge is different. We, too, need a hierarchical framework for the areas of the world we study, from the largest to the smallest. But our classification scheme is horizontal, not vertical. It is *spatial*. Our equivalent of the biologists' overarching kingdoms (of plants and animals) is the Earth's natural partitioning into landmasses and oceans. The next level is the division of the inhabited landmasses into geographic realms based on human as well as natural properties.

The Criteria for Realms

In any classification system, criteria are the key. Not all animals are mammals; the criteria for inclusion in that biological class are more specific and restrictive. A dolphin may look and act like a fish, but both anatomically and functionally dolphins belong to the class of mammals.

Geographic realms are based on sets of spatial criteria. First, they are the largest units into which the inhabited world can be divided. The criteria on which such a broad regionalization is based include both physical (that is, natural) and human (or social) yardsticks. South America is a geographic realm, for example, because physically it is a continent and culturally it is dominated by a set of social norms. The realm called South Asia, on the other hand, lies on a Eurasian landmass shared by several other geographic realms; high mountains, wide deserts, and dense forests combine with a distinctive social fabric to create this well-defined realm centered on India.

Second, geographic realms are the result of the interaction of human societies and natural environments, a *functional* interaction revealed by farms, mines, fishing ports, transport routes, dams, bridges, villages, and countless other features that mark the landscape. According to this criterion, Antarctica is a continent but not a geographic realm.

Third, geographic realms must represent the most comprehensive and encompassing definition of the great clusters of humankind in the world today. China lies at the heart of such a cluster, as does India. Africa constitutes a geographic realm from the southern margin of the Sahara (an Arabic word for "desert") to the Cape of Good Hope and from its Atlantic to its Indian Ocean shores.

Figure G-2 displays the 12 world geographic realms based on these criteria. As we will show in more detail later, waters, deserts, and mountains as well as cultural and political shifts mark the borders of these realms. We will discuss the position of these boundaries as we examine each realm. For the moment, keep in mind the following:

4 • Where geographic realms meet, **transition zones**, not sharp boundaries, mark their contacts.

We need only remind ourselves of the border zone between the geographic realm in which most of us live, North America, and the adjacent realm of Middle Amer-

"From the observation platform atop the Seoul Tower one would be able to see into North Korea except for the range of hills in the background: the capital lies in the shadow of the DMZ (demilitarized zone), relic of one of the hot conflicts of the Cold War. The vulnerable Seoul-Inchon metropolitan area ranks among the world's largest, its population approaching 20 million. I asked my colleague, a professor of geography at Seoul National University, why the central business district, the cluster of buildings in the center of this photo, does not seem to reflect the economic power of this city (note the sector of traditional buildings in the foreground, some with blue roofs, the culture's favorite color). 'Turn around and look across the [Han] River,' he said. 'That's the new Seoul, and there the skyline matches that of Singapore, Beijing or Tokyo.' Height restrictions, disputes over land ownership, and congestion are among the factors that slowed growth in this part of Seoul and caused many companies to build in the Samsung area. Looking in that direction, you cannot miss the large U.S. military facility, right in the heart of the urban area, on some of the most valuable real estate and right next to an upscale shopping district. In the local press, the debate over the presence of American forces dominated the letters page day after day."

ica. The line on Figure G-2 coincides with the boundary between Mexico and the United States, crosses the Gulf of Mexico, and then separates Florida from Cuba and the Bahamas. But Hispanic influences are strong in North America north of this boundary, and the U.S. economic influence is strong south of it. The line, therefore, represents an ever-changing zone of regional interaction. Again, there are many ties between South Florida and the Bahamas, but the Bahamas resemble a Caribbean more than a North American society.

In Africa, the transition zone from Subsaharan to North Africa is so wide and well defined that we have put it on the world map; elsewhere, transition zones tend to be narrower and less easily represented. In these early

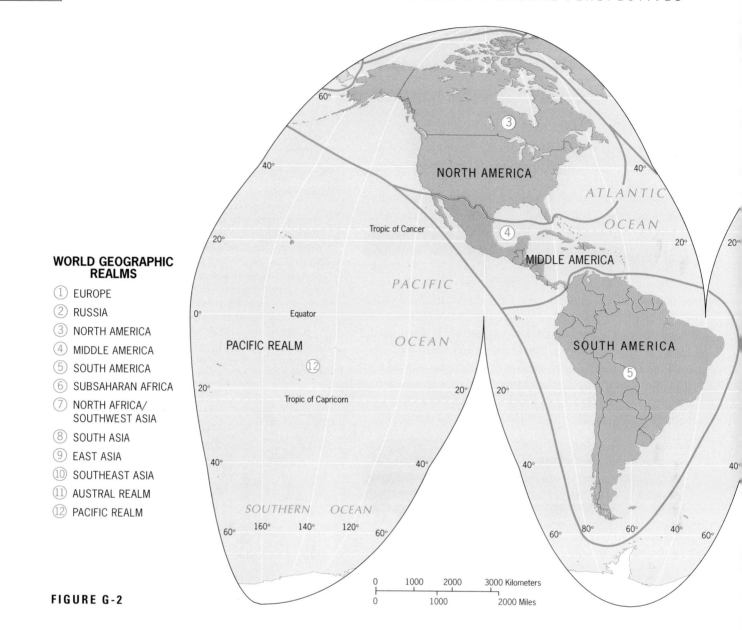

WORLD GEOGRAPHIC REALMS

1. EUROPE
2. RUSSIA
3. NORTH AMERICA
4. MIDDLE AMERICA
5. SOUTH AMERICA
6. SUBSAHARAN AFRICA
7. NORTH AFRICA/ SOUTHWEST ASIA
8. SOUTH ASIA
9. EAST ASIA
10. SOUTHEAST ASIA
11. AUSTRAL REALM
12. PACIFIC REALM

FIGURE G-2

years of the twenty-first century, such countries as Belarus (between Europe and Russia) and Kazakhstan (between Russia and Muslim Southwest Asia) lie in inter-realm transition zones. Remember, over much (though not all) of their length, borders between realms are zones of regional change.

5 • Geographic realms **change** over time.

Had we drawn Figure G-2 before Columbus made his voyages (1492–1504), the map would have looked different: Amerindian states and peoples would have determined the boundaries in the Americas; Australia and New Guinea would have constituted one realm; and New Zealand would have been part of the Pacific Realm. The colonization, Europeanization, and Westernization of the world changed that map dramatically. During the four decades after World War II relatively little change

took place, but since 1985 far-reaching realignments have again been occurring.

As we try to envisage what a **New World Order** **6** will look like on the map, note that the 12 geographic realms can be divided into two groups: (1) those dominated by one major political entity, in terms of territory or population or both (North America/United States, Middle America/Mexico, South America/Brazil, South Asia/India, East Asia/China, Southeast Asia/Indonesia as well as Russia and Australia), and (2) those contain many countries but no dominant state (Europe, North Africa/Southwest Asia, Subsaharan Africa, and the Pacific Realm). For several decades two major powers, the United States and the former Soviet Union, dominated the world. Will a multipolar world arise from the rubble of that bipolar world? We will address this question in the pages that follow.

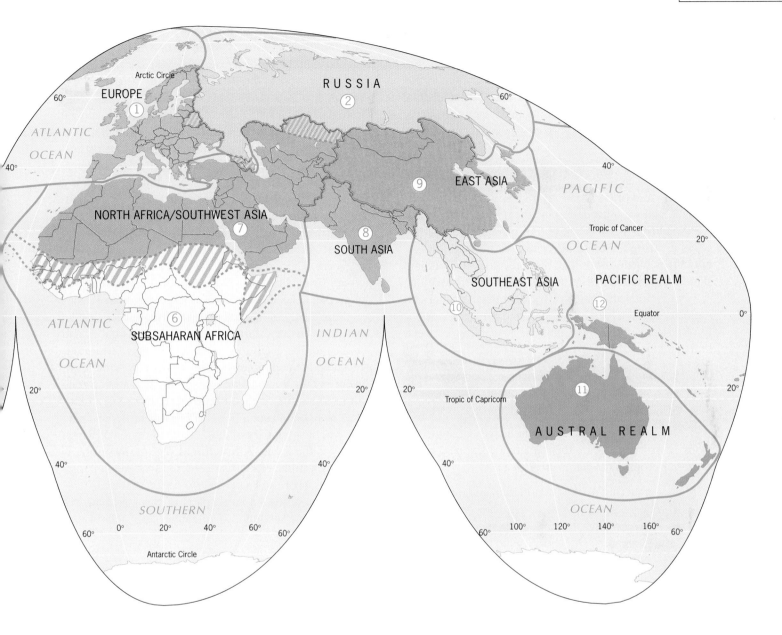

The Criteria for Regions

The spatial division of the world into geographic realms establishes a broad global framework, but for our purposes a more refined level of spatial classification is needed. This brings us to an important organizing concept in geography: the **regional concept**. To continue the analogy with biological taxonomy, we now go from phylum to order. To establish regions within geographic realms, we need more specific criteria.

Let us use the North American realm to demonstrate the regional idea. When we refer to a part of the United States or Canada (e.g., the South, the Midwest, or the Prairie Provinces), we employ a regional concept—not scientifically but as part of everyday communication. We reveal our *perception* of local or distant space as well as our mental image of the region we are describing.

But what exactly is the Midwest? How would you draw this region on the North American map? Regions are easy to imagine and describe, but they can be difficult to outline on a map. One way to define the Midwest is to use the borders of States: certain States are part of this region, others are not. You could also use agriculture as the chief criterion: the Midwest is where corn and/or soybeans occupy a certain percentage of the farmland. Each method results in a different delimitation; a Midwest based on States is different from a Midwest based on farm production. Therein lies an important principle: regions are scientific devices that allow us to make spatial generalizations, and they are based on artificial criteria we establish to help us construct them. If you were studying the geography behind politics, then a Midwest region defined by State boundaries would make sense. If you were studying agricultural distributions, you would need a different definition.

Given these different dimensions of the same region, we can identify properties that all regions have in common. To begin with, all regions have *area*. This observation would seem obvious, but there is more to this idea than meets the eye. Regions may be intellectual constructs, but they are not abstractions: they exist in the real world, and they occupy space on the Earth's surface.

It follows that regions have *boundaries*. Occasionally, nature itself draws sharp dividing lines, for example, along the crest of a mountain range or the margin of a forest. More often, **regional boundaries** are not self-evident, and we must determine them using criteria that we establish for that purpose. For example, to define a citrus-growing agricultural region, we may decide that only areas where more than 50 percent of all farmland stands under citrus trees qualify to be part of that region.

All regions also have **location**. Often the name of a region contains a locational clue, as in Amazon Basin or Indochina (a region of Southeast Asia lying between India and China). Geographers refer to the **absolute location** of a place or region by providing the latitudinal and longitudinal extent of the region with respect to the Earth's grid coordinates. A far more practical measure is a region's **relative location**, that is, its location with reference to other regions. Again, the names of some regions reveal aspects of their relative locations, as in *Eastern* Europe and *Equatorial* Africa.

• Many regions are marked by a certain homogeneity.

Homogeneity or sameness may lie in a region's human (cultural) properties, its physical (natural) characteristics, or both. Siberia, a vast region of northeastern Russia, is marked by a sparse human population that resides in widely scattered, small settlements of similar form; frigid climates; extensive areas of permafrost (permanently frozen subsoil); and cold-adapted vegetation. This dominant uniformity makes it one of Russia's natural and cultural regions, extending from the Ural Mountains in the west to the Pacific Ocean in the east. When regions display a measurable and often visible internal homogeneity, they are called **formal regions**. But not all formal regions are visibly uniform. For example, a region may be delimited by the area in which, say, 90 percent of the people or more speak a particular language. This cannot be seen in the landscape, but the region is a reality, and we can use this criterion to draw its boundaries accurately. It, too, is a formal region.

Other regions are marked *not* by their internal sameness, but by their functional integration—that is, the way they work. These regions are defined as **spatial systems** and are formed by the areal extent of the activities that define them. Take the case of a large city with its surrounding zone of suburbs, urban-fringe countryside, satellite towns, and farms. The city supplies goods and services to this encircling zone, and it buys farm products and other commodities from it. The city is the heart, the *core* of this region, and we call the surrounding zone of interaction the city's **hinterland**. But the city's influence wanes on the outer periphery of that hinterland, and there lies the boundary of the functional region of which the city is the focus. A **functional region**, therefore, is forged by a structured, urban-centered system of interaction. It has a core and a periphery. As we shall see, core-periphery contrasts in some parts of the world are becoming strong enough to endanger the stability of countries.

• Global interconnections shape regions.

There was a time in global history when world regions existed in isolation or near-complete insularity.

Interconnections in the spatial system of the Corn Belt: Garden City, Iowa. Roads and grain elevators underscore the linkages between the town and its surroundings. The spatial homogeneity of the region is quite evident from the agricultural landscape beyond (evenly dispersed farmsteads, repetition of field size and land uses). Also visible is the signature geometry of the Township-and-Range system of land division which, beginning in 1785, superimposed a rectangular grid of 6-mile by 6-mile townships on all of the U.S. States opened for settlement after that date. Within that geographical matrix, townships were subdivided into 36 square-mile (640-acre) properties that were further split into 160-acre (and sometimes smaller) farms. As the photo shows, this land division scheme dominates every aspect of the cultural landscape: not only the rectangular fields and the north-south/east-west layout of roads and railways in the countryside, but also the configuration and spacing of streets, buildings, and properties in Garden City itself.

Pre-Columbian America, precolonial Africa, and pre-imperial Australia contained cultural regions untouched by external influences. Contact between European and East Asian societies was maintained by intermediaries and produced no significant cultural exchanges. Today, the opposite is true: no world region exists in isolation. Every geographic region displays a mix of local traditions and foreign influences, the result of centuries of colonialism, trade, migration, cultural pervasion, economic invasion, geopolitical intervention, and, more recently, media penetration (radio, television) and computer linkages. Such interactions blur regional boundaries and also have the effect of diluting the strength of local traditions. In Australia, little remains of the once-dominant cultures; in the Americas, native societies are mere remnants of major civilizations. Africa was never depopulated as Australia and the Americas were, but cultural institutions were overwhelmed by successive waves of intrusion. Even in populous and culturally durable Asia, external influences transformed the most traditional societies. World regions today, therefore, display a variable mix of local and global forces.

REGIONS AT SCALE

In this book we examine the geography of the world by dividing it into regions and using regional concepts as we proceed. These concepts form the building blocks from which the regional framework emerges.

Regions come in many sizes. The Russian formal region of Siberia is huge. By comparison, a functional region formed by a city and its hinterland is small. It is always possible to identify smaller regions within larger ones. And this brings us to the geographic concept of **scale**.

The map is the geographer's strongest ally. It does for geography what taxonomic (and other) classification systems do for the other sciences. Maps display enormous amounts of information. They can suggest spatial relationships, they pose questions for researchers, and they often suggest where the answers may lie. Maps represent the surface of the Earth at various levels of generalization. A map's scale is the ratio of the distance between two locations on a map and the actual distance between these two locations on the Earth's surface. That ratio is expressed as a fraction.

In Figure G-3, note how that fraction grows larger as the level of detail increases—and the area displayed shrinks. Map A shows almost all of North America at the small scale of 1:103,000,000. The scale of map B is almost twice as large but shows only the northeast quadrant of the realm. At 1:24,000,000 (map C) only the Canadian province of Quebec can mostly be shown; at 1:1,000,000

(map D) you can see details of the urban area of Montreal at a scale about 100 times as large as map A.

In a book that surveys the major regions of each world realm, we will obviously have to operate at relatively small scales. Occasionally, we will enlarge our scale to focus on an issue in somewhat greater detail, but mostly our view will be macroscopic and general.

THE PHYSICAL SETTING

This book focuses on the geographic realms and regions produced by human activity over thousands of years. But we should never forget the natural environments in which all this activity took place because we can still recognize the role of these environments in how people make their living. Certain areas of the world, for example, presented opportunities for plant and animal domestication that other areas did not. The people who happened to live in those favored areas learned to grow wheat, rice, or root crops, and to domesticate oxen, goats, or llamas. We can still discern those early "patterns of opportunity" on the map in the twenty-first century. From such opportunities came adaptation and invention, and thus arose villages, towns, cities, and states. But people living in more difficult environments found it much harder to achieve this organization. Take tropical Africa. There, the human communities could not domesticate any of the many species of wildlife, from gazelles and zebras to giraffes and buffalo. Wild animals were a threat, not an opportunity. Eventually humans domesticated only one African animal, the guinea fowl. Early African peoples faced environmental disadvantages that persist today. The modern map carries many imprints of the past.

Planet Earth presents a jumble of **natural landscapes** ranging from rugged mountain chains to smooth coastal plains (Fig. G-1). As we investigate each geographic realm, we will note the natural landscapes that form its physical base because these landscapes continue to play a role in their formation. Mountain ranges form barriers to movement, but they also protect peoples against enemies. River basins in Asia still contain some of the world's largest population concentrations because they are well-watered and have fertile soils, but river basins in tropical Africa and South America support no such numbers. Every geographic realm has its distinctive combination of natural landscapes.

Continents and Plates

The study of natural landscapes is part of **physical geography**. Nearly a century ago a physical geographer, Alfred Wegener, used spatial analysis to explain the details he

EFFECT OF SCALE

FIGURE G-3

saw on maps like Figure G-1. He studied the outlines of the continents and, marshalling vast evidence, theorized that the Earth's landmasses are pieces of a giant supercontinent that existed hundreds of millions of years ago. He named that postulated supercontinent *Pangaea*, and Africa lay at the heart of it. When North and South America split away from Eurasia and Africa, the Atlantic Ocean opened—but you can still see how the opposite coastlines fit together.

Wegener's hypothesis of **continental drift** seemed to explain much of what Figure G-1 shows, including the global distribution of major mountain ranges. As South America "drifted" westward, away from Africa, the rocks of its leading edge crumpled like folds of an accordion, creating the Andes Mountains. On the opposite side of Africa, Australia moved eastward, so that its major mountain chain now lies near its eastern margin. Africa itself moved little, which is why no major linear mountain range formed there.

Wegener's hypothesis, based on spatial evidence, pointed the way for the research of others. Eventually geologists, building on the geographic evidence, formulated the notion of plate tectonics. The continental landmasses, they reasoned, rest upon great slabs of crust called **tectonic plates**. These plates are mobile, moving slowly a few centimeters (an inch or so) a year, propelled by heat-driven convection cells in the molten rock below.

Where these giant plates collide, the plate that consists of lighter (less heavy) rock rides up over the weightier plate, which is pushed downward in a process called **subduction**. Such gigantic collisions create the mountain chains that girdle the Earth, causing earthquakes and volcanic eruptions in the process. In Figure G-4, note that the Pacific Ocean is encircled by a nearly continuous plate-collision zone called the **Pacific Ring of Fire**. Disastrous earthquakes, volcanic eruptions, landslides, and other threats are facts of daily life here (Fig. G-5).

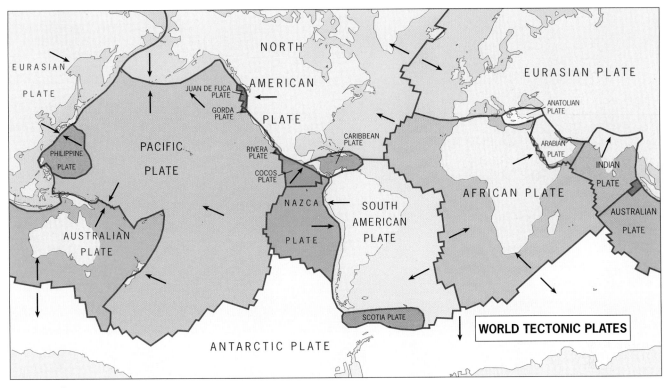

FIGURE G-4

"The land of New Zealand's South Island dropped off precipitously under the waters of the Tasman Sea here, along the southwestern coast. As we approached the shore we crossed the contact zone between the Pacific Plate, on which most of the South Island lies, and the Australian Plate (see Fig. G-4). Subduction along this zone creates some of the world's most spectacular scenery as the crust is pushed precipitously downward, to be recycled in the mantle below . . . diagonally across the Pacific from New Zealand lies volcanically and seismically active Alaska. Mount Edgecumbe, about 16 miles (26 km) west of Sitka, dominates the landscape here. I took the day and walked across the bridge toward the old volcano, chatting with locals as I went. Edgecumbe is listed as an extinct volcano, but Sitkans have their doubts. Tlingit oral history tells of a violent eruption that buried the whole countryside under a deep blanket of ash. 'The earth tremors around here tell me that it's a little early to call it extinct,' said one resident. 'A few centuries is nothing when it comes to geologic activity. I'd say the old lady is dormant.' The scars of erosion, outlined by the summer snow in the picture, suggest a lengthy dormancy, but the crater rim still rises nearly 3300 feet (1000 m) above the water."

FIGURE G-5

The complex and varied natural landscapes of the continents result not only from geologic forces but also from processes of rock *weathering* and *erosion*. Rocks vary greatly in their resistance to these processes, resulting in highly diverse and often spectacular scenery. Rivers, glaciers, coastal waves, even the wind all sculpt the surface, creating a mosaic of natural landscapes that is unique to the solar system.

Climate

It would be impossible to discuss the world's geographic realms and regions without reference to their **climates**. Many of humanity's inventions were motivated by the need to overcome problems posed by climate: clothing against cold, irrigation against drought, air conditioning against heat. When we think of some distant region, the first question we are likely to ask is, "What is its climate?"

A more appropriate question would be "What is its climate *today*?" Global climates change, sometimes rapidly. Archeological evidence tells us that farms, villages, and towns that sprang up in well-watered, fertile areas of Southwest Asia were overtaken by **desertification**. Dependable rainfall cycles gave way to intermittent droughts, river levels dropped, wells failed. Eventually the expanding desert claimed what had once been thriving farmlands and bustling settlements. Some settlements were buried under advancing dunes.

Over the longer term, too, climate is subject to wide swings. Little more than 12,000 years ago, an eyeblink in geologic history, giant icesheets covered virtually all of

Canada and much of the U.S. Midwest (Fig. G-6) as far south as the Ohio River. Most of Europe was covered from the Alps to Scandinavia. This was only the latest phase of **glaciation**, a period of lowered temperatures, ice surges, and dropping sea levels that has prevailed over the past 12 million years. It all started with a cooling trend, perhaps as long as 20 million years ago, the beginning of a global **ice age**. Eventually this gradual cooling gave way to massive advances of icesheets and valley glaciers, marking the beginning of what geologists call the Pleistocene epoch less than 3 million years ago. Ever since, frigid glaciations have been followed by relatively warm spells called **interglaciations**. During the current interglaciation, named the *Holocene* and already over 10,000 years in duration, human civilizations have evolved from the smallest and simplest communities to the populous countries and megacities of today.

Climatic Regions

Climatic variability also occurs in the short term, and we should not be surprised that climates continue to change. Unless we prepare for it, these dynamics can produce social and political dislocation. Thus we need to understand the climatic environments of the world's geographic realms and regions. Our technological advances notwithstanding, climate still plays a key role in the lives and livelihoods of billions of people. A perturbation in the system—such as an El Niño event or a drought—can have widespread repercussions on economies and even politics.

EXTENT OF GLACIATION DURING THE PLEISTOCENE

Land areas covered by glaciers

Present-day glaciers

FIGURE G-6

FROM THE FIELD NOTES

"Walking the hilly countryside anywhere in Indonesia leaves you in no doubt about the properties of *Af* climate. The sweltering sun, the hot, humid air, the daily afternoon rains, and the lack of relief even when the sun goes down—the still atmosphere lies like a heavy blanket on the countryside to make this a challenging field trip. Deep, fertile soils form rapidly here on the volcanic rocks, and the entire landscape is green, most of it now draped by rice fields and terraced paddies. The farmer whose paddies I photographed here told me that his land produces three crops per year. Consider this: the island of Jawa, about the size of Louisiana, has a population of more than 125 million—its growth made possible by this combination of equatorial circumstances . . . Alaska, almost a dozen times as large as Jawa, has a population under three-quarters of a million. Here climates range from *Cfc* to *E*, soils are thin and take thousands of years to develop, and the air is arctic. We sailed slowly into Glacier Bay, in awe of the spectacular, unspoiled scenery, and turned into a bay filled with ice floes, some of them serving as rafts for sleeping seals. Calving (breaking up) into the bay was the Grand Pacific Glacier, with evidence of its recent recession all around. Less than 300 years ago, all of Glacier Bay was filled with ice; today you can sail miles to the Johns Hopkins (shown here) and Margerie tidewater glaciers' current outer edges. Global warming in action—and a reminder that any map of climate is a still picture of a changing world."

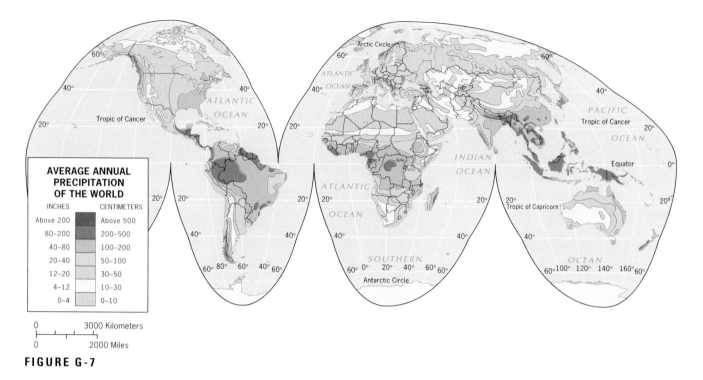

FIGURE G-7

It is no easier to generalize about climates than it is about natural landscapes. A regional climate is the average of countless weather observations among which measures of precipitation and temperature are key. Also taken into account are seasonal variations and extremes. In terms of precipitation, the Earth displays enormous spatial variation (Fig. G-7). Equatorial and tropical areas tend to be well-watered because the **hydrologic cycle**—the system that carries evaporated ocean water (leaving the salt behind) over the landmasses where it falls as rain or snow—is most efficient in the high heat and humidity of the low latitudes. But in Africa this equatorial bounty soon gives way to subtropical aridity. From over 80 inches (200 cm) of rain in The Congo near the equator, the annual total drops to 4 inches (10 cm) or less in the Sahara to the north and the Kalahari-Namib to the south. In higher latitudes, only certain favored areas (notably Western Europe) receive ample precipitation.

Average temperatures decline from equatorial toward polar latitudes, but the gradation is irregular, affected by elevation (mountains have a cooling effect) and location relative to coastlines (interiors of continents are hotter in summer, cooler in winter). The size, shape, and terrain of the landmasses complicate the map of average temperatures. Ocean currents also influence the pattern. Warm waters offshore, combined with onshore winds, give most of Europe a warmth that is not experienced at similar latitudes elsewhere.

How can we forge regional coherence out of such a jumble of data? The effort to create a relatively simple, small-scale world map of climates has continued for nearly a century. Figure G-8 is based on the system Wladimir Köppen devised and Rudolf Geiger then modified. This system, which has the advantage of simplicity, is represented by a set of letter symbols. The first (capital) letter is the critical one: the *A* climates are humid and tropical, the *B* climates are arid, the *C* climates are mild and humid, the *D* climates show increasing extremes of seasonal heat and cold, and the *E* climates reflect the frigid conditions at and near the poles.

Figure G-8 merits your attention because familiarity with it will help you understand much of what follows in this book. The map has practical utility, too. Although it depicts **climatic regions**, daily weather in each color-coded region is relatively standard. If, for example, you are familiar with the weather in the large area mapped as *Cfa* in the southeastern United States, you will feel at home in Uruguay (South America), Kwazulu-Natal (South Africa), New South Wales (Australia), and Fujian Province (China). Let us look at the world's climatic regions in some detail.

Humid Equatorial (*A*) Climates

The humid equatorial, or tropical, climates are characterized by high temperatures all year and by heavy precipitation. In the *Af* subtype, the rainfall arrives in substantial amounts every month; but in the *Am* areas, the arrival of the annual wet *monsoon* (the Arabic word for "season" [see p. 378]) marks a sudden enormous increase in precipitation. The *Af* subtype is named after the vegetation that develops there—the tropical rainforest. The *Am* subtype, prevailing in part of peninsular India, in a coastal area of West Africa, and in sections of Southeast Asia, is appropriately referred to as the monsoon climate. A third tropical climate, the savanna (*Aw*), has a wider daily and annual temperature range and a more strongly seasonal distribution of rainfall. As Figure G-7 indicates, savanna rainfall totals tend to be lower than that in the rainforest zone, and savanna seasonality is often expressed in a "double maximum." Each year produces two periods of increased rainfall separated by pronounced dry spells. In many savanna zones, inhabitants refer to the "long rains" and the "short rains" to identify those seasons; a persistent problem is the unpredictability of the rain's arrival. Savanna soils are not among the most fertile, and when the rains fail hunger looms. Savanna regions are far more densely peopled than rainforest areas, and millions of residents of the savanna subsist on what they cultivate. Rainfall variability is their principal environmental problem.

Dry (*B*) Climates

Dry climates occur in both lower and higher latitudes. The difference between the *BW* (true desert) and the moister *BS* (semiarid steppe) varies but may be taken to lie at about 10 inches (25 cm) of annual precipitation. Parts of the central Sahara in North Africa receive less than 4 inches (10 cm) of rainfall. Most of the world's arid areas have an enormous daily temperature range, especially in subtropical deserts. In the Sahara, there are recorded instances of a maximum daytime shade temperature of over 120°F (49°C) followed by a nighttime low of 48°F (9°C). Soils in these arid areas tend to be thin and poorly developed; soil scientists have an appropriate name for them—aridisols.

Humid Temperate (*C*) Climates

As the map shows, almost all these mid-latitude climate areas lie just beyond the Tropics of Cancer and Capricorn (23½° North and South latitude, respectively). This is the prevailing climate in the southeastern United States from Kentucky to central Florida, on North America's west coast, in most of Europe and the Mediterranean, in southern Brazil and northern Argentina, in coastal South Africa, in eastern Australia, and in eastern China and southern Japan. None of these areas suffers climatic extremes or severity, but the winters can be cold, especially away from water bodies that moderate temperatures. These areas lie midway between the winterless equatorial climates and the summerless polar zones. Fertile and productive soils have developed under this regime, as we will note in our discussion of the North American and European realms.

The humid temperate climates range from moist, as along the densely forested coasts of Oregon, Washington, and British Columbia, to relatively dry, as in the so-called Mediterranean (dry-summer) areas that include not only coastal southern Europe and northwestern Africa but also the southwestern tips of Australia and Africa, central Chile, and Southern California. In these Mediterranean environments, the scrubby, moisture-preserving vegetation creates a natural landscape different from that of richly green Western Europe.

Humid Cold (*D*) Climates

The humid cold (or "snow") climates may be called the continental climates, for they seem to develop in the interior of large landmasses, as in the heart of Eurasia or North America. No equivalent land areas at similar latitudes exist in the Southern Hemisphere; consequently, no *D* climates occur there.

Great annual temperature ranges mark these humid continental climates, and cold winters and relatively cool summers are the rule. In a *Dfa* climate, for instance, the warmest summer month (July) may average as high as 70°F (21°C), but the coldest month (January) might average only 12°F (−11°C). Total precipitation, much of it snow, is not high, ranging from over 30 inches (75 cm) to a steppe-like 10 inches (25 cm). Compensating for this paucity of precipitation are cool temperatures that inhibit the loss of moisture from evaporation and evapotranspiration (moisture loss to the atmosphere from soils and plants).

Some of the world's most productive soils lie in areas under humid cold climates, including the U.S. Midwest, parts of southern Russia and Ukraine, and Northeast China. The winter dormancy (when all water is frozen) and the accumulation of plant debris during the fall balance the soil-forming and enriching processes. The soil differentiates into well-defined, nutrient-rich layers, and substantial organic humus accumulates. Even where the annual precipitation is light, this environment sustains extensive coniferous forests.

Cold Polar (*E*) and Highland (*H*) Climates

Cold polar (*E*) climates are differentiated into true icecap conditions, where permanent ice and snow keep vegetation from gaining a foothold, and the tundra, which may have average temperatures above freezing up to four months of the year. Like rainforest, savanna, and steppe, the term *tundra* is vegetative as well as climatic, and the boundary between the *D* and *E* climates in Figure G-8 corresponds closely to that between the northern coniferous forests and the tundra.

Finally, the *H* climates—unclassified highlands mapped in gray (Fig. G-8)—resemble the *E* climates.

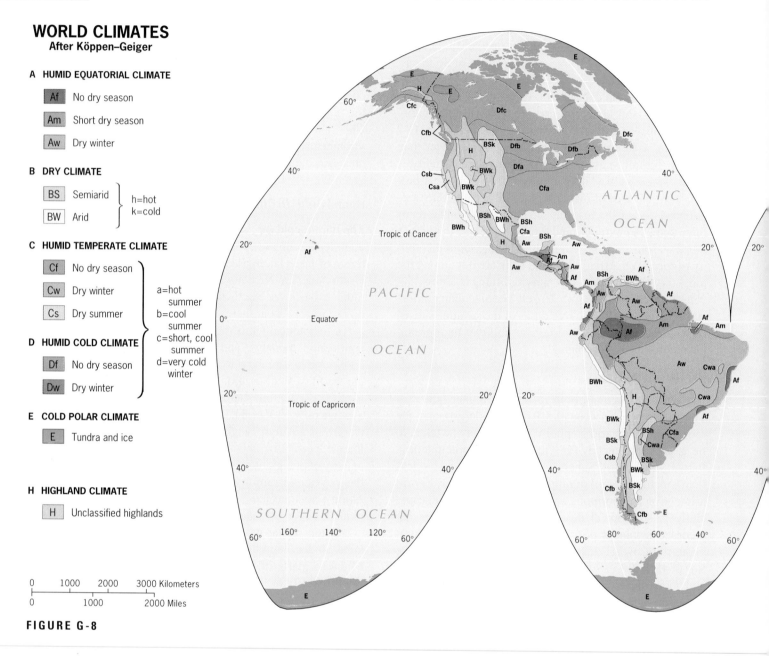

WORLD CLIMATES
After Köppen–Geiger

A HUMID EQUATORIAL CLIMATE

Af	No dry season
Am	Short dry season
Aw	Dry winter

B DRY CLIMATE

| BS | Semiarid | h=hot |
| BW | Arid | k=cold |

C HUMID TEMPERATE CLIMATE

Cf	No dry season
Cw	Dry winter
Cs	Dry summer

a=hot summer
b=cool summer
c=short, cool summer
d=very cold winter

D HUMID COLD CLIMATE

| Df | No dry season |
| Dw | Dry winter |

E COLD POLAR CLIMATE

| E | Tundra and ice |

H HIGHLAND CLIMATE

| H | Unclassified highlands |

```
0      1000    2000    3000 Kilometers
0           1000         2000 Miles
```

FIGURE G-8

High elevations and the complex topography of major mountain systems often produce near-Arctic climates above the tree line, even in the lowest latitudes such as the equatorial section of the high Andes of South America.

Let us not forget an important qualification concerning Figure G-8: this is a still-picture of a changing scene, a single frame from an ongoing film. Climate is still changing, and less than a century from now climatologists are likely to be modifying the climate maps to reflect new data. Who knows: we may have to redraw even those familiar coastlines. Environmental change is a never-ending challenge.

▮ REGIONS AND CULTURES

Whenever we explore a geographic realm or region, we should assess the physical stage that forms its base—its total physical geography or **physiography**. Still, the realms and regions we discuss in the chapters that follow are defined primarily by human-geographic criteria, and one criterion, culture, is especially significant. Therefore, we should carefully consider the culture concept in a regional context.

When anthropologists define **culture**, they tend to concentrate on abstractions: learning, knowledge and its transmission, and behavior. Geographers are most inter-

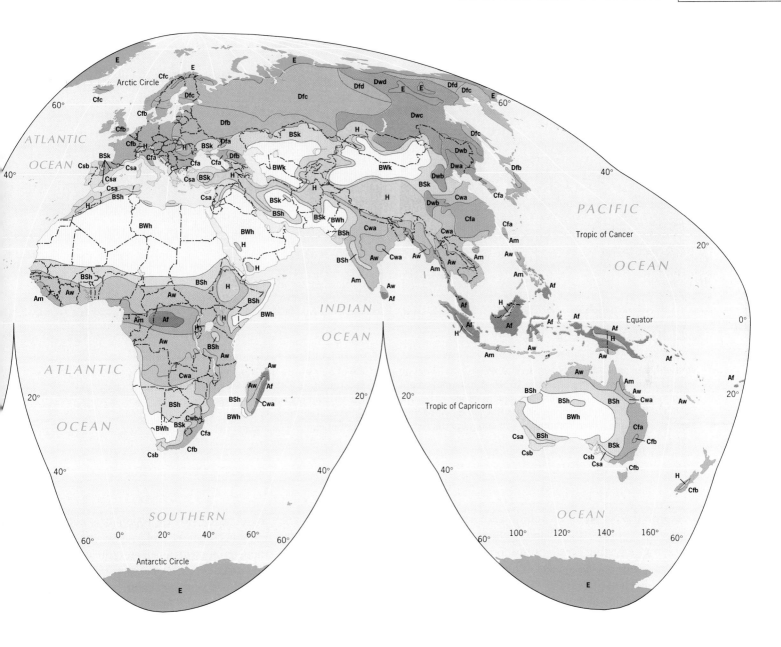

ested in how culture and the patterns of behavior associated with it are imprinted on the landscape. Thus we will examine how members of a society perceive and exploit their available resources, how they maximize the opportunities and adapt to the limitations of their natural environment, and how they organize their portion of the Earth. Some human works remain etched on the Earth's surface for a long time: the Egyptian pyramids, the Great Wall of China, and Roman roads and bridges are still in place millennia after they were built. Over time, regions take on dominant qualities that collectively constitute a

32 **regional character**, a personality, a distinct atmosphere. Regional character is a crucial criterion in ascer-

taining how we divide the human world into major geographic realms and regions.

Cultural Landscapes

As we noted, geographers are particularly concerned with the impress of culture on the Earth's physical surface. Culture gives visible character to a region in many ways. Often a single scene in a photograph or picture can reveal, in general terms, where the photo was taken. The architecture, forms of transportation, goods being carried, and clothing of the people (all these are part of culture) help us guess the region in the photo. We can do

FROM THE FIELD NOTES

"The Atlantic-coast city of Bergen, Norway displayed the Norse cultural landscape more comprehensively, it seemed, than any other Norwegian city, even Oslo. The high-relief site of Bergen creates great vistas, but also long shadows; windows are large to let in maximum light. Red-tiled roofs are pitched steeply to enhance runoff and inhibit snow accumulation; streets are narrow and houses clustered, conserving warmth . . . The coastal village of Mengkabong on the Borneo coast of the South China Sea represents a cultural landscape seen all along the island's shores, a stilt village of the Bajau, a fishing people. Houses and canoes are built of wood as they have been for centuries. But we could see some evidence of modernization: windows filling wall openings, water piped in from a nearby well."

this because the people of any culture are active agents of change; they transform the land they occupy by building structures, creating lines of transport and communication, parceling out fields, and tilling the soil (among countless other activities).

The composite of human imprints on the Earth's surface is called the **cultural landscape**, a term that came into general use in geography during the 1920s. Carl Ortwin Sauer (a professor of geography at the University of California, Berkeley) developed a school of cultural geography that focused on the concept of cultural landscape. In a paper titled "Recent Developments in Cultural Geography" (1927), Sauer proposed his most straightforward definition of the cultural landscape: *the forms superimposed on the physical landscape by the activities of man.* Such forms result from cultural processes—causal forces that shape cultural patterns—that unfold over a long time and involve the cumulative influences of successive occupants.

When it comes to mapping a cultural landscape, its material, tangible properties are the easiest to observe and record. Take, for instance, the urban "townscape"— a prominent element of the overall cultural landscape— and compare a major U.S. city with, say, a major Japan-

ese city. Photographs of these two metropolitan scenes would reveal the differences quickly, of course, but so would maps. The American city with its square-grid layout of the **central business district (CBD)** and its widely dispersed, now heavily urbanized suburbs contrasts sharply with the clustered, space-conserving Japanese metropolis. In a rural example, the spatially lavish subdivision and ownership patterns of American farmland look unmistakably different from the traditional African countryside, with its irregular, often tiny patches of land surrounding a village. Still, the totality of a cultural landscape can never be captured by a photograph or map because the personality of a region involves more than its prevailing spatial organization. One must also include the region's visual appearance, its noises and odors, the shared experiences of its inhabitants, and even their pace of life.

Culture and Ethnicity

Language, religion, and other cultural traditions often are durable and persistent. Culture is not necessarily based on **ethnicity**, however, and as we study the world's human-geographic regions, we should be aware that peoples of different ethnic stocks can achieve a common cultural landscape, while people of the same ethnic background can be divided along cultural lines.

The recent events in the former Eastern European country of Yugoslavia are a case in point. As the old Yugoslavia broke up in the early 1990s, its component parts were first affected and then engulfed by what was often described as "ethnic" conflict. When the crisis reached Bosnia-Herzegovina in the heart of Yugoslavia, three groups fought a bitter civil war. These groups were identified as Bosnian Muslims, Serbs, and Croats—but in fact all were ethnic Slavs (Yugoslavia means "Land of the South Slavs"). What distanced them from one another, and what kept the conflict going, was cultural tradition, not ethnicity. The Bosnian Muslims and the Serbs had developed different communities in different parts of the former Yugoslavia. The Muslims were descendants of Slavs who, a century ago or more, had been converted by the Turks (who once ruled here) from Christianity to Islam. Each of these groups feared domination by the others, and so South Slav turned against South Slav in what was, in truth, a culture-based conflict.

Even though the post–Cold War world is still young, it has already witnessed numerous intraregional conflicts; in later chapters, we will explain some of these conflicts in geographic context. Not all of them have ethnicity at their roots. Culture is a great unifier; it can also be a powerful divider.

REALMS OF POPULATION

Earlier we noted that population numbers by themselves do not define geographic realms or regions. Population distributions, and the functioning society that gives them common ground, are more significant criteria. That is why we can identify one geographic realm (the Austral) with just under 24 million people and another (East Asia) with more than 1.5 billion inhabitants. Neither population numbers nor territorial size alone can delimit a geographic realm. Nevertheless, the map of world population distribution (Fig. G-9) suggests the relative location of several of the world's geographic realms, based on the strong clustering of population in certain areas. Before we examine these clusters in some detail, remember that the Earth's human population now totals over 6 billion—6 thousand million people confined to the landmasses that constitute less than 30 percent of our planet's surface, much of which is arid desert, rugged mountain terrain, or frigid tundra. (Remember that Figure G-9 is an early twenty-first-century still-picture of an ever-changing scene: the rapid growth of humankind continues.) After thousands of years of slow growth, world population during the nineteenth and twentieth centuries expanded at an increasing rate. That rate has recently been slowing down somewhat, but consider this: it took about 17 centuries after the birth of Christ for the world to add 250 million people to its numbers; now we are adding 250 million about every *four years*.

Demographic (population-related) issues will arise repeatedly in later chapters as we survey the world's most crowded realms. For the moment, we will confine ourselves to the overall global situation. For that purpose let us compare the map of world **population distribution** to the demographic data shown in Table G-1 (pp. 33–37). This table provides information about the total populations in each geographic realm as well as their individual countries. Also, if you compare Figure G-9 to the maps of terrain (Fig. G-1) and climate (Fig. G-8), you will note that some of the largest population concentrations lie in the fertile basins of major rivers (China's Huang He and Chang Jiang, India's Ganges). We live in a modern world, but old ways of life still prescribe where hundreds of millions on this Earth live.

Major Population Clusters

The world's greatest population cluster, *East Asia*, lies centered on China and includes the Pacific-facing Asian coastal zone from the Korean Peninsula to Vietnam. The map indicates that the number of people per unit area— the **population density**—tends to decline from the coastal zone toward the interior. Note, however, the ribbon-like extensions marked *A* and *B* (Fig. G-9). As the map of world natural landscapes (Fig. G-1) confirms, these are populations concentrated in the valleys of China's major rivers, the Huang He (Yellow) and the Chang Jiang (Yangzi), respectively. Here in East Asia, most people are farmers, not city dwellers. There are great cities in China (such as Beijing and Shanghai), but their total population still is far outnumbered by the farmers—those who live and work on the land and whose crops of rice and wheat feed not only themselves but also the people in those cities.

The *South Asia* population cluster lies centered on India and includes the populous neighboring countries of Bangladesh and Pakistan. This huge agglomeration of humanity focuses on the broad plain of the lower Ganges River (*C* in Fig. G-9). This cluster is nearly as large as that of East Asia and at present growth rates will overtake East Asia in less than 20 years. As in East Asia, most people are farmers, but in South Asia pressure on the land is even greater, and farming is less efficient. As we note in Chapter 8, the population issue looms large in this realm.

FIGURE G-9

The third-ranking population cluster, *Europe*, also lies on the world's biggest landmass but at the opposite end from China. The European cluster, including western Russia, counts over 700 million inhabitants, which puts it in a class with the two larger Eurasian concentrations—but there the similarity ends. In Europe, the key to the linear, east-west orientation of the axis of population (*D* in Fig. G-9) is not a fertile river basin but a zone of raw materials for industry. Europe is among the world's most highly urbanized and industrialized realms, its human agglomeration sustained by industries and offices rather than paddies and pastures.

The three world population concentrations just discussed (East Asia, South Asia, and Europe) account for more than 3.5 billion of the world's 6.4 billion people. No other cluster comes close to these numbers. The next-ranking cluster, *Eastern North America*, is only about one-quarter the size of the smallest of the Eurasian concentrations. As in Europe, the population in this area is concentrated in major metropolitan complexes; the rural areas are now relatively sparsely settled. The heart of the North American cluster lies in the urban complex that lines the U.S. northeastern seaboard from Boston to Washington, D.C., and includes New York, Philadelphia, and Baltimore—the great multimetropolitan agglomeration that urban geographers refer to as **Megalopolis**.

The quantitative dominance of Eurasia's population clusters is revealed dramatically by a special map trans-

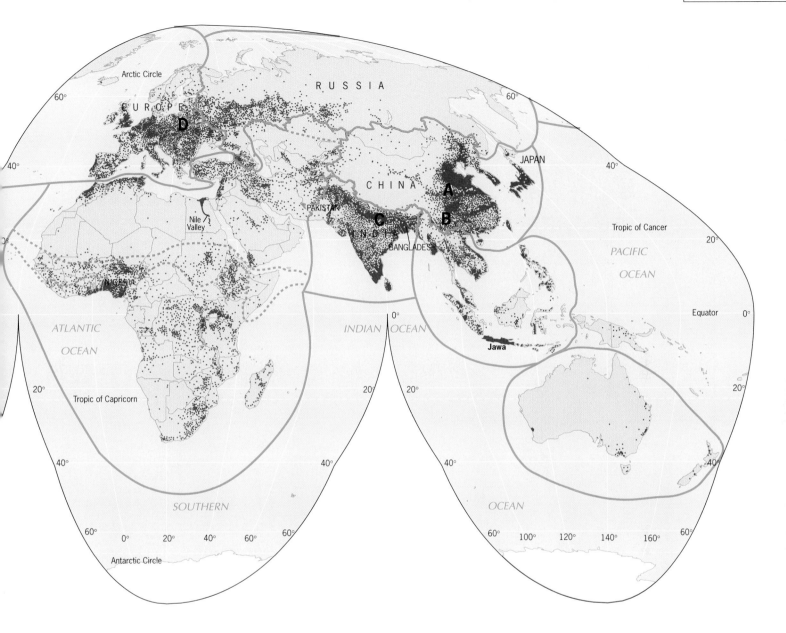

formation called a **cartogram** (Fig. G-10). Here the map of the world is redrawn, so that the area of each country reflects *not* territory but population numbers. China and India stand out because, unlike the European population cluster, their populations are not fragmented among many national political entities.

Neither Figure G-9 nor Figure G-10 can tell us how the populations they exhibit are changing. As we will see, varying rates of population growth mark the world's realms and regions; some countries' populations are growing rapidly, while populations in other countries are actually shrinking. Another key factor is **urbanization**, the percentage of the population living in cities and towns. Not only is there much geographic variation in

the level of urbanization among realms and regions, but some realms are urbanizing much more rapidly than others. Urbanization is a defining property of geographic regions, and we will explain why they differ so markedly in this respect.

REALMS, REGIONS, AND STATES

Our analysis of the world's regional geography requires data, and it is crucial to know the origin of these data. Unfortunately, we do not have a uniformly-sized grid to

CARTOGRAM OF THE WORLD'S NATIONAL POPULATIONS, 2004

1 Guatemala
2 El Salvador
3 Honduras
4 Nicaragua
5 Costa Rica
6 Panama
7 Jamaica
8 Haiti
9 Dominican Republic
10 Puerto Rico
11 Trinidad and Tobago
12 Guyana
13 Suriname
14 Ecuador
15 Bolivia

16 Paraguay
17 Uruguay
18 Chile
19 Guadeloupe
20 Barbados
21 Netherlands Antilles
22 Bahamas
23 Martinique
24 Norway
25 Sweden
26 Finland
27 Austria
28 Switzerland
29 Portugal
30 Czech Republic

31 Slovakia
32 Slovenia
33 Croatia
34 Bosnia
35 Serbia-Montenegro
36 Hungary
37 Macedonia
38 Albania
39 Bulgaria
40 Greece
41 Cyprus
42 Lebanon
43 Israel
44 Jordan
45 United Arab Emirates

46 Yemen
47 Saudi Arabia
48 Oman
49 Libya
50 Niger
51 Chad
52 Tunisia
53 Central African Republic
54 Senegal
55 Mauritania
56 Mali
57 Guinea
58 Gambia
59 Guinea-Bissau
60 Sierra Leone

61 Liberia
62 Ivory Coast
63 Burkina Faso
64 Togo
65 Benin
66 Equatorial Guinea
67 Cameroon
68 Gabon
69 Congo
70 Rwanda
71 Burundi
72 Zambia
73 Zimbabwe
74 Malawi
75 Angola

76 Namibia
77 Botswana
78 Moçambique
79 Lesotho
80 Swaziland
81 Comoros
82 Madagascar
83 Mauritius
84 Djibouti
85 Eritrea
86 Mongolia
87 Afghanistan
88 Bhutan
89 Maldives
90 Laos

91 Cambodia
92 Malaysia
93 Singapore
94 Brunei
95 Papua New Guinea
96 New Zealand
97 Belarus
98 Estonia
99 Latvia
100 Lithuania
101 Moldova
102 Georgia
103 Armenia
104 Azerbaijan
105 Turkmenistan
106 Tajikistan
107 Kyrgyzstan

FIGURE G-10

superimpose over the globe: we must depend on the world's 180-plus countries to report vital information. Irregular as the boundary framework shown on the world map (see Front Endpaper or Fig. G-11) may be, it is all we have. Fortunately, all large and populous countries are subdivided into provinces, States, or other internal entities, and their governments provide information on each of these subdivisions when they conduct their census.

Although we often refer to the political entities on the world map as *countries*, the appropriate geographic term for them is *states*. To avoid confusion with subdivisions of the same name, we capitalize internal States (as in the preceding paragraph). Thus we refer to the State of California and the State of Queensland (Australia), but to the state of Japan and the state of Argentina.

The **state** as a political, social, and economic entity has been developing for thousands of years, ever since agricultural surpluses made possible the growth of large and powerful towns that could command hinterlands and control peoples far beyond their walls. But the modern state is a relatively recent phenomenon. The boundary framework we see on the world political map today substantially came about during the nineteenth century, and the independence of dozens of former colonies (which made them states as well) occurred during the twentieth century. Today, the **European state model**—a clearly and legally defined territory inhabited by a population governed from a capital city by a representative government—prevails in the aftermath of the collapse of colonial and communist empires.

As Figures G-2 and G-11 suggest, geographic realms are mostly assemblages of states, and the borders between realms frequently coincide with the boundaries between countries—for example, between North America and Middle America along the U.S.-Mexico boundary. But a realm boundary can also cut *across* a state, as does the one between Subsaharan Africa and the Muslim-dominated realm of North Africa/Southwest Asia. Here the boundary takes on the properties of a wide transition zone, but it still divides states such as Chad and Sudan. The transformation of the margins of the former Soviet Union is creating similar cross-country transitions. Newly independent states such as Belarus (between Europe and Russia) and Kazakhstan (between Russia and Muslim Southwest Asia) lie in zones of regional change.

Most often, however, geographic realms consist of groups of states whose boundaries also mark the limits of the realms. Look at Southeast Asia, for instance. Its northern border coincides with the political boundary that separates China (a realm practically unto itself) from Vietnam, Laos, and Myanmar (Burma). The boundary between Myanmar and Bangladesh (which is part of the South

Asian realm) defines its western border. Here, the state boundary framework helps delimit geographic realms.

The global boundary framework is even more useful in delimiting regions *within* geographic realms. We shall discuss regional divisions every time we introduce a geographic realm, but an example is appropriate here. In the Middle American realm, we recognize four regions. Two of these lie on the mainland: Mexico, the giant of the realm, and Central America, which consists of the seven comparatively small countries located between Mexico and the Panama-Colombia border (which is the boundary with the South American realm). Central America often is misdefined in news reports; the correct regional definition is based on the politico-geographical framework.

To our earlier criteria of physiography, population distribution, and cultural geography, therefore, we now add political geography as a determinant of world-scale geographic regions. In doing so, we should be aware that the global boundary framework continues to change and that boundaries are created (as in 1993 between the Czech Republic and Slovakia) as well as eliminated (e.g., between former West and East Germany in 1990). But the overall system, much of it resulting from colonial and imperial expansionism, has endured, despite the predictions of some geographers that the "boundaries of imperialism" would be replaced by newly negotiated ones in the postcolonial period.

Toward the end of this book, we will discuss a recent, ominous development in boundary-making: the extension of boundaries onto and into the oceans and seas. This process has been consuming the last of the Earth's open frontiers, with uncertain consequences.

PATTERNS OF DEVELOPMENT

Finally, we turn to economic geography for criteria that allow us to place countries, regions, and realms in regional groupings based on their level of **development**. **Economic geography** focuses on spatial aspects of how people make a living; thus it deals with patterns of production, distribution, and consumption of goods and services. As with all else in this world, these patterns exhibit considerable variation. Individual states report their imports and exports, farm and factory production, and many other economic data to the United Nations and other international agencies. From such information we can determine the comparative economic well-being of the world's countries.

The World Bank, one of the agencies that monitor economic conditions, classifies countries into four groups: (1) high-income, (2) upper-middle-income,

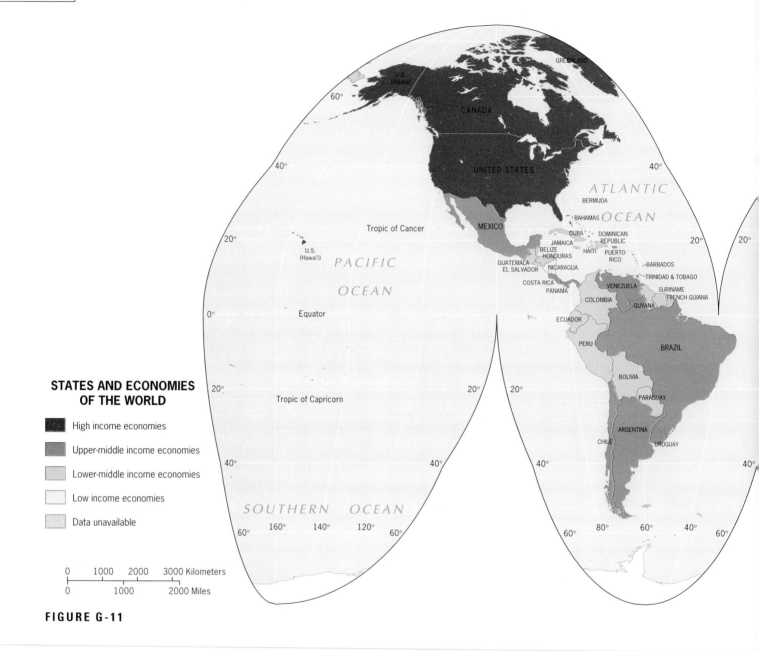

STATES AND ECONOMIES OF THE WORLD

■ High income economies

■ Upper-middle income economies

■ Lower-middle income economies

□ Low income economies

▨ Data unavailable

0   1000   2000   3000 Kilometers

0   1000   2000 Miles

FIGURE G-11

(3) lower-middle-income, and (4) low-income countries. These groupings display regional clustering when mapped (Fig. G-11). The higher-income economies are concentrated in Europe, North America, and along the western Pacific Rim, most notably in Japan and Australia. The low-income countries dominate in Africa and parts of Asia.

As Figure G-11 shows, certain geographic-realm boundaries can be discerned on this economic map, including those between North America and Middle America, as well as Australia and Southeast Asia. Within the great geographic realms, you can see some distinct regional boundaries: between Western and Eastern Europe, between Brazil and Andean South America (the region to the west), and between the oil-rich Arabian

Peninsula and the countries of the Middle East to its northwest.

Figure G-11 reveals the level of economic development among the world's countries—based, we should remind ourselves, on the political framework as the reporting grid. Until recently, economists and economic geographers used this basis to divide the world into developed countries (DCs) and underdeveloped countries (UDCs). But that distinction has lost much of its relevance, for reasons Figure G-11 cannot show. Today, the world has **core areas** where the richer countries are clustered. But many countries, whatever their level of income, display cores of development (where they often resemble the richest societies) as well as **peripheries** of poverty and underdevelopment (see box titled "Core-

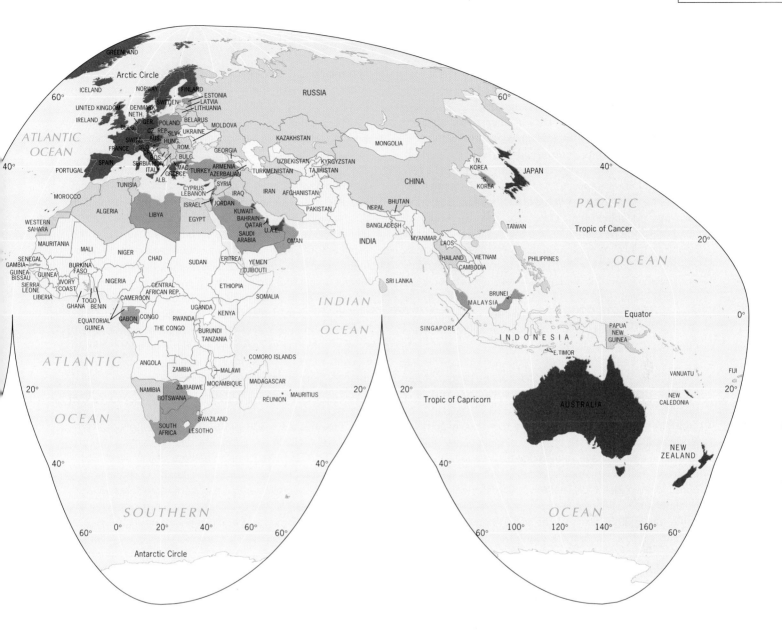

Periphery Regional Relationships"). Even comparatively wealthy countries still have areas of severe underdevelopment. Other countries, mapped as low-income economies, have given rise to bustling, skyscrapered cities whose streets are jammed with luxury automobiles and shops selling opulent goods. As the rich cores get richer and the poor peripheries get poorer, the averages reported as "national" economic statistics lose much of their geographic meaning. Perhaps an index of **regional disparity** within each economy would give us a truer picture.

It is therefore no longer appropriate to divide the world geographically into developed and underdeveloped realms. East Asia is dominated by lower-income economies, but it includes Japan, one of the world's richest states. Europe is one of the world's highest-income realms, but not in Albania, Moldova, or Ukraine.

Not long ago, countries also were categorized as being part of a First World (capitalist), Second World (socialist/communist), Third World (underdeveloped), Fourth World (severely underdeveloped), or even Fifth World (poorest). This classification, too, has lost most of its significance in the twenty-first century. The most meaningful distinction we can make is based on the notion of **advantage**: some countries have it and many others do not. Such advantage takes many forms: geographic location (to be on the coast is generally better than to be landlocked!), raw materials, government, political stability, productive skills, and much more. As the gap between the advantaged and the disadvantaged

Core-Periphery Regional Relationships

*O*urs is a world of uneven development. The advantages lie with the developed countries (DCs), and economic indicators show that the gap between the DCs and many underdeveloped countries (UDCs) is growing. How has this situation become so extreme? And what perpetuates it?

Regional contrasts in development have been attributed to many factors. Climate has been blamed (as a conditioner of human capacity), as has cultural heritage (e.g., resistance to innovation), colonial exploitation, and, more recently, neocolonialism. Obviously, the distribution of accessible natural resources and the territorial location of countries play a major part. Certain areas have always had more opportunities for interaction and exchange than others. Isolation from the mainstreams of change has always been a disadvantage. These factors, to a greater or lesser degree depending on location, have been important.

One hallmark of national spatial organization is the evolution of core areas, foci of human activity that function as the leading regions of control and change. Today we tend to associate this regional concept with a country's heartland—its largest population cluster, most productive and influential region, the area possessing the greatest centrality and accessibility that usually contains a powerful capital city.

This is nothing new. Archeologists have unearthed evidence of the earliest states, proving that city-dominated cores formed their foci. These were small urban centers, but later they were recast as city-states, culture hearths, seats of empires, and sources of modern technological revolutions. But such cores would not have developed without contributions from their surrounding areas. One of the earliest developments in ancient cities was the creation of organized armed forces to help rulers secure taxes and tribute from the people living in the countryside. Thus core and periphery (margin) became functionally tied together: the core had requirements, and the periphery supplied them.

Have modern times ended this core-periphery relationship? The answer not only is no—the situation has worsened. You can see core (have) and periphery (have-not) relationships in every country in the world. In many African countries (Kenya, Senegal, and Zimbabwe, to name examples in three different regions), the former colonial capital now lies at the focus of the national core area. Its tall buildings and government centers symbolize the concentration of privilege, power, and control. Here are the trappings of development, the paved roads and streetlights, hospitals and schools, corporate headquarters and businesses, industries and markets. Farms producing high-priced goods, especially perishables, for the city dwellers who can afford to buy them surround the capital. But farther out into the countryside, life changes. In the periphery, the core area is distant, even ominous— always a concern. How much will markets in the core pay for crops? How much will goods made there, goods that the countryside needs, cost? How high will taxes be? This core-periphery system keeps regional development contrasts high. The core may appear to be a thriving metropolis and look like a city in a developed country. But in the periphery underdevelopment reigns.

Looking at the world from this perspective, we see that core-periphery systems encompass not just individual countries but the entire globe. Today the whole of Western Europe, North America, and Japan function as core areas. Therefore, what we said about that powerful city-region in an otherwise underdeveloped country also applies, at the macroscale, to these global core areas. Power, money, influence, and decision-making organizations are concentrated in Europe, North America, and Japan.

Imagine that you, as a geographer, are asked to locate a factory (say, a textile plant) to sell products in core-area markets. Because salaries and wages in the core area are high, you look for an alternative site in the periphery, where wages are much lower. Raw materials also can be bought locally. So you advise the corporation to establish its factory in a certain country in the periphery, thereby initiating a relationship between core-based business and periphery-based producers. Now, however, the corporate management is concerned about political stability in that periphery country. Soon you are talking to govern-

states widens (and it *is* widening), the stability of the political world is at risk, and with it the advantages of the "haves" over the "have nots."

As we proceed with our investigation of the world's geographic realms, the concept of development will be examined in a regional context and from several viewpoints. Figure G-11 reflects events that began long before the Industrial Revolution in the late eighteenth and nineteenth centuries: Europe, even by the middle of the eighteenth

century, had laid the foundations for its colonial expansion. The Industrial Revolution then magnified Europe's demands for raw materials, and its manufactures increased its imperial control. Western countries thereby gained an enormous head start, while colonial dependencies continued to supply raw materials and to consume Western industrial products. Thus was born a system of international exchange and capital flow that changed little when the colonial period ended. Developing countries, well aware of ment representatives about guarantees, and the economic connection also becomes a political one.

Is all this necessarily disadvantageous to the countries in the periphery, the mainly underdeveloped countries, the countries of the disadvantaged world? After all, the core-area corporations and businesses make investments, stimulate economies, and put people to work in the countries of the periphery. However, a dependency relationship develops that soon diminishes any such advantage. The profits earned return mostly to the core area and benefit the periphery only minimally. When you see that ultra-modern high-rise hotel towering above the sandy beaches of a Caribbean island and watch its staff at work, you observe both sides of it. An investment was made, and jobs were created. But the hotel's profits go back to the multinational corporation's bank account—in New York, Tokyo, or London.

Another problem concerns the sociopolitical effects. That textile factory we mentioned needs a manager and other administrative personnel. These people tend to be (or become) part of the elite, the "upper class" in that periphery country, a kind of cadre of the world core area in the disadvantaged realms. Because they represent core-area interests, these people have divided loyalties: what is best for core-area enterprises is not always best for their home country. Such examples of core-periphery interactions abound, and the entire world is now more interconnected than many of us realize. Superimposed on the traditional cultural landscapes of the disadvantaged countries is a growing global network of interaction that reaches into even the smallest village store in the remotest part of a UDC. The numerous geographic realms and regions we will come to recognize in this book are, above all, enmeshed in a worldwide economic-geographic system—a system tightly oriented to the world's core areas.

Thus the countries of the periphery find themselves locked within a global economic system over which they have no control. Even countries that possess commodities in high demand in the core (such as the oil of the OPEC states) have difficulty converting their temporary advantage into longer-term parity with core-area powers. Countries whose income depends heavily on the export of raw materials such as strategic minerals (or single crops like bananas or sugar) are at the mercy of those who do the buying. But there are other reasons for the depressing condition of many periphery countries. Like the contemporary city, the world core area constantly siphons off the skilled and professional people from the periphery. How many doctors, engineers, and teachers from India are working in England and the United States? And how badly India now needs such trained people! Every loss is magnified—but oppressive political circumstances in periphery countries contribute to this "brain drain."

The continuing underdevelopment of periphery countries also involves traditional cultural attitudes and values. The introduction of core-area tentacles into outlying regions where traditional culture is strong may cause a reaction, perhaps a resurgence of fundamentalism. This further restricts economic change. The invasions of modern ways can be unsettling. Do not forget that the long traditions of societies in the periphery have generated large bureaucracies that are threatened by the change development entails. Occasionally, the tangible presence of an external (core) installation such as an office building, airline agency, or retail establishment is violently attacked. Such incidents symbolize ideological opposition to the intrusion they represent.

The countries of the periphery, therefore, face severe problems. They are pawns in a global economic game whose rules they cannot change. Their internal problems are intensified by the aggressive involvement of core-area interests. The way they use resources (such as whether to produce food for local consumption or to produce export crops) is strongly affected by foreign interference. They suffer far more from environmental degradation, overpopulation, and mismanagement than core-area countries do. Their inherited disadvantages have grown, not lessened, over time. The widening gaps that result between enriching cores and persistently impoverished peripheries threaten the future of the world.

their predicament, accused the developed world of perpetuating its long-term advantage through **neocolonialism**—the entrenchment of the old system under a new guise.

Symptoms of Underdevelopment

Although it is no longer appropriate to divide the world into developed and underdeveloped geographic realms, make no mistake: there still *are* regions within realms, and countries within regions, that suffer from underdevelopment. As we will discover during our global journey, underdevelopment takes many forms, displays many symptoms, and has many causes. To grasp some of the symptoms, see Table G-1 and compare, say, Bangladesh or Moçambique with Japan or Canada. Population growth rates in the periphery are higher, life expectancies shorter, urbanization rates lower, incomes smaller. As we will find, the poorer countries suffer from high infant and child

FROM THE FIELD NOTES

"February 1, 2003. A long-held hope came true today: thanks to a Brazilian intermediary I was allowed to enter and spend a day in two of Rio de Janeiro's hillslope *favelas*, an eight-hour walk through one into the other. Here live millions of the city's poor, in areas often ruled by drug lords and their gangs, with minimal or no public services, amid squalor and stench, in discomfort and danger. And yet life in the older *favelas* has become more comfortable as shacks are replaced by more permanent structures, electricity is sometimes available, water supply, however haphazard, is improved, and an informal economy brings goods and services to the residents. I stood in the doorway of a resident's single-room dwelling for this overview of an urban landscape in transition: satellite-television disks symbolize the change going on here. The often blue cisterns catch rainwater; walls are made of rough brick and roofs of corrugated iron or asbestos sheeting. No roads or automobile access, so people walk to the nearest road at the bottom of the hill. Locals told me of their hope that they will some day have legal rights to the space they occupy. During his campaign for president of Brazil, Lula da Silva suggested that long-term inhabitants should be awarded title, and in 2003 his government approved the notion. It will be complicated: as the photo shows, people live quite literally on top of one another, and mapping the chaos will not be simple (but will be made possible with Geographic Information Systems). This would allow the government to tax residents, but it would also allow residents to obtain loans based on the value their *favela* properties, and bring millions of Brazilians into the formal economy. The hardships I saw on this excursion were often dreadful, but you could sense the hope for and anticipation of a better future."

mortality rates, poor overall health and sanitation, inefficient farming, insufficient diets, overcrowded cities, and many other ills. Many countries in the periphery are trapped in a global economic system in which the export of unfinished or partially finished raw materials is their only source of outside income.

The Specter of Debt

All of us are aware of the dangers of going too deeply into debt. Borrow too much money, and payments of interest and principal may leave too little for routine needs such as food and clothing, not to mention repairs and replacement of equipment. Individuals, families, villages, and cities must devise budgets and balance incomes against expenditures.

So it is with countries. National governments, like individuals and families, must sometimes borrow to make ends meet. If a drought curtails domestic food production, a government may borrow against future oil or ore exports to pay for urgently needed staples. Should domestic harvests fail in the following year as well, further borrowing will be necessary. This means that the government's future budgets will be burdened by ever-higher "debt retirement" payments, limiting its ability to spend on schools, roads, and clinics. Should the decision be made to raise taxes to help make the payments, the result may be a slowdown in the economy. The cycle of debt is hard to escape.

All too often it is not just a food emergency or some other unavoidable problem, but a government's mismanagement that leads to excessive debt. In the postcolonial period, many governments of former colonies sought shortcuts to boost their economies, building dams, factories, and ports that did not justify their huge investments. Of course the governments and corporations of the former colonial powers were all too willing to do the building, making large profits and even lending the new governments money to build still more.

Soon, many of the civilian governments that took over from the colonial powers, weakened by such errors and losing the confidence of their citizens, fell prey to military takeovers. And the military dictatorships that replaced them needed weapons to maintain their hold on power—weapons readily sold for credit by the arms manufacturers in the wealthy core countries. During the Cold War, dictatorial regimes in the periphery became close allies of both superpowers, and proxy wars were fought in Asia, Africa, and the Americas. By the time the Cold War ended and representative government began to return to some (though not all) of these countries, they were so deeply in debt to the rich core countries that they had no prospect of recovery.

But make no mistake: rich countries also mismanage their budgets and run up serious external debts. A dramatic case, as we will see in more detail in Chapter 5, involves Argentina. A century ago, Argentina was one of the world's ten richest countries, ahead of France and

Germany. But since then its economy has lurched from one crisis to the next, a sequence of booms and busts that had political as well as economic consequences. Just a decade ago, Argentina enjoyed one of its boom times, its economy achieving robust growth, held up to the region and the globalizing world as a model of productive free-market reforms. Then mismanagement, corruption, and bad luck in the form of financial crises in Mexico and neighboring Brazil combined to bring the economy to a halt. By the middle of 2002, Argentina's debt amounted to U.S. $155 billion—more than $4000 per person in the entire population.

The foreign debt of a country is usually measured against its Gross Domestic Product (GDP), the total value of all goods and services produced within the country by a state's economy during a year. (The gross *national* product, or GNP, adds citizens' income from foreign investments and other external sources.) Argentina's foreign debt in late 2002 was 55 percent of its GDP, but numerous countries had even higher debt ratios (Nigeria, 84 percent; Russia, 64 percent, Zimbabwe, 59 percent; Malaysia, 57 percent). Argentina's most recent debacle became notorious because of its enormous assets and potential, a combination of advantages of which other debt-ridden countries can only dream.

The issue of national debt can have serious geographic implications. We live in an era of globalization and regionalization—of ever-intensifying interactions in many spheres, notably the economic. When a country as substantial and important as Argentina encounters a debt crisis, free-trade accords are at risk and neighbors fear infection. That is one reason why, when NAFTA-neighbor Mexico faced a fiscal crisis in 1995, the United States acted quickly to help the Mexican economy over its plight. Argentines also note that Brazil got more help than they did when the Brazilian economy stalled and devaluation of its currency took place. But Brazil has nearly half of South America's population and is the cornerstone of another regional free-trade organization (see Chapter 5). Meanwhile, Argentina languishes as its model fails.

National debt also plays a key role in the economic geography of Europe. As we will note in Chapter 1, the countries that constitute the European Union have agreed to limit their public debt to 3 percent of GDP—but not all of them are making it, leading to some serious disharmony. In 2002, when 12 of the European Union's 15 member-states replaced their own currencies with the new Euro under the terms of the European Monetary Union, several countries expressed their unhappiness over the inclusion of Italy, whose financial affairs were never in order and whose ability to achieve a 3 percent debt ceiling was in doubt. In early 2003, it was clear that not just Italy but several others among those complainers were also unable to achieve the 3 percent limit, including one of the loudest critics, Germany. Now the European Union is set for rapid expansion, and the debt issue will roil the debate (see Chapter 1).

In the international arena, the debt issue arouses strong passions. Poorer countries appeal to the richer ones to cancel their debts; the richer countries want the poorer ones to reform their economies, change their spending practices, and democratize their governments. Meanwhile, the global economic system sustains the gap between creditors and debtors.

Globalization

In August 1999 a Frenchman named José Bové, leader of the so-called Peasant Federation, became a national hero. He achieved his stardom because a group of his sympathizers destroyed a McDonald's restaurant being built in the town of Millau; when Mr. Bové came to trial ten months later, some 40,000 supporters rallied to his cause. To Mr. Bové, McDonald's is a symbol of **globalization**, a process he views as the Americanization of France's traditions, and trashing it was a matter of cultural self-defense. In November 1999 he was in Seattle, working to disrupt a meeting of the World Trade Organization (WTO). As globalization proceeds, resistance to it mounts, not just in France but even in the United States itself.

Why does an apparently beneficial process arouse such heated passions? Globalization breaks down barriers to international trade, stimulates commerce, brings jobs to remote places, and promotes social, cultural, political, and other kinds of exchanges. High-tech workers in India are employed by computer firms based in California. Japanese cars are assembled in Thailand. American shoes are made in China. Thousands of McDonald's restaurants serve their familiar (and standard) menus to customers from Tokyo to Tel Aviv.

Opponents of globalization argue that the negatives far outweigh the positives. Those high-tech workers in India are paid a fraction of what their Californian counterparts earn, so that many of them will want to leave India for better pay abroad at a time when the country has great need of their skills. The wages and working conditions for the Chinese shoe-factory workers are far below those acceptable to American labor. And as for those McDonald's restaurants, they promote the consumption of junk food over better traditional fare. Opposition to globalization in the United States often centers on the loss of American jobs when corporations move their factories to foreign countries where labor is cheaper.

In a way, the current globalization process is a revolution—but it is not the first of its kind. The first "globalization revolution" occurred during the nineteenth and early twentieth centuries, when Europe's colonial expansion spread ideas, inventions, products, and habits around the world. Colonialism transformed the world as the European powers built cities, transport networks, dams, irrigation systems, power plants, and other facilities, often with devastating impact on local traditions, cultures, and economies. From goods to games (soap to soccer) people in much of the world started doing similar things. The largest of all colonial empires, that of Britain, made English a worldwide language, a key element in the current, second globalization process.

The present globalization is even more revolutionary than the colonial phase because it is driven by more modern, higher-speed communications. When the British colonizers planned the construction of their ornate Victorian government and public buildings in (then) Bombay, the architectural drawings had to be prepared in London and sent by boat to India. When the Chinese government in the 1980s decided to create a Manhattan-like commercial district on the riverfront in Shanghai, the plans were drawn in the United States, Japan, and Western Europe and transmitted to Shanghai via the Internet. One container ship carrying products from China to the American market hauls more cargo than a hundred colonial-era boats.

And, as the pages that follow will show frequently, the world's national-political boundaries are becoming increasingly porous. Economic alliances enable manufacturers to send raw materials and finished products across borders that once inhibited such exchanges. Groups of countries forge unions whose acronyms (NAFTA, Mercosur) stand for freer trade. The ultimate goal of the World Trade Organization is to lower the remaining trade barriers the world over, boosting not just regional commerce but also global trade.

As with all revolutions, the overall consequences of the present globalization process are uncertain. Critics underscore that one of its outcomes is a growing gap between rich and poor, a polarization of wealth that will destabilize the world. Core-periphery contrasts are intensified, not lessened, by globalization as the poor in peripheral societies are exploited by core-based corporations. Proponents argue that, as with the Industrial Revolution, it will take time for the benefits to spread—but that globalization's ultimate effects will be advantageous to all.

Indeed, the world is functionally shrinking, and we will find evidence for this throughout this book. But the "global village" still retains its distinctive neighborhoods, and two revolutionary globalizations have failed to erase their particular properties. In the chapters that follow we use the vehicle of geography to visit and investigate them.

THE REGIONAL FRAMEWORK

At the beginning of this Introduction, we outlined a map of the great geographic realms of the world (Fig. G-2). We then addressed the task of dividing these realms into regions, and we used criteria ranging from physical geography to economic geography. The result is Figure G-12. Before we begin our survey, here is a summary of the 12 geographic realms and their regional components.

Europe (1)

Territorially small and politically fragmented, Europe remains disproportionately influential in global affairs. A core geographic realm, Europe has five regions: Western Europe, the British Isles, Northern (Nordic) Europe, Mediterranean Europe, and Eastern Europe.

Russia (2)

Territorially enormous and politically unified, Russia was the dominant force in the former Soviet Union that disbanded in 1991. Undergoing a difficult transition from dictatorship to democracy and from communism to capitalism, Russia is geographically complex and changing. We define four regions: the Russian Core, the Eastern Frontier, Siberia, and the Far East.

North America (3)

Another realm in the global core, North America consists of the United States and Canada. We identify nine regions: the North American Core, the Maritime Northeast, French Canada, the Continental Interior, the South, the Southwest, the Western Frontier, the Northern Frontier, and the Pacific Hinge.

Middle America (4)

Nowhere in the world is the contrast between core and periphery as sharply demarcated as it is between North and Middle America. This small, fragmented realm clearly divides into four regions: Mexico, Central America, and the Greater Antilles and Lesser Antilles of the Caribbean.

South America (5)

The continent of South America also defines a geographic realm in which Iberian (Spanish and Portuguese) influences dominate the cultural geography but Amerindian imprints survive. We recognize four regions: the North, composed of Caribbean-facing states; the Andean West, with its strong Amerindian influences; the Southern Cone; and Brazil, the realm's giant.

Subsaharan Africa (6)

Between the African Transition Zone in the north and the southernmost Cape of South Africa lies Subsaharan Africa. The realm consists of five regions: Southern Africa, East Africa, Equatorial Africa, West Africa, and the African Transition Zone.

North Africa/Southwest Asia (7)

This vast geographic realm has several names, extending as it does from North Africa into Southwest and, indeed, Central Asia. Some geographers call it *Naswasia* or *Afrasia*. There are six regions: Egypt and the Lower Nile Basin plus the Maghreb in North Africa; the Middle East, the Arabian Peninsula, the Empire States, and Turkestan in Southwest Asia.

South Asia (8)

Physiographically one of the most clearly defined geographic realms, South Asia has a complex cultural geography. It consists of five regions: India at the center; Pakistan to the west; Bangladesh to the east; the mountainous North; and the peninsular South, which includes island Sri Lanka.

East Asia (9)

The vast East Asian geographic realm extends from the deserts of Central Asia to the tropical coasts of the South China Sea and from Japan to Xizang (Tibet). We identify five regions: China Proper, including North Korea; Xizang (Tibet) in the southwest; desert Xinjiang in the west; Mongolia in the north; and the Jakota Triangle (Japan, South Korea, and Taiwan) in the east.

Southeast Asia (10)

Southeast Asia is a varied mosaic of natural landscapes, cultures, and economies. Influenced by India, China, Europe, and the United States, it includes dozens of religions and hundreds of languages plus economies representing both core and periphery. Physically, Southeast Asia consists of a peninsular mainland and an arc consisting of thousands of islands. The two regions (Mainland and Insular) are based on this distinction.

Austral Realm (11)

Australia and its neighbor New Zealand form the Austral geographic realm by virtue of continental dimensions, insular separation, and dominantly Western cultural heritage. The four regions are defined by physical as well as cultural geography: a highly urbanized, two-part core and a vast, desert-dominated interior in Australia; and

FROM THE FIELD NOTES

"It was an equatorial day here today, in hot and humid Singapore. Walked the three miles to the Sultan Mosque this morning and observed the activity arising from the Friday prayers, then sat in the shade of the large ficus tree on the corner of Arab Street and watched the busy pedestrian and vehicular traffic. An Indian man saw me taking notes and sat down beside me. For some time he said nothing, then asked whether I was writing a novel. 'If only I could!' I answered, and explained that I was noting locations and impressions for photos and text in a geography book. He said that he was Hindu but liked living in this area because some of what he called "Old Singapore" still survived. 'No problems between Hindus and Muslims here," he said. 'The police maintain strict security here around the Mosques, especially after September 2001. The government tolerates all religions, but it doesn't tolerate religious conflict.' He offered to show me the Hindu shrine where he and his family joined others from Singapore's Hindu community. As we walked down Arab Street, he pointed skyward. 'That's the future,' he said. 'The symbol of globalization. Do you mention it in your book? It's in the papers every day, supposedly a good thing for Singapore. Well, maybe. But a lot of history has been lost to make space for what you see here.' I took the photo, and promised to mention his point."

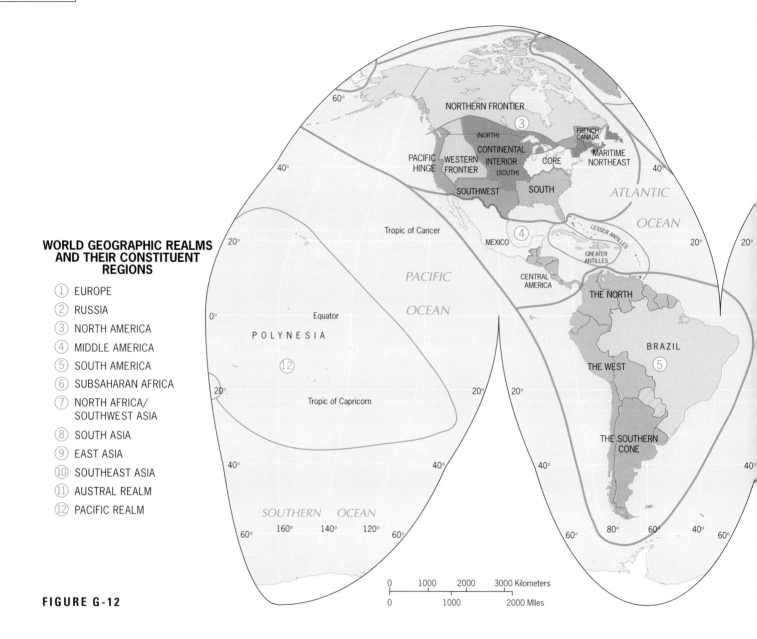

WORLD GEOGRAPHIC REALMS AND THEIR CONSTITUENT REGIONS

1 EUROPE
2 RUSSIA
3 NORTH AMERICA
4 MIDDLE AMERICA
5 SOUTH AMERICA
6 SUBSAHARAN AFRICA
7 NORTH AFRICA/ SOUTHWEST ASIA
8 SOUTH ASIA
9 EAST ASIA
10 SOUTHEAST ASIA
11 AUSTRAL REALM
12 PACIFIC REALM

FIGURE G-12

two main islands in New Zealand that exhibit considerable geographic contrast.

Pacific Realm (12)

The vast Pacific Ocean, larger than all the landmasses combined, contains tens of thousands of islands large and small. Dominant cultural criteria warrant three regions: Melanesia, Micronesia, and Polynesia.

THE PERSPECTIVE OF GEOGRAPHY

As this introductory chapter demonstrates, our world regional survey is no mere description of places and areas. We have combined the study of realms and regions with a look at geography's ideas and concepts—the notions, generalizations, and basic theories that make the discipline what it is. We continue this method in the chapters ahead so that we will become better acquainted with the world and with geography. By now you are aware that geography is a wide-ranging, multifaceted discipline. It is often described as a social science, but that is only half the story: in fact, geography straddles the divide between the social and the physical (natural) sciences. Many of the ideas and concepts you will encounter have to do with the many interactions between human societies and natural environments.

Regional geography allows us to view the world in 51 an all-encompassing way. As we have seen, regional geography borrows information from many sources to create an overall image of our divided world. Those sources are not random. They form topical or **systematic geo-** 52

graphy. Research in the systematic fields of geography makes our world-scale generalizations possible. As Figure G-13 shows, these systematic fields relate closely to those of other disciplines. Cultural geography, for example, is allied with anthropology; it is the spatial perspective that distinguishes cultural geography. Economic geography focuses on the spatial dimensions of economic activity; political geography concentrates on the spatial imprints of political behavior. Other systematic fields include historical, medical, behavioral, environmental, agricultural, and coastal geography. We will also draw on information from biogeography, marine geography, population geography, and climatology (as we did earlier in this chapter).

These systematic fields of geography are so named because their approach is global, not regional. Take the geographic study of cities, urban geography. Urban-

ization is a worldwide process, and urban geographers can identify certain human activities that all cities in the world exhibit in one form or another. But cities also display regional properties. The model Japanese city is quite distinct from, say, the African city. Regional geography, therefore, borrows from the systematic field of urban geography, but it injects this regional perspective.

In the following chapters we call upon these systematic fields to give us a better understanding of the world's realms and regions. As a result, you will gain insights into the discipline of geography as well as the regions we investigate. This will prove that geography is a relevant and practical discipline when it comes to comprehending, and coping with, our fast-changing world (see box titled "What Do Geographers Do?" and Appendix B for a discussion of career opportunities).

**THE RELATIONSHIP BETWEEN
REGIONAL AND SYSTEMATIC GEOGRAPHY**

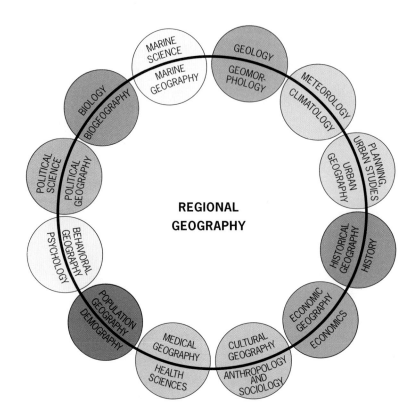

FIGURE G-13

What Do Geographers Do?

A systematic spatial perspective and an interest in regional study are the unifying themes and enthusiasms of geography. Geography's practitioners include physical geographers, whose principal interests are the study of geomorphology (land surfaces), research on climate and weather, vegetation and soils, and the management of water and other natural resources. There also are geographers whose research and teaching concentrate on the ecological interrelationships between the physical and human worlds. They study the impact of humankind on our globe's natural environments and the influences of the environment (including such artificial contents as air and water pollution) on human individuals and societies.

Other geographers are regional specialists, who often concentrate their work for governments, planning agencies, and multinational corporations on a particular region of the world. Still other geographers—who now consti-

tute the largest group of practitioners—are devoted to topical or systematic subfields such as urban geography, economic geography, and cultural geography (see Fig. G-13). They perform numerous tasks associated with the identification and resolution (through policy-making and planning) of spatial problems in their specialized areas. And, as in the past, there remain many geographers who combine their fascination for spatial questions with technical know-how.

Computerized cartography, geographic information systems, remote sensing, and even environmental engineering are among the specializations listed by the 10,000-plus professional geographers of North America. In Appendix B, you will find considerable information on the discipline, how one trains to become a geographer, and the many exciting career options that are open to the young professional.

	Land Area (sq mi)	Population 2004 (Millions)	Population 2025 (Millions)	Population Density		Birth Rate	Death Rate	Natural Increase %	Doubling Time (years)	Infant Mortality Per 1000 (births)	Life Expectancy		Percent Urban Pop	Per Capita GNP ($US)
				Arithmetic	Physiological						Male (years)	Female (years)		
World	**51,789,601**	**6,365.1**	**7,859.0**	**123**	**1,286**	**21**	**9**	**1.2**	**58**	**54**	**65**	**69**	**47**	**$7,140**
Europe	**2,284,860**	**583.2**	**588.7**	**255**	**901**	**10**	**11**	**−0.1**		**8**	**70**	**78**	**73**	**$16,150**
Albania	11,100	3.2	4.1	286	1,426	17	5	1.2	58	12	72	76	46	$3,600
Austria	32,378	8.2	8.4	253	1,512	9	9	0.0		5	75	81	54	$26,330
Belarus	80,154	9.8	9.4	122	422	9	14	−0.5		9	63	75	70	$7,550
Belgium	11,787	10.3	10.8	876	3,644	11	10	0.1	700	5	75	82	97	$27,470
Bosnia	19,741	3.4	3.6	174	1,243	12	8	0.4	175	7	70	77	54	$6,050
Bulgaria	42,822	7.7	6.6	180	421	9	14	−0.5		13	68	75	69	$5,560
Croatia	21,830	4.3	4.1	196	944	10	12	−0.2		7	70	77	54	$7,960
Cyprus	3,571	0.9	1.0	255	2,104	12	7	0.5	140	5	75	80	66	$20,780
Czech Republic	30,448	10.3	10.3	337	840	9	11	−0.2		4	72	78	77	$13,780
Denmark	16,637	5.4	5.9	325	550	12	11	0.1	700	5	75	79	85	$27,250
Estonia	17,413	1.4	1.2	80	340	9	14	−0.5		9	65	76	69	$9,340
Finland	130,560	5.2	5.3	40	554	11	10	0.1	700	4	74	81	61	$24,570
France	212,934	60.0	64.2	282	856	13	9	0.4	175	5	76	83	74	$24,420
Germany	137,830	82.4	78.1	598	1,851	9	10	−0.1		4	75	81	86	$24,920
Greece	50,950	11.0	10.4	216	1,163	10	10	0.0		6	76	81	59	$16,860
Hungary	35,919	10.0	9.2	280	551	10	13	−0.3		9	67	76	64	$11,990
Iceland	39,768	0.3	0.3	8		15	6	0.9	78	3	78	81	93	$28,710
Ireland	27,135	3.8	4.5	142	1,112	14	8	0.6	117	6	74	79	58	$25,520
Italy	116,320	58.1	57.5	499	1,651	9	9	0.0		5	77	83	90	$23,470
Latvia	24,942	2.3	2.2	91	351	8	14	−0.6		11	65	76	68	$7,070
Liechtenstein	62	0.1	0.04	489		12	7	0.5	140	8			23	
Lithuania	25,174	3.5	3.5	138	398	9	12	−0.3		8	68	78	67	$6,980
Luxembourg	999	0.5	0.6	505	2,100	13	9	0.4	175	5	75	81	88	$45,470
Macedonia	9,927	2.0	2.2	204	861	15	9	0.6	117	12	70	75	59	$5,020
Malta	124	0.4	0.4	3,245		11	8	0.3	233	6	74	80	91	$16,530
Moldova	13,012	4.3	4.5	330	638	9	10	−0.1		18	64	71	46	$2,230
Netherlands	15,768	16.2	17.7	1,029	4,955	13	9	0.4	175	5	76	81	62	$25,850
Norway	125,050	4.5	5.0	36	1,273	13	10	0.3	233	4	76	81	74	$29,630
Poland	124,087	38.6	38.6	311	699	10	10	0.0		8	70	78	62	$9,000
Portugal	35,514	10.4	9.7	294	1,131	12	10	0.2	350	6	73	80	48	$16,990
Romania	92,042	22.3	20.6	242	612	10	12	−0.2		18	67	74	55	$6,360
Serbia-Montenegro	39,448	10.7	10.7	272	1,329	12	11	0.1	700	13	70	75	52	$5,950
Slovakia	18,923	5.4	5.2	285	937	10	10	0.0		9	69	77	57	$11,040
Slovenia	7,819	2.0	2.0	256	2,137	9	9	0.0		5	72	79	50	$17,310
Spain	195,363	41.4	44.3	212	715	10	9	0.1	700	5	76	83	64	$19,260
Sweden	173,730	8.9	9.1	51	799	10	11	−0.1		3	77	82	84	$23,970
Switzerland	15,942	7.3	7.6	460	4,790	10	8	0.2	350	5	77	83	68	$30,450
Ukraine	233,089	47.5	45.1	204	366	8	15	−0.7		12	62	74	67	$3,700
United Kingdom	94,548	60.3	64.8	638	2,586	11	10	0.1	700	6	75	80	90	$23,550

| | Land Area (sq mi) | Population 2004 (Millions) | Population 2025 (Millions) | Population Density | | Birth Rate | Death Rate | Natural Increase % | Doubling Time (years) | Infant Mortality Per 1000 (births) | Life Expectancy | | Percent Urban Pop | Per Capita GNP ($US) |
				Arithmetic	Physiological						Male (years)	Female (years)		
Russia	**6,592,819**	**143.8**	**130.2**	**22**	**274**	**9**	**16**	**−0.7**		**15**	**59**	**72**	**73**	**$8,010**
Armenia	11,506	3.8	3.7	332	2,059	8	6	0.2	350	16	70	74	67	$2,580
Azerbaijan	33,436	8.3	10.2	249	1,386	14	6	0.8	88	13	69	75	51	$2,740
Georgia	3,571	4.4	3.6	1,232	182	9	9	0.0		18	69	77	56	$2,680
North America	**7,699,508**	**322.2**	**382.0**	**42**	**358**	**14**	**9**	**0.5**	**140**	**6**	**74**	**80**	**75**	**$33,410**
Canada	3,849,670	31.6	36.0	8	164	11	7	0.4	175	5	76	81	78	$27,170
United States	3,717,796	290.9	346.0	78	412	15	9	0.6	117	7	74	80	75	$34,100
Middle America	**1,048,105**	**184.2**	**233.0**	**176**	**1,445**	**26**	**6**	**2.0**	**35**	**31**	**70**	**75**	**66**	
Antigua and Barbuda	170	0.1	0.1	607	2,867	22	6	1.6	44	17	68	73	37	$10,000
Bahamas	5,359	0.3	0.4	57	7,894	18	5	1.3	54	16	70	75	84	$16,400
Barbados	166	0.3	0.3	1,833	4,111	15	8	0.7	100	13	70	76	38	$15,020
Belize	8,865	0.3	0.4	35	1,784	29	6	2.3	30	21	70	74	49	$5,240
Costa Rica	19,730	4.0	5.2	204	3,413	21	4	1.7	41	11	75	79	45	$7,980
Cuba	42,803	11.4	11.8	267	1,111	12	7	0.5	140	6	74	78	75	
Dominica	290	0.1	0.1	350	3,763	16	8	0.8	88	16	71	76	71	
Dominican Rep.	18,815	9.2	12.1	488	2,336	26	5	2.1	33	47	67	71	61	$5,710
El Salvador	8,124	6.9	9.3	850	3,198	30	7	2.3	30	30	67	73	58	$4,410
Grenada	131	0.1	0.1	782	6,828	19	7	1.2	58	14			34	$6,960
Guadeloupe	660	0.5	0.5	774	5,215	17	6	1.1	64	8	73	80	48	
Guatemala	42,042	12.8	19.8	305	2,548	36	7	2.9	24	41	63	69	39	$3,770
Haiti	10,714	7.4	9.6	687	3,471	33	15	1.8	39	80	48	51	35	$1,470
Honduras	43,278	7.1	9.6	163	1,091	33	6	2.7	26	42	64	68	46	$2,400
Jamaica	4,243	2.7	3.3	631	4,555	20	5	1.5	47	24	73	77	50	$3,440
Martinique	425	0.4	0.4	956	12,701	14	6	0.8	88	7	76	82	93	
Mexico	756,052	106.0	131.7	140	1,199	26	5	2.1	33	25	73	78	74	$8,790
Netherlands Antilles	309	0.2	0.2	658	6,774	14	6	0.8	88	12	73	79	70	
Nicaragua	50,193	5.7	8.6	114	1,355	34	5	2.9	24	40	66	71	57	$2,080
Panama	29,158	3.0	3.8	103	1,499	23	4	1.9	37	17	72	77	62	$5,680
Puerto Rico	3,456	4.0	4.1	1,147	29,137	15	7	0.8	88	11	71	80	71	
Saint Lucia	239	0.2	0.2	857	12,802	18	6	1.2	58	13	69	74	30	$5,400
St. Vincent and the Grenadines	151	0.1	0.1	678	5,121	20	8	1.2	58	22	71	74	44	$5,210
Trinidad and Tobago	1,981	1.3	1.4	664	4,385	14	8	0.6	117	17	68	73	72	$8,220
South America	**6,898,579**	**365.4**	**463.0**	**53**	**1,049**	**22**	**6**	**1.6**	**44**	**29**	**67**	**74**	**79**	**$7,070**
Argentina	1,073,514	37.3	47.2	35	392	19	8	1.1	64	18	70	77	90	$12,050
Bolivia	424,162	9.2	13.2	22	1,100	32	9	2.3	30	61	61	64	64	$2,360
Brazil	3,300,154	178.0	219.0	54	1,090	19	7	1.2	58	33	65	73	81	$7,300
Chile	292,135	16.0	19.5	55	1,105	18	6	1.2	58	12	73	79	86	$9,100
Colombia	439,734	45.2	59.7	103	2,819	22	6	1.6	44	21	68	74	71	$6,060
Ecuador	109,483	13.6	18.5	124	2,117	28	6	2.2	32	30	68	73	61	$2,910
French Guiana	34,749	0.2	0.3	6	613	26	5	2.1	33	17	72	79	79	
Guyana	83,000	0.8	0.7	10	543	24	8	1.6	44	56	59	67	36	$3,670

	Land Area (sq mi)	Population 2004 (Millions)	Population 2025 (Millions)	Population Density Arithmetic	Population Density Physiological	Birth Rate	Death Rate	Natural Increase %	Doubling Time (years)	Infant Mortality Per 1000 (births)	Life Expectancy Male (years)	Life Expectancy Female (years)	Percent Urban Pop	Per Capita GNP ($US)
Paraguay	157,046	6.3	10.1	40	686	31	5	2.6	27	37	69	73	54	$4,450
Peru	496,224	27.7	35.7	56	1,870	26	7	1.9	37	33	66	71	72	$4,660
Suriname	63,039	0.4	0.5	7	687	24	7	1.7	41	26	68	74	69	$3,480
Uruguay	68,498	3.4	3.8	50	728	16	10	0.6	117	14	71	79	92	$8,880
Venezuela	352,143	26.1	34.8	74	1,913	24	5	1.9	37	20	71	77	87	$5,740
Subsaharan Africa	**8,366,638**	**683.8**	**1,025.0**	**82**	**1,360**	**39**	**17**	**2.2**	**32**	**88**	**49**	**50**	**30**	**$1,540**
Angola	481,351	13.4	28.2	28	1,394	48	20	2.8	25	122	44	47	32	$1,180
Benin	43,483	7.0	12.0	161	1,259	41	12	2.9	24	85	53	56	39	$980
Botswana	224,606	1.6	1.2	7	372	31	22	0.9	78	60	39	40	49	$7,170
Burkina Faso	105,792	13.4	21.6	126	974	47	17	3.0	23	105	46	47	15	$970
Burundi	10,745	7.0	12.4	651	1,607	43	21	2.2	32	116	40	41	8	$580
Cameroon	183,568	17.0	24.7	93	3,157	37	12	2.5	28	77	54	56	48	$1,590
Cape Verde Is.	1,556	0.5	0.7	341	3,014	37	7	3.0	23	31	66	72	53	$4,760
Central African Republic	240,533	3.7	4.9	16	519	38	18	2.0	35	98	42	46	39	$1,160
Chad	495,753	9.6	18.2	19	659	49	16	3.3	21	103	49	53	21	$870
Comoros Is.	861	0.6	1.1	746	2,040	47	12	3.5	20	86	54	59	29	$1,590
Congo	132,046	3.4	6.3	26	2,574	44	14	3.0	23	72	49	53	41	$570
Congo, The	905,531	58.7	106.0	65	2,234	46	15	3.1	23	102	47	51	29	$680
Djibouti	8,958	0.7	0.8	81		39	19	2.0	35	117	42	44	83	$690
Equatorial Guinea	10,830	0.5	0.9	49	975	43	17	2.6	27	108	49	53	37	$5,600
Eritrea	45,045	4.8	8.3	106	1,022	43	12	3.1	23	77	53	58	16	$960
Ethiopia	426,371	71.1	117.6	167	1,535	40	15	2.5	28	97	51	53	15	$660
Gabon	103,347	1.2	1.4	12	1,245	32	16	1.6	44	57	49	51	73	$5,360
Gambia	4,363	1.6	2.7	364	2,262	42	13	2.9	24	82	51	55	37	$1,620
Ghana	92,100	21.1	26.5	229	2,000	32	10	2.2	32	56	56	59	37	$1,910
Guinea	94,927	8.9	14.1	93	4,668	45	18	2.7	26	119	47	48	26	$1,930
Guinea-Bissau	13,946	1.4	2.2	98	1,139	45	20	2.5	28	126	43	46	22	$710
Ivory Coast	124,502	16.8	25.6	135	1,710	36	16	2.0	35	95	44	47	46	$1,500
Kenya	224,081	32.4	33.3	144	2,103	34	14	2.0	35	74	47	49	20	$1,010
Lesotho	11,718	2.3	2.4	195	1,771	33	15	1.8	39	84	50	52	16	$2,590
Liberia	43,000	3.5	6.0	82	9,448	49	17	3.2	22	139	49	52	45	$3,950
Madagascar	226,656	17.6	30.8	78	1,958	34	14	2.0	35	74	47	49	20	$820
Malawi	45,745	11.4	12.8	250	926	46	22	2.4	29	104	37	38	20	$600
Mali	478,838	12.0	21.6	25	1,272	49	19	3.0	23	113	46	48	26	$780
Mauritania	395,954	2.7	5.1	7	683	34	14	2.0	35	74	53	55	55	$1,630
Mauritius	788	1.2	1.4	1,550	3,117	16	7	0.9	78	14	68	75	43	$9,940
Moçambique	309,494	20.4	20.6	66	1,684	43	23	2.0	35	135	38	37	28	$800
Namibia	318,259	1.9	2.0	6	583	35	20	1.5	47	72	44	41	27	$6,410
Niger	489,189	12.4	25.7	25	847	55	20	3.5	20	123	45	46	17	$740
Nigeria	356,668	129.0	198.1	371	1,981	41	17	2.4	29	75	52	52	36	$800
Réunion	969	0.7	0.9	744	4,242	20	5	1.5	47	8	70	79	73	
Rwanda	10,170	7.7	8.0	759	2,320	42	21	2.1	33	107	39	40	5	$930

	Land Area (sq mi)	Population 2004 (Millions)	Population 2025 (Millions)	Population Density Arithmetic	Population Density Physiological	Birth Rate	Death Rate	Natural Increase %	Doubling Time (years)	Infant Mortality Per 1000 (births)	Life Expectancy Male (years)	Life Expectancy Female (years)	Percent Urban Pop	Per Capita GNP ($US)
São Tomé and Principe	371	0.2	0.3	577	35,708	43	8	3.5	20	50	64	67	44	
Senegal	75,954	10.4	16.5	137	1,169	38	12	2.6	27	68	52	55	43	$1,480
Seychelles	174	0.1	0.1	587	25,553	18	7	1.1	64	10	67	73	63	$7,350
Sierra Leone	27,699	5.9	10.6	212	3,028	49	25	2.4	29	153	38	40	37	$480
Somalia	246,201	8.3	14.9	34	1,705	48	19	2.9	24	126	45	48	28	
South Africa	471,444	44.5	35.1	94	943	25	15	1.0	70	45	50	52	54	$9,160
Swaziland	6,703	1.1	1.4	171	1,579	41	20	2.1	33	109	40	41	25	$4,600
Tanzania	364,900	39.2	59.8	108	3,834	40	13	2.7	26	99	51	53	22	$520
Togo	21,927	5.6	7.6	256	703	40	11	2.9	24	80	53	57	31	$1,410
Uganda	93,066	26.2	48.0	282	1,359	48	18	3.0	23	88	42	44	16	$1,210
Zambia	290,583	10.4	14.3	36	518	42	22	2.0	35	95	37	37	38	$750
Zimbabwe	105,873	12.5	10.3	118	1,197	29	20	0.9	78	65	39	36	32	$2,550
North Africa/ SW Asia	**7,459,064**	**530.6**	**730.0**	**71**	**978**	**27**	**8**	**1.9**	**37**	**48**	**65**	**68**	**53**	
Afghanistan	251,772	29.2	45.9	116	965	43	19	2.4	29	154	46	44	22	
Algeria	919,591	32.5	43.0	35	1,180	23	5	1.8	39	54	68	71	49	$5,040
Bahrain	266	0.7	1.7	2,733	242,284	22	3	1.9	37	9	73	75	87	$14,410
Egypt	386,660	74.1	96.1	192	9,638	27	7	2.0	35	44	65	68	43	$3,670
Iran	630,575	67.1	81.3	106	1,061	17	6	1.1	64	32	68	70	66	$5,910
Iraq	169,236	24.8	41.2	147	1,223	35	10	2.5	28	103	56	59	68	
Israel	8,131	6.8	9.3	836	5,000	21	6	1.5	47	5	76	81	91	$19,330
Jordan	34,444	5.5	8.7	161	4,043	28	5	2.3	30	31	69	71	79	$3,950
Kazakhstan	1,049,151	14.9	14.7	14	121	15	10	0.5	140	20	60	71	56	$5,490
Kuwait	6,880	2.4	3.9	354		32	3	2.9	24	9	74	78	100	$18,690
Kyrgyzstan	76,641	5.1	6.5	67	989	20	7	1.3	54	23	65	72	35	$2,540
Lebanon	4,015	4.4	5.4	1,101	5,263	21	7	1.4	50	33	72	75	88	$4,550
Libya	679,359	5.7	8.3	8	833	28	4	2.4	29	30	73	77	86	
Morocco	172,413	30.8	40.5	179	852	25	6	1.9	37	50	67	71	55	$3,450
Oman	82,031	2.8	5.1	34		33	4	2.9	24	17	72	75	72	
Palestinian Territories	2,417	3.8	7.4	1,554		40	4	3.6	19	26	71	74	57	
Qatar	4,247	0.6	0.8	149	14,717	31	4	2.7	26	12	70	75	91	
Saudi Arabia	829,996	25.4	40.9	31	1,531	35	6	2.9	24	19	71	73	83	$11,390
Sudan	967,494	34.2	49.6	35	745	36	12	2.4	29	82	55	57	27	$1,520
Syria	71,498	18.1	26.5	253	909	31	6	2.5	28	24	70	70	50	$3,340
Tajikistan	55,251	6.5	7.8	117	1,992	19	4	1.5	47	19	66	71	27	$1,090
Tunisia	63,170	10.0	11.6	159	879	17	6	1.1	64	26	70	74	63	$6,070
Turkey	299,158	69.3	85.0	232	729	22	7	1.5	47	35	67	72	66	$7,030
Turkmenistan	188,456	5.8	7.2	31	1,058	19	5	1.4	50	25	63	70	44	$3,800
United Arab Emirates	32,278	3.6	4.5	112		17	2	1.5	47	19	72	77	78	$19,410
Uzbekistan	172,741	26.3	37.2	152	1,826	22	5	1.7	41	20	68	73	38	$2,360
Western Sahara	97,344	0.3	0.4	3		46	17	2.9	24	140	46	48	95	
Yemen	203,849	19.8	39.6	97	3,245	44	11	3.3	21	75	57	61	26	$770

	Land Area (sq mi)	Population 2004 (Millions)	Population 2025 (Millions)	Population Density		Birth Rate	Death Rate	Natural Increase %	Doubling Time (years)	Infant Mortality Per 1000 (births)	Life Expectancy		Percent Urban Pop	Per Capita GNP ($US)
				Arithmetic	Physiological						Male (years)	Female (years)		
South Asia	**1,732,734**	**1,419.5**	**1,843.0**	**819**	**1,837**	**27**	**9**	**1.8**	**39**	**69**	**62**	**63**	**28**	**$2,213**
Bangladesh	55,598	139.5	177.8	2,510	3,800	30	8	2.2	32	66	59	59	23	$1,590
Bhutan	18,147	0.9	1.4	52	2,598	34	9	2.5	28	61	66	66	16	$1,440
India	1,269,340	1,085.5	1,363.0	855	1,688	26	9	1.7	41	68	62	64	28	$2,340
Maldives	116	0.3	0.5	2,685	31,151	23	4	1.9	37	37	67	66	27	$4,240
Nepal	56,826	24.9	36.1	438	2,770	31	11	2.0	35	64	58	57	11	$1,370
Pakistan	307,375	149.6	242.1	487	1,862	30	9	2.1	33	86	63	63	33	$1,860
Sri Lanka	25,332	18.7	22.1	740	5,355	18	6	1.2	58	17	70	74	30	$3,460
East Asia	**4,546,050**	**1,530.2**	**1,690.0**	**337**	**3,735**	**13**	**7**	**0.6**	**117**	**31**	**70**	**74**	**44**	**$6,280**
China	3,696,521	1,306.0	1,463.7	353	3,524	13	6	0.7	100	31	70	73	38	$3,920
Japan	145,869	127.4	121.1	873	7,965	8	8	0.0		3	78	85	78	$27,080
Korea, North	46,541	23.6	26.4	506	3,621	18	10	0.8	88	42	62	67	59	
Korea, South	38,324	49.2	50.5	1,283	6,793	13	5	0.8	88	8	72	80	79	$17,300
Mongolia	604,826	2.5	3.3	4	409	23	8	1.5	47	37	61	65	57	$1,760
Taiwan	13,969	22.7	25.3	1,627	6,764	11	6	0.5	140	6	73	78	77	$15,100
Southeast Asia	**1,735,448**	**552.2**	**706.0**	**318**	**2,377**	**22**	**7**	**1.5**	**47**	**41**	**65**	**70**	**36**	**$3,450**
Brunei	2,228	0.4	0.5	186	20,767	22	3	1.9	37	15	71	76	67	$24,910
Cambodia	69,900	12.7	18.4	182	1,439	28	11	1.7	41	95	54	58	16	$1,440
East Timor	5,741	0.8	1.2	143		29	15	1.4	50	135	47	48	8	$480
Indonesia	735,355	224.0	281.9	305	3,176	22	6	1.6	44	46	66	70	39	$2,830
Laos	91,429	5.8	8.6	63	2,153	36	13	2.3	30	104	52	55	17	$1,540
Malaysia	127,317	25.3	35.6	199	6,655	23	4	1.9	37	8	70	75	57	$8,330
Myanmar/Burma	261,228	50.3	60.2	192	1,393	25	12	1.3	54	90	54	59	27	$960
Philippines	115,830	83.9	129.9	724	3,836	29	5	2.4	29	26	65	71	47	$4,220
Singapore	239	4.3	8.0	17,856	1,066,867	12	4	0.8	88	2	76	80	100	$24,910
Thailand	198,116	63.6	72.1	321	948	14	6	0.8	88	20	70	75	31	$6,320
Vietnam	128,066	81.9	104.1	640	3,835	19	5	1.4	50	30	67	70	24	$2,000
Austral Realm	**3,093,340**	**23.9**	**27.8**	**8**		**13**	**7**	**0.6**	**114**	**5**	**77**	**82**	**84**	**$23,894**
Australia	2,988,888	19.9	23.2	7	106	13	7	0.6	117	5	77	82	85	$24,970
New Zealand	104,452	4.0	4.6	38	22	14	7	0.7	100	5	76	81	77	$18,530
Pacific Realm	**213,401**	**8.5**	**12.2**	**40**	**6,928**	**8**	**1**	**0.6**	**110**	**10**	**14**	**14**	**24**	**$628**
Federated States of Micronesia	270	0.1	0.2	389		31	6	2.5	28	45	65	67	27	$3,050
Fiji	7,054	0.9	1.0	132	1,316	25	6	1.9	37	20	65	69	46	$4,480
French Polynesia	1,544	0.2	0.3	134	14,747	21	5	1.6	44	8	69	74	53	
Guam	212	0.2	0.2	982	9,458	24	4	2.0	35	9	75	80	38	$9,010
Marshall Islands	69	0.1	0.1	1,559		42	5	3.7	19	37	66	69	65	$3,080
New Caledonia	7,174	0.2	0.3	29	2,902	21	6	1.5	47	7	70	76	71	$18,870
Papua New Guinea	178,703	5.2	8.0	29		34	11	2.3	30	77	56	58	15	$2,180
Samoa	1,097	0.2	0.2	191	1,003	30	6	2.4	29	25	65	72	21	$5,050
Solomon Islands	11,158	0.5	0.9	48	4,950	41	7	3.4	21	25	67	68	13	$1,710
Vanuatu	4,707	0.2	0.4	45	2,257	36	6	3.0	23	45	67	66	21	$2,960

Relief

Meters		Feet
3050		10 000
1525		5000
610		2000
305		1000
152.5		500
0	Sea Level	0
152.5		500
1525		5000
3050		10 000

Below Sea Level

Scale 1: 16 000 000; one inch to 250 miles. Conic Projection
Elevations and depressions are given in feet

Longitude West of Greenwich   Longitude East of Greenwich

Continued on pages 230-231

| 0 | 50 | 100 | 200 | 300 | 400 | 500 Miles |

| 0 | 100 | 200 | 400 | 600 | 800 Kilometers |

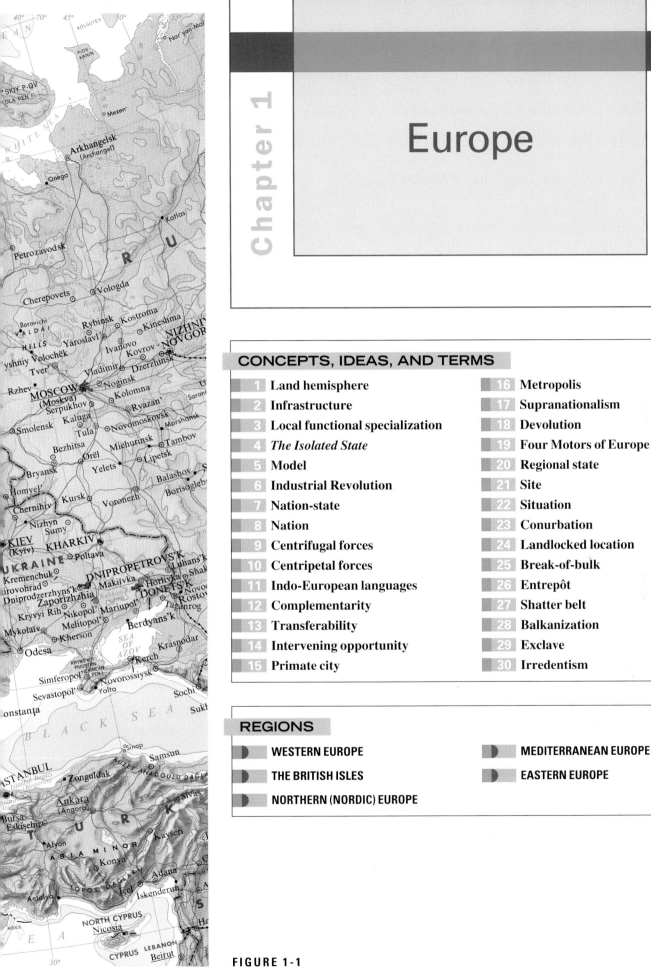

Chapter 1

Europe

CONCEPTS, IDEAS, AND TERMS

1 Land hemisphere	16 Metropolis
2 Infrastructure	17 Supranationalism
3 Local functional specialization	18 Devolution
4 *The Isolated State*	19 Four Motors of Europe
5 Model	20 Regional state
6 Industrial Revolution	21 Site
7 Nation-state	22 Situation
8 Nation	23 Conurbation
9 Centrifugal forces	24 Landlocked location
10 Centripetal forces	25 Break-of-bulk
11 Indo-European languages	26 Entrepôt
12 Complementarity	27 Shatter belt
13 Transferability	28 Balkanization
14 Intervening opportunity	29 Exclave
15 Primate city	30 Irredentism

REGIONS

WESTERN EUROPE	MEDITERRANEAN EUROPE
THE BRITISH ISLES	EASTERN EUROPE
NORTHERN (NORDIC) EUROPE	

FIGURE 1-1

*I*T IS APPROPRIATE to begin our investigation of the world's geographic realms in Europe because over the past five centuries Europe and Europeans have influenced and changed the rest of the world more than any other realm or people has done. European empires spanned the globe and transformed societies far and near. European colonialism propelled the first wave of globalization. Millions of Europeans migrated from their homelands to the Old World as well as the New, changing (and sometimes nearly obliterating) traditional communities and creating new societies from Australia to North America. Colonial power and economic incentive combined to impel the movement of millions of imperial subjects from their ancestral homes to distant lands: Africans to the Americas, Indians to Africa, Chinese to Southeast Asia, Malays to South Africa's Cape, Native Americans from east to west. In agriculture, industry, politics, and other spheres, Europe generated revolutions—and then exported those revolutions across the world, thereby consolidating the European advantage.

But throughout much of that 500-year period of European hegemony, Europe also was a cauldron of conflict. Religious, territorial, and political disputes precipitated bitter wars that even spilled over into the colonies. And during the twentieth century, Europe twice plunged the world into war. The terrible, unprecedented toll of World War I (1914–1918) was not enough to stave off World War II (1939–1945), which ended with the first-ever use of nuclear weapons in Japan. In the aftermath of that war, Europe's weakened powers lost most of their colonial possessions and a new rivalry emerged: an ideological Cold War between the communist Soviet Union and the capitalist United States. This Cold War lowered an Iron Curtain across the heart of Europe, leaving most of the east under Soviet control and most of the west in the American camp. Western Europe proved resilient, overcoming the destruction of war and the loss of colonial power to regain economic strength. Meanwhile, the Soviet communist experiment failed at home and abroad,

MAJOR GEOGRAPHIC QUALITIES OF

Europe

1. The European realm lies on the western extremity of the Eurasian landmass, a locale of maximum efficiency for contact with the rest of the world.

2. Europe's lingering and resurgent world influence results largely from advantages accrued over centuries of global political and economic domination.

3. The European natural environment displays a wide range of topographic, climatic, and soil conditions and is endowed with many industrial resources.

4. Europe is marked by strong internal regional differentiation (cultural as well as physical), exhibits a high degree of functional specialization, and provides multiple exchange opportunities.

5. European economies are dominated by manufacturing, and the level of productivity has been high; levels of development generally decline from west to east.

6. Europe's nation-states emerged from durable power cores that formed the headquarters of world colonial empires. A number of those states are now plagued by internal separatist movements.

7. Europe's rapidly aging population is generally well off, highly urbanized, well educated, and enjoys long life expectancies.

8. A growing number of European countries are experiencing population declines; in many of these countries, the natural decrease is partially offset by immigration.

9. Europe has made significant progress toward international economic integration and, to a lesser extent, political coordination.

and in 1990 the last vestiges of the Iron Curtain were lifted. Since then, a massive effort has been underway to reintegrate and reunify Europe from the Atlantic coast to the Russian border, the key geographic story of this chapter.

DEFINING THE REALM

As FIGURE 1-1 SHOWS, Europe is a realm of peninsulas and islands on the western margin of the world's largest landmass, Eurasia. It is a realm of 583 million people and 39 countries, but it is territorially quite small. Yet despite its modest proportions it has had—and continues to have—a major impact on world affairs. For many centuries Europe has been a hearth of achievement, innovation, and invention.

The European realm is bounded on the west, north, and south by Atlantic, Arctic, and Mediterranean waters, respectively. But where is Europe's eastern limit? Some scholars place it at the Ural Mountains, deep inside Russia, thereby recognizing a "European" Russia and, presumably, an "Asian" one as well. Our regional definition places Europe's eastern boundary between Russia and its numerous European neighbors to the west. This definition

EUROPE'S CLIMATES

B DRY CLIMATE

| BS | Semiarid |
| BW | Arid |

h=hot
k=cold

C HUMID TEMPERATE CLIMATE

Cfa	
Cfb	No dry season
Cfc	

| Csa | Dry summer |
| Csb | (Mediterranean) |

a=hot summer   c=short, cool summer
b=cool summer   d=very cold winter

D HUMID COLD CLIMATE

Dfa	
Dfb	No dry season
Dfc	

E COLD POLAR CLIMATE

| E | Tundra and ice |

H HIGHLAND CLIMATE

| H | Undifferentiated highlands |

0   200   400   600   800 Kilometers
0   100   200   300   400   500 Miles

FIGURE 1-2

RELATIVE LOCATION: EUROPE IN THE LAND HEMISPHERE

FIGURE 1-3

Azimuthal equidistant projection centered on Hamburg, Germany

is based on several geographic factors, including European-Russian contrasts in territorial dimensions, population size, cultural properties, and historic aspects, all discernible on the maps in Chapters 1 and 2.

Europe's peoples have benefited from a large and varied store of raw materials. Whenever the opportunity or need arose, the realm proved to contain what was required. Early on, these requirements included cultivable soils, rich fishing waters, and wild animals that could be domesticated; in addition, extensive forests provided wood for houses and boats. Later, mineral fuels and ores propelled industrialization.

From the balmy shores of the Mediterranean Sea to the icy peaks of the Alps, and from the moist woodlands and moors of the Atlantic fringe to the semiarid prairies north of the Black Sea, Europe presents an almost infinite range of natural environments (Fig. 1-2). Compare Western Europe and eastern North America on Figure G-8 (pp. 14–15) and you will see the moderating influence of the warm ocean current known as the North Atlantic Drift and its onshore windflow.

The European realm is home to peoples of numerous cultural-linguistic stocks, including not only Latins, Germanics, and Slavs but also minorities such as Finns, Magyars (Hungarians), Basques, and Celts. This diversity of ancestries continues to be an asset as well as a lia-

bility. It has generated not only interaction and exchange, but also conflict and war.

Europe also has outstanding locational advantages. Its *relative location*, at the heart of the **land hemisphere**, creates maximum efficiency for contact with the rest of the world (Fig. 1-3). A "peninsula of peninsulas," Europe is nowhere far from the ocean and its avenues of seaborne trade and conquest. Hundreds of miles of navigable rivers, augmented by an unmatched system of canals, open the interior of Europe to its neighboring seas and to the waterways of the world.

Also consider the scale of the maps of Europe in this chapter. Europe is a realm of moderate distances and close proximities. Short distances and large cultural differences make for intense interaction, the constant circulation of goods and ideas. That has been the hallmark of Europe's geography for over a millennium.

LANDSCAPES AND OPPORTUNITIES

Europe's area may be small, but its landscapes are varied and complex. Regionally, we identify four broad units: the Central Uplands, the southern Alpine Mountains, the Western Uplands, and the North European Lowland (Fig. 1-4).

FIGURE 1-4

The *Central Uplands* form the heart of Europe. It is a region of hills and low plateaus loaded with raw materials whose farm villages grew into towns and cities when the Industrial Revolution transformed this realm.

The *Alpine Mountains*, a highland region named after the Alps, extend from the Pyrenees on the French-Spanish border to the Balkan Mountains near the Black Sea, and include Italy's Appennines and Eastern Europe's Carpathians.

The *Western Uplands*, geologically older, lower, and more stable than the Alpine Mountains, extend from Scandinavia through western Britain and Ireland to the heart of the Iberian Peninsula.

The *North European Lowland* extends in a lengthy arc from southeast Britain and central France across Germany and Denmark into Poland and Ukraine, from where it continues well into Russia. Also known as the Great European Plain, this has been an avenue for human

FROM THE FIELD NOTES

"I got aboard a large ship in Dover, England and sailed up the Seine River to Rouen, France. What a lesson in regional geography: from the highest deck I had a panoramic view over the landscape of what I had learned to call the Paris Basin. Every square inch of the flat to undulating countryside of this part of the North European Lowland seemed to be in productive use in one of the most meticulously organized, parklike rural landscapes I had ever seen. Clusters of trees signaled villages; no road was without its line of poplars or oaks. The Seine River, narrow but surprisingly deep given the size of the ship, traverses its basin in wide meanders, which must have caused the land planners a headache. Still, the French pattern of 'long lots,' giving as many farmers as possible some frontage on a waterway and then extending the parcel of soil deep inland, had some regularity even here (you can also see it in Canada's Quebec). Two of those long lots, one fallow and the other under pasture, can be seen in the foreground."

migration time after time, so that complex cultural and economic mosaics developed here together with a jig-saw-like political map. As Figure 1-4 shows, many of Europe's major rivers and connecting waterways serve this populous region, where many of Europe's leading cities (London, Paris, Amsterdam, Copenhagen, Berlin, Warsaw) are located.

EUROPE'S PREMODERN HERITAGE

Modern Europe was peopled in the wake of the Pleistocene's most recent glacial retreat and global warming—a gradual warming that caused tundra to give way to deciduous forest and ice-filled valleys to turn into grassy vales. On Mediterranean shores, Europe witnessed the rise of its first great civilizations, on the islands and peninsulas of Greece and later in what is today Italy.

Ancient Greece lay exposed to influences radiating from the advanced civilizations of Mesopotamia and the Nile Valley, and in their fragmented habitat the Greeks laid the foundations of European civilization. Their achievements in political science, philosophy, the arts, and other spheres have endured for 25 centuries. But the ancient Greeks never managed to unify their domain, and their persistent conflicts proved fatal when the Romans challenged them from the west. By 147 BC, the last of the sovereign Greek intercity leagues (alliances) had fallen to the Roman conquerors.

The center of civilization and power now shifted to Rome in present-day Italy. Borrowing from Greek culture, the Romans created an empire that stretched from Britain to the Persian Gulf and from the Black Sea to Egypt; they made the Mediterranean Sea a Roman lake carrying armies to distant shores and goods to imperial Rome. With an urban population that probably exceeded 1 million, Rome was the first metropolitan-scale urban center in Europe.

The Romans founded numerous other cities throughout their empire and linked them to the capital through a vast system of highway and water routes, facilitating political control and enabling economic growth in their provinces. It was an unparalleled **infrastructure**, much **2** of which long outlasted the empire itself.

Roman rule brought disparate, isolated peoples into the imperial political and economic sphere. By guiding (and often forcing) these groups to produce particular goods or materials, they launched Europe down a road for which it would become famous: **local functional** **3** **specialization**. The workers on Elba, a Mediterranean island, mined iron ore. Those near Cartagena in Spain produced silver and lead. Certain farmers were taught irrigation to produce specialty crops. Others raised livestock for meat or wool. The *production of particular goods by particular people in particular places* became and remained a hallmark of the realm.

The Romans also spread their language across the empire, setting the stage for the emergence of the *Romance* languages; they disseminated Christianity; and

Several cultures made lasting contributions to the architecture of Iberia: the Romans, the Arabs, and the Spanish and Portuguese themselves. The city of Segovia, capital of the Autonomous Community of Castile-Leon, was already more than six centuries old when it was taken by the Romans in about 80 BC; the Moors took it early in the eighth century AD, and the Spanish freed it in 1079. Today, the city's Old Town, encircled by ancient walls, lies centered on the *Plaza Mayor*; nearby lies the *Plaza Azoguejo*, another busy square. Crossing the *Plaza Azoguejo* is one of the best-preserved Roman aqueducts (water-carriers) still in existence, known locally as *El Puente*. As the background of this picture (behind the red-tiled roofs) shows, the occupiers of Segovia (and other towns), whether they were Romans, Arabs, or Spaniards, built their structures on the highest and most defensible hill in the area. This caused a water-supply problem, which the Romans solved by building aqueducts that brought water from far upstream to the center.

they established durable systems of education, administration, and commerce. But when their empire collapsed in the fifth century, disorder ensued, and massive migrations soon brought Germanic and Slavic peoples to their present positions on the European stage. Capitalizing on Europe's weakness, the Arab-Berber Moors from North Africa, energized by Islam, conquered most of Iberia and penetrated France. Later the Ottoman Turks invaded Eastern Europe and reached the outskirts of Vienna.

Europe's revival—its *Renaissance*—did not begin until the fifteenth century. After a thousand years of feudal turmoil marking the "Dark" and "Middle" Ages, powerful monarchies began to lay the foundations of modern states. The discovery of continents and riches across the oceans opened a new era of *mercantilism*, the competitive accumulation of wealth chiefly in the form of gold and silver. Best placed for this competition were the kingdoms of Western Europe. Europe was on its way to colonial expansion and world domination.

THE REVOLUTIONS OF MODERNIZING EUROPE

Even as Europe's rising powers reached for world domination overseas, they fought with each other in Europe itself. Powerful monarchies and land-owning ("landed") aristocracies had their status and privilege challenged by ever-wealthier merchants and businesspeople. Demands for political recognition grew; cities mushroomed with

the development of industries; the markets for farm products burgeoned; and Europe's population, more or less stable at about 100 million since the sixteenth century, began to increase.

The Agrarian Revolution

We know Europe as the focus of the Industrial Revolution, but before this momentous development occurred, another revolution was already in progress: the *agrarian revolution*. Port cities and capital cities thrived and expanded, and their growing markets created economic opportunities for farmers. This led to revolutionary changes in land ownership and agricultural methods. Improved farm practices, better equipment, superior storage facilities, and more efficient transport to the urban markets marked a revolution in the countryside. The colonial merchants brought back new crops (the American potato soon became a European staple), and market prices rose, drawing more and more farmers into the economy.

The transformation of Europe's farmlands reshaped its economic geography, producing new patterns of land use and market links. The economic geographer Johann Heinrich von Thünen (1783–1850), himself an estate farmer who had studied these changes for several decades, published his results in 1826 in a pioneering work entitled ***The Isolated State***. 4

Von Thünen used information from his own farmstead to build what today we call a **model** (an idealized 5 representation of reality that demonstrates its most important properties) of the location of productive activities in Europe's farmlands. Since a model is an abstraction that must always involve assumptions, von Thünen postulated a self-contained area (hence the "isolation") with a single market center, flat and uninterrupted land without impediments to cultivation or transportation. In such a situation, transport costs would be directly proportional to distance.

Von Thünen's model revealed four zones or rings of land use encircling the market center (Fig. 1-5). Innermost and directly adjacent to the market would lie a zone of intensive farming and dairying, yielding the most perishable and highest-priced products. Immediately beyond lay a zone of forest used for timber and firewood (still a priority in von Thünen's time). Next there would be a ring of field crops, for example, grains or potatoes. A fourth zone would contain pastures and livestock. Beyond lay wilderness, from where the costs of transport to market would become prohibitive.

In many ways, von Thünen's model was the first analysis in a field that would eventually become known as *location theory*. Von Thünen knew, of course, that the real Europe did not present the conditions postulated in his model. But it did demonstrate the economic-geographic forces that shaped the new Europe, which is why it is still being discussed today. More than a century after the publication of *The Isolated State*, geographers

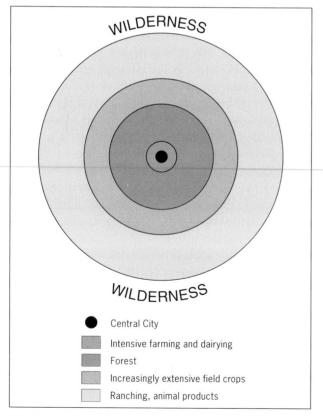

- ● Central City
- Intensive farming and dairying
- Forest
- Increasingly extensive field crops
- Ranching, animal products

FIGURE 1-5

Samuel van Valkenburg and Colbert Held produced a map of twentieth-century European agricultural intensity, revealing a striking, ring-like concentricity focused on the vast urbanized area lining the North Sea—now the dominant market for a realmwide, macroscale "Thünian" agricultural system (Fig. 1-6).

The Industrial Revolution

The term **Industrial Revolution** suggests that an agrarian Europe was suddenly swept up in wholesale industrialization that changed the realm in a few decades. In reality, seventeenth- and eighteenth-century Europe had been industrializing in many spheres, long before the chain of events known as the Industrial Revolution began. From the textiles of England and Flanders to the iron farm implements of Saxony (in present-day Germany), from Scandinavian furniture to French linens, Europe had already entered a new era of local functional specialization. It would therefore be more appropriate to call what happened next the period of Europe's *industrial intensification*.

In the 1780s, James Watt and others devised a steam-driven engine, which was soon adopted for numerous industrial uses. At about the same time, coal (converted into carbon-rich coke) was recognized as a vastly superior substitute for charcoal in smelting iron. These momentous innovations had a rapid effect. The power loom revolutionized the weaving industry. Iron smelters, long dependent on Europe's dwindling forests for fuel, could now be concentrated near coalfields. Engines could move locomotives as well as power looms. Ocean shipping entered a new age.

Britain had an enormous advantage, for the Industrial Revolution occurred when British influence reigned worldwide and the significant innovations were achieved in Britain itself. The British controlled the flow of raw materials, they held a monopoly over products that were in global demand, and they alone possessed the skills necessary to make the machines that manufactured the products. Soon the fruits of the Industrial Revolution were being exported, and the modern industrial spatial organization of Europe began to take shape. In Britain, manufacturing regions, densely populated and heavily urbanized, developed near coalfields in the English Midlands, at Newcastle to the northeast, in southern Wales, and along Scotland's Clyde River around Glasgow.

In mainland Europe, a belt of major coalfields extends from west to east, roughly along the southern margins of the North European Lowland, due eastward from southern England across northern France and Belgium, Germany (the Ruhr), western Bohemia in the Czech Republic, Silesia in southern Poland, and the Donets Basin

FIGURE 1-6

VON THÜNEN RINGS: EUROPE

Intensity of Production Based on Index of Yields of 8 Leading Crops.

(Donbas) in eastern Ukraine. Iron ore is found in a broadly similar belt, and the industrial map of Europe reflects the resulting concentrations of economic activity (Fig. 1-7). Another set of manufacturing regions emerged in and near the growing urban centers of Europe, as the same map demonstrates. London—already Europe's leading urban focus and Britain's richest domestic market—was typical of these developments. Many local industries were established here, taking advantage of the large supply of labor, the ready availability of capital, and the proximity of so great a number of potential buyers. Although the Industrial Revolution thrust other places into prominence, London did not lose its primacy: industries in and around the British capital multiplied.

The industrial transformation of Europe, like the agrarian revolution, became the focus of geographic research. One of the leaders in this area was the economic geographer Alfred Weber (1868–1958), who published a spatial analysis of the process titled *Concerning the Location of Industries* (1909). Unlike von Thünen, Weber focused on activities that take place at particular points rather than over large areas. His model, therefore, represented the factors of industrial location, the clustering or dispersal of places of intense manufacturing activity.

One of Weber's most interesting conclusions has to do with what he called *agglomerative* (concentrating) and *deglomerative* (dispersive) forces. It is often advantageous for certain industries to cluster together, sharing equipment, transport facilities, labor skills, and other assets of urban areas. This is what made London (as well as Paris and other cities that were not situated on rich deposits of industrial raw materials) attractive to many manufacturing plants that could benefit from agglomeration and from the large markets that these cities anchored. As Weber found, however, excessive agglomeration may lead to disadvantages such as competition for increasingly expensive space, congestion, overburdening of infrastructure, and environmental pollution. Manufacturers may then move away, and deglomerative forces will increase.

The Industrial Revolution spread eastward from Britain onto the European mainland throughout the middle and late nineteenth century (see inset, Fig. 1-7). Population skyrocketed, emigration mushroomed, industrializing cities burst at the seams. European states already had acquired colonial empires before this revolution started; now colonialism gave Europe an unprecedented advantage in its dominance over the rest of the world.

**EUROPE:
INDUSTRIALIZATION**

— European Core Boundary

▬ Major Industrial District

▬ Major Coalfield

⚒ Minor Coal Deposit

● Iron Ore Deposit

⠿ Canal

| 0 | 250 | 500 Kilometers |
| 0 | 100 | 200 | 300 Miles |

**SPREAD OF THE
INDUSTRIAL
REVOLUTION**

| 0 | 250 | 500 Kilometers |
| 0 | 250 | 500 Miles |

MIDLANDS
1840s

RUHR 1860s

1850s

SAXONY

SILESIA
1870s

1880s

DONBAS

FIGURE 1-7

Political Revolutions

Revolution in a third sphere—the political—had been going on in Europe even longer than the agrarian or industrial revolutions. Europe's *political revolution* took many different forms and affected diverse peoples and countries, but in general it headed toward parliamentary representation and democracy. Most dramatic was the French Revolution (1789–1795), but political transformation had come much earlier to the Netherlands, Britain, and Scandinavian countries. Other parts of Europe remained under the control of authoritarian (dictatorial) regimes headed by monarchs or despots. Europe's patchy political revolution lasted into the twentieth century, and by then *nationalism* (national spirit, pride, patriotism) had become a powerful force in European politics.

When you look at the political map of Europe, the question that arises is how did so small a geographic realm come to be divided into so many political entities? Europe's map is a legacy of its feudal and royal periods, when powerful kings, barons, dukes, and other rulers, rich enough to fund armies and powerful enough to exact taxes and tribute from their domains, created bounded territories in which they reigned supreme. Royal marriages, alliances, and conquests actually simplified Europe's political map. In the early nineteenth century there still were 39 German states; Germany as we know it today did not emerge until the 1870s.

Europe's political revolution produced a form of political-territorial organization known as the **nation-state**, a state embodied by its culturally distinctive population. But what is a nation-state and what is not? The term **nation** has multiple meanings. In one sense it refers to a people with a single language, a common history, a similar ethnic background. In the sense of *nationality* it relates to legal membership in the state, that is, citizenship. Very few states today are so homogeneous culturally that the culture is conterminous with the state. Europe's prominent nation-states of a century ago—France, Spain, the United Kingdom, Italy—have become multicultural societies, their nations defined more by an intangible "national spirit" and emotional commitment than by cultural or ethnic homogeneity. Today, Poland, Hungary, and Sweden are among the few states that still satisfy the definition of the nation-state in Europe.

Mercantilism and colonialism empowered the states of Western Europe; the United Kingdom (Britain) was the superpower of its day. But all countries, even Europe's nation-states in their heyday, are subject to divisive stresses. Political geographers use the term **centrifugal forces** to identify and measure the strength of such division, which may result from religious, racial, linguistic, political, or regional factors. In the United States, racial issues form a centrifugal force; during the Vietnam (Indochina) War (1964–1975), politics created strong and dangerous disunity.

Centrifugal forces are measured against **centripetal forces**, the binding, unifying glue of the state. General satisfaction with the system of government and administration, legal institutions, and other functions of the state (notably including its treatment of minorities) can ensure stability and continuity when centrifugal forces threaten. In the recent case of Yugoslavia, the centrifugal forces unleashed after the end of the Cold War exceeded the weak centripetal forces in that relatively young state, and it disintegrated.

Europe's political revolution continues. Today a growing group of European states is trying to create a realmwide union, what might some day become a European superstate.

CONTEMPORARY EUROPE

Europe has been a regional laboratory of political revolution and evolution, and some of its nation-states were among the first of their kind to emerge on the world stage. Enriched and empowered by colonialism, European states competed and fought with each other, but Europe's nations survived and prospered. Strong cultural identities and historic durability gave European peoples a confidence that continues to mark the realm today, long after Europe's global empires collapsed.

The European realm exhibits only limited geographic homogeneity, which is a challenge that confronts those leading states that want to create a more unified European system. As Figure 1-8 shows, most Europeans speak **Indo-European languages**, but in fact Europe remains a veritable Tower of Babel. Not only are many of those Indo-European languages not mutually understandable, but peoples such as the Hungarians and the Finns have other linguistic sources. English has become Europe's *lingua franca*, but generally with declining effectiveness from west to east.

Christian religious traditions, another factor that might serve as a unifier, have instead been the source of endless conflict. Shared Christian values, for example, have done little to bring peace to Northern Ireland, where longstanding sectarian strife continues. Religion and politics remain closely connected, and some political parties still have religious names, such as Germany's Christian Democratic Union and the Dutch Katholieke Volkspartij (Catholic People's Party).

LANGUAGES OF EUROPE

0   200   400   600 Kilometers

0   100   200   300 Miles

MAJOR INDO-EUROPEAN BRANCHES	OTHER INDO-EUROPEAN BRANCHES	URALIC LANGUAGE FAMILY	

MAJOR INDO-EUROPEAN BRANCHES

GERMANIC GROUP

WESTERN GERMANIC

1 Dutch
2 German
3 Frysian
4 English

NORTHERN GERMANIC

5 Danish
6 Swedish
7 Norwegian
8 Icelandic
9 Faeroese

ROMANCE GROUP

10 Portuguese
11 Spanish
12 Catalan
13 Provençal
14 French
15 Italian
16 Rhaeto-Romansch
17 Romanian
18 Corsican-Italian
19 Sardinian-Italian
20 Walloon

SLAVIC GROUP

WEST SLAVONIC

21 Polish
22 Slovak
23 Czech
24 Lusatian

EAST SLAVONIC

25 Russian
26 Ukrainian
27 Belarussian

SOUTH SLAVONIC

28 Slovene
29 Serbo-Croatian
30 Macedonian
31 Bulgarian

OTHER INDO-EUROPEAN BRANCHES

CELTIC GROUP

BRITANNIC

32 Breton
33 Welsh

GAELISH

34 Irish Gaelic
35 Scots Gaelic

BALTIC GROUP

36 Latvian
37 Lithuanian

HELLENIC

38 Greek

THRACIAN/ILLYRIAN GROUP

39 Albanian

INDO-IRANIAN GROUP

40 Romani (dispersed)

URALIC LANGUAGE FAMILY

FINNO-UGRIC GROUP

41 Finnish
42 Karelian
43 Saami
44 Estonian
45 Hungarian
46 Komi

SAMOYEDIC GROUP

47 Samoyedic

ALTAIC LANGUAGE FAMILY

TURKIC GROUP

48 Turkish

OTHER LANGUAGES

BASQUE

49 Basque

Areas with significant concentrations of other languages (usually adjacent national languages)

Boundary between languages

FIGURE 1-8

A TGV (*train à grande vitesse*) is poised to depart from Paris's Montparnasse Station on its 400-mile (650-km) run to Bordeaux in southwestern France, a trip scheduled to be completed in just 2 hours and 55 minutes at an average speed of 137 miles (220 km) per hour. The French and the Japanese have long been world leaders in fast-train technology, and it is now a primary European objective to link all parts of the European Union through a high-speed rail network. This includes offshore Britain, which is linked to the continent by the tunnel beneath the English Channel that opened a decade ago.

The name *European* is sometimes taken to refer not just to someone residing in Europe, but as a racial reference, describing a common ancestry. But here again, Europe's purported homogeneity is more apparent than real. In terms of physical characteristics, Europeans, from Swede to Spaniard and from Scot to Sicilian, are as varied as any of the world's other geographic realms.

Spatial Interaction

If not culture, what does unify Europe? The answer lies in this realm's outstanding opportunities for productive contact and profitable interaction. The ancient Romans knew it, but they centered their system on the imperial capital. Modern Europeans have seized the same opportunities to create a regionwide structure of *spatial interaction* that links regions, countries, and places in countless ways. The American geographer Edward Ullman conceptualized this process around three operating principles: (1) complementarity, (2) transferability, and (3) intervening opportunity.

Complementarity occurs when one area has a surplus of a commodity required by another area. The mere existence of a particular resource or product is no guarantee of trade: it must be needed elsewhere. One of Europe's countless examples (at various levels of scale) involves Italy, leader among Mediterranean countries in economic development but lacking adequate coal supplies. Italy imports coal from Western Europe, and in turn Italy exports to Western Europe its citrus fruits, olives, and grapes—which are in high demand on Western European markets. This is a case of *double comple-*

mentarity, and even the physical barrier of the Alps has not restricted this two-way trade by rail and road.

Transferability refers to the ease with which a commodity is transported between producer and consumer. Sheer distance, in terms of cost and time, may make it economically impractical to transfer a product. This is not a problem in modestly sized Europe, where distances are short and transport systems efficient.

Intervening opportunity, the third of Ullman's spatial interaction principles, holds that potential trade between two places, even if they are in a position of complementarity and do not have problems of transferability, will develop only in the absence of a closer, intervening source of supply. Using our current example, if a major coal reserve were discovered in southern Switzerland, Italy would avail itself of that (intervening) opportunity, reducing or eliminating its imports from Western Europe.

Europe's internal spatial interaction is facilitated by what in many respects is the world's most effective network of railroads, highways, waterways, and air routes. This network is continuously improving as tunnels, bridges, high-speed rail lines, and augmented airports are built. The continent's burgeoning cities and their environs, however, are increasingly troubled by severe congestion.

An Urbanized Realm

Overall, 73 percent of Europe's population resides in cities and towns, but this average is exceeded by far in the west and not attained in the east.

Large cities are the crucibles of their nations' cultures. In his 1939 study of the pivotal role of great cities in the development of national cultures, American geographer Mark Jefferson postulated the law of the **primate city**, which stated that "a country's leading city is always disproportionately large and exceptionally expressive of national capacity and feeling." Though imprecise, this "law" can readily be demonstrated using European examples. Certainly Paris personifies France in countless ways, and nothing in England rivals London. In both of these primate cities, the culture and history of a nation and empire are indelibly etched in the urban landscape. Similarly, Vienna is a microcosm of Austria, Warsaw is the heart of Poland, Stockholm typifies Sweden, and Athens is Greece. Today, each of these (together with the other primate cities of Europe) sits atop a hierarchy of urban centers that has captured the lion's share of its national population growth since World War II.

Primate cities tend to be old, and in general the European cityscape looks quite different from the American. Seemingly haphazard inner-city street systems impede traffic; central cities may be picturesque, but they are also cramped. The urban layout of the London region (Fig. 1-9) reveals much about the internal spatial structure of the European **metropolis** (the central city and its suburban ring). The metropolitan area remains focused on the large city at its center, especially the downtown *central business district (CBD)*, which is the oldest part of the urban agglomeration and contains the region's largest concentration of business, government, and shopping facilities as well as its wealthiest and most prestigious residences. Wide residential sectors radiate outward from the CBD across the rest of the central city, each one home to a particular income group. Beyond the central city lies a sizeable suburban ring, but residential densities are much higher here than in the United States because the European tradition is one of setting aside recreational spaces (in "greenbelts") and living in apartments rather than in detached single-family houses. There is also a greater reliance on public transportation, which further concentrates the suburban development pattern. That has allowed many nonresidential activities to suburbanize as well, and today ultramodern outlying business centers increasingly compete with the CBD in many parts of urban Europe (see photo p. 53).

A Changing Population Composition

When a population urbanizes, average family size declines, and so does the overall rate of natural increase. There was a time when Europe's population was (in the terminology of population geographers) exploding,

FIGURE 1-9

METROPOLITAN LONDON

Built-up area — Road
Greenbelt — Railroad
Central Business District

0 5 10 15 20 Kilometers
0 5 10 Miles

Booming *suburban downtowns* (see Chapter 3) are transforming the U.S. metropolitan landscape, and such "edge cities" are now multiplying in Western Europe as well. The biggest and most important of these complexes is *La Défense*, located just west of Paris, which has grown so robustly that it is now continental Europe's single largest business district, whose annual transactions exceed the GNP of the Netherlands! Besides its economic prowess, the cubic Grande Arche that forms the centerpiece of *La Défense* is an architectural jewel that is the western anchor of Paris's "Historic Axis" which includes the *Arc de Triomphe* (top center of photo), *Champs Elysées, Place de la Concorde*, and the *Louvre*.

sending millions to the New World and the colonies and still growing at home. But today Europe's indigenous population, unlike most of the rest of the world's, is actually shrinking. To keep a population from declining, the (statistically) average woman must bear 2.1 children. For Europe as a whole, that figure was 1.4 in 2002. Six European countries, including Spain and the Czech Republic, recorded figures below 1.3—the lowest ever seen in any human population.

Such *negative population growth* poses serious challenges for any nation. When the population pyramid becomes top-heavy, the number of workers whose taxes pay for the social services of the aged goes down, leading to reduced pensions and dwindling funds for health care. Governments that impose tax increases endanger the business climate; their options are limited. Europe, and especially Western Europe, is experiencing a *population implosion* that will be a formidable challenge in decades to come.

Meanwhile, *immigration* is partially offsetting the losses European countries face. Millions of Turks (mainly to Germany), Algerians (France), Moroccans (Spain), West Africans (Britain), and Indonesians (Netherlands) are changing the social fabric of what once were unicultural nation-states.

EUROPE'S TRANSFORMATION

At the end of World War II, much of Europe lay shattered, its cities and towns devastated, its infrastructure wrecked, its economies decimated. The Soviet Union had taken control over the bulk of Eastern Europe, and communist parties seemed poised to dominate the political life of major Western European countries.

In 1947 U.S. Secretary of State George C. Marshall proposed a European Recovery Program designed to counter all this dislocation and to create stable political conditions in which democracy would survive. Over the next four years, the United States provided about $13 billion in assistance to Europe (almost $100 billion in today's money). Because the Soviet Union refused U.S. aid and forced its Eastern European satellites to do the same, the Marshall Plan applied solely to 16 European countries, including defeated (West) Germany and Turkey.

European Unification

The Marshall Plan did far more than stimulate European economies. It showed European leaders that their countries needed a joint economic-administrative structure not only to coordinate the financial assistance, but also to ease the flow of resources and products across Europe's mosaic of boundaries, to lower restrictive trade tariffs, and to seek ways to effect political cooperation.

For all these needs Europe's governments had some guidelines. While in exile in Britain, three small countries—Belgium, the Netherlands, and Luxembourg—had been discussing an association of this kind even before the end of the war. There, in 1944, they formulated and signed the Benelux Agreement, intended to achieve total economic integration. When the Marshall Plan was launched, the Benelux precedent helped speed the creation of the Organization for European Economic Cooperation (OEEC), which was established to coordinate the investment of America's aid (see box titled "Supranationalism in Europe").

Soon the economic steps led to greater political cooperation as well. In 1949 the participating governments created the Council of Europe, the beginnings of what was to become a European Parliament meeting in Strasbourg, France. Europe was embarked on still another political revolution—the formation of a multinational union involving a growing number of European states. Geographers define **supranationalism** as the voluntary **17** association in economic, political, or cultural spheres of

---

Supranationalism in Europe

1944	Benelux Agreement signed.
1947	Marshall Plan created (effective 1948–1952).
1948	Organization for European Economic Cooperation (OEEC) established.
1949	Council of Europe created.
1951	European Coal and Steel Community (ECSC) Agreement signed (effective 1952).
1957	Treaty of Rome signed, establishing European Economic Community (EEC) (effective 1958), also known as the Common Market and "The Six." European Atomic Energy Community (EURATOM) Treaty signed (effective 1958).
1959	European Free Trade Association (EFTA) Treaty signed (effective 1960).
1961	United Kingdom, Ireland, Denmark, and Norway apply for EEC membership.
1963	France vetoes UK EEC membership; Ireland, Denmark, and Norway withdraw applications.
1965	EEC-ECSC-EURATOM Merger Treaty signed (effective 1967).
1967	European Community (EC) inaugurated.
1968	All customs duties removed for intra-EC trade; common external tariff established.
1973	United Kingdom, Denmark, and Ireland admitted as members of EC, creating "The Nine." Norway rejects membership in the EC by referendum.
1979	First general elections for a European Parliament held; new 410-member legislature meets in Strasbourg. European Monetary System established.
1981	Greece admitted as tenth member of EC.
1985	Greenland, acting independently of Denmark, withdraws from EC.
1986	Spain and Portugal admitted as the eleventh and twelfth members of EC. Single European Act ratified, targeting a functioning European Union in the 1990s.
1987	Turkey and Morocco make first application to join EC. Morocco is rejected; Turkey is told that discussions will continue.
1990	Charter of Paris signed by 34 members of the Conference on Security and Cooperation in Europe (CSCE). Former East Germany, as part of newly reunified Germany, incorporated into EC.
1991	Maastricht meeting charts European Union (EU) course for the 1990s.
1993	Single European Market goes into effect. Modified European Union Treaty ratified, transforming EC into EU.
1995	Austria, Finland, and Sweden admitted into EU, enlarging its membership from 12 to 15 countries.
1999	European Monetary Union (EMU) goes into effect. Member-states commit themselves to limit their external debt to 3 percent of GDP.
2001	Denmark's voters reject EMU participation by 53 to 47 percent.
2002	The euro is introduced as preexisting national currencies disappear in 12 countries.
2004	Historic expansion of EU from 15 to 25 countries with the admission of Cyprus, the Czech Republic, Estonia, Hungary, Latvia, Lithuania, Malta, Poland, Slovakia, and Slovenia.

three or more independent states willing to yield some measure of sovereignty for their mutual benefit. In later chapters we will encounter other supranational organizations, including the North American Free Trade Agreement (NAFTA), but none has reached the plateau achieved by the European Union (EU).

In Europe, the key initiatives arose from the Marshall Plan and lay in the economic arena; political integration came more haltingly. Through the Treaty of Rome signed in 1957, six countries joined to become the European Economic Community (EEC), also called the "Common Market." In 1973 the United Kingdom, Ireland, and Denmark joined, and the renamed European Community (EC) had nine members. As Figure 1-10 shows, membership reached 15 countries in 1995, after the organization had been renamed yet one more time to become the European Union (EU).

The European Union is not just a paper organization for bankers and manufacturers. It has a major impact on the daily lives of its member countries' citizens in countless ways. (You will see one of these ways when you arrive at an EU airport and find that EU-passport holders move through inspection at a fast pace, while non-EU citizens wait in long immigration lines.) Some of the

FIGURE 1-10

FROM THE FIELD NOTES

"No place in Europe displays the flag of the European Union as liberally as does the Dutch city of Maastricht, where the European Union Treaty of 1991 was signed. The flag, seen here at City Hall, shows twelve yellow stars (representing the signatories to the Treaty) against a blue background. Although three additional states joined the EU in 1995 [and 10 more in 2004], the flag will remain as is, and will not, as the U.S. flag does, change to reflect changing times."

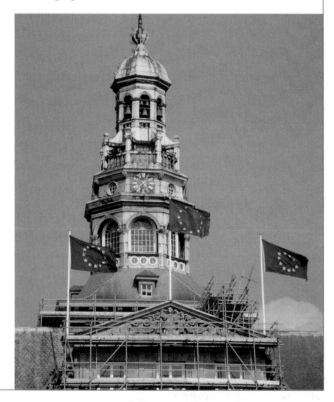

EU's conditions are not universally popular: farmers in particular repeatedly object to EU regulations on agriculture. One major plan is to replace all national currencies such as France's franc and Germany's mark with the euro, but by late 2003 three of the then-15 EU members, including the United Kingdom, had not joined the euro movement. Still, use of the euro took effect in 12 EU countries in 2002, and the old currencies disappeared.

Expansion in 2004

In December 2002, the European Commission, the executive body of the European Union, announced that it had given final approval for the admission of ten new members, raising the total of member-states to 25 and adding 75 million people to the EU's population. As Figure 1-10 shows, the ten new member-states are the three Baltic republics of Estonia, Latvia, and Lithuania; the two components of the former Czechoslovakia, now the Czech Republic and Slovakia; two cornerstones of the region, Poland and Hungary; the first fragment of the former Yugoslavia to join the EU, Slovenia; and two Mediterranean island-states, Cyprus and Malta.

Not included in this round of admissions are two countries with lagging economies that front the Black Sea, Romania and Bulgaria, and the only Muslim aspirant, Turkey. The Commission indicated that Romania and Bulgaria might be admitted in 2007, but Turkey's prospects were less favorable.

The 2004 expansion is not without serious complications. The accession treaty required tough negotiations over the rules governing the entry of the ten, the most difficult of which involved agriculture. This issue has already soured relations among the original 15 members, some of whom (notably the United Kingdom, Germany, and the Netherlands) want to phase out direct support payments to farmers, while others, led by France and Spain, like the system as it is. In effect, this pits the more urbanized and industrialized members against those with larger agricultural sectors, and all ten of the new members are in the latter category. How can the enlarged EU afford to pay subsidies to millions of eastern farmers? The current plan is to pay the farmers in the new member-states only 25 percent of the subsidies received by the farmers in the original 15 countries, and to increase these payments to parity for all by 2013. In return, the new members will have to submit to production quotas in order to avoid undue competition on western markets from cheap eastern produce. Not surprisingly, the new members object to this requirement, arguing that their farm industry is key to their prosperity and that constraining it would damage their economic prospects.

The huge EU budget is difficult to unravel or understand, but one thing is clear: the richer countries subsidize the poorer, thus reducing the economic disparities that are Europe's legacy as well as the world's. All member-states, including the newest members, must pay a membership fee to the EU; the richer economies pay far more than the poorer ones, but even the poorer ones must contribute. From the funds thus collected, all members receive subsidies, but the poorer countries, such as Spain, Portugal, and Greece, receive far more than they contribute. These funds are used to build infrastructure and stimulate economic development; in general, the poorer countries have used them to help their most poverty-stricken provinces. Richer members always grumbled about the system, but they have also paid their share. Now, however, ten new members, all of them poor by existing definitions, join the Union—and become eligible for similar subventions. Portugal and Spain will find themselves among the richer members of the newly enlarged EU, and they will lose much of their advantage. Indeed, they may have to help pay for the subsidies to the new members.

The new members, meanwhile, will be faced with membership fees well before receiving any subventions from Brussels (EU's headquarters city). It will take time for the funds they receive to start having an impact, but their obligations to the EU will come due the moment they join. Add to this the conditions attached to farm subsidies, and you can see why not all the voters in the new member-states are enthusiastic over their incorporation (see the Issue Box titled "Europe: How Desirable Is Economic and Political Union?").

Europe: How Desirable Is Economic and Political Union?

IN SUPPORT OF UNION

"As a public servant in the French government I have been personally affected by the events of the past half-century. I am in favor of European unification because Europe is a realm of comparatively small countries, all of which will benefit from economic and political union. Europe's history is bedeviled by its fragmentation—its borders and barriers and favored cores and disadvantaged peripheries, its different economies and diverse currencies, its numerous languages and varied cultures. Now we have the opportunity and determination to overcome these divisions and to create an entity whose sum is greater than its parts, a supranational Europe that will benefit all of its members.

Regional **ISSUE**

"In the process we will not only establish a common currency, eliminate trade tariffs and customs barriers among our members, and enable citizens of one member-state to live and work in any of the others. We will also make European laws that will supersede the old "national" laws and create regulations that will be adhered to by all members on matters ranging from subsidies to help small farms survive to rules regarding environmental protection. In order to assist the poorer members of this European Union, the richer members will send billions of euros to the needy ones, paid for from the taxes levied on their citizens. This has already lifted Spain and Ireland from poverty to prosperity. But to ensure that "national" economies do not get out of line, there will be rules to govern their fiscal policies, including a debt limit of 3 percent of GDP.

"The EU will require sacrifices from people and governments, but the rewards will make these worth it. Our European Parliament, with representatives from all member countries, is the harbinger of a truly united Europe. The European Commission deals with the practical problems as we move toward unification, including the momentous EU enlargement of 2004. Less than a half-century after the first significant steps were taken, the EU has 25 member-states including former wartime enemies, reformed communist economies, and newly independent entities. Rather than struggling alone, these member-states will form part of an increasingly prosperous and economically powerful whole.

"I hope that administrative coordination and economic union will be followed by political unification. The ultimate goal of this great movement should be the creation of a federal United States of Europe, a worthy competitor and countervailing force in a world dominated by another federation, the United States of America."

IN OPPOSITION TO THE EUROPEAN UNION

"I am a taxi driver in Vienna, Austria, and here's something to consider: the bureaucrats, not the workers, are the great proponents of this European Union plan. Let me ask you this: how many member countries' voters have had the chance to tell their governments whether they want to join the EU and submit to all those confounded regulations? It all happened before we realized the implications. At least the Norwegians had common sense: they stayed out of it, because they didn't want some international body to tell them what to do with their oil and gas and their fishing industry. But I can tell you this: if the workers of Europe had had the chance to vote on each step in this bureaucratic plot, we'd have ended it long ago.

"So what are we getting for it? Higher prices for everything, higher taxes to pay for those poor people in southern Italy and needy Portugal, costly environmental regulations, and—here's the worst of it—a flood of cheap labor and an uncontrolled influx of immigrants. Notice that this new 2004 expansion wasn't voted on by the Germans (they're too small to matter much, but Germany is the largest EU member), and do you know why? Because the German government knew well and good that the people wouldn't support this ridiculous scheme. This whole European Union program is concocted by the elite, and we have to accept it whether we like it or not.

"Does make me wonder, though. Do these EU-planners have any notion of the real Europe they're dealing with? They talk about how those "young democracies" joining in 2004 will have their democratic systems and their market reforms locked into place by being part of the Union, but I have friends who try to do business in Poland and Slovakia, and they tell me that there may be democracy, but all they experience is corruption, from top to bottom. I'll bet that those subsidies we'll be paying for to help these "democracies" get on their feet will wind up in Swiss bank accounts. By the way, the Swiss, you'll notice, aren't EU members either. Wonder why?

"A United States of Europe? Not as far as I am concerned. I am Austrian, and I am and will always be a nationalist as far as my country is concerned. I speak German, but already English is becoming the major EU language, though the Brits were smart enough not to join the euro. I'm afraid that these European Commission bureaucrats have their heads in the clouds, which is why they can't see the real Europe they're putting in a straitjacket."

The 2004 expansion is complicated by still another problem. As we note in detail later, Cyprus remains a divided island, with a Greek south and a Turkish north separated by a sharp cultural and economic discontinuity. The Greeks of the south have been the negotiators for Cypriot membership in the EU, claiming to represent the whole island; but the Turks of the north agree to no such thing, especially in view of the EU's treatment of Turkey itself. The admission to membership of only the southern, Greek sector angers Turks both in Cyprus and in Turkey, and this geographic dilemma is a challenge for all involved.

Despite these (and other) problems, what has already been achieved in historically fractious, culturally fragmented Europe is unprecedented. In the west, only Norway, Iceland, and Switzerland remain outside the EU; and in the east, after the accession of Romania and Bulgaria, probably before 2010, only the fragments of the former Yugoslavia, Albania, Ukraine, Moldova, and Belarus are nonmembers. If Turkey can qualify in the next 10 to 15 years, the EU will take another momentous step in its supranational quest.

As our Issue Box statements underscore, not all Europeans are in favor of their Union. Still, the influence of the European Parliament, headquartered in Strasbourg, is growing; border formalities and other obstructions to movement and circulation in Europe are diminishing; evidence of the EU's support system is everywhere; and monetary union has transformed economic life. As of 2004, the 25 member-states of the EU represent more than 450 million people who constitute one of the world's richest markets and who produce more than 40 percent of all exports. Some European leaders continue to dream of a United States of Europe, a federal superpower that will constitute a major force in world affairs as the twenty-first century unfolds.

Centrifugal Forces

Europe has been described as a realm of geographic contradictions. Its current progress toward unification should be seen against a history of conflict, division, and self-destruction. Will supranationalism finally overcome the centrifugal forces that have so long and so frequently afflicted this part of the world?

Even as Europe's states have been working to join forces in the EU, many of those same states are confronting severe centrifugal stresses. The term **devolution** has come into use to describe the powerful centrifugal forces whereby regions or peoples within a state, through negotiation or active rebellion, demand and gain political strength and sometimes autonomy at the expense of the center. Most states exhibit some level of internal regionalism, but the process of devolution is set into motion when a key centripetal, binding force—the nationally accepted idea of what a country stands for—erodes to the point that a regional drive for autonomy, or for outright secession, is launched.

As Figure 1-11 shows, devolution affects numerous European countries. States large and small, young and old, EU members and non-EU members must deal with the problem. Even the long-stable United Kingdom is affected. England, the historic core area of the British Isles, dominates the UK in terms of population as well as political and economic power. The country's three other entities—Scotland, Wales, and Northern Ireland—were acquired over several centuries and attached to England (hence the "United" Kingdom). But neither time nor representative democratic government was enough to eliminate all latent regionalism in these three components of the UK. During the 1960s and 1970s, the British government confronted a virtual civil war in Northern Ireland and rising tides of nationalism in Scotland and Wales. In 1997 the government in London gave the Scots and Welsh the opportunity to vote for greater autonomy in new regional parliaments that would have limited but significant powers over local affairs. The Scots voted overwhelmingly in favor, and the Welsh by a slim majority—and thus a major devolutionary step was taken in one of Europe's oldest, most durable, and most unified states.

Even while this devolutionary process was ongoing, the British government still joined the European Union on behalf not only of the English, but also of the Scots, Welsh, and Northern Irelanders seeking a new relationship with London. As the map shows, the UK is not alone in such contradiction. Spain faces severe devolutionary forces in its Basque area and lesser ones in Catalonia and Galicia; France contends with a secessionist movement on its island of Corsica; Belgium is driven by Flemish-Walloon separatism; Italy confronts devolutionary pressures in South Tyrol and Lombardy. In recent decades Eastern Europe has been a cauldron of devolution as Yugoslavia and Czechoslovakia collapsed, Moldova fragmented, and Ukraine suffered from the stresses of its historical (Russian-penetrated) geography.

Political devolution is not the only centrifugal force to buffet European states. As the European Union materialized, its freedoms (in the form of money flows, labor movements, and transferabilities) led to the emergence of powerful urban regions as hubs of economic power and influence, in some ways beyond the control of their national governments. Examples include the Rhône-Alpes region in France, centered on Lyon; Lombardy in Italy, focused on Milan; Catalonia in Spain, anchored by

FIGURE 1-11

Barcelona; and Baden-Württemberg in Germany, cen-
19 tered on Stuttgart. This group, known as the **Four Motors of Europe**, bypasses not only their national governments in dealing with each other but even extends their
business channels to span the world. The Japanese economist Kenichi Ohmae calls these (and other) economic
20 powerhouses **regional states**, entities that defy old borders and are shaped by the globalizing economy of
which they have become a part.

Like devolutionary forces, these emerging regional
states are changing the map of Europe. In some ways,

they are beginning to supplant the old framework that
arose from the nation-states of an earlier period. Elsewhere, economic development on both sides of international boundaries is having a similar effect, creating
Euroregions that foster more localized, cross-border
cooperation on a continent once rigidly partitioned.

All these developments underscore how far-reaching Europe's current transformation is. The Marshall
Plan jump-started it; good economic times sustained it;
and the end of the Cold War stimulated it. But Europe's
gains remain clouded by violent discord in parts of the

United Kingdom, Spain, France, and the Balkans, by devolutionary forces elsewhere, and by uncertainties over EU expansion and further political integration.

The Muslim Presence

Immigration has been changing the composition of Europe's population since World War II. Eastern Europe has included clusters of Muslim converts since the days of the Ottoman Empire, but many more Muslims have immigrated to other parts of Europe in the postwar period from Turkey, Morocco, Algeria, and other Islamic countries. In 2003 an estimated 28 million documented and undocumented Muslim residents lived in European countries, representing as much as 70 percent of the population in Albania, 43 percent in Bosnia, 30 percent in Macedonia (all, along with Kosovo, small populations with Ottoman histories) and between 3 and 10 percent in such large states as France, Germany, and the United Kingdom, where the numbers are larger and the introduction of Islam more recent (Fig. 1-12). France's Muslim population, largely North African in origin, approaches 5 million; Germany's, mainly Kurds from Turkey, is estimated to be nearly 3.5 million; Britain's 2.1 million Muslims, coming from Pakistan and Bangladesh as well as, more recently, from Somalia, Iraq, Iran, and Afghanistan, are more diverse.

The vast majority of these immigrants are intensely devout, politically aware, and culturally insular. They continue to arrive in a Europe where native populations are stagnant or declining, where religious institutions are weakening, where secularism is rapidly rising, where political positions often appear to be anti-Islamic, and where cultural norms are incompatible with Muslim traditions. Many young men, unskilled and uncompetitive in an EU where unemployment is already high, get involved in petty crime or the drug trade, are harassed by the police, and turn to their faith for solace and reassurance. Unlike most other immigrant groups, Muslim communities also tend to resist assimilation, making Islam the essence of their identity. In Britain alone there are more than 1500 mosques, in themselves a transformation of local cultural landscapes.

In truth, European governments have not done enough to foster the very integration they see Muslims rejecting. The French assumed that their North African immigrants would aspire to assimilation as Muslim children went to public schools from their urban public-housing projects; instead they got into disputes over the dress codes of Muslim girls. The Germans for decades would not award German citizenship to children of immigrant parents born on German soil. The British found their modest efforts at integration stymied by the Muslim practice of importing brides from Islamic countries, which had the effect of perpetuating segregation. In many European cities, the housing projects that form the living space of Muslim residents (by now many of them born and retired there) are dreadful, barren, impoverished environments unseen by Europeans whose images of Paris, Manchester, or Frankfurt are far different.

Under such circumstances, it might surprise outside observers that the architects of the European Union would want to enlarge the EU to include a dominantly Muslim country, Turkey. The accession of Turkey to the EU is likely to be long delayed (it will probably come after Romania and Bulgaria have been admitted), but the first steps are being taken. Turkey is trying to meet certain standards required of all members, notably in the human-rights arena, where the Turks still have a long way to go; the European Commission is working to persuade all of its members, including long-dubious Greece, that Turkish membership would be an asset to the organization. Critics of the initiative say that, once again, this proves that the bureaucratic elite of Europe is out of touch with the people in general, but EU leaders point to the advantages Turkey's incorporation would bring.

First and foremost, it would prove that the "Christian Club" is capable of encompassing a Muslim country, proving its capacity to embrace cultures even more divergent than those of Europe itself. Second, Turkey's membership could lead to the solution of one of Europe's most intractable problems, the political division of Cyprus (see p. 86). Third, Turkey might well become a bridge between Europe and energy-rich Central Asia, where the Turks have historic influence. Fourth, Turkey could become helpful in the long-range stabilization of parts of the Balkans where Christian and Muslim communities still are at odds. And fifth, the several million people of Turkish and Turkish-Kurdish descent who live and work in Germany and other EU countries would, with Turkish accession, cease being outsiders.

In the near future, however, the EU's relationships with its Muslim minorities will remain difficult. Muslim extremism in Europe is not widespread, but Europe was part of the stage for the events of 9/11 and the "War on Terror" has reached into its cities. Governments of EU members find themselves having to consider the impact of their foreign-policy decisions, relating to Islamic issues, on the large Muslim communities in their midst. This has already caused discord between the EU and the United States, notably during and after the 2002 elections in Germany. Once again, Europe faces a cultural challenge with major implications.

EUROPE: MUSLIMS AS PERCENTAGE OF NATIONAL POPULATION, 2004

- Less than 1%
- 1–4%
- 4–7%
- 7–12%
- 12–25%
- More than 25%

0   250   500 Kilometers
0   100   200   300 Miles

Longitude East of Greenwich

FIGURE 1-12

REGIONS OF THE REALM

SO COMPLEX ARE Europe's physical and human geographies that it proves to be one of the most difficult geographic realms when it comes to regionalization. Europe may not be large territorially, but it incorporates enormous environmental, cultural, political, and economic diversity. Europe extends from moist Atlantic shores to continental Russian borders and from Arctic cold to Mediterranean warmth. Small as Europe is, the realm contains nearly 40 countries.

To better comprehend Europe's geographic diversity, we group its countries into five regions based on their proximity, environmental similarities, historical associations, cultural congruities, social parallels, and economic linkages. On such bases, we map (1) Western Europe, (2) the British Isles, (3) Northern (Nordic) Europe, (4) Mediterranean Europe, and (5) Eastern Europe (Fig. 1-13).

Over the past decade, Europeans have begun to distinguish between two parts of Eastern Europe. The westernmost countries in this region, bordering the Baltic Sea and Western Europe, refer to themselves as "Central Europe" as distinct from the "real" Eastern Europe consisting, in their perspective, of Belarus, Ukraine, Moldova, Romania, Bulgaria, Albania, and the fragments of the former Yugoslavia except Slovenia (Fig. 1-13). Note that none of this "real" Eastern Europe is poised to join the European Union; Central Europe, essentially, consists of the ten countries joining in 2004.

The Central Europe concept is undoubtedly temporary, arising in part to confirm the contrasts between the EU entrants and the nonentrants. We will, therefore, retain the framework shown in Figure 1-13, which is based on the formal-region concept discussed on page 6, in order to simplify a complex realm. But it is indeed possible to approach the problem of regional classification in other ways.´

The first part of this chapter focused on Europe's ongoing integration, of which the European Union is only the most prominent manifestation. As the divisive effects of European boundaries diminish and cross-border interconnections increase, a new regionalization of Europe is emerging. Using this functional-region concept, we would recognize a European core, consisting essentially of regions (1) through (4) on Figure 1-13 and a periphery, mostly in region (5). Embedded in this functional framework are the "Four Motors," "Euroregions," and other spatial expressions of Europe's growing interconnections.

The framework in Figure 1-13 suits our purposes better because it provides insight into both Europe's enduring formal regions and its changing functional regions. For all its operational integration, Europe still is a realm whose differences in natural environments, economic opportunities, historical experiences, and relative locations have produced societies and countries with particular, often contrasting cultural traditions and outlooks. These contrasts make Europe's political map what it is: one of the most tortuously fragmented in the world. And as we noted earlier, devolutionary forces are at work to fracture it even further. The outcome of Europe's contest between centripetal and centrifugal forces is far from certain.

MAJOR CITIES OF THE REALM

City	Population* (in millions)
Amsterdam, Netherlands	1.1
Athens, Greece	3.1
Barcelona, Spain	2.7
Berlin, Germany	3.3
Brussels, Belgium	1.1
Frankfurt, Germany	3.7
London, UK	7.6
Lyon, France	1.4
Madrid, Spain	4.0
Milan, Italy	4.3
Paris, France	9.8
Prague, Czech Republic	1.2
Rome, Italy	2.6
Stuttgart, Germany	2.7
Vienna, Austria	2.1
Warsaw, Poland	2.3

*Based on 2004 estimates.

▶ WESTERN EUROPE

Western Europe is the heart of the realm, the hub of its economic power, and the focus of its unifying drive. Germany and France dominate it, flanked by the three Benelux countries in the northwest and the three Alpine states in the east (Fig. 1-14). With a combined population of some 185 million representing eight of the world's richest economies, Western Europe is a powerful force not only in Europe but also on the international stage.

— European
1 Western Eu
2 British Isles
3 Northern (Nordic) Europe
4 Mediterranean Europe
5 Eastern Europe

0   250   500 Kilometers
0 100 200 300 Miles

Longitude East of Greenwich

FIGURE 1-13

FIGURE 1-14

Dominant Germany

Germany today is Europe's most populous country, its most powerful economy, and its strongest force for union. German leaders ceaselessly promote notions of greater European unity, and Germany has been the chief proponent of a Europe unified both politically and economically.

Twice during the twentieth century Germany plunged Europe and the world into war, until, finally, in 1945, the defeated and devastated German state was divided into two parts, West and East (see the notation in red on Fig. 1-15). Its eastern boundaries were also changed, leaving the industrial district of Silesia in newly-defined

Poland, that of Saxony in communist-ruled and Soviet-controlled East Germany, and the Ruhr in West Germany (Fig. 1-14). Aware that these were the industrial centers that had enabled Nazi Germany to seek world domination through war, the victorious allies laid out this new boundary framework to make sure this would not happen again.

In the aftermath of World War II, Soviet and Allied administration of East and West Germany differed. Soviet rule in East Germany was established on the Russian-communist model and, given the extreme hardships the USSR had suffered at German hands during the war, harshly punitive. The American-led authority in West Germany was less strict and aimed more at reha-

FIGURE 1-15

bilitation. When the Marshall Plan was instituted, West Germany was included, and its economy recovered rapidly. Meanwhile, West Germany was reorganized politically into a modern federal state along democratic foundations.

West Germany's economy thrived. Between 1949 and 1964 its GNP tripled while industrial output rose 60 percent. It absorbed millions of German-speaking refugees from Eastern Europe (and many escapees from communist East Germany as well). Since unemployment was virtually nonexistent, hundreds of thousands of Turkish and other foreign workers arrived to take jobs Germans could not fill or did not want.

Simultaneously, West Germany's political leaders participated enthusiastically in the OEEC and in the negotiations that led to the six-member Common Market. Geography worked in West Germany's favor: it had common borders with all but one of the EEC memberstates. Its transport infrastructure, rapidly rebuilt, was second to none in the realm. More than compensating for its loss of Saxony and Silesia were the expanding Ruhr (in the hinterland of the Dutch port of Rotterdam) and the newly emerging industrial complexes centered on Hamburg in the north, Frankfurt (the leading financial center as well) in the center, and Stuttgart in the south. West Germany exported huge quantities of iron, steel, motor vehicles, machinery, textiles, and farm products.

No economy grows without setbacks and slowdowns, however, and Germany experienced such problems in the 1970s and 1980s, when energy shortages, declining competitiveness on world markets, lagging modernization (notably in the aging Ruhr), and social dilemmas involving rising unemployment, an aging population, high taxation, and a backlash against foreign resident workers roiled West German society. And then, quite suddenly, the collapse of the communist Soviet Union opened the door to reunification with East Germany.

In 1990 West Germany had a population of about 62 million and East Germany 17 million. Communist misrule in the East had yielded outdated factories, crumbling infrastructures, polluted environments, drab cities, inefficient farming, and inadequate legal and other institutions. Reunification was more a rescue than a merger, and the cost to West Germany was enormous. When the West German government imposed sales-tax increases and an income-tax surcharge on its citizens, a sizeable percentage of Westerners doubted the wisdom of reunification. It was projected that it would take decades to reconstruct Virginia-sized East Germany: ten years later, exports from the former East still contributed only about 7 percent of the national total.

Once again, however, Germany proved its capacity to recover. Helped by the weakness of the new currency, the euro, German exports rebounded on world markets. Moreover, conditions in the former East Germany began to improve, helped by the relocation of the national capital from Bonn to Berlin. Various statistical indicators that had taken a beating from the merger (lower life expectancies, reduced urbanization, smaller GNP) began to recover as heavy investment in the East took effect. Now in the second decade of its reunified status, Germany's dominance in European affairs has resumed.

The Federal Republic

Before reunification, West Germany functioned as a federal state consisting of ten States (*Länder*) (Fig. 1-15). East Germany was divided under communist rule into 15 districts including East Berlin. Upon reunification, East Germany was reorganized into six new States based on traditional provinces within its borders. Figure 1-15 makes a key point: regional disparity in terms of income per person remains a serious problem between the former East and West. Note that five of former East Germany's six

States (Berlin being the sole exception) are in the two lowest income categories, while *none* of the ten former West German States is in these lowest categories. Nevertheless, the gap is narrowing, and with 82 million inhabitants, over 7 million foreigners, and more than 4 million ethnic Germans born outside the country, Germany is again exerting its dominance over a mainland Europe in which it has no peer.

France

German dominance in the European Union is a constant concern in the other leading Western European country. The French and the Germans have been rivals in Europe for centuries. France (population: 60 million) is an old state, by most measures the oldest in Western Europe. Germany is a young country, created in 1871 after a loose association of German-speaking states had fought a successful war against . . . the French.

Territorially, France is much larger than Germany, and the map suggests that France has a superior relative location, with coastlines on the Mediterranean Sea, the Atlantic Ocean, and, at Calais, even a window on the North Sea. But France does not have any good natural harbors, and oceangoing ships cannot navigate its rivers and other waterways far inland. France has no equivalent to Rotterdam either internally or externally.

The map of Western Europe (Fig. 1-14) reveals a significant demographic contrast between France and Germany. France has one dominant city, Paris, at the heart of the Paris Basin, France's core area. No other city in France comes close to Paris in terms of population or centrality: Paris has 9.8 million residents, whereas its closest rival, Lyon, has only 1.4 million. Germany has no city to match Paris, but it does have a number of cities with populations between 1 and 5 million. And as Table G-1 shows, Germany is much more highly urbanized overall than France.

Why should Paris, without major raw materials nearby, have grown so large? Whenever geographers investigate the evolution of a city, they focus on two important locational qualities: its **site** (the physical attributes of the place it occupies) and its **situation** (its location relative to surrounding areas of productive capacity, other cities and towns, barriers to access and movement, and other aspects of the greater regional framework in which it lies).

The site of the original settlement at Paris lay on an island in the Seine River, a defensible place where the river was often crossed. This island, the *Île de la Cité*, was a Roman outpost 2000 years ago; for centuries its security ensured continuity. Eventually the island became overcrowded, and the city expanded along the banks of the river (Fig. 1-16, left map).

SITE

SITE AND SITUATION OF PARIS, FRANCE

- ▪▪▪ 12th Century Wall
- ═══ Wall of 1840s
- ■ Forests, Parks
- □ Built-up area

0 2 4 6 8 10 Kilometers
0 1 2 3 4 5 Miles

SITUATION

FIGURE 1-16

FROM THE FIELD NOTES

"Because the Seine has a meandering course in an often steep-sided valley, large ships can sail upstream only as far as Rouen, not to the French capital of Paris. At the river's mouth, the port of Le Havre is no great natural harbor, so that the lower Seine is lined with smaller riverside ports handling such commodities as oil, coal, and (as in the photo at the left) scrap iron . . . Beyond and to the southeast of Paris lies Bourgogne (Burgundy), famous for its vineyards and wines and known also for crops such as wheat, rye, mustard, and sunflowers. From the top of an old church I had a superb view of the rural landscape of Burgundy, with its nucleated settlements and its highly fragmented fields. The diversity of colors signifies the variety of crops even in this small area, but farming here is quite efficient. Farmers, who often own several separate parcels of land, use downsized mechanized equipment; in the Burgundy tradition they live in their clustered villages where they have vegetable gardens and keep their few livestock."

Soon the settlement's advantageous situation stimulated its growth and prosperity. Its fertile agricultural hinterland thrived, and, as an enlarging market, Paris's focality increased steadily. The Seine River is joined near Paris by several navigable tributaries (the Oise, Marne, and Yonne). When canals extended these waterways even farther, Paris was linked to the Loire Valley, the Rhône-Saône Basin, Lorraine (an industrial area), and the northern border with Belgium. When Napoleon reorganized France and built a radial system of roads—followed later by railroads—that focused on Paris from all parts of the country, the city's primacy was assured

(Fig. 1-16, right map). The only disadvantage in Paris's situation lies in its seaward access: oceangoing ships can sail up the Seine River only as far as Rouen.

Paris, in accordance with Weber's agglomeration principle, grew into one of Europe's greatest cities. French industrial development was less spectacular, but French agriculture remained Europe's most productive and varied, exploiting the country's wide range of soils and climates. Today France's economic geography is marked by new high-tech industries. It is a leading producer of high-speed trains, aircraft, fiber-optic communications systems, and space-related technologies. It also is the world leader in

AMONG THE REALM'S GREAT CITIES . . . **PARIS**

*I*f the greatness of a city were to be measured solely by its number of inhabitants, Paris (9.8 million) would not even rank in the world's top 20. But if greatness is measured by a city's historic heritage, cultural content, and international influence, Paris has no peer. Old Paris, near the *Île de la Cité* that housed the original village where Paris began and carries the eight-century-old Notre Dame Cathedral, contains an unparalleled assemblage of architectural and artistic landmarks old and new. The *Arc de Triomphe*, erected by Napoleon in 1806 (though not completed until 1836), commemorates the emperor's victories and stands as a monument to French neoclassical architecture, overlooking one of the world's most famous streets, the *Champs Elysées*, which leads to the grandest of city squares, the *Place de la Concorde*, and on to the magnificent palace-turned-museum, the *Louvre*.

Even the Eiffel Tower, built for the 1889 International Exposition over the objections of Parisians who regarded it as ugly and unsafe, became a treasure. From its beautiful Seine river bridges to its palaces and parks, Paris embodies French culture and tradition. It is perhaps the ultimate primate city in the world.

As the capital of a globe-girdling empire, Paris was the hearth from which radiated the cultural forces of Francophone assimilation, transforming much of North, West, and Equatorial Africa, Madagascar, Indochina, and many smaller colonies into societies on the French model. Distant cities such as Dakar, Abidjan, Brazzaville, and Saigon acquired a Parisian atmosphere. France, meanwhile, spent heavily to keep Paris, especially Old Paris, well maintained—not just as a relic of history, but as a functioning, vibrant center, an example to which other cities can aspire.

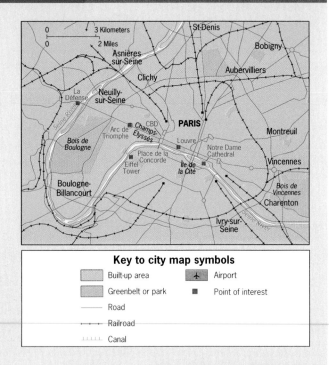

Key to city map symbols

Built-up area		✈	Airport
Greenbelt or park		■	Point of interest
—— Road			
•—•—• Railroad			
⊔⊔⊔⊔ Canal			

Today, Old Paris is ringed by a new and different Paris. Stand on top of the *Arc de Triomphe* and turn around from the *Champs Elysées*, and the tree-lined avenue gives way to *La Défense*, an ultramodern high-rise complex that is one of Europe's leading business districts (see photo, p. 53). But from atop the Eiffel Tower you can see as far as 50 miles (80 km) and discern a Paris visitors rarely experience: grimy, aging industrial quarters, and poor, crowded neighborhoods where discontent and unemployment fester—and where Muslim immigrants cluster in a world apart from the splendor of the old city.

FIGURE 1-17

nuclear power, which currently supplies more than 75 percent of its electricity and thereby reduces its dependence on foreign oil.

When Napoleon reorganized France in the early 1800s, he broke up the country's large traditional subregions and established more than 80 small *départements* (additions and subdivisions later increased this number to 96). Each *département* had representation in Paris, but the power was concentrated in the capital, not in the individual *départements*. France became a highly centralized state and remained so for nearly two centuries (see

inset map, Fig. 1-17). Only the island *département* of Corsica produced a rebel movement, whose violent opposition to French rule continued for decades and even touched the mainland.

Today, France is decentralizing. A new subnational framework of 22 historically significant provinces, groupings of *départements* called *regions* (Fig. 1-17), has been established to accommodate the devolutionary forces felt throughout Europe and, indeed, throughout the world. These regions, though still represented in the Paris government, have substantial autonomy in such areas as taxation, borrowing, and development spending. The cities that anchor them benefit because they are the seats of governing regional councils that can attract investment, not only within France but also from abroad.

Lyon, France's second city and headquarters of the region named Rhône-Alpes, has become a focus for growth industries and multinational firms. This region is evolving into a self-standing economic powerhouse that is becoming a driving force in the European economy; indeed, it is one of the Four Motors of Europe (see p. 59) with its own international business connections to countries as far away as China and Chile.

France has one of the world's most productive and most diversified economies, based in one of humanity's richest cultures. Protecting that cultural heritage in a unifying Europe and a globalizing world is an imperative that causes occasional differences with Germany over the future of the European Union.

Benelux

Three countries are crowded into the northwest corner of Western Europe: Belgium, the Netherlands, and tiny Luxembourg, collectively referred to by their first syllables (*Be-Ne-Lux*). The major differences that evolved between the agriculturally productive Netherlands and the industrially developed Belgium yielded a double complementarity that led to the 1944 Benelux customs union.

The Benelux countries are among the most densely populated on Earth. For centuries the Dutch have been expanding their living space—not by warring with their neighbors but by wresting land from the sea. The greatest project so far, the draining of almost the entire Zuider Zee (Southern Sea), began in 1932 and continues. In the province of Zeeland, islands are being connected by dikes and the water is being pumped out, adding more *polders* (reclaimed lands) to the total national territory.

Geographically, **Belgium** is marked by a cultural fault line that extends diagonally across the state, separating the Flemish in the northwest (55 percent of the total population of 10.3 million) from the Walloons in the southeast (33 percent). Devolution threatens here, but Belgium's great asset is its capital, Brussels, a headquarters of the European Union.

The regional geography of the **Netherlands** (population: 16.2 million) is noted for the *Randstad*, a triangular urban core area anchored by the constitutional capital, Amsterdam, Europe's largest port, Rotterdam, and the seat of government, The Hague. This **conurbation**, as geographers call large multimetropolitan complexes formed by the coalescence of two or more urban areas, forms a ring-shaped complex that encircles a still-rural center (*rand* means edge or margin).

Luxembourg, small as it is, has the distinction of recording by far the highest per-capita GNP in all of Europe (Table G-1). Financial, service, tourist, and information-technology industries make this country, still run by a "Grand Duke," the most prosperous ministate in the realm.

The Alpine States

Switzerland, Austria, and the *microstate* of Liechtenstein on their border share an absence of coasts and the mountainous topography of the Alps—and little else. Austria speaks one language; the Swiss speak German in the north, French in the west, Italian in the southeast, and

FROM THE FIELD NOTES

"Hundreds of bicycles are parked under the overhang of an office building in Utrecht, Netherlands. We watched the commuters arrive in the morning, filling the rows of bike stalls, then leave after work. From executives in business attire to office workers in jeans, all came to work by bicycle. Imagine the size of the parking lot required if these people had driven to work; imagine also the amount of gasoline saved, and pollution avoided. The continued use of the bicycle in the automobile era also has had an effect in limiting the sprawl of Dutch (and other European) cities. In the often cold and rainy weather of the Netherlands, the shorter the bike trip the better, so there is less attraction to living in a distant suburb."

even a bit of Rhaeto-Romansch in the remote central highlands (Fig. 1-8). Austria has a large primate city; multicultural Switzerland does not. Austria has a substantial range of domestic raw materials; Switzerland does not. Austria is twice the size of Switzerland and has a larger population, but far more trade crosses the Swiss Alps between Western and Mediterranean Europe than crosses Austria.

Switzerland, not Austria, is in most ways the leading state in Alpine Western Europe (Table G-1). Mountainous terrain and **landlocked location** can constitute crucial barriers to economic development, tending to inhibit the dissemination of ideas and innovations, obstruct circulation, constrain farming, and divide cultures. That is why Switzerland is such an important lesson in human geography. Through the skillful maximization of their opportunities (including the transfer needs of their neighbors), the Swiss have transformed their seemingly restrictive environment into a prosperous state. They used the waters cascading from their mountains to generate hydroelectric power in order to develop highly specialized industries. Swiss farmers perfected ways to optimize the productivity of mountain pastures and valley soils. Swiss leaders converted their country's isolation into stability, security, and neutrality, making it a world banking giant, a global magnet for money. Zürich, in the German sector, is the financial center; Geneva, in the French sector, is one of the world's most international cities. The Swiss feel that they do not need to join the EU, and they have not done so.

Austria, which joined the EU in 1995, is a remnant of the Austro-Hungarian Empire and has a historical geography that is far more reminiscent of unstable Eastern Europe than of Switzerland. Even Austria's physical geography seems to demand that the country look eastward: it is at its widest, lowest, and most productive in the east, where the Danube links it to Hungary, its old ally in the anti-Muslim wars of the past.

Vienna, by far the Alpine subregion's largest city, also lies on the country's eastern perimeter. One of the world's most expressive primate cities with magnificent architecture and monumental art, Vienna today is Western Europe's easternmost city, situated on the doorstep of fast-changing Eastern Europe. But Austria's cultural landscapes and economic and political standards are those of Western Europe, a nation-state of the European core.

▶ THE BRITISH ISLES

Off the coast of mainland Western Europe lie two major islands, surrounded by a constellation of tiny ones, that constitute the British Isles, a discrete region of the European realm (Fig. 1-18). The larger of the two major islands, which also lies nearest to the mainland (a mere 21 miles, or 34 km, at the closest point), is the island called *Britain*; its smaller neighbor to the west is *Ireland*.

The names attached to these islands and the countries they encompass are the source of some confusion. They still are called the British Isles, even though British dominance over most of Ireland ended in 1921. The nation-state that occupies Britain and a corner of northeastern Ireland is officially called the United Kingdom of Great Britain and Northern Ireland—United Kingdom for short and UK by abbreviation. But this country often is referred to simply as Britain, and its people are known as the British. The nation-state of Ireland officially is the Republic of Ireland (*Eire* in Irish Gaelic), but it does not include the whole island of Ireland.

How convenient it would be if physical and political geography coincided! Unfortunately, the two do not. During the long British occupation of Ireland, which is overwhelmingly Catholic, many Protestants from northern Britain settled in northeastern Ireland. In 1921, when British domination ended, the Irish were set free—except in that corner in the north, where London kept control to protect the area's Protestant settlers. That is why the country to this day is officially known as the United Kingdom of Great Britain and Northern Ireland.

Northern Ireland (Fig. 1-18) was home not only to Protestants from Britain, but also to a substantial population of Irish Catholics who found themselves on the wrong side of the border when Ireland was liberated. Ever since, conflict has intermittently engulfed Northern Ireland and spilled over into Britain and even into Western Europe. It has been and remains one of Europe's costliest struggles.

Although all of Britain lies in the United Kingdom, political divisions exist here as well. England is the largest of these units, the center of power from which the rest of the region was originally brought under unified control. The English conquered Wales in the Middle Ages, and Scotland's link to England, cemented when a Scottish king ascended the English throne in 1603, was ratified by the Act of Union of 1707. Thus England, Wales, Scotland, and Northern Ireland became the United Kingdom.

The British Isles form a distinct region of Europe for several reasons. Britain's insularity provided centuries of security from turbulent Europe, protecting the evolving British nation as it achieved a system of parliamentary government that had no peer in the Western world. Having united the Welsh, Scots, and Irish, the British set out to forge what would become the world's largest colonial empire. An era of mercantilism and domestic manufacturing (the latter based on water power from

FIGURE 1-18

streams flowing off the Pennines, Britain's mountain backbone) foreshadowed the momentous Industrial Revolution, which transformed Britain—and much of the world. British cities became synonyms for specialized products as the smokestacks of factories rose like forests over the urban scene. London on the Thames River anchored an English core area that mushroomed into the headquarters of a global political, financial, and cultural empire. As recently as World War II, the narrow

English Channel ensured the United Kingdom's impregnability against German invasion, giving the British time to organize their war machine. When the United Kingdom emerged from that conflict as a leading power among the victorious allies, it seemed that its superpower role in the postwar era was assured.

Two unanticipated developments changed that prospect: the worldwide collapse of colonial empires and the rapid resurgence of mainland Europe. Always ambivalent about the EC and EU, and with its first membership application vetoed by the French in 1963, Britain (admitted in 1973) has worked to restrain moves toward tighter integration. When most member-states adopted the new euro in favor of their national currencies, the British kept their pound sterling and delayed their participation in the European Monetary Union (EMU). As for a federalized Europe, to Britain this is out of the question. In this as in other respects, Britain's historic, insular standoffishness continues.

The United Kingdom

As we noted earlier, the British Isles as a region consists of two political entities: the United Kingdom and Ireland. The UK, with an area about the size of Oregon and a population of just over 60 million, is by European standards quite a large country. Based on a combination of physiographic, historical, cultural, economic, and politi-

cal criteria, the United Kingdom can be divided into five subregions (numbered in Fig. 1-18):

1. *Southern England.* Centered on the gigantic London metropolitan area, this is a the UK's most affluent subregion, with one-third of the country's population. Financial, communications, engineering, and energy-related industries cluster in this economically and politically dominant area. London exemplifies the momentum of long-term agglomeration; today this subregion benefits anew not only from its superior links to the European mainland but also as one of the three leading "world cities" on the global scene.

2. *Northern England.* As the map suggests, this subregion should really be called Northern, Central, and Western England. Its center, appropriately called the Midlands, was the focus of the Industrial Revolution and the spectacular rise of Manchester, the source; Liverpool, the great port; Birmingham, unmatched industrial city; and

FROM THE FIELD NOTES

"From a ferry on the Mersey River, the skyline of downtown Liverpool seemed to mirror the troubles of this city, the nucleus of a metropolitan area with almost one million inhabitants. The dominant buildings in the CBD remain those built during Liverpool's heyday a century ago, when Liverpool's docks outnumbered even London's, and the city thrived on its trade with America and Africa as Merseyside's manufacturing industries prospered. After World War II, Liverpool declined as its port functions dwindled, its industries weakened, and unemployment soared; this view reveals no modern, glass-sided buildings towering over the older cityscape, symbols of investment and confidence in the future. Once the apex of what was called the *Liverpool Triangle* linking this port to West Africa and eastern America, the Merseyside area now lies remote from the centers of British and European action. Some investment has been made in waterfront redevelopment, but the city's stagnation is evident throughout its historic center (where memories of the famed rock group, the Beatles, have developed into a minor industry). 'There may be ambivalence about the European Union in other parts of Britain,' said my colleague of the Department of Geography at the University of Liverpool as we walked the streets, 'but not here. At least we can see more tangible results of EU investment, for which Liverpool is eligible.' But what will lift Merseyside out of its doldrums?"

AMONG THE REALM'S GREAT CITIES . . . LONDON

*S*ail westward up the meandering Thames River toward the heart of London, and be prepared to be disappointed. London does not overpower or overwhelm with spectacular skylines or beckoning beauty. It is, rather, an amalgam of towns—Chelsea, Chiswick, Dulwich, Hampstead, Islington—each with its own social character and urban landscape. Some of these towns come into view from the same river the Romans sailed 2000 years ago: Silvertown and its waterfront urban renewal; Greenwich with its famed Observatory; Thamesmead, the model modern commuter community. Others somehow retain their identity in the vast metropolis that remains (Fig. 1-9), in many ways Europe's most civilized and cosmopolitan city. And each contributes to the whole in its own way: every part of London, it seems, has its memories of empire, its memorials to heroes, its monuments to wartime courage.

Along the banks of the Thames, London displays the heritage of state and empire: the Tower and the Tower Bridge, the Houses of Parliament (officially known as the Palace of Westminster), the Royal Festival Hall. Step ashore, and you find London to be a memorable mix of the historic and (often architecturally ugly) modern, of the obsolete and the efficient, of the poor and the prosperous. Public transportation, by world standards, is excellent, even though facilities are aging rapidly; traffic, however, often is chaotic, gridlocked by narrow streets. Recreational and cultural amenities are second to none. London seems to stand with one foot in the twenty-first century and the other in the nineteenth. Its airports are ultramodern. But when the Eurostar TGV train from Paris emerges from the Channel Tunnel, it has to slow down for the final hour in its

three-hour run because the tracks into London's Waterloo Station still have not been upgraded to carry it at its normal high speed.

London remains one of the world's most livable metropolises for its size (7.6 million) largely because of farsighted urban planning that created and maintained, around the central city, a so-called Greenbelt set aside for recreation, farming, and other nonresidential, noncommercial uses (Fig. 1-9). Although London's growth eroded this Greenbelt, in places leaving only "green wedges," the design preserved crucial open space in and around the city, channeling suburbanization toward a zone at least 25 miles (40 km) from the center. The map reveals the design's continuing impact.

---

many other manufacturing cities and towns. But here obsolescence has overtaken what was once ultramodern, and the North now suffers from Rustbelt conditions and high unemployment among immigrants from South Asia, Africa, and the Caribbean.

3. *Wales.* This nearly rectangular, rugged territory was a refuge for ancient Celtic peoples, and in its western counties more than half the inhabitants still speak Welsh. Because of the high-quality coal reserves in its southern tier, Wales, too, was engulfed by the Industrial Revolution, and Cardiff, the capital, was once the world's leading coal exporter. But the fortunes of Wales also declined, and many Welsh emigrated. Among the nearly 3 million who remained, however, the flame of Welsh nationalism survived,

and in 1997 the voters approved the establishment of a Welsh Assembly to administer public services in Wales, a first devolutionary step.

4. *Scotland.* Nearly twice as large as the Netherlands and with a population about the size of Denmark's, Scotland is a major component of the United Kingdom. As Figure 1-18 suggests, most of Scotland's more than 5 million people live in the Scottish Lowlands anchored by Edinburgh, the capital, in the east and Glasgow in the west. Attracted there by the labor demands of the Industrial Revolution (coal and iron reserves lay in the area), the Scots developed a worldclass shipbuilding industry. Decline and obsolescence were followed by high-tech development, notably in the hinterland of Glasgow, and Scottish participation

in the exploitation of oil and gas reserves under the North Sea (Fig. 1-18), which transformed the eastern ports of Aberdeen and Leith (Edinburgh). But many Scots feel that they are disadvantaged within the UK and should play a major role in the EU. Therefore, when the British government put the option of a Scottish parliament before the voters in 1997, 74 percent approved. Many Scots hope that total independence lies in their future.

5. *Northern Ireland.* Prospects of devolution in Scotland pale before the devastation caused by political and sectarian conflict in Northern Ireland. With a population of 1.7 million occupying the northeastern one-sixth of the island of Ireland, here we observe the troubled legacy of British colonial rule. About 55 percent of the people in Northern Ireland trace their ancestry to Scotland or England and are Protestants; some 44 percent are Roman Catholics, who share their Catholicism with virtually the entire population of the Irish Republic on the other side of the border. Although Figure 1-18 suggests that there are majority areas of Protestants and Catholics in Northern Ireland, no clear separation exists; mostly they live in clusters throughout the territory, including walled-off neighborhoods in the major cities of Belfast and Londonderry. Partition is no solution to a conflict that has raged for over three decades at a cost of thousands of lives; Catholics accuse London as well as the local Protestant-dominated administration of discrimination, whereas Protestants accuse Catholics of seeking union with the Republic of Ireland. In the late 1990s another mediation effort produced a complicated settlement creating a Northern Ireland Assembly to which powers will be devolved from London, but the violence continued.

Republic of Ireland

What Northern Ireland is missing through its damaging conflicts is shown by the Irish Republic itself: a solid service-based economy, the fastest-growing in all of Europe over the past decade. The capital, Dublin, reflects what is happening all over the country: expanding cities and towns (with heightened real-estate prices), mushrooming industrial parks, bustling traffic (for which the Irish have more colorful adjectives), and construction projects everywhere.

For much of this, the Irish have their membership in the EU to thank. When Ireland joined, the country's economy was sluggish. But Ireland was eligible for the subsidies given by the EU to its poorer members, and

FROM THE FIELD NOTES

"I had heard of the high-technology complex developing near the Dublin (Ireland) airport, and wanted to get a first-hand look. It wasn't easy. Entry into the Clonshaugh Industrial Estate, as it is called, is tightly controlled (left). Once inside, I was refused entry onto the grounds of a major computer manufacturer because I had no prior appointment (the company has since gone out of business here). Numerous high-tech facilities and fulfillment operations reflected the burgeoning economy of Ireland at the time. Meanwhile, land values were soaring in Dublin (and elsewhere), and the old city was taking on the trappings of globalization. Not everyone was favorably disposed toward this development (right), and some warned that the boom that gave Ireland the name of "Celtic Tiger" would not last forever." (By 2002, those who had their doubts turned out to be correct. The Irish economy slowed, and a difficult adjustment is under way.)

this stimulated an economic boom. Improvements in infrastructure made the country attractive to global companies, and for the first time ever, workers of Irish descent began to return from foreign places to take jobs at home. Non-Irish immigrants were also arriving, posing some novel problems for what has always been a closely knit, rather isolated society.

The Republic of Ireland fought itself free from British colonial rule just three generations ago. Its cool, moist climate had earlier led to the adoption of the American potato as the staple crop, but excessive rain and a blight in the late 1840s caused famine and cost over 1 million lives. Another two million Irish emigrated.

Independence in 1921 did not bring economic prosperity, and Ireland stagnated until the 1990s. Then, European telecommunications service industries saw labor and locational opportunities in Ireland, and soon Ireland was Europe's leading call center for the realm's rapidly expanding toll-free telephone market. Other service industries followed suit, benefiting from Ireland's well-educated but not highly paid labor pool. By the late 1990s Ireland was booming, with its economy producing annual surpluses and its image sparkling as the *Celtic Tiger*. That expansion came to a halt in 2002, a victim of the worldwide technology bust, and the country's economic prospects today remain uncertain.

◗ NORTHERN (NORDIC) EUROPE

North of Europe's Western European core area lies a disconnected group of six countries that exemplify core-periphery contrasts. Northern Europe is a region of difficult environments: generally cold climates, poorly developed soils, limited mineral resources, long distances. Together, the six countries in this region—Sweden, Norway, Denmark, Finland, Estonia, and Iceland—contain just over 25 million inhabitants, which is a lower total population than that of Benelux and only one-seventh that of Western Europe. The overall land area, on the other hand, is almost the size of the entire European core. Here in peripheral *Norden*, as the people call their northerly domain, national core areas lie in the south: note the location of the capitals of Helsinki, Stockholm, and Oslo at approximately the same latitude (Fig. 1-19).

Northern Europe's peripheral situation is more than environmental. As viewed from Europe's core area, Norden is on the way to nowhere. No major shipping lanes lead from Western Europe past Norden to other productive areas of the world, limiting interaction of the sort that ties the British Isles to the mainland. Moreover, except for relatively small Denmark and Estonia, all of Norden lies separated from the European mainland by

water. At all levels of spatial generalization, isolation is pervasive in Norden.

Northern Europe's remoteness, isolation, and environmental severity also have had positive effects for this region. The countries of the Scandinavian Peninsula lay removed from the wars of mainland Europe (although Norway was overrun by Nazi Germany during World War II). The three major languages—Danish, Swedish, and Norwegian—are mutually intelligible, which creates one of the criteria delimiting this region. Another regional criterion is the overwhelming adherence to the same Lutheran church in each of the Scandinavian countries (Norway, Sweden, and Denmark), Iceland, and Finland. Furthermore, democratic and representative governments emerged early, and individual rights and social welfare have long been carefully protected. Women participate more fully in government and politics here than in any other region of the world.

Sweden is the largest Nordic country in terms of both population and territory. Most Swedes live south of the zone where the evergreen forest yields to mixed deciduous and needleleaf forests, in what is climatically the most moderate part of the country (Fig. 1-19). Here lie the capital, core area, and, as Figure 1-7 shows, the main industrial districts; here, too, are the main agricultural areas that benefit from the lower relief, better soils, and milder climate.

Sweden long exported raw or semifinished materials to industrial countries, but today the Swedes are making finished products themselves, including automobiles, electronics, stainless steel, furniture, and glassware. Much of this production is based on local resources, including a major iron ore reserve at Kiruna in the far north (there is a steel mill at Luleå). Swedish manufacturing, in contrast to that of several Western European countries, is based in dozens of small and medium-sized towns specializing in particular products. Energy-poor Sweden was a pioneer in the development of nuclear power, but a national debate over the risks involved has reversed that course.

Norway does not need a nuclear power industry to supply its energy needs. It has found its economic opportunities on, in, and beneath the sea. Norway's fishing industry, now augmented by highly efficient fish farms, long has been a cornerstone of the economy, and its merchant marine spans the world. But since the 1970s, Norway's economic life has been transformed by the bounty of oil and natural gas discovered in its sector of the North Sea (Fig. 1-19). In 2000, Norway ranked as the second-biggest exporter of oil in the world.

With its limited patches of cultivable soil, high relief, extensive forests, frigid north, and spectacularly fjorded coastline, Norway has nothing to compare to Sweden's agricultural or industrial development. Its

FIGURE 1-19

cities, from the capital Oslo and the North Sea port of Bergen to the historic national focus of Trondheim as well as Arctic Hammerfest, lie on the coast and have difficult overland connections. Norway has been described as a necklace, its beads linked by the thinnest of strands. But this has not constrained national development. Norway in 2000 had the second-lowest unemployment rate in Europe (after tiny Luxembourg). In terms of income per capita, Norway is the third-richest country in the world.

Norwegians have a strong national consciousness and a spirit of independence. In 1994, when Sweden and Finland voted to join the European Union, the Norwegians again said no. They did not want to trade their economic independence for the regulations of a larger, even possibly safer, Europe.

Denmark, territorially small by Scandinavian standards, has a population of 5.4 million, second largest in the Nordic region after Sweden. It consists of the Jutland Peninsula and several islands to the east at the gateway to the Baltic Sea; it is on one of these islands, Sjaelland, that the capital of Copenhagen is located. Copenhagen, the "Singapore of the Baltic," has long been a port that collects, stores, and transships large quantities of goods. This **break-of-bulk** function exists because many oceangoing vessels cannot enter the shallow Baltic Sea, making the city an **entrepôt** where transfer facilities and activities prevail. The completion of the Øresund bridge-tunnel link to southern Sweden enhanced Copenhagen's situation in 2000 (Fig. 1-19).

Denmark remains a kingdom, and in centuries past Danish influence spread far beyond its present confines. Remnants of that period now challenge Denmark's governance. Greenland came under Danish rule after union with Norway (1380) and remained a Danish domain when that union ended (1814). In 1953 Greenland's status changed from colony to province, and in 1979 the 60,000 inhabitants were given home rule with an Inuit name: *Kalaallit Nunaat*. They promptly exercised their rights by withdrawing from the European Union, of which they had become a part when Denmark joined. Another restive dependency is the Faeroe Islands, located between Scotland and Iceland. These 17 small islands and their 45,000 inhabitants were awarded self-government in 1948, complete with their own flag and currency, but even this was not enough to defuse demands for total independence. A referendum in mid-2001 confirmed that even Denmark is not immune from Europe's devolutionary forces (Fig. 1-11).

Finland, territorially almost as large as Germany, has only 5.2 million residents, most of them concentrated in the triangle formed by the capital, Helsinki, the textile-producing center, Tampere, and the shipbuilding center, Turku (Fig. 1-19). A land of evergreen forests and glacial lakes, Finland has an economy that has long been sustained by wood and wood product exports. But the Finns, being a skillful and productive people, have developed a diversified economy in which the manufacture of machinery and telecommunications equipment as well as the growing of staple crops are important.

As in Norway and Sweden, environmental challenges and relative location have created Nordic cultural landscapes in Finland, but the Finns are not a Scandinavian people; their linguistic and historic links are instead with the Estonians across the Gulf of Finland. As we will see in Chapter 2, ethnic groups speaking Finno-Ugric languages are widely dispersed across what is today western Russia.

Estonia, northernmost of the three "Baltic states," is part of Nordic Europe by virtue of its ethnic and linguistic ties to Finland. But during the period of Soviet control from 1940 to 1991, Estonia's demographic structure changed drastically: today about 30 percent of its 1.4 million inhabitants are Russians, most of whom came there as colonizers.

After a difficult period of adjustment, Estonia today is forging ahead of its Baltic neighbors (see Table G-1) and catching up with its Nordic counterparts. Busy traffic links Tallinn, the capital, with Helsinki, and a new free-trade zone at Muuga Harbor facilitates commerce with Russia.

Iceland, the volcanic, glacier-studded island in the frigid waters of the North Atlantic just south of the Arctic Circle, is the sixth country in this region. Inhabited by people with Scandinavian ancestries (population: 290,000), Iceland and its small neighboring archipelago, the Westermann Islands, are of special scientific interest because they lie on the Mid-Atlantic Ridge, where the Eurasian and North American tectonic plates of the Earth's crust are diverging and new land can be seen forming (see Fig. G-4).

Iceland's population is almost totally urban, and the capital, Reykjavik, contains about half the country's inhabitants. The nation's economic geography is almost entirely oriented to the surrounding waters, whose seafood harvests give Iceland one of the world's highest standards of living—but at the risk of overfishing. Disputes over fishing grounds and fish quotas have intensified in recent decades; the Icelanders argue that, unlike the Norwegians or the British, they have little or no alternative economic opportunity.

▶ MEDITERRANEAN EUROPE

South of Europe's core lie the six countries that comprise the Mediterranean region: Italy, Spain, Portugal, Greece, and the island countries of Cyprus and Malta (Fig. 1-20). Like Northern Europe, this is a region of peninsulas, and again it is discontinuous. It lies separated from the European core, and core-periphery contrasts in some areas are quite sharp. A degree of continuity, dating from Greco-Roman times, marks the region's languages and religion, lifeways, and cultural landscapes. As Figure 1-2 shows, natural environments in this region are dominated by a climatic regime that bears its very name—Mediterranean. Dry, hot summers are the norm, so that moisture is often in short supply during the growing season, and specially adapted plants mark local agriculture.

In terms of raw materials, Southern (Mediterranean) Europe is not nearly as well endowed as the European core, as reflected by the map of industrial complexes (Fig. 1-7). Only northern Italy and northern Spain (the former through massive imports of coal and iron ore) have become part of the core area. Furthermore, the region has been largely deforested, and the limited and highly seasonal water supply constrains Southern Europe's hydroelectric opportunities.

A key contrast between Northern and Southern Europe lies in their populations. Although territorially smaller than Northern Europe, Mediterranean Europe has nearly five times as many people (122 million). Population distribution continues to reflect the agricultural bases of the preindustrial era, with large concentrations in coastal lowlands and fertile river basins, although the growth of major industrial centers, notably in northern Italy and northern Spain, has superimposed a new mosaic. Still, urbanization in Southern Europe is far below that in Western or Northern Europe, or in the British Isles (Portugal's 48 percent is one of the lowest in the realm). Other data indicate that living standards in Mediterranean Europe also lag behind (Table G-1). This was, and remains, one of the challenges the European Union must confront as its leaders strive for equality among the now 25 members.

Italy

Centrally located in the Mediterranean region, most populous of the Mediterranean states, best connected to the European core, and economically most advanced is Italy (58.1 million), a charter member of Europe's Common Market.

AMONG THE REALM'S GREAT CITIES . . . ROME

From a high vantage point, Rome seems to consist of an endless sea of tiled roofs, above which rise numerous white, ochre, and gray domes of various sizes; in the distance, the urban perimeter is marked by highrises fading in the urban haze. This historic city lives amid its past as perhaps no other as busy traffic encircles the Colosseum, the Forum, the Pantheon, and other legacies of Europe's greatest empire.

Founded about 3000 years ago at an island crossing point on the Tiber River about 15 miles (24 km) from the sea, Rome had a high, defensible site. A millennium later, with a population some scholars estimate as high as 1 million, it was the capital of a Roman domain that extended from Britain to the head of the Persian Gulf and from the shores of the Black Sea to North Africa. Rome's emperors endowed the city with magnificent, marble-faced, columned public buildings, baths, stadiums, obelisks, arches, and statuary; when Rome became a Christian city, the domes of churches and chapels added to its luster.

It is almost inconceivable that such a city could collapse, but that is what happened after the center of Roman power shifted eastward to Constantinople (now Istanbul). By the end of the sixth century, Rome probably had fewer than 50,000 inhabitants, and in the thirteenth, a mere 30,000. Papal rule and a Renaissance revival lay ahead, but in 1870, when Rome became the capital of newly united Italy, it still had a population of only 200,000.

Now began a growth cycle that eclipsed all previous records. As Italy's political, religious, and cultural focus (though not an industrial center to match), Rome grew to 1 million by 1930, to 2 million by 1960, and subsequently to 2.6 million where it has leveled off today. The religious enclave of Vatican City, Roman Catholicism's headquarters, makes Rome a twin capital; the Vatican functions as an independent entity and has a global influence that Italy cannot equal.

Rome today remains a city whose economy is dominated by service industries: national and local government, finance and banking, insurance, retailing, and tourism employ three-quarters of the labor force. The new city sprawls far beyond the old, walled, traffic-choked center where the Roman past and the Italian future come face to face.

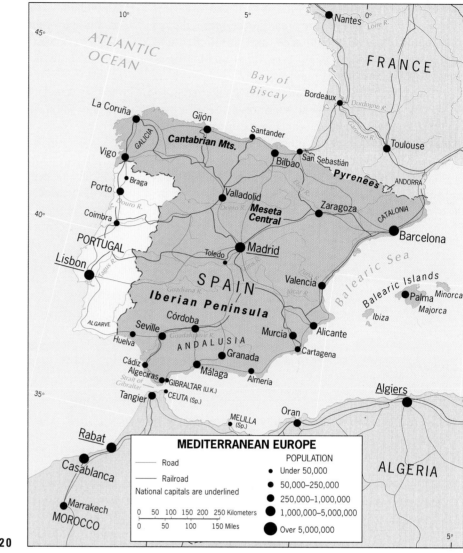

FIGURE 1-20

MEDITERRANEAN EUROPE

Road
Railroad
National capitals are underlined

POPULATION
• Under 50,000
• 50,000–250,000
● 250,000–1,000,000
● 1,000,000–5,000,000
● Over 5,000,000

0 50 100 150 200 250 Kilometers
0 50 100 150 Miles

Administratively, Italy is organized into 20 regions, many with historic roots dating back centuries (Fig. 1-21). Several of these regions have become powerful economic entities centered on major cities, such as Lombardy (Milan) and Piedmont (Turin); others are historic hearths of Italian culture, including Tuscany (Florence) and Veneto (Venice). These regions in the northern half of Italy stand in strong social, economic, and political contrast to such southern regions as Calabria (the "toe" of the Italian "boot") and Italy's two major Mediterranean islands, Sicily and Sardinia. Not surprisingly, Italy is often described as two countries—a progressive north and a stagnant south, or *Mezzogiorno*.

North and south are bound by the ancient headquarters, Rome, which lies astride the narrow transition zone between Italy's contrasting halves. This zone is referred to as the *Ancona Line*, after the city on the Adriatic coast where the zone reaches the other side of the peninsula (Fig. 1-21). Whereas Rome remains Italy's capital and cultural focus, the functional core area of Italy has shifted northward into Lombardy in the basin of the Po River. Here lies Southern Europe's leading manufacturing complex, in which a large, skilled labor force and ample hydroelectric power from Alpine and Appennine slopes combine with a host of imported raw materials to produce a wide range of machinery and precision equipment. The Milan–Turin–Genoa triangle exports appliances, instruments, automobiles, ships, and many specialized products. Meanwhile, the Po Basin, lying on the margins of the region's dominant Mediterranean climatic regime, enjoys a more even pattern of rainfall distribution throughout the year, making it a productive agricultural zone as well.

Metropolitan Milan embodies the new, modern Italy. Not only is Milan (at 4.3 million) Italy's largest city and leading manufacturing center—making Lombardy one of Europe's Four Motors—but it also is the country's financial and service-industry headquarters. Today the Milan area, a cornerstone of the European core, has just 9 percent of Italy's population but accounts for one-third of the entire country's national income.

As in Germany, the lowest-income regions of Italy lie concentrated in one part of the country: the Mezzogiorno in the south. But Italy's north-south disparity continues to grow, which has led to taxpayers' revolts over the subsidies the state pays to the poorer southern regions (a separatist movement has even espoused an independent *Padania* in the north [see Fig. 1-11]). In truth, the south receives the bulk of Italy's illegal immigrants, whether from Africa across the Mediterranean or from the former Yugoslavia and Albania across the Adriatic. Sicily and the Mezzogiorno exemplify the problems of a periphery.

Iberia

At the western end of Southern Europe lies the Iberian Peninsula, separated from France and Western Europe by the rugged Pyrenees and from North Africa by the narrow Strait of Gibraltar (Fig. 1-20). Spain (population: 41.4 million) occupies most of this compact Mediterranean landmass. Portugal lies in its southwestern corner.

Imperial Romans, Muslim Moors, and Catholic kings left their imprints on Iberia, notably the boundary between Spain and Portugal dating from the twelfth century. The golden age of colonialism was followed by dictatorial rule and economic stagnation.

FIGURE 1-21

Today both Spain and Portugal are democracies, and their economies are doing well, helped in the case of Portugal by EU subsidies. As Figure 1-22 shows, **Spain** followed the leads of Germany and France and decentralized its administrative structure, creating 17 regions called Autonomous Communities (ACs). Each AC has its own parliament and administration that control planning, public works, cultural affairs, education, environmental matters, and even, to some extent, international commerce. Each AC can negotiate its own degree of autonomy with the central government in Madrid. This new system, however, has not been enough to defuse Spain's most problematic devolutionary issue, which involves

the Basques in the AC mapped as Basque Country (Fig. 1-22, especially the top inset map).

Particularly noteworthy is Catalonia, the triangular AC in Spain's northeastern corner, adjacent to France. Centered on prosperous, productive Barcelona, Catalonia is Spain's leading industrial area, an AC with a population of 6.5 million. It is imbued with a fierce nationalism, endowed with its own language and culture, and economically it is one of the Four Motors of Europe. Note that while most of Spain's industrial raw materials lie in the northwest, its major industrial development has taken place in the northeast, where innovations and skills propel a high-technology-driven regional economy. In

FIGURE 1-22

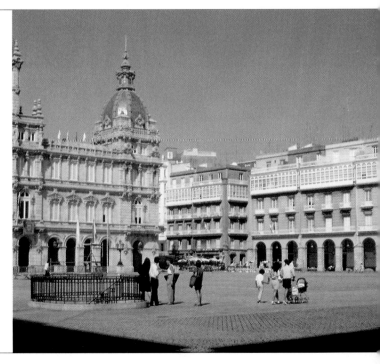

"Every major Spanish city has its *plaza*, a central square flanked by its most impressive public and religious buildings. They tend to look the same, but the details tend to have local or regional significance. This one, in the city of La Coruña in the northwestern corner of Spain, displayed a feature I had seen along many streets: to the classic architecture had been added small, white-framed and windowed balconies called *miradores*. I assumed that this innovation had some environmental cause, and when I asked about it I was told that La Coruña is the windiest city in Spain, and that the specially glazed windows were installed to cope with this. High-velocity winds, of course, are a feature of Mediterranean climates; as the map shows, La Coruña lies especially exposed. 'But please realize,' said the local teacher I had asked about this, 'that there's more to La Coruña than interesting baroque architecture and *miradores*. This was a Roman port, then held by the Moors, then the Portuguese, now Madrid.' Now Madrid? 'Yes, this is Galicia, not just another division of Spain, and we're not only closer to Western Europe than any other part of Iberia, but we're on the direct route between Western Europe and eastern South America. We want more autonomy, more identity.' Here we go again, I thought as I took my notes. Identity. Autonomy. Devolution."

recent years, Catalonia—with 6 percent of Spain's territory and 16 percent of its population—has annually produced 25 percent of all Spanish exports and nearly 40 percent of its industrial exports. Such economic strength translates into political power, and in Spain the issue of Catalonian separatism is never far from the surface.

Ironically, Catalan nationalism was spurred by the very EU subsidies that were intended to redress imbalances in member-states. Spain received hundreds of millions of euros from Brussels, and the central government decided to spend a substantial part of this fund to build a high-speed train link between Madrid and Seville in Andalusia, the poor, southern AC on the Mediterranean coast (Fig. 1-22). This, the government hoped, would stimulate development where it was most needed. But public opinion in Catalonia held that Barcelona, not Seville, should have been rewarded with that modern rail link, and Catalan nationalism got another boost.

As Figure 1-22 indicates, Spain's capital and largest city, Madrid, lies near the geographic center of the state. It also lies along an economic-geographic divide. Catalonia and Madrid are Spain's most prosperous ACs; the contiguous group of five ACs between the Basque Country and Valencia rank next. Tourism (notably along the Mediterranean coast) and winegrowing (especially in La Rioja) contribute importantly here. Ranking below this cluster of ACs are the four such units of the northwest: Galicia, Castile-Leon, and industrialized Asturias and Cantabria. Incomes in these regions are below the national average because industrial obsolescence, dwindling raw material sources, and emigration have plagued their economies. Worst off, however, are the three large ACs to the south of Madrid: Extremadura, Castile-La Mancha, and especially Andalusia. Drought, inadequate land reform, scarce resources, and remoteness from Spain's fast-growing northeast are among the factors that inhibit development here. Even as EU members seek to reduce the differences among themselves, individual countries face the geographic consequences of focused growth that deepens internal divisions.

Although Spain and the United Kingdom are both EU members committed to cooperative resolution of territorial problems, the two countries are embroiled in a dispute over a sliver of land at the southern tip of Iberia: legendary Gibraltar (see lower inset map, Fig. 1-22). "The Rock" was ceded by Spain to the British in perpetuity in 1713 (though not the neck of the peninsula linking it to the mainland, which the British later occupied) and has been a British colony ever since. The 30,000 residents are used to British institutions, legal rules, and schools. Successive Spanish governments have demanded that Gibraltar be returned to Madrid's rule, but the colony's residents are against it. And under their 1969 constitution, Gibraltarians have the right to vote on any transfer of sovereignty.

Now the British and Spanish governments are trying to reach an agreement under which they would share the administration of Gibraltar for an indefinite time; both are EU members, so little would have to change. The advantages would be many: economic development is being slowed by the ongoing dispute, border checks would be lifted, EU benefits would flow. But Gibraltarians have their doubts and want to put the issue to a referendum. Spain will not accept the idea of a referendum, and the matter is far from resolved.

Spain's refusal to allow, and abide by, a referendum in Gibraltar stands in remarkable contrast to its demand for a referendum in its own outposts, two small exclaves on the coast of Morocco, Ceuta and Melilla (Fig. 1-22). Morocco has been demanding the return of these two small cities, but Spain has refused on the grounds that the local residents do not want this. The matter has long simmered quietly but came to international attention in 2002 when a small detachment of Moroccan soldiers seized the island of Perejil, an uninhabited, Spanish possession off the Moroccan coast also wanted by Morocco. In the diplomatic tensions that followed, the entire question of Spain's holdings in North Africa (and its anti-Moroccan stance in the larger matter of Western Sahara) exposed some contradictions Madrid had preferred to conceal.

The state of **Portugal** (population: 10.4 million), a comparatively poor country that has benefited enormously from its admission to the EU, occupies the southwestern corner of the Iberian Peninsula. One rule of EU membership is that the richer members assist the poorer ones; Portugal shows the results in a massive renovation project in its capital, Lisbon, as well as in the modernization of surface transport routes.

Unlike Spain, which has major population clusters on its interior plateau as well as its coastal lowlands, Portugal has a population that is concentrated along and near the Atlantic coast. Lisbon and the second city, Porto, are coastal cities; the best farmlands lie in the moister western and northern zones of the country. But the farms are small and inefficient, and although Portugal remains dominantly rural, it must import as much as half of its foodstuffs. Exporting textiles, wines, cork, and fish, and running up an annual deficit, the indebted Portuguese economy remains a far cry from those of other European countries of similar dimensions.

Greece and Cyprus

At the eastern end of the region we define as Southern Europe lies **Greece**, an outlier of both Mediterranean Europe and the European Union. Greece's land boundaries are with Turkey, Bulgaria, Macedonia, and Albania; as Figure 1-20 reveals, it also owns islands just offshore from mainland Turkey. Altogether, the Greek archipelago numbers some 2000 islands ranging in size from Crete (3218 square miles [8335 sq km]) to small specks of land in the Cyclades. In addition, Greeks represent the great majority on the now-divided island of Cyprus, the southern part of which aspires to EU membership.

Ancient Greece was a cradle of Western civilization, and later it was absorbed by the expanding Roman Empire. For some 350 years beginning in the mid-fifteenth century, Greece was under the sway of the Ottoman Turks. Greece regained independence in 1827, but not until nearly a century later, through a series of Balkan wars, did it acquire its present boundaries. During World War II, Nazi Germany occupied and ravaged the country, and in the postwar period the Greeks have quarreled with the Turks, the Albanians, and the newly independent Macedonians. Today, Greece finds itself between the Muslim world of Southwest Asia and the Muslim communities of Eastern Europe; it is still the only noncontiguous mainland EU member even after the accession of ten more states in 2004.

Volatile as Greece's surroundings are, Greece itself is a country on the move, an EU success story to rival Ireland's. Political upheavals in the 1970s and economic stagnation in the 1980s are all but forgotten in the new century: now Greece, its economy booming, is described as the locomotive for the Balkans, a beacon for the EU in a crucial part of the world. Infrastructure improvements focused on the Athens urban area, where more than one-third of the population is concentrated, include new subways, a new beltway, and a new airport. Meanwhile, Greece is host for the 2004 Summer Olympics, which entails another infusion of capital investment that will make a permanent mark on the capital. Longstanding concerns over a potential domestic terrorist threat against the Olympic Games

AMONG THE REALM'S GREAT CITIES . . . ATHENS

Take a map of Greece. Draw a line to encompass its land area as well as all the islands. Now find the geographic center of this terrestrial and maritime territory, and you will find a major city nearby. That city is Athens, capital of modern Greece, culture hearth of ancient Greece, source of Western civilization.

Athens today is only part of a greater metropolis that sprawls across, and extends beyond, a mountain-encircled, arid basin that opens southward to the Bay of Phaleron, an inlet of the Aegean Sea. Here lies Piraeus, Greece's largest port, linked by rail and road to the adjacent capital. Other towns, from Keratsinion in the west to Agios Dimitrios in the east, form part of an urban area that not only lies at, but *is*, the heart of modern Greece. With a population of 3.1 million, most of the country's major industries, its densest transport network, and a multitude of service functions ranging from government to tourism, Greater Athens is a well-defined national core area.

Arrive at the new Eleftherios Venizelos Airport east of Athens and drive into the city, and you can navigate on the famed Acropolis, the 500-foot (150-m) high hill that was ancient Athens' sanctuary and citadel, crowned by one of humanity's greatest historic treasures, the Parthenon, temple to the goddess Athena. It is only one among many relics of the grandeur of Athens 25 centuries ago, now engulfed by mainly low-rise urbanization and all too often obscured by smog generated by factories and vehicles and trapped in the basin. Conservationists decry the damage air pollution has inflicted on the city's remaining historic structures, but in truth the greatest destruction was wrought by the Ottoman Turks, who plundered the country

during their occupation of it, and the British and Germans, who carted away priceless treasures to museums and markets in Western Europe.

Travel westward from the Middle East or Turkey, and Athens will present itself as the first European city on your route. Travel eastward from the heart of Europe, however, and your impression of its cultural landscape may differ: behind the modern avenues and shopping streets are bazaar-like alleys and markets where the atmosphere is unlike that of any other in this geographic realm. And beyond the margins of Greater Athens lies a Greece that lags far behind this bustling, productive coreland.

was defused in 2002 through a series of arrests. Given the location of the 2004 Summer Games, the Greek organizers obviously faced additional security challenges, but the breakup of the left-wing terrorist movement called November 17th was a key achievement.

Greece will confront serious challenges in the post–2004 period. With the accession of ten Eastern European members to the EU, Greece will lose much of its EU subsidy to these poorer countries. Greece's democratic institutions still need strengthening, and corruption remains a serious problem. Educational institutions need modernization. And while Greece is on better terms today with its fractious neighbors and invests in development projects in Macedonia and Bulgaria (for example, an oil pipeline from Salonika to a Greek-owned refinery in Skopje, Macedonia), it will take skillful diplomacy to navigate the shoals of historic discord.

Modern Greece is a nation of 11 million centered on historic Athens, one of the realm's great cities. With its port of Piraeus, metropolitan Athens contains about 40 percent of the Greek population, making it one of Europe's most congested and polluted urban areas. Athens is the quintessential primate city; the monumental architecture of ancient Greece still dominates its cultural landscape. The Acropolis and other prominent landmarks attract a steady stream of visitors; tourism is one of Greece's leading sources of foreign revenues, and Athens is only the beginning of what the country has to offer.

Deforestation, soil erosion, and variable rainfall make farming difficult in much of Greece, but the country remains strongly agrarian and is self-sufficient in sta-ple foods. Greece remains the poorest EU country in terms of GNP and suffers from a huge and persistent trade deficit.

Cyprus lies in the far northeast corner of the Mediterranean Sea, much closer to Turkey than to Greece (Fig. 1-13), but it is peopled dominantly by Greeks rather than Turks. In 1571 the Turks conquered Cyprus, then ruled by Venice, and controlled it until 1878 when the British took over. Most of the island's Turks arrived during the Ottoman period; the Greeks have been there longest.

When the British were ready to give Cyprus independence after World War II, the 80-percent-Greek majority mostly preferred union with Greece. Ethnic conflict followed, but in 1960 the British granted Cyprus independence under a constitution that prescribed majority rule but guaranteed minority rights.

This fragile order broke down in 1974, and civil war engulfed the island. Turkey sent in troops and massive dislocation followed, resulting in the partition of Cyprus into northern Turkish and southern Greek sectors separated by the "Green Line" (Fig. 1-20, inset map). In 1983 the 40 percent of Cyprus under Turkish control, with about 100,000 inhabitants (and some 30,000 Turkish soldiers), declared itself the independent Turkish Republic of Northern Cyprus. Only Turkey recognizes this ministate (which now contains a population of 190,000); the international community recognizes the government on the Greek side as legitimate. With about 900,000 residents, a relatively prosperous economy based on agriculture and tourism, and strong links with Europe, the Greek-side government succeeded in placing Cyprus

FROM THE FIELD NOTES

"For a few days in early 2003, the island ministate of Malta held the attention of Europe and, indeed, the world. It was the first of the ten states joining the European Union in 2004 to perform one of the requirements for entry: a referendum on the question among the voters. The multicultural Maltese are very active politically, and vigorous campaigns for as well as against EU membership crested with large and vocal meetings. Polls suggested that the vote would be close, and EU leaders worried: what if the very first country to test its will were to reject the initiative? Might this lead other candidates to do the same, threatening the whole EU expansion plan? But Malta's government, which favored incorporation, persuaded an albeit slender majority that Malta would benefit from EU membership, and the yes vote carried the day. Valletta, the capital (shown here) could celebrate."

among the ten countries joining the EU in 2004, even though the *de facto* partition of the island continues.

Late in 2002, a ray of hope appeared in the form of a map drawn at the United Nations that would alter the partition line in such a way as to award the Greek majority a larger share of Cyprus' territory (see inset map, Fig. 1-20), in return for which the Greek majority would abandon claims to all other areas in the north, including the Karpas Peninsula. On this basis, a federal system would bestow significant autonomy on each of the two parts of what would become a reunified state, all of which would join the EU in 2004. Public demonstrations and polls on both sides of the partition line indicated majority approval of this plan, but the head of the Turkish Republic of Northern Cyprus would not endorse it. Nevertheless, in another important development, the two governments agreed to open the gates along the Green Line so that Greeks and Turks could cross over and visit their former homes, and tens of thousands took that opportunity on an April day in 2003.

Should the new Cyprus plan succeed, the Turkish north, as part of the new state, would become the first Islamic territory to join the EU. As such it could be a vanguard of Turkish-Greek cooperation elsewhere, and of the entry into the European Union of Turkey itself.

The Mediterranean region contains one other ministate, **Malta**, located south of Sicily. Malta is a small archipelago of three inhabited and two uninhabited islands with a population of just under 400,000. An ancient crossroads and culturally rich with Arab, Phoenician, Italian, and British infusions, Malta became a British dependency and served British shipping and its military. It suffered terribly during the World War II bombings, but despite limited natural resources recovered during the postwar period. Today Malta has a booming tourist industry and a relatively high standard of living, an economy meriting membership in the EU.

▶ EASTERN EUROPE

As Figure 1-13 shows, Eastern Europe is not only territorially the largest region in the European realm: it also contains more countries (17) than any other European region. Almost all of Eastern Europe lies outside the core, and the problems of the periphery affect many of its countries. From the North European Lowland in Poland to the rugged highlands of the south, this is a region of physiographic, cultural, and political fragmentation. Open plains, major rivers, strategic mountains, isolated valleys, and crucial corridors all have influenced Eastern Europe's tumultuous migrations, epic battles, foreign invasions, and imperial episodes. Illyrians, Slavs, Turks, Hungarians, and other peoples converged on this region from near and far. Ethnic and cultural differences have kept them in chronic conflict.

Geographers call this region a **shatter belt**, a zone [27] of persistent splintering and fracturing. Geographic terminology uses several expressions to describe the breakup of established order, and these tend to have their roots in this part of the world. One of them is **balkanization**. [28] The southern half of Eastern Europe is referred to as the Balkans or Balkan Peninsula, after the name of a mountain range in Bulgaria. Balkanization denotes the recurrent division and fragmentation of this part of Eastern Europe, and it is now applied to any place where such processes take place. A more recent term is *ethnic cleansing*—the forcible ouster of entire populations from their homelands by a stronger power bent on taking their territories. The term may be new, but the process is as old as Eastern Europe itself.

Each episode in the historical geography of Eastern Europe has left its legacy in the cultural landscape. Twenty centuries ago the Roman Empire ruled much of it (Romania is a cartographic reminder of that period); during the past half-century, the Soviet Empire controlled almost all of it. In the intervening two millennia, Christian Orthodox church doctrines spread from the southeast, and Roman Catholicism advanced from the northwest. Turkish (Ottoman) Muslims invaded and created an empire that reached the environs of Vienna. By the time the Austro-Hungarian Empire ousted the Turks, millions of Eastern Europeans had been converted to Islam. Albania today remains a dominantly Muslim country. In the twentieth century, Eastern Europe became a battleground between superpowers, and the complicated map reflects the results through 1991 (Fig. 1-23).

The subsequent collapse of the Soviet Union freed several of Russia's neighbors, and with only one exception— Belarus—these countries turned their gaze from Moscow to the west, specifically to the European Union and its economic promise. This changed the map of Eastern Europe, repositioning the realm boundary eastward and adding five countries to the region (Fig. 1-24): Latvia, Lithuania, Belarus, Moldova, and Ukraine. Meanwhile, however, two of Eastern Europe's established states fell apart: Czechoslovakia peacefully, Yugoslavia violently.*

---

*Between 1919 and 1991, the name *Yugoslavia* referred to the multicultural state shown in Figure 1-23. When civil war destroyed the old order, parts of Yugoslavia became independent states, including Slovenia, Croatia, Bosnia, and Macedonia. The largest component, dominated by the Serbs, kept the name *Yugoslavia*. But in 2002 the then-Yugoslav government announced that it would agree to a name change, calling their country Serbia-Montenegro after its two main components. (The future of Kosovo, under United Nations administration but technically still a part of Serbia-Montenegro, remained unresolved in late 2003.)

FIGURE 1-23

The Geographic Framework

To clarify this complex regional geography, we group Eastern Europe's 17 countries by geographic subregion:

1. Countries Facing the Baltic Sea
2. The Landlocked Center
3. Countries Facing the Black Sea
4. Countries Facing the Adriatic Sea

Countries Facing the Baltic Sea

Poland (population: 39 million) dominates this subregion, which also includes Lithuania and Latvia and, by virtue of relative location, Belarus (although Belarus does not possess a Baltic Sea coast). Wedged between coastal Poland and Lithuania is the Russian exclave of Kaliningrad, which is not functionally a part of Eastern Europe (see Chapter 2).

FIGURE 1-24

Figure 1-23 displays Poland's most recent boundary shifts: in the aftermath of World War II, the whole country shifted westward, losing land to the (then) USSR in the east and gaining it at the expense of Germany to the west. Situated on the North European Lowland, Poland traditionally was an agrarian country, but during the Soviet-communist period Silesia (once part of Germany) became its industrial heartland, and Katowice, Wroclaw, and Krakow grew into major industrial cities amid some of the world's worst environmental pollution. The Soviets invested far less in agriculture, collectivizing farms

without modernizing the technology and leaving post-Soviet farming in a decrepit state.

Lithuania (3.5 million) and Latvia (2.3 million) face the Baltic Sea north of Poland. **Lithuania**, remnant of the Grand Duchy of Lithuania that once dominated Eastern Europe from the Baltic to the Black Sea, is centered on the interior capital of Vilnius. When Kaliningrad became a Russian **exclave** as a World War II prize, Lithuania was left with only about 50 miles (80 km) of Baltic coastline and a small port (Klaipeda) that was not even connected to Vilnius by rail. But since Lithuania (like

Latvia and Estonia) fell under Soviet domination, that seemed to matter little as its trade was with the Soviet Union, not the outside world.

Latvia, on the other hand, is centered on the Baltic port of Riga, but Latvia experienced far more Russian immigration than Lithuania did: today Latvians make up only about 54 percent of the population. Its economy was geared almost totally to Moscow, and after the half-century of Soviet domination over the Baltics (1940–1991) ended, Latvia was left with huge economic problems and few options. Nonetheless, Latvia, along with its Baltic neighbors, Lithuania and Estonia, were among the ten countries approved for inclusion in the European Union in 2004.

Belarus has no Baltic coast, but it borders all three of the coastal states in this subregion. About 80 percent of its nearly 10 million people are Belarussians ("White" Russians), a West Slavic people. Only some 13 percent are (East Slavic) Russians. Devastated during World War II, Belarus became one of Moscow's most loyal satellites, and the Soviets made Mensk (Minsk), the capital, into a large industrial center. But in the post-Soviet era, Belarus has lagged badly, and its government has retained powers reminiscent of the communist era. Unlike its neighbors, Belarus has no interest in joining the EU. On the contrary, its overtures have been toward Moscow: it seeks to rejoin Russia in some formal way.

The Landlocked Center

Three states comprise this subregion: the Czech Republic, Slovakia, and Hungary. Until 1993, the first two formed the state of Czechoslovakia, but their "velvet divorce" broke it up. The Czechs got the better of the deal; Slovakia is this subregion's poorest country by far.

The **Czech Republic** (10.3 million) centers on Bohemia, the mountain-enclosed core area that contains the historic capital, Prague. This province always has been cosmopolitan and Western in its exposure, outlook, development, and linkages. Prague lies in the basin of the Elbe River, Bohemia's traditional outlet through northern Germany to the North Sea. It is a classic primate city, its cultural landscape faithful to Czech traditions (see photo, p. 92); but it also is an industrial center. The encircling mountains contain many valleys with small towns that specialize, Swiss-style, in fabricating high-quality goods. In Eastern Europe the Czechs always were the leaders in technology and engineering; even during the communist period their products found markets in foreign countries near and far.

As Figure 1-24 shows, the Czech Republic incorporates an eastern province named Moravia, linked to Poland's Silesia through the Moravian Gate between the Sudeten and Carpathian Mountains. Here the Soviets brought their industrialization drive, making Ostrava and Brno important manufacturing cities but leaving behind obsolescence and environmental problems.

The separation of the Czech and Slovak societies created a virtual nation-state in the Czech Republic, with only a small minority of several hundred thousand Roma, or gypsies, among the Czechs. The treatment of this minority has become an issue in the discussions leading to Czech membership in the EU.

As Figure 1-25 shows, **Slovakia** inherited Czechoslovakia's largest minority, the ethnic Hungarian com-

FIGURE 1-25

munity, which now constitutes about 11 percent of the country's 5.4 million inhabitants and is concentrated in the southern zone along the Danube River. The country's division resulted from ethnic and cultural factors, but the Slovaks also did not share the Czechs' enthusiasm for post-Soviet economic reforms. The Hungarian minority found itself at odds with the new government in Bratislava, but the issues between them have been shelved in the interest of the 2004 accession to the EU, which both sides want.

Hungary also approaches the status of nation-state, but there are Hungarians not only in Slovakia but also in Serbia-Montenegro, Austria, and Romania (Fig. 1-25),

all remnants of a time when Hungary ruled much of this region. Hungarian governments have repeatedly expressed support for these ethnic cohorts in neighboring countries, a practice referred to as **irredentism**. The term derives from a nineteenth-century campaign by Italy to incorporate an Italian-speaking area of Austria, calling it *Italia Irredenta* (Unredeemed Italy).

The Hungarians (Magyars) moved into the middle Danube River Basin more than a thousand years ago from an Asian source; they have neither Slavic nor Germanic roots. They converted their fertile lowland into a thriving state while retaining their cultural and linguistic identity, eventually forging an imperial power in the region. The

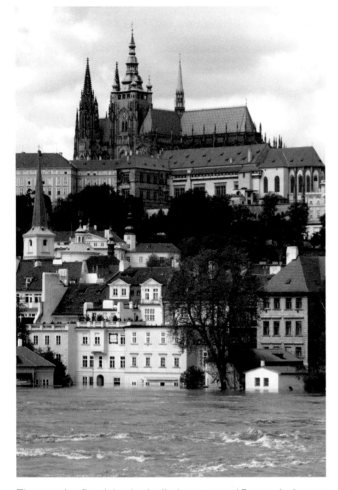

The massive flood that bedeviled east-central Europe in August 2002 was at its worst in the upper basin of the Elbe River. One of the cities hit hardest was the Czech capital, Prague, where the Vltava River overflowed its banks to an extent not seen in more than two centuries. In this photo, taken at the height of the disaster, much of the Old City (Mala Strana) district is inundated, which necessitated the evacuation of more than 40,000 residents. Looming above it all are the spires of St. Vitus Cathedral, located at the heart of the Prague Castle complex that is the Czech Republic's seat of government and most revered national monument.

twin-cities capital astride the Danube, Buda and Pest (better known as Budapest), is a primate city nearly ten times the size of the next largest town in Hungary—reflecting the continuing rural character of the country.

Hungary's inclusion in the group of ten states acceding to the EU in 2004 is especially important, but Hungary is an economically divided country. Look again at Figure 1-24, on which you could draw a line from north to south, just east of Budapest, and discern two Hungaries: a relatively prosperous west, where foreign investment has produced a progressive economic geography, and a much poorer, much more agrarian (and less populous) east, which is more like rural Poland and Romania. With a population of 10 million and a western

sector that shows the way, Hungary should stand among those new EU members that prove the wisdom of the 2004 expansion.

Countries Facing the Black Sea

Four countries form Eastern Europe's Black Sea quadrant: Ukraine, Moldova, Romania, and Bulgaria (Fig. 1-24). All except Moldova have coastlines on the Black Sea, but none of their core areas or capital cities lies on the coast. This reflects an inward orientation that characterizes the subregion as a whole.

Ukraine is Eastern Europe's most populous country (47.5 million); territorially, it is also the largest state in the entire realm. Its capital, Kiev (Kyyiv), is a major historic, cultural, and political focus. Briefly independent before the communist takeover in Russia, Ukraine regained its sovereignty as a much-changed country in 1991. Once a land of farmers tilling its famously fertile soils, Ukraine emerged from the Soviet period with a huge industrial complex in its east—and with a large (over 20 percent) Russian minority. Ukraine's boundaries also changed during the Soviet era. In 1954 a Soviet dictator capriciously transferred the entire Crimea Peninsula, including its Russian inhabitants, to Ukraine as a reward for its productivity.

The Dnieper River forms a useful geographic reference to comprehend Ukraine's spatial division (Fig. 1-24). To its west lies agrarian, rural, mainly Roman Catholic Ukraine; in its great southern bend and eastward lies industrial, urban, Russified (and Russian Orthodox) Ukraine. Soviet planners made eastern Ukraine a communist Ruhr based on local coal and iron ore, making the Donets Basin (*Donbas* for short) a key industrial complex. Meanwhile, the Russian Soviet Republic supplied Ukraine with oil and gas.

Ukraine is a critically important country for Europe but is in trouble today. Political mismanagement and corruption, a faltering economy, falling social indicators, and rising crime are causing many Ukrainians to call for a reunion with Russia and a return to the "good old days." And yet this country has access to international shipping lanes, a large resource base, massive farm production, educated and skilled labor, and a large domestic market. A legacy of dependence is taking its toll.

Moldova, Ukraine's small and impoverished neighbor, was a Romanian province seized by the Soviets in 1940 and made into a (deliberately landlocked) "Soviet Republic" of the USSR. Romanians remain in the majority among it 4.3 million people, but most of the Russians and Ukrainians (each about 13 percent) have moved to the small strip of land between the Dniester River and the Ukrainian border, proclaiming there a "Republic of Transdniestra." Devolutionary forces form only one of

Moldova's problems; its farm economy is declining steadily, and by many measures this is now Europe's poorest country.

The situation in **Romania** is not much better; it is actually more serious because more than 22 million people are affected. Once known for its occasional defiance of Moscow and for its energy exports, Romania has a stagnant economy and a political system that is still far less democratic than EU membership would require. Its oil and gas reserves are nearly exhausted, agricultural output is declining, a subsistence economy is spreading, and the country is deeply in debt. Many talented Romanians are leaving in search of opportunity elsewhere.

Romania's drab and decaying capital, Bucharest, and its surrounding core area lie in the interior linked by rail to the Black Sea port of Constanta. Once known as Eastern Europe's most civilized society, its capital the Paris of the region, Romania today is the basket case of the Balkan Peninsula.

Across the Danube lies Romania's neighbor, **Bulgaria**, southernmost of this subregion's countries. The rugged Balkan Mountains form Bulgaria's physiographic backbone, separating the Danube and Maritsa basins. As the map shows, Bulgaria has five neighbors, several of which are in political turmoil.

The Bulgarian state emerged in 1878, when the Russian czar's armies drove the Turks out of this area. The Slavic Bulgars, who form 86 percent of the population of 7.7 million, were (unlike the Romanians) loyal allies of Moscow during the Soviet period. But they did not treat their Turkish minority, about 9 percent of the population, very kindly, closing mosques, prohibiting use of the Turkish language, and forcing Turkish families to adopt Slavic names. Conditions for the remaining Turks improved somewhat after the end of the Soviet period.

Bulgaria has a Black Sea coast and an outlet, the port of Varna, but the country does not generate much external trade; the capital, Sofia, lies near the border with Serbia-Montenegro. Bulgaria needs economic reform, but like the rest of this subregion, its prospects are not bright.

Countries Facing the Adriatic Sea

As recently as 1990, only two Eastern European countries fronted the Adriatic Sea: Yugoslavia and Albania (Fig. 1-23). Albania survives, but the former Yugoslavia has splintered into five countries—and its disintegration may not be over.

We turn first to the former Yugoslavia ("Land of the South Slavs"), a country extending from Austria to Greece, thrown together in 1918 after World War I, containing 7 major and 17 smaller ethnic and cultural groups (Fig. 1-25). The Slovenes and Croats in the north were Roman Catholics; the Serbs in the south adhered to the

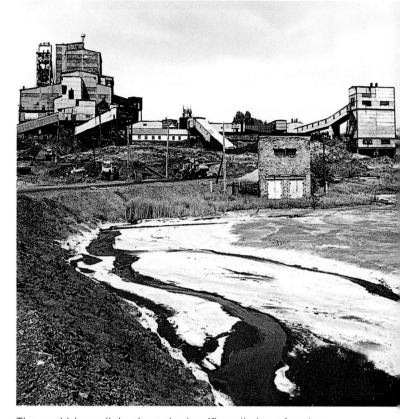

The world knew little about the horrific pollution of underground water supplies, streams, soils, and air during the Soviet Union's rapid industrialization. In the former communist colonies as well as Russia itself, this was environmental degradation on a huge scale. Since the end of communist rule, successful efforts have been made to mitigate the problem, but the task is massive—and help must come from countries that have their own record of pollution. This coal mine stands on a bend in the Mius River in Ukraine; the air smells foul, pollutants seep into the soil through piles of waste, and an ugly scum covers the water. Multiply this scene several thousand times, and you have an idea what the countries of the former Soviet Union confront.

Serbian Orthodox Church. Several million Muslims also formed part of the cultural mosaic. Two alphabets were in use in separate regions of the country. At first, the Royal House of Serbia dominated Yugoslavia; after 1945, a communist dictatorship personified by World War II hero Marshal Tito held Yugoslavia together. But when the communist system collapsed in Eastern Europe and the Soviet Union disintegrated, so did the Yugoslav state.

Communist social planners laid the groundwork for the disaster that befell Yugoslavia. They divided the country into six internal "republics" based on the Soviet model, each dominated (except Bosnia) by one major group. Since all these "republics" inevitably incorporated minorities, the state guaranteed their rights, albeit through autocratic means.

When the communist system collapsed, individual "republics" proclaimed their independence—that is, the majorities in these entities did so. What remained of the machine of state, still dominated by the Serbs, tried to halt this disintegration. That effort soon failed, and new countries named Slovenia, Croatia, Bosnia, Macedonia, and, by subtraction, Serbia appeared on the map. Minorities in these new states, however, objected and, in Croatia and Bosnia, rose against those who were advocating statehood. The result was a catastrophic conflict just when grandiose Euro-unification schemes were under way. Europe, which twice during the twentieth century had plunged the world into war and had vowed that genocide would "never again" cloud its horizons, failed this first test of its declarations. EU members, led by Germany, quickly accorded official recognition to the post-Yugoslavian states, even before the concerns of the frightened minorities could be considered. Then, European powers stood by as more than 250,000 people were killed, perhaps a million more were injured, ethnic cleansing and mass executions depopulated entire areas, refugees streamed into neighboring countries, and historic treasures were demolished.

As we will discover, the devolutionary process continues in the twenty-first century. Today, the still-evolving map emerging from the former Yugoslavia consists of five countries: Slovenia, Croatia, Bosnia, Macedonia, and Serbia-Montenegro. Devolutionary pressures in the last of these five have wrested the province of Kosovo from Serbian control and continue to affect the province of Montenegro.

Slovenia was the first "republic" to secede and proclaim independence. Ethnically the most homogeneous but territorially small with only 2 million of the former Yugoslavia's 23 million citizens, Slovenia lay farthest from the Serbian power core and seized its opportunity. This Alpine-mountain state, with the most productive economy among the republics even before independence, today has this subregion's highest per-capita GNP by far. It achieved approval for inclusion among the ten states admitted to the EU in 2004.

Croatia's crescent-shaped territory, with prongs along the Hungarian border and along the Adriatic coast, only partially reflects the distribution of Croats in the former Yugoslavia (Fig. 1-25). About 85 percent of Croatia's 4.3 million people are Croats, but another 800,000 live in a broad zone of southern Bosnia. Croatia's Serb minority, about 12 percent when independence came, has dwindled under Croatian pressure to under 5 percent. Recent democratic reforms have improved Croatia's prospects.

Bosnia was the cauldron of calamity. Multicultural, effectively landlocked, and situated between the Croatian state to the west and the Serbian stronghold to the east,

Bosnia fell victim to disastrous conflict among Serbs, Croats, and Muslims (the Muslims constituted about 50 percent of the population). In 1995, a U.S. diplomatic initiative resulted in a truce that partitioned Bosnia (red line in Fig. 1-26), making the country a fragile federation.

Macedonia was the southernmost "republic" of the former Yugoslavia and became a state with 2 million inhabitants. Over 60 percent of these inhabitants are Macedonian Slavs, about 30 percent are Muslim Albanians, and the remainder are Turks, Serbs, and Roma (gypsies). Small, landlocked, poor, and powerless, Macedonia first faced the ire of Greece, which argued that the name *Macedonia* was Greek property; then confronted a massive flow of refugees entering from war-torn Kosovo; and most recently coped with an Albanian rebel movement in its northwestern corner, where its main minority is concentrated.

But it is in **Serbia-Montenegro** where the major problems of this subregion are now focused and where the unfinished devolutionary process continues. This geographic name now refers to what is left of the domain once ruled or dominated by the Serbs: *Serbia*, centered on the old Serb capital of Belgrade on the Danube River; *Vojvodina* in the north, part of which was once Hungarian territory and where a Hungarian minority numbers over 350,000; *Montenegro* on the Adriatic coast, once strongly supportive of the Serbs but recently trying to distance itself from Belgrade; and *Kosovo* in the south, one of the historic Muslim strongholds in the subregion, now under the administration of the North Atlantic Treaty Organization (NATO) following NATO's ouster of the repressive Serb authorities.

The total population of this country-in-transition is nominally 10.7 million, but that of Serbia is just over 6 million, whereas Kosovo has about 2 million, Vojvodina some 1.7 million, and Montenegro 650,000. The two key politico-geographical problems involve the future of Kosovo and of Montenegro. When Serbian rule over Kosovo was ended by aerial bombardment and NATO's rule was instituted, there was no clear plan for the territory's future. Many Muslims expect eventual independence, but the new government in Belgrade may demand something short of that. And in Montenegro a slender majority want independence from Belgrade as well—another development the Serbs may not be willing to accept. The devolutionary process in former Yugoslavia has not yet run its course.

It is appropriate to save our discussion of **Albania** for last because even in this turbulent region Albania is unusual. It is the only dominantly Muslim state in Europe; some 70 percent of its 3.2 million people adhere to Islam. Albania also rivals Moldova as the poorest country in Europe (its per-capita GNP is slightly higher), ranking lowest on several indices of well-being. It has by far the

FIGURE 1-26

fastest rate of population growth in the realm. Most Albanians subsist on livestock herding and farming on the one-fifth of this mountainous, earthquake-prone country that can be used for agriculture. Thousands of Albanians have tried to reach Italy across the Adriatic Sea.

Kosovar and Macedonian Muslims look to Albania for support, but the Tirane government cannot embark on irredentist campaigns. Albania has its own cultural divide between the poverty-stricken Gegs in the north and the somewhat better-off Tosks in the south. National unity is its primary goal.

With 583 million inhabitants in 39 countries, including some of the world's highest-income economies, a politically stable and economically integrated Europe could become a superpower in this century. But Europe's political geography is anything but stable, as devolutionary forces and cultural conflict continue to trouble the realm. Moreover, economic integration so far involves only 25 countries, a process that will become more difficult as the European Union confronts applications from marginally qualifying states in the east. Europe always has been a realm of revolutionary change, and it remains so today.

The upheaval in the Balkans continues to require a substantial military presence in several parts of what was once Yugoslavia. In Kosovo, more than 30,000 NATO peacekeeping troops remained in place in mid-2003, the remnant of an even larger contingent that arrived in 1999. Here, after an attempted bombing in April 2003, Danish members of that multinational force guard a crucial bridge crossing in a narrow valley 25 miles (40 km) northwest of Pristina.

Scale 1:20 000 000; one inch to 315 miles.
Lambert's Azimuthal, Equal Area Projection
Elevations and depressions are given in feet

FIGURE 2-1

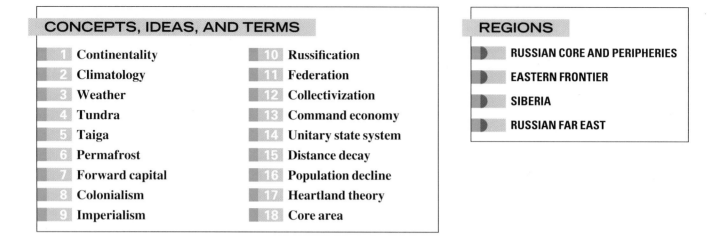

CONCEPTS, IDEAS, AND TERMS

1 Continentality
2 Climatology
3 Weather
4 Tundra
5 Taiga
6 Permafrost
7 Forward capital
8 Colonialism
9 Imperialism
10 Russification
11 Federation
12 Collectivization
13 Command economy
14 Unitary state system
15 Distance decay
16 Population decline
17 Heartland theory
18 Core area

REGIONS

▶ RUSSIAN CORE AND PERIPHERIES
▶ EASTERN FRONTIER
▶ SIBERIA
▶ RUSSIAN FAR EAST

On October 23, 2002, a large theater audience in Moscow was enjoying a musical about shenanigans in the country's remote, wild northeast when, suddenly, heavily armed people in combat fatigues appeared on the stage and herded the startled performers into the wings. Pointing their guns at the audience, they announced that all 700-plus theatergoers were hostages, that their survival would depend on the negotiations about to take place, and that the theater was already boobytrapped with bombs and mines. The hostage-takers, 41 men and women, also had bombs strapped to their bodies, and they were prepared to die for their cause.

What cause could generate an act like this? The audience knew the moment their captors appeared on the stage. The cause was an independent Chechnya, and the hostage-takers were Chechen rebels. All were Muslims, and the women wore long, black robes and hair-covering shawls. Their demand was an immediate cessation of Russian military activity in their homeland and the withdrawal of all Russian forces from the Republic of Chechnya. It was only the latest episode in a geopolitical struggle that had already brought death and destruction to Moscow before.

The disastrous outcome of this attack brought the parties no closer to a resolution. No negotiations developed; conditions in the theater deteriorated rapidly. When the captors announced that they would start killing hostages on the morning of the third day and shots were fired in the theater during the night, Russian antiterror units pumped what turned out to be a lethal gas into the hall, immobilizing all. Storming the auditorium, they killed all the rebels and rescued the hostages—but at a high cost. By the time the theater was cleared, 128 of the captives were dead from the effects of the gas. And none of the Chechens was alive to be questioned about the planning, logistics, and funding of the operation.

The lethal assault and its tragic aftermath reminded Russians as well as the world that, more than a decade after the collapse of the Soviet Union, the Russian state is still an empire and that not all the parts of the empire submit willingly to Moscow's rule. Indeed, Chechnya, located in Russia's mountainous periphery between the Black and the Caspian seas, is not the only entity in the still-evolving Russian state that has shown signs of revolt. But, as we will see later in this chapter, Russia has largely succeeded in bringing its numerous nationalities into its fold, including several with Islamic imprints. In Chechnya, however, it has failed, and nothing—from accommodation to repression—has worked. And now a territory the size of Connecticut with a population barely over 1 million is threatening what is territorially the largest state in the world with a population of more than 140 million through the weapon of terrorism. It is a mirror not only on Russia but on the world of the present.

DEFINING THE REALM

The Russian state constitutes a world geographic realm because of it territorial size, relative location, and substantial population. Turn back one page and consider the map: Russia borders Norway as well as North Korea and ten countries in between; it is about three times as large as the contiguous United States; and it extends across 11 time zones from its eastern frontier (the volcano-studded Kamchatka Peninsula) to its western port of St. Petersburg.

MAJOR GEOGRAPHIC QUALITIES OF

Russia

1. Russia is the largest territorial state in the world. Its area is nearly twice as large as that of the next-ranking country (Canada).

2. Russia is the northernmost large and populous country in the world; much of it is cold and/or dry. Extensive rugged mountain zones separate Russia from warmer subtropical air, and the country lies open to Arctic air-masses.

3. Russia was one of the world's major colonial powers. Under the czars, the Russians forged the world's largest contiguous empire; the Soviet rulers who succeeded the czars took over and expanded this empire.

4. For so large an area, Russia's population of under 145 million is comparatively small. The population remains heavily concentrated in the westernmost one-fifth of the country.

5. Development in Russia is concentrated west of the Ural Mountains; here lie the major cities, leading industrial regions, densest transport networks, and most productive farming areas. National integration and economic development east of the Urals extend mainly along a narrow corridor that stretches from the southern Urals region to the southern Far East around Vladivostok.

6. Russia is a multicultural state with a complex domestic political geography. Twenty-one internal Republics, originally based on ethnic clusters, function as politico-geographical entities.

7. Its large territorial size notwithstanding, Russia suffers from land encirclement within Eurasia; it has few good and suitably located ports.

8. Regions long part of the Russian and Soviet empires are realigning themselves in the postcommunist era. Eastern Europe and the heavily Muslim Southwest Asia realm are encroaching on Russia's imperial borders.

9. The failure of the Soviet communist system left Russia in economic disarray. Many of the long-term components described in this chapter (food-producing areas, railroad links, pipeline connections) broke down in the transition to the postcommunist order.

10. Russia long has been a source of raw materials but not a manufacturer of export products, except weaponry. Few Russian (or Soviet) automobiles, televisions, cameras, or other consumer goods reach world markets.

The western half of Russia is bisected by the north-south trending Ural Mountains, which extend from the island of Novaya Zemlya in the Arctic Ocean to the border with Kazakhstan. The vast majority of all Russians live to the west of the Urals, which has given rise to the false notion of a "European" Russia and, consequently, a non-European Russia farther east. But the cultural landscapes of Russia continue from Moscow to Novosibirsk to Vladivostok on the Pacific coast. The Urals bound the Siberian Lowland, but they do not limit "European" Russia.

FROM THE FIELD NOTES

"Walking the streets of St. Petersburg in 1994, more than 30 years after our first trip to the city, brought home the magnitude of the changes after 1991. Not only Nevsky Prospekt, the city's main street seen here, but other streets as well, are crowded with cars and pedestrians. Advertisements are everywhere. Small makeshift stalls selling goods ranging from books to family memorabilia line the sidewalks. Luxury goods and services have made their appearance, and the newly wealthy shop at recently opened, swank stores. Also in evidence are the newly destitute and desperate; just down the street from this scene a little girl stood playing the violin all day, her father leaning against the wall nearby, waiting to collect the coins passersby gave her. Crime has escalated as well, and extortion threatens many new small businesses. Like other Russian cities, St. Petersburg (*Peter* as locals call it) is in post-Soviet ferment."

Russia's predominantly Slavic population of 143.8 million may not match those of China or India, but it is much larger than that of any European country. The demographic as well as the political map confirms Russia's status as a geographic realm. As we will see, environmental, historic, and ideological factors combine to strengthen Russia's geographic identity.

Although the geographic realm under discussion is dominated by the giant Russian state, it also includes three small countries in the area (known as *Transcaucasia*) between the Black Sea and the Caspian Sea: Georgia, Armenia, and Azerbaijan. For centuries Russia was a powerful empire whose influence reached far beyond its borders into Transcaucasia, Turkestan, Mongolia, and even China. Today, Russia retains strong links with all three of these Transcaucasian republics, placing the realm boundary along the borders with Turkey and Iran (Fig. G-2).

Russia is a land of vast distances, remote frontiers, bitter cold, and isolated outposts. Its core area lies west of the Urals, centered on Moscow. To the north, south, and east lie distant, problematic peripheries where economies falter and political systems fail. During the early twentieth century, Russia embarked on an immense social experiment with the goal of an egalitarian communist state that would be a model for the rest of the world. The experiment failed, but not before Russia and its Eurasian empire had been transformed at gigantic cultural, economic, and political cost. The geographic story of this chapter centers on Russia's postcommunist reorganization.

RUSSIAN ROOTS

The name *Russia* evokes cultural-geographic images of a stormy past: terrifying czars, conquering Cossacks, Byzantine bishops, rousing revolutionaries, clashing cultures. Russians repulsed the Tatar (Mongol) hordes, forged a powerful state, colonized a vast contiguous empire, defeated Napoleon, adopted communism, and, when the communist system failed, lost most of their imperial domain. Today, Russia seeks a new place in the globalizing world.

Precommunist Russia may be described as a culture of extremes. It was a culture of strong nationalism, resistance to change, and despotic rule. Enormous wealth was concentrated in a small elite. Powerful rulers and bejeweled aristocrats perpetuated their privileges at the expense of millions of peasants and serfs who lived in dreadful poverty. The Industrial Revolution arrived late in Russia, and a middle class was slow to develop. Yet the Russian nation gained the loyalty of many of its citizens, and its writers and artists were among the world's greatest. Authors such as Tolstoy and Dostoyevsky chronicled the plight of the poor; composers celebrated the indomitable Russian people, and, as Tchaikovsky did in his *1812 Overture*, commemorated their victories over foreign foes.

Under the czars, Russia grew from nation into empire. The czars' insatiable demands for wealth, territory, and power sent Russian armies across the plains of Siberia, through the deserts of interior Asia, and into the mountains along Russia's rim. Russian pioneers ventured even farther, entering Alaska, traveling down the Pacific coast of North America, and planting the Russian flag near San Francisco in 1812. But as Russia's empire expanded, its internal weaknesses gnawed at the power of the czars. Peasants rebelled. Unpaid (and poorly fed) armies mutinied. When the czars tried to initiate reforms, the aristocracy objected. The empire at the beginning of the twentieth century was ripe for revolution, which began in 1905 (see box titled "The Soviet Union, 1924–1991").

The last czar, Nicholas II, was overthrown in 1917, and civil war followed. The victorious communists led by V. I. Lenin soon swept away much of the Russia of the past. The Russian flag disappeared. The czar and his family were executed. The old capital of Russia, St. Petersburg, was renamed Leningrad in honor of the revolutionary leader. Moscow, in the interior of the country, was chosen as the new capital for a country with a new name, the *Soviet Union*. Eventually, this Union consisted of 15 political entities, each a Soviet Socialist Republic (SSR). Russia was just one of these republics, and the name *Russia* disappeared from the international map.

But on the Soviet map, Russia was the giant, the dominant republic, the Slavic center. Not for nothing was the communist revolution known as the Russian Revolution. The other republics of the Soviet Union were for minorities the czars had colonized or for countries that fell under Soviet sway later, but none could begin to match Russia. The Soviet Empire was the legacy of czarist expansionism, and the new communist rulers were Russian first and foremost (with the exception of Josef Stalin, who was a Georgian). Russians moved by the millions to the non-Russian republics, where the Russification of the empire proceeded, just as the British and French and Dutch and Portuguese were also moving to their colonies in large numbers. The Soviet Union was a Russian colonial empire, and like all colonial empires it was doomed to failure.

Officially, the Union of Soviet Socialist Republics endured from 1924 to 1991. When the USSR collapsed, the republics were set free and world maps had to be redrawn. Now Russia and its former colonies had to adjust to the realities of a changing world order. It proved to be a difficult, sometimes desperate journey.

The Soviet Union, 1924–1991

*F*or 67 years Russia was the cornerstone of the *Soviet Union*, the so-called Union of Soviet Socialist Republics (USSR). The Soviet Union was the product of the Revolution of 1917, when more than a decade of rebellion against the rule of Czar Nicholas II led to his abdication. Russian revolutionary groups were called soviets ("councils"), and they had been active since the first workers' uprising in 1905. In that year, thousands of Russian workers marched on the czar's palace in St. Petersburg in protest, and soldiers fired on them. Hundreds were killed and wounded. Russia descended into chaos.

A coalition of military and professional men forced the czar's abdication in 1917. Then Russia was ruled briefly by a provisional government. In November 1917, the country held its first democratic election—and, as it turned out, its last for more than 70 years.

The provisional government allowed the exiled activists in the Bolshevik camp to return to Russia (there were divisions among the revolutionaries): Lenin from Switzerland, Trotsky from New York, and Stalin from internal exile in Siberia. In the ensuing political struggle, Lenin's Bolsheviks gained control over the revolutionary soviets, and this ushered in the era of communism. In 1924, the new communist empire was formally renamed the Union of Soviet Socialist Republics, or Soviet Union in shorthand.

Lenin the organizer died in 1924 and was succeeded by Stalin the tyrant, and many of the peoples under Moscow's control suffered unimaginably. In pursuit of communist reconstruction, Stalin and his adherents starved millions of Ukrainian peasants to death, forcibly relocated entire ethnic groups (including a people known as the Chechens), and exterminated "uncooperative" or "disloyal" peoples. The full extent of these horrors may never be known.

On December 25, 1991, the inevitable occurred: the Soviet Union ceased to exist, its economy a shambles, its political system shattered, the communist experiment a failure. The last Soviet president, Mikhail Gorbachev, resigned, and the Soviet hammer-and-sickle flag flying atop the Kremlin was lowered for the last time and immediately replaced by the white, red, and blue Russian tricolor. A new and turbulent era had begun, but Soviet structures and systems will long cast their shadows over transforming Russia.

At first, leaders in Russia and some of the SSRs hoped that a new voluntary association among these republics would continue the links forged during communist times, but the Commonwealth of Independent States (CIS) they created never amounted to much more than an acronym. The SSRs soon went their own ways (Table 2-1); only Belarus has sought to rejoin Russia in a formal union.

Loosening ties with Russia has been easier for some former SSRs than others. The forces of economic geography cannot be redirected overnight and, as Table 2-1 indicates, sizeable Russian minorities remain in several of the former colonies. Most of the westernmost SSRs are reorienting themselves toward Europe, with three joining the EU in 2004. The five former Soviet Central Asian republics now are part of a distinct geographic region, Turkestan, where Islam is reviving and links to the Muslim world are intensifying. But in Transcaucasia in the southwest, as we shall presently see, the situation is more complicated.

RUSSIA'S PHYSICAL ENVIRONMENTS

The historical geography of Russia is the story of Slavic expansion from its populous western heartland across interior Eurasia to the east, and into the mountains and deserts of the south. This eastward march was hampered not only by vast distances but also by harsh natural conditions. As the northernmost populous country on Earth, Russia has virtually no natural barriers against the onslaught of Arctic air. Moscow lies farther north than Edmonton, Canada, and St. Petersburg lies at latitude 60° North—the latitude of the southern tip of Greenland. Winters are long, dark, and bitterly cold in most of Russia; summers are short and growing seasons limited. Many a Siberian frontier outpost was doomed by cold, snow, and hunger.

To make matters worse, precipitation totals range from modest to minimal because the warm, moist air carried across Europe from the North Atlantic Ocean loses much of its warmth and moisture by the time it reaches Russia. Figures G-7 and G-8 reveal the consequences. Russia's climatic **continentality** (inland climatic environment remote from moderating and moistening maritime influences) is expressed by its prevailing *Dfb* and *Dfc* conditions. Compare the Russian map to that of North America, and you note that, except for a small corner off the Black Sea, Russia's climatic conditions resemble those of the U.S. Upper Midwest and Canada. Along its entire north, Russia has a zone of *E* climates, the most frigid on the planet. In these Arctic latitudes originate the polar airmasses that dominate its environments.

Table 2-1
WHAT IS HAPPENING TO THE OTHER 14 EX-SOVIET REPUBLICS?

Name	Population (Millions)	Percent Russians	Religion	Official Language	Status
Armenia	3.8	2	Armenian Orthodox (Christian) 48%	Armenian	Armenia is in conflict with neighboring Azerbaijan over the exclave of Nagorno-Karabakh. Public opinion favors closer association with Russia.
Azerbaijan	8.3	6	Shi'ite Muslim 93%	Azerbaijani	Azerbaijan is seeking settlement with Armenia over the Nagorno-Karabakh exclave, and trying to reduce dependence on Russia for oil-export routes.
Belarus	9.8	14	Belarussian Orthodox, Roman Catholic 49%	Belarussian; Russian	Belarus persuaded Russia to sign a "union" agreement in 1997. Political repression, human-rights violations, stagnating economy.
Estonia	1.4	29	Estonian Orthodox, Estonian Lutheran 34%	Estonian	After a period of discrimination against the Russian minority, conditions are improving along with a booming economy. Relations with Nordic neighbors and Russia strengthening. Joined European Union (EU) in 2004.
Georgia	4.4	6	Georgian, Russian Orthodox 47%; Sunni Muslim 11%	Georgian	Georgia has unfriendly relations with Russia and accuses Russia of encouraging separatist elements in the province of Abkhazia. Russian military bases on Georgian soil, and Russian border closures are seen as threats to Georgian sovereignty.
Kazakhstan	14.9	35	Sunni Muslim 47%; Russian Orthodox 10%	Kazakh	Kazakhstan has difficult relations with Russia over treatment of Russians in the northern oblasts, over the use of Cossack troops to patrol the northern border, over Russian oil-company rights in the Caspian area, and over Russian weapons-testing sites inside Kazakhstan.
Kyrgyzstan	5.1	16	Sunni Muslim 71%	Kyrgyz; Russian	An increasingly authoritarian government is impairing the democratization process by imprisoning critics and stifling the press. The Kyrgyz Communist Party has demanded a Belarus-type "union" with Russia.
Latvia	2.3	32	Christian Churches 40%	Latvian	Relations between the government and the large Russian minority have been difficult but improving; naturalization slow. Joined EU in 2004.
Lithuania	3.5	8	Roman Catholic 79%	Lithuanian	Relations with Russia improved in 1998 with the signing of an agreement settling a boundary dispute. Joined EU in 2004.
Moldova	4.3	13	Eastern Orthodox 49%	Romanian	Relations with Russia have suffered from the attempted breakaway by the pro-Russian trans-Dniester region. Economy in trouble.
Tajikistan	6.5	7	Sunni Muslim 80%	Tajik	Civil war has pitted the ruling post-communist regime against United Tajik Opposition (UTO) led by the Islamic Renaissance Party.
Turkmenistan	5.8	6	Sunni Muslim 87%	Turkmen	Turkmenistan's government accuses Russia of impeding the flow of natural gas to its customers; an alternative pipeline via Iran is in the works. Problems continue with Azerbaijan over Caspian Sea oil and gas reserves.
Ukraine	47.5	22	Ukrainian Orthodox 61%	Ukrainian	A large and potentially powerful country, Ukraine has settled several disputes with Russia including the rights of its large Russian minority and the disposition of the Soviet Black Sea fleet as well as two Black Sea ports. Pro-Russian sentiment in Ukraine appears to be rising.
Uzbekistan	26.3	6	Sunni Muslim 88%	Uzbek	Repeated setbacks to the post-Soviet democratization process include stifling of opposition and press. Fear of Islamic fundamentalism has produced repressive laws and disputes with Russia as well as the United States and European countries.

By studying Russia's climates we can begin to understand what the map of population distribution (Fig. G-9) shows. **Climatology** is a field of geography that investigates not only the distribution of climatic conditions over the Earth's surface but also the processes that cause this spatial arrangement. The Earth's atmosphere traps heat received in the form of radiation from the Sun, but this *greenhouse effect* varies in space—and over time. As we noted in the Introduction, much of Russia was in the grip of a glaciation until the onset of the warmer Holocene. Today, a natural warming cycle is in progress, augmented by human action. This enhanced global warming threatens some societies (for example, those with extensive coastal plains that would be inundated by rising sea levels), but Russia may be one of the winners in this process of environmental change. Warmer temperatures and increased moisture may expand the frontiers of settlement shown in Figure G-9.

Climate and weather (there is a distinction: *climate* is a long-term average, whereas **weather** refers to existing atmospheric conditions at a given place and time) have always challenged Russia's farmers. Conditions are most favorable in the west, but even there temperature extremes, variable and undependable rainfall, and short growing seasons make farming difficult. During the Soviet period, fertile and productive Ukraine supplied much of Russia's food needs, but even then Russia often had to import grain. Soviet rulers wanted to reduce their country's dependence on imported food, and their communist planners built major irrigation projects to increase crop yields in the colonized republics of Central Asia. As we will see in Chapter 7, some of these attempts to overcome nature's limitations spelled disaster for the local people.

Physiographic Regions

To assess the physiography of this vast country, it is useful to refer to Figures 2-1 and 2-2 simultaneously. Note how mountains and deserts encircle Russia: the Caucasus in the southwest ⑧; the Central Asian Ranges in the center ⑦; the Eastern Highlands facing the Pacific from the Bering Sea to the East Sea (Sea of Japan) ⑥. The Kamchatka Peninsula has a string of active volcanoes in one of the world's most earthquake-prone zones (Fig. G-5). Warm subtropical air thus has little opportunity to penetrate Russia, while cold Arctic air sweeps southward without impediment. Russia's Arctic north is a gently sloping lowland broken only by the Urals and Eastern Highlands.

Russia's vast and complex physical stage can be divided into eight physiographic regions, each of which, at a larger scale, can be subdivided into smaller units. In the Siberian region, one criterion for such subdivision

FIGURE 2-2

would be the vegetation. The Russian language has given us two terms to describe this vegetation: **tundra**, the treeless plain along the Arctic shore where mosses, lichens, and some grasses survive, and **taiga**, the mostly coniferous forests that begin to the south of where the tundra ends and extend over vast reaches of Siberia (which means the "sleeping land" in Russian).

The Russian Plain ① is the eastward continuation of the North European Lowland, and here the Russian state formed its first core area. Travel north from Moscow at its heart, and the countryside soon is covered by needleleaf forests like those of Canada; to the south lie the grain fields of southern Russia and, beyond, those of Ukraine. Note the Kola Peninsula and Barents Sea in the far north: warm water from the North Atlantic comes around northern Norway and keeps the port of Murmansk ice free most of the year. The Russian Plain is bounded on the east by the Ural Mountains ②; though not a high range, it is topographically prominent because it separates two extensive plains. The Urals Range is more than 2000 miles (3200 km) long and reaches from the shores of the Kara Sea to the border with Kazakhstan. It is not a barrier to east-west transportation, and its southern end is densely populated. Here the Urals yield minerals and fossil fuels.

East of the Urals lies Siberia. The West Siberian Plain ③ has been described as the world's largest unbroken lowland; this is the vast basin of the Ob and Irtysh rivers. Over the last 1000 miles (1600 km) of its course to the Arctic Ocean, the Ob falls less than 300 feet (90 m). In Figure 2-2, note the dashed line that extends from the Finnish border to the East Siberian Sea and offsets the Arctic Lowland. North of the line, water in the ground is permanently frozen; this **permafrost** creates another obstacle to permanent settlement. Looking again at the West Siberian Plain, we see that the north is permafrost-ridden and the central zone is marshy. The south, however, has such major cities as Omsk and Novosibirsk, within the corridor of the Trans-Siberian Railroad (see photo below).

East of the West Siberian Plain the country begins to rise, first into the central Siberian Plateau ④, another sparsely settled, remote, permafrost-affected region. Here winters are long and cold, and summers are short; the area remains barely touched by human activity. Beyond the Yakutsk Basin ⑤, the terrain becomes mountainous and the relief high. The Eastern Highlands ⑥ are a jumbled mass of ranges and ridges, precipitous valleys, and volcanic mountains. Lake Baykal lies in a trough that is over 5000 feet (1500 m) deep—the deepest rift lake in the world (see Chapter 6). On the Kamchatka Peninsula, volcanic Mount Klyuchevskaya reaches nearly 15,600 feet (4750 m).

The northern part of region ⑥ is Russia's most inhospitable zone, but southward along the Pacific coast the climate is less severe. Nonetheless, this is a true frontier region. The forests provide opportunities for lumbering, a fur trade exists, and there are gold and diamond deposits. To help develop this promising area, the Soviets constructed the new Baykal-Amur Mainline (BAM) Railroad, a 2200-mile-long (3540 km) route that roughly parallels the aging Trans-Siberian line to the south (see Fig. 2-6).

Mountains also mark the southern margins of much of Russia. The Central Asian Ranges ⑦, from the Kazakh border in the west to Lake Baykal in the east, contain

The Trans-Siberian Express travels through the taiga on its way from Moscow to Vladivostok. This service is a crucial element in Russia's cohesion, but passenger and freight volumes are down, costs are up, and reliability is suffering.

many glaciers whose annual meltwaters send alluvium-laden streams to enrich farmlands at lower elevations. The Caucasus ⑧, in the land corridor between the Black and Caspian seas, form an extension of Europe's Alpine Mountains and exhibit a similarly high *relief* (range of elevations). Here Russia's southern border is sharply defined by *topography* (surface configuration).

As our physiographic maps suggests, the more habitable terrain in Russia becomes latitudinally narrower from west to east. Beyond the southern Urals, the zone of settlement in Russia becomes discontinuous (in Soviet times it extended into northern Kazakhstan, where a large Russian population still lives). Figure 2-1 reveals that isolated towns did develop in Russia's vast eastern reaches, even in Siberia, but the tenuous ribbon of settlement does not widen again until it reaches the country's Far Eastern Pacific Rim.

EVOLUTION OF THE RUSSIAN STATE

Four centuries ago, when European kingdoms were sending fleets to distant shores to search for riches and capture colonies, there was little indication that the largest of all empires would one day center on a city in the forests halfway between Sweden and the Black Sea. The plains of Eurasia south of the taiga had seen waves of migrants sweep from east to west: Scythians, Sarmatians, Goths, Huns, and others came, fought, settled, and were absorbed or driven off. Eventually, the Slavs, heirs to these Eurasian infusions, emerged as the dominant culture in what is today Ukraine, south of Russia and north of the Black Sea.

From this base on fertile, productive soils, the Slavs expanded their domain, making Kiev (Kyyiv) their southern headquarters and establishing Novgorod on Lake Ilmen in the north. Each was the center of a state known as a *Rus*, and both formed key links on a trade route between the German-speaking Hanseatic ports of the Baltic Sea and the trading centers of the Mediterranean. During the eleventh and twelfth centuries, the Kievan Rus and the Novgorod Rus united to form a large and comparatively prosperous state astride forest and steppe (short-grass prairie).

The Mongol Invasion

Prosperity attracts attention, and far to the east, north of China, the Mongol Empire had been building under Genghis Khan. In the thirteenth century, the Mongol (Tatar) "hordes" rode their horses into the Kievan Rus,

and the state fell. By mid-century, Slavs were fleeing into the forests, where the Mongol horsemen had less success than on the open plains of Ukraine. Here the Slavs reorganized, setting up several Russes. Still threatened by the Tatars, the ruling princes paid tribute to the Mongols in exchange for peace.

Moscow, deep in the forest on the Moscow River, was one of these Russes. Its site was remote and defensible. Over time, Moscow established trade and political links with Novgorod and became the focus of a growing area of Slavic settlement. When the Mongols, worried about Moscow's expanding influence, attacked near the end of the fourteenth century and were repulsed, Moscow emerged as the unchallenged center of the Slavic Russes. Soon its ruler, now called a Grand Duke, took control of Novgorod, and the stage was set for further growth.

Grand Duchy of Muscovy

Moscow's geographic advantages again influenced events. Its centrality, defensibility, links to Slavic settlements including Novgorod, and open frontiers to the north and west, where no enemies threatened, gave Moscow the potential to expand at the cost of its old enemies, the Mongols. Many of these invaders had settled in the basin of the Volga River and elsewhere, where the city of Kazan had become a major center of Islam, the Tatars' dominant religion. During the rule of Ivan the Terrible (1547–1584), the Grand Duchy of Muscovy became a major military power and imperial state. Its rulers called themselves czars and claimed to be the heirs of the Byzantine emperors. Now began the expansion of the Russian domain (Fig. 2-3), marked by the defeat of the Tatars at Kazan and the destruction of hundreds of mosques.

The Cossacks

This eastward expansion of Russia was spearheaded by a relatively small group of seminomadic peoples who came to be known as Cossacks and whose original home was in present-day Ukraine. Opportunists and pioneers, they sought the riches of the eastern frontier, chiefly fur-bearing animals, as early as the sixteenth century. By the mid-seventeenth century they reached the Pacific Ocean, defeating Tatars in their path and consolidating their gains by constructing *ostrogs* (strategic fortified waystations) along river courses. Before the eastward expansion halted in 1812 (Fig. 2-3), the Russians had moved across the Bering Strait to Alaska and down the western coast of North America into what is now northern California (see box titled "Russians in North America").

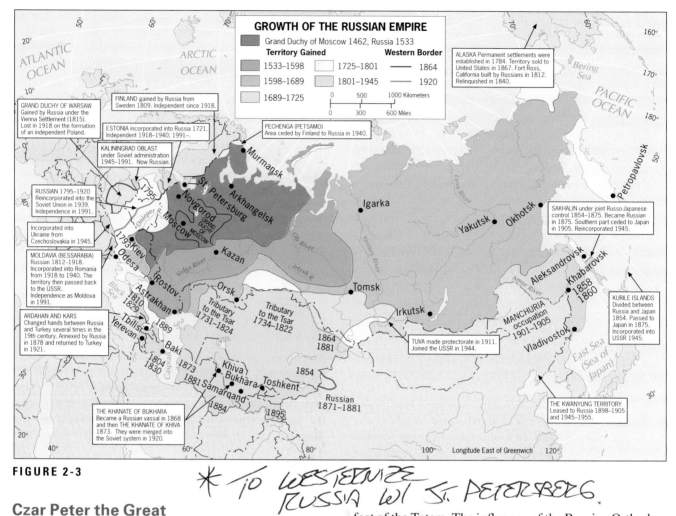

FIGURE 2-3

To WESTERNIZE RUSSIA W/ St. PETERSBURG.

Czar Peter the Great

When Peter the Great became czar (he ruled from 1682 to 1725), Moscow already lay at the center of a great empire—great, at least, in terms of the territories it controlled. The Islamic threat had been ended with the defeat of the Tatars. The influence of the Russian Orthodox Church was represented by its distinctive religious architecture and powerful bishops.

Peter consolidated Russia's gains and hoped to make a modern, European-style state out of his loosely knit coun-

---

Russians in North America

The first white settlers in Alaska were Russians, not Western Europeans, and they came across Siberia and the Bering Strait, not across the Atlantic and North America. Russian hunters of the sea otter, valued for its high-priced pelt, established their first Alaskan settlement at Kodiak Island in 1784. Moving south along the North American coast, the Russians founded additional villages and forts to protect their tenuous holdings until they reached as far as the area just north of San Francisco Bay, where they built Fort Ross in 1812.

But the Russian settlements were isolated and vulnerable. European fur traders began to put pressure on their Russian competitors, and St. Petersburg found the distant settlements a burden and a risk. In any case, American, British, and Canadian hunters were decimating the sea otter population, and profits declined. When U.S. Secretary of State William Seward offered to purchase Russia's holdings in 1867, St. Petersburg quickly agreed—for $7.2 million. Thus Alaska, including its lengthy southward coastal extension—the Alaskan "panhandle"—became U.S. territory and, eventually, the forty-ninth State. Although Seward was ridiculed for his decision—Alaska was called "Seward's Folly" and "Seward's Icebox"—he was vindicated when gold was discovered there in the 1890s. The twentieth century proved the wisdom of Seward's action, strategically as well as economically. At Prudhoe Bay off Alaska's northern Arctic slope, large oil reserves were tapped in the 1970s and are still being exploited. And like Siberia, Alaska probably contains yet-unknown riches.

try. He built St. Petersburg as a **forward capital** on the doorstep of Swedish-held Finland, fortified it with major military installations, and made it Russia's leading port.

Czar Peter the Great, an extraordinary leader, was in many ways the founder of modern Russia. In his desire to remake Russia—to pull it from the forests of the interior to the waters of the west, to open it to outside influences, and to relocate its population—he left no stone unturned. Prominent merchant families were forced to move from other cities to St. Petersburg. Ships and wagons entering the city had to bring building stones as an entry toll. The czar himself, aware that to become a major power Russia had to be strong at sea as well as on land, went to the Netherlands to work as a laborer in the famed Dutch shipyards to learn the most efficient method for building ships. Meanwhile, the czar's forces continued to conquer people and territory: Estonia was incorporated in 1721, widening Russia's window to the west, and major expansion soon occurred south of the city of Tomsk (Fig. 2-3).

Czarina Catherine the Great

Under Czarina Catherine the Great, who ruled from 1760 to 1796, Russia's empire in the Black Sea area grew at the expense of the Ottoman Turks. The Crimea Peninsula, the port city of Odesa (Odessa), and the whole northern coastal zone of the Black Sea fell under Russian control. Also during this period, the Russians made a fateful move: they penetrated the area between the Black and Caspian seas, the mountainous Caucasus with dozens of ethnic and cultural groups, many of which were Islamized. The cities of Tbilisi (now in Georgia), Baki (Baku in Azerbaijan), and Yerevan (Armenia) were captured. Eventually, the Russian push toward an Indian Ocean outlet was halted by the British, who held sway in Persia (modern Iran), and also by the Turks.

Meanwhile, Russian colonists entered Alaska, founding their first American settlement at Kodiak Island in 1784 and moving southward, establishing numerous forts to protect their tenuous holdings against Native American peoples. Eventually, they reached nearly as far south as San Francisco Bay where, in the fateful year 1812, they erected Fort Ross.

Catherine the Great had made Russia a colonial power, but the Russians eventually gave up on their North American outposts. The sea-otter pelts that had attracted the early pioneers were running out, European and white American hunters were cutting into the profits, and Native American resistance was growing. When U.S. Secretary of State William Seward offered to purchase Russia's Alaskan holdings in 1867, the Russian government quickly agreed—for $7.2 million. Thus Alaska and its Russian-held panhandle became U.S. territory and, in 1959, the forty-ninth State.

FROM THE FIELD NOTES

"Not only the city of St. Petersburg itself, but also its surrounding suburbs display the architectural and artistic splendor of czarist Russia. The czars built opulent palaces in these outlying districts (then some distance from the built-up center), among which the Catherine Palace, begun in 1717 and completed in 1723 followed by several expansions, was especially majestic. During my first visit in 1994 the palace, parts of which had been deliberately destroyed by the Germans during World War II, was still being restored; large black and white photographs in the hallways showed what the Nazis had done and chronicled the progress of the repairs during communist and post-communist years. A return visit in 2000 revealed the wealth of sculptural decoration on the magnificent exterior (left) and the interior detail of a set of rooms called the "golden suite" of which the Amber Room (right) exemplifies eighteenth-century Russian Baroque at its height."

A Russian Empire

Although Russia withdrew from North America, Russian expansionism during the nineteenth century continued in Eurasia. While extending their empire southward, the Russians also took on the Poles, old enemies to the west, and succeeded in taking most of what is today the Polish state, including the capital of Warsaw. To the northwest, Russia took over Finland from the Swedes in 1809. During most of the nineteenth century, however, the Russians were preoccupied with Central Asia—the region between the Caspian Sea and western China—where Toshkent (Tashkent) and Samarqand (Samarkand) came under St. Petersburg's control (Fig. 2-3). The Russians here were still bothered by raids of nomadic horsemen, and they sought to establish their authority over the Central Asian steppe country as far as the edges of the high mountains that lay to the south. Thus Russia gained many Muslim subjects, for this was Islamic Asia they were penetrating. Under czarist rule, these people retained some autonomy. Much farther to the east, a combination of Japanese expansionism and a decline of Chinese influence led Russia to annex from China several provinces to the east of the Amur River. Soon thereafter, in 1860, the port of Vladivostok on the Pacific was founded.

Now began the events that were to lead to the first involuntary halt to the Russian drive for territory. As Figure 2-3 shows, the most direct route from western Russia to the port of Vladivostok lay across northeastern China, the territory then still called Manchuria. The Russians had begun construction of the Trans-Siberian Railroad in 1892, and they wanted China to permit the track to cross Manchuria. But the Chinese resisted. Then, taking advantage of the Boxer Rebellion in China in 1900 (see Chapter 9), Russian forces occupied Manchuria so that railroad construction might proceed.

This move, however, threatened Japanese interests in this area, and the Japanese confronted the Russians in the Russo-Japanese War of 1904–1905. Not only was Russia defeated and forced out of Manchuria: Japan even took possession of the southern part of Sakhalin Island, which they named Karafuto and retained until the end of World War II.

THE COLONIAL LEGACY

Thus Russia, like Britain, France, and other European powers, expanded through **colonialism**. Yet whereas the other European powers expanded overseas, Russian influence traveled overland into Central Asia, Siberia, China, and the Pacific coastlands of the Far East. What emerged was not the greatest empire but the largest territorially contiguous empire in the world. At the time of the Japanese war, the Russian czar controlled more than 8.5 million square miles (22 million sq km), just a tiny fraction less than the area of the Soviet Union after the 1917 Revolution. Thus the communist empire, to a large extent, was the legacy of St. Petersburg and European Russia, not the product of Moscow and the socialist revolution.

The czars embarked on their imperial conquests in part because of Russia's relative location: Russia always lacked warm-water ports. Had the Revolution not intervened, their southward push might have reached the Persian Gulf or even the Mediterranean Sea. Czar Peter the Great envisaged a Russia open to trading with the entire world; he developed St. Petersburg on the Baltic Sea into Russia's leading port. But in truth, Russia's historical geography is one of remoteness from the mainstreams of change and progress, as well as one of self-imposed isolation. Not even a string of warm-water ports would have been likely to transform Russia into an outward-looking, trading state. The czars' objectives were primarily strategic, not economic.

An Imperial, Multinational State

Centuries of Russian expansionism did not confine itself to empty land or unclaimed frontiers. The Russian state became an imperial power that annexed and incorporated many nationalities and cultures. This was done by employing force of arms, by overthrowing uncooperative rulers, by annexing territory, and by stoking the fires of ethnic conflict. By the time the ruthless Russian regime began to face revolution among its own people, czarist Russia was a hearth of **imperialism**, and its empire contained peoples representing more than 100 nationalities. The winners in the ensuing revolutionary struggle—the communists who forged the Soviet Union—did not liberate these subjugated peoples. Rather, they changed the empire's framework, binding the peoples colonized by the czars into a new system that would in theory give them autonomy and identity. In practice, it doomed those peoples to bondage and, in some cases, extinction.

When the Soviet system failed and the Soviet Socialist Republics became independent states, Russia was left without the empire that had taken centuries to build and consolidate—and that contained crucial agricultural and mineral resources. No longer did Moscow control the farms of Ukraine and the oil and natural gas reserves of Central Asia. But look again at Figure 2-3 and you will see that, even without its European and Central Asian colonies, Russia remains an empire. Russia lost the "republics" on its periphery, but Moscow still rules over a domain that extends from the borders of Finland to North Korea. Inside that domain Russians are in the overwhelming majority, but many subjugated nationalities, from Tatars to Yakuts, still inhabit ancestral homelands. Accommodating these many indigenous peoples is one of the challenges facing the Russian Federation today.

On the basis of physiographic, ethnic, historic, and cultural criteria, therefore, Russia constitutes a geographic realm, although transition zones rather than sharp boundaries mark its limits in some areas. Encircled by mountains and deserts, ruled by climatic continentality, unified by a dominant culture, and unmatched on Earth in terms of dimensions, Russia stands apart—from Europe to the west, China to the east, Central Asia to the south. But Russia, as we will see, is a society in transition. The realm's boundaries are unstable, still changing. To the west, Belarus at the opening of the twenty-first century was redirecting its interests from Europe toward Moscow. In the Caucasus, Armenia and Georgia effectively were in the Russian realm's orbit. In Central Asia, millions of Russians live across the border in northern Kazakhstan.

In the 1990s, Russia reorganized in the aftermath of the collapse of the Soviet Union. This reorganization cannot be understood without reference to the seven decades of Soviet communist rule that went before. We turn next to this crucial topic.

THE SOVIET LEGACY

The era of communism may have ended in the Soviet Empire, but its effects on Russia's political and economic geography will long remain. Seventy years of centralized planning and implementation cannot be erased overnight; regional reorganization toward a market economy cannot be accomplished in a day.

While the world of capitalism celebrates the failure of the communist system in the former Soviet realm, it should remember why communism found such fertile ground in the Russia of the 1910s and 1920s. In those days Russia was infamous for the wretched serfdom of its peasants, the cruel exploitation of its workers, the excesses of its nobility, and the ostentatious palaces and riches of the czars. Ripples from the Western European Industrial Revolution introduced a new age of misery for those laboring in factories. There were workers' strikes and ugly retributions, but when the czars finally tried to better the lot of the poor, it was too little too late. There was no democracy, and the people had no way to express

FROM THE FIELD NOTES

"June 25, 1964. We left Moscow this morning after our first week in the Soviet Union, on the road to Kiev. Much has been seen and learned after we arrived in Leningrad from Helsinki (also by bus) last Sunday. Some interesting contradictions: in a country that launched sputnik nearly a decade ago, nothing mechanical seems to work, from elevators to vehicles. The arts are vibrant, from the ballet in the Kirov Theatre to the Moscow Symphony. But much of what is historic here seems to be in disrepair, and in every city and town you pass, rows and rows of drab, gray apartment buildings stand amid uncleared rubble, without green spaces, playgrounds, trees. My colleagues here talk about the good fortune of the residents of these buildings: as part of the planning of the "socialist city," their apartments are always close to their place of work, so that they can walk or ride the bus a short distance rather than having to commute for hours. Our bus broke down as we entered the old part of Tula, and soon we learned that it would be several hours before we could expect to reboard. We had stopped right next to a row of what must have been attractive shops and apartments a half century ago; everything decorative seems to have been removed, and nothing has been done to paint, replace rotting wood, or fix roofs. I walked up the street a mile or so, and there was the "new" Tula—the Soviet imprint on the architecture of the place. This scene could have been photographed on the outskirts of Leningrad, Moscow, or any other urban center we had seen. What will life be like in the Soviet city of the distant future?"

or channel their grievances. Europe's democratic revolution passed Russia by, and its economic revolution touched the czars' domain only slightly. Most Russians, and tens of millions of non-Russians under the czars' control, faced exploitation, corruption, starvation, and harsh subjugation. When the people began to rebel in 1905, there was no hint of what lay in store; even after the full-scale Revolution of 1917, Russia's political future hung in the balance.

The Russian Revolution was no unified uprising. There were factions and cliques; the Bolsheviks ("Majority") took their ideological lead from Lenin, while the Mensheviks ("Minority") saw a different, more liberal future for their country. The so-called "Red" army factions fought against the "Whites," while both battled the forces of the czar. The country stopped functioning; the people suffered terrible deprivations in the countryside and in the cities. Most Russians (and other nationalities within the empire as well) were ready for radical change.

That change came when the Revolution succeeded and the Bolsheviks bested the Mensheviks, most of whom were exiled. In 1918, the capital was moved from Petrograd (as St. Petersburg had been renamed in 1914, to remove its German appellation) to Moscow. This was a symbolic move, the opposite of the forward-capital principle: Moscow lay deep in the Russian interior, not even on a major navigable waterway (let alone a coast), amid the same forests that much earlier had protected the Russians from their enemies. The new Soviet Union would look inward, and the communist system would achieve with Soviet resources and labor the goals that had for so long eluded the country. The chief political and economic architect of this effort was the revolutionary leader who prevailed in the power struggle: V. I. Lenin (born Vladimir Ilyich Ulyanov).

The Political Framework

Russia's great expansion had brought many nationalities under czarist control; now the revolutionary government sought to organize this heterogeneous ethnic mosaic into a smoothly functioning state. The czars had conquered, but they had done little to bring Russian culture to the peoples they ruled. The Georgians, Armenians, Tatars, and residents of the Muslim states of Central Asia were among dozens of individual cultural, linguistic, and religious groups that had not been "Russified." In 1917, however, the Russians themselves constituted only about one-half of the population of the entire country. Thus it was impossible to establish a Russian state instantly over this vast political region, and these diverse national groups had to be accommodated.

FIGURE 2-4

The question of the nationalities became a major issue in the young Soviet state after 1917. Lenin, who brought the philosophy of Karl Marx to Russia, talked from the beginning about the "right of self-determination for the nationalities." The first response by many of Russia's subject peoples was to proclaim independent republics, as they did in Ukraine, Georgia, Armenia, Azerbaijan, and even in Central Asia. But Lenin had no intention of permitting the Russian state to break up. In 1923, when his blueprint for the new Soviet Union went into effect, the last of these briefly independent units was fully absorbed into the sphere of the Moscow regime. Ukraine, for example, declared itself independent in 1917 and managed to sustain this initiative until 1919. But in that year the Bolsheviks set up a provisional government in Kiev, the Ukrainian capital, thereby ensuring the incorporation of the country into Lenin's Soviet framework.

The Bolsheviks' political framework for the Soviet Union was based on the ethnic identities of its many incorporated peoples. Given the size and cultural complexity of the empire, it was impossible to allocate territory of equal political standing to all the nationalities; the communists controlled the destinies of well over 100 peoples, both large nations and small isolated groups. It was decided to divide the vast realm into Soviet Socialist Republics (SSRs), each of which was delimited to correspond broadly to one of the major nationalities. At the time, Russians constituted about half of the developing Soviet Union's population, and, as Figure 2-4 shows, they also were (and still are) the most widely dispersed ethnic group in the realm. The Russian Republic, therefore, was by far the largest designated SSR, comprising just under 77 percent of total Soviet territory.

Within the SSRs, smaller minorities were assigned political units of lesser rank. These were called Autonomous Soviet Socialist Republics (ASSRs), which in effect were republics within republics; other areas were designated Autonomous Regions or other nationality-based units. It was a complicated, cumbersome, often poorly designed framework, but in 1924 it was launched officially under the banner of the Union of Soviet Socialist Republics (USSR).

Eventually, the Soviet Union came to consist of 15 SSRs (shown in Fig. 2-5), including not only the original

FIGURE 2-5

republics of 1924 but also such later acquisitions as Moldova (formerly Moldavia in Romania), Estonia, Latvia, and Lithuania. The internal political layout often was changed, sometimes at the whim of the communist empire's dictators. But no communist apartheid-like system of segregation could accommodate the shifting multinational mosaic of the Soviet realm. The republics quarreled among themselves over boundaries and territory. Demographic changes, migrations, war, and economic factors soon made much of the layout of the 1920s obsolete. Moreover, the communist planners made it Soviet policy to relocate entire peoples from their homelands in order to better fit the grand design, and to reward or punish—sometimes capriciously. The overall effect, however, was to move minority peoples eastward and to replace them with Russians. This **Russification** of the Soviet Empire produced substantial ethnic Russian minorities in all the non-Russian republics.

The Soviet planners called their system a **federation**. We focus in more detail on this geographic concept in Chapter 11, but we note here that *federalism* involves the sharing of power between a country's central government and its political subdivisions (provinces, States, or, in the Soviet case, Socialist Republics). Study the map of the former Soviet Union (Fig. 2-5), and an interesting geographic corollary emerges: every one of the 15 Soviet Republics had a boundary with a non-Soviet neighbor. Not one was spatially locked within the others. This seemed to give geographic substance to the notion that any Republic was free to leave the USSR if it so desired. Reality, of course, was different. Moscow's control over the republics made the Soviet Union a federation in theory only.

The centerpiece of the tightly controlled Soviet "federation" was the Russian Republic. With half the vast state's population, the capital city, the realm's core area, and over three-quarters of the Soviet Union's territory, Russia was the empire's nucleus. In other republics, "Soviet" often was simply equated with "Russian"—it was the reality with which the lesser republics lived. Russians came to the other republics to teach (Russian was taught in the colonial schools), to organize (and often dominate) the local Communist Party, and to implement Moscow's economic decisions. This was colonialism, but somehow the communist disguise—how could socialists, as the communists called themselves, be colonialists?—and the contiguous spatial nature of the empire made it appear to the rest of the world as something else. Indeed, on the world stage the Soviet Union became a champion of oppressed peoples, a force in the decolonization process. It was an astonishing contradiction that would, in time, be fully exposed.

The Soviet Economic Framework

The geopolitical changes that resulted from the founding of the Soviet Union were accompanied by a gigantic economic experiment: the conversion of the empire from a czarist autocracy with a capitalist veneer to communism. From the early 1920s onward, the country's economy would be centrally planned—the communist leadership in Moscow would make all decisions regarding economic planning and development. Soviet planners had two principal objectives: (1) to accelerate industrialization and (2) to **collectivize** agriculture. To accomplish these objectives, the entire country was mobilized, with a national planning commission (Gosplan) at the helm. For the first time ever on such a scale, and for the first time in accordance with Marxist-Leninist principles, an entire country was organized to work toward national goals prescribed by a central government.

The Soviet planners believed that agriculture could be made more productive by organizing it into huge state-run enterprises. The holdings of large landowners were expropriated, private farms were taken away from the farmers, and the land was consolidated into collective farms. Initially, all such land was meant to be part of a *sovkhoz*, literally a grain-and-meat factory in which agricultural efficiency, through maximum mechanization and minimum labor requirements, would be at its peak. But many farmers opposed the Soviets and tried to sabotage the program in various ways, hoping to retain their land.

The farmers and peasants who obstructed the communists' grand design suffered a dreadful fate; their tragedy is one of the factors that nurtured anti-Russian nationalism in the now-independent republics. In the 1930s, for instance, Stalin confiscated Ukraine's agricultural output and then ordered part of the border between the Russian and Ukrainian republics sealed—thereby leading to a famine that killed millions of farmers and their families. In the Soviet Union under communist totalitarianism, the ends justified the means, and untold hardship came to millions who had already suffered under the czars. In his book *Lenin's Tomb*, David Remnick estimates that between 30 and 60 million people lost their lives from imposed starvation, political purges, Siberian exile, and forced relocation. It was an incalculable human tragedy, but the secretive character of Soviet officialdom made it possible to hide it from the world.

Collectivization of agriculture did not increase farm productivity as the Soviet planners had hoped, even as millions of acres of land were brought into cultivation through ambitious (and environmentally disastrous) irrigation schemes. The largest of these ventures was the Virgin and Idle Lands Program launched in the 1950s in Kazakhstan, turning pasturelands into wheatfields. Pesticides poisoned the groundwater, thousands of people became ill—and still the Soviets had to import food from abroad.

The Soviet planners hoped that collectivized and mechanized farming would free hundreds of thousands of workers to labor in factories. Industrialization was the prime objective of the regime, and here the results were superior. Productivity rose rapidly, and when World War II engulfed the empire in 1941, the Soviet manufacturing sector was able to produce the equipment and weapons needed to repel the German invaders.

Yet even in this context, the Soviet grand design entailed liabilities for the future. The USSR practiced a **command economy**, in which state planners assigned the production of particular manufactures to particular places, often disregarding the rules of economic geography. For example, the manufacture of railroad cars might be assigned (as indeed it was) to a factory in Latvia. No other factory anywhere else would be permitted to produce this equipment—even if supplies of raw materials would make it cheaper to build them near, say, Volgograd 1200 miles away. Despite an expanded and improved transport network (Fig. 2-6), such practices made manufacturing in the USSR extremely expensive, and the absence of competition made managers complacent and workers less productive then they could be.

Of course, the Soviet planners never imagined that their experiment would fail and that a market-driven economy would replace their command economy. When that happened, the transition was predictably difficult; indeed, it is far from over, and it is putting severe strains on the now more democratic state.

RUSSIA'S CHANGING POLITICAL GEOGRAPHY

When the USSR dissolved in 1991, Russia's former empire devolved into 14 independent countries, and Russia itself was a changed nation. Russians now made up about 83 percent of the population of under 150 million, a far higher proportion than in the days of the Soviet Union. But numerous minority peoples remained under Moscow's new flag, and millions of Russians found themselves under new governments in the former Republics.

Soviet planners had created a complicated administrative structure for their "Russian Soviet Federative Socialist Republic," and Russia's postcommunist leaders had to use this framework to make their country function. In 1992, most of Russia's internal "Republics," Autonomous Regions, Oblasts, and Krays (all of them components of the administrative hierarchy) signed a document known as the Russian Federation Treaty, committing them to cooperate in the new federal system. At first a few units refused to sign, including Tatarstan, scene of Ivan the Terrible's brutal conquest more than four centuries ago, and a Republic in the Caucasus periphery, then known as Chechenya-Ingushetia, where Muslim rebels waged a campaign for independence. As the map shows, Chechenya-Ingushetia split into two separate Republics whose names were now spelled Chechnya and Ingushetia (Fig. 2-7). Eventually, only Chechnya refused to sign the Russian Federation Treaty, and subsequent Russian military intervention led to a prolonged and violent conflict, with disastrous consequences for Chechnya's people and infrastructure (the capital, Groznyy, was completely destroyed [photo, p. 119]). The Chechnya war continues today (see the Issue Box entitled "Chechnya" on p. 118) and is a disaster for Russia's government as well.

The Federal Framework of Russia

The spatial framework of the still-evolving Russian Federation is as complex as that of the Russian Federative Socialist Republic of communist times. As the twenty-first century opened, the Federation consisted of 89 entities: 2 Autonomous Federal Cities, 21 Republics, 11 Autonomous Regions (Okrugs), 49 Provinces (Oblasts), and 6 Territories (Krays). Moscow and St. Petersburg are the two Autonomous Federal Cities. The 21 Republics, recognized to accommodate substantial ethnic minorities in the population, lie in several clusters (Fig. 2-7).

Russia's post–1991 governments have faced complicated administrative problems. A multinational, multicultural state that had been accustomed to authoritarian rule and government control over virtually everything—from factory production to everyday life—now had to be governed in a new way. Democratization of the political system, transition to a market economy, sale of state-owned industries (privatization), and liberation of the press and other media had to take place in an orderly way, or the country would, literally, fall apart.

Russia's leaders knew that their options were limited. They could continue to hold as much power as possible at the center, making decisions in Moscow that would apply to all the Republics, Regions, and other subdivisions of the state. (Such a **unitary state system**, with its centralized government and administration, marked authoritarian kingdoms of the past and serves totalitarian dictatorships of the present.) Or they could share power with the Republics and Regions, allowing elected regional leaders to come to Moscow to represent the interests of their people. This is the federal system Russia chose as the only way to accommodate the country's economic and cultural diversity.

FIGURE 2-6

OCEAN

East
Siberian
Sea

UNITED STATES

Bering
Sea

North
Land

90°   100°   110°   120°   130°   140°   150°   160°   170°   80°   70°   60°
180°   170°

Laptev
Sea

New Siberian Is.

Wrangel I.

Kolyma R.

Ust Nern

Magadan

Kamchatka
Peninsula

Aldan R.

Okhotsk

Sea
of
Okhotsk

Petropavlovsk-
Kamchatskiy

50°

Vilyuy R.

Lena R.

Yakutsk

Lesser Tunguska R.

Suntar

Sakhalin

F E D E R A T I O N

Aleksandrovsk

Angara R.

Baykal-Amur Mainline Railroad

Amur R.

Vanino

Krasnoyarsk

Bratsk

Komsomolsk

Kurile Is.

Tayshet

Lena R.

Trans-Siberian Railroad

Khabarovsk

Lake
Baykal

Chita

Amur R.

Blagoveshchensk

Ussuri R.

Irkutsk

Ulan-Ude

L. Khanka

Vladivostok

40°

Harbin

Nakhodka

Ulaanbaatar

Changchun

East Sea
(Sea of
Japan)

140°

MONGOLIA

CHINA

Shenyang

NORTH
KOREA

JAPAN

Beijing

120°

Pyongyang

SOUTH
KOREA
Seoul

100°   110°   130°

RUSSIA AND ITS INTERNAL DIVISIONS

Russian administrative units

Autonomous Regions within administrative units

Peripheries in transition

Each internal republic is colored separately

National capital is underlined

| 0 | 200 | 400 | 600 | 800 Kilometers |
| 0 | 100 | 200 | 300 | 400 | 500 Miles |

RUSSIAN ADMINISTRATIVE UNITS
(named after their capitals)

1. Astrakhan	20. Pskov
2. Belgorod	21. Rostov
3. Bryansk	22. Ryazan
4. Chelyabinsk	23. St. Petersburg
5. Ivanovo	24. Samara
6. Kaluga	25. Saratov
7. Kemerovo	26. Smolensk
8. Kostroma	27. Stavropol
9. Krasnodar	28. Tambov
10. Kurgan	29. Tomsk
11. Kursk	30. Tula
12. Lipetsk	31. Tver
13. Moscow	32. Ulyanovsk
14. Novgorod	33. Vladimir
15. Novosibirsk	34. Volgograd
16. Omsk	35. Voronezh
17. Orenburg	36. Vyatka
18. Orël	37. Yaroslavl
19. Penza	

Longitude East of Greenwich

FIGURE 2-7

CHECHNYA

Regional ISSUE

IN SUPPORT OF RUSSIAN CONTROL OVER CHECHNYA

"As a policeman here in Moscow, I strongly support the government and the army in their efforts to establish control over the criminal elements trying to take control of Chechnya. Let us remember what happened there. When the Russian government back in 1991 had to take charge of our country and its many components, there was lots of opposition. The Tatars (also Muslims, like the Chechens) talked about establishing an independent state in their republic. The Kalmyks, the Udmurts, the Bashkirs, the Chuvash and many others proclaimed that they wanted everything ranging from autonomy to independence. There were even Russians demanding it in places like Yekaterinburg and Primorskiy. But the government under President Yeltsin negotiated all these claims and gave those people reasons to want to stay under the Russian flag. Except the Chechens. Nothing was going to satisfy them.

"And the Chechens were even given their own republic after the changeover from Soviet to Russian administration. In 1991 they still shared their territory with the Ingush, who are also Muslims, in the so-called Chechen-Ingush Republic. So what did the Chechens do? They installed their separatist leader as ruler of the republic and started fighting with the Ingush minority. To help solve this crisis, Russia's government divided the republic's territory into two, the larger, richer part including the capital and the oilfields for the Chechens and the remainder for the Ingush.

"But it wasn't good enough for the Chechens. They attacked the Russians living in Chechnya, causing the army to move in to protect Russians and Russian interests. They broke the truce that followed. Then they mounted a full-scale war from their mountain hideouts and caused us hideous casualties. They didn't care that their capital was totally destroyed; these Muslim warlords fought among themselves even as they fought us. Next they started using terror to get their way. Remember the autumn of 1999, when they blew up apartment buildings in Moscow, causing more than 300 dead? And how about what they did in Dagestan, where they took over a bunch of villages and declared an "independent Chechen republic"? Not to mention that hospital full of doctors, nurses, and patients, whom they took hostage, eventually killing hundreds of innocent victims. These are people to whom we should entrust the government of an independent country on our borders? Never.

"So don't criticize us when it comes to our strategies to deal with these barbarians. Foreign governments say that the Russian army does terrible things to captured Chechens, but what about the treatment of Russian soldiers by Chechen rebels? War is brutal, and this is a war for the soil of Mother Russia.

"If we give up in Chechnya, other minorities will get ideas about independence too, and that would be the end of the nation. There is nothing to negotiate. Russia must and will prevail."

WHY CHECHNYA DESERVES INDEPENDENCE

"My grandparents were born in what is today the Russian colony of Chechnya, and they died a horrible death somewhere in Soviet Central Asia. I am here to avenge their deaths and to punish the Russians for what they did to my people.

"Russians seem to think that we are fighting for independence just because we want it. Why aren't we like all those other minorities that have come to terms with their lives under Moscow's heel? Well, in our case there is more to it. We fought the Russian imperialists to a standstill less than two centuries ago, when the czars' armies colonized Islamic peoples from the Caucasus to Central Asia. Yes, we were eventually defeated, but those Russians never really penetrated our mountain hideouts. Then came the Soviets, who thought they did us a big favor by creating one of those "Autonomous Republics" for us along with the poor Ingush. But look at the map. They combined our traditional homeland with a stretch of flatland to the north, which was full of Russian farmers and oilmen. Do you think Groznyy was a Chechen town? Think again. Or look for a mosque on those photographs of the 'good old days.'

"But we might have put up with it all except for what happened during the war between the Soviets and the Germans, World War II. Some of us Chechens were happy to see the Germans do to the Soviets what the Soviets had done to us, but Josef Stalin accused us all of collaborating with the enemy. In a few short months he packed all 500,000 of my people on trains and sent them to exile in Kazakhstan. Thousands died on the trains and were simply thrown onto the tracks. Many more, probably about 125,000, perished in the desert. I don't know where or when my grandparents died, but my parents survived.

"We got our homeland back because the Soviet dictator Khrushchev reinstated our Republic in 1957, and the survivors straggled back. But nobody ever forgot what was done to us. We're not talking of a few hundred hostages here. We're talking about the killing of a quarter of a nation. Still, we tried to restart our society under those atheist Soviets, keeping our Islamic practices quiet and private. But then the Soviet Empire collapsed and we got our chance. In 1991 we declared our independence from the new Russia, and at first it seemed that the Russians would be sensible and approve.

"But before long the Russians changed their minds and tried to intervene in our internal affairs. We managed to defeat the Russian army once again, and the war devastated Groznyy, but the Russians would not even consider independence. We got assistance from Muslims elsewhere and money from many sources, but we can see that Russia will not give us what we want. So we turn to any means we can to wound the Russian bear, and we will chase it out of Chechnya with the help of Allah."

Vote your opinion at www.wiley.com/college/deblij

Chechnya was the only republic that rejected incorporation in the Russian Federation, precipitating a series of conflicts that killed tens of thousands of soldiers and civilians, devastated the territory's infrastructure, destroyed the capital, Groznyy, almost totally, brought terrorism to Moscow, caused political crises in the Russian government, and defied resolution for more than a decade. Today, Muslim rebels still are holed up in the (southern) Caucasus sector of the tiny territory while as many as 80,000 Russian military are deployed to assert government control over the capital and the northern two-thirds of Chechnya. In Groznyy, survivors continue to inhabit parts of shattered buildings, basements, shelters, and huts built from the rubble of war, carrying on with life but risking death from Russian guns and rebel bombs.

In a *federal system*, the national government usually is responsible for matters such as defense, foreign policy, and foreign trade. The Regions (or provinces, States, or other subdivisions) retain authority over affairs ranging from education to transportation. A federal system does not create unity out of diversity, but it does allow diverse components of the state to coexist, their common interests represented by the national government and their regional interests by their local administrations. Some countries owe their survival as coherent states to their federal frameworks. India and Australia are cases in point.

But to maintain a generally acceptable balance of power between the center and the Regions (or States) is difficult. Disputes over "States' rights" continue to roil the American political scene more than two centuries after the Constitution was adopted. Russia, which also owes its continuity to a grand federal experiment, is barely more than ten years old. Power shifts continue, and the framework of the Russian Federation will undoubtedly change over time.

Another problem arises from Russia's sheer size. Geographers refer to the principle of **distance decay** to explain how increasing distances between places tend to reduce interactions among them. Because Russia is the world's largest country, distance forms a significant ingredient in the relationships between the capital and the outlying areas. Furthermore, Moscow lies in the far west of the giant country, half a world away from the shores of the Pacific. Not surprisingly, one of the most obstreperous Regions has been Primorskiy, the Region of Vladivostok. In 1997, the Russian government sought to dismiss the elected governor of Primorskiy because, among other things, he refused to implement federal decisions and would not forward the Region's tax revenues to Moscow.

Still another problem is reflected by Figure 2-7: the enormous size variation among the Republics (and to a lesser extent, the Regions). Whereas the (territorially) smallest Republics are concentrated in the Russian core area, the largest lie far to the east, where Sakha is nearly a thousand times as large as Ingushetiya. On the other hand, the populations of the huge eastern Republics are tiny compared to those of the smaller ones in the core. Such diversity spells administrative difficulty.

But perhaps the most serious current problem for Russia is the growing social disharmony between Moscow and the subnational political entities. Almost wherever one goes in Russia today, Moscow is disliked and often berated by angry locals. The capital is seen as the privileged playground for those who have benefited most from the post-Soviet transition—bureaucrats and hangers-on whose economic policies have driven down standards of living, whose greed and corruption have hurt the economy, whose actions in Chechnya have been a disaster, who have allowed foreigners (prominently the United States) to erode Russian power and prestige, who fail to pay the wages of workers toiling in industries still owned by the state, and

**RUSSIA: THE NEW FEDERAL
DISTRICTS PROCLAIMED
MAY 13, 2000 AND THEIR CAPITALS**

FIGURE 2-8

who do not represent the Russian people. Complaints about the capital are not uncommon in free countries, but in Russia the mistrust between capital and subordinate areas has become so serious that it constitutes a key challenge for the government.

In 2000, the Putin administration moved to diminish the influence of the Regions by creating a new spatial framework that combines the 89 Regions, Republics, and other entities into seven new administrative units—not to enhance their influence in Moscow but to increase Moscow's authority over them (Fig. 2-8). Each of these new federal administrative districts has its capital city, which will become the conduit for Moscow's "guidance," as the official plan puts it. This framework will counter some powerful vested interests in the Republics and Regions, however, and it will reverse a trend toward ever-greater local self-determination. It is far from certain that the system will work.

CHANGING SOCIAL GEOGRAPHIES

Russia's difficulties are underscored by the country's population data. The population of 144 million (2004) is shrinking—not because of the factors we cited in the case of Western Europe but because of the dislocation, turmoil, fears, and uncertainties arising from the post-Soviet transition. This **population decline** now amounts to about 1 million per year. Male life expectancy has dropped from 66 in the mid-1960s to 59 in the early 2000s; it is also declining for women. Among males, alcoholism, suicide, and other manifestations of social disorder drive down life expectancy. In the general population, drug abuse, heavy smoking, and poor diets are to blame. The incidence of diseases is rising. Tuberculosis is taking a heavy toll; AIDS cases are multiplying rapidly; cancer rates are rising. Russia confronts a health crisis that threatens to overshadow all its other problems.

Coupled with this distress is the decline of national institutions. The armed forces are in disrepair. Major universities lack modern facilities. Leading industries are failing; the energy industry is in the grip of powerful magnates who flout the rules. Legal systems, banking operations, and property regulations are not sufficiently developed to ensure orderly procedures. Corruption and cronyism are rife, and some observers describe organized crime as Russia's most successful industry.

Yet Russia remains a formidable nuclear power with a vast arsenal of weapons. Russia's quest for social and political recovery and stability is a matter of global interest.

Symbol of the new era in Russia: a Coca Cola sign and a bank logo tower over traffic and pedestrians on Novi Arbat street near the Kremlin in Moscow. The city's skyline still does not reflect its political or economic power or its rank as one of the world's most populous centers, but Moscow's ambience has been transformed since 1990.

RUSSIA'S PROSPECTS

When Peter the Great began to reorient his country toward Europe and the outside world, he envisaged a Russia with warm-water ports, a nation no longer encircled by Swedes, Lithuanians, Poles, and Turks, and a force in European affairs. Core-periphery relationships always have been crucial to Russia, and they remain so today.

Heartlands and Rimlands

Russia's relative location has long been the subject of study and conjecture by geographers. Exactly a century ago, the British geographer Sir Halford Mackinder (1861–1947) argued in a still-discussed article entitled "The Geographical Pivot of History" that western Russia and Eastern Europe enjoyed a combination of natural protection and resource wealth that would someday propel its occupants to world power. The protected core area, he reasoned, overshadowed the exposed periphery. Eventually, this pivotal interior region of Eurasia, which he later called the *heartland*, would become a stage for world domination.

Mackinder's **heartland theory** was published when Russia was a weak, economically backward society ripe for revolution, but Mackinder stuck to his guns. When Russia, now in control of Eastern Europe and with a vast colonial empire, emerged from World War II as a superpower, Mackinder's conjectures seemed prophetic.

But not all geographers agreed. Probably the first scholar to use the term *rimland* (today *Pacific Rim* is part of everyday language) was Nicholas Spykman, who in 1944 countered Mackinder by calculating that Eurasia's periphery, not its core, held the key to global power. Spykman foresaw the rise of rimland states as superpowers and viewed Japan's emergence to wealth and power as just the beginning of that process.

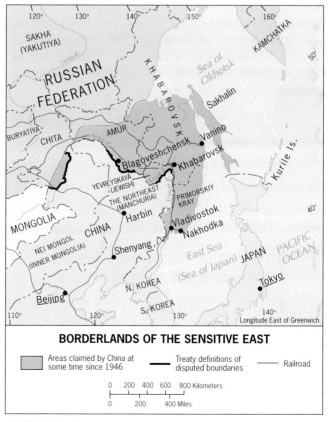

BORDERLANDS OF THE SENSITIVE EAST

Areas claimed by China at some time since 1946 | Treaty definitions of disputed boundaries | Railroad

0 200 400 600 800 Kilometers

0 200 400 Miles

FIGURE 2-9

Russia's Role in the World

Even as Russia's leaders seek to address the enormous internal problems that beset this giant country, Moscow also tries to maintain a prominent international role. Russia faces an encroaching NATO in Europe, a strong Muslim presence in Central Asia, a rising China along its eastern flanks. During the 1990s, the Russians settled more than a dozen lingering border disputes with China (some remain unresolved). But for the future, the larger question is likely to involve China's "lost" territories. As Figure 2-9 shows, Russia's eastern territory just touches North Korea, virtually landlocking China's historic in-dustrial heartland. Russians took these lands from China more than a century ago, and China has a long memory.

Russia in the early twenty-first century faced several other issues with geographic overtones: (1) relations with the expanding European Union and NATO; (2) owner-ship of oil and gas reserves in the Caspian Sea Basin be-yond its land borders; (3) the Russian diaspora in north-ern Kazakhstan and elsewhere in its former domain; (4) Chechnya's resistance to incorporation into the Russian Federation; and (5) projection of continuing Russian power and influence in the world despite the country's in-ternal disarray.

REGIONS OF THE REALM

SO VAST IS RUSSIA, so varied its physiography, and so diverse its cultural landscape that regionalization re-quires a small-scale perspective and a high level of gen-eralization. Figure 2-10 outlines a four-region frame-work: the Russian Core and its Peripheries west of the Urals, the Eastern Frontier, Siberia, and the Far East. As we will see, each of these massive regions contains major subregions.

RUSSIAN CORE AND PERIPHERIES

18 The heartland of a state is its **core area**. Here much of the population is concentrated, and here lie its leading cities, major industries, densest transport networks, most inten-sively cultivated lands, and other key components of the country. Core areas of long standing strongly reflect the imprints of culture and history. The Russian core area, broadly defined, extends from the western border of the Russian realm to the Ural Mountains in the east (Fig. 2-10). This is the Russia of Moscow and St. Petersburg, of the Volga River and its industrial cities.

Central Industrial Region

At the heart of the Russian Core lies the Central Industrial Region (Fig. 2-11). The precise definition of this subre-gion varies, for all regional definitions are subject to de-bate. Some geographers prefer to call this the Moscow Region, thereby emphasizing that for over 250 miles (400 km) in all directions from the capital, everything is ori-ented toward this historic focus of the state. As both Fig-ures 2-1 and 2-6 show, Moscow has maintained its deci-sive centrality: roads and railroads converge in all directions from Ukraine in the south; from Mensk (Be-larus) and the rest of Eastern Europe in the west; from St. Petersburg and the Baltic coast in the northwest; from

Nizhniy Novgorod (formerly Gorkiy) and the Urals in the east; from the cities and waterways of the Volga Basin in the southeast (a canal links Moscow to the Volga, Rus-sia's most important navigable river); and even to the subarctic northern periphery that faces the Barents Sea (see box titled "Facing the Barents Sea").

Moscow (population: 8.1 million) is the focus of an area that includes some 50 million inhabitants (more than one-third of the country's total population), many of them concentrated in such major cities as Nizhniy Novgorod, the automobile-producing "Soviet Detroit"; Yaroslavl, the tire-producing center; Ivanovo, the heart of the textile industry; and Tula, the mining and metallurgical center where lignite (brown coal) deposits are worked.

St. Petersburg (the former Leningrad) remains Rus-sia's second city, with a population of 4.5 million. Under the czars, St. Petersburg was the focus of Russian political and cultural life, and Moscow was a distant

MAJOR CITIES OF THE REALM	
City	Population* (in millions)
Baki (Baku), Azerbaijan	2.1
Irkutsk, Russia	0.7
Kazan, Russia	1.1
Moscow, Russia	8.1
Nizhniy Novgorod, Russia	1.3
Novosibirsk, Russia	1.3
St. Petersburg, Russia	4.5
Tbilisi, Georgia	1.4
Vladivostok, Russia	0.8
Volgograd, Russia	1.0
Yekaterinburg, Russia	1.2
Yerevan, Armenia	1.5

*Based on 2004 estimates.

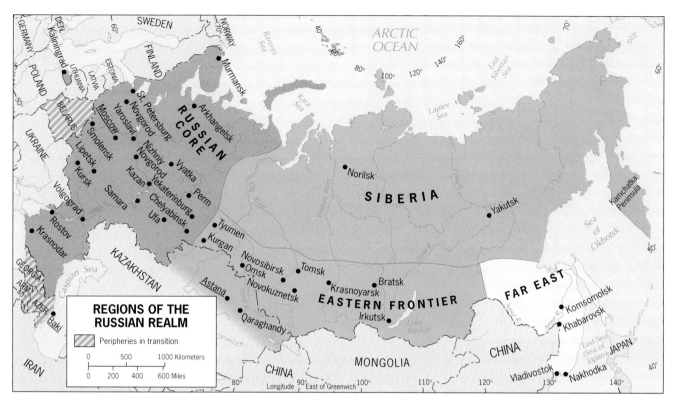

REGIONS OF THE RUSSIAN REALM

Peripheries in transition

0	500	1000 Kilometers
0	200 400	600 Miles

FIGURE 2-10

RUSSIA'S MANUFACTURING REGIONS

Manufacturing regions

Railroad

0	500	1000 Kilometers
0	200 400	600 Miles

FIGURE 2-11

Facing the Barents Sea

*N*orth of the latitude of St. Petersburg, the region we have named the Russian Core takes on Siberian properties. However, the Russian presence is much stronger in this remote northern periphery than it is in Siberia proper. Two substantial cities, Murmansk and Arkhangelsk, are road- and rail-connected outposts in the shadow of the Arctic Circle.

Murmansk lies on the Kola Peninsula not far from the border with Finland (Fig. 2-6). In its hinterland lie a variety of mineral deposits, but Murmansk is particularly important as a naval base. During World War II, Allied ships brought supplies to Murmansk; the city's remoteness shielded it from German occupation. After the war, it became a base for nuclear submarines. This city is also an important fishing port and a container facility for cargo ships.

Arkhangelsk is located near the mouth of the Northern Dvina River where it reaches an arm of the White Sea (Fig. 2-6). Ivan the Terrible chose its site during Muscovy's early expansion; the czar wanted to make this the key port on a route to maritime Europe. Yet Arkhangelsk, mainly a port for lumber shipments, has a shorter ice-free season than Murmansk, whose port can be kept open with the help of the warm North Atlantic Drift ocean current (and by icebreakers if necessary).

Nothing in Siberia east of the Urals rivals either of these cities—yet. But their existence and growth prove that Siberian barriers to settlement can be overcome.

AMONG THE REALM'S GREAT CITIES . . . **MOSCOW**

*I*n the vastness of Russia's territory, Moscow, capital of the Federation, seems to lie far from the center, close to its western margin. But in terms of Russia's population distribution, Moscow's centrality is second to none among the country's cities. Moscow (8.1 million) lies at the heart of Russia's primary core area and at the focus of its circulation systems.

On the banks of the meandering Moscow River, Moscow's skyline of onion-shaped church domes and modern buildings rises from the forested flat Russian Plain like a giant oasis in a verdant setting. Archeological evidence points to ancient settlement of the site, but Moscow enters recorded history only in the middle of the twelfth century. Forest and river provided defenses against Tatar raids, and when a Muscovy force defeated a Tatar army in the late fourteenth century, Moscow's primacy was assured. A huge brick Kremlin (citadel, fortress), with walls more than a mile in length and with 18 towers, was built to ensure the city's security. From this base Ivan the Terrible expanded Muscovy's domain and laid the foundations for Russia's vast empire.

The triangular Kremlin and the enormous square in front of it (Red Square of revolutionary times), flanked by St. Basil's Cathedral and overlooking the Moscow River, is the center of old Moscow and is still the heart of the city. From here, avenues and streets radiate in all directions to the Garden Ring and beyond (see photo, p. 121). Until the 1970s, the Moscow Circular Motorway encircled most of the built-up area, but today the metropolis sprawls far outside this beltway. For all its size, Moscow never developed a world-class downtown skyline. Communist policy was to create a "socialist city" in which neighborhoods—consisting

of apartment buildings, workplaces, schools, hospitals, shops, and other amenities—would be clustered so as to obviate long commutes into a high-rise city center.

Moscow may not be known for its architectural appeal, but the city contains noteworthy historic as well as modern structures, including the Cathedral of the Archangel and the towers of Moscow State University.

second city. Today, however, St. Petersburg has none of Moscow's locational advantages, at least not with respect to the domestic market. It lies well outside the Central Industrial Region near the northwestern corner of the country, 400 miles (650 km) from Moscow. Neither is it better off than Moscow in terms of resources: fuels, metals, and foodstuffs must all be brought in, mostly from far away. The former Soviet emphasis on self-sufficiency even reduced St. Petersburg's asset of being on the Baltic coast because some raw materials could have been imported much more cheaply across the Baltic Sea from foreign sources than from domestic sites in distant Central Asia (only bauxite deposits lie nearby, at Tikhvin).

Yet St. Petersburg was at the vanguard of the Industrial Revolution in Russia, and its specialization and skills have remained important. Today, the city and its immediate environs contribute about 10 percent of the country's manufacturing, much of it through fabricating high-quality machinery.

Povolzhye: The Volga Region

A second region lying within the Russian Core is the *Povolzhye*, the Russian name for an area that extends along the middle and lower valley of the Volga River. It would be appropriate to call this the Volga Region, for that greatest of Russia's rivers is its lifeline and most of the cities that lie in the Povolzhye are on its banks (Fig. 2-11). In the 1950s, a canal was completed to link the lower Volga with the lower Don River (and thereby the Black Sea).

AMONG THE REALM'S GREAT CITIES . . . ST. PETERSBURG

*C*zar Peter the Great and his architects transformed the islands of the Neva Delta, at the head of the Gulf of Finland, into the Venice of the North, its palaces, churches, waterfront façades, bridges, and monuments giving St. Petersburg a European look unlike that of any other city in Russia. Having driven out the Swedes, Peter laid the foundations of the Peter and Paul Fortress on the bank of the wide Neva River in 1703, and the city was declared the capital of Russia in 1712.

Peter's "window on Europe," St. Petersburg (named after the saint, not the czar) was to become a capital to match Paris, Rome, and London. During the eighteenth century, St. Petersburg acquired a magnificent skyline dominated by tall, thin spires and ornate cupolas, and graced by baroque and classical architecture. The Imperial Winter Palace and the adjoining Hermitage Museum at the heart of the city are among a host of surviving architectural treasures.

Revolution and war lay in St. Petersburg's future. In 1917, the Russian Revolution began in the city (then named Petrograd), and following the communist victory it lost its functions as a capital to Moscow and its name to Lenin. As Leningrad, it suffered through the 872-day Nazi siege during World War II, holding out heroically through endless bombardment and starvation that took nearly 1 million lives and severely damaged many of its buildings.

The communist period witnessed the neglect and destruction of some of Leningrad's most beautiful churches, the intrusion and addition of crude monuments and bleak apartment complexes, and the rapid industrialization and growth of the city (which now totals 4.5 million). Immediately after the collapse of the Soviet Union, it regained its original name and a new era opened. The Russian Orthodox Church revived, building restoration went forward (see photo, p. 107), and tourism boomed. But the transformation from communist to capitalist ways has been accompanied by social problems that include crime and poverty.

The Volga River was an important historic route in old Russia, but for a long time neighboring regions overshadowed it. The Moscow area and Ukraine were far ahead in industry and agriculture. The Industrial Revolution that came late in the nineteenth century to the Moscow Region did not have much effect in the Povolzhye. Its major function remained the transit of foodstuffs and raw materials to and from other regions.

This transport function is still important, but the Povolzhye has changed. First, World War II brought furious development because the Volga River, located east of Ukraine, was far from the German armies that invaded from the west. Second, in the postwar era the Volga-Urals Region for some time was the largest known source of petroleum and natural gas in the entire Soviet Union. From near Volgograd (formerly Stalingrad) in the southwest to Perm on the Urals' flank in the northeast lies a belt of major oilfields (Fig. 2-12).

Third, the transport system has been greatly expanded. The Volga–Don Canal directly connects the Volga waterway to the Black Sea; the Moscow Canal extends the northern navigability of this river system into the heart of the Central Industrial Region; and the Mariinsk canals provide a link to the Baltic Sea. Today, the Volga Region's population exceeds 25 million, and the cities of Samara (formerly Kuybyshev), Volgograd, Kazan, and Saratov all have populations between 1.0 and 1.3 million. Manufacturing has also expanded into the middle Volga Basin, emphasizing more specialized engineering industries. The huge Fiat-built auto assembly plant in Tolyatti, for example, is one of the world's largest of its kind.

The Urals Region

The Ural Mountains form the eastern limit of the Russian Core. They are not particularly high; in the north they consist of a single range, but southward they broaden into a hilly zone. Nowhere are they an obstacle to east-west transportation. An enormous storehouse of metallic mineral resources located in and near the Urals has made this area a natural place for industrial development. Today, the Urals Region, well connected to the Volga and Central Industrial Region, extends from Serov in the north to Orsk in the south (Fig. 2-11).

The Central Industrial, Volga, and Urals regions form the anchors of the Russian core area. For decades they have been spatially expanding toward one another, their interactions ever more intensive. These regions of the Russian Core stand in sharp contrast to the compara-

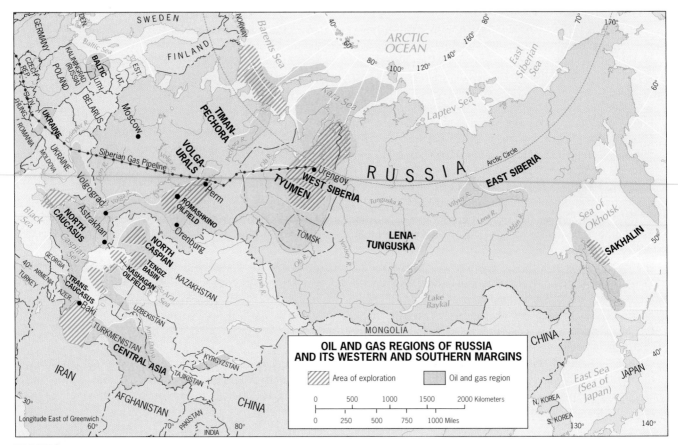

OIL AND GAS REGIONS OF RUSSIA AND ITS WESTERN AND SOUTHERN MARGINS

Area of exploration   Oil and gas region

FIGURE 2-12

tively less developed, forested, Arctic north and the remote upland to the south between the Black and Caspian seas. Thus even within this Russian coreland, frontiers still await growth and development.

The Internal Southern Periphery

We turn now to an area that has generated some of the most difficult politico-geographical problems the new Russia has faced since the collapse of the Soviet Union. Between the Black Sea to the west and the Caspian Sea to the east, and with the Caucasus Mountains to the south, Moscow seeks to stabilize and assert its authority over a tier of minority "republics," holdovers from the Soviet era (Fig. 2-13).

A combination of geographic factors causes the problems Moscow confronts in this peripheral zone. Here the Russians (and later the Soviets) met and stalled the advance of Islam. Here the cultural mosaic is as intricate as anywhere in the Federation: Dagestan, the southernmost republic fronting the Caspian Sea, has 2 million inhabitants comprising some 30 ethnic groups speaking about 80 languages. And here Russia and its Caspian holdings penetrate the reservoirs of oil and natural gas which this huge, energy-rich basin contains. Groznyy, the capital of Chechnya which adjoins Dagestan to the west, was a crucial oil-industry center, pipeline junction, and service hub during the Soviet era.

But Chechnya also contains a sizeable Muslim population, and fiercely independent Muslim Chechens used Caucasus mountain hideouts to resist Russian colonization during the nineteenth century. Accused of collaboration with the Nazis during World War II, on Stalin's orders the Soviets exiled the entire Chechen population to Central Asia, with much loss of life. Rehabilitated and allowed to return by Premier Nikita Khrushchev in the 1950s, the Chechens seized their opportunity in 1991 after the collapse of the Soviet Union and fought the Russian army to a stalemate. But Moscow never granted Chechnya independence (one quarter of the population of 1.2 million was Russian), and an uneasy standoff continued. Meanwhile, Chechen hit-and-run attacks on targets in neighboring Republics continued, and in 1999 three apartment buildings in Moscow were bombed, resulting in 230 deaths. Russian authorities blamed Chechnyan terrorists for these bombings and ordered a full-scale attack on Chechnya's Muslim holdouts. Groznyy, already severely damaged in the earlier conflict, was totally devastated and the rebels were driven into the mountains, but the Russian armed forces, still taking substantial losses, were unable to establish unchallenged control of all of the Republic.

Russia's costly problems in Chechnya illustrate the fractious nature of its internal southern periphery, from Buddhist Kalmykiya to also-Muslim Ingushetiya. Here the still-evolving federal framework will be put to the test for many years to come.

The External Southern Periphery

Beyond Russia's tier of internal republics lies still another periphery, legally outside the Russian sphere but closely tied to it nevertheless. This is *Transcaucasia*, historically a battleground for Christians and Muslims, Russians and Turks, Armenians and Persians. Today it is a subregion containing three former Soviet Socialist Republics: landlocked Armenia, coastal Georgia on the Black Sea, and Azerbaijan on the land-encircled Caspian Sea. Although these three countries are independent states today, they are so closely bound up with Russia that they are, functionally, part of the Russian realm (Fig. 2-13).

Armenia

As Figure 2-1 shows, landlocked Armenia (population: 3.8 million) occupies some of the most rugged and mountainous terrain in the earthquake-prone Transcaucasus. The Armenians are an embattled people who adopted Christianity 17 centuries ago and for more than a millennium sought to secure their ancient homeland here on the margins of the Muslim world. During World War I, the Ottoman Turks massacred much of the Christian Armenian minority and drove the survivors from eastern Anatolia and what is now Iraq into the Transcaucasus. At the end of that war in 1918, an independent Armenia arose, but its autonomy lasted only two years. In 1920, Armenia was taken over by the Soviets; in 1936, it became one of the 15 constituent republics of the Soviet Union. The collapse of the Soviet Empire gave Armenia what it had lost three generations earlier: independence.

Or so it seemed. Soon afterward, the Armenians found themselves at war with neighboring Azerbaijan over the fate of some 150,000 Armenians living in Nagorno-Karabakh, a pocket of territory surrounded by Azerbaijan. Such a separated territory is called an *exclave*, and this one had been created by Soviet sociopolitical planners who, while acknowledging the cultural (Christian) distinctiveness of this cluster of Armenians, nevertheless gave (Muslim) Azerbaijan jurisdiction over it.

That was a recipe for trouble: the arrangement was made to work under authoritarian Soviet rule, but once this rule ended, the Christian Armenians encircled by Muslim Azerbaijan felt insecure and appealed to Armenia for help. In the ensuing conflict, Armenian troops entered Azerbaijan and gained control over the exclave, even ousting Azerbaijanis from the zone between the main body of Armenia and Nagorno-Karabakh (Fig. 2-13). The international community, however, has not recognized

**SOUTHERN RUSSIA:
INTERIOR AND EXTERIOR PERIPHERIES**

POPULATION

- Under 50,000
- 50,000-250,000
- 250,000-1,000,000
- 1,000,000-5,000,000
- Over 5,000,000

National capitals are underlined

———— Railroad
———— Road
·········· Canal
•—•—• Oil pipeline
○—○—○ Proposed oil pipeline

FIGURE 2-13

Armenia's occupation, and officially the territory remains a part of Azerbaijan. After the turn of the twenty-first century, the matter still remained unresolved.

Georgia

Of the three former Soviet Republics in Transcaucasia, only Georgia has a Black Sea coast and thus an outlet to the wider world. Smaller than South Carolina, Georgia is a country of high mountains and fertile valleys. Its social and political geographies are complicated. The population of 4.4 million is more than 70 percent Georgian but also includes Armenians (8 percent), Russians (6 percent), Ossetians (3 percent), and Abkhazians (2 percent). The Georgian Orthodox Church dominates the religious community, but about 10 percent of the people are Muslims, most of them concentrated in Ajaria in the southwest.

Unlike Armenia and Azerbaijan, Georgia has no exclaves, but its political geography is problematic nonetheless (Fig. 2-13). Within Georgia's borders lie three minority-based autonomous entities: the Abkhazian and Ajarian Autonomous Republics, and the South Ossetian Autonomous Region.

Sakartvelos, as the Georgians call their country, has a long and turbulent history. Tbilisi, the capital for 15 centuries, lay at the core of an empire around the turn of the thirteenth century, but the Mongol invasion ended this era. Next, the Christian Georgians found themselves in the path of wars between Islamic Turks and Persians. Turning northward for protection, the Georgians were annexed by the Russians, who were looking for warm-water ports, in 1800. Like other peoples overpowered by the czars, the Georgians took advantage of the Russian Revolution to reassert their independence; but the Soviets reincorporated Georgia in 1921 and proclaimed a Georgian Soviet Socialist Republic in 1936. Josef Stalin, the communist dictator who succeeded Lenin, was a Georgian.

Georgia is renowned for its scenic beauty, warm and favorable climates, agriculture (especially tea), timber, manganese, and other products. Georgian wines, tobacco, and citrus fruits are much in demand. The diversified economy could support a viable state.

Unfortunately, Georgia's political geography is loaded with centrifugal forces. After Georgia declared its independence in 1991, factional fighting destroyed its first elected government. But worse was to come. In the Autonomous Region of South Ossetia, a movement for union with (Russian) North Ossetia created havoc; Abkhazia proclaimed independence; and Muslim-infused Ajaria asserted itself by ignoring a government preoccupied with problems elsewhere. The price of all this instability was Russian intervention, and today Georgia remains a dysfunctional state.

Azerbaijan

Azerbaijan is the name of an independent state *and* of a province in neighboring Iran. The Azeris (short for Azerbaijanis) on both sides of the border have the same ancestry: they are a Turkish people divided between the (then) Russian and Persian empires by a treaty signed in 1828. By that time, the Azeris had become Shi'ite Muslims, and when the Soviet communists laid out their grand design for the USSR, they awarded the Azeris their own republic. On the Persian side, the Azeris were assimilated into the Persian Empire, and their domain became a province. Today the former Soviet Socialist Republic is the independent state of Azerbaijan (population: 8.3 million), and the 10 million Azeris to the south live in the northwesternmost Iranian province.

During the brief transition to independence and at the height of their war with the Armenians, the dominantly Muslim Azeris tended to look southward, toward Iran. But geographic realities dictate a more practical orientation. Azerbaijan possesses huge reserves of oil and natural gas; under the Soviets it was one of Moscow's chief regional sources of fuels. The center of the oil industry is Baki (Baku), the capital on the shore of the Caspian Sea—but the Caspian Sea is a lake. To export its oil, Azerbaijan needs pipelines, but those of Soviet vintage link Baki to Russia's Black Sea terminal of Novorossiysk. During the mid-1990s, when Azerbaijan's leaders announced plans to build new pipelines via Georgia or Iran, Russia objected and even meddled in Azeri politics to stymie such plans.

A look at the map (Fig. 2-13) shows Azerbaijan's options. A pipeline could be routed from Baki through Georgia to the Black Sea coast near Batumi, avoiding Russian territory altogether. Another route could run through Georgia and across Turkey to the Mediterranean terminal at Ceyhan. Still another option would run the pipeline across Iran to a terminal on the Persian Gulf. Moscow objects to such alternatives, but Azerbaijan now has powerful international allies. American, French, British, and Japanese oil companies are developing Azerbaijan's offshore Caspian reserves, and the U.S. government prefers the Turkish route.

All these alternate routes are in the future, however; for the present, Azerbaijan's oil must flow across Russian territory, and Azerbaijan must cooperate with Moscow. Given its Muslim roots and its location, the time may come when Azerbaijan turns toward the Islamic world. For the moment, however, this country is inextricably bound up with its neighbors in Transcaucasia and the north.

THE EASTERN FRONTIER

From the eastern flanks of the Ural Mountains to the headwaters of the Amur River, and from the latitude of Tyumen to the northern zone of neighboring Kazakhstan, lies Russia's vast Eastern Frontier Region, product of a gigantic experiment in the eastward extension of the Russian Core (Fig. 2-10). As the maps of cities and surface communications suggest, this eastern frontier is more densely peopled and more fully developed in the west than in the east. At the longitude of Lake Baykal, settlement has become linear, marked by ribbons and clusters along the east-west railroads. Two subregions dominate the geography: the Kuznetsk Basin in the west and the Lake Baykal area in the east.

The Kuznetsk Basin (Kuzbas)

Some 900 miles (1450 km) east of the Urals lies another of Russia's primary regions of heavy manufacturing resulting from the communist period's national planning: the Kuznetsk Basin, or *Kuzbas* (Fig. 2-11). In the 1930s, it was opened up as a supplier of raw materials (especially coal) to the Urals, but that function became less important as local industrialization accelerated. The original plan was to move coal from the Kuzbas west to the Urals and allow the returning trains to carry iron ore east to the coalfields. However, good-quality iron ore deposits were subsequently discovered near the Kuznetsk Basin itself. As the new resource-based Kuzbas industries grew, so did its urban centers. The leading city, located just outside the region, is Novosibirsk, which stands at the intersection of the Trans-Siberian Railroad and the Ob River as the symbol of Russian enterprise in the vast eastern interior. To the northeast lies Tomsk, one of the oldest Russian towns in all of Siberia, founded in the seventeenth century and now caught up in the modern development of the Kuzbas Region. Southeast of Novosibirsk lies Novokuznetsk, a city that produces steel for the region's machine and metal-working plants and aluminum products from Urals bauxite.

The Lake Baykal Area (Baykaliya)

East of the Kuzbas, development becomes more insular, and distance becomes a stronger adversary. North of the Tyva Republic and eastward around Lake Baykal, larger and smaller settlements cluster along the two railroads to the Pacific coast (Fig. 2-11). West of the lake, these rail corridors lie in the headwater zone of the Yenisey River and its tributaries. A number of dams and hydroelectric projects serve the valley of the Angara River, particularly the city of Bratsk. Mining, lumbering, and some farming sustain life here, but isolation dominates it. The city of Irkutsk, near the southern end of Lake Baykal, is the principal service center for a vast Siberian region to the north and for a lengthy east-west stretch of southeastern Russia.

Beyond Lake Baykal, the Eastern Frontier really lives up to its name: this is southern Russia's most rugged, remote, forbidding country. Settlements are rare, many being mere camps. The Buryat Republic (Fig. 2-7) is part of this zone; the territory bordering it to the east was taken from China by the czars and may become an issue in the future. Where the Russian-Chinese boundary turns southward, along the Amur River, the region called the Eastern Frontier ends and Russia's Far East begins.

SIBERIA

Before we assess the potential of Russia's Pacific Rim, we should remember that the ribbons of settlement just discussed hug the southern perimeter of this giant country, avoiding the vast Siberian region to the north (Fig. 2-10). Siberia extends from the Ural Mountains to the Kamchatka Peninsula—a vast, bleak, frigid, forbidding land. Larger than the conterminous United States but inhabited by only an estimated 15 million people, Siberia quintessentially symbolizes the Russian environmental plight: vast distances, cold temperatures worsened by strong Arctic winds, difficult terrain, poor soils, and limited options for survival.

But Siberia also has resources. From the days of the first Russian explorers and Cossack adventurers, Siberia's riches have beckoned. Gold, diamonds, and other precious minerals were found. Later, metallic ores including iron and bauxite were discovered. Still more recently, the Siberian interior proved to contain sizeable quantities of oil and natural gas (Fig. 2-12) and began to contribute significantly to Russia's energy supply.

As the physiographic map (Fig. 2-2) shows, major rivers—the Ob, Yenisey, and Lena—flow gently northward across Siberia and the Arctic Lowland into the Arctic Ocean. Hydroelectric power development in the basins of these rivers has generated electricity used to extract and refine local ores, and run the lumber mills that have been set up to exploit the vast Siberian forests.

The human geography of Siberia is fragmented, and much of the region is virtually uninhabited (Fig. 2-4). Ribbons of Russian settlement have developed; the Yenisey River, for instance, can be traced on this map of Soviet peoples (a series of small settlements north of Krasnoyarsk), and the upper Lena Valley is similarly fringed by ethnic Russian settlement. Yet hundreds of miles of empty territory separate these ribbons and other islands of habitation.

The political geography of eastern Siberia is marked by the growing identity of Sakha (the Yakut Republic). As additional resources are discovered here (including oil and natural gas), this Republic, centered on the capital, Yakutsk, will become more important.

Siberia, Russia's freezer, is stocked with goods that may become mainstays of future national development. Already, precious metals and mineral fuels are bolstering the Russian economy. In time, we may expect Siberian resources to play a growing role in the economic development of the Eastern Frontier and the Russian Far East as well. One step in that process was already taken during Soviet times: the completion of the BAM (Baykal-Amur Mainline) Railroad in the 1980s. This route, lying north of and parallel to the old Trans-Siberian Railroad, extends 2200 miles (3540 km) eastward from Tayshet (near the important center of Krasnoyarsk) directly to the Far East city of Komsomolsk (Fig. 2-6). In the post-Soviet era, the BAM Railroad has been beset by equipment breakdowns and workers' strikes. Nonetheless, it is a key element of the infrastructure that will serve the Eastern Frontier's economic growth in the twenty-first century.

THE RUSSIAN FAR EAST

Imagine this: a country with 5000 miles (8000 km) of Pacific coastline, two major ports, interior cities nearby, huge reserves of resources ranging from minerals to fuels to timber, directly across from one of the world's largest economies—all this at a time when the Asian Pacific Rim was the world's fastest-growing economic region. Would not that country have burgeoning cities, busy harbors, growing industries, and expanding trade?

In the Russian Far East (Fig. 2-10), the answer is—no. Activity in the port of Vladivostok is a shadow of what it was during the Soviet era, when it was the communists' key naval base. The nearby container terminal at Nakhodka suffers from breakdowns and inefficiencies. The railroad to western Russia carries just a fraction of the trade it did during the 1970s and 1980s. Cross-border trade with China is minimal. Trade with Japan is inconsequential. The region's cities are grimy, drab, moribund. Utilities are shut off for hours at a time because of fuel shortages and system breakdowns. Outdated factories are closed down, their workers dismissed. Political relations with Moscow are poor. There is potential here, but little of it has been realized.

As a region, the Russian Far East consists of two parts: the mainland area extending from Vladivostok to the Stanovoy Mountains and the large island of Sakhalin (Figs. 2-1, 2-14). This is cold country: icebreakers have to keep the ports of Vladivostok and Nakhodka open throughout the winter. Winters here are long and bitterly cold; summers are brief and cool. Although the population is small (about 7 million), food must be imported because not much can be grown. Most of the region is rugged, forested, and remote. Vladivostok, Khabarovsk, and Komsomolsk are the only cities of any size. Nakhodka and the newer railroad terminal at Vanino are smaller towns; the population of the whole island of Sakhalin is about 700,000 (on an island the size of Caribbean Hispaniola [16.6 million]). Offshore lie productive fishing grounds. Most important of all are the huge oil and gas reserves recently discovered on and around Sakhalin: a new era may be dawning as Russian-partnered foreign companies move in to exploit them.

Russia's oil and natural gas reserves are key to the overall health of the Russian economy. Oil and gas are Russia's most valuable exports by far, and Europe's dependence on Russian energy supplies grows continuously. This photograph shows a small segment of what has been described as the biggest work site in the world: the laying of the Trans-Siberian gas pipeline from near the shore of the Arctic Ocean to the border of the Czech Republic, where it is being connected to the gas pipeline network leading to Western Europe. Here are some statistics: the pipeline shown here is nearly five feet (1.42 m) in diameter. It will lie across permafrost and through trenches for a total of almost 2800 miles (4450 km) from the Arctic to Eastern Europe. It will be capable of sending 34 billion cubic meters of gas from source to consumers.

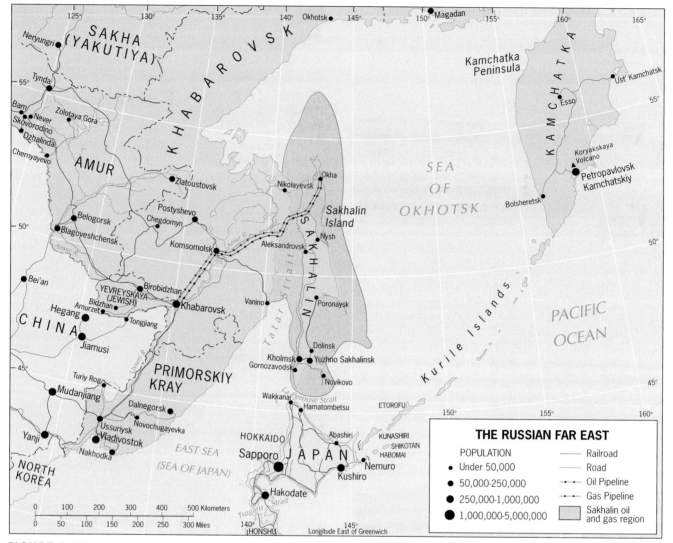

FIGURE 2-14

The Soviet regime rewarded people willing to move to this region with housing and subsidies. The communists, like the czars before them, realized the importance of this frontier (Vladivostok means "We Own the East"), and they used every possible incentive to develop it and link it ever closer to Russia's distant western core. Freight rates on the Trans-Siberian Railroad, for example, were about 10 percent of their real costs; the trains were always loaded in both directions. Vladivostok was a military base and a city closed to foreigners, and Moscow invested heavily in its infrastructure. Komsomolsk in the north and Khabarovsk near the region's center were endowed with state-owned industries using local resources: iron ore from Komsomolsk, oil from Sakhalin, timber from the ubiquitous forests. The steel, chemical, and furniture industries sent their products westward by train, and they received food and other consumer goods from the Russian heartland.

For several reasons, the post-Soviet transition has been especially difficult here in the Far East. The new economic order has canceled the region's communist-era advantages: the Trans-Siberian Railroad now must charge the real cost of transporting products from the Eastern Frontier to the Russian Core. State-subsidized industries must compete on market principles, their subsidies having ended. The decline of Russia's armed forces has hit Vladivostok hard. The fleet lies rusting in port; service industries have lost their military markets; the shipbuilding industry has no government contracts. Coal miners in the Bureya River Valley (a tributary of the Amur) go unpaid for months and go on strike; coal-fired power plants do not receive fuel shipments, and cities and towns go dark.

Locals put much of the blame for their region's failure on Moscow, and with reason. As Figure 2-7 shows, the Far East contains only five administrative regions: Primorskiy Kray, Khabarovsk Kray, Amur Oblast, Sakhalin Oblast, and Yevreyskaya, originally the Jewish Autonomous Region. This does not add up to much po-

FROM THE FIELD NOTES

"Standing in an elevated doorway in the center of Vladivostok, you can see some of the vestiges of the Soviet period: the omnipresent, once-dominant, GUM department store behind the blue seal on the left, and the communist hammer-and-sickle on top of the defunct hotel on the right. But in the much more colorful garb of the people, and the private cars in the street, you see reflections of the new era. Once a city closed to foreigners, Vladivostok now throbs with visitors—and has become a major point of entry for contraband goods."

litical clout, which is the way Moscow appears to want it. Compare Figures 2-8 and 2-14 and you will note that the city of Khabarovsk, not the Primorskiy Kray capital of Vladivostok, has been made the headquarters of the Far Eastern Region as defined by Moscow.

For all its stagnation, Russia's Far East will figure prominently in Russian (and probably world) affairs.

Here Russia meets China on land and Japan at sea (an unresolved issue between Russia and Japan involves several Kurile islands). Here lie vast resources ranging from Sakhalin's fuels to Siberia's lumber, Sakha's gold to Khabarovsk's metals. Here Russia has a foothold on the Pacific Rim, a window on the ocean on whose shores the world is being transformed.

The Kamchatka Peninsula constitutes one of the Pacific Rim's most geologically active zones, with more than 125 volcanoes of which more than 20 are active. Sudden massive eruptions can endanger airliners on the northern Pacific route between North America and East Asia, which skirts Kamchatka. The region's largest city, Petropavlovsk-Kamchatskiy, lies at the foot of the 11,400-foot (3456-m) high Koryakskaya Volcano near the Pacific coast in southeastern Kamchatka. Soviet-era apartment buildings dominate the cultural landscape of this bleak fishing port founded during Russia's eighteenth-century eastward expansion. Note that, in this *Dfc* climate, the countryside is much greener (with deciduous as well as coniferous vegetation) than it is toward the north and the Siberian interior.

FIGURE 3-1

Relief

Meters		Feet
3050		10 000
1525		5000
610		2000
305		1000
0	Sea Level	0
		Below
152.5	500	Sea Level
1525	5000	
3050	10 000	
6100	20 000	

A-520000-76 -5-5-18
COPYRIGHT BY
RAND McNALLY & COMPANY
MADE IN U.S.A.

Scale 1:40 000 000: one inch to 630 miles. Lambert's Azimuthal Equal Area Projection
Elevations and depressions are given in feet

Longitude   West   100°   of   Greenwich

| 0 | 200 | 400 | 600 | 800 | 1000 Miles |
| 0 | 400 | 800 | 1200 | 1600 Kilometers |

North America

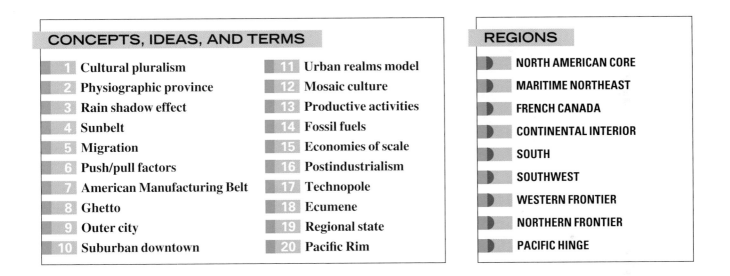

NORTH AMERICA AS a geographic name means different things to different people. To physical geographers, North America refers to the landmass that extends from Alaska to Panama. As Figure G-4 shows, the North American continent is the dominant feature of the North American tectonic plate, which reaches almost—but not quite—to Panama. To human geographers, North America (sometimes inappropriately called "Anglo" America) signifies a geographic realm consisting of two large countries with much in common, Canada and the United States. We therefore recognize three realms in the Americas: (1) *North America*, propelled by the United States; (2) *Middle America*, anchored by Mexico; and (3) *South America*, dominated by Brazil.

DEFINING THE REALM

DEFINED IN THE context of human geography, North America is constituted by two of the world's most highly advanced countries by virtually every measure of human development. Blessed by an almost endless range of natural resources and bonded by trade as well as culture, Canada and the United States are locked in a mutually productive embrace that is reflected by the statistics. In an average year of the recent past, 85 percent of Canadian exports went to the United States and 74 percent of Canada's imports came from its southern neighbor. For the United States, Canada is its leading export market and its number one source of imports.

Both countries also rank among the world's most highly urbanized, with more than 75 percent of their populations concentrated in cities that have evolved into metropolitan agglomerations. The old core cities are now

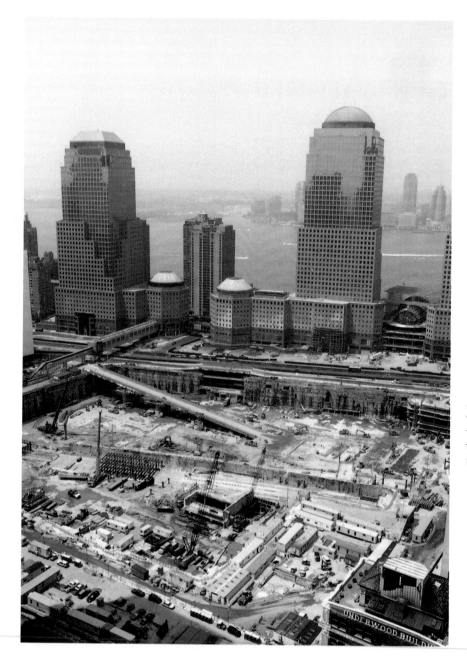

The destruction of the twin towers of the World Trade Center on September 11, 2001, left a gaping hole not only in the skyline of New York City but in the ground as well. In a remarkable operation that began before the smoke had cleared, the rubble of terrorism's aftermath was removed in less than nine months, as shown in this photograph taken on Memorial Day 2002. But the success of the terrorists' deadly onslaught set in motion a chain of events that is changing the world as well as the United States. American forces invaded Afghanistan and ousted the regime that had given refuge to Islamic militants. Claiming a terrorist connection among other justifications, the United States led an armed attack on Iraq and overthrew its dictator. American soldiers found themselves headed for Pakistan, Djibouti, the Philippines, and other destinations in a new "War on Terror." Meanwhile, U.S. relations with long-term allies such as France and Canada were strained by disagreements over this campaign even as economic stresses attributable to post–9/11 uncertainties buffeted America as a whole and New York City in particular. Those who watched the dreadful events at this site unfold, and who predicted that the world and the nation would never be the same, were right.

encircled by rings of suburban development, which draw an ever-greater proportion of the metropolitan population and activity base into their newly urbanized environments. Nothing symbolizes the North American city as strongly as the skyscrapered panoramas of New York, Chicago, and Toronto or the vast, beltway-connected suburban expanses of Los Angeles, Washington, and Houston.

North Americans also are the most mobile people in the world. Commuters stream into and out of suburban business centers and central-city downtowns by the millions each working day; most of them drive cars, whose numbers have multiplied more than six times faster than the human population since 1970. Moreover, each year nearly one out of every six individuals changes his or her residence.

Although Canada and the United States share many historical, cultural, and economic qualities, they also differ in significant ways, as can be seen on the map. The United States, somewhat smaller territorially than Canada, occupies the heart of the North American continent and, as a result, encompasses a greater environmental range. The U.S. population is dispersed across most of the country, forming major concentrations along both the (north-south-trending) Atlantic and Pacific coasts. The overwhelming majority of Canadians, however, live in an interrupted east-west corridor that lies across south-

MAJOR GEOGRAPHIC QUALITIES OF

North America

1. North America encompasses two of the world's biggest states territorially (Canada is the second largest in size; the United States is third).

2. Both Canada and the United States are federal states, but their systems differ. Canada's is adapted from the British parliamentary system and is divided into ten provinces and three territories. The United States separates its executive and legislative branches of government, and it consists of 50 States, the Commonwealth of Puerto Rico, and a number of island territories under U.S. jurisdiction in the Caribbean Sea and the Pacific Ocean.

3. Both Canada and the United States are plural societies. Although ethnicity is increasingly important, Canada's pluralism is most strongly expressed in regional bilingualism. In the United States, major divisions occur along racial lines.

4. A large number of Quebec's French-speaking citizens supports a movement that seeks independence for the province. The movement's high-water mark was reached in the 1995 referendum in which (minority) non-French-speakers were the difference in the narrow defeat of separation. The prospects for a break-up of the Canadian state have diminished significantly since 2000.

5. North America's population, not large by international standards, is the world's most highly urbanized and mobile. Largely propelled by a continuing wave of immigration, the realm's population total is expected to grow by more than 40 percent over the next half-century.

6. By world standards, this is a rich realm where high incomes and high rates of consumption prevail. North America possesses a highly diversified resource base, but nonrenewable fuel and mineral deposits are consumed prodigiously.

7. North America is home to one of the world's great manufacturing complexes. The realm's industrialization generated its unparalleled urban growth, but a new postindustrial society and economy are rapidly maturing in both countries.

8. The two countries heavily depend on each other for supplies of critical raw materials (e.g., Canada is the leading source of U.S. energy imports) and have long been each other's chief trading partners. Today, the North American Free Trade Agreement (NAFTA), which also includes Mexico, is linking all three economies ever more tightly as barriers to international trade and investment flow are steadily being dismantled.

9. North Americans are the world's most mobile people. Although plagued by recurrent congestion problems, the realm's networks of highways, commercial air routes, and cutting-edge telecommunications are the most efficient on Earth.

---

ern Canada, mainly within 200 miles (320 km) of the U.S. border. The United States also encompasses North America's northwestern extension, Alaska. (Offshore Hawai'i, however, belongs in the Pacific Realm.)

Demographic contrasts between the United States and Canada are substantial. The United States is the world's third most populous country with 290.9 million in 2004; Canada ranks 35th with 31.6 million. Comparatively small as its population may be, however, Canada has varied cultures and traditions ranging from multilingualism (French has equal status with English) to ethnic diversity that, as we will note later, have strong spatial expression. Thus Canada shares with the United States a characteristic feature of this realm: the prevalence of **cultural pluralism**. In the United States, this is exhibited by a range of ethnic backgrounds from European and African to Asian and Native American. And while the civil rights movement and other initiatives have elimi-

nated *de jure* (legal) segregation in the United States, ethnic clustering and *de facto* (real-world) self-segregation continue to mark the social landscape. In part, this is due to a troubling aspect of this wealthiest of all realms: poverty and deprivation still afflict a substantial minority as income gaps between the successful rich and the struggling poor continue to widen.

NORTH AMERICA'S PHYSICAL GEOGRAPHY

Before we examine the human geography of the United States and Canada more closely, we need to consider the physical setting in which they are rooted. The North American continent extends from the Arctic Ocean to Panama (Fig. 3-1), but we will confine ourselves here to

The photo at the left, capped by the tapering Library Tower, shows the heart of the central business district (CBD) of Los Angeles. The central city's skyscraper-dominated skyline originated during the era of industrial urbanization that peaked in the mid-twentieth century. But we now live in the twenty-first century, and the postindustrial transformation of urban America is all but complete. The landscape of the latter is symbolized by the low-rise world headquarters of IBM, completed in 1997, the centerpiece of a forested suburban office campus located in Armonk, New York about 30 miles north of Times Square. This is actually IBM's second headquarters facility on the site, the company having first moved out of Manhattan to this suburban Westchester County community more than 40 years ago. These scenes represent the most far-reaching change in the history of the American city, shaped by the forces of the past four decades that turned it inside-out as well as replacing its single-centered with a multi-centered spatial structure.

the territory north of Mexico—a geographic realm that still stretches from the near-tropical latitudes of southern Florida and Texas to subpolar Alaska and Canada's far-flung northern periphery. The remainder of the North American continent comprises a separate realm, *Middle America*, which is covered in Chapter 4.

Physiography

North America's physiography is characterized by its clear, well-defined division into physically homogeneous regions called **physiographic provinces**. Each region is marked by considerable uniformity in relief, climate, vegetation, soils, and other environmental conditions, resulting in a scenic sameness that comes readily to mind. For example, we identify such regions when we refer to the Rocky Mountains, the Great Plains, and the Appalachian Highlands. However, not all the physiographic provinces of North America are so easily delineated.

Figure 3-2 maps the complete layout of the continent's physiography and includes a cross-sectional terrain profile along the 40th parallel. The most obvious aspect of this map of North America's physiographic provinces is the north-south alignment of the continent's great mountain backbone, the Rocky Mountains, whose rugged topography dominates the western segment of the continent from Alaska to New Mexico. The major feature of eastern North America is another, much lower chain of mountain ranges called the Appalachian Highlands, which also trend approximately north-south and extend from Canada's Atlantic Provinces to Alabama. The orientation of the Rockies and Appalachians is important because, unlike Europe's Alps, they do not form a topographic barrier to polar or tropical airmasses flowing southward or northward, respectively, across the continent's interior.

Between the Rocky Mountains and the Appalachians lie North America's vast interior plains, which extend from the Mackenzie Delta on the Arctic Ocean to the Gulf of Mexico. We can subdivide these into several provinces: (1) the great Canadian Shield, which is the geologic core area containing North America's oldest rocks; (2) the Interior Lowlands, covered largely by glacial debris laid down by ice, meltwater, and wind during the Pleistocene glaciation; and (3) the Great Plains, the extensive sedimentary surface that rises slowly westward toward the

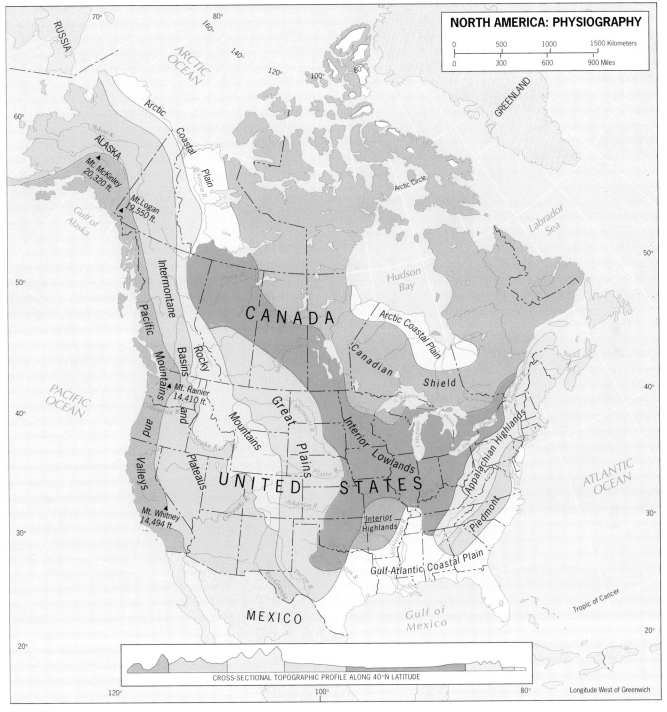

FIGURE 3-2

Rocky Mountains. Along the southern margin, these interior plainlands merge into the Gulf-Atlantic Coastal Plain, which extends from southern Texas along the seaward margin of the Appalachian Highlands and the neighboring Piedmont until it ends at Long Island just to the east of New York City.

On the western side of the Rocky Mountains lies the zone of Intermontane Basins and Plateaus. Within the conterminous United States, this physiographic province includes: (1) the Colorado Plateau in the south, with its thick sediments and spectacular Grand Canyon; (2) the lava-covered Columbia Plateau in the north, which forms the watershed of the Columbia River; and (3) the central Basin-and-Range country (Great Basin) of Nevada and Utah, which contains several extinct lakes from the glacial period as well as the surviving Great Salt Lake.

What geographers call the southern Rocky Mountains is one of the world's most clearly demarcated *physiographic* provinces, wedged between the Great Plains to the east and the Intermontane Basins and Plateaus to the west, and displaying a distinctive natural landscape of high relief and spectacular scenery. Glacial erosion carved this topography and sculpted valleys such as the Aspen Valley shown here; forests drape countrysides and snow covers the peaks. Gold finds once drew settlers here, but today the leading industry is tourism. The slopes and facilities of Aspen are world-renowned.

This province is called *intermontane* because of its position between the Rocky Mountains to the east and the Pacific coast mountain system to the west.

From the Alaskan Peninsula to Southern California, the west coast of North America is dominated by an almost unbroken corridor of high mountain ranges that originated from the contact between the North American and Pacific Plates (Fig. G-4). The major components of this coastal mountain belt include California's Sierra Nevada, the Cascades of Oregon and Washington, and the long chain of highland massifs that line the British Columbia and southern Alaska coasts. Three broad valleys—which contain dense populations—are the only noteworthy interruptions: California's Central (San Joaquin-Sacramento) Valley; the Cowlitz-Puget Sound lowland of Washington State, which extends southward into western Oregon's Willamette Valley; and the lower Fraser Valley, which slices through southern British Columbia's coast range.

Climate

The world climate map (Fig. G-8) clearly depicts the various climatic regimes and regions of North America. In general, temperature varies latitudinally—the farther north one goes, the cooler it gets. Regional land-and-water-heating differentials, however, distort this broad pattern. Because land surfaces heat and cool far more rapidly than water bodies, yearly temperature ranges are much larger where *continentality* (interior remoteness from the sea) is greatest.

Precipitation generally tends to decline toward the west (except for the Pacific coastal strip itself) as a result of the **rain shadow effect**. This occurs because Pacific airmasses, driven by prevailing winds, carry their moisture onshore but soon collide with the Sierra Nevada-Cascades wall, forcing them to rise—and cool—in order to crest those mountain ranges. Such cooling is accompanied by major condensation and precipitation, so that by the time these airmasses descend and warm along the eastern slopes to begin their journey across the continent's interior, they have already deposited much of their moisture. Thus the mountains produce a downwind "shadow" of dryness, which is reinforced whenever eastward-moving air must surmount other ranges farther inland, especially the massive Rockies. Indeed, this semiarid (and in places truly arid) environment extends so deeply into the central United States that a broad division can be made between Arid (western) and Humid (eastern) America, which face each other along a fuzzy boundary that is best viewed as a wide transition zone. Although the separating criterion of 20 inches (50 cm) of annual precipitation is easily mapped (see Fig. G-7), that generally north-south *isohyet* (the line connecting all places receiving exactly 20 inches per year) can and does swing widely across the drought-prone Great Plains from year to year because highly variable warm-season rains from the Gulf of Mexico come and go in unpredictable fashion.

On the other hand, precipitation in Humid America is far more regular. The prevailing westerly winds (blowing from west to east—winds are always named for the direction *from* which they come), which normally come up dry for the large zone west of the 100th meridian, pick up considerable moisture over the Interior Lowlands, and distribute it throughout eastern North America. A large number of storms develop here on the highly active weather front between tropical Gulf air to the south and polar air to the north. Even if major storms do not materialize, local weather disturbances created by sharply contrasting temperature differences are always a danger. There are more tornadoes (nature's most violent weather) in the central United States each year than anywhere else on Earth. And in winter, the northern half of this region receives large amounts of snow, particularly around the Great Lakes.

Figure G-8 shows the absence of humid temperate (*C*) climates from Canada (except along the narrow Pacific coastal zone) and the prevalence of cold in Canadian environments. East of the Rocky Mountains, Canada's most *moderate* climates correspond to the *coldest* of the United States. Nonetheless, southern Canada shares the environmental conditions that mark the Upper Midwest and Great Lakes areas of the United States, so that agricultural productivity in the Prairie Provinces and in Ontario is substantial. Canada is a leading food exporter (chiefly wheat), as is the United States, despite its comparatively short growing season.

The broad environmental partitioning into Humid and Arid America is also reflected in the distribution of the realm's soils and vegetation. For farming purposes there is usually sufficient soil moisture to support crops where annual precipitation exceeds the critical 20 inches; where the yearly total is less, soils may still be fertile (especially in the Great Plains), but irrigation is often necessary to achieve their full agricultural potential. As for vegetation, the Humid/Arid America dichotomy is again a valid generalization: the natural vegetation of areas receiving more than 20 inches of water annually is *forest*, whereas the drier climates give rise to a *grassland* cover.

Hydrography (Surface Water)

Surface water patterns in North America are dominated by the two major drainage systems that lie between the Rockies and the Appalachians: (1) the five Great Lakes (Superior, Michigan, Huron, Erie, and Ontario) that drain into the St. Lawrence River and (2) the mighty Mississippi-Missouri river network, fed by such major tributaries as the Ohio, Tennessee, and Arkansas rivers. Both are products of the last episode of Pleistocene glaciation, and together they amount to nothing less than the best natural inland waterway system in the world. Human intervention has further enhanced this network of navigability, mainly through the building of canals that link the two systems as well as the St. Lawrence Seaway.

Elsewhere, the northern east coast of the continent is well served by a number of short rivers leading inland from the Atlantic. In fact, many of the major northeastern seaboard cities of the United States—such as Washington, D.C., Baltimore, and Philadelphia—are located at the waterfalls that marked the limit to tidewater navigation (hence their designation as *Fall Line cities*). Rivers in the Southeast and west of the Rockies at first offered little practical value because of their orientation and the difficulty of navigating them. In the far west, however, the Colorado and Columbia rivers have become supremely important as suppliers of drinking and irrigation water as well as hydroelectric power.

INDIGENOUS NORTH AMERICA

When the first Europeans set foot on North American soil, the continent was occupied by millions of people whose ancestors had reached the Americas from Asia, via Alaska and probably also across the Pacific, more than 13,000 years before (and possibly as long as 30,000 years ago). In search of Asia, the Europeans misnamed them "Indians," but the historic affinities of these earliest Americans were with the peoples of eastern and northeastern Asia, not India. In North America these *Native Americans* or *First Nations*—as they are now called in the United States and Canada, respectively—had organized themselves into hundreds of nations with a rich mosaic of languages and a great diversity of cultures (Fig. 3-3). Farmers grew crops the Europeans had never seen; other nations depended chiefly on fishing, herding, hunting, or some combination of these. Elaborate houses, efficient watercraft, effective weaponry, decorative clothing, and wide-ranging art forms distinguished the aboriginal nations. Certain nations had formulated sophisticated health and medical practices; ceremonial life was complex and highly developed; and political institutions were mature and elaborate.

The eastern nations were the first to bear the brunt of the European invasion. By the end of the eighteenth century, ruthless, land-hungry settlers had driven most of the Native American peoples living along the Atlantic and Gulf coasts from their homes and lands, beginning a westward push that was to devastate indigenous society. The U.S. Congress in 1789 proclaimed that "Indian . . . land and property shall never be taken from them without their consent," but in fact this is just what happened. One of the sorriest episodes in American history involved the removal of the eastern Cherokee, Chickasaw, Choctaw, Creek, and Seminole from their homelands in forced marches a thousand miles westward to Oklahoma. One-fourth of the entire Cherokee population

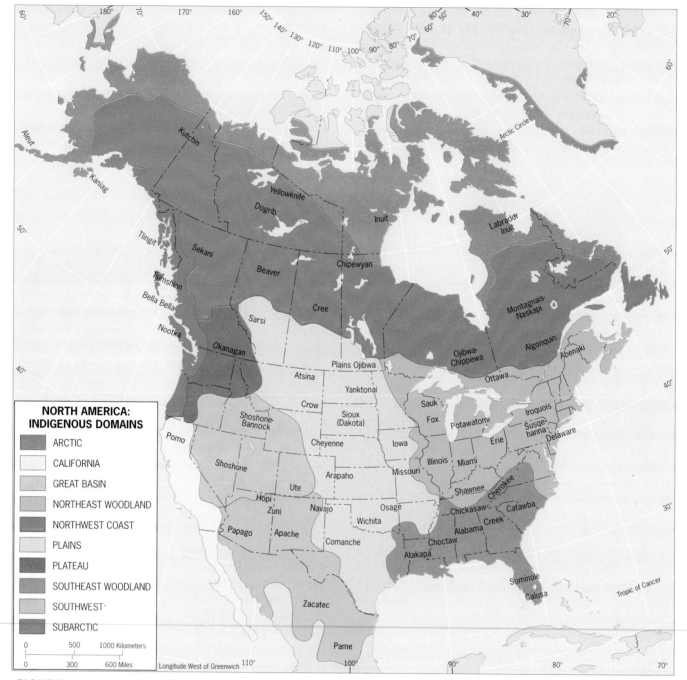

FIGURE 3-3

died along the way from exposure, starvation, and disease, and the others fared little better. Again, Congress approved treaties that would at least protect the native peoples of the Plains (Fig. 3-3) and those farther to the west, but after the mid-nineteenth century the white settlers ignored those guarantees as well. A half-century of war left what remained of North America's nations with about 4 percent of U.S. territory in the form of mostly impoverished reservations.

In what is today Canada, too, the comparatively small First Nations population was overwhelmed by the numbers and power of European settlers, and decimated by the diseases they introduced. Efforts at restitution and recognition of First Nations rights, however, have gone farther in Canada than in the United States.

As we noted earlier, two large countries, with geographic similarities as well as differences, constitute the North American realm. Canada and the United States share physiographic provinces as well as border-straddling economic and cultural subregions (for example, large-scale grain cultivation in the Great Plains). It would be possible to base the regionalization of this realm on contrasts between the two states; but as we will see in the second part of this chapter, quite a different set of regions emerges from our geographic analysis. To fully appreciate this regional framework, we must examine in some detail the changing human geography of the United States and Canada individually.

THE UNITED STATES

As we note throughout this book, the administrative components of states are assuming ever-greater importance in the scheme of things. In Europe, all the major states have reorganized their political-economic systems, forged regions from smaller units, and devolved power to the provinces. As a result, it is increasingly important not just to know Spain but also Catalonia, not just France but also

Rhône-Alpes, not just Italy but also Lombardy. So it is in the United States. We should have a clear mental map of the functional layout of this federation, which is why it may be helpful to take a few moments to review Figure 3-4.

Population in Time and Space

The current population distribution of the United States is shown in Figure 3-5. It is important to note that this map is the latest "still" in a motion picture, one that has been unreeling for nearly four centuries since the founding of the first permanent European settlements on the northeastern coast. Slowly at first, then with accelerating speed after 1800, as one major transportation breakthrough followed another, Americans (and Canadians) took charge of their remarkable continent and pushed the settlement frontier westward to the Pacific. The swiftness of this expansion

FIGURE 3-4

FIGURE 3-5

was dramatic, but Americans have long been the world's most mobile people. In fact, migrations continue to redistribute population in the United States today, perhaps the most significant being the persistent drift of people and livelihoods toward the South and West (the so-called **Sunbelt**).

To understand the contemporary population map, we need to review the major forces that have shaped, and con-

tinue to shape, the distribution of Americans and their activities. Since its earliest days, the United States has attracted a steady influx of immigrants who were rapidly assimilated into the societal mainstream (see box titled "The Migration Process"). Within the country, people have sorted themselves out to maximize their proximity to existing economic opportunities, and they have shown little resistance to relocating as the nation's evolving economic

The Migration Process

The United States as well as Canada is the product of **migration** (a change in residential location intended to be permanent). After tens of thousands of years of native settlement, European explorers reached the shores of North America beginning with Columbus in 1492 (and probably centuries before that). The first permanent colonies were established in the early 1600s, and from them evolved the modern United States of America. The Europeanization of North America doomed the continent's aboriginal societies, but this was only one of many areas around the world where local cultures and foreign invaders came face to face. Between 1835 and 1935, perhaps as many as 75 million Europeans departed for distant shores—most of them bound for the Americas (Fig. 3-6). Some sought religious freedom; others escaped poverty and famine; still others simply hoped for a better life.

Studies of the *migration decision* indicate that migration flows vary in size with (1) the perceived degree of difference between one's home, or source, and the destination; (2) the effectiveness of the information flow, that is, the news about the destination that migrants sent to those who stayed behind waiting to decide; and (3) the distance between the source and the destination (shorter moves attract many more migrants than longer ones). More than a century ago, the British social scientist Ernst Georg Ravenstein studied the migration process, and many of his conclusions remain valid today. For example, every migration stream from source to destination produces a counter-stream of returning migrants who cannot adjust, are unsuccessful, or are otherwise persuaded to go back home. Studies of migration also conclude that several factors are at work in the migration process. **Push factors** motivate people to move away; **pull factors** attract them to new destinations. To those early Europeans, the United States was a new frontier, a place where one might acquire a piece of land and some livestock. The opportunities were reported to be unlimited.

That perception has never changed, and immigration continues to significantly shape the human-geographic complexion of the United States. Today's inmigrants account for more than one-third of the country's annual population growth. Important details about this latest stage of the international migration process are discussed on pages 153–154.

HUMAN MIGRATIONS IN MODERN TIMES

— European emigration

— Migrations by other peoples (generated by European intervention)

FIGURE 3-6

geography has successively favored different sets of places over time.

During the past century these transformations spawned a number of major migrations: (1) the still-continuing westward shift but now with that southerly, Sunbelt deflection; (2) the rapid growth of metropolitan areas, first triggered by the late-nineteenth-century Industrial Revolution, which since the 1960s has been largely rechanneled from the central cities to the suburban ring; (3) the movement of African Americans from the rural south to the urban north, which since the 1970s has become a stronger return flow, particularly among middle-income blacks; and (4) the influx of immigrants from outside the United States, mostly European before 1960 but now dominated by a new wave from Middle America and eastern Asia that is directed, respectively, toward the States near the southern border and the urban areas of the Pacific coast.

Let us now look more closely at the historical geography of these changing population patterns, considering first the initial rural influence and then the decisive impacts of industrial urbanization.

Colonial and Nineteenth-Century Development

The current spatial distribution of the U.S. population is rooted in the colonial era of the seventeenth and eighteenth centuries that was dominated by England and France. The French sought mainly to organize a lucrative fur-trading network, while the English established settlements along the coast of what is today the northeastern U.S. seaboard. These British colonies quickly became differentiated in their local economies, a diversity that was to endure and later shape American cultural geography. The northern colony of New England (Massachusetts Bay and environs) specialized in commerce; the southern Chesapeake Bay colony (Tidewater Virginia and Maryland) emphasized the plantation farming

of tobacco; the Middle Atlantic area lying in between (southeastern New York, New Jersey, eastern Pennsylvania) was home to a number of smaller, independent-farmer colonies.

These neighboring colonies soon thrived and yearned to expand, but the British government responded by closing the inland frontier and tightening economic controls. By 1783 this move had led to colonial unification, British defeat in the Revolutionary War, and independence for the newly formed United States of America. The western frontier of the fledgling nation now swung open, and the zone north of the Ohio River was promptly settled following the discovery that the soils (and climate) of the Interior Lowlands were more favorable for farming than those of the Atlantic Coastal Plain and Piedmont. This triggered the rapid growth of trans-Appalachian agriculture and the widening of seaboard-interior trading ties, and the new interregional complementarities signified that U.S. spatial organization was assuming national-scale proportions.

By the time the westward-moving frontier swept across the Mississippi Valley in the 1820s, the three former seaboard colonies (**A**, **B**, and **C** in Fig. 3-7) had be-

FIGURE 3-7

come separate *culture hearths*—primary source areas and innovation centers from which migrants carried cultural traditions into the central United States (as the arrows in Fig. 3-7 show). The northern half of this vast interior space soon became well unified as its infrastructure steadily improved following the introduction of the railroad in the 1830s. The American South, however, did not wish to integrate itself economically with the North, preferring to export tobacco and cotton from its plantations to overseas markets; its insistence on preserving slavery to support this system soon led the South into secession, disastrous Civil War (1861–1865), and a dismal aftermath that took a full century to overcome.

The second half of the nineteenth century saw the frontier cross the western United States, and by 1869 agriculturally booming California was linked to the rest of the nation by transcontinental railroad (these same steel tracks also opened up the bypassed, semiarid Great Plains). When the American frontier closed in the 1890s, today's rural settlement pattern was firmly in place, anchored to a set of enduring national agricultural regions (discussed later in this chapter). By then, however, the exodus of rural Americans toward the burgeoning cities had begun in response to the Industrial Revolution that had taken hold after 1870.

Post-1900 Industrial Urbanization

The Industrial Revolution occurred almost a century later in the United States than in Europe, but when it finally did cross the Atlantic in the 1870s, it took hold so successfully and advanced so robustly that only 50 years later America was surpassing Europe as the world's mightiest industrial power. The impact of industrial urbanization occurred simultaneously at two levels of spatial generalization. At the national or *macroscale*, a system of new cities swiftly emerged, specializing in the collection, processing, and distribution of raw materials and manufactured goods, linked together by an efficient web of railroad lines. Within that urban network, at the local or *microscale*, individual cities prospered in their new roles as manufacturing centers, generating an internal structure that still forms the geographic framework of most of the central cities of America's large metropolitan areas. We now examine the urban trend at both of these scales.

Evolution of the U.S. Urban System The rise of the national urban system in the late nineteenth century was based on the traditional external role of cities: providing goods and services for their hinterlands in exchange for raw materials. Because people (both as laborers and as consumers), commercial activities, investment capital, and transport facilities were already agglomerated in existing (preindustrial) cities, the emerging industrialization movement tended to favor such locations. Their growing incomes, in turn, permitted industrially intensifying cities to invest in a bigger local infrastructure of private and public services as well as housing—and thereby convert each round of industrial expansion into a new stage of urban development.

The national urban system had actually been in the process of formation long before its flowering after 1870. Its evolution over the past two centuries has been studied by John Borchert, who identified **five epochs of metropolitan evolution** based on transportation technology and industrial energy. (1) *The Sail-Wagon Epoch* (1790–1830), marked by primitive overland and waterway circulation; the leading cities were the northeastern ports, which were more heavily oriented to the European overseas trade than to their barely accessible western hinterlands. (2) *The Iron Horse Epoch* (1830–1870), dominated by the arrival and spread of the steam-powered railroad; a nationwide transport system had now been forged, and the national urban system began to take shape, with New York emerging as the primate city by 1850. (3) *The Steel-Rail Epoch* (1870–1920), which spanned the Industrial Revolution and saw the full establishment of the national metropolitan system; the new forces shaping growth were the increasing scale of manufacturing, the rise of the steel and auto industries in Midwestern cities, and the introduction of steel rails that enabled trains to travel faster and haul heavier cargoes. (4) *The Auto-Air-Amenity Epoch* (1920–1970), which encompassed the later stage of U.S. industrial urbanization and the maturation of the national urban hierarchy; its key elements were the automobile and the airplane, the expansion of white-collar services jobs, and the growing locational pull of *amenities* (pleasant environments) that increasingly stimulated the urbanization of the suburbs and selected Sunbelt locales. (5) *The Satellite-Electronic-Jet Propulsion Epoch* (1970-), shaped by the newest advancements in information management, computer technologies, global communications, and intercontinental travel; it favors globally oriented metropolises (now often called *world cities*), particularly those along the Pacific and Atlantic coasts that function as international gateways.

Industrialization and the accompanying growth of the urban system reconfigured the realm's economic landscape. The most notable regional transformation was the emergence of the North American Core, or **American Manufacturing Belt**, which contained the lion's share of industrial activity in both the United States and Canada. As Figure 3-8 shows, the geographic form of the Core Region—which includes southern Ontario—was a near-rectangle whose four corners were Boston, Milwaukee, St. Louis, and Baltimore. However, because manufacturing is such a spatially concentrated activity, less than

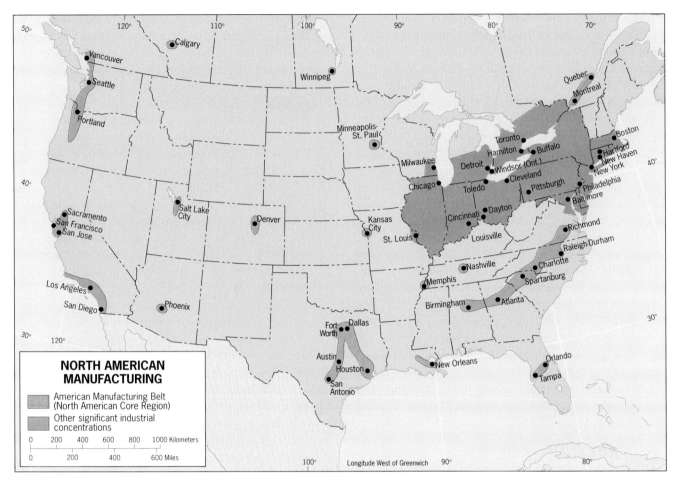

FIGURE 3-8

1 percent of the territory of the Belt is devoted to industrial land use: most of its factories are tightly clustered into a dozen districts centered on the cities mapped in Figure 3-8.

At the subregional scale, as transportation breakthroughs permitted progressive urban decentralization and *megalopolitan growth*, the expanding peripheries of major cities soon coalesced to form a number of conurbations. The most important of these by far is the *Atlantic Seaboard Megalopolis* (Fig. 3-9), the 600-mile (1000-km) urbanized northeastern coastal strip extending from southern Maine to Virginia that contains metropolitan Boston, New York, Philadelphia, Baltimore, and Washington. This was the economic heartland of the Core; the seat of U.S. government, business, and culture; and the trans-Atlantic trading interface between much of North America and Europe. Six other primary conurbations have also emerged: *Lower Great Lakes* (Chicago–Detroit–Cleveland–Pittsburgh), *Piedmont* (Atlanta–Charlotte–Raleigh/Durham), *Florida* (Jacksonville–Tampa–Orlando–Miami), *Texas* (Houston–Dallas/Ft. Worth–San Antonio), *California* (San Diego–Los Angeles–San Francisco), and the *Pacific Northwest*

(Portland–Seattle–Vancouver). Note that the last spills across the border into Canada, which has also spawned its own nationally predominant conurbation—*Main Street* (Windsor–Toronto–Montreal–Quebec City).

The Changing Structure of the U.S. Metropolis
The internal structure of the metropolis reflected the same mixture of forces that shaped the national urban system, especially transportation technology. Rails—in this case, lighter street-rail lines—once again shaped spatial organization as horse-drawn trolleys were succeeded by electric streetcars in the late nineteenth century. The mass introduction of the automobile after World War I changed all that, and America steadily turned from building compact cities to the widely dispersed metropolises of the post-World War II highway era. By 1970, the new intraurban expressway network had equalized location costs throughout the metropolis, setting the stage for suburbia to swiftly transform itself from a residential preserve into a complete outer city with amenities and new prestige that proved highly attractive to the business world. As the newly urbanized

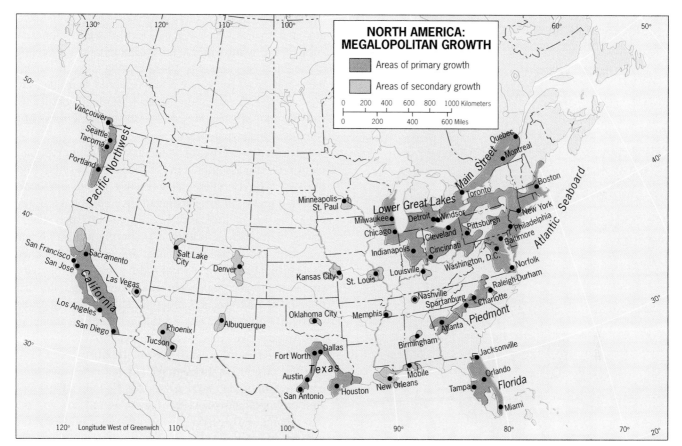

FIGURE 3-9

suburbs increasingly captured major economic activities, many large cities saw their status diminish to that of coequal. Their once thriving central business districts (CBDs) were now all but reduced to serving the less affluent populations that increasingly dominated the central city's neighborhoods.

This growth process was conceptualized into a four-stage model by John Adams, who identified the four **eras of intraurban structural evolution** (see Fig. 3-10). Stage I, prior to 1888, was the *Walking-Horsecar Era*, which produced a compact pedestrian city in which everything had to be within a 30-minute walk, a layout only slightly augmented when horse-drawn trolleys began to operate after 1850. The 1888 invention of the electric traction motor launched Stage II, the *Electric Streetcar Era* (1888–1920); higher speeds enabled the 30-minute travel radius and the urbanized area to expand considerably along new outlying trolley corridors; in the older core city, the CBD, industrial, and residential land uses differentiated into their modern form. Stage III, the *Recreational Automobile Era* (1920–1945), was marked by the initial impact of cars and highways that steadily improved the accessibility of the outer metropolitan ring, thereby launching a wave of mass suburbanization that

further extended the urban frontier. During this era, the still-dominant central city experienced its economic peak and the partitioning of its residential space into neighborhoods sharply defined by income, ethnicity, and race. Stage IV, the *Freeway Era* (under way since 1945), saw the full impact of automobiles, with the metropolis turning inside-out as expressways pushed suburban development more than 30 miles (50 km) from the CBD.

The social geography of the evolving industrial metropolis has been marked by the development of a residential mosaic that exhibited the congregating of ever-more-specialized groups. The electric streetcar, which introduced "mass" transit that every urbanite could afford, allowed the heterogeneous, immigrant-dominated city population to sort itself into ethnically uniform neighborhoods. When the United States sharply curtailed foreign immigration in the 1920s, industrial managers discovered the large African American population of the rural South—increasingly unemployed there as cotton-related agriculture declined—and began to recruit these workers by the thousands for the factories of Manufacturing Belt cities. This influx had an immediate impact on the social geography of the industrial city because whites were unwilling to share their living space with the racially different

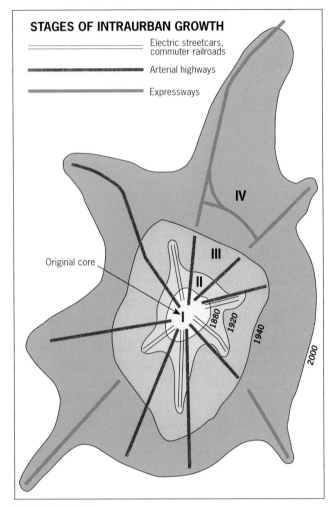

STAGES OF INTRAURBAN GROWTH

———— Electric streetcars, commuter railroads

———— Arterial highways

———— Expressways

Original core

I II III IV

1880 1920 1940 2000

FIGURE 3-10

entertainment facilities, and even major league sports stadiums and arenas, which together formed burgeoning new **10** **suburban downtowns** that are an automobile-age version of the CBD (see photo, p. 151). As suburban downtowns flourish, they attract tens of thousands of local residents to organize their lives around them—offering workplaces, shopping, leisure activities, and all the other elements of a complete urban environment.

These newest spatial elements of the contemporary metropolis are assembled in the model displayed in Figure 3-11. The rise of the outer city has produced a *multicentered* metropolis consisting of the traditional CBD as well as a set of increasingly coequal suburban downtowns, with each activity center serving a discrete and self-sufficient surrounding area. James Vance defined **11** these tributary areas as **urban realms**, recognizing in his studies that each such realm maintains a separate, distinct economic, social, and political significance and strength. Figure 3-12 applies the urban realms model to Los Angeles; we could easily draw a similar regionalization scheme for other large U.S. metropolises.

The position of the central city within the new multinodal metropolis of realms is eroding. No longer the dominant metropolitanwide center for goods and services, the CBD increasingly serves the less affluent residents of the innermost realm and those working downtown. As manufacturing employment declined precipitously, many large cities adapted successfully by shifting toward service industries. Accompanying this switch is downtown commercial revitalization, but in many cities for each shining new skyscraper that goes up several old commercial buildings are abandoned. Residential reinvestment has also occurred in many downtown-area neighborhoods but

newcomers. The result was the involuntary segregation of these newest migrants, who were channeled into geographically separate, all-black areas. By the 1950s, these **8** mostly inner-city areas became large expanding **ghettos**, speeding the departure of many white central-city communities and reinforcing the trend toward a racially divided urban society.

The huge suburban component of the intraurban residential mosaic is not home just to the more affluent residents of the metropolis: over the past three decades it received so massive an infusion of nonresidential activities **9** that it was transformed into a full-fledged **outer city**. As its ties to the central city loosened, the outer city's growing independence was accelerated by the rise of major new suburban nuclei (particularly near key freeway interchanges) to serve the new local economies. These multipurpose activity nodes often developed around large regional shopping centers, whose prestigious images attracted scores of industrial parks, office campuses and highrises, hotels, restaurants,

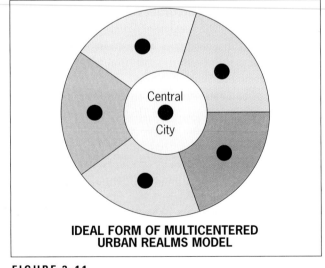

Central City

IDEAL FORM OF MULTICENTERED URBAN REALMS MODEL

FIGURE 3-11

FROM THE FIELD NOTES

"Monitoring the urbanization of U.S. suburbs for the past three decades has brought us to Tyson's Corner, Virginia on many a field trip and data-gathering foray. It is now hard to recall from this late-1990s view that only 40 years ago this place was merely a near-rural crossroads. But as nearby Washington, D.C. steadily decentralized, *Tyson's* capitalized on its unparalleled regional accessibility (its Capital Beltway location at the intersection with the radial Dulles Airport Toll Road) to attract a seemingly endless parade of high-level retail facilities, office complexes, and a plethora of supporting commercial services. We would rank it today among the three largest and most influential suburban downtowns (or 'edge cities') in North America."

FIGURE 3-12

This photo of a portion of northern Manhattan island was taken in Fort Lee, New Jersey (see map p. 170), looking eastward from atop the Palisades escarpment that lines the west bank of the Hudson River. Located at the base of the Palisades in the foreground is a cluster of upscale Japanese restaurants, groceries, department stores, and electronics and myriad other retailers. This luxurious shopping and entertainment complex is Yaohan Plaza–New York, which serves the tens of thousands of Japanese who work for the New York branch offices of their global corporations. Most live in high-income suburbs north and west of New York City, for which Fort Lee is especially convenient because three expressways and the George Washington Bridge (just off the left of the photo) converge there. The slice of Manhattan seen across the river, extending from about 160th Street on the left to 140th Street on the right, is a totally different social world. This is not the sky-scrapered, international-class CBD south of 59th Street, but the highly stressed inner city of the southern end of Washington Heights. The towers that form the skyline backdrop at the upper right are in the heart of the adjacent neighborhood of Harlem. These highrises are not commercial buildings but the deteriorating—and thoroughly depressing—low-income housing projects that grew out of the sterile urban renewal programs of the post-World War II era.

usually requires the displacement of established lower-income residents, an emotional issue that has sparked many conflicts. Beyond the CBD zone, the vast inner city remains the problem-ridden domain of low- and moderate-income people, with most forced to reside in ghettos.

Cultural Geography

In the United States, over the past two centuries, the contributions of a wide spectrum of immigrant groups have shaped—and continue to shape—a rich and varied cultural complex. Great numbers of these newcomers were willing to set aside their original cultural baggage in favor of assimilation into the emerging culture of their adopted homeland, which itself was a hybrid nurtured by constant infusions of new influences. For most upwardly mobile immigrants, this plunge into the much-touted "melting pot" promised a ticket for acceptance into mainstream American society.

American Cultural Bases

As the hybrid culture of the United States matured, it built on a set of powerful values and beliefs: (1) love of newness; (2) desire to be near nature; (3) freedom to move; (4) individualism; (5) societal acceptance; (6) aggressive pursuit of goals; and (7) a firm sense of destiny. Brian Berry has discerned these cultural traits in the behavior of people throughout the evolution of urban America. A "rural ideal" has prevailed throughout U.S. history and is still expressed in a strong bias against residing in cities. When industrialization made urban living unavoidable, those able to afford

it soon moved to the emerging suburbs (*newness*) where a form of country life (*close to nature*) was possible in a semiurban setting. The fragmented metropolitan residential mosaic, composed of myriad slightly different neighborhoods, encouraged frequent *unencumbered mobility* as middle-class life revolved around the *individual* nuclear family's *aggressive pursuit* of its aspirations for *acceptance into the next higher stratum of society*. To most Americans these accomplishments confirmed that they could attain their goals through hard work and perseverance and that they had the ability to realize their *destiny* by achieving the "American Dream" of homeownership, affluence, and total satisfaction.

Language and Religion

Although linguistic variations play a far more important role in Canada, more than one-eighth of the U.S. population speaks a primary language other than English. Differences in English usage are also evident at the subnational level in the United States, where regional variations (*dialects*) can still be noted despite the recent trend toward a truly national society. The South and New England immediately come to mind as areas that still possess distinctive accents.

North America's Christian-dominated kaleidoscope of religious faiths contains important spatial variations. Many major Protestant denominations are clustered in particular regions, with Southern Baptists localized in the southeastern quadrant of the United States, Lutherans in the Upper Midwest and northern Great Plains, and Mormons west of the Rockies but focused on Utah.

Roman Catholics are most visibly concentrated in Manufacturing Belt metropolises, New England, and the Mexican borderland zone. Judaism is the nation's most highly agglomerated major religious group, whose largest congregations are clustered in the cities and suburbs of Megalopolis, Southern California, South Florida, and the Midwest.

Ethnic Patterns

Ethnicity (national ancestry) has always played a key role in American cultural geography. Today, whites of European background no longer dominate the increasingly diverse U.S. ethnic tapestry, with ethnics of color and non-European origin comprising almost one-third of the population—a proportion that demographers predict will rise to 50 percent by mid-century. In the late 1990s, Hispanic Americans surpassed African Americans to become the nation's largest minority, and today they account for 13 percent of the U.S. total.

The spatial distribution of the four largest ethnic minorities is mapped in Figure 3-13. Hispanics are regionally clustered, with about 60 percent residing in California, Texas, New Mexico, and South Florida; nonetheless,

population geographers expect this rapidly expanding minority to continue to disperse across the country, especially in the West and along the eastern seaboard. African Americans are regionally concentrated as well, most notably in the South, reflecting the legacy of slavery and the plantation economy; blacks are also a major presence in the central cities of the Manufacturing Belt and west coast (most urban areas do not register prominently on these small-scale maps). Asian Americans are the most agglomerated of the leading ethnic minorities, their distribution dominated by urban-based clusters along the Pacific coast (entry points for the large number who are recent immigrants). The main concentrations of Native Americans are found in the West, where they largely occupy tribal lands on reservations ceded by the federal government; many also reside in communities widely scattered across the nation.

The changing ethnic complexion of the United States has long been influenced by immigration, and that still holds true. At the end of the 1990s, nearly 700,000 legal immigrants annually entered the country. The source areas, however, have changed dramatically over the past half-century. During the 1950s, just over 50 percent came

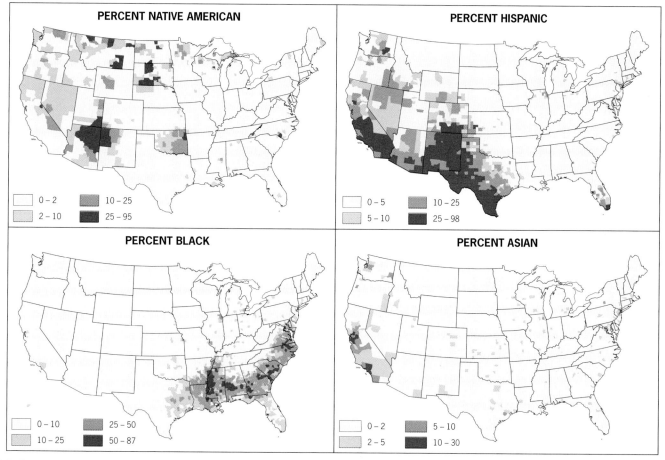

FIGURE 3-13

from Europe, 25 percent from Middle and South America, and 15 percent from Canada. Today, almost 50 percent come from Middle and South America, 30 percent from Asia, and only about 15 percent from Europe and Canada. Given this recent and continuing influx of immigrants, it is not surprising that the 2000 census reported that more than 28 million residents had been born in other countries—the highest total in U.S. history (see Issue Box titled "Immigrants: How Many Can North America Accommodate?").

Immigrants: How Many Can North America Accommodate?

IMMIGRATION BRINGS BENEFITS—THE MORE THE MERRIER!

"THE UNITED STATES and Canada are nations of immigrants. What would have happened if our forebears had closed the door to America after they arrived and stopped the Irish, the Italians, the Eastern Europeans, and so many other nationalities from entering this country? Now we're arguing over Latinos, Asians, Russians, Muslims, you name it. Fact is, newcomers have always been viewed negatively by most of those who came before them. When Irish Catholics began arriving in the 1830s the Protestants already here accused them of assigning their loyalty to some Italian pope rather than to their new country, but Irish Catholics soon proved to be pretty good Americans. Sound familiar? Muslims can be very good Americans too. It just takes time, longer for some immigrant groups than others. But don't you see that America's immigrants have always been the engine of growth? They become part of the world's most dynamic economy and make it more dynamic still.

"My ancestors came from Holland in the 1800s, and the head of the family was an architect from Rotterdam. I work here in western Michigan as an urban planner. People who want to limit immigration seem to think that only the least-educated workers flood into the United States and Canada, depriving the less-skilled among us of jobs and causing hardship for citizens. But in fact America attracts skilled and highly educated as well as unskilled immigrants, and they all make contributions. The highly educated foreigners, including doctors and technologically skilled workers, are quickly absorbed into the workforce; you're very likely to have been treated by a physician from India or a dentist from South Africa. The unskilled workers take jobs we're not willing to perform at the wages offered. Things have changed! A few decades ago, American youngsters on summer break flooded the job market in search of temporary employment in hotels, department stores, and restaurants. Now they're vacationing in Europe or trekking in Costa Rica, and the managers of those establishments bring in temporary workers from Jamaica and Romania.

"And our own population is aging, which is why we need the infusion of younger people immigration brings with it. We don't want to become like Japan or some European countries, where they won't have the younger working people to pay the taxes needed to support the social security system. I agree with opponents of immigration on only one point: what we need is legal immigration, so that the new arrivals will get housed and schooled, and illegal immigration must be curbed. Oherwise, we need more, not fewer, immigrants."

LIMIT IMMIGRATION NOW

"The percentage of recent immigrants in the U.S. population is the highest it has been in 70 years, and in Canada in 60 years. America is adding the population of San Diego every year, over and above the natural increase, and not counting illegal immigration. This can't go on. By 2010, 15 percent of the U.S. population will consist of recent immigrants. One-third of them will not have a high-school diploma. They will need housing, education, medical treatment, and other social services that put a huge strain on the budgets of the States they enter. The jobs they're looking for often aren't there, and then they start displacing working Americans by accepting lower wages. It's easy for the elite to pontificate about how great immigration is for the American melting pot, but they're not the ones affected on a daily basis. Immigration is a problem for the working people. We see company jobs disappearing across the border to Mexico, and at the same time we have Mexicans arriving here by the hundreds of thousands, legally and illegally, and more jobs are taken away.

"And don't talk to me about how immigration now will pay social security bills later. I know a thing or two about this because I'm an accountant here in Los Angeles, and I can calculate as well as the next guy in Washington. Those fiscal planners seem to forget that immigrants grow older and will need social security too. And as for that supposed slowdown in the aging of our population because immigrants are so young and have so many children, over the past 20 years the average age in the United States has dropped by four months. So much for that nonsense. What's needed is a revamping of the tax structure, so those fat cats who rob corporations and then let them go under will at least have paid their fair share into the national kitty. There'll be plenty of money to fund social services for the aged. We don't need unskilled immigrants to pay those bills.

"And I'm against this notion of amnesty for illegal immigrants being talked about these days. All that would do is to attract more people to try to make it across our borders. I heard President Fox of Mexico propose opening the U.S.-Mexican border the way they're opening borders in the European Union. Can you imagine what would happen? What our two countries really need is a policy that deters illegal movement across that border, which will save lives as well as jobs, and a system that will confine immigration to legal channels. These days, that's not just a social or economic issue; it's a security matter as well."

Regional **ISSUE**

Although still overwhelmingly directed at urban areas, the geography of immigrant destinations is changing. The once-dominant Manufacturing Belt States have slipped to less than 25 percent, with most of the newcomers now heading for either metropolitan New York or Chicago. Not unexpectedly, with the ongoing growth of Asian- and Hispanic American populations (Fig. 3-13), the Sunbelt States have taken up the slack, with California, Texas, and Florida attracting more than 50 percent of all U.S. immigrants.

The Emerging Mosaic Culture

American cultural geography continues to evolve. What is now taking place is a new fragmentation into the emerging nationwide **mosaic culture**, an increasingly heterogeneous complex of separate, uniform "tiles" that cater to more specialized groups than ever before. No longer based solely on such broad divisions as income, race, and ethnicity, today's residential communities of interest are also forming along the dimensions of age, occupational status, and especially lifestyle. Their success and steady proliferation reflect an obvious satisfaction on the part of most Americans. Yet such balkanization—fueled by people choosing to interact only with others exactly like themselves—threatens the survival of important democratic values that have prevailed throughout the evolution of U.S. society.

The Changing Geography of Economic Activity

The economic geography of the United States today is the product of all of the foregoing, as bountiful environmental, human, and technological resources have cumulatively blended together to create one of the world's most advanced economies. Perhaps the greatest triumph was overcoming the tyranny of distance, as people and activities were organized into a continentwide spatial economy that took maximum advantage of agricultural, industrial, and urban development opportunities. Yet, despite these past achievements, American economic geography at the outset of the twenty-first century is once again in the throes of restructuring as the transition is being completed from industrial to postindustrial society.

Major Components of the Spatial Economy

Economic geography is mainly (though not exclusively) concerned with the locational analysis of **productive activities**. Four major sets may be identified:

- **Primary activity:** the extractive sector of the economy in which workers and the environment come into direct contact, especially in *mining* and *agriculture*.

- **Secondary activity:** the *manufacturing* sector in which raw materials are transformed into finished industrial products.
- **Tertiary activity:** the *services* sector, including a wide range of activities from retailing to finance to education to routine office-based jobs.
- **Quaternary activity:** today's dominant sector, involving the collection, processing, and manipulation of *information*; a subset, sometimes referred to as **quinary activity**, is the managerial activity associated with decision-making in large organizations.

Historically, each of these activities has successively dominated the American labor force for a time over the past 200 years, with the quaternary sector now dominant. Agriculture dominated until the late nineteenth century, giving way to manufacturing by 1900. The steady growth of services after 1920 finally surpassed manufacturing in the 1950s but now shares a dwindling portion of the limelight with the still-rising quaternary sector. The approximate breakdown by major sector of employment in the U.S. labor force today is agriculture, 2 percent; manufacturing, 15 percent; services, 18 percent; and quaternary, 65 percent (with about 10 percent in the quinary sector). We now proceed to review these major productive components of the spatial economy in the following coverage of resource use, agriculture, manufacturing, and the postindustrial revolution.

Resource Use

The United States (and Canada) was blessed with abundant deposits of mineral and energy resources. Fortunately, these resources were usually concentrated in sufficient quantities to make long-term extraction an economically feasible proposition, and most of the richest raw material sites are still the scene of major drilling or mining operations. Moreover, the continental and offshore mineral/fuel storehouse may yet contain significant undiscovered resources for future exploitation.

Mineral Resources North America's rich mineral deposits are localized in three zones: the Canadian Shield north of the Great Lakes, the Appalachian Highlands, and scattered areas throughout the mountain ranges of the West. The Shield's most noteworthy minerals are iron ore, nickel, gold, uranium, and copper. Besides vast deposits of soft (bituminous) coal, the Appalachian region also contains hard (anthracite) coal in northeastern Pennsylvania and iron ore in central Alabama. The western mountain zone contains significant deposits of coal, copper, lead, zinc, molybdenum, uranium, silver, and gold.

Fossil Fuel Energy Resources The realm's most strategically important resources are its petroleum (oil), natural gas, and coal supplies—the **fossil fuels**, so named because they were formed by the geologic compression and transformation of plant and tiny animal organisms that lived hundreds of millions of years ago. These energy supplies are mapped in Figure 3-14, which reveals abundant deposits and far-flung distribution networks.

The leading *oil*-production areas of the United States are located along and offshore from the Texas–Louisiana Gulf Coast; in the Midcontinent district, extending through western Texas–Oklahoma–eastern Kansas; and along Alaska's central North Slope facing the Arctic Ocean. (Canada's major oilfields lie in a wide crescent curving southeastward from northern Alberta to southern Manitoba and beneath the waters around Newfoundland in the extreme east.) The distribution of *natural gas* deposits generally resembles the geography of oilfields because petro-

FIGURE 3-14

leum and energy gas are usually found in similar geologic formations (the floors of ancient shallow seas). The realm's *coal* reserves rank among the greatest on Earth, and the U.S. portion alone contains at least a 400-year supply; the main producing coalfields are found in Appalachia, the northern U.S. Great Plains/southern Alberta, and southern Illinois/western Kentucky.

Agriculture

Despite the post–1900 emphasis on developing the non-primary sectors of the spatial economy, agriculture remains an important element in America's human geography. Because it is the most space-consuming economic activity, vast expanses of the U.S. (and Canadian) landscape are clothed with fields of grain. In addition, great herds of livestock are sustained by pastures and fodder crops because this wealthy realm can afford the luxury of feeding animals from its farmlands to meet huge demands for red meat in its diet. The growing application of high-technology mechanization to farming has steadily increased both the volume and value of total agricultural production. It also has been accompanied by a sharp reduction in the number of those actively engaged in agri-

culture, and today only slightly more than 1.5 percent of the U.S. population still lives on farms.

The regionalization of U.S. agricultural production is shown in Figure 3-15, its spatial organization developed largely within the framework of the *von Thünen model* (see pp. 45–46). As in Europe (Fig. 1-6), the early-nineteenth-century, original-scale model of town and hinterland expanded outward (driven by constantly improving transportation technology) from a locally "isolated state" to encompass the entire continent by 1900. As this macrogeographic structure formed, the greatly enlarged original Thünian production zones (Fig. 1-5) were modified: (1) the first ring was now differentiated into an inner fruit/vegetable zone and a surrounding belt of dairying; (2) the forestry ring was displaced to the outermost limits of the regional system because railroads could now transport wood quite cheaply; (3) the field crops ring subdivided into an inner mixed crop-and-livestock ring to produce meat (the Corn Belt, as it came to be called) and an outer zone that specialized in the mass production of wheat grains (the Wheat Belts); and (4) the ranching area remained in its outermost position, a grazing zone that supplied young animals to be fattened in

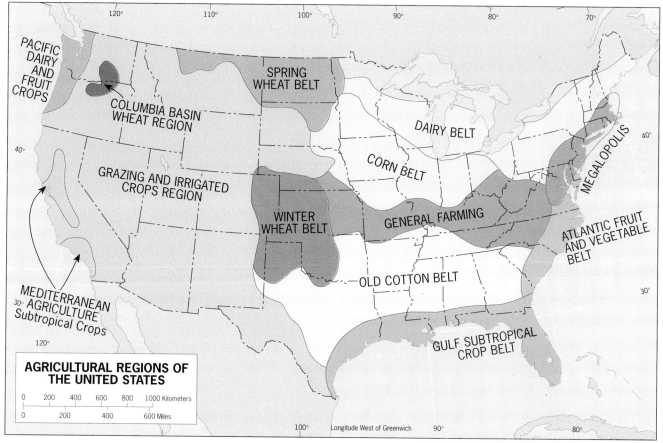

FIGURE 3-15

the meat-producing Corn Belt as well as supporting an indigenous sheep-raising industry. The "supercity" anchoring this macro-Thünian regional system was the northeastern Megalopolis, well on its way toward coalescence and already the dominant food market and transport focus of the entire country.

Although the circular rings of the model are not apparent in Figure 3-15, many spatial regularities can be observed (remember that von Thünen, too, applied his model to the real world and thereby distorted the theoretically ideal pattern). Most significant is the sequence of farming regions as distance from the national market increased, especially westward from Megalopolis toward central California, which was the main directional thrust of the historic inland penetration of the United States. The Atlantic Fruit and Vegetable Belt, Dairy Belt, Corn Belt, Wheat Belts, and Grazing Region are indeed consistent with the model's logical structure, each zone successively farther inland astride the main transcontinental routeway. We can attribute deviations from the scheme to irregularities in the environment or to unique conditions. For example, the nearly year-round growing seasons of California and the Gulf Coast-Florida region permit those distant areas (with the help of efficient refrigerated transport) to produce fruits and vegetables in competition with the regional system's (winter-dormant) innermost zone.

Manufacturing

The geography of North America's industrial production has long been dominated by the Manufacturing Belt (Fig. 3-8). The emergence of this region was propelled by (1) superior access to the Megalopolis national market that formed its eastern edge and (2) proximity to industrial resources, particularly iron ore and coal for the pivotal steel industry that arose in its western half. We noted earlier that manufacturers had a strong locational affinity for cities, and the internal structure of the Belt became organized around a dozen urban-industrial districts interlinked by a dense transportation network. As these industrial centers expanded, they swiftly achieved **economies of scale**, savings accruing from large-scale production in which the cost of manufacturing a single item was further reduced as individual companies mechanized assembly lines, specialized their workforces, and purchased raw materials in massive quantities.

This efficient production pattern served the nation well throughout the remainder of the industrial age. Because of the effects of *historical inertia*—the need to continue using hugely expensive manufacturing facilities for their full, multiple-decade lifetimes to cover initial investments—the Manufacturing Belt is not going to disappear anytime soon. But as its aging facilities are retired, a process well under way, the distribution of American industry will change. As transportation costs equalize among U.S. regions, as energy costs now favor the south-central oil- and gas-producing States, as high-technology manufacturing advances reduce the need for lesser skilled labor, and as locational decision-making intensifies its attachment to noneconomic factors, industrial management has increasingly demonstrated its willingness to relocate to regions it perceives as more desirable in the South and West.

Parts of the Manufacturing Belt have been resisting this trend, particularly the industrial Midwest, whose prospects have greatly improved since the "Rustbelt" days of the 1970s and 1980s. Shedding that label in recent years, Midwestern manufacturers are successfully reinventing their operations by rooting out inefficiencies, investing in cutting-edge factories and technologies, and not only beating back foreign competition in the United States but significantly expanding their exports. Nonetheless, there is a downside to this progress because as high-tech manufacturing proliferates, older factory workers discover that the demand for their lower-tech skills shrinks accordingly.

As the U.S. manufacturing map continues to change, the economic lives of countless communities are affected. If the 1990s were a sign of things of come, entrepreneurial restructuring will also play an increasingly important role because corporate mergers and acquisitions often result in the geographic reshuffling of thousands of jobs. Another recent trend likely to intensify is the economic warfare that occurs when employers seek to locate major new facilities and deliberately pit cities, counties, and even States against one another in order to get the best deal. Often exhorted by politicians seeking reelection, communities offer ever greater incentive packages that include tax breaks, labor retraining, and infrastructural improvements. However, many such competitions can fail to produce a decent rate of return. For example, when Alabama recently went to extreme lengths to lure Mercedes-Benz to build a new assembly plant near Tuscaloosa, it wound up costing the State government $200,000 per expected job. (The cost to Tennessee's government to attract the Saturn automaking complex a few years earlier had been one-tenth that figure.) Moreover, these business strategies can be worked in reverse; not long ago, New York City offered a $60 million incentive package to keep two Wall Street brokerage firms from moving to the suburbs, and Disneyland got its city and State to spend hundreds of millions on local road improvements after the company threatened to cancel expansion plans and perhaps even move the theme park out of Southern California.

The Postindustrial Revolution

16 The signs of **postindustrialism** are visible throughout the United States (and much of Canada) today; they are popularly grouped under such labels as "the computer age" or "the new economy." The term *postindustrial* by itself, of course, tells us mainly what the theme of the American economy no longer is. Yet many social scientists also use the term to refer to a set of societal traits that signal an historic break with the past (for example, work experiences now focus mainly on person-to-person interaction rather than person-product or person-environment contact). Many of the urban spatial expressions of the ongoing societal transformation have already been highlighted; here we focus on some broader economic-geographical patterns.

High-technology, white-collar, office-based activities are the leading growth industries of the postindustrial economy. Most are relatively footloose and are therefore responsive to such noneconomic locational forces as geographic prestige, local amenities, and proximity to recreational opportunities. Northern California's *Silicon Valley*—the world's leading center for computer research and development and the headquarters of the U.S. microprocessor industry—epitomizes the blend of locational qualities that attract a critical mass of high-tech companies to a given locality. These include: (1) a world-class research university (Stanford); (2) technological know-how in the form of a large pool of highly educated, highly skilled labor; (3) close proximity to a cosmopolitan urban center (San Francisco); (4) abundant venture capital; (5) a local economic climate and entrepreneurial culture that supports risk-taking and forgives failure by both smaller and big companies; (6) a locally based network of global business linkages; and (7) a high-amenity environment in the form of top-quality housing, pleasant weather, scenic countrysides, and year-round recreational opportunities.

The development of Silicon Valley is so significant to the new postindustrial era that regional-planning scholars Manuel Castells and Peter Hall have concep-**17** tualized it as the first **technopole**. Technopoles are planned technoindustrial complexes that innovate, promote, and manufacture the hardware and software products of the new informational economy. On the landscape of the outer suburban city, where almost all of these complexes are located, the signature of a technopole is a low-density cluster of ultramodern, low-rise buildings laid out as a campus—an image as symbolic of today's spatial economy as the smoke-belching factory was of the industrial age a century ago. From Silicon Valley, technopoles have spread in all directions—from San Diego in southwesternmost California to the Route 128 corridor around Boston and from North Carolina's Research Triangle to the lakeside suburbs of Seattle—and many of these technopoles will be noted in the regional section that concludes this chapter. These high-tech complexes are also becoming a global phenomenon as they spring up in other geographic realms. They often arise as joint ventures between government and the private sector; among the local names they go by are science city (Japan; Russia), technopolis (France; Japan), science park (Taiwan)—and even Silicon Glen (Scotland).

■ CANADA

Like the United States, Canada is a federal state, but it is organized differently. Canada is divided administratively into ten provinces and three territories (Fig. 3-16). The provinces—where 99.7 percent of all Canadians live—range in territorial size from tiny, Delaware-sized Prince Edward Island to sprawling Quebec, more than twice the area of Texas. Beginning in the east, the four Atlantic Provinces are Nova Scotia, New Brunswick, Prince Edward Island, and Newfoundland and Labrador. To their west lie Quebec and Ontario, Canada's two biggest provinces. Most of western Canada (which is what the Canadians call everything west of Lake Superior) is covered by the three Prairie Provinces—Manitoba, Saskatchewan, and Alberta. In the far west, beyond the Canadian Rockies and facing the Pacific, lies the tenth province, British Columbia.

The three territories—Yukon, the Northwest Territories, and Nunavut—together occupy a massive area half the size of Australia but are inhabited by only about 100,000 people. Nunavut is the newest addition to Canada's political map and deserves special mention. Created in 1999, this new territory is the outcome of a major aboriginal land claim agreement between the *Inuit* people (formerly called Eskimos) and the federal government, and encompasses all of Canada's eastern Arctic. Nunavut covers fully one-fifth of Canada's land, an area larger than that of any other province or territory. Since 80 percent of its 28,000 residents are Inuit, Nunavut—which means "our land"—is the first territory to have a substantial degree of native self-government (in addition to outright ownership of an area half the size of Texas).

In population size (in descending order), Ontario (12.0 million) and Quebec (7.5 million) are again the leaders; British Columbia ranks third with 4.3 million; next come the three Prairie Provinces with a combined total of 5.3 million; the Atlantic Provinces are the four smallest, together containing 2.4 million. The sparsely populated territories in the far north contain barely

FIGURE 3-16

100,000 residents. Canada's total population of 31.6 million is only slightly larger than one-tenth the size of the U.S. population. Spatially, as noted at the outset of the chapter, Canadians overwhelmingly reside within 200 miles (320 km) of the U.S. border.

Though comparatively small, Canada's population is divided by culture and tradition, and this division has a pronounced regional expression. Just under 60 percent of Canada's citizens speak English as their mother tongue, 23 percent speak French, and the remaining 17 percent other languages; about 18 percent of the population speak both English and French (of these, 43 percent of those who speak French as a first language are bilingual, but only 9 percent of the English speakers). The spatial clustering of more than 90 percent of the country's *Francophones* (French-speakers) in Quebec accentuates this Canadian social division along ethnic and linguistic lines. More than 80 percent of this province's

population is French Canadian, and Quebec is the historic, traditional, and emotional focus of French culture in Canada. Over the past few decades a strong nationalist movement emerged in Quebec to demand outright separation from the rest of Canada. That movement crested in 1995, when independence was narrowly defeated in a landmark referendum. Even though support for secession has waned since then, this ethnolinguistic division continues to be the cornerstone of Canada's cultural pluralism, and the issue could still pose a threat to national unity in the future.

Population in Time and Space

The map showing the distribution of Canada's population (Fig. 3-5, p. 144) reveals that only about one-eighth of this enormous country can be classified as its **ecumene**—the inhabitable zone of permanent settlement. As Figure

3-16 indicates, the Canadian ecumene is dominated by a discontinuous strip of population clusters that lines the southern border. We can identify four such clusters on the map, the largest by far being Main Street (home to more than six out of every ten Canadians). As noted earlier (Fig. 3-9), *Main Street* is the conurbation that stretches across southernmost Quebec and Ontario, from Quebec City on the lower St. Lawrence River southwest through Montreal and Toronto to Windsor on the Detroit River. The three lesser clusters are: (1) the Saint John-Halifax crescent in central New Brunswick and Nova Scotia; (2) the prairies of southern Alberta, Saskatchewan, and Manitoba; and (3) the southwestern corner of British Columbia, focused on Canada's third-largest metropolis, Vancouver.

Pre-Twentieth-Century Canada

Canada's population map evolved more slowly than that of the United States. As Figure 3-7 shows, compared to European expansion in the United States, penetration of the Canadian interior lagged several decades behind. In 1850, for example, when the eastern United States and much of the Pacific coast had already been settled, Canada's frontier had only reached Lake Huron. In terms of political geography, Canada did not unify before the last third of the nineteenth century—and then only reluctantly because of fears the United States was about to expand in a northerly direction. Moreover, the Canadian push to the west was delayed by major physical obstacles, including the barren Canadian Shield north and west of the Great Lakes, and the rugged highlands separating the Great Plains from the Pacific (which lacked convenient mountain passes).

The evolution of modern Canada—as well as its contemporary cultural geography—is also deeply rooted in the bicultural division discussed above. Its origin lies in the fact that it was the French, not the British, who were the first European colonizers of present-day Canada, beginning in the 1530s. *New France*, during the seventeenth century, grew to encompass the St. Lawrence Basin, the Great Lakes region, and the Mississippi Valley. A series of wars between the English and French began in the 1680s, ending with France's defeat and the cession of New France to Britain in 1763. By the time London took control of its new possession, the French had made considerable progress in their North American domain. French laws, the French land-tenure system, and the Roman Catholic Church prevailed, and substantial settlements (including Montreal on the St. Lawrence) had been established. The British, anxious to avoid a war of suppression and preoccupied with problems in their other American colonies, gave former French Quebec—the territory extending from the Great Lakes to the mouth of the St. Lawrence—the right to retain its legal and land-tenure systems as well as freedom of religion.

After the American War for Independence, London was left with a region it called British North America (the name *Canada*—derived from an aboriginal word meaning settlement—was not yet in use), but whose cultural imprint still was decidedly French. The Revolutionary War drove many thousands of English refugees northward, and soon difficulties arose between them and the French in British North America. In 1791, heeding appeals by these new settlers, the British Parliament divided Quebec into two provinces: Upper Canada, the region upstream from Montreal centered on the north shore of Lake Ontario, and Lower Canada, the valley of the St. Lawrence. Upper and Lower Canada became, respectively, the provinces of Ontario and Quebec (Fig. 3-16). By parliamentary plan, Ontario would become English-speaking and Quebec would remain French-speaking.

This earliest cultural division did not work well, and in 1840 the British Parliament tried again. This time it reunited the two provinces through the Act of Union under which Upper and Lower Canada would have equal representation in the provincial legislature. That, too, failed, and efforts to find a better system finally led to the 1867 British North America Act that established the Canadian federation (consisting initially of Upper and Lower Canada, New Brunswick, and Nova Scotia, and later, between 1870 and 1999, to be joined by the other provinces and territories). Under the 1867 Act, Ontario and Quebec were again separated, but this time Quebec was given important guarantees: the French civil code was left unchanged, and the French language was protected in Parliament and in the courts.

Canada Since 1900

By the close of the nineteenth century, the fledgling Canadian federation was making major strides toward regional development and the spatial integration of a continentwide economy. In 1886, the transcontinental Canadian Pacific Railway had finally been completed to Vancouver, a condition of British Columbia's entry into the federation 15 years earlier. This new lifeline soon spawned the settlement of not only the far west but also the fertile Prairie Provinces, whose wheat-raising economy expanded steadily as immigrants from the east and abroad poured in during the early twentieth century. Industrialization also began to stir, and by 1920 Canadian manufacturing had surpassed agriculture as the leading source of national income. As noted earlier (Fig. 3-8), the dominant zone of industrial activity is the Toronto–Hamilton–Windsor corridor of southern Ontario (which also functions as one of the 12 districts of the American Manufacturing Belt). That has been the case since the Industrial Revolution took hold in Canada during World

This thriving highrise CBD anchors a metropolis that is located where the Great Plains meet the Rocky Mountains. No, this is *not* Denver, Colorado—it is Alberta's Calgary. Calgary started out as a fort on the western edge of Canada's Great Plains; the settlement got a boost when the Grand Pacific Railway reached it in 1883. Later a second railroad, followed by the Trans-Canada Highway, further improved its cross-continental linkages. But Calgary's biggest shot in the arm was the discovery of oil- and gasfields in its hinterland, which put petroleum refining and petrochemical industries ahead of flour milling and meat packing as industrial mainstays. Today Calgary also is the service center for a large region, with a high-tech sector spearheading its modernization.

War I (1914–1918), when the country heavily supported Britain's war effort on the European mainland.

As industrial intensification proceeded, it was accompanied by the expected parallel maturation of the national urban system. Along the lines of Borchert's five epochs of American metropolitan evolution, Maurice Yeates has constructed a similar multistage model of urban-system development that divides the telescoped Canadian experience into three eras. The initial *Frontier-Staples Era* (prior to 1935) encompasses the century-long transition from a frontier-mercantile economy to one oriented to staples (production of raw materials and agricultural goods for export), with increasing manufacturing activity in the budding industrial heartland. By 1930, Montreal and Toronto, reflecting their different cultural constituencies, had emerged as the two leading cities atop the national urban hierarchy (thus Canada has no single primate city).

Next came the *Era of Industrial Capitalism* (1935–1975), during which Canada achieved U.S.-style prosperity. However, most of this development—involving the massive growth of manufacturing, tertiary activities, and urbanization—took place after 1950 because the Great Depression lasted until World War II, which was followed by a lengthy recession. A major stimulus was the investment of U.S. corporations in Canadian branch-plant construction, especially in the automobile industry in Ontario near the automakers' Detroit-area headquarters. In western Canada, the rapid growth of oil and natural gas production fueled Alberta's urban development, but new agricultural technologies reduced farm labor needs and sparked a rural outmigration in neighboring Saskatchewan and Manitoba. The postwar period also saw the ascent of Main Street, which on less than 2 percent of Canada's land quickly came to contain more than 60 percent of its people, contributed two-thirds of its national income, and had nearly 75 percent of its manufacturing jobs.

The third stage, ongoing since 1975, is the *Era of Global Capitalism*, signifying the rise of additional foreign investment from the Asian Pacific Rim and Europe. This, of course, is also the era of transformation into a postindustrial economy and society, and in the process Canada is experiencing many of the same upheavals as the United States. Interestingly, while U.S. urban growth has recently leveled off at 75 percent of the total population, Canada still continues to urbanize (reaching 78 percent in the early 2000s). Most of this development is occurring in the form of new suburbanization. This represents a departure from the recent past because the pre-1990 Canadian metropolis had experienced far less automobile-generated deconcentration than the United States, with the central city retaining much of its middle-income population and economic base. But today Canada's large cities are turning inside-out, and the new intraurban geography is increasingly symbolized by the suburban downtowns that anchor the ultramodern business complexes along the Highway 401 freeway north of Toronto and Alberta's West Edmonton mega-mall, the world's biggest shopping center.

Cultural/Political Geography

The historic cleavage between Canada's French- and English-speakers, supposedly resolved by the British North America Act, has resurfaced in the past three decades to dominate the country's cultural and political geography. By the time the Canadian federation observed its centennial in 1967, it had become evident that Quebecers regarded themselves as second-class citizens; they believed that bilingualism meant that French-speakers had to learn English but not vice versa; and they perceived that Quebec was not getting its fair share of the country's wealth. Since the 1960s, the intensity of ethnic feelings in Quebec has risen in surges despite the federal government's efforts to satisfy the province's demands. During the 1970s, while a separatist political party came to power in Quebec, a new federal constitution was drawn up in Ottawa. In 1980, Quebec's voters solidly rejected sovereignty when given that choice in a referendum. But the new constitution did *not* satisfy the Quebecers, and throughout the 1980s and early 1990s the Ottawa government struggled unsuccessfully to devise a plan, acceptable to all the provinces, that would keep Quebec in the Canadian federation.

By 1995, with Canada's interest in constitutional reform exhausted, a second referendum on Quebec's sovereignty could no longer be put off. With the reenergized separatist party again leading the way, the Francophone-dominated electorate almost approved independence (which attracted an astonishing 49.4 percent of the vote), an outcome now regarded by Canadians as their country's "near-death experience." Despite calls by many separatists for a follow-up vote that might turn narrow defeat into victory, plans for a third referendum have now been shelved because public opinion polls in Quebec have shown a steady erosion of support for secession since 1995. Several reasons lie behind that changing attitude: (1) the implementation of provincial laws that firmly established the use of French and the primacy of Quebecois culture; (2) the increased bilinguality of Quebec's Anglophones (English speakers), which has calmed Francophone fears concerning assimilation into Canada's English-speaking mainstream; (3) the arrival of a new wave of immigrants (some Francophone, some Anglophone) from Asia, Africa, Eastern Europe, and the Caribbean, who have accelerated the weakening of language barriers by settling in both French-speaking and English-speaking neighborhoods; and (4) the landslide defeat of the ruling separatist Parti Quebecois in the provincial election of April 2003.

Seizing the initiative outside Quebec after 1995, the federal government responded by asking Canada's Supreme Court to review the constitutionality of the province's separation, and in 1998 the court ruled that Quebec had no right to unilateral secession under international law. However, the ruling also declared that Canada is not indivisible; therefore, should a clear majority of Quebec's electorate vote to secede, the Ottawa government and the other provinces would be obliged to negotiate the terms of separation as if the matter were a

new amendment to Canada's constitution. Armed by this legal interpretation, the federal government moved to pass legislation that formalizes the court's rulings—and makes the process of secession for a breakaway province far more cumbersome. First, the lower house of the Canadian Parliament will determine the clarity of any secessionist question approved in a provincial referendum (both the 1980 and 1995 votes in Quebec were based on vaguely worded questions that would have received less support had their language been more explicit and hard line). Second, that body is also empowered to decide what constitutes a "clear majority" of voters, which is likely to be well beyond a total of 50-percent-plus-one. And third, the legislation lays the groundwork for the court-mandated negotiations over the terms of separation, in which a breakaway province would have to reach settlement with Ottawa and the nine other provinces on such issues as borders, guaranteeing the rights of linguistic and aboriginal minorities, and paying off its share of the Canadian national debt. Should the Quebec situation ever advance to this stage, it is easy to envision that certain provinces (particularly British Columbia and Alberta) would continue their longstanding hard line and insist that Quebec pay a very high price for its independence. Given this likelihood, and all that has gone into the search for a resolution to the Quebec crisis since 1980, negotiations might well end in an impasse—with unknown consequences for the future of northern North America.

Among the issues that a departing Quebec would have to negotiate, most significant from a geographic standpoint is that the French linguistic region does *not* coincide with the province's territorial boundaries. As can be seen in the map on page 168, "French Canada" is mainly confined to the area centered on the middle and lower St. Lawrence Valley, with some spillover into northern New Brunswick. Moreover, as the distribution of the "no" vote on separation in the 1995 referendum strongly suggests (Fig. 3-17; areas colored pink), there are numerous non-French communities located within the Francophone region—especially along the Ontario and U.S. borders as well as in metropolitan Montreal. During the late 1990s, dozens of English-speaking municipalities in this southern periphery of Quebec spearheaded a partitionist movement by declaring their intention to stay in Canada, arguing that they have the same right to secede from Quebec that Quebec has to secede from Canada. Even though this claim was vehemently opposed by the separatist provincial government then in power (which realized that a dismembered independent Quebec would be less attractive to voters in any future referendum), opinion polls showed that perhaps as many as half the French speakers agreed with it.

The Cree and Inuit of northern Quebec held their own plebiscites and voted to remain in Canada.

Cree

Inuit

DIVIDED QUEBEC: THE 1995 REFERENDUM VOTE ON SEPARATION FROM CANADA

National capital is underlined.
Scale varies in this perspective.

Yes to separation
No to separation
⚒ Forestry (pulp and paper)
▬ Hydroelectric station
⚓ Manufacturing center
✕ Mining center
▲ First Nation community
✈ Major airport
— Major highway
— Passenger railroad
— — Power transmission line

Chicoutimi

Jonquière

Quebec

Montreal

Pontiac

FIGURE 3-17

On the wider Canadian scene, the culture-based Quebec crisis has stirred the ethnic feelings of the country's 1.2 million native (First Nations and Inuit) peoples. Their assertions have also received a sympathetic hearing in Ottawa, and in recent years breakthroughs were achieved with the creation of Nunavut and a treaty for limited tribal self-government in northern British Columbia. A leading concern among these peoples is that their aboriginal rights be protected by the federal government against the provinces. This is especially true for the First Nations of Quebec's northern frontier, the Cree, whose historic domain covers more than half of the province of Quebec as it appears on current maps. Administration of the Cree was assigned to Quebec's government in 1912, a responsibility that a move toward independence might well invalidate. In any case, the Cree would also likely be empowered to seek independence. This would leave the French-speaking remnant of Quebec with only about 45 percent of the province's present

territory (which would be further diminished if the Anglophone partitionist movement succeeded). As Figure 3-16 shows, the territory of the Cree is no unproductive wilderness: it contains vital facilities of the James Bay Hydroelectric Project, a massive scheme of dikes, dams, and artificial lakes that has transformed much of northwestern Quebec and generates electrical power for a huge market within and outside the province.

Finally, the events of the past quarter-century have increasingly impacted Canada's national political landscape. Regionalism has also intensified in the west, whose leaders (some even to the point of threatening the region's own secession) opposed federal concessions to Quebec and insisted that the equal treatment of all ten provinces is a basic principle that precludes the designation of special status for anyone. The latest federal elections have clearly revealed the emerging fault lines that surround Quebec and set the western provinces (British Columbia, Alberta, and Saskatchewan) off from the rest of the country. Even the remaining Ontario-led center and the eastern bloc of Atlantic Provinces voted divergently, and today the politico-geographical hypothesis of "Four Canadas" may be on its way toward becoming reality. Thus, with its national unity under pressure, Canada today confronts the coalescing forces of *devolution* that threaten to transform the new fault lines into permanent fractures.

Economic Geography

As in the United States, the growth of Canada's spatial economy has been supported by a diversified, high-quality *resource base*. We noted earlier that the Canadian Shield is endowed with rich mineral deposits and that oil and natural gas are extracted in sizeable quantities in Alberta to the southwest of the Shield. Canada has long been a leading *agricultural* producer and exporter, especially of wheat and other grains from its breadbasket

in the Prairie Provinces; new technologies keep productivity high, but labor requirements continue to diminish and farm workers now account for only about 2 percent of the national workforce. Postindustrialization has caused substantial employment decline in the *manufacturing sector*, with Southern Ontario's industrial heartland being the most adversely affected area. On the other hand, Canada's robust *tertiary and quaternary sectors* (which today employ more than 70 percent of the total workforce) are creating a host of new economic opportunities. Fortunately, Southern Ontario is benefiting from this postindustrial transformation because the country's leading high-technology, research-and-development complex has grown up around Waterloo and Guelph, a pair of university towns about 60 miles (100 km) west of Toronto.

Canada's economic future is also going to be strongly affected by the continuing development of its trading relationships. The landmark United States-Canada Free Trade Agreement, signed in 1989, initiated a now-completed

AMONG THE REALM'S GREAT CITIES . . . **TORONTO**

*T*oronto, the capital of Ontario and Canada's largest metropolis (5.1 million), is the historic heart of English-speaking Canada. The landscape of much of its center is dominated by exquisite Victorian-era architecture and surrounds a healthy downtown that is one of Canada's leading economic centers. Landmarks abound in this CBD, including the SkyDome stadium, the famous City Hall with its facing pair of curved highrises, and mast-like CN Tower, which at 1815 feet (553 m) is the world's tallest freestanding structure. Toronto also is a major port and industrial complex, with miles of facilities lining the shoreline of Lake Ontario.

Livability is one of the first labels Torontonians apply to their city, which has retained more of its middle class than central cities of its size in the United States. *Diversity* is another leading characteristic because this is North America's richest urban ethnic mosaic. Among the largest of more than a dozen thriving ethnic communities are those dominated by Italians, Portuguese, Chinese, Greeks, and Ukrainians; overall, Toronto now includes residents from 169 countries who speak more than 100 languages; and the immigrant inflow continues strongly, with those born outside Canada constituting more than 40 percent of the city's current population. *Vibrancy* is yet another hallmark of this remarkable city, whose resiliency was tested but not broken by the SARS-virus outbreak in the spring of 2003.

Toronto has worked well in recent decades, thanks to a metropolitan government structure that fostered central city–suburban cooperation. But that relationship is becoming stressed as the outer city gains a critical mass of population, economic activity, and political clout, while the central city was recently forced to accommodate an unwanted amalgamation with five of its nearest suburban neighbors.

phasing out of all tariffs and investment restrictions between the two countries, whose annual cross-border flow of goods and services is the largest in the international trade arena. (In the early 2000s more than four-fifths of Canada's exports went to the United States, from which it also derived about three-quarters of its imports.) In 1994 these economic linkages were further tightened by the implementation of the North American Free Trade Agreement (NAFTA), which consolidated the gains of the 1989 pact and opened major new opportunities for both countries by adding Mexico to the trading partnership (see box titled "North American Free Trade Agreement [NAFTA]").

Because these free-trade agreements increasingly impact the Canadian spatial economy, they are likely to weaken domestic east-west linkages and strengthen international north-south ties. Since many local cross-border linkages built on geographical and historical commonalities are already well developed, they can be expected to intensify in the future: the Atlantic Provinces with neighboring New England; Quebec with New York State; Ontario with Michigan and adjacent Midwestern States; the Prairie Provinces with the Upper Midwest; and, above all, British Columbia with the (U.S.) Pacific Northwest. Such functional reorientations, of course, constitute yet another set of powerful devolutionary

North American Free Trade Agreement (NAFTA)

*T*he economies of Canada, the United States, and Mexico formalized their interactions under the *North American Free Trade Agreement (NAFTA)*, which took effect in 1994. This economic alliance has forged the world's largest trading bloc, a (U.S.) $8-trillion-plus market of 428 million consumers (surpassing the European Union's 382 million). Between now and its completion date of 2009, NAFTA will integrate its constituent economies through a number of steps. Thousands of individual tariffs, quotas, and import licenses have been or are scheduled to be rolled back to eliminate most trade barriers for agricultural and manufactured goods as well as tertiary and quaternary services. Moreover, restrictions on the flow of investment capital across international borders are being lifted, and regulations concerning foreign ownership of productive facilities continue to be liberalized.

The architects of NAFTA cite the opportunities the maturing pact will bring. Business investment will surge as unimpeded exports flow throughout the vast new marketplace. At the same time, intensified competition will give consumers access to a wider variety of quality products at reduced prices. Canada—which joined NAFTA mainly to safeguard the concessions it received in its 1989 Free Trade Agreement with the United States—faced the most skeptical citizenry at the outset, but those doubts evaporated as Canadian exports (and new jobs) under NAFTA have risen substantially since 1993. Trade with the United States, already heightened under the bilateral 1989 agreement, has propelled much of this growth, especially the southward export of goods whose tariffs had been lowered.

This progress notwithstanding, NAFTA's evolution continues to be criticized. Organized labor has objected to the redistribution and losses of jobs in a number of industries, dislocations which it perceives stem largely from the implementation of the trade agreement. Environmental activists complain that the pact's architects and component governments have not adequately addressed their concerns. And a number of social scientists and public policy planners argue that NAFTA should be doing much more to reduce regional income disparities and uneven development within its trade area.

As NAFTA realizes its goals, it will require each member country to concentrate on the production of those goods and services for which it has a comparative advantage (in research and development, technological skills, managerial know-how, and the like). As workers move from less productive to more productive industries, the geography of employment will be restructured—a process that would be occurring anyway as the three countries continue to adapt to continental and global economic change.

NAFTA's success has attracted the intense interest of all the other countries in the Western Hemisphere. Although most of them are involved in their own regional efforts aimed at greater economic integration (as we shall see in Chapters 4 and 5), the great majority desire to become part of an enlarged NAFTA—which triggered its own expansion by inviting Chile to join in 1995 (an initiative so far thwarted by the U.S. Congress). To that end, ambitious plans were launched at summit meetings in the mid-1990s to create the *Free Trade Area of the Americas (FTAA)*, a hemisphere-encompassing free-trade zone that would establish a single market of nearly 900 million consumers. If FTAA is implemented by its target date of 2005, it should accelerate the completion of NAFTA—as well as NAFTA's transformation into a much larger supranational organization.

forces confronting the Ottawa government because most of these potential economic fault lines coincide with those that politically demarcate the "Four Canadas."

The rising importance of this framework of transnational regions straddling the U.S.-Canadian border recalls Kenichi Ohmae's **regional state** concept introduced in Chapter 1. A regional state is a "natural economic zone" that defies old borders, and it is shaped by the global economy of which it is a part; its leaders deal directly with foreign partners and negotiate the best terms they can with the national governments under which they operate. Writing about Canada recently, Ohmae identified both a Pacific Northwest (the Seattle–Vancouver axis) and a Great Lakes regional state (the remarkably intertwined Ontario–Michigan industrial complex), and warned that the manner in which Ottawa's leaders dealt with these new economic entities was critical to the survival of the Canadian state. Today his warning is more meaningful than ever because the Pacific Northwest and Great Lakes have been joined by most of the other areas listed above. Indeed, these growing international interactions are increasingly evident in the overall regional configuration of North America, to which we now turn in the concluding part of this chapter.

REGIONS OF THE REALM

THE ONGOING TRANSFORMATION of North America's human geography is fully reflected in its internal regional organization. As new forces uproot and redistribute people and activities, old locational rules no longer apply. However, the varied character of the realm's physical, cultural, and economic landscapes ensures that meaningful regional differences will persist. We will now examine the current areal arrangement of the United States and Canada within a framework of nine regions (Fig. 3-18).

Although manufacturing remains highly important within the transformed American economy, productivity and obsolescence problems during the late twentieth century increased production costs in the Core Region, thereby erasing many of its historical competitive advantages over newly emerging industrial areas in the southern and western United States. In parts of the Core (especially its eastern portion), employment dropped sharply

THE NORTH AMERICAN CORE

The Core Region (Fig. 3-18)—synonymous with the American Manufacturing Belt—was introduced earlier in this chapter. Serving as the historic workshop for the linked spatial economies of the United States and Canada, this region was the unquestioned leader and centerpiece during the century between the Civil War and the close of the industrial age (1865–1970). The rise of postindustrialism over the past three decades has diminished that linchpin regional role, and strengthening challengers to the south and west will continue to siphon key functions away from the Core as this century unfolds. But make no mistake: this is still the geographic heart of North America. Here one finds the largest city and federal capital of both countries. Here, too, are the leading financial markets, corporate headquarters, media centers, cultural facilities, research complexes, and the busiest airports and intercity expressways. Moreover, both the U.S. and Canadian portions of the Core still contain more than one-third of their respective national populations.

MAJOR CITIES OF THE REALM	
City	Population* (in millions)
Atlanta	4.6
Boston	6.0
Chicago	9.4
Dallas–Ft. Worth	5.7
Denver	2.8
Detroit	5.6
Houston	5.0
Los Angeles	17.0
Montreal, Canada	3.5
New York	21.4
Ottawa, Canada	1.1
Philadelphia	6.3
San Diego	3.0
San Francisco	7.3
Seattle	3.8
Toronto, Canada	5.1
Vancouver, Canada	2.2
Washington, D.C.	5.2

*Based on 2004 estimates.

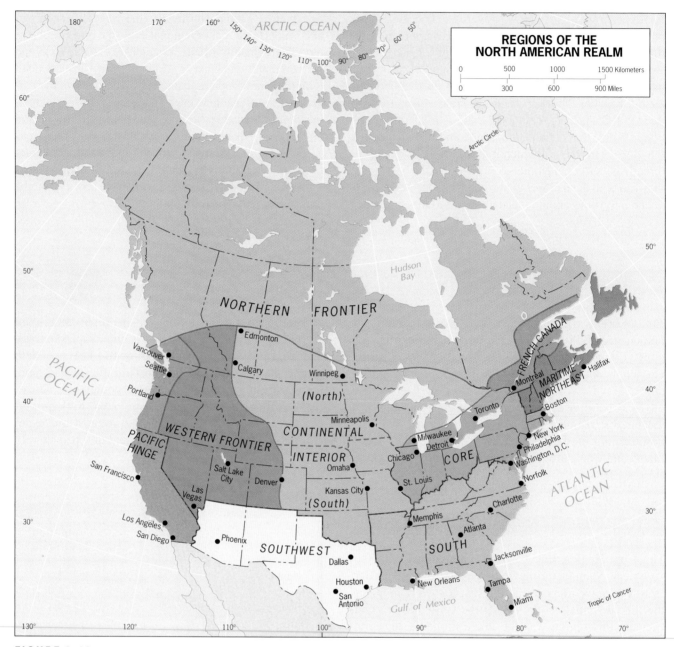

FIGURE 3-18

in the smokestack industries as factories closed or relocated, accelerating the economic decline of numerous mill towns, blue-collar suburbs, and inner central cities. But other areas fought back, none more successfully than the Midwest portion of the Manufacturing Belt, which reinvented itself by aggressively pursuing the high-tech upgrading of its aged industrial base and becoming far more competitive in the international marketplace.

The reconcentration of the pivotal automobile industry since the 1980s is a prime example of the booming Midwest's new manufacturing geography, which

foreign as well as domestic carmakers have shaped. A key role was played by Japanese-owned Toyota, which decided to challenge the "Big Three" (GM, Ford, and [then] Chrysler) on its home turf. Toyota's investments were directed at the Core's less-developed southwestern border zone, where a 500-mile (800-km)-long corridor of Interstate-64 between St. Louis and northern West Virginia became the axis of a self-sufficient, state-of-the-art automaking complex centered on the Lexington, Kentucky area. Detroit's Big Three were forced to keep pace and successfully met the challenge by modernizing fabrication technologies and dispersing certain opera-

AMONG THE REALM'S GREAT CITIES . . . **CHICAGO**

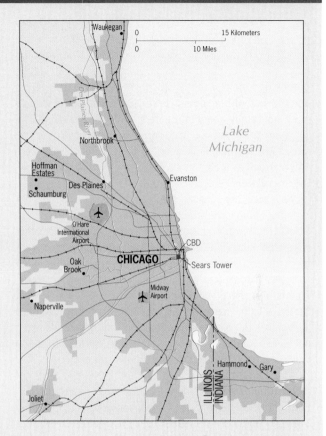

*C*hicago lies near the southern end of Lake Michigan not far from the geographic center of the contermi- nous United States. Its centrality and crossroads location were evident to its earliest indigenous settlers, who devel- oped the site as a portage where canoes could be hauled from the lake to the headwaters of a nearby stream that led to the Mississippi River. Centuries later, when the modern U.S. spatial economy emerged, Chicago became the lead- ing hub on the continental transport network. As a freight- rail node, it still reigns supreme; and even though most long-distance passengers switched to jet planes a genera- tion ago, Chicago remains the quintessential North Ameri- can hub city, with O'Hare International consistently rank- ing among the world's busiest airports.

Poet Carl Sandburg described Chicago as "the city of big shoulders," underscoring its personality and prowess as the leading U.S. manufacturing center during the cen- tury following the Industrial Revolution of the 1870s and 1880s. Among the myriad products that emanated from Chicago's industrial crucible were the first steel-frame skyscraper, elevated railway, refrigerated boxcar, cooking range, electric iron, nuclear reactor, and window envelope. Chicago also became a major commercial center, spawn- ing a skyscraper-dominated central business district sec- ond only to New York's in size and influence.

Chicago today (metro population: 9.4 million) is trying hard to make the transition to the postindustrial age. Its di- versified economy has proven to be a precious asset in this quest, and the city is determined to reinvent itself as a first- order service center. The new forces it must contend with, however, are symbolized in the fate of the emptying Sears Tower, the city's (and the nation's) tallest building: not long ago Sears & Roebuck abandoned the structure for a new headquarters campus along Interstate-90 in the far north-

western suburb of Hoffman Estates. This shift acknowl- edges the rise of Chicagoland's thriving outer city, one of the country's largest and most successful. Around O'Hare, a huge and still-expanding suburban office/industrial-park complex has truly become Chicago's "Second City," under- scoring that major airports themselves will increasingly function as urban growth magnets in the ever more global- ized spatial economy of the twenty-first century.

---

tions away from their traditional base in Michigan to locations elsewhere in the Midwest and as far away as Mexico.

Postindustrial development has also spawned a num- ber of new (or recycled) growth centers in the Core. The major metropolitan complexes of the northeastern Megalopolis—already the scene of much quaternary and quinary economic activity—are adjusting fairly well. The Boston area, well endowed with research fa- cilities, continues to attract innovative high-tech busi- nesses, mostly toward the Route 128 freeway corridor that girdles the central city (North America's pioneering

technopole of the 1950s). New York City remains the national leader in finance and advertising, and is home to significant elements of the broadcast and print media; but even here the city increasingly shares its business leadership with its huge outer suburban city.

The Core Region metropolis that has gained the most from postindustrialization is Washington, D.C. As the information-producing and managerial sectors have blossomed, and as the U.S. federal government ex- tended its ties to the private sector, Washington and its surrounding outer cities in the affluent Maryland and Virginia suburbs (interconnected by the 66-mile [105-

AMONG THE REALM'S GREAT CITIES . . . **NEW YORK**

*N*ew York is much more than the largest city of the North American realm. It is one of the most famous places on Earth; it is the hemisphere's gateway to Europe and the rest of the Old World; it is a tourist mecca; it is the seat of the United Nations; it is one of the globe's most important financial centers—a true "world city" in every sense of that overused term. Unfortunately, the terrorists who planned and executed the attacks of 9/11 were only too well aware of New York's symbolic importance on the international stage.

New York City consists of five boroughs, centered by the island of Manhattan, which contains the CBD south of 59th Street (the southern border of Central Park). Here is a sky-scrapered landscape unequaled anywhere, studded with more fabled landmarks, streets, squares, and commercial facilities than any other cities except London and Paris. This is also the cultural and media capital of the United States, which means the city's influence constantly radiates across the planet thanks to New York-based television networks, newspapers and magazines, book publishers, fashion and design leaders, and artistic trendsetters.

At the metropolitan scale, New York forms the center of a vast urban region, 150 miles (250 km) square (population: 21.4 million), that sprawls across parts of three States in the heart of Megalopolis. That outer city has become a giant in its own right with its population of more than 13 million, massive business complexes, and flourishing suburban downtowns. Thus, despite its global connections, New York is caught in the currents that affect U.S. central cities today. Its ghettos of disadvantaged populations grow steadily; its aging port and industrial base continue to decline; its corporate community is increasingly stressed by the high cost of doing business in ultra-

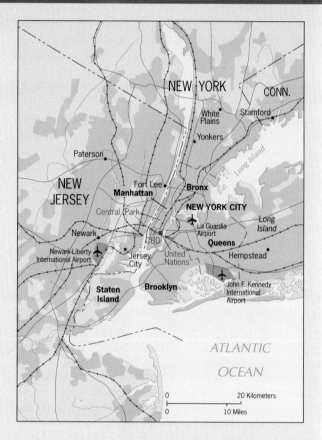

expensive Manhattan, a situation exacerbated by heightened expenses for security in the post–9/11 era.

New York has not yet found the solutions but is struggling to overcome these problems by reinventing itself in order to remain a leading economic player in the twenty-first century. The staggering events of September 2001 notwithstanding, that effort continues unabated today.

---

km] Beltway encircling the capital city) has amassed an enormous complex of office, research, high-tech, lobbying, and consulting firms. The most important development has occurred along the outer-suburban expressway corridor between Tyson's Corner on the Beltway (see photo, p. 151) and Dulles International, one of the world's fastest-growing airports. Since the mid-1990s, a large number of major telecommunications and Internet companies have been attracted to the Dulles area, one of the country's leading fiber-optic cable hubs thanks to its local cluster of federal intelligence agencies as well as the corporate headquarters of America Online. This growth is now fanning out from Dulles

across much of northern Virginia's sector of the Washington metropolis, which is well on its way to becoming one of the realm's foremost technopoles. Moreover, some are already calling this activity complex the "technological capital" of the United States because more technical workers are employed here than in either Silicon Valley or suburban Boston's Route 128 agglomeration. This economic transformation is being felt throughout the Washington region: the area's approximately 350,000 federal government workers are today far outnumbered by the more than 500,000 employed by computer-service, telecom, Internet, aerospace, and biotech companies.

▶ THE MARITIME NORTHEAST

The Maritime Northeast consists of upper New England and the neighboring Atlantic Provinces of easternmost Canada (Fig. 3-18). New England, one of the realm's historic culture hearths, has retained a strong regional identity for almost 400 years. The urbanized southern half of New England has been the northeastern anchor of the Core since the mid-nineteenth century—where it must be regionally classified. However, the six New England States (Maine, New Hampshire, Vermont, Rhode Island, Massachusetts, and Connecticut) still share many common characteristics. Besides this overlap with the Manufacturing Belt (and Megalopolis) in its south, the Maritime Northeast region also extends northeastward across the Canadian border to encompass almost all of the four Atlantic Provinces of New Brunswick, Nova Scotia, Prince Edward Island, and Newfoundland and Labrador.

A long association based on economic and cultural similarities has tied northern New England to Atlantic Canada. Both have a strong maritime orientation, are rural in character, possess difficult environments with limited land resources, and were historically bypassed in favor of more fertile inland areas. Thus economic growth in upper New England has always lagged behind that of the rest of the realm. Development has centered on primary activities, mainly fishing the (once) rich offshore banks of the nearby North Atlantic, forestry in the uplands, and farming in the few fertile valleys available. Recreation and tourism have boosted the regional economy in recent times, with scenic coasts and mountains attracting millions from the neighboring Core Region. The growth of skiing has also helped, extending the tourist season through the harsh winter months.

Since 1980, New England's roller-coaster economy has experienced spectacular prosperity, hard times, and, most recently, cautious recovery. Weaning itself from its overreliance on the activities that brought on the troubles of a decade ago, New England is developing a more diversified economic base built around telecommunications, financial services, health care, and biotechnology. Nonetheless, the economic revival has not taken hold in much of northern New England. Most of the benefits of the recovery are concentrated in the zone closest to metropolitan Boston, not just its immediate suburbs focused on the again-booming Route 128 corridor, but a wide swath of fringe areas that spill over into the southern parts of Maine and particularly New Hampshire. New England's continued growth, however, is by no means assured: apart from the uneven distribution of recovery, the region's costs of doing business are higher than those in the South and Southwest, and the continuing outflow of highly skilled labor raises concerns about the size and quality of the future workforce.

The Atlantic Provinces also experienced hard economic times in the 1990s. Most adversely affected was the groundfish industry as offshore stocks of flounder, haddock, and cod became severely depleted through overfishing. Here, too, disagreements over Canada's maritime fishing boundaries led to the seizure of vessels and heightened tensions with other nations, including the United States. New opportunities, never easy to come by, are most promising in the provinces peripheral to the more heavily populated Nova Scotia and New Brunswick. In tiny Prince Edward Island, tourism was given a boost by the 1997 opening of an 8-mile (13-km) bridge across Northumberland Strait, the island's first direct link to the mainland since it joined the Canadian federation in 1873.

Economic prospects are even brighter in remote Newfoundland and Labrador, Canada's poorest province, which was the tenth (and last) province to enter the federation just over 50 years ago. Major offshore oil deposits (see Fig. 3-14), centered on the Grand Banks about 200 miles (320 km) east of the capital, St. John's, were discovered in the 1990s just as government-mandated fisheries restrictions were inflicting disaster on the seafood industry. The construction and opening of seabed drilling platforms and coastal support facilities have been swift, additional oil as well as natural gas deposits are being discovered closer to the coast and onshore, and prospects look good that oil can transform the economy of Newfoundland and Labrador along the lines of Norway's over the past quarter-century. (Prospects are also brightening in Nova Scotia, where offshore gas deposits now promise a similar bonanza.)

▶ FRENCH CANADA

Francophone Canada constitutes the effectively settled, southern portion of Quebec, which straddles the central and lower St. Lawrence Valley from where that river crosses the Ontario-Quebec border just upstream from Montreal to its mouth in the Gulf of St. Lawrence. Also included is a sizeable concentration of French speakers, known as the Acadians, who reside just beyond Quebec's provincial boundary in neighboring New Brunswick (Fig. 3-18). The Old World charm of Quebec's cities (see photo, p. 172) is matched by an equally unique rural settlement landscape introduced by the French: narrow rectangular farms, known as *long lots*, are laid out in sequence perpendicular to the St. Lawrence, other rivers, and the roads that parallel them, thus allowing each farm access to an adjacent routeway.

An unmistakably French cultural landscape in North America: the Rue Saint-Louis in the capital of Quebec Province, Quebec. Not an English sign in sight in this center of French/Quebecois culture. The green spires in the background belong to Le Chateau Frontenac, a grand hotel built on the site of the old Fort St. Louis (the Count of Frontenac was a prominent governor of what was then known as New France). French-speaking, Roman Catholic Quebec lies at the core of French Canada.

The economy of French Canada, however, is no longer rural (although dairying remains a leading agricultural pursuit) and exhibits urbanization rates similar to those of the rest of the country. Industrialization is widespread, supported by cheap hydroelectric energy generated at huge dams in northern Quebec, but relatively little of the region's manufacturing could be classified as high-tech. Tertiary and postindustrial commercial activities are concentrated around Montreal, and tourism and recreation are also important to the regional economy. But the health of these sectors—now improving as the independence movement has subsided—is tied to the resolution of Quebec's political status within Canada.

This uncertainty reminds us that, over the past third of a century, Quebec's heightened nationalism has had a major impact on the province that extends well beyond the economic sphere. During the 1980s, the provincial government enacted new laws that strengthened the French language and culture throughout Quebec. With English domination ended, the French Canadians then channeled their energies into developing Quebec's economy. A byproduct of that effort was the rapid urbanization of young families and a loosening of their ties to Roman Catholicism. The outcome was a sharp decline in the provincial birth rate, which the Quebec government tried to offset by encouraging (with cash payments) large families and Francophone immigration (mostly from former colonies of France around the world).

As noted earlier, French Canada also includes Acadia, Canada's largest cluster of Francophones outside Quebec. Here in northernmost New Brunswick, the approximately 250,000 French-speakers constitute one-third of the province's population. The Acadians, however, not only shun the notion of independence for themselves but actively promote all efforts to keep Quebec within the Canadian federation. Unlike the Quebecers, the Acadians have devoted their energies in recent years to accommodation with the Anglophone community. Today, a new relationship (known as "cohabitation") has been worked out between the two groups, based as much on mutual respect for each other's languages as the strict equality of a new, government-mandated system of bilingualism.

THE CONTINENTAL INTERIOR

The Continental Interior extends across the center of both the conterminous United States and the southern tier of Canada (Fig. 3-18). With few exceptions—most notably in the region's northeastern corner where the barren but mineral-rich Canadian Shield meets the Great Lakes—agriculture is the predominant feature of the landscape. Whereas the innermost fruit-and-vegetable and dairying belts in the U.S. macro-Thünian regional system (Fig. 3-15) are in competition with other economic activities in the North American Core to the east, by the time one reaches the Mississippi Valley, meat and grain production prevail for much of the next thousand miles (1600 km) west to the base of the Rocky Moun-

AMONG THE REALM'S GREAT CITIES . . . MONTREAL

*M*ontreal lies on a large triangular island in the St. Lawrence River, its historic core laid out along the base of what Montrealers call "the mountain"—flat-topped Mount Royal, from which the city takes its name. Culturally, Montreal is the second largest French-speaking city in the world (after Paris), and therein lie its opportunities and challenges. Without a doubt, this is still one of the realm's most cosmopolitan cities, a strong European flavor pervading its stately CBD, its lively neighborhoods, and its bustling street life. To escape the long, harsh winters here, much of downtown Montreal has become an underground city with miles of handsome passageways lined with shops, restaurants, and movie theaters, all of it served by one of the world's most modern and attractive subway systems.

Yet, despite the sparkle and continental atmosphere, Montreal has not been a happy place in recent years. With the escalation of tensions between Francophone and Anglophone Canada over the past three decades, it is understandable that the confrontation should overshadow life in Quebec's largest metropolis (3.5 million), and the city continues to be marked by divisions along ethnolinguistic lines. Spatially, this polarization is still evident along the Boulevard St. Laurent, the longstanding boundary between the French-speakers in Montreal's East End (who constitute about 75 percent of the city's population) and the English-speakers who cluster in the West End (the remnant of a larger Anglophone community that was reduced by 40 percent when the French-supremacy language laws were implemented). These days, however, things are improving as linguistically-mixed communities proliferate throughout Montreal, driven by an influx of immigrants from outside North America who settle in both English- and French-speaking neighborhoods.

Montreal's economy has also suffered. During the 1990s, hundreds of Anglophone firms (and thousands of their workers) fled to Ontario and points west, leaving in their wake a commercial real estate tailspin, rising unemployment, and a noticeable decline in the city's prosperity. Today, with support for secession at a low ebb in Quebec, Montreal is steadily progressing in its comeback. Tourists have returned in droves; a construction boom is reshaping parts of the CBD; new foreign-trading ties, especially with the United States, are reviving the business community; and vibrant suburbs are leading the metropolis to reinvent itself as a hub for information technology, telecommunications, and biopharmaceutical companies.

---

tains. Because the eastern half of the Continental Interior lies in Humid America and closer to the national food market on the northeastern seaboard, mixed crop-and-livestock farming wins out over less competitive wheat raising, which is relegated to the fertile but semiarid environment of the central and western Great Plains on the dry side of the 100th meridian. The latter area also contains Canada's agricultural heartland north of the 49th-parallel border. Although these fertile Prairie Provinces are less subject to serious drought than the U.S. high plains to the south, they are situated at a more northerly latitude and thus have a shorter growing season.

The distribution of North American corn and wheat farming is mapped in Figure 3-19, and the edges of the most productive zones of these two leading Continental

Interior crops coincide with several of the region's boundaries. As we saw earlier in Figure 3-15, the Corn Belt is most heavily focused on Iowa and part of neighboring Illinois, while the Spring and Winter Wheat Belts are respectively centered on North Dakota and western Kansas. This separation of the Wheat Belts is also the basis for subdividing the Continental Interior region, as the dashed line in Figure 3-18 indicates. The North subregion encompasses the Upper Midwest, the northern Great Plains, and the cultivable southern portions of Canada's three Prairie Provinces; the Spring Wheat Belt is its leading agricultural component and is so named because crops are planted in the spring and harvested in late summer or early autumn. The South subregion contains a major part of the Corn Belt in its eastern

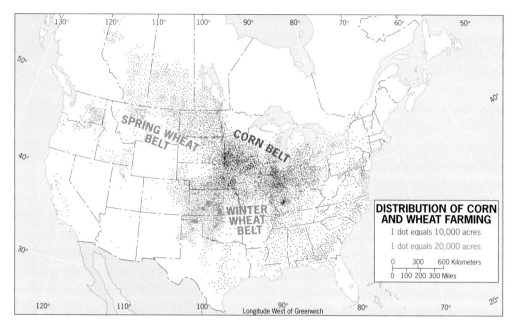

DISTRIBUTION OF CORN AND WHEAT FARMING
1 dot equals 10,000 acres
1 dot equals 20,000 acres

0   300   600 Kilometers
0 100 200 300 Miles

FIGURE 3-19

half and all of the Winter Wheat Belt in its western. The Corn Belt should probably be renamed the *Corn-Soybean Belt* because it also produces huge soybean crops that are grown in rotation with corn (see box titled "Geography of the Mighty Soybean"). The Winter Wheat Belt is centered in the southern Great Plains, and its name is derived from the practice of planting wheat in the fall and harvest-

ing it by late spring; during its infancy, winter wheat can withstand the milder winter temperatures south of 40°N latitude. The great advantage of this growth cycle is that the crop is harvested before the hot, dry summer takes hold in this semiarid environment.

Throughout the Continental Interior, economic activity is markedly oriented toward farming. Its leading

Geography of the Mighty Soybean

Over the past century, no crop in this realm has expanded faster than the soybean. This hardy legume was one of the earliest plants domesticated, and for nearly 5000 years the Chinese have revered it as a grain vital to life itself. The advantages of soybeans are numerous. They are the world's cheapest and most efficient source of protein, with a level of concentration double that of meat or fish. They are easy to cultivate, adapting to a wide range of environmental conditions, and they are less vulnerable to pests and diseases than other temperate-climate food crops. Economically, crop yields are comparatively high, and the demand for soybeans has steadily risen across the world in recent decades.

The dramatic increase of soybean production in the United States has been propelled not only by these advantages but also by this crop's particular suitability to the Corn Belt. It was quickly discovered that soybeans were an ideal crop to rotate with corn because they recondition the soil with nitrogen and other nutrients needed by corn. In terms of cultivation, soybeans also proved highly profitable to grow on the large, flat fields of the Corn Belt (which continues to account for about 80 percent of all U.S. soybean production) where their farm-machinery

requirements were identical to those already in place for raising corn. Moreover, clever marketing constantly widened the versatility of soybeans beyond their use as a nutritious livestock feed, and today they are used in a variety of human foods (ranging from vegetable oil to a plethora of low-calorie products) as well as such industrial goods as fertilizers, paints, and insect sprays.

All of the above also contributed to steadily rising foreign demand, and in many recent years soybeans have ranked as the leading U.S. export commodity. Other producing countries have recognized this opportunity as well, and intensifying global competition after 1990 has resulted in a leveling off of America's share of the world market. Nonetheless, the United States still produces roughly half of the world's annual soybean output. As we note in Chapter 5, most of its competition (another 33 percent of the global supply) comes from Brazil and Argentina in mid-latitude South America, but China is constantly expanding its acreage in a bid to become a major producer. Soybeans will continue to play a key role in North American and international agriculture as the twenty-first century unfolds, buoyed in no small part by this grain's important potential in the struggle to ease the global food shortage.

metropolises—Kansas City, Minneapolis–St. Paul, Winnipeg, Omaha, and even Denver—are major processing and marketing points for pork and beef packing, flour milling, and soybean, sunflower, and canola oil production (all of which are increasingly exported). Although the family farm still dominates the rural landscape, the aggressive incursion of large-scale corporate farming now threatens that way of life. This is particularly true for hog raising in the western Corn Belt, where big companies are taking control of the entire production process in their ultramodern, factory-like complexes that gobble up huge swaths of farmland and disgorge appalling quantities of animal wastes that pollute air and groundwater for miles around. (A major new area for such mass-scale hog raising rapidly materialized during the late 1990s in the region's southwest corner, centered on the panhandle of far western Oklahoma.) As individual agricultural opportunities decline, especially in the less intensively farmed Great Plains to the west, the unrelenting exodus of younger people leaves behind an ever more elderly population widely dispersed across a constellation of stressed rural communities struggling to survive the depopulation trend.

Despite the key role that agriculture plays in the regional economy, the nonfarming sectors of the Continental Interior continue to develop and diversify, especially in and around major metropolitan centers. Minnesota has been the most successful in promoting such growth, capitalizing on the reputations of that State's stable economy, highly educated workforce, track record as a product innovator, and "north woods" amenities. In the urban areas of the eastern Great Plains, telemarketing has become a major new pursuit. Omaha, Nebraska is known as the "toll-free capital" of the United States because many credit-card, reservations, and catalog-order companies are concentrated there; Gateway 2000, the computer sales giant, has singlehandedly put North Sioux City, South Dakota on the map; and Sprint, a global telecommunications leader, has built its new headquarters just southwest of Kansas City. Moreover, the development of energy resources has spurred the growth of the region's northwest; southern Alberta, most notably, is attracting investments from all over the United States to expand production of its massive deposits of natural gas—today's clean-burning fuel of choice for home heating and electrical power generation.

▶ THE SOUTH

The American South occupies the realm's southeastern corner, extending from the lush Bluegrass Basin of northern Kentucky to the swampy bayous of Louisiana's Gulf Coast, and from the knobby hills of West Virginia to the flat sandy islands off the southernmost extremity of peninsular Florida (Fig. 3-18). Of the realm's nine regions, none has undergone more overall change during the past half-century. Choosing to pursue its own sectional interests practically from the advent of nationhood, the South experienced an even deeper economic and cultural isolation from the rest of the United States during the long and bitter aftermath of the ruinous Civil War. For over a century it languished in economic stagnation, but the 1970s finally witnessed a reassessment of the nation's perception of the region that launched a still-ongoing wave of growth and change unprecedented in Southern history.

Propelled by the forces that created the Sunbelt phenomenon, people and activities streamed into the urban South. Cities such as Atlanta, Charlotte, Miami, and Tampa became booming metropolises practically overnight, and conurbations swiftly formed in such places as southern and central Florida, the Carolina Piedmont, and the western Gulf Coast from Houston (just across the Texas border) through New Orleans to Mobile, Alabama. This "bulldozer revolution" was matched in the most favored rural areas by an agricultural renaissance that stressed such higher-value commodities as soybeans, winter wheat, poultry, and wood products. On the social front, institutionalized racial segregation was dismantled more than three decades ago.

Yet for all the growth that has taken place since 1970, the South remains a region beset by many economic problems because the geography of its development has been decidedly uneven. Although several metropolitan and certain farming areas have benefited, many others containing sizeable populations have not, and the juxtaposition of progress and backwardness is frequently encountered in the Southern landscape. Not surprisingly, the gap between rich and poor is wider here than in any other U.S. region. This economic disparity also carries racial dimensions: compared to whites, blacks have much higher poverty rates and rank well behind in median family income and achievement of middle-class status. Within this checkerboard spatial framework of development, the South's new affluence is largely concentrated in its central-city CBDs and burgeoning suburban rings. At the other end of the income spectrum, its far more widely dispersed struggling areas include less favored agricultural zones (such as the chronically depressed bottomlands that line the lower Mississippi) and myriad rural towns whose only jobs are tied to small-scale plants in locally dying industries such as food processing and garmentmaking. And at the regional scale, yet another geographic pattern of inequality emerges: except for centrally located Atlanta and the Interstate-85 "Boombelt" corridor extending from there northeastward into central North Carolina, a significant proportion of the South's recent economic growth has taken place on its periphery in

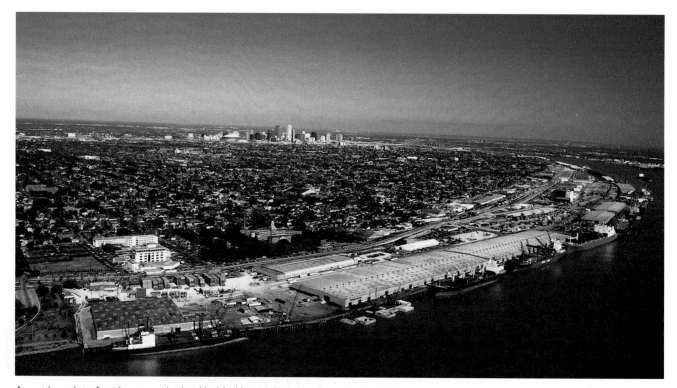

A port is an interface between the land behind it and the water body in front of it, and some of the world's most productive ports lie at the mouths of major rivers—such as Rotterdam at the North Sea outlet of Western Europe's Rhine. New Orleans, located at the apex of the Mississippi Delta, has benefited from such a favored location since the early nineteenth century. Today, in the era of NAFTA, as the products of the U.S. heartland increasingly enter the international trade arena, southern Louisiana senses a heightened opportunity to tap this goods flow by convincing shippers to export by water via the Mississippi. In order to attract this business, huge investments have recently been made to modernize and expand the Port of New Orleans. Some of the results are visible here along this stretch of the great river just above the Crescent City itself, part of a huge new port complex that will extend more than 20 miles (32 km) upstream from downtown New Orleans (seen in the middle distance).

the Washington suburbs of northern Virginia, on the western Gulf Coast, and along Florida's central (Interstate-4) and southern-coast corridors.

The transformation of the South is also changing the complexion of its demographic mosaic (as was demonstrated so dramatically in Florida in the aftermath of the 2000 presidential election). Racially and economically, the composition of the region's population continues to be reshaped by the in-migration of whites and blacks from the North, with both groups exhibiting higher income levels than their resident Southern counterparts. Ethnically, the rapidly increasing presence of Hispanics and Asians is intensifying the South's cultural heterogeneity—and heightening ethnic tensions as low-skilled newcomers increase the competition for a shrinking number of blue-collar jobs. In age structure, too, the region is diversifying as the influx of retirees steadily increases, not just in Florida but also in Georgia, Virginia, and the Carolinas. Finally, we should remember that the massive construction of new urban landscapes since 1980 has provided ample opportunity for creating fragmented residential environments that typify today's mosaic culture. Thus the filling of the tiles of the Southern population mosaic by an ever more diverse array of social and lifestyle groups is another signal that this once outcast region is completing its convergence with the rest of the country.

This 30-year transition period has ended with the South not only fully absorbed into the *national* economy and society, but also becoming an important player on the *international* stage. During the 1990s, European and Japanese companies were active investors in the region, especially the carmakers who built entire manufacturing complexes. In terms of relative location, the most obvious foreign business opportunities to develop in this era of NAFTA lie to the south in Middle and South America. South Florida has been in the vanguard here for a generation, and Miami appropriately bills itself as "the gateway to Latin America." Lately, other parts of the South have indicated an interest in undercutting this monopoly: Atlanta has opened a number of new nonstop air routes, and New Orleans is seeking to enlarge its share of the lucrative seaborne trade by completing the world's highest technology and longest linear port (24 miles/38 km).

And in Memphis plans are proceeding to make this metropolis a global distribution hub, based on the huge airshipment facilities recently built there by Federal Express and United Parcel Service.

THE SOUTHWEST

Before the late twentieth century, most geographers did not even identify a distinct southwestern region, classifying the U.S.-Mexican borderland zone into separate southward extensions of the Continental Interior and the mountain/plateau country to its west. Today, however, this steadily developing area—constituted by the States of Texas, New Mexico, and Arizona—has earned its place on the regional map of North America (Fig. 3-18). The still-emerging Southwest is also unique in the United States because it is a bicultural regional complex. Atop the crest of the Sunbelt wave, in-migrating Anglo-Americans join a rapidly expanding Mexican American population anchored to the sizeable long-time resident Hispanic population, which traces its roots to the Spanish colonial territory that once extended from Texas's Gulf Coast to central California. In fact, when we add the large Native American population, it is not inappropriate to recognize the Southwest as a *tricultural* region.

Recent rapid development in Texas, Arizona, and New Mexico is built on a three-pronged foundation: (1) availability of huge amounts of *electricity* to power air conditioners through the long, brutally hot summers; (2) sufficient *water* to supply everything from swimming pools to irrigated crops, so that large numbers may reside in this dry environment; and (3) the *automobile*, so that

affluent newcomers may spread themselves out at much-desired low densities. The first and third of these have been rather easily attained because the eastern half of the Southwest is abundantly endowed with oil and natural gas (Fig. 3-14). The future of water supplies, however, is far more problematical. For instance, Phoenix and Tucson could not keep growing at their current swift pace without the massive new Central Arizona Project (CAP) Canal, which transports Colorado River water across hundreds of miles of desert.

The economic growth of the Southwest in recent years has been led by Texas, which during the 1990s weaned itself away from dependence on its oil and gas industries (whose share of the State's GNP is now only about one-quarter of the 1980 level). Following a serious oil-induced recession in the 1980s, the Lone Star State's mix of economic activities was restructured and diversified. Success was not long in coming, and today the eastern Texas triangle formed by Dallas–Ft. Worth, Houston, and San Antonio has become one of the world's most productive postindustrial complexes, specializing in business-information and health-care services and high-technology manufacturing. The metropolis centered on the State capital—Austin, which lies near the heart of the triangle—has become a technopole that is second only to Silicon Valley as a leader in computer research and development as well as a producer of semiconductors and PCs. To the north at the triangle's apex, the Dallas-Ft. Worth (DFW) Metroplex is keeping pace with its own expanding cluster of high-tech industries focused on the *Telecom Corridor* technopole just north of Dallas (the world's largest concentration of telecommunications firms). As Figure 3-20 shows, the geography of employment is increasingly oriented to

FIGURE 3-20

booming suburban freeway corridors, a spatial pattern intensifying as well in metropolitan Austin, Houston, and San Antonio. The DFW Metroplex is also the leading Texas command post for international trade and NAFTA-spawned development, and it is becoming the northern anchor of a new transnational growth corridor that extends for more than 500 miles (800 km) southward from Dallas through Austin and San Antonio to Monterrey 150 miles (250 km) inside Mexico.

The western flank of the Southwest—whose development is mostly concentrated in southern Arizona—is unique. Equaled only by parts of southern Israel's Negev, this is the most technologically transformed desert environment on Earth. But while the Negev is devoted to agriculture, Arizona's far more intensive development is marked by vast urbanized landscapes focused on two of the largest western U.S. cities, Phoenix and Tucson. During most of the 1990s, metropolitan Phoenix was expanding into the surrounding desert at the rate of an acre per hour, and it still grows at the astonishing yearly rate of roughly 4 percent. Although the forces of the slow-growth movement are gaining momentum, Arizona's conservative political and business leaders have been loath to restrict development. This debate may soon become moot, however, because everyone agrees that retrievable water supplies—the State's lifeblood and key to its future—are reaching their limit.

THE WESTERN FRONTIER

To the north of the western half of the Southwest lies the realm's newest region: the swiftly growing, economically booming Western Frontier. In its east-west extent, this mountain-studded plateau region stretches between (and includes) the Rocky Mountains and the Sierra Nevada–Cascades chain; its longer north-south axis reaches from northern Arizona's Grand Canyon country to the edge of the subarctic snowforest west of the Canadian Rockies (Fig. 3-18). The region's heartland encompasses Utah, Nevada, Idaho, and western Colorado, and it is here that the old intermountain West is most actively reinventing itself. Until recently, such an opportunity would have been unthinkable because of this region's remoteness, dryness, and sparse population. But the Western Frontier's geographic prospects are changing significantly as the postindustrial national economy matures, advances in communications and transportation technologies eliminate interregional differences in the costs of doing business, and amenities become the leading locational preference of employers. Undoubtedly, proximity to the increasingly expensive and overcrowded Pacific coast is also a reinforcing factor: well over 1 million disenchanted Californians have moved into the Western Frontier since 1990, lured by its sunny climate, wide-open spaces, spectacular scenery, proliferating jobs, and lower cost of living.

The label *Western Frontier* further emphasizes that the transformation under way is being driven by the influx of people and activities from outside the region. Indeed, the populations of the four States of its heartland all rank among the five fastest-growing in the United States since 1990, with the more than 2 million migrants who arrived during the past decade already constituting more than 20 percent of the Western Frontier's population. In the economic-geographic sphere, the region's identity is being reforged as traditional resource-based industries (mining, timber, livestock grazing) decline, while in-migrating and new indigenous companies create thousands of high-tech manufacturing and specialized services jobs. The spatial pattern of this development is strongly focused on the region's mushrooming urban areas and creates an unusual frontier: instead of a single line of advancing settlement, a dispersed set of population clusters is simultaneously expanding around large- and middle-sized cities, high-amenity zones, and interstate freeway corridors.

The hottest growth area in all of the Western Frontier—and one of the fastest-growing metropolises in the United States—is the Las Vegas Valley of southernmost Nevada. This boom was triggered by Las Vegas's burgeoning recreation industry (which attracts some 30 million tourists annually), but it is also the product of new economic development dominated by the influx of high-tech and professional services firms that specialize in computers, finance, and health care. Business entrepreneurs have flocked here, and the new employment opportunities sparked a local population explosion, spearheaded by those relocating Californians; relatively cheap land, modest housing prices, low taxes, and the sunny climate further fueled this migration.

Other urban areas are also experiencing accelerated growth, particularly the region's two largest metropolises, centered on Denver and Salt Lake City (which have spawned conurbations extending more than 50 miles north and south of the central city). In both cases the leading amenity is the Rocky Mountains, which soar majestically above the urbanizing flatlands and foothills. Metropolitan Denver has expanded rapidly along (and into) central Colorado's Front Range, building a diverse new, globally oriented economic base dominated by telecommunications, computer software, and financial services. On a somewhat smaller scale, Salt Lake City has followed in Denver's footsteps on the Rockies' western flank and now centers a thriving conurbation that

There is something quite remarkable about the ongoing urbanization of arid valleys in the high desert of the Western Frontier, and nowhere is this phenomenon more starkly on view than in the Las Vegas Valley in Nevada's southern-most extremity. This particular municipality is Henderson, not only Las Vegas's largest sub-urb but also the fastest-growing urban settle-ment in the entire United States between 1990 and 2000. Many affluent retirees have also been attracted to this immediate area, and to-gether with their younger counterparts have shaped a classic mosaic-culture urban fabric dominated by a low-density checkerboard of self-contained, socially homogeneous com-munities. Architectural critics and environmen-talists, of course, would have no trouble in de-claring the filling of this landscape to be one of the most blatant examples of urban sprawl on view anywhere in North America.

lines northern Utah's Wasatch Front from Ogden in the north to Provo in the south. In fact, the budding techno-pole of *Software Valley*, the 40-mile (65-km) corridor linking Salt Lake City and Provo, contains one of the world's largest concentrations of information technology companies, built up around two universities, a highly edu-cated workforce, and a postindustrial complex specializ-ing in the production of word-processing and computer-simulation software.

As much as 90 percent of the Western Frontier's development is confined to its widely dispersed urban areas, but the effects of this rapid growth are rippling deeply into the region's vast rural zones. Some of the problems—such as the steadily rising incidence of wildfires as zones bordering forests become ever more populated—are caused by the spillover effects of urban-ization on adjacent areas. But many others result from the collision of lifestyles as higher-income, recreation-seeking city dwellers invade a countryside whose long-time, nonaffluent residents are already struggling to pre-serve a way of life based on declining activities such as mining and extensive agriculture. In the valleys of the Col-orado Rockies, ranchers are further squeezed by the up-wardly spiralling land values set off by the massive growth of ski resorts—and usually are forced to sell and relocate their operations to less favorable land in even more remote

areas. In the small towns of northern Idaho and western Montana, social tensions are inflamed as local right-wing militias increasingly encounter more liberal transplanted Californians. And seemingly everywhere, the plans of developers are blocked by environmentalists—who in the 1990s often received the support of the federal gov-ernment (the region's biggest landowner). The surpris-ingly rapid emergence of a new preservation ethic, which increasingly threatens environmentally insensitive land and resource developers, is part of a larger anti-growth movement that is certain to intensify. Throughout this re-gion today, the buzzwords are "managing growth without restricting economic opportunity"—which usually trans-lates into bringing opposing interests together to work out compromises.

Whatever our new century brings, the growth boom of the 1990s that created the realm's newest region has forever altered the economic-geographic and demographic com-plexion of much of the old intermountain West. Now re-born as the Western Frontier, this no-longer-remote region is putting the often disastrous boom/bust cycles of its min-ing- and grazing-dominated past behind it and is making major strides toward establishing itself as a globally con-nected, technological powerhouse on the inland doorstep of North America's burgeoning sector of the Pacific Rim (see box titled "The Pacific Rim Connection," p. 181).

THE NORTHERN FRONTIER

The northern half of the realm, lying poleward of roughly 52°N latitude, constitutes the Northern Frontier region (Fig. 3-18). This is by far North America's biggest regional subdivision, covering almost 90 percent of Canada and all of the U.S. State of Alaska. Indeed, if this huge territory were a separate country, it would rank as the world's fifth-largest in size. The latitudinal extent of this region leaves no doubt as to why it is designated as *northern*. In fact, after northern Russia (whose dimensions of continentality are even more pronounced), this is our planet's second greatest expanse of harsh cold environment affecting an inhabited area. As with the Arctic fringes of Eurasia, very few physical barriers in northern North America ameliorate the masses of supercold air, which throughout the long winter plunge directly southward from the ice-covered Arctic Ocean. Not surprisingly, most of the Northern Frontier's people and activities are concentrated in the milder, heavily forested southern half of the region.

Unquestionably, this region is also a *frontier*—and far more classically so than the Western Frontier just discussed. The latter's newly activated, dispersed, urbanized development pattern and dramatic population influx are largely attributable to its improved economic-geographic position astride a national axis, within which significant east-west interaction was already taking place. The Northern Frontier, on the other hand, represents a thrust into undeveloped country in an entirely new direction, one marked by the leisurely, long-term advance of a line across parts of Canada's near- and middle-northern periphery that absorbs the areas it crosses into the national spatial economy. Boom and bust cycles since the late nineteenth century have controlled the rate of movement of this development line, which frequently has ebbed backward as much as it has flowed forward. In general, the still sparsely populated Northern Frontier is mostly a region of recent European settlement based on the extraction of newly discovered resources. As we shall see, however, First Nations and other indigenous peoples, who have always regarded this region as a homeland, are today playing a growing role in the affairs of the Northern Frontier and will have much to say about shaping its future.

As noted earlier, the Canadian Shield, which covers the eastern half of the Northern Frontier (Fig. 3-2), is one of the world's richest storehouses of mineral resources. Most of the Shield's developing areas are focused on mining complexes that extract such high-demand metallic ores as nickel, uranium, copper, gold, silver, lead, and zinc. The leading zones of production are located along the Shield's southern margins, but since 1980 the search for additional deposits, the expansion of road and rail networks, and the opening of new mines have all advanced steadily northward. The Yukon and Northwest Territories have proven bountiful, with gold- and diamond-mining paving the way; and at the opposite, eastern edge of the Shield, near Voisey's Bay on the central Labrador coast of Newfoundland and Labrador, the largest body of high-grade nickel ores ever discovered is being opened up for full-scale production. Besides metallic minerals, the Northern Frontier is endowed with an abundance of other resources (which include fossil fuels, hydroelectric power opportunities, timber, and fisheries). These are dominated by substantial oil and gas reserves that are concentrated in a wide zone east of the Canadian Rockies between the Yukon and the U.S. border (Fig. 3-14). Most intriguing are the Athabasca Tar Sands of northeastern Alberta, which contain more crude oil than all of Saudi Arabia but deposited as a tar-like substance (bitumen) in near-surface sand formations. Recent technological breakthroughs are enabling this energy supply to be extracted ever more profitably, output is steadily rising, and by 2010 the Tar Sands are expected to become Canada's dominant oil-producing zone.

The productive activities of the Northern Frontier are linked together within a far-flung network of mines, oil- and gasfields, pulp mills, and hydropower stations that have spawned hundreds of settlements and thousands of miles of interconnecting transport and communications facilities. Although dominant, this commercial economy constitutes just one component of a dual regional economy. The other component, as Robert Bone points out in his book *The Geography of the Canadian North*, consists of the native economy that "evolved from a subsistence hunting economy to one in which wage employment, transfer payments, and trapping provide cash income while hunting provides country food." This native economy, which adheres to the traditional values of the indigenous peoples (First Nations), has long been tied to the commercial economy as a source of wages. It has also been heavily land-based, and in recent years First Nations throughout the region have become more assertive and filed legal actions to recover lands that Europeans took in the past without negotiating treaties. Both the courts and the federal government have been sympathetic, and the creation of Nunavut in 1999 (see map, p. 160) has been the most dramatic breakthrough to date.

The Northern Frontier also contains Alaska, whose geographic opportunities and challenges differ somewhat from those of the Canadian North. Besides being the largest U.S. State, Alaska earns a sizeable income from oil production on its northeastern Arctic slope, and the south-central coast around Anchorage increasingly benefits from its near-midpoint location on the Pacific Rim air route between the "Lower 48" States and East Asia. These advantages have propelled a growth spurt of more than 100 percent since 1970, and today Alaska's total of just over 650,000 accounts for 30 percent of the Northern Frontier's population on less than one-sixth of its land area. Here, too, metallic minerals have long drawn migrants northward. But, unlike Canada to the east, Alaska's recent boom was triggered by the opening of its vast oilfields on the North Slope, one of the hemisphere's leading energy sources since the 800-mile Trans-Alaska Pipeline began operating in 1977 (Fig. 3-14). Dwindling supplies at the main field at Prudhoe Bay are now forcing producers to turn to the huge additional reserves of the Arctic slope, but strong opposition from nature preservationist groups has slowed their plans. To the east lies the Arctic National Wildlife Refuge, where the Bush administration wants to begin drilling. To the west lies the National Petroleum Reserve-Alaska, which the federal government partially opened in 1999 to oil leasing by drillers, who insist that their newly gained sensitivity paired with the latest technologies will minimize the impact of production on the tundra's fragile ecosystems. Another major project that faces a similar battle is the proposal to construct a natural-gas pipeline to parallel the oil pipeline (huge supplies of liquified gas are a byproduct of oil drilling). As noted earlier, demand for clean-burning natural gas by electricity producers is rising all across North America.

◗ THE PACIFIC HINGE

The Pacific coastlands of the conterminous United States and southwesternmost Canada, which comprise the realm's ninth region (Fig. 3-18), have been a powerful lure to migrants since the Oregon Trail was pioneered more than 150 years ago. Unlike the remainder of western North America south of 50°N latitude, the strip of land between the Sierra Nevada—Cascade mountain wall and the sea receives adequate moisture. It also possesses a far more hospitable environment, with generally delightful weather south of San Francisco, highly productive farmlands in California's Central Valley, and such scenic glories as the Big Sur coast, Washington State's Olympic Mountains, and the spectacular waters surrounding San Francisco, San Diego, Seattle, and Vancouver. Most of the major development here took place during the post–World War II era, accommodating enormous population and economic growth along this outermost edge of the conterminous United States. But with the arrival of the twenty-first century, it is clear that, in terms of its economic geography, the West Coast no longer represents an end but rather a beginning—a gateway to an abundance of growing opportunities that in recent times have blossomed on many of the distant shores encircling the Pacific Basin. That is why we use the term *hinge*, because now this region increasingly forms an interface between North America and the booming, still-emerging *Pacific Rim* (see box titled "The Pacific Rim Connection").

Fifty-plus years of unrelenting growth have taken their toll, and whereas the West Coast States and British Columbia are now savoring the prospects of their Pacific Rim location, they must also confront the less pleasant consequences of regional maturity. This is especially true in California, where the massive development of America's most populated and multiethnic State (36 million in 2004, 14 percent larger than all of Canada) has been overwhelmingly concentrated in the teeming conurbation extending south from San Francisco through San Jose, the San Joaquin Valley, the Los Angeles Basin, and the southwestern coast into San Diego at the Mexican border. Environmental hazards bedevil this entire corridor, including inland droughts, coastal-zone flooding, mudslides, brushfires, and particularly earthquakes—with the ominous San Andreas Fault practically the axis of megalopolitan coalescence. To all this, humans have added their own abuses of California's fragile habitat, from overuse of water supplies (requiring vast aqueduct systems to import water from hundreds of miles away) to the chronic air pollution of the Los Angeles-San Diego coastal strip.

The Pacific Rim Connection

$\mathcal{T}$oday we continue to witness the rise of a far-flung region born of a string of economic miracles on the shores of the Pacific Ocean. It is a still-discontinuous region, led by its foci on the Pacific's western margins: Japan, coastal China, South Korea, Taiwan, Thailand, Malaysia, Singapore. The regional term **Pacific Rim** has come into use to describe this dramatic development, which over the past quarter-century has redrawn the map of the Pacific periphery—not only in East and Southeast Asia but also in the United States and Canada, and even in such Southern Hemisphere locales as Australia and South America's Chile.

The Pacific Rim also is a superb example of a functional region, with economic activity in the form of capital flows, raw-material movements, and trade linkages generating urbanization, industrialization, and labor migration. In the process, human landscapes from Sydney to Santiago are being transformed within the 20,000-mile (32,000-km) corridor that girdles the globe's largest body of water. We will continue to discuss this important regional development and its spatial impacts, in later chapters, wherever the Pacific Rim intersects Pacific-bordering geographic realms.

Research by physical geographers indicates that the western third of the conterminous United States has the country's highest air pollution potential. Urban areas such as Los Angeles and Denver became notorious in the late twentieth century for their frequent smog outbreaks, although both have dramatically improved local air quality since the 1980s. Today, among the worst offenders are Seattle and San Diego, which grew explosively over the past two decades. In this photo Seattle's landmark Space Needle immediately gives away the city's identity, but the nearby skyline of its CBD is shrouded by the murky, late-afternoon smog. Nonetheless, on days when brisk winds blow onshore from Puget Sound to the right, the veil of smog is (temporarily) flushed away and the famous snowcapped cone of Mount Rainier high atop the Cascades, almost 60 miles (100 km) southeast of here, can still be glimpsed through the remaining haze.

These challenges notwithstanding, Californians over the past half-century have built one of the realm's most productive economic machines. Indeed, if the Golden State were an independent country, its economy would today rank as the world's sixth largest. The road to success, however, has often proven to be a bumpy one, as was the case in Southern California during the 1990s. This subregion began the decade struggling as the national recession combined with the unexpected end of the Cold War to deliver a serious blow to the Los Angeles-area economy, which at the time was heavily dependent on its aerospace industry driven by federal government defense expenditures. By the late 1990s, a turnaround was underway as Southern California successfully diversified its economy and developed such new growth sectors as the entertainment industry, foreign trade, and the nation's largest single metropolitan manufacturing complex—whose products range from high-tech medical equipment to the clothing churned out by the low-tech factories of a burgeoning, immigrant-staffed garment industry that continues to outpace New York City's.

The northern portion of the region encompasses Northern California and the Pacific Northwest, the latter extending northward from Oregon's Willamette Valley through the Cowlitz-Puget Sound lowland of western Washington State into the adjoining coastal zone of British Columbia beyond the Canadian border. During the 1990s Northern California, focused on the San Francisco Bay Area, experienced a much smoother path than the southern part of the State. Here an unprecedented period of economic expansion continues, led by Silicon Valley. Within that technopole, no community stands to gain more from this boom

than Palo Alto, which is situated at the largest switching point on the Internet that handles more than half of its global traffic. As Internet commerce grows explosively, companies are flocking to Palo Alto to take advantage of its unparalleled access to cyberspace. This is further facilitated by the city's state-of-the-art local fiber-optic network. What is taking shape here is nothing less than an infrastructural hub that will become one of the cornerstones of the so-called New Economy.

Continuing up the coast, the Pacific Northwest is also deeply involved in the pioneering and production of computer technology, especially in the technopole centered on suburban Seattle's Redmond (Microsoft's headquarters). Originally built on timber and fishing—primary activities that still survive here despite human pressures on the environment—this subregion found its impetus for industrialization in the massive Columbia River dam projects of the 1930s and 1950s, which generated the cheap hydroelectricity that attracted aluminum and aircraft manufacturers. Although Boeing is still a major employer, the recent influx of affluent newcomers is fueling the growth of software and Internet companies that is spearheading Seattle's transformation into the prototype metropolis of the New Economy. Nonetheless, what appear to be short-term economic problems have arisen in the early twenty-first century as a result of the downturn in the global computer and airline industries.

North of the Canadian border, the Pacific Hinge's northernmost subregion stands out from the others not only because of its British Columbia location but also because its economic core, Vancouver, has become the most Asianized metropolis in North America. Vancouver

AMONG THE REALM'S GREAT CITIES . . . **LOS ANGELES**

*I*t is hard to think of another city that has been in more headlines in the past several years than Los Angeles. Much of this publicity has been negative. But despite its earthquakes, mudslides, brushfires, pollution alerts, showcase trials, and riots, L.A.'s glamorous image has hardly been dented. This is compelling testimony to the city's resilience and vibrancy, and to its unchallenged position as the Western world's entertainment capital.

The plane-window view during the descent into Los Angeles International Airport, almost always across the heart of the metropolis, gives a good feel for the immensity of this urban landscape. It not only fills the huge natural amphitheater known as the Los Angeles Basin, but it also oozes into adjoining coastal strips, mountain-fringed valleys and foothills, and even the margins of the Mojave Desert more than 50 miles (80 km) inland. This quintessential spread city, of course, could only have materialized in the automobile age. In fact, most of it was built rapidly over the past 60 years, propelled by the swift expansion of a high-speed freeway network unsurpassed anywhere in metropolitan America. In the process, Greater Los Angeles became so widely dispersed that today it has reorganized within a geographic framework of six sprawling urban realms (see Fig. 3-12).

The metropolis as a whole is home to 17 million Southern Californians, constitutes North America's second largest urban agglomeration, and forms the southern anchor of the huge California conurbation that lines the Pacific coast from

San Francisco southward to the Mexican border (Fig. 3-9). It also is the Pacific Hinge's leading trade, manufacturing, and financial center. In the global arena, Los Angeles is the eastern Pacific Rim's biggest city and the origin of the largest number of transoceanic flights and sailings to Asia.

---

lies closer to East Asia on the air and sea routes than the cities to its south, and it got a head start in forging trade and investment linkages to the western Pacific Rim thanks to its large and growing Asian population. Today, ethnic Chinese residents constitute fully 20 percent of the metropolitan population, whose total Asian component now exceeds a remarkable 35 percent. Although a new east-west hybrid culture is trying to establish itself, frictions intensified during the late 1990s as wealthy Asians (many of them expatriates of now-Chinese Hong Kong) took over neighborhoods and replaced traditional homes with much larger houses, surrounded them with walls and fences where none had existed, and cut down most of the trees. More serious is the growing belief by the majority Anglos that the newcomers do not really want to assimilate, preferring instead to avoid the use of English and live, shop, and transact business exclusively within their own communities.

With North America ever more tightly enmeshed in the global economy, nowhere are the realm's international linkages more apparent today than in the Pacific Hinge. From the northern end of this region to the Mexican border,

its geographic advantages as a gateway to the Pacific Rim are providing opportunities undreamed of barely a decade ago. Asianizing British Columbia now exports more than 40 percent of its goods to the countries of the western Pacific Rim (a proportion far higher than in any other Canadian province or U.S. State). The rest of the Vancouver-Seattle-Portland corridor is following suit, led by its high-technology industries. In California, the foreign trade sector of the State's gigantic economy has tripled since 1990, and today more than 15 percent of its jobs are export-dependent. Both Northern and Southern California are expected to thrive as these trends continue, with metropolitan Los Angeles becoming the financial, manufacturing, and trading capital of the eastern Pacific Rim as well as its transport hub (L.A. overtook New York in the mid-1990s as the leading U.S. seaport in the value of goods passing through it). Thus it is abundantly clear today that the mid-latitude West Coast no longer serves as North America's back door, but forms the hinges of a new front door to the Pacific arena thrown wide open to foster international interactions of every kind.

FIGURE 4-1

Scale 1:16 000 000; one inch to 250 miles. Polyconic Projection
Elevations and depressions are given in feet

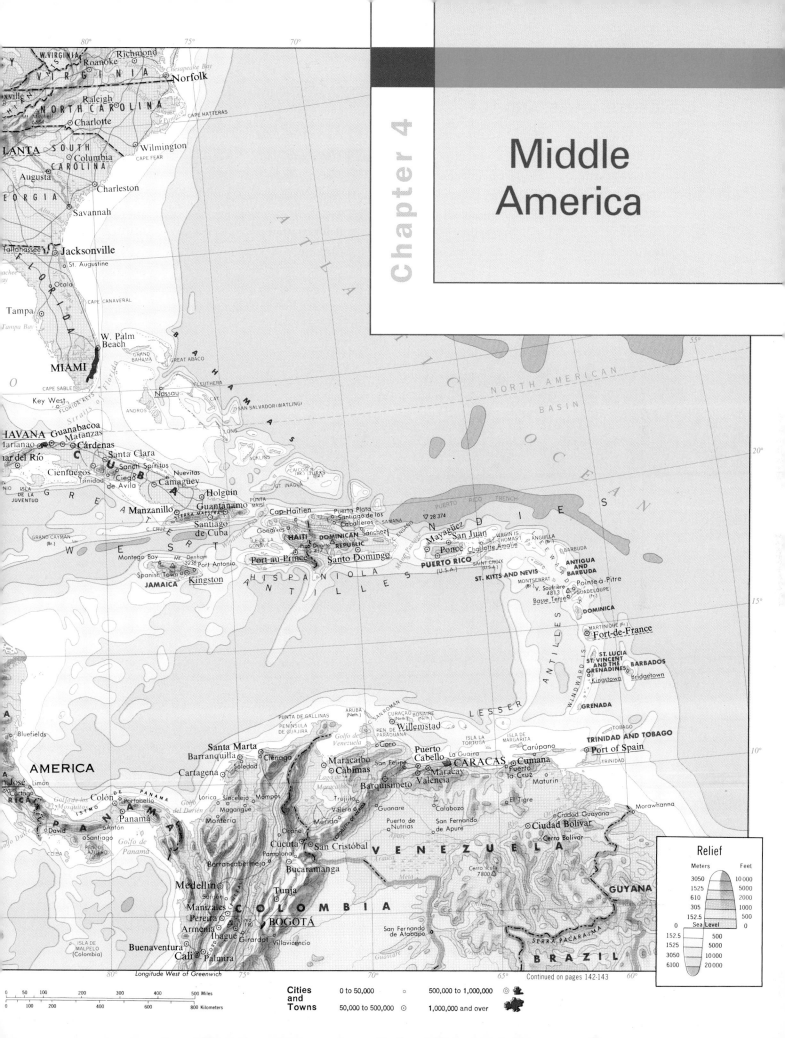

Chapter 4

Middle America

Continued on pages 142-143

CONCEPTS, IDEAS, AND TERMS

1. Land bridge
2. Archipelago
3. Culture hearth
4. Mainland-Rimland framework
5. Mestizo
6. Hacienda
7. Plantation
8. Acculturation
9. Transculturation
10. Maquiladora
11. Dry canal
12. Altitudinal zonation
13. *Tierra caliente*
14. *Tierra templada*
15. *Tierra fría*
16. *Tierra helada*
17. *Tierra nevada*
18. Tropical deforestation
19. Mulatto

REGIONS

MEXICO

CENTRAL AMERICA

CARIBBEAN BASIN

THE GREATER ANTILLES

THE LESSER ANTILLES

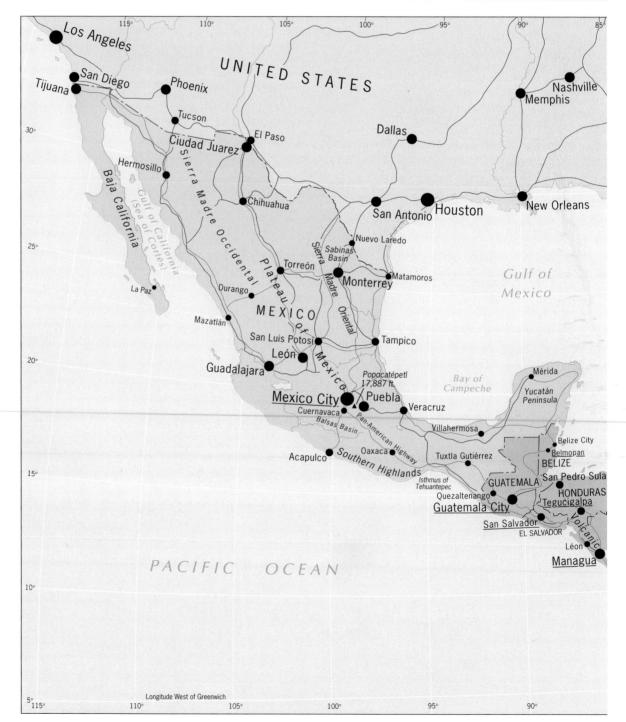

IDDLE AMERICA IS a realm of vivid contrasts, turbulent history, continuing political turmoil, and an uncertain future. The realm's diversity, visible in Figure 4-1, encompasses all the lands and islands between the United States to the north and South America to the south. Its regions are mapped in Figure 4-2 and include: (1) the substantial landmass of *Mexico*; (2) the narrowing strip of land to its southeast that constitutes *Central America*; and (3) the many large and small islands—respectively known as the *Greater Antilles* and *Lesser Antilles*—of the Caribbean Sea to the east.

Middle America is a realm of soaring volcanoes and forested plains, of mountainous islands and flat coral cays. Moist tropical winds sweep in from the east, watering windward (wind-facing) coasts while leaving leeward (wind-protected) areas dry. Soils vary from fertile volcanic to desert barren. Spectacular scenery abounds, and tourism is one of the realm's leading industries.

FIGURE 4-2

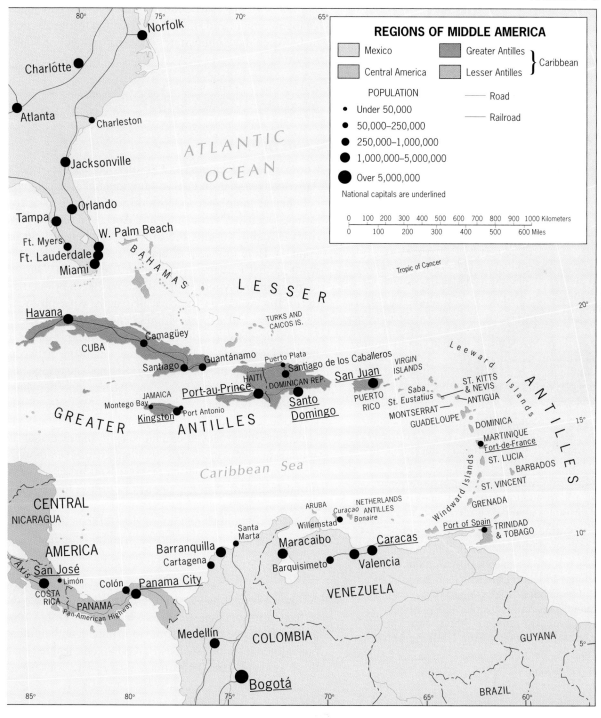

DEFINING THE REALM

Is MIDDLE AMERICA a discrete geographic realm? Some geographers combine Middle and South America into a realm they call "Latin" America, citing the dominant Spanish (and in Brazil, Portuguese) heritage and the prevalence of Roman Catholicism. But these criteria apply more strongly to South America. In Middle America, large populations exhibit African and Asian as well as European ancestries. And nowhere in South America has native, Amerindian culture contributed to modern civilization as strongly as it has in Mexico. The Caribbean Basin is a patchwork of independent states, territories in political transition, and residual colonial dependencies. The Dominican Republic speaks Spanish, adjacent Haiti uses French, Dutch is spoken in Aruba, while English is spoken in Jamaica. Middle America thus gives vivid definition to concepts of cultural-geographical pluralism.

Middle America is occasionally called *Central America*, but that name actually refers to a region within the Middle American realm (Fig. 4-2). Central America comprises the republics that occupy the strip of mainland between Mexico and Panama: Guatemala, Belize, Honduras, El Salvador, Nicaragua, and Costa Rica. Although Panama itself is regarded here as belonging to Central America, it should be noted that many Central Americans still do not consider Panama to be part of their region because for most of its history that country was a part of South America's Colombia. Again, as we define it, *Middle America* includes all the mainland and island countries and territories that lie between the United States and the continent of South America.

PHYSIOGRAPHY

Land Bridge

Compared to the landmass of South America, the Middle American realm is divided and fragmented. Even its funnel-shaped mainland, a 3800-mile (6000-km) connection between the North and South American realms, narrows to a slim 40-mile (65-km)-wide ribbon of land in Panama. Here this strip of land, or *isthmus*, bends eastward so that Panama's orientation is east-west. On the map, mainland Middle America looks like a bridge between the Americas, and this is exactly what physical geographers call such an isthmian link: a **land bridge**. ▪

If you examine a globe, you can see other present and former land bridges: Egypt's Sinai Peninsula between Asia and Africa, the (now-broken) Bering land bridge between northeasternmost Asia and Alaska, and the shallow waters between New Guinea and Australia. Such land bridges, though temporary surface features in geologic time, have played crucial roles in the dispersal of animals and humans across the planet. But even though mainland Middle America forms a land bridge, its internal fragmentation has always inhibited movement. Mountain ranges, swampy coastlands, and dense rainforests make contact and interaction difficult, especially where they all combine in most of Central America.

Island Chains

As shown in Figure 4-2, the approximately 7000 islands of the Caribbean Sea stretch in a lengthy arc from Cuba and the Bahamas east and south to Trinidad, with numerous outliers outside (such as Barbados) and inside (e.g., the Cayman Islands) the main chain. The four large

MAJOR GEOGRAPHIC QUALITIES OF

Middle America

1. Middle America is a fragmented realm that consists of all the mainland countries from Mexico to Panama and all the islands of the Caribbean Basin to the east.

2. Middle America's mainland constitutes a crucial barrier between Atlantic and Pacific waters. In physiographic terms, this is a land bridge that connects the continental landmasses of North and South America.

3. Middle America is a realm of intense cultural and political fragmentation. The political geography defies unification efforts, but countries and regions are beginning to work together to solve mutual problems.

4. Middle America's cultural geography is complex. African influences dominate the Caribbean, whereas Spanish and Amerindian traditions survive on the mainland.

5. The realm contains the Americas' least-developed territories. New economic opportunities may help alleviate Middle America's endemic poverty.

6. In terms of area, population, and economic potential, Mexico dominates the realm.

7. Mexico is reforming its economy and has experienced major industrial growth. Its hopes for continuing this development are tied to overcoming its remaining economic problems and to expanding trade with the United States and Canada under the North American Free Trade Agreement (NAFTA).

In terms of natural hazards, Middle America is the most dangerous realm of all. Hurricanes, volcanic eruptions, earthquakes, and landslides threaten virtually every corner of this part of the world. In January 2001, two of these hazards lethally combined to devastate the Las Colinas neighborhood of Santa Tecla, a suburb of the city of San Salvador, capital of the Central American country of El Salvador. A severe earthquake (magnitude 7.6 on the Richter Scale) rocked much of that country, causing only moderate damage overall, but a few places took the brunt of the seismic forces unleashed by the temblor. One was the geologically unstable hillside above Las Colinas, which in a matter of seconds produced this massive landslide that buried hundreds of homes and claimed more than half of the 687 Salvadorians who perished in the quake that day.

islands—Cuba, Hispaniola (containing Haiti and the Dominican Republic), Puerto Rico, and Jamaica—are called the *Greater Antilles*. All the remaining smaller islands are called the *Lesser Antilles*.

Once again, the map hints at the physiography. The entire Antillean **archipelago** (island chain) consists of the crests and tops of mountain chains that rise from the floor of the Caribbean, the result of collisions between the Caribbean Plate and its neighbors (Fig. G-4). Some of these crests are relatively stable, but elsewhere they contain active volcanoes, and almost everywhere in this realm, earthquakes are an ever-present danger—in the islands as well as on the mainland. Add the seasonal risk of hurricanes spawned and nurtured by the realm's abundant tropical waters, and the Middle American environment easily ranks among the world's most hazardous.

LEGACY OF MESOAMERICA

Mainland Middle America was the scene of the emergence of a major ancient civilization. Here lay one of the world's true **culture hearths** (see Fig. 7-3), a source area from which new ideas radiated and whose population could expand and make significant material and intellectual progress. Agricultural specialization, urbanization, and transport networks developed, and writing, science, art, and other spheres of achievement saw major advances. Anthropologists refer to the Middle American culture hearth as *Mesoamerica*, which extended southeast from the vicinity of present-day Mexico City to central Nicaragua. Its development is especially remarkable because it occurred in

very different geographic environments, each presenting obstacles that had to be overcome in order to unify and integrate large areas. First, in the low-lying tropical plains of what is now northern Guatemala, Belize, and Mexico's Yucatán Peninsula, and perhaps simultaneously in Guatemala's highlands to the south, the Maya civilization arose more than 3000 years ago. Later, far to the northwest on the high plateau in central Mexico, the Aztecs founded a major civilization centered on the largest city ever to exist in pre-Columbian times.

The Lowland Maya

The Maya civilization is the only one in the world that ever arose in the lowland tropics. Its great cities, with their stone pyramids and massive temples, still yield archeological information today. Maya culture reached its zenith from the third to the tenth centuries AD.

The Maya civilization, forged by a series of city-states, unified an area larger than any of the modern Middle American countries except Mexico. Its population probably totaled between 2 and 3 million; certain Maya languages are used in the area to this day. The Maya city-states were marked by dynastic rule that functioned alongside a powerful religious hierarchy, and the great cities that now lie in ruins were primarily ceremonial centers. We also know that Maya culture produced skilled artists and scientists, and that these people achieved a great deal in agriculture and trade. They grew cotton, created a rudimentary textile industry, and exported cotton cloth by seagoing canoes to other parts of Middle America in return for valuable raw materials.

"We spent Monday and Tuesday upriver at Lamanai, a huge, still mostly overgrown Maya site deep in the forest of Belize. On Wednesday we drove from Belize City to Altun Ha, which represents a very different picture. Settled around 200 B.C., Altun Ha flourished as a Classic Period center between A.D. 300 and 900, when it was a thriving trade and redistribution center for the Caribbean merchant canoe traffic and served as an entrepôt for the interior land trails, some of them leading all the way to Teotihuacán. Altun Ha has an area of about 2.5 square miles (6+ sq km), with the main structures, one of which is shown here, arranged around two plazas at its core. I climbed to the top of this one to get a perspective, and sat down to have my sandwich lunch, imagining what this place must have looked like as a bustling trade and ceremonial center when the Roman Empire still thrived, but a more urgent matter intruded. A five-inch tarantula emerged from a wide crack in the sun-baked platform, and I noticed it only when it was about two feet away, apparently attracted by the crumbs and a small piece of salami. A somewhat hurried departure put an end to my historical-geographical ruminations."

The Highland Aztecs

In what is today the intermontane highland zone of Mexico, significant cultural developments were also taking place. Here, just north of present-day Mexico City, lay Teotihuacán, the first true urban center in the Western Hemisphere, which prospered for nearly seven centuries after its founding around the beginning of the Christian era. Later the Toltecs migrated into this area from the north, conquered and absorbed the local Amerindian peoples, and formed a powerful state. When, about 300 years later, the Toltec state itself was penetrated by new elements from the north, it was already in decay; but its technology was readily adopted and developed by the conquering Aztecs.

The Aztec state, the pinnacle of organization and power in pre-Columbian Middle America, is thought to have originated in the early fourteenth century with the founding of a settlement on an island in a lake that lay in the *Valley of Mexico* (the area surrounding what is now Mexico City). This urban complex, a functioning city as well as a ceremonial center, named Tenochtitlán, was soon to become the greatest city in the Americas and the capital of a large, powerful state. The Aztecs soon gained control over the entire Valley of Mexico, a pivotal 30-by-40-mile (50-by-65-km) mountain-encircled basin that is still the heart of the modern state of Mexico. Both elevation and interior location moderate its climate; for a tropical area, it is quite dry and very cool. The basin's lakes formed a valuable means of internal communication, and the Aztecs built canals to connect several of them. This fostered a busy canoe traffic, bringing agricultural produce to the cities, and tribute was paid by their many subjects to the headquarters of the ruling nobility.

With its heartland consolidated, the new Aztec state soon began to conquer territories to the east and south. The Aztecs' expansion of their empire was driven by their desire to subjugate peoples and towns in order to extract taxes and tribute. As Aztec influence spread throughout Middle America, the state grew ever richer, its population mushroomed, and its cities thrived and expanded.

The Aztecs produced a wide range of impressive accomplishments, although they were better borrowers and refiners than they were innovators. They developed irrigation systems, and they built elaborate walls to terrace slopes where soil erosion threatened. Indeed, the greatest contributions of Mesoamerica's Amerindians surely came from the agricultural sphere. Corn (maize), the sweet potato, various kinds of beans, the tomato, squash, cacao beans (the raw material of chocolate), and tobacco are just a few of the crops that grew in Mesoamerica when the Europeans first made contact.

COLLISION OF CULTURES

We in the Western world all too often believe that history began when the Europeans arrived in some area of the world and that the Europeans brought such superior power to the other continents that whatever existed there previously had little significance. Middle America confirms this misperception: the great, feared Aztec state apparently fell before a relatively small band of Spanish invaders in an incredibly short period of time (1519–1521). But let us not lose sight of a few facts. At first, the Aztecs believed the Spaniards were "White Gods" whose arrival had been predicted by Aztec prophecy. Hernán Cortés, for all his 508 soldiers, did not singlehandedly overthrow this powerful empire: he ignited a rebellion by Amerindian peoples who had fallen under Aztec domination and had seen their relatives carried off for human sacrifice to Aztec gods. Led by Cortés with his horses and guns, these peoples rose against their Aztec oppressors and joined the band of Spaniards headed toward Tenochtitlán (where thousands of them would die in combat against Aztec defenders).

Effects of the Conquest

Spain's defeat of Middle America's dominant indigenous state opened the door to Spanish penetration and supremacy. Throughout the realm, the confrontation between Hispanic and native cultures spelled disaster for the Amerindians: a catastrophic decline in population (perhaps as high as 90 percent), rapid deforestation, pressure on vegetation from grazing animals, substitution of Spanish wheat for maize (corn) on cropland, and the concentration of Amerindians into newly built towns.

The Spaniards were ruthless colonizers but not more so than other European powers that subjugated other cultures. True, the Spaniards first enslaved the Amerindians and were determined to destroy the strength of indigenous society. But biology accomplished what ruthlessness could not have achieved in so short a time: diseases introduced by the Spaniards and slaves imported from Africa decimated millions of Amerindians.

Middle America's cultural landscape—its great cities, its terraced fields, its dispersed aboriginal villages—was thus drastically modified. Unlike the Amerindians, who had used stone as their main building material, the Spaniards employed great quantities of wood and used charcoal for heating, cooking, and smelting metal. The onslaught on the forests was immediate, and rings of deforestation swiftly expanded around the colonizers' towns. The Spaniards also introduced large numbers of cattle and sheep, and people and livestock now had to compete for available food (requiring the opening of vast areas of marginal land that further disrupted the region's food-production balance). Moreover, the Spaniards introduced their own crops (notably wheat) and farming equipment, and soon large fields of wheat began to encroach upon the small plots of corn that the natives cultivated.

The Spaniards' most far-reaching cultural changes derived from their traditions as town dwellers. To facilitate domination, the Amerindians were moved off their land into nucleated villages and towns that the Spaniards established and laid out. In these settlements, the Spaniards could exercise the kind of rule and administration to which they were accustomed (Fig. 4-3). The internal focus of each Spanish town was the central *plaza* or market square, around which both the local church and government buildings were located. The surrounding street pattern was deliberately laid out in *gridiron* form, so that any insurrections by the resettled

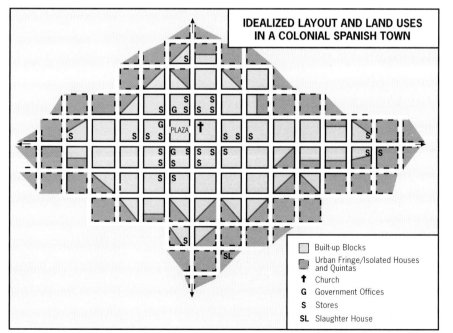

IDEALIZED LAYOUT AND LAND USES IN A COLONIAL SPANISH TOWN

Built-up Blocks
Urban Fringe/Isolated Houses and Quintas
† Church
G Government Offices
S Stores
SL Slaughter House

FIGURE 4-3

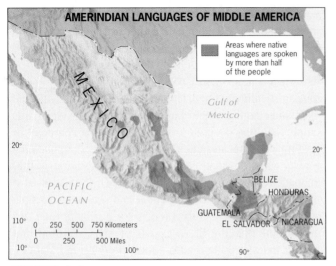

FIGURE 4-4

Amerindians could be contained by having a small military force seal off the affected blocks and then root out the troublemakers. Each town was located near what was thought to be good agricultural land (which

was often not so good), so that the Amerindians could go out each day and work in the fields. Packed tightly into these towns and villages, they came face to face with Spanish culture. Here they learned the Europeans' Roman Catholic religion and Spanish language, and they paid their taxes and tribute to a new master. Nonetheless, the nucleated indigenous village survived under colonial (and later postcolonial) administration and is still a key feature of remote Amerindian areas in southeastern Mexico and inner Guatemala, where to this day native languages prevail over Spanish (Fig. 4-4).

Once the indigenous population was conquered and resettled, the Spaniards were able to pursue another primary goal in their New World territory: the exploitation of its wealth for their own benefit. Lucrative trade, commercial agriculture, livestock ranching, and especially mining were avenues to affluence. And wherever the Spaniards ruled—in towns, farms, mines, or indigenous villages—the Roman Catholic Church was the supreme cultural force transforming Amerindian society. With Jesuits and soldiers working together to advance the

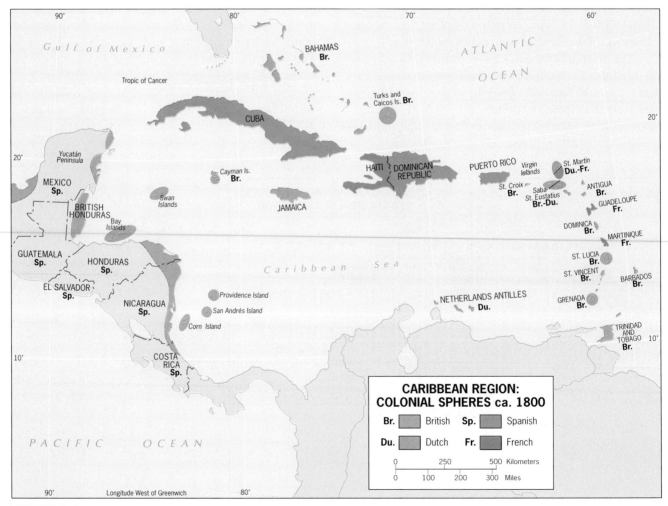

FIGURE 4-5

frontiers of New Spain, in the wake of conquest it was the church that controlled, pacified, organized, and acculturated the Amerindian peoples.

MAINLAND AND RIMLAND

In Middle America outside Mexico, only Panama, with its twin attractions of interoceanic transit and gold deposits, became an early focus of Spanish activity. From there, following the Pacific side of the isthmus, Spanish influence radiated northwestward through Central America and into Mexico. The major arena of international competition in Middle America, however, lay not on the Pacific side but on the islands and coasts of the Caribbean Sea. Here the British gained a foothold on the mainland, controlling a narrow coastal strip that extended southeast from Yucatán to what is now Costa Rica. As the colonial-era map (Fig. 4-5) shows, in the Caribbean the Spaniards faced the British, French, and Dutch, all interested in the lucrative sugar trade, all searching for instant wealth, and all seeking to expand their empires.

Later, after centuries of European colonial rivalry in the Caribbean Basin, the United States entered the picture and made its influence felt in the coastal areas of the mainland, not through colonial conquest but through the introduction of widespread, large-scale, banana plantation agriculture. The effects of these plantations were as far-reaching as the impact of colonialism on the Caribbean islands. Because the diseases the Europeans had introduced were most rampant in these hot, humid lowlands (as well as the Caribbean islands to the east), the Amerindian population that survived was too small to provide a sufficient workforce. This labor shortage was remedied through the generation of the trans-Atlantic slave trade from Africa that transformed the Caribbean Basin's demography (see map, p. 279).

These contrasts between the Middle American highlands and the coastal areas and Caribbean islands were conceptualized by John Augelli into the **Mainland-Rimland framework** (Fig. 4-6). Augelli recognized (1) a Euro-Amerindian **Mainland**, which consisted of continental Middle America from Mexico to Panama, with the exception of the Caribbean coastal belt from mid-Yucatán southeastward; and (2) a Euro-African **Rimland**, which included this coastal zone as well as the islands of the Caribbean. The terms *Euro-Amerindian* and *Euro-African* underscore the cultural heritage of each region. On the Mainland, European (Spanish) and Amerindian influences are paramount and also include **mestizo** sectors where the

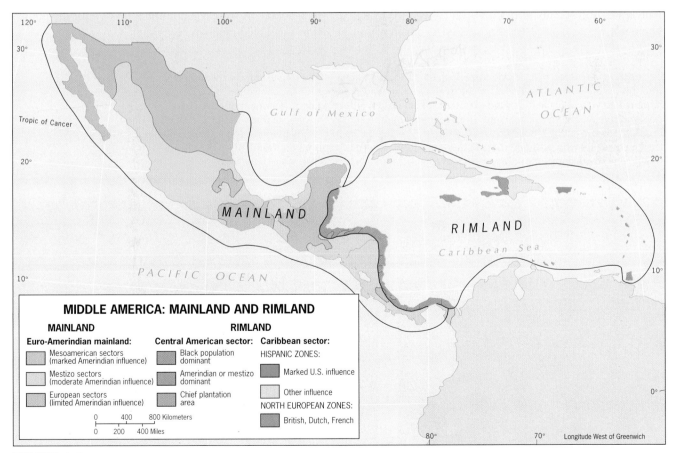

FIGURE 4-6

two ancestries mixed. In the Rimland, the heritage is European and African.

As Figure 4-6 shows, the Mainland is subdivided into several areas based on the strength of the Amerindian legacy. The Rimland is also subdivided, with the most obvious division being the one between the mainland-coastal plantation zone and the islands. Note, too, that the islands themselves can be classified according to their cultural heritage (Figs. 4-5, 4-6).

Supplementing these contrasts are regional differences in outlook and orientation. The Rimland was an area of sugar and banana plantations, of high accessibility, of seaward exposure, and of maximum cultural contact and mixture. The Mainland, being farther removed from these contacts, was an area of greater isolation. The Rimland was the region of the great *plantation*, and its commercial economy was therefore susceptible to fluctuating world markets and tied to overseas investment capital. The Mainland was the region of the *hacienda*, which was more self-sufficient and less dependent on external markets.

The Hacienda

This contrast between plantation and hacienda land tenure in itself constitutes strong evidence for the Rimland-Mainland division. The hacienda was a Spanish institution, but the modern plantation, Augelli argued, was the concept of Europeans of more northerly origin. In the **hacienda**, Spanish landowners possessed a domain whose

productivity they might never push to its limits: the very possession of such a vast estate brought with it social prestige and a comfortable lifestyle. Native workers lived on the land—which may once have been *their* land—and had plots where they could grow their own subsistence crops. All this is written as though it is mostly in the past, but the legacy of the hacienda, with its inefficient use of land and labor, is still visible throughout mainland Middle America.

The Plantation

The **plantation** was conceived as something entirely different from the hacienda. Robert West and John Augelli list five characteristics of Middle American plantations that illustrate the differences between hacienda and plantation: (1) plantations are located in the humid tropical coastal lowlands; (2) plantations produce for export almost exclusively—usually a single crop; (3) capital and skills are often imported so that foreign ownership and an outflow of profits occur; (4) labor is seasonal—needed in large numbers mainly during the harvest period—and such labor has been imported because of the scarcity of Amerindian workers; and (5) with its "factory-in-the-field" operation, the plantation is more efficient in its use of land and labor than the hacienda. The objective was not self-sufficiency but profit, and wealth rather than social prestige is a dominant motive for the plantation's establishment and operation.

Contrasting land uses in the Middle American Rimland and Mainland give rise to some very different rural cultural landscapes. Huge stretches of the realm's best land continue to be controlled by (often absentee) landowners whose haciendas yield export or luxury crops, or foreign corporations that raise fruits for transport and sale on their home markets. The banana plantation shown here lies near the Caribbean coast of Honduras and is owned by United Brands Company. The vast fields of banana plants stand in strong contrast to the lone peasant who ekes out a bare subsistence from small cultivable plots of land, often in high-relief countryside where grazing some goats or other livestock is the only way to use most of the land.

Legacy of Land-Tenure Traditions

During the twentieth century, both traditional systems of land tenure changed a great deal. In the Rimland, massive U.S. corporate investment in the Caribbean coastal belt of Guatemala, Honduras, Nicaragua, Costa Rica, and Panama transformed that area and brought a new concept of plantation agriculture to the region. On the Mainland, the hacienda system came under increasing pressure from national governments that viewed it as an economic, political, and social liability. Indeed, some haciendas have now been parceled out to small landholders, and others have been pressed into greater specialization and productivity. In Mexico, hacienda land has been placed in *ejidos*, where it is communally owned by groups of families (though privatization is now underway). With its remnants continuing to break down today, both the hacienda and the plantation have for centuries contributed to the different social and economic directions that gave the Mainland and Rimland their respective and distinct regional personalities.

POLITICAL DIFFERENTIATION

Today continental Middle America is fragmented into eight different countries, all but one of which have Hispanic origins. The largest of them all is Mexico—the giant of Middle America—whose 737,000 square miles (1,910,000 sq km) constitute more than 70 percent of the realm's entire land area (the Caribbean Basin included), and whose 106 million people outnumber those of all the other countries and islands of Middle America combined.

The cultural variety in Caribbean Middle America is much greater. Here Hispanic-influenced Cuba dominates: its area is almost as large as that of all the other islands put together, and its population of 11.4 million is well ahead of the next-ranking country—the Dominican Republic (9.2 million), also of Spanish heritage. As we noted, however, the Caribbean is hardly an arena of exclusive Hispanic cultural heritage: for example, Cuba's southern neighbor, Jamaica (population 2.7 million, mostly black), has a legacy of British involvement, while to the east in Haiti (7.4 million, overwhelmingly black) the strongest imprints have been African and French. The Lesser Antilles also exhibit great cultural diversity. There are the (once Danish) U.S. Virgin Islands; French Guadeloupe and Martinique; a group of British-influenced islands, including Barbados, St. Lucia, and Trinidad and Tobago; and the Dutch St. Maarten (shared with the French) and A-B-C islands of the Netherlands Antilles—Aruba, Bonaire, and Curaçao—off the northwestern Venezuelan coast.

Independence movements stirred Middle America at an early stage. On the Mainland, revolts against Spanish authority (beginning in 1810) achieved independence for Mexico by 1821, and for the Central American republics by the end of the 1820s. The colonial map of the Caribbean was a patchwork of holdings by British, Dutch, French, Spanish, and Danish interests (Fig. 4-5). The United States, concerned over European designs in the realm, proclaimed the Monroe Doctrine in 1823 to deter any European power from reasserting its authority in newly independent republics or from further expanding its existing domains. By the end of the nineteenth century, the United States itself had become a major force in Middle America. The Spanish-American War of 1898 made Cuba independent and put Puerto Rico under the U.S. flag; soon afterward, the Americans were in Panama constructing the Panama Canal. Meanwhile, with U.S. corporations driving a boom based on huge banana plantations, the Central American republics had become colonies of the United States in all but name.

Independence came to the Caribbean Basin in fits and starts. Afro-Caribbean Jamaica as well as Trinidad and Tobago, where the British had brought a large South Asian population, attained full sovereignty from the United Kingdom in 1962; other British islands (among them Barbados, St. Vincent, and Dominica) became independent later. France, however, retains Martinique and Guadeloupe as overseas *départements* of the French Republic, and the Dutch islands are at various stages of autonomy.

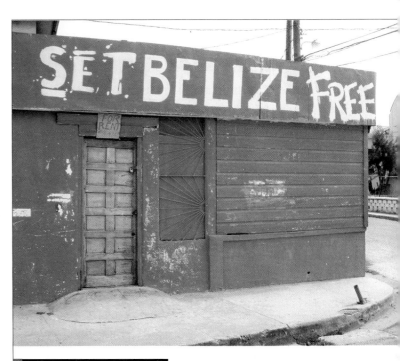

FROM THE FIELD NOTES

"Evidently the battle was won."

REGIONS OF THE REALM

MIDDLE AMERICA CONSISTS of four geographic regions: (1) Mexico, the giant of the realm in every respect; (2) Central America, the string of seven small republics occupying the land bridge to South America; (3) the four large islands that constitute the Greater Antilles of the Caribbean; and (4) the numerous islands of the Caribbean's Lesser Antilles (Fig. 4-2).

▶ MEXICO

By virtue of its physical size, large population, cultural identity, resource base and economy, and relative location, the state of Mexico by itself constitutes a geographic region in this multifaceted realm. The U.S-Mexican boundary on the map crosses the continent from the Pacific to the Gulf, but Mexican cultural influences penetrate deeply into the southwestern States and American impacts reach far into Mexico. To Mexicans the border is a reminder of territory lost to the United States in historic conflicts; to Americans it is a symbol of economic contrasts and illegal immigration. Along the Mexican side of this border, the effects of NAFTA (the North American Free Trade Agreement among Canada, the United States, and Mexico) are transforming Mexico's economic geography.

MAJOR CITIES OF THE REALM	
City	Population* (in millions)
Ciudad Juárez, Mexico	1.5
Guadalajara, Mexico	3.9
Guatemala City, Guatemala	3.9
Havana, Cuba	2.3
Managua, Nicaragua	1.2
Mexico City, Mexico	27.6
Monterrey, Mexico	3.7
Panama City, Panama	1.3
Port-au-Prince, Haiti	2.1
Puebla, Mexico	1.9
San José, Costa Rica	1.1
San Juan, Puerto Rico	1.5
San Salvador, El Salvador	1.5
Santo Domingo, Dominican Rep.	2.9
Tegucigalpa, Honduras	1.1
Tijuana, Mexico	1.8

*Based on 2004 estimates.

Mexico is a giant not just in regional but also in global terms. Territorially, it is more than three times as large as Europe's largest country; a few years ago Mexico became the eleventh country in the world to count more than 100 million people (larger, again, than any European state). Mexico's population grew rapidly during the last three decades of the twentieth century, doubling in just 28 years; but demographers have recently noted a sharp drop in fertility, and they are predicting that Mexico's population (currently 106 million) will stop growing altogether by about 2050. This will have enormous implications for the country's economy, and it will reduce the cross-border migration that is of so much concern at present.

Physiography

The physiography of Mexico is reminiscent of that of the western United States, although its environments are more tropical. Figure 4-7 shows several prominent features: the elongated Baja (Lower) California Peninsula in the northwest, the far eastern Yucatán Peninsula, and the Isthmus of Tehuantepec in the southeast where the Mexican landmass tapers to its narrowest extent. Here in the southeast, Mexico most resembles Central America physiographically; a mountain backbone forms the isthmus, curves southeast into Guatemala, and extends northwest toward Mexico City. Shortly before reaching the capital, this mountain range divides into two chains, the Sierra Madre Occidental in the west and Sierra Madre Oriental in the east (Figs. 4-1, 4-7). These diverging ranges frame the funnel-shaped Mexican heartland, the center of which consists of the rugged, extensive Plateau of Mexico (the Valley of Mexico lies near its southeastern end). As Figure G-8 reveals, Mexico's climates are marked by dryness, particularly in the broad, mountain-flanked north. Most of the better-watered areas lie in the southern half of the country where the major population concentrations have developed.

Population Patterns

The distribution of population across Mexico's 31 internal States is shown in Figure 4-8. The largest concentration, containing more than half the Mexican people, extends across the densely populated "waist" of the country from Veracruz State on the eastern Gulf coast to Jalisco State on the Pacific. The center of this corridor is dominated by the most populous State, Mexico (**3** on the map), at whose heart lies the Federal District of Mexico City (**9**). In the dry and rugged terrain to the north of this central corridor lie Mexico's least-populated States. Southern Mexico also exhibits

FIGURE 4-7

a sparsely peopled periphery in the hot and humid lowlands of the Yucatán Peninsula, but here most of the highlands of the continental spine contain sizeable populations.

Another major feature of Mexico's population map is urbanization, driven by the *pull* of the cities (with their perceived opportunities for upward mobility) in tandem with the *push* of the economically stagnant countryside. Today, 74 percent of Mexicans reside in towns and cities, a surprisingly high proportion for a developing country. Undoubtedly, these numbers are affected by the explosive recent growth of Mexico City, which now totals 27.6 million (making it the largest urban concentration on Earth) and is home to 26 percent of the national popula-

tion. Among the other leading cities are Guadalajara, Puebla, and León in the central population corridor, and Monterrey, Ciudad Juárez, and Tijuana in the northern U.S. border zone (Fig. 4-8). Urbanization rates at the other end of Mexico, however, are at their lowest in those remote uplands where Amerindian society has been least touched by modernization.

Nationally, the Amerindian imprint on Mexican culture remains quite strong. Today, 60 percent of all Mexicans are mestizos, 20 percent are predominantly Amerindian, and about 10 percent are full-blooded Amerindians; only 9 percent are Europeans. Certainly the Mexican Amerindian has been Europeanized, but the Amerindianization of

AMONG THE REALM'S GREAT CITIES . . .   MEXICO CITY

*M*iddle America has only one great metropolis: Mexico City. With 27.6 million inhabitants, Mexico City is home to just over one-fourth of Mexico's population and grows by about 300,000 each year. Even more significantly, Mexico City has surpassed Tokyo, Japan (current population: 26.7 million) to become the world's largest urban agglomeration.

Lakes and canals marked this site when the Aztecs built their city of Tenochtitlán here seven centuries ago. The conquering Spaniards made it their headquarters, and following independence the Mexicans made it their capital. Centrally positioned and well connected to the rest of the country, Mexico City, hub of the national core area, became the quintessential primate city.

Vivid social contrasts mark the cityscape. Historic plazas, magnificent palaces, churches, villas, superb museums, ultramodern skyscrapers, and luxury shops fill the city center. Beyond lies a zone of comfortable middle-class and struggling, but stable, working-class neighborhoods. Outside this belt, however, lies a ring of more than 500 slums and countless, even poorer *ciudades perdidas*—the "lost cities" where newly arrived peasants live in miserable poverty and squalor. (These squatter settlements contain no less than *one-third* of the metropolitan area's population.) Mexico City's more affluent residents have also been plagued by problems in recent years as the country's social and political order came close to unraveling. Rampant crime remains a serious concern, much of it associated with corrupt police as the control of the country's long-time ruling party (the PRI) weakened and collapsed.

Environmental crises parallel the social problems. Local surface waters have long since dried up, and groundwater supplies are approaching depletion; to meet demand, the metropolis must now import a third of its water by pipeline from across the mountains (with about 40 percent lost through leakages in the city's crumbling waterpipe network). Air pollution here is probably the world's worst as more than 4 million motor vehicles and 40,000 factories

churn out smog that in Mexico City's thin, high-altitude air sometimes reaches *100 times* the acceptably safe level. And add to all this a set of geologic hazards: severe land subsidence as underground water supplies are overdrawn; the ever-present threat of earthquakes that can wreak havoc on the city's unstable surface (the last big one occurred in 1985); and the risk of volcanic activity as nearby Mount Popocatépetl may be ending centuries of dormancy.

In spite of it all, the great city continues to beckon, and every year more than 100,000 of the desperate and the dislocated arrive with hope—and little else.

modern Mexican society is so powerful that it would be inappropriate here to speak of one-way, European-dominated **acculturation**. Instead, what took place in Mexico is **transculturation**—the two-way exchange of culture traits between societies in close contact. In the southeastern periphery (Fig. 4-4), several hundred thousand Mexicans still speak only an Amerindian language, and millions more still use these languages in everyday conversation even though they also speak Mexican Spanish. The latter has been strongly shaped by Amerindian

influences, as have Mexican modes of dress, foods and cuisine, sculpture and painting, architectural styles, and folkways. This fusion of heritages, which makes Mexico unique, is the product of an upheaval that began to reshape the country nearly a century ago.

Revolution and Its Aftermath

Modern Mexico was forged in a revolution that began in 1910 and set into motion events that are still unfolding

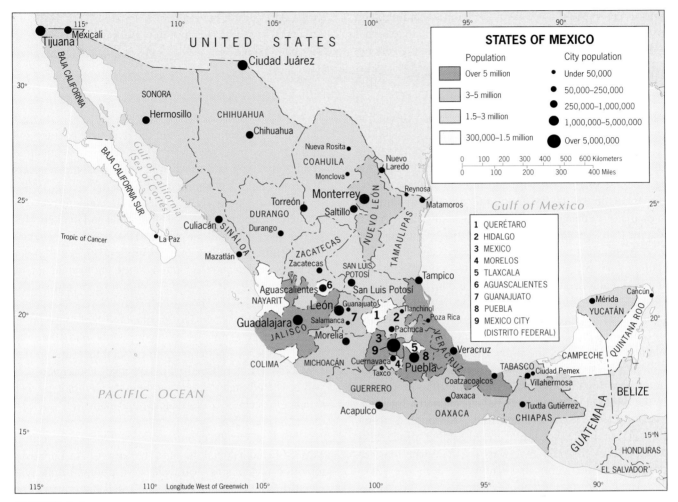

FIGURE 4-8

today. At its heart, this revolution was about the redistribution of land, an issue that had not been resolved after Mexico freed itself from Spanish colonial control in the early nineteenth century. As late as 1900, more than 8000 haciendas blanketed virtually all of Mexico's good farmland, and about 95 percent of all rural families owned no land whatsoever and toiled as *peones* (landless, constantly indebted serfs) on the haciendas. The triumphant revolution produced a new constitution in 1917 that launched a program of expropriation and parceling out of the haciendas to rural communities.

Since 1917, more than half the cultivated land of Mexico has been redistributed, mostly to peasant communities consisting of 20 families or more. On such farmlands (known as *ejidos*) the government holds title to the land, and use rights are parceled out to villages and then individuals for cultivation. Most of these *ejidos* lie in central and southern Mexico, where Amerindian agricultural traditions survived, but the newly created landholdings were too finely fragmented to provide a real development opportunity. Despite an understandable temporary decline in agricultural productivity during the transition, the miracle is that land reform was carried off without major dislocation. Although considerable poverty persisted in the countryside, it was recognized that Mexico, alone among Middle America's countries with large Amerindian populations, had made significant strides toward solving the land question. Just how much farther Mexico has to go, however, became evident in 1994 when the issue resurfaced at the heart of a rebellion that broke out in southeasternmost Chiapas State.

Chiapas, the poorest of the 31 States, lies wedged against the Guatemalan border and is more reminiscent of Central America than Mexico. The complexion of its rapidly growing population is heavily Amerindian and dominated by families of peasant farmers who eke out a precarious existence by cultivating tiny plots of land in the rainforested hills. For centuries, the better valley soils have been incorporated into the estates of the large landholders, a system that endures here virtually unaffected by the land redistribution that reshaped so much of rural Mexico. In response to this lack of development, during the 1980s the Mexican government had pledged to introduce new services and programs in Chiapas, but its main effort was an ineffective coffee-raising scheme and a feeble attempt to privatize Amerindian lands. This only intensified the longstanding bitterness of the Chiapans, and a radical group of Mayan peasant farmers began to organize to resume the historic struggle of Amerindian *peones* to gain land and fair treatment.

On January 1, 1994, this organization, now calling itself the Zapatista National Liberation Army (ZNLA), ignited a guerrilla war with coordinated attacks on several Chiapan towns. By timing this uprising to coincide with the birth date of NAFTA and by taking the name of Zapata (the legendary leader of the 1910 revolution), the ZNLA ensured maximum publicity for its agenda, which included land reform, access to greater economic opportunity, heightened cultural identity, and local autonomy. The Mexican military reacted to this rebellion with a heavy hand but quickly called off its offensive when the international press exposed a number of human rights violations. Ever since, the army has maintained a major Chiapan presence—but always leaving sufficient maneuvering room for the unrepentant guerrillas to regroup in the mountains, where they have remained a formidable political, if not a military, force. Although negotiations soon replaced the initial armed conflict, peace talks collapsed in 1996 and gave way to a stalemate that has dragged into the twenty-first century.

The Chiapas conflict raises a number of broader issues that have important implications for Mexico's future. First, it emphasizes that certain areas of the country do not participate in the ongoing development thrust (in spatial terms, the Chiapas situation is a classic case of core-periphery confrontation). A second issue, of course, is devolution. The ZNLA demand for "autonomy" is modeled after Spain's Autonomous Communities (see p. 82) and involves the decentralization of powers from the federal to the State government, which allows the States more local control, particularly over cultural affairs. A third issue is especially vital to Mexico's social geography: the ZNLA crusade elicited wide sympathy among indigenous populations in other southern States and could well spark a nationwide civil rights movement for Amerindians. As in Caribbean societies (as we shall see), darkness of skin color is directly related to a person's social status, a linkage that in Mexico further extends to an individual's degree of "Indianness." Clearly, large numbers of predominantly Amerindian Mexicans (a group now totaling 20 percent of the national population) made social gains during the past century. But the remaining 10 percent are full-blooded Amerindians, peoples whose ancestral homes are widely distributed throughout Mexico (Fig. 4-9). They choose to preserve their pre-Hispanic cultural traditions, having been shunned by a mainstream society whose discrimination continues to render them second-class citizens.

The Changing Geography of Economic Activity

During the last two decades of the twentieth century and in the first years of the twenty-first, Mexico's economic geography experienced change, and in some respects progress—though not without setbacks. In the early 1990s, the implementation of NAFTA led to an economic boom as Mexico became part of a free-trade zone and market comprising over 400 million people. This boom transformed urban landscapes along the 1936-mile (3115-km) border between Mexico and the United States, but it could not, of course, immediately close the economic gap between the two sides. And like all booms, Mexico's NAFTA surge entailed risks. When Mexico encountered budgetary problems and an economic recession during the mid-1990s, its NAFTA role was not enough to cushion the blow. However, NAFTA membership did have a key consequence: it was in United States' interest to help its free-trade partner weather the economic storm, and huge loans were made to make this possible.

Then, just as the economy seemed to be recovering, Mexico faced a social crisis in the form of the Zapatista rebellion in Chiapas, followed by a political upheaval when the political party that had ruled the country for more than seven consecutive decades was ousted in the 2000 election. Mexico's new president, Vicente Fox, was encouraged by signals from his American counterpart, George W. Bush, that Mexico would have priority standing among U.S. foreign-policy concerns. President Fox immediately suggested that NAFTA should be more than a free-trade program: immigration across the U.S.-Mexican border should be liberalized (on the EU model), so that the problem of illegal migration and sometimes deadly border-policing would be mitigated. The U.S. administration's reaction to this idea was not immediately favorable, but then the events of September 11, 2001 intervened and Mexico's standing among American priorities suffered a setback.

Nonetheless, Mexico is in progressive transition in many spheres: its democratic institutions are strengthening; its economy is more robust than it was at the time of NAFTA's inception; its once-rapid population growth is declining; its social fabric (notably relations with Amerindian minorities) is improving. The familiar problems of a country in transition, such as unchecked urbanization, inadequate infrastructure, corruption, and violent crime, will continue to afflict Mexico for decades to come. But, as the following discussion confirms, Mexico

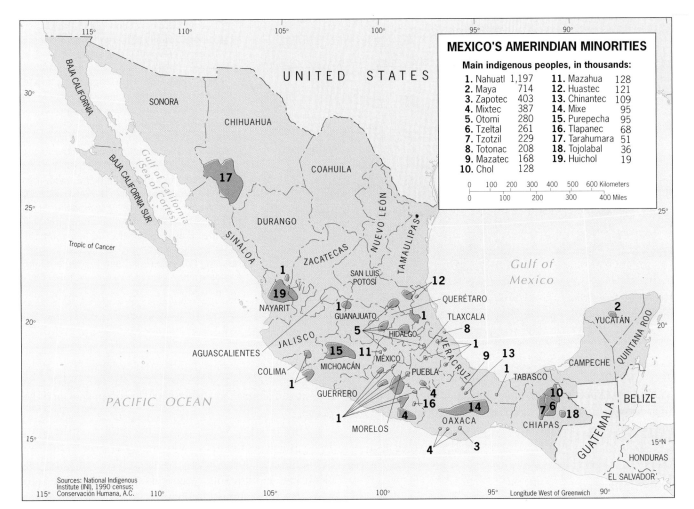

MEXICO'S AMERINDIAN MINORITIES

Main indigenous peoples, in thousands:

1.	Nahuatl	1,197	11.	Mazahua	128
2.	Maya	714	12.	Huastec	121
3.	Zapotec	403	13.	Chinantec	109
4.	Mixtec	387	14.	Mixe	95
5.	Otomi	280	15.	Purepecha	95
6.	Tzeltal	261	16.	Tlapanec	68
7.	Tzotzil	229	17.	Tarahumara	51
8.	Totonac	208	18.	Tojolabal	36
9.	Mazatec	168	19.	Huichol	19
10.	Chol	128			

Sources: National Indigenous Institute (INI), 1990 census; Conservación Humana, A.C.

FIGURE 4-9

is a far stronger economy and society today than it was just one generation ago.

Agriculture

Although traditional subsistence agriculture and the output of the inefficient *ejidos* have not changed a great deal in the poorer areas of rural Mexico, larger-scale commercial agriculture has diversified during the past three decades and made major gains with respect to both domestic and export markets. The country's arid northern tier has led the way as major irrigation projects have been built on streams flowing down from the interior highlands. Along the booming northwest coast of the mainland, which lies within a day's drive of Southern California, mechanized large-scale cotton production now supplies an increasingly profitable export trade. Here, too, wheat and winter vegetables are grown, with fruit and vegetable cultivation attracting foreign investors.

Energy Resources

Although Mexico's metal mining industries are less important today than they once were, since 1970 the country has enjoyed the advantages and suffered the problems of being a major petroleum producer. Huge oilfields centered on the southern Gulf coast's Bay of Campeche around Villahermosa in Tabasco State brought Mexico abundant revenues when the world oil price was high in the late 1970s and serious economic difficulties when the price fell after 1980. These discoveries of massive oil and natural gas reserves have made Mexico self-sufficient in energy, adding to already substantial reserves located in oilfields along the Gulf coast to the northwest and western Yucatán to the northeast (Fig. 4-7). The high petroleum prices of the 1970s stimulated the beginnings of Mexico's economic-geographic transformation, but the oil crash of the 1980s thwarted its momentum for a decade. This crash occurred because the government found itself

The border between Mexico and the United States occasionally provides some stunning contrasts. Here, looking westward, we observe the opposing economic geographies that mark the border landscape of the Imperial Valley east of the urban area formed by Mexicali, Mexico and Calexico, California. The larger town of Mexicali (on the left) sprawls eastward along the border, with a cluster of *maquiladora* assembly plants, visible in the left foreground, surrounded by high-density, poor-quality housing. On the U.S. side, lush irrigated croplands blanket the otherwise dry countryside, fed by the All-American Canal that taps the waters of the Colorado River before they cross into Mexico. At this point, the canal swings northward to bypass border-hugging Calexico (right rear) before rejoining the international boundary beyond the town's western edge.

without the expected oil revenues it needed to pay off the huge foreign loans it had taken to finance domestic development programs. Such a disaster could recur if a future oil boom again tempts Mexican leaders to take such risks.

Industrialization

Manufacturing is the centerpiece of Mexico's latest development episode, but this economic sector actually got its start more than a century ago. Blessed with a wide range of raw materials (Fig. 4-7), the country began to industrialize in 1903 with the completion of an iron and steel plant in the northern city of Monterrey. A second steel complex was built at nearby Monclova in the 1950s, a period that saw the spreading of factories across many parts of central Mexico. Since that time, the industrial sector has grown steadily.

The most significant recent development in Mexico's manufacturing geography is the growth of maquiladora plants in the U.S. border zone. **Maquiladoras** are factories that assemble imported, duty-free components and raw materials into finished products. At least 80 percent of these goods are then reexported to the United States. Although foreign owners benefit from Mexico's lower wage rates, this industrial system also offers many economic advantages for Mexico. These advantages include the expansion of job opportunities, increased foreign investment, and the transfer of new technologies into the country. Unfortunately, there is a downside too: factory workers are usually exploited through long work weeks, receive minimal fringe benefits, and most can only afford to reside in

the squalid slums and shacktowns that often surround the plants.

Shortly after NAFTA went into effect, in the mid-1990s, there were about 1800 industrial plants employing some 500,000 workers to assemble such goods as electronic equipment, electrical appliances, and auto parts. By 2000, almost 4000 maquiladoras with over 1.2 million employees were operating all along the northern border zone in the urban areas mapped in red in Figure 4-10, accounting for nearly one-third of Mexico's industrial jobs and 45 percent of its total exports.

Since peaking in 2000, however, this expansion has reversed, raising concerns about the future of Mexico's border-zone development. In 2002, economic geographers reported that, for the first time since the inception of NAFTA, U.S. imports from Mexico had declined from the previous year (while U.S. imports from China showed no sign of any slowdown). Some of this decline could be attributed to the downturn of the U.S. economy and to the increasing strength of Mexico's peso, making exports more expensive and thus less competitive. But the underlying reasons were more serious. Many American and other foreign corporations had moved their factories to Mexico because labor there was much cheaper: maquiladora workers earn between U.S. $2.00 and $2.50 an hour, and benefits are far less costly. But the cost of labor in China ranges from U.S. $0.35 to $1.00 per hour. If proximity to the U.S. market is not an overriding factor, as it is for bulky products like automobiles, companies can save money by moving their factories to Asia. And that is exactly what took place during 2000 and

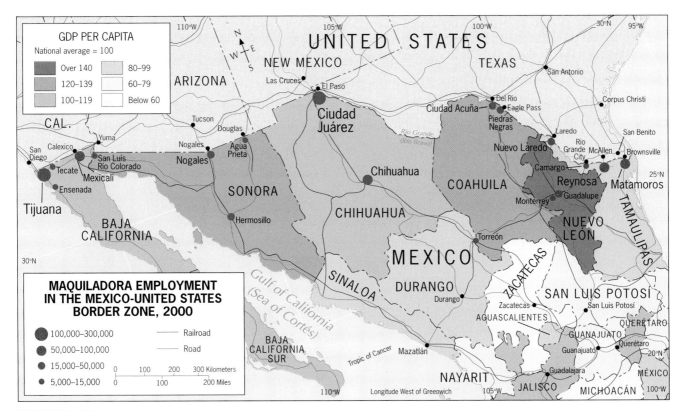

FIGURE 4-10

2001: the Dutch electronics firm Philips, for example, abandoned its plant in Ciudad Juárez and relocated that operation to Suzhou, China, putting nearly 1000 Mexican employees out of work. And the Japanese company Canon closed its facility in Tijuana and transferred operations to Southeast Asia. By some estimates, maquiladora-based industries have lost 300,000 jobs since the beginning of this decade.

Obviously, the maquiladora complex cannot compete on wages alone. There must be other advantages for companies to locate—and stay—in Mexico. Certainly, Mexico's wages remain far below those in the United States, and factories will continue to be opened even as others close. And the cost of transporting heavy, bulky products like cars and refrigerators across the Pacific Ocean will keep many plants firmly based in Mexico. But Mexico has work to do in other areas. Education is one: there is a persistent shortage of Mexican managers and skilled technical workers. Infrastructure urgently needs improvement: add general transportation woes to the tie-ups along the U.S.-Mexican border, and the maquiladoras' advantages dwindle. Bureaucratic red tape, crime, corruption, and persistently high labor turnover all combine to cause corporations to be on the lookout for alternative locations. The maquiladora industries were the most dynamic of the 1990s, rivalling the momentous

economic developments occurring on the Asian Pacific Rim (see Chapter 9). The current warning signs should cause the Mexican government to act to protect this vital sector of its growing economy.

Fortunately, the government had already been capitalizing on the success of the maquiladoras by promoting industrial growth elsewhere in the country. What development planners desire most is that industrial firms create complete manufacturing complexes within the heart of Mexico rather than limit their investments to assembly plants that hug the U.S. border. A number of multinational corporations have undertaken such ventures (most notably U.S. automakers and several of their Japanese and German competitors), and government intiatives continue in order to lure others.

To speed the opening of the rest of the country, the government is prioritizing efforts aimed at upgrading Mexico's infrastructure to international standards, particularly its telecommunications, transportation, and electrical-power networks. Individual megaprojects to boost regional development are also being pursued, most ambitiously the **dry canal** across the 150-mile (250-km) Tehuantepec isthmus to move containerized goods between the Pacific port of Salina Cruz and the Gulf port of Coatzacoalcos. This would not only be an overland trade route to rival the aging Panama Canal, but also an ultramodern road/rail corridor lined with factories that specialize in the manufacturing and assembly of goods in transit between the affluent countries of the Asian Pacific Rim and the North Atlantic Basin.

Uneven Regional Development

Despite such efforts, an economic divide is deepening between the southern and northern halves of Mexico. South of the capital, technological and social development lags far behind the rest of the country as traditional productive activities (low-output farming, mining, logging) continue to dominate. North of Mexico City, growth corridors are stirring that increasingly exhibit the landscape of the global economy in the form of new manufacturing facilities, technology training academies, and even the beginnings of a

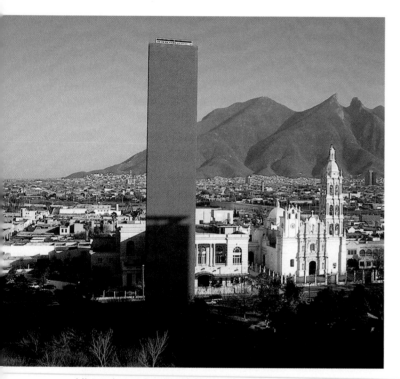

Likened to a glowing hot bar of freshly cast steel plucked from a furnace, the *Faro del Comercio* (Lighthouse of Commerce) shoots skyward in the historic heart of Monterrey. Saddle Mountain in the background is a reminder that this northeastern Mexican metropolis is situated at a strategic point where the desert mets the great Sierra Madre Oriental range, in the heart of an area so well endowed with raw materials that Mexico's iron and steel industry was born here a century ago. Ever since, Monterrey has been an economic leader, and its new landmark signifies that no city more boldly confronts the Mexican future. Today, building on the opportunities offered by NAFTA, Monterrey is busy forging a role for itself on the international circuit. Capitalizing on its function as the southern anchor of the booming growth corridor that stretches northward to Texas' Dallas–Ft. Worth Metroplex, Monterrey has built a new downtown business complex that includes a convention center booked solidly with trade shows at least two years ahead. And in 1999, the city even hosted the National League's season-opening baseball game.

"Silicon Valley of Mexico" on the edge of Guadalajara. To capture the forces transforming northern Mexico, we focus on one of its leading urban centers, Monterrey.

Burgeoning Monterrey—150 miles (250 km) inside Mexico, yet close enough to the Texas border to have benefited from all the recent development trends—is frequently singled out as a model of what a future Mexican growth center should be. Among this city's assets are a highly educated and well-paid labor force, a stable international business community, and a thriving high-technology complex of ultramodern industrial facilities that has attracted blue-chip multinational companies. In addition, a new expressway link to the Rio Grande is helping to forge an international growth corridor between Monterrey and Dallas-Fort Worth, which is becoming a primary axis of cross-border trade with Texas. For Mexico as a whole, however, there may also be a negative side to Monterrey's success. This city and its surrounding State of Nuevo León have now become so prosperous (note its standout GDP in Fig. 4-10) that there is widening talk of secession. Similar to the devolutionary movement in northern Italy's Lombardy (p. 81), affluent Nuevo León—which receives back less than 20 centavos in services for every tax peso (100 centavos) it sends to Mexico City—has increasingly complained that the federal government tries to develop the impoverished South by taking more money from the industrial North.

NAFTA and Continuing Challenges

The launching of the North American Free Trade zone in 1994 was supposed to mark an economic turning point for Mexico. Instead, NAFTA's early years were plagued by an unending string of crises that included the Chiapas rebellion, monetary devaluation, economic recession, crime waves associated with uncontrolled drug trafficking, and political scandals and transitions. Nonetheless, trade with the United States increased substantially: more than 85 percent of Mexican exports now go to the United States, which is also the source for about 75 percent of all imported goods. At the same time, Mexico has become a leading U.S. trade partner, in 1997 dislodging Japan as the United States' second-largest export market after Canada.

NAFTA in some respects is modeled on the European Union, although it is much less integrative than its European counterpart. When the EU was established, it was the first time that what we would today call high-income countries entered into a set of reciprocal agreements with middle-income countries. The membership terms of the EU required that the richer countries subsidize the economies of the poorer ones, and hundreds of billions of dollars have flowed from Brussels to Lisbon, Madrid, Dublin, and Athens. No similar commitment exists between Canada and the United States on the one hand and Mexico on the other, and until the EU's 2004 expansion into Eastern Europe no other free-trade agreement linked economies as high-income as the U.S. and Canadian to one as comparatively low-income as the Mexican. But consider this: in the Seventh Edition of this book (1994) we reported Mexico's GNP as U.S. $2490 per person; in this edition, ten years later, we note that it stands at U.S. $8790. This average conceals severe regional disparities—as we noted, Mexico's northern States are far more affluent than those in the south—but it does underscore the country's overall progress.

As Mexico strives to join the ranks of the world's advantaged states, the movements of its inhabitants indicate that the country is still a wobbly giant. About one million Mexicans, lured by the promise of jobs in manufacturing and agriculture, annually migrate to the U.S. border zone. Most migrants would prefer to keep right on going north, and hundreds of thousands do manage to cross the U.S. boundary each year. Large numbers of frustrated Mexicans also attempt to leave the country illegally, but are intercepted and deported by U.S. authorities. Undoubtedly, many more get through, which has triggered heightened public support in the United States for tougher measures to curb this flow.

Mexico today is at a crossroads. Its most recent national election installed a regime that was expected to produce new initiatives aimed at reducing the divisiveness of the U.S. border, devolving power to the States, resolving the Chiapas conflict, and continuing the advancement of the economy. Expectations remain high throughout the country, but the results so far have been modest.

THE CENTRAL AMERICAN REPUBLICS

Crowded onto the narrow segment of the Middle American land bridge between Mexico and the South American continent are the seven countries of Central America (Fig. 4-11). Territorially, they are all quite small; their population sizes range from Guatemala's 12.8 million down to Belize's 295,000. Physiographically, the land bridge consists of a highland belt flanked by coastal lowlands on both the Caribbean and Pacific sides (Fig. 4-1). These highlands are studded with volcanoes, and local areas of fertile volcanic soils are scattered throughout them. From earliest times, the region's inhabitants have been concentrated in this upland zone, where tropical temperatures are moderated by elevation and rainfall is sufficient to support a variety of crops.

Altitudinal Zonation of Environments

Continental Middle America and the western margin of South America are areas of high relief and strong environmental contrasts. Even though temperate intermontane basins and valleys are most favored, people also cluster in hot tropical lowlands as well as high plateaus just below the snow line in South America's Andes Mountains. In each of these zones, distinct local climates, soils, vegetation, crops, domestic animals, and modes of life prevail. Such **altitudinal zones** (diagrammed in Fig. 4-12) are known by specific names as if they were regions with distinguishing properties—as, in reality, they are.

The lowest vertical zone, from sea level to 2500 feet (about 750 m), is known as the ***tierra caliente***, the "hot land" of the coastal plains and low-lying interior basins where tropical agriculture predominates. Above this zone lie the tropical highlands containing Middle and South America's largest population clusters, the ***tierra templada*** of temperate land reaching up to about 6000 feet (1800 m). Temperatures here are cooler; prominent among the commercial crops is coffee, while corn (maize) and wheat are the staple grains. Still higher, from about 6000 feet to nearly 12,000 feet (3600 m), is the ***tierra fría***, the cold country of the higher Andes where hardy crops such as potatoes and barley are mainstays. Above the tree line, which marks the upper limit of the *tierra fría*, lies the ***tierra helada***; this fourth altitudinal zone, extending from about 12,000 to 15,000 feet (3600 to 4500 m), is so cold and barren that it can support only the grazing of sheep and other hardy livestock. The highest zone of all is the ***tierra nevada***, a zone of permanent snow and ice associated with the loftiest Andean peaks.

Population Geography

The distribution of population within Central America, apart from its concentration in the uplands, also exhibits greater densities toward the Pacific than toward the Caribbean coastlands (Fig. G-9). El Salvador, Belize, and (to some degree) Panama are exceptions to the rule that people in continental Middle America are concentrated in the *tierra templada* zone. El Salvador, in particular, is mostly tropical *tierra caliente*, and the majority of its people are crowded onto the intermontane plains lying less than 2500 feet (750 m) above sea level.

By contrast, the Caribbean coastal lowlands—hot, wet, and awash in infertile soils—support comparatively few people. In the region's most populous republic, Guatemala, the heartland also has long been in the southern highlands. Although the large majority of Costa Rica's population is clustered in the Central Valley around San José, the Pacific lowlands have been the scene of major in-migration since banana plantations were established there.

FIGURE 4-11

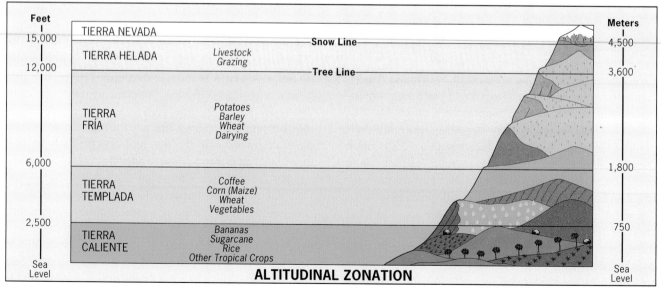

FIGURE 4-12

Even in Panama there is a strong Pacific orientation, with more than half of all Panamanians living in the southwestern lowlands and on adjoining mountain slopes.

Middle America's smaller republics face the same problems as the less developed parts of Mexico, and they also share many of the difficulties that confront the poorer Caribbean islands. No present or future challenge, however, is greater than Central America's overpopulation. The region's population explosion began a half-century ago, expanding from a base of about 9 million people in 1950 to 40 million in 2004. Unlike Mexico, which has markedly reduced its rate of natural increase since 1980, Central America (except for Costa Rica and Panama) is on a course that will see a doubling of today's population to 80 million by 2030. This amounts to nothing less than an onrushing demographic catastrophe in a region already unable to cope with most of its social, economic, and natural resource problems.

Emergence From a Turbulent Era

Devastating social inequities, repressive governments, brutal external interference, and tyrannies by armies and police have afflicted most of Central America for much of its modern history. The roots of these conditions are old and deep, beginning during the colonial period and enduring almost continuously until the 1990s. Some of the republics, notably Costa Rica and Belize, have fared somewhat better than others, but the overall picture is dismal.

Part of the problem lies in Central America's cultural history. In precolonial times this was a comparatively densely populated region, home to ancient and well-organized Maya and other Amerindian societies with strong cultural traditions. They long resisted Spanish colonization and, though decimated, survived in substantial numbers in their traditional strongholds. Later, when the colonies established by their conquerors became independent states, they found themselves oppressed in new ways, and resisted again. Struggles between poor, isolated, traditional communities and invading, richer, better-armed mestizo groups became endemic to the region.

Another part of the problem lay in Central America's rugged, forested physiography. The mountainous, volcanically active, forested land bridge between North and South America contains numerous isolated and inaccessible locales. Here the Amerindian communities had some advantage, but so did their oppressors—who concealed from the outside world the often terrible retribution meted out to the "rebels."

Still another element in Central America's conflict-filled historical geography is the United States. From the days of the Monroe Doctrine until the recent past, Amer-ican governments have regarded Middle America as their sphere of influence, supporting "friendly" regimes against rebels or guerrillas and turning a blind eye to the excesses of the ruling regimes. In 1954, the United States encouraged and assisted a coup against the first democratically elected leader of Guatemala, who appeared to be leaning toward Moscow; this resulted in several decades of civil war and internal terror from which the country has yet to recover. When communism did gain a Middle American foothold on the island of Cuba, Washington's determination to preclude anything similar on the mainland caused it to support such autocratic rulers as the Somoza family in Nicaragua. The predictable Sandinista revolution against Somoza led to a U.S.-sponsored counterrevolution, and Nicaragua suffered anew. In El Salvador, the American government also supported a regime whose excesses only came to light during the 1990s.

The U.S. role in Central America has been the dominant external factor, but in recent years that role has begun to change. Today the American government supports democratic government in the region; it has negotiated the transfer of the Panama Canal to the Panamanian government; it has given some encouragement to those trying to expose the injustices of the past; and it has even helped in the prosecution of Central American war criminals living in retirement in the United States. It is probably too early to state that a new era of U.S.-Central American relations has opened, but change is clearly under way.

No discussion of Central America's prospects would be complete without reference to another circumstance: natural disasters. This is a region bedeviled by earthquakes, volcanic eruptions, and hurricanes, which batter infrastructures, kill thousands, and hamper economic development (see photo, p. 189). In 1998, Honduras lost hundreds of bridges and tens of thousands of houses during a direct hit by Hurricane Mitch; it will take years for the country just to get back to where it was prior to this onslaught. Countless instances of nature's wrath mark the landscapes of countries already disadvantaged in this turbulent region.

The Republics

Guatemala, northwesternmost of Central America's republics, has more land neighbors than any other. Straight-line boundaries across the tropical forest mark much of the border with Mexico, creating the box-like region of Petén between Chiapas State on the west and Belize on the east. To the southeast lie Honduras and El Salvador (Fig. 4-11). This heart of the ancient Maya Empire, which remains strongly infused by Amerindian culture and tradition, has just a small window on the Caribbean but a longer Pacific coastline. Guatemala was still part of Mexico when the Mexicans threw off the Spanish yoke, and although independent from Spain after 1821, it did not become a separate republic until 1838. Mestizos, not the Amerindian majority, secured the country's independence.

Most populous of the seven republics with 12.8 million inhabitants (mestizos are in the majority with 56 percent, Amerindians 44 percent), Guatemala has seen much conflict. Repressive regimes made deals with U.S. and other foreign economic interests that stimulated development, but at a high social cost. Over the past half-century, military regimes have dominated political life. The deepening split between the wretchedly poor Amerindians and the better-off mestizos, who here call themselves *ladinos*, generated a civil war that started in 1960 and has since claimed more than 200,000 lives as well as 50,000 "disappearances." An overwhelming number of the victims have been of Mayan descent; the mestizos control the government, army, and land-tenure system.

Because the insurgents were able to take refuge in Guatemala's remote, rainforested plains on the Mexican border, the war was unwinnable for the ladinos based in the core area of the southern highlands. Rising international outrage over nearly 40,000 human-rights violations, perpetrated by the armed forces and their paramilitary death squads, finally forced the government to the negotiating table, and in 1996 a peace agreement was concluded. For more than a year the political violence ceased, and initial steps toward stability were taken. But in 1998 the hopeful new atmosphere was shattered by the assassination of one of the country's leading human-rights advocates, Bishop Juan José Gerardi, an event that continues to cast a deep shadow over the country.

The tragedy of Guatemala is that its economic geography has considerable potential but has long been shackled by the unending internal conflicts that have kept the income level of 80 percent of the population below the poverty line. The country's mineral wealth includes nickel in the highlands and oil in the lower-lying north. Agriculturally, soils are fertile and moisture is ample over highland areas large enough to produce a wide range of crops including excellent coffee. Ironically, this promising upland zone is now an area of outmigration for a horde of mostly indigenous, land-hungry peasants who are streaming northward into the infertile tropical rainforests of the Petén. The push factors in this migration are the increasing pressures on traditional subsistence farmlands propelled by the realm's highest birth rate, the return of tens of thousands of Amerindians from Mexican exile following the civil war, and ineffective land reform. The pull factors are the (unjustified) beliefs that the land is rich and job opportunities are plentiful, plus a knowledge that the government uses the Petén as a

social safety valve and allows squatter settlements while failing to enforce conservation regulations that are supposed to protect the area's environmentally sensitive woodlands.

Belize, strictly speaking, is not a Central American republic in the same tradition as the other six. Until 1981, this country, a wedge of land between northern Guatemala, Mexico's Yucatán Peninsula, and the Caribbean (Fig. 4-11), was a dependency of the United Kingdom known as British Honduras. Slightly larger than Massachusetts and with a minuscule population of only 295,000 (many of African descent), Belize has been more reminiscent of a Caribbean island than of a continental Middle American state. Today, all that is changing as the demographic complexion of Belize is being reshaped. Thousands of residents of African descent have recently emigrated (many went to the United States) and were replaced by tens of thousands of Spanish-speaking immigrants. Most of the latter are refugees from strife in nearby Guatemala, El Salvador, and Honduras, and their proportion of the Belizean population has risen from 33 to nearly 50 percent since 1980. Within the next few years the newcomers will be in the majority, Spanish will become the *lingua franca*, and Belize's cultural geography will exhibit an expansion of the Mainland at the expense of the Rimland.

The Belizean transformation extends to the economic sphere as well. No longer just an exporter of sugar and bananas, Belize is producing new commercial crops, and its seafood-processing and clothing industries have become major revenue earners. Also important is tourism, which annually lures more than 150,000 vacationers to the country's Mayan ruins, resorts, and newly legalized casinos; a growing speciality is ecotourism, based on the natural attractions of the country's near-pristine environment. Belize is also known as a center for *offshore banking*—a financial haven for foreign companies and individuals who want to avoid paying taxes in their home countries.

Honduras is a country on hold as it struggles to rebuild its battered infrastructure and economy. We noted earlier that hurricanes rank among the most dangerous of Middle America's natural hazards, and Honduras had the misfortune of taking the brunt of the region's worst in 1998. That tropical cyclone, named Mitch, rampaged all across northern Central America and proved to be the costliest disaster in the modern history of the Western Hemisphere. Honduras was hit the hardest, and the consequences were catastrophic as massive floods and mudslides were unleashed across the country, killing 9200 people, demolishing more than 150,000 homes, destroying 21,000 miles of roadway and 335 bridges, and rendering 2 million homeless. Also devastated was the critical agricultural sector that employed two-thirds of Honduras's

The text recounts the details of the horrifying impact of Hurricane Mitch on Honduras in late 1998. Yet, despite this devastating blow to their already struggling country, the resilience of the Hondurans amazed international disaster-response teams. The gargantuan reconstruction effort began almost immediately, particularly at the grass-roots level. Here we see a new housing project nearing completion adjacent to the town of Choluteca in southernmost Honduras, just inland from the Gulf of Fonseca, the country's small window on Central America's Pacific coastline. This contribution to the post-Mitch rural rebuilding movement was a gift of Caritas, a French-based multinational consortium of more than 150 Roman Catholic relief and social-service organizations.

labor force, accounted for nearly a third of its gross domestic product, and earned more than 70 percent of its foreign revenues.

With 7.1 million inhabitants, about 90 percent mestizo, bedeviled Honduras still has years to go even to restore what was already the third poorest economy in the Americas (after Haiti and Nicaragua). Agriculture, livestock, forestry, and limited mining formed the mainstays of the pre–1998 economy, with the familiar Central American products—bananas, coffee, shellfish, and apparel—earning most of the external income. There has also been some growth of maquiladora-type light industry around San Pedro Sula near the northwestern coast, dominated by a complex of sweatshops that finish clothing designed and mostly produced elsewhere. Foreign investors, however, have been wary of risking additional funds because of the country's reputation as one of the world's most corrupt (U.S.-bound drug smuggling flourishes here). Nonetheless, Honduras has a democratically elected government, although the military wields considerable power.

Honduras, in direct contrast to Guatemala, has a lengthy Caribbean coastline and a small window on the Pacific (Fig. 4-11). The country also occupies a critical place in the political geography of Central America, flanked as it is by Nicaragua, El Salvador, and Guatemala—all of whom are continuing to grapple with the aftermath of years of internal conflict and, most recently, natural disaster. The road back to economic viability is an arduous one, but once

traversed will still leave four out of five Hondurans deeply mired in poverty and the country with little overall improvement in its development prospects.

El Salvador is Central America's smallest country territorially, smaller even than Belize, but with a population about 25 times as large (6.9 million) it is the most densely peopled. Again, like Belize, it is one of only two continental republics that lack coastlines on both the Caribbean and Pacific sides (Fig. 4-11). El Salvador adjoins the Pacific in a narrow coastal plain backed by a chain of volcanic mountains, behind which lies the country's heartland. Unlike neighboring Guatemala, El Salvador has a quite homogeneous population (94 percent mestizo and just 5 percent Amerindian). Yet ethnic homogeneity has not translated into social or economic equality or even opportunity. Whereas other Central American countries were called banana republics, El Salvador was a coffee republic, and the coffee was produced on the huge landholdings of a few landowners and on the backs of a subjugated peasant labor force. The military supported this system and repeatedly suppressed violent and desperate peasant uprisings.

From 1980 to 1992, El Salvador was torn by a devastating civil war that was worsened by outside arms supplies from the United States (supporting the government) and Nicaragua (aiding the Marxist rebel forces). But ever since the negotiated end to that war, efforts have been under way to prevent a recurrence because El Salvador is having difficulty overcoming its legacy of searing inequality. The civil war did have one positive result: affluent citizens who left the country and did well in the United States and elsewhere send substantial funds back home, which now provide the largest single source of foreign revenues. This has helped stimulate such industries as apparel and footwear manufacturing as well as food processing. But a major stumbling block to revitalization of the agricultural sector has again been land reform, and El Salvador's future still hangs in the balance.

Nicaragua is best approached by reexamining the map (Fig. 4-11), which underscores the country's pivotal position in the heart of Central America. The Pacific coast follows a southeasterly direction, but the Caribbean coast is oriented north-south so that Nicaragua forms a triangle of land with its lakeside capital, Managua, located in a valley on the mountainous, earthquake-prone, Pacific side (the country's core area has always been located here). The Caribbean side, where the uplands yield to a coastal plain of rainforest, savanna, and swampland, has for centuries been home to Amerindian peoples such as the Miskito, who have been remote from the focus of national life.

Until the end of the 1970s, Nicaragua was the typical Central American republic, ruled by a dictatorial government and exploited by a wealthy land-owning minority, its export agriculture dominated by huge plantations owned by foreign corporations. It was a situation ripe for insurgency, and in 1979 leftist rebels overthrew the government; but the new regime quickly produced its own excesses, resulting in civil war for most of the 1980s. The conflict ended in 1990, and more democratic governments have since been voted into office. Nonetheless, the former rebels continue to wield influence in Nicaraguan society, and the country remains too divided to resolve most economic and political issues.

Nicaragua's economy has been a leading casualty of this turmoil, and for the past two decades it has ranked as continental Middle America's poorest. Still mired in the difficult aftermath of Hurricane Mitch, Nicaragua is struggling to rebuild much of its infrastructure. Thus it faces yet another challenge piled upon its already formidable burdens. Agricultural recovery is a leading priority: not only are commercial-farm products major foreign income earners, but if farming opportunities are not promptly restored, emigration from the countryside to the badly overcrowded towns and cities will accelerate and swamp prospects for urban economic development. Moreover, in the months before Mitch, the Nicaraguans had finally resolved property-ownership issues stemming from land redistribution during the civil war. With more than 200,000 families and at least one-third of the country's arable farmland affected, continued national stability necessitates that this vital social reform get back on track.

When "normal" conditions return, Nicaragua will still possess few comparative advantages in the competition for foreign investment capital against the region's more prosperous republics. Among the more promising development possibilities is the construction of a *dry canal* across the flatlands of southern Nicaragua, with high-speed trains ferrying freight-filled containers between Caribbean and Pacific ports that can be directly loaded onto ultramodern ships too large to fit through the isthmus's aging Panama Canal. Perhaps Middle America's land bridge can support additional interocean transit corridors, but here Nicaragua will confront competitors who are proposing such projects in Mexico (as we saw), Colombia, and even in Panama itself. Economic opportunity, however, must also be weighed against looming demographic disaster: unless Nicaragua can reduce the very high natural increase rate in its population of 5.7 million, the country's living standards will not have a chance to improve.

Costa Rica underscores what was said about Middle America's endless variety and diversity because it differs significantly from its neighbors and from the norms of Central America as well. Bordered by two volatile countries (Nicaragua to the north and Panama to the east), Costa Rica is a nation with an old democratic tradition and, in this cauldron, no standing army for the past half-century! Although the country's Hispanic imprint is similar to that found elsewhere on the Mainland, its early independence, its good fortune to lie remote from regional strife, and its leisurely pace of settlement allowed Costa Rica the luxury of concentrating on its economic development. Perhaps most important, internal political stability has prevailed over much of the past 175-odd years.

Like its neighbors, Costa Rica, is divided into environmental zones that parallel the coasts. The most densely settled is the central highland zone, lying in the cooler *tierra templada*, whose heartland is the Valle Central (Central Valley), a fertile basin that contains the country's main coffee-growing area and the leading population cluster focused on San José—the virtually slumless capital city that is the most cosmopolitan urban center between Mexico City and the primate cities of northern South America. To the east of the highlands are the hot and rainy Caribbean lowlands, a sparsely populated segment of Rimland where many plantations have been abandoned and replaced by subsistence farmers. Between 1930 and 1960, the U.S.-based United Fruit Company shifted most of the country's banana plantations from this crop-disease-ridden coastal plain to Costa Rica's third zone—the plains and gentle slopes of the Pacific coastlands. This move gave the Pacific zone a major boost in economic growth, and it is now an area of diversifying and expanding commercial agriculture.

The long-term development of Costa Rica's economy has given it the region's highest standard of living, literacy rate, and life expectancy. Agriculture continues to dominate (with bananas, coffee, seafood, and tropical fruits the leading exports), and tourism is expanding steadily as the country now attracts more than a million foreign visitors a year. But the most impressive economic gains are being generated by a single new industrial complex that the Intel Corporation opened just outside San José in 1998. Here, the world's leading maker of semiconductors has built one of its largest assembly/test facilities in which silicon chips manufactured at fabrication plants are transformed into the processors that constitute the brains of personal computers.

These advances notwithstanding, the country's veneer of development cannot mask serious problems. In terms of social structure, about one-quarter of the population of 4 million is trapped in an unending cycle of poverty, and the huge gap between the poor and the affluent is constantly widening. In the environmental sphere, recent progress has come at the expense of the tropical forest: even though the rate of woodland destruction is now slowing, it is too late to avert an ecological disaster because more than 75 percent of Costa Rica's rainforest has vanished (see box titled "Tropical Deforestation").

Politically, Costa Rica remains quite stable. Despite its proximity to the region's trouble spots, it has resisted involvement because the overwhelming majority of its peace-loving people prefer the country to maintain its neutrality as "the Switzerland of Central America."

Panama owes its existence to the idea of a canal connecting the Atlantic and Pacific oceans to avoid the lengthy circumnavigation of South America. In the 1880s, when Panama was still an extension of neighboring Colombia, a French company tried and failed to build such a waterway here. By the turn of the twentieth century, U.S. interest in a Panama canal rose sharply, and the United States in 1903 proposed a treaty that would permit a renewed effort at construction across Colombia's Panamanian isthmus. When the Colombian Senate refused to go along, Panamanians rebelled, and the United States supported this uprising by preventing Colombian forces from intervening. The Panamanians, at the behest of the United States, declared their independence from Colombia, and the new republic immediately granted the United States rights to the Canal Zone, averaging about 10 miles (16 km) in width and just over 50 miles (80 km) in length.

Soon canal construction commenced, and this time the project succeeded as American technology and medical advances triumphed over a formidable set of obstacles. The Panama Canal (see inset map, Fig. 4-11) was opened in 1914, a symbol of U.S. power and influence in Middle America. The Canal Zone was held by the United States under a treaty that granted it "all the rights, powers, and authority" in the area "as if it were the sovereign of the territory." Such language might suggest that the United States held rights over the Canal Zone in perpetuity, but the treaty nowhere stated specifically that Panama permanently yielded its own sovereignty in that transit corridor. In the 1970s, as the canal was transferring more than 14,000 ships per year (that number is now only slightly lower, but the cargo tonnage is up significantly) and generating hundreds of millions of dollars in tolls, Panama sought to terminate U.S. control in the Canal Zone. Delicate negotiations began. In 1977, an agreement was reached on a staged withdrawal by the United States from the territory, first from the Canal Zone and then from the Panama Canal itself (a process completed on December 31, 1999).

Panama today reflects some of the usual geographic features of the Central American republics. Its population of 3 million is more than two-thirds mestizo and also contains substantial Amerindian, white, and black minorities. Spanish is the official language, but English is also widely used. Ribbon-like and oriented east-west, Panama's topography is mountainous and hilly. Eastern Panama, especially Darien Province adjoining Colombia, is densely forested, and here is the only remaining gap in the intercontinental Pan American Highway. Most of the rural population lives in the uplands west of the canal; there, Panama produces bananas, shrimps and other seafood, sugarcane, coffee, and rice. Much of the urban population is concentrated in the vicinity of the waterway, anchored by the cities at each end of the canal, Colón and Panama City.

The Panama Canal, despite its age and inability to accommodate the largest 20 percent of the vessels in today's global merchant fleet, remains the country's focus, its lifeline, and—with a significant proportion of the world's cargoes moving through the waterway each year—its future. To help assure that future astride this crucial international trading artery, the canal's new Panamanian owners are busily pursuing opportunities they hope will transform their trading hub into a full-fledged Central American economic tiger. Their initial task, now successfully completed, was to demonstrate to the international community that they are fully capable of operating and maintaining the canal without sharply increasing its tolls. They have also begun to attract major new foreign investments, which are being channeled into a number of maritime, railroad, and manufacturing projects.

Near the northern end of the Panama Canal lies the city of Colón, site of the Colón Free Zone, a huge trading entrepôt designed to transfer and distribute goods bound for South America. It is augmented by the Manzanillo

Tropical Deforestation

*B*efore the Europeans arrived, two-thirds of continental Middle America was covered by tropical rainforests. The clearing and destruction of this precious woodland resource to make way for expanding settlement frontiers and the exploitation of new economic opportunities began in the sixteenth century during the Spanish colonial era, and the practice has continued systematically ever since. In recent decades, however, the pace of **tropical deforestation** in Central America has accelerated alarmingly, and since 1950 almost 90 percent of the region's forests have been decimated. Now, about 3 million acres of Central American and Mexican woodland disappear each year, an area equivalent to one-third the size of Belgium. El Salvador has already lost 99 percent of its forests, and most of the six other republics will soon reach that stage.

The causes of tropical deforestation are related to the persistent economic and demographic problems of disadvantaged countries. In Central America, the leading cause has been the need to clear rural lands for cattle pasture as many countries, especially Costa Rica, became meat producers and exporters. The price of this environmental degradation has been enormous, although some gains have been recorded. Because tropical soils are so nutrient-poor, newly deforested areas can function as pastures for only a few years at most. These fields are then abandoned for other freshly cut lands and quickly become the ravaged landscape seen in the photo in this box. Without the protection of trees, local soil erosion and flooding immediately become problems, affecting still-productive nearby areas (a sequence of events that reached catastrophic dimensions all across Honduras and northern Nicaragua when Hurricane Mitch struck in 1998). A second cause of deforestation is the rapid logging of tropical woodlands as the timber industry increasingly turns from the exhausted forests of the mid-latitudes to harvest the rich tree resources of the equatorial zones, responding to accelerating global demands for housing, paper, and furniture. The third major contributing factor is related to the region's population explosion: as more and more peasants are required to extract a subsistence from inferior lands, they have no choice but to cut down the remaining forest for both firewood and additional crop-raising space, and their intrusion prevents the trees from regenerating (Haiti is the extreme example of this denudation process).

Although deforestation is a depressing event, tropical pastoralists, farmers, and timber producers do not consider it to be life-threatening, and perhaps it even seems to offer some short-term economic advantages. Why, then, should there be such an outcry from the scientific community? And why should the World Resources Institute call this "the world's most pressing land-use problem"? The answer is that unless immediate large-scale action is taken, by the middle of this century the world's tropical rainforests will be reduced to two disappearing patches—the western Amazon Basin of northern South America and the middle Congo Basin of Equatorial Africa.

The tropical forest, therefore, must be a very important part of our natural world, and indeed it is. Biologically, the rainforest is by far the richest, most diversified arena of life on our planet: even though it covers only about (a shrinking) 3 percent of the Earth's land area, it contains about three-quarters of all plant and animal species. Its loss would cause not only the extinction of millions of species, but also what ecologist Norman Myers calls "the death of birth" because the evolutionary process that produces new species would be terminated. Because tropical rainforests already yield countless valuable medicinal, food, and industrial products, many potential disease-combatting drugs or new crop varieties to feed undernourished millions will be irretrievably lost if they are allowed to disappear.

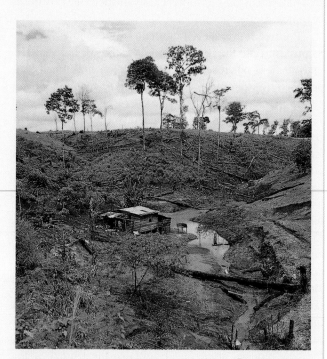

This scene in Costa Rica shows how badly the land can be scarred in the wake of deforestation. Without roots to bind the soil, tropical rains swiftly erode the unprotected topsoil.

FROM THE FIELD NOTES

"The Panama Canal remains an engineering marvel 90 years after it opened in August 1914. The parallel lock chambers each are 1000 feet long and 110 feet wide, permitting vessels as large as the Queen Elizabeth II to cross the isthmus. Ships are raised by a series of locks to Gatún Lake, 85 feet above sea level. We watched as tugs helped guide the QEII into the Gatún Locks, a series of three locks leading to Gatún Lake, on the Atlantic side. A container ship behind the QEII is sailing up the dredged channel leading from the Limón Bay entrance. The lock gates are 65 feet wide and seven feet thick, and range in height from 47 to 82 feet. The motors that move them are recessed in the walls of the lock chambers. Once inside the locks, the ships are pulled by powerful locomotives called *mules* that ride on rails that ascend and descend the system. It was still early morning, and a major fire, probably a forest fire, was burning near the city of Colón, where land clearing was in progress. This was the beginning of one of the most fascinating days ever."

International Terminal, an ultramodern port facility capable of transshipping more than 1000 containers a day. By 2002 China had become the third-largest user of the Canal (after the United States and Japan) and accounted for more than 20 percent of the cargo entering the Colón Free Zone.

The Panama Canal has political as well as economic implications for Panama. For many years, Panama has recognized Taiwan as an independent entity and has sponsored resolutions to get Taiwan readmitted to the United Nations (where it was ousted in favor of China in 1971). In return, Taiwan has spent hundreds of millions of dollars in Panama in the form of investment and direct aid. But now, China's growing presence in the Canal and the Colón Free Zone are causing a conundrum. Panama has awarded a contract to a Hong Kong-based (and thus Chinese) firm to operate and modernize the ports at both ends of the Canal, resulting in investments approaching $500 million. China is lobbying Panama for a switch in recognition, using this commitment as leverage. But the Panamanians are cautious. Having just ousted the United States, they are not eager to be dominated by another giant. It is more profitable to play the "two Chinas" off against each other.

Panama also feels the effects of its location adjacent to rebel-plagued, drug-infested, and unstable Colombia. For many years Panama City, Central America's only coastal capital city, was a financial center that handled more than just the funds generated by the Canal. The link between the Colón Free Zone and Panama City also had a money-laundering dimension, and the Miami-like skyline of the capital reflected that. In the 1990s Colombia was the Colón Free Zone's biggest customer, but when Panama began a crackdown on money laundering, its share of trade dropped significantly. In October 2000 *The Economist* quoted a businessman in the Free Zone as stating that before 1999, perhaps 90 percent of what was sold to Colombia was contraband; after the crackdown it was down to 70 percent.

Fortunately for Panama, its border with Colombia in the Darien is short (170 mi/270 km) and densely forested, occasionally providing hideouts for rebel bands and traffickers but creating a buffer between the main theaters of rebel activity and the Panamanian core area. Colombia's campaign against drug production, as we note in Chapter 5, focuses on other parts of the country and plagues its other neighbors far more. But Colombia's chronic difficulties are having another effect: well-to-do Colombians, looking for a relatively safe place to live and invest, are moving to Panama in growing numbers. Panama welcomes them, an ironic revival of linkage to a country on which the Panamanians turned their backs a century ago.

◗ THE CARIBBEAN BASIN

As Figure 4-2 reveals, the Caribbean Basin, Middle America's island region, consists of a broad arc of numerous islands extending from the western tip of Cuba to the southern coast of Trinidad. The four larger islands or *Greater Antilles* (Cuba, Hispaniola, Jamaica, and Puerto

Rico) are clustered in the western half of this arc. The smaller islands or *Lesser Antilles* extend to the east in a crescent-shaped zone from the Virgin Islands to Trinidad and Tobago. (Breaking this tectonic-plate-related regularity are the Bahamas and the Turks and Caicos, north of the Greater Antilles, and numerous other islands too small to appear on a map at the scale of Fig. 4-2.)

On these islands, whose combined land area constitutes only 9 percent of Middle America, lie 33 states and several other entities. (Europe's colonial flags have not totally disappeared from this region, and the U.S. flag flies over Puerto Rico.) The populations of these states and territories, however, comprise 21 percent of the entire geographic realm, making this the most densely peopled part of the Americas.

Economic and Social Patterns

This is also one of the poorest parts of the world. Environmental, demographic, political, and economic circumstances combine to impede development in the Caribbean region at almost every turn. Cuba should be the Caribbean's leader, but its dictatorial regime of the 1950s collapsed against a socialist revolution that turned the island into a communist outpost—and made it an adversary of the United States with calamitous consequences for its people. Haiti should be a magnet for tourists and a source of textiles and farm products, but it is the Western Hemisphere's poorest country from which desperate people risk their lives in flimsy boats to reach the mainland. Jamaica, with its fertile fields, scenic coasts, and bauxite mines, should have a far higher GNP than Table G-1 records. But the Caribbean's farmers do not have free and open access to American markets, and European farmers are protected against Caribbean products by subsidies and tariffs. The Caribbean region is scenically beautiful, but it is a difficult place to make a living.

It was not always thus. Untold wealth flowed to the European colonists who invaded this region, enslaved and eventually obliterated the indigenous Amerindian (Carib) population, and brought to these islands in bondage the Africans to work on the sugar plantations that made them rich. By the time competition from elsewhere ended the near-monopoly Caribbean sugar had enjoyed on European markets, the seeds of disaster had been sown.

As the sugar trade collapsed, millions were pushed into a life of subsistence, malnutrition, and even hunger. Population growth now created ever-greater pressure on the land, but Caribbean islanders had few emigration options in their fragmented, mountainous habitats. The American market for sugar allowed some island countries to revive their sugar exports, including the Dominican Republic and Jamaica (Cuba's exports went to the Soviet Union, but when the USSR disintegrated Cuba's sugar industry was doomed). Some agricultural diversification occurred in the Lesser Antilles where bananas, spices, and sea-island cotton replaced sugar, but farming is a risky profession in the Caribbean. Foreign markets are undependable and producers are trapped in a disadvantageous international economic order they cannot change.

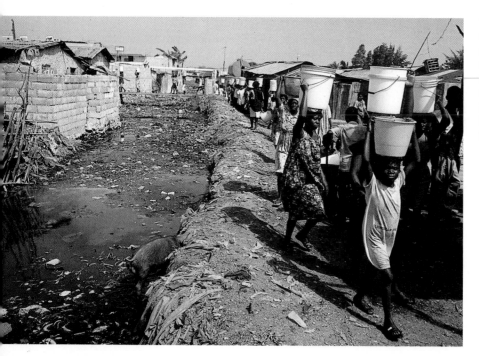

The Caribbean is a region of sharp, often searing contrasts. Haiti is the Western Hemisphere's poorest country, its city slums (here in the capital, Port-au-Prince) the most desperate of human environments. Tourism and wealthy waterfront living create another face of the region, generating landscapes of conspicuous consumption—but producing some of the limited economic opportunities in the Caribbean Basin.

As a result, the majority of those who remain in the rural areas eke out a precarious living from small plots of ground. Mired in poverty and subject to disease, they do not have the power to change their lives. Even if all the better land now devoted to export crops were assigned to subsistence farmers, the picture would improve very little. Farming methods are dated, equipment is in short supply, and land inheritance customs, where these apply, have the effect of subdividing peasant families' plots excessively, reducing productivity still more. And soil erosion is a constant threat. For example, Haiti's land has become so ravaged by agricultural overuse and deforestation that almost the entire country is now an agricultural wasteland.

Not surprisingly, many farmers and their families simply abandon their land and leave for towns and cities, hoping to do better there. Although the Caribbean region is less urbanized, overall, than other parts of the Americas, it is far more urbanized than, say, Africa or the Pacific Realm. Nearly two-thirds of the Caribbean's population now live in cities such as Santo Domingo in the Dominican Republic, Havana in Cuba, Port-au-Prince in Haiti, and San Juan in Puerto Rico. Many of the region's cities reflect the poverty of the migrants who were driven to seek refuge there. Port-au-Prince has some of the world's worst slums, and desolate squatter settlements ring cities such as Kingston, Jamaica, and smaller towns.

The human geography of the Caribbean region still carries imprints of the cultures of Subsaharan Africa. And the legacy of European domination also lingers. The historical geography of Cuba, Hispaniola, and Puerto Rico is suffused with Hispanic culture; Haiti and Jamaica carry stronger African legacies. But the reality of this ethnic diversity is that European lineages still hold the advantage. Hispanics tend to be in the best positions in the Greater Antilles; people who have mixed European-African ancestries, and who are described as **mulatto**, rank next. The largest part of this social pyramid is also the least advantaged: the Afro-Caribbean majority. In virtually all societies of the Caribbean, the minorities hold disproportionate power and exert discrepant influence. In Haiti, the mulatto minority accounts for perhaps 5 percent of the population but has long held most of the power. In the adjacent Dominican Republic, the pyramid of power puts Hispanics (16 percent) at the top, the mixed sector (73 percent) in the middle, and the Afro-Caribbean minority (11 percent) at the bottom. Historic advantage has a way of perpetuating itself.

The composition of the population of the islands is further complicated by the presence of Asians from both China and India. During the nineteenth century, the emancipation of slaves and ensuing local labor shortages brought some far-reaching solutions. Some 100,000 Chinese emigrated to Cuba as indentured laborers, and Jamaica, Guadeloupe, and especially Trinidad saw nearly 250,000 South Asians arrive for similar purposes. To the African-modified forms of English and French heard in the Caribbean, therefore, can be added several Asian languages. The ethnic and cultural variety of the plural societies of Caribbean America is indeed endless.

Countries in many realms and regions of the world today are seeking ways to strengthen their economies through international cooperation, and the Caribbean region is no exception. Aware of the negative impact NAFTA was likely to have on their access to the U.S. market, 25 Caribbean countries in 1994 formed the Association of Caribbean States (ACS). This group includes Spanish-speaking Cuba and the Dominican Republic as well as former British colonies and even mainland Central American states as well as Colombia and Venezuela. In fact, trade among ACS members is very limited; their joint objective is access to major international markets. But in this arena, the ACS countries have actually lost ground both in the United States (where Mexico has the advantage under NAFTA rules) and the European Union, where farm subsidies make it difficult for ACS countries to compete.

The Greater Antilles

The four islands of the Greater Antilles contain five political entities: Cuba, Haiti, the Dominican Republic, Jamaica, and Puerto Rico (Fig. 4-2). Haiti and the Dominican Republic share the island of Hispaniola.

Cuba, the largest Caribbean island-state in terms of both territory (43,000 square miles/111,000 sq km) and population (11.4 million), lies only 90 miles (145 km) from the southern the tip of Florida (Fig. 4-13). Havana, the now-dilapidated capital, lies almost directly across from the Florida Keys on the northwest coast of the elongated island. As the map shows, sugar long was Cuba's economic mainstay; the plantations, once the property of rich landowners, extend all across the territory. But in a few years this map will show that sugarcane is losing its position as the leading Cuban foreign-exchange earner. Mills are being closed down, and the cane fields are being cleared for other crops and for pastures. Cuba's other economic opportunities are considerable: there are three mountainous areas, of which the southeastern chain, the Sierra Maestra, is the highest and most extensive. These highlands create considerable environmental diversity as reflected by extensive, timber-producing tropical forests and varied soils on which crops ranging from tobacco to subtropical and tropical fruits are grown. Rice and beans are the staples (but Cuba cannot meet its needs and must import food); the savannas of the center and west

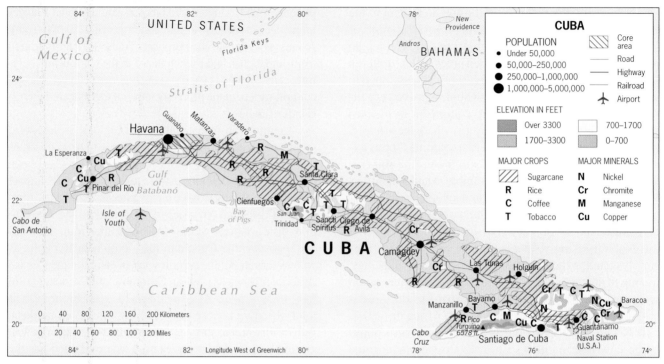

FIGURE 4-13

support livestock. Cuba has limited mineral reserves and no oil, but its nickel deposits are extensive and have been mined for a century.

Like several Central American republics, Cuba was strongly influenced by U.S. power throughout its modern history. American forces got involved on Cuba's side in the war of independence against Spain, and when Cuba became independent, in 1899, the U.S. military restored order and helped build infrastructure. Fifty years later, when Fulgencio Batista was dictator, Havana was an American playground and the regime could be assured of Washington's noninterference while Cuba's economy thrived on gambling, graft, tourism, and the exploitation of some of the region's poorest workers.

Such were the inequalities in Cuba's social sphere that the country was ripe for revolution, and it came in the 1950s. While Cuba's military was supplied with U.S. weapons and the Batista regime had Washington's approval, there was no American intervention as the insurgency gathered momentum. Eventually Fidel Castro overthrew the Batista dictatorship and soon aligned himself with the Soviet Union. Communism had gained a foothold on the doorstep of the United States.

This transition precipitated the first of many waves of emigration by Cubans to the United States, mainly Florida. In 1961, Washington backed an incursion by forces drawn from these exiles—the Bay of Pigs invasion—which failed disastrously. In the following year, the Soviet decision to place missiles in Cuba aimed at the United States provoked a crisis that brought the world close to nuclear war.

In the decades that followed, Cuba became dependent on Soviet markets for its products and for Soviet aid for its social experiment. Cuban sugar had a privileged place on Soviet markets; Soviet oil came to Cuba at bargain prices. The Castro regime made use of its revenues to encourage communist insurrections in Central America and in Africa, and to build impressive educational and medical programs at home. But Cuba was not immune to the failures and inefficiencies of communist systems that were stultifying Eastern Europe: its economy stagnated, repression of opponents intensified, and Cubans took to risking their lives on small boats to reach the shores of the United States. When the Soviet Union collapsed in 1991 and Cuba's sugar markets and cheap energy were no more, it was time for the communist regime to liberalize and set a new course.

In this effort, Castro and his clique have proven to be a failure, although the American policy of isolating Cuba and prohibiting trade has limited his options. Without liberalizing political life in Cuba, Castro did allow some free-market reforms that stimulated tourism, one of the island's great opportunities, and permitted growth in some sectors of the economy, notably light manufacturing and telecommunications. But when some restiveness arose among the people, including a muted demand for free and open elections, he backtracked and closed the door that had been slightly ajar.

Poverty, crumbling infrastructure, crowded slums, and unemployment mark the Cuban cultural landscape, but Cuba's regime still has support among the general population. During the Castro period, much was done to bring the Afro-Cuban population into the mainstream through education and health provision. Cubans point to Guatemala, Nicaragua, and El Salvador and ask whether those countries are better off than they are under Castro. But in the United States, the view is different: Cuba, exiles and locals agree, could be the shining star of the Caribbean, its people free, its tourist economy booming, its products flowing to American markets. "The future [economic tiger] of the Caribbean," suggested a Cuban geographer recently. Much will have to change for that prediction to come true.

Jamaica lies across the deep Cayman Trench from southern Cuba, and a cultural gulf separates these two countries as well. Jamaica, a former British dependency, has an almost entirely Afro-Caribbean population. As a member of the British Commonwealth, Jamaica still recognizes the British monarch as the chief of state, represented by a governor-general. The effective head of government in this democratic country, however, is the prime minister. English remains the official language here, and British customs still linger.

Smaller than Connecticut and with 2.7 million people, Jamaica's GNP has steadily declined over the past few decades despite its relatively slow population growth. Tourism has become the largest source of income, but the markets for bauxite (aluminum ore), of which Jamaica is a major exporter, have dwindled. And like other Caribbean countries, Jamaica has trouble making money from its sugar exports. Jamaican farmers also produce crops ranging from bananas to tobacco, but the country faces the disadvantages on world markets common to those in the periphery. Meanwhile, Jamaica must import all of its oil and much of its food, because the densely populated coastal flatlands suffer from overuse and shrinking harvests.

The capital, Kingston on the south coast, reflects Jamaica's economic struggle. Almost none of the hundreds of thousands of tourists who visit the country's beaches, explore its Cockpit Country of (karst) limestone towers and caverns, or populate the cruise ships calling at Montego Bay or other points along the north coast get even a glimpse of what life is like for the ordinary Jamaican (see the Issue Box entitled "The Role of the Tourist Industry in Middle American Economies").

Haiti, the poorest state in the Western Hemisphere by virtually any measure, occupies the western part of the island of Hispaniola, directly across the Windward Passage from eastern Cuba (Fig. 4-14). Arawaks, not Caribs, formed the dominant indigenous population here, but the Spanish colonists who first took Hispaniola killed many, worked most of the survivors to death on their plantations, and left the others to die of the diseases they brought with them. French pirates established themselves in coastal coves along Hispaniola's west coast even as Spanish activity focused on the east, and just before the eighteenth century opened France formalized the colonial status of "Saint Dominique." During the following century, French colonists established vast plantations in the valleys of the northern mountains and in the central plain, laid out elaborate irrigation systems, built dams, and brought in a large number of Africans in bondage to work the fields. Prosperity made the colonists rich, but the African workers suffered terribly. They rebelled and, in a momentous victory over their European oppressors, established the independent republic of Haiti (resurrecting the original Arawak name for it) in 1804.

Defeating the French turned out to be easier than running the first Afro-Caribbean republic in the hemisphere. Strife among Haitian groups destabilized the country; the agricultural system the French had designed fell apart for lack of maintenance; conflict arose between Haiti and its neighbor, the Hispanic Dominican Republic. The United States took an interest in Haiti's affairs and became the country's dominant commercial partner, but its operations were thwarted by the continuing round of revolts and killings. In 1915 U.S. armed forces occupied Haiti and controlled it until 1934, but following their departure the republic fell into another round of dictatorial rule, "presidents for life," private military-police controls, political murder, corruption, and economic failure. Toward the end of the twentieth century the United States sought to assist Haiti toward some form of democratic government, alienating the mulatto minority that, for nearly two centuries, had held power. A reformist outsider, Jean-Bertrand Aristide, was elected president in 1990, but a year later he was forced into exile by the old oligarchy, and once again Haiti was under military rule. In 1994 the United States threatened an invasion unless the democratically elected president was allowed to return, and Aristide came back to form a new government.

These events occurred against a backdrop of economic and social disaster. Today Haiti's GNP is only about half that of Jamaica and lower than that of many of Africa's poorest countries; foreign aid makes possible most of the country's limited public expenditures. Health conditions are dreadful: malnutrition is common, diseases ranging from malaria to tuberculosis are rife, hospital beds and doctors are in short supply. Conditions in and around the capital, Port-au-Prince, are among the worst in the world (see photo, p. 214)—600 miles (1000 km) from the United States. Countless desperate Haitians have sought to escape to America, and many have succeeded; a "Little Haiti" is

The Role of the Tourist Industry in Middle American Economies

IN SUPPORT OF THE TOURIST INDUSTRY

"As THE GENERAL MANAGER of a small hotel in St. Maarten, on the Dutch side of the island, I can tell you that without tourists, we would be in deep trouble economically. Mass tourism, plain and simple, has come to the rescue in the Caribbean and in other countries of Middle America. Look at the numbers. Here it's the only industry that is growing, and it already is the largest worldwide industry. For some of the smaller countries of the Caribbean, this is not just the leading industry but the only one producing external revenues. Whether it's hotel patrons or cruise-ship passengers, tourists spend money, create jobs, fill airplanes that give us a link to the outside world, require infrastructure that's good not just for them but for locals too. We've got better roads, better telephone service, more items in our stores. All this comes from tourism. I employ 24 people, most of whom would be looking for nonexistent jobs if it weren't for the tourist industry.

"And it isn't just us here in the Caribbean. Look at Belize. I just read that tourism there, based on their coral reefs, Mayan ruins, and inland waterways, brought in U.S. $100 million last year, in a country with a total population of only about 300,000! That sure beats sugar and bananas. In Jamaica, they tell me, one in every three workers has a job in tourism. And the truth is, there's still plenty of room for the tourist industry to expand in Middle America. Those Americans and Europeans can close off their markets against our products, but they can't stop their citizens from getting away from their awful weather by coming to this tropical paradise.

"Here's another good thing about tourism. It's a clean industry. It digs no mine shafts, doesn't pollute the atmosphere, doesn't cause diseases, doesn't poison villagers, isn't subject to graft and corruption the way some other industries are.

"Last but not least, tourism is educational. Travel heightens knowledge and awareness. There's always a minority of tourists who just come to lie on the beach or spend all their time in some cruise-ship bar, but most of the travelers we see in my hotel are interested in the place they're visiting. They want to know why this island is divided between the Dutch and the French, they ask about coral reefs and volcanoes, and some even want to practice their French on the other side of the border (don't worry, no formalities, just drive across and start talking). Tourism's the best thing that happened to this part of the world, and other parts too, and I hope we'll never see a slowdown."

CRITICAL OF THE TOURIST INDUSTRY

"You won't get much support for tourism from some of us teaching at this college in Puerto Rico, no matter how important some economists say tourism is for the Caribbean. Yes, tourism is an important source of income for some countries, like Kenya with its wildlife and Nepal with its mountains, but for many countries that income from tourism does not constitute a real and fundamental benefit to the local economies. Much of it may in fact result from the diversion to tourist consumption of scarce commodities such as food, clean water, and electricity. More of it has to be reinvested in the construction of airport, cruise-port, overland transport, and other tourist-serving amenities. And as for items in demand by tourists, have you noticed that places with many tourists are also places where prices are high?

"Sure, our government people like tourism. Some of them have a stake in those gleaming hotels where they can share the pleasures of the wealthy. But what those glass-enclosed towers represent is globalization, powerful multinational corporations colluding with the government to limit the opportunities of local entrepreneurs. Planeloads and busloads of tourists come through on prearranged (and prepaid) tour promotions that isolate those visitors from local society.

"You geographers talk about cultural landscapes. Well, picture this: luxury liners gliding past poverty-stricken villages, luxury hotels towering over muddy slums, restaurants serving caviar when, down the street, children suffer from malnutrition. If the tourist industry offered real prospects for economic progress in poorer countries, such circumstances might be viewed as the temporary, unfortunate byproducts of the upward struggle. Unfortunately, the evidence indicates otherwise. Name me a tourism-dependent economy where the gap between the rich and poor has narrowed.

"As for the educational effect of tourism, spare me the argument. Have you sat through any of those "culture" shows staged by the big hotels? What you see there is the debasing of local culture as it adapts to visitors' tastes. Ask hotel workers how they really feel about their jobs, and you'll hear many say that they find their work dehumanizing because expatriate managers demand displays of friendliness and servitude that locals find insulting to sustain.

"I've heard it said that tourism doesn't pollute. Well, the Alaskans certainly don't agree—they sued a major cruise line on that issue and won. Not very long ago, cruise-ship crews routinely threw garbage-filled plastic bags overboard. That seems to have stopped, but I'm sure you've heard of the trash left by mountainclimbers in Nepal, the damage done by off-road vehicles in the wildlife parks of Kenya, the coral reefs injured by divers off Florida. Tourism is here to stay, but it is no panacea."

Vote your opinion at www.wiley.com/college/deblij

HAITI AND DOMINICAN REPUBLIC

POPULATION
- Under 50,000
- 50,000–250,000
- 250,000–1,000,000
- 1,000,000–5,000,000

— Road — Railroad ✈ Airport

ELEVATION IN FEET
- Over 3300
- 1700–3300
- 700–1700
- 0–700

MAJOR CROPS
- **B** Bananas
- **S** Sugarcane
- **R** Rice
- **C** Coffee
- **F** Fish/Shellfish
- **L** Livestock

MAJOR MINERALS
- **Al** Bauxite
- **Ni** Nickel
- **Au** Gold
- **Na** Salt
- **Gp** Gypsum
- **Cu** Copper

FIGURE 4-14

part of Miami's ethnic mosaic. But unlike Cubans, who are regarded as legitimate refugees because they escaped from the political repression of communism, Haitians are routinely turned back because their claim to asylum is economic, not political.

The **Dominican Republic** has a larger share of the island of Hispaniola than Haiti (Fig. 4-14) in terms of both territory and population, and it is economically much better off despite its own searing social inequalities. Its GNP is more than four times that of Haiti, and the country's economy recorded the fastest growth rate in all of Middle America between 1995 and 2003. Fly along the north-south border between the two countries, and you see a crucial difference: to the west, Haiti's hills and plains are treeless and gulleyed, its soils eroded and its streams silt-laden. To the east, forests drape the countryside and streams run clear.

Indigenous Caribs inhabited this eastern part of Hispaniola, and when the European colonists arrived on the island, they were in the process of driving the Arawaks westward. Spanish colonists made this a prosperous colony, but then Mexico and Peru diverted Spanish attention from Hispaniola and the territory was ceded to France. But Hispanic culture lingered, and after the revolution in Haiti the French gave eastern Hispaniola back to Spain. After the Dominican Republic declared its independence in 1821, Haitian forces invaded it and occupied the republic until 1844, creating an historic animosity that still lingers.

The mountainous Dominican Republic has a wide range of natural environments and a far stronger resource base than Haiti. Nickel, gold, and bauxite have long been exported along with sugar, bananas, coffee, and cocoa, but tourism (the great opportunity lost to Haiti) is the leading industry. A long period of dictatorial rule punctuated by revolutions and U.S. military intervention ended in 1978 with the first peaceful transfer of power following a democratic election. The election of 2000, bringing President Hipólito Mejía into office, proved the robustness of Dominican democracy.

Political stability has brought rich economic rewards. GDP in 2001 grew 10 percent over the previous year, a remarkable surge; the economy is being stimulated by the establishment of a telecommunications sector in 43 free-trade zones where high-tech industries flourish. The government, however, recognizes that the gap between rich and poor (generally, Hispanics and Afro-Caribbeans, respectively) must be closed: an estimated 44 percent of the republic's citizens still live below the poverty line, most of them engaged in subsistence farming. Such poverty is seen not only in the rural, noncommercial countryside but also in the slums of the capital, Santo Domingo, and the smaller cities and towns in which nearly two-thirds of all inhabitants are concentrated. Even in its comparatively bright spots, the Caribbean region has a long way to go to overcome its obstacles.

Puerto Rico is the largest U.S. domain in Middle America, the easternmost and smallest island of the Greater Antilles (Fig. 4-15). This 3400-square-mile (8800-sq-km) island, with a population of 4 million, is larger than Delaware and more populous than Oregon. It fell to the United States more than a century ago during the Spanish-American War of 1898. Since the Puerto Ricans had been struggling for some time to free themselves from Spanish control, this transfer of power was, in their view, only a change from one colonial power to another. As a result, the first half-century of U.S. administration was difficult, and it was not until 1948 that Puerto Ricans were permitted to elect their own governor.

When the island's voters approved the creation of a Commonwealth in a 1952 referendum, Washington and San Juan, the two seats of government, entered into a complicated arrangement. Puerto Ricans are U.S. citizens but pay no federal taxes on local incomes. The Puerto Rican Federal Relations Act governs the island under the terms of its own constitution and awards it considerable autonomy. Puerto Rico receives large annual subsidies from Washington, totaling about U.S. $9 billion per year at the end of the 1990s, which is roughly equivalent to all U.S. foreign aid to the rest of the world combined.

Despite these apparent advantages in the poverty-mired Caribbean, Puerto Rico has not thrived under U.S. administration. Long dependent on a single-crop economy (sugar), the island's industrialization during the 1950s and 1960s was based on its comparatively cheap labor, tax breaks for corporations, political stability, and special access to the U.S. market. As a result, pharmaceuticals, electronic equipment, and apparel top today's list of exports, not sugar or bananas. But this industrialization failed to stem a tide of emigration that carried more than 1 million Puerto Ricans to New York City alone. The same wages that favored corporations kept many Puerto Ricans poor or unemployed. Behind the impressive waterfront skyline of San Juan with its high-rise hotels and tourist attractions lies a landscape of economic malaise. Today, approximately 60 percent of all Puerto Ricans continue to live below the poverty line.

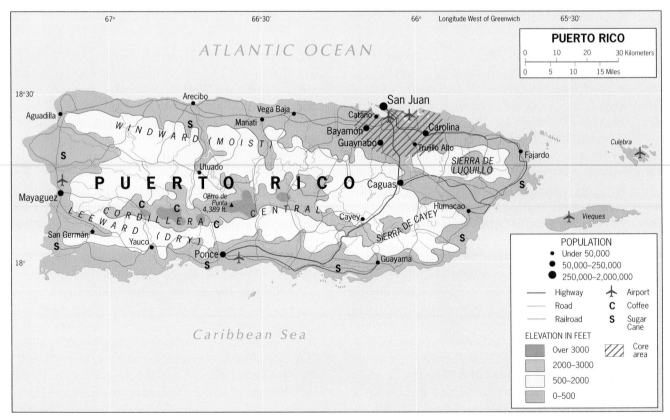

FIGURE 4-15

In his recent book, *Puerto Rico: The Trials of the Oldest Colony in the World*, José Trias Monge leaves no doubt as to the consequences of what he considers to be Washington's failed governance of its largest dependency. Puerto Rico has become one of the major points of U.S. entry for the international drug trade; its crime rates are much higher than those in the United States; corruption is rife; and its political debate is bad tempered. The island's newspapers reflect a degree of polarization that should raise warning flags for the future.

During the 1990s, Puerto Ricans were twice asked to vote on their political future in referenda authorized by the U.S. Congress. On each occasion the island's electorate rejected both independence and joining the United States as the fifty-first State, choosing instead to continue their Commonwealth status. In the first referendum in 1993, the status quo option narrowly defeated Statehood 48.4 to 46.2 percent, and there were subsequent indications that support for joining the United States was on the rise. But the 1998 referendum revealed only a minimal shift in opinion as the Commonwealth vote again bested Statehood, this time by 50.2 to 46.5 percent. Those preferring independence, however, lost ground as the 1993 vote of 4.4 percent shrank to 2.5 percent five years later.

The outcome of the latest referendum will not silence those who, like Trias Monge, argue that any status short of outright sovereignty will serve only to perpetuate Puerto Rico's standing as an American dependency—branding the United States as a colonial power no matter how lofty and democratic its intentions. Thus Puerto Rico will continue to pose a formidable challenge to American statecraft well into the future.

The Lesser Antilles

As Figure 4-2 shows, the Greater Antilles are flanked by two clusters of islands: the extensive Bahamas-Turks/Caicos archipelago to the north and the Lesser Antilles to the east and south. The Bahamas, the former British colony that is now the closest Caribbean neighbor to the United States, alone consists of nearly 3000 coral islands, most of them rocky, barren, and uninhabited, but about 700 carrying vegetation, of which some 30 are inhabited. Centrally positioned New Providence carries most of the country's 300,000 inhabitants and contains the capital, Nassau, a leading tourist attraction.

The Lesser Antilles are grouped geographically into the Leeward Islands and the Windward Islands, a (climatologically incorrect) reference to the prevailing airflows in this tropical area. The Leeward Islands extend from the U.S. Virgin Islands to the French dependencies of Guadeloupe and Martinique, and the Windward Islands from St. Lucia to the Netherlands Antilles off the Venezuelan coast (Fig. 4-2). It would be impractical to detail the individual geographic characteristics of each of the Lesser Antilles, but we should note that these countries and territories share the environmental risks of this region: earthquakes, volcanic eruptions, and hurricanes; that they face to varying degrees similar economic problems in the form of limited domestic resources, overpopulation, soil deterioration, land fragmentation, and market limitations; that tourism has become the leading industry for many; that political status ranges from complete sovereignty to continuing dependency; and that cultural diversity is strong not only between islands but within them as well. In economic terms, the GNP figures given (where available) in Table G-1 may look encouraging, but these numbers conceal the social reality in virtually every entity: the gap between the fortunate few who are well-off and the great majority who are poor remains enormous.

Under such circumstances one looks for hopeful signs, and one of these signs comes from **Trinidad and Tobago**, the two-island republic at the southern end of the Lesser Antilles. This country (population: 1.3 million) has embarked on a natural-gas-driven industrialization boom that could turn this two-island country into an economic tiger. Trinidad has long been an oil producer, but lower world prices and dwindling supplies in recent years forced a reexamination of its natural gas deposits to help counter the downturn. That quickly resulted in the discovery of major new supplies, and this cheap and abundant fuel has sparked a local gas-production boom as well as an influx of energy, chemical, and steel companies from Western Europe, Canada, and even India. Many of the new industrial facilities have agglomerated at the ultramodern Point Lisas Industrial Estate outside the capital, Port of Spain, and they have propelled Trinidad to become the world's leading exporter of ammonia and methanol. Natural gas is also an efficient fuel for the manufacturing of metals, and steelmakers as well as aluminum refiners have been attracted to locate here. With Trinidad lying only a few miles from the Venezuelan coast of South America, it is also a sea-lane crossroads that is well connected to the vast, near-coastal supplies of iron ore and bauxite that are mined in nearby countries, particularly the Brazilian Amazon.

Middle America is a physically, culturally, and economically fragmented and diverse geographic realm that defies generalization. Given its strong Amerindian presence, its North American infusions, and its lingering Western European traditions, this most certainly is not "Latin" America.

FIGURE 5-1

Chapter 5

South America

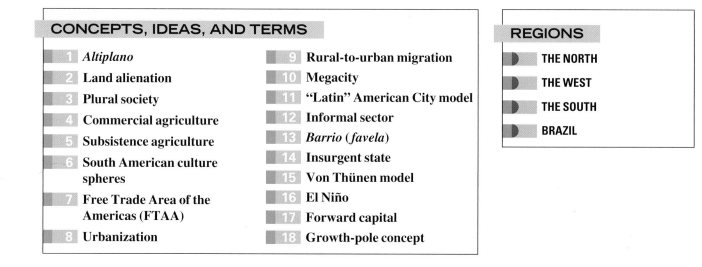

CONCEPTS, IDEAS, AND TERMS

1 *Altiplano*
2 Land alienation
3 Plural society
4 Commercial agriculture
5 Subsistence agriculture
6 South American culture spheres
7 Free Trade Area of the Americas (FTAA)
8 Urbanization
9 Rural-to-urban migration
10 Megacity
11 "Latin" American City model
12 Informal sector
13 *Barrio* (*favela*)
14 Insurgent state
15 Von Thünen model
16 El Niño
17 Forward capital
18 Growth-pole concept

REGIONS

THE NORTH
THE WEST
THE SOUTH
BRAZIL

O F ALL THE continents, South America has the most familiar shape—a giant triangle connected by mainland Middle America's tenuous land bridge to its sister continent in the north. South America also lies not only south but mostly east of its northern counterpart. Lima, the capital of Peru—one of the continent's westernmost cities—lies farther east than Miami, Florida. Thus South America juts out much more prominently into the Atlantic Ocean toward southern Europe and Africa than does North America. But lying so far eastward means that South America's western flank faces a much wider Pacific Ocean, with the distance from Peru to Australia nearly twice that from California to Japan.

As if to reaffirm South America's northward and eastward orientation, the western margins of the continent are rimmed by one of the world's longest and highest mountain ranges, the Andes, a gigantic wall that ex-

tends unbroken from Tierra del Fuego near the southern tip of the triangle to Venezuela in the far north (Fig. 5-1). The other major physiographic feature of South America dominates its central north—the Amazon Basin; this vast humid-tropical amphitheater is drained by the mighty Amazon, which is fed by several major tributaries. Much of the remainder of the continent can be classified as plateau, with the most important components being the Brazilian Highlands that cover most of Brazil southeast of the Amazon Basin, the Guiana Highlands located north of the lower Amazon Basin, and the cold Patagonian plateau that blankets the southern third of Argentina. Figure 5-1 also reveals two other noteworthy river basins beyond Amazonia: the Paraná-Paraguay Basin of south-central South America, and the Orinoco Basin in the far north that drains interior Colombia and Venezuela.

DEFINING THE REALM

Long characterized by regional disparities, political turmoil, and developmental inertia, South America has entered a new era of opportunity. Its major countries, heretofore accustomed to going their separate ways, are discovering the benefits of forging closer individual and multinational ties. More democratic forms of government now mark the political landscape. New transport routes are opening settlement frontiers in many once-remote parts of the continent. The perception now taking hold is that, finally, things are improving and this realm is at the threshold of a period of unprecedented economic growth.

Such optimism, however, must be tempered by the recognition that serious challenges remain to be overcome. As certain areas of South America make real progress, too many others continue to be plagued by infrastructure shortcomings, inefficiency and corruption, and endless rounds of no-gain, boom/bust cycles. Most importantly, throughout the realm, the gulf between the rich and the poor is steadily widening, with the richest 20 percent of the population controlling *70 percent* of South America's wealth, while the poorest 20 percent controls only 2 percent. These numbers reveal the greatest gap between affluence and poverty to be found in any geographic realm—and show that South America is still very much a continent of stupendous contrasts.

MAJOR GEOGRAPHIC QUALITIES OF

South America

1. South America's physiography is dominated by the Andes Mountains in the west and the Amazon Basin in the central north. Much of the remainder is plateau country.

2. Half of the realm's area and half of its population are concentrated in one country—Brazil.

3. South America's population remains concentrated along the continent's periphery. Most of the interior is sparsely peopled, but sections of it are now undergoing significant development.

4. Interconnections among the states of the realm are improving. Economic integration has become a major force, particularly in southern South America.

5. Regional economic contrasts and disparities, both in the realm as a whole and within individual countries, are strong.

6. Cultural pluralism exists in almost all of the realm's countries and is often expressed regionally.

7. Rapid urban growth continues to mark much of the South American realm, and the urbanization level overall now exceeds the levels of both the United States and Europe.

THE HUMAN SEQUENCE

Although modern South America's largest populations are situated in the east and north, during the height of the Inca Empire the Andes Mountains contained the most densely peopled and best organized state on the continent. Although the origins of Inca civilization are still shrouded in mystery, it has become generally accepted that the Incas were descendants of ancient peoples who came to South America via the Middle American land bridge. Thus for thousands of years before the Europeans arrived in the sixteenth century, indigenous Amerindian societies had been developing in South America.

The Inca Empire

About one thousand years ago, a number of regional cultures thrived in Andean valleys and basins and at places along the Pacific coast. By A.D. 1300, the Incas had established themselves in the intermontane basin of Cuzco (Fig. 5-2). With their hearth consolidated, they were now ready to begin forging the greatest pre-European empire in the Americas by steadily conquering and extending their authority over the peoples of coastal Peru and other Andean basins. When the Inca civilization is compared to that of ancient Mesopotamia, Egypt, and the Mexican Aztec Empire, it quickly becomes clear that this civilization was an unusual achievement. Everywhere else, rivers and waterways provided avenues for the circulation of goods and ideas. Here, however, an empire was forged from a series of elongated basins (called ***altiplanos***) in the high Andes, created when mountain valleys between parallel and converging ranges filled with erosional materials from surrounding uplands. These *altiplanos* are often separated by some of the world's most rugged terrain, with high snowcapped mountains alternating with precipitous canyons.

More impressive than the Incas' military victories was their subsequent capacity to integrate the peoples and regions of the Andean domain into a stable and efficiently functioning state. The odds would seem to have

INDIGENOUS AND COLONIAL DOMAINS OF SOUTH AMERICA

- Peoples of the Caribbean fringe
- Andean peoples
- Peoples of the tropical forest
- Marginal peoples
- ← Route of Spanish penetration
- ← Route of Portuguese penetration
- — Colonial boundaries

FIGURE 5-2

been against them because as they progressed their domain became ever more elongated, making effective control much more difficult. The Incas, however, were expert road and bridge builders, colonizers, and administrators, and in an incredibly short time they unified their new territories that stretched from Colombia southward to central Chile (brown zone, Fig. 5-2).

The Incas themselves were always in a minority in this huge state, and their position became one of a ruling elite in a rigidly class-structured society. A bureaucracy of Inca administrators strictly controlled the life of the empire's subjects, and the state was so highly centralized that a takeover at the top was enough to gain power over the entire empire—as the Spaniards quickly proved in the 1530s. The Inca Empire disintegrated abruptly under the impact of the Spanish invaders, but it left behind spectacular ruins such as those at Peru's Machu Picchu. It also bequeathed a legacy of social values that have remained a part of Amerindian life in the Andes to this day and still contribute to fundamental divisions between the Hispanic and Amerindian population in this part of South America.

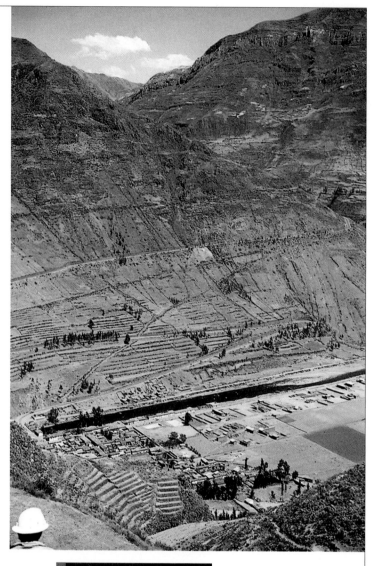

FROM THE FIELD NOTES

"From this high vantage point I got a good perspective of a valley near Pisac in the Peruvian Andes, not far from Cuzco. This was part of the Incan domain when the Spaniards arrived to overthrow the empire, but the terraces you can see actually predate the Inca period. Human occupation in these rugged mountains is very old, and undoubtedly the physiography here changed over time. Today these slopes are arid and barren, and only a few hardy trees survive; the stream in the valley bottom is all the water in sight. But when the terrace builders transformed these slopes, the climate may have been more moist, the countryside greener."

The Iberian Invaders

In South America as in Middle America, the location of indigenous peoples largely determined the direction of the thrusts of European invasion. The Incas, like Mex-ico's Maya and Aztec peoples, had accumulated gold and silver at their headquarters, possessed productive farmlands, and constituted a ready labor force. Not long after the defeat of the Aztecs in 1521, Francisco Pizarro sailed southward along the continent's northwestern coast, learned of the existence of the Inca Empire, and withdrew to Spain to organize its overthrow. He returned to the Peruvian coast in 1531 with 183 men and two dozen horses, and the events that followed are well known. In 1533, his party rode victorious into Cuzco.

At first, the Spaniards kept the Incan imperial structure intact by permitting the crowning of an emperor who was under their control. But soon the breakdown of the old order began. The new order that gradually emerged in western South America placed the indigenous peoples in serfdom to the Spaniards. Great haciendas were formed by **land alienation** (the takeover of former Amerindian lands), taxes were instituted, and a forced-labor system was introduced to maximize the profits of exploitation.

Lima, the west coast headquarters of the Spanish conquerors, soon became one of the richest cities in the world, its wealth based on the exploitation of vast Andean silver deposits. The city also served as the capital of the viceroyalty of Peru, as the Spanish authorities quickly integrated the new possession into their colonial empire (Fig. 5-2). Subsequently, when Colombia and Venezuela came under Spanish control and, later, when Spanish settlement expanded in what is now Argentina and Uruguay, two additional viceroyalties were added to the map: New Granada and La Plata.

Meanwhile, another vanguard of the Iberian invasion was penetrating the east-central part of the continent, the coastlands of present-day Brazil. This area had become a Portuguese sphere of influence because Spain and Portugal had signed a treaty in 1494 to recognize a north-south line 370 leagues west of the Cape Verde Islands as the boundary between their New World spheres of influence. This border ran approximately along the meridian of 50°W longitude, thereby cutting off a sizeable triangle of eastern South America for Portugal's exploitation (Fig. 5-2). But a brief look at the political map of South America (Fig. 5-3) shows that this treaty did not limit Portuguese colonial territory to the east of the 50th meridian. Instead, Brazil's boundaries were bent far inland to include almost the entire Amazon Basin, and the country came to be only slightly smaller in territorial size than all the other South American countries combined. This westward thrust was the work of many Brazilian elements, particularly the *Paulistas*, the settlers of São Paulo who needed Amerindian slave labor to run their plantations.

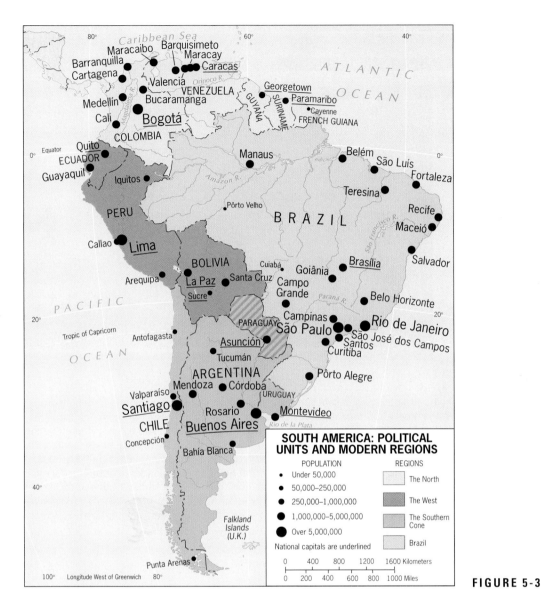

SOUTH AMERICA: POLITICAL
UNITS AND MODERN REGIONS

POPULATION
- Under 50,000
- 50,000–250,000
- 250,000–1,000,000
- 1,000,000–5,000,000
- Over 5,000,000

National capitals are underlined

REGIONS
- The North
- The West
- The Southern Cone
- Brazil

0   400   800   1200   1600 Kilometers

0   200   400   600   800   1000 Miles

FIGURE 5-3

The Africans

As Figure 5-2 shows, the Spaniards initially got very much the better of the territorial partitioning of South America—not just in land quality but also in the size of the aboriginal labor force. When the Portuguese began to develop their territory, they turned to the same lucrative activity that their Spanish rivals had pursued in the Caribbean—the plantation cultivation of sugar for the European market. And they, too, found their labor force in the same source region, as millions of Africans were brought in slavery to the tropical Brazilian coast north of Rio de Janeiro (see map, p. 279). Not surprisingly, Brazil now has South America's largest black population, which is still heavily concentrated in the country's poverty-stricken north-eastern States. With blacks of "pure" and mixed ancestry

today accounting for 44 percent of Brazil's population of 178 million, the Africans decidedly constitute the third major immigration of foreign peoples into South America.

Longstanding Isolation

Despite their adjacent location on the same continent, their common language and cultural heritage, and their shared national problems, the countries that arose out of South America's Spanish viceroyalties (as well as Brazil) have existed in a considerable degree of isolation from one another. Distance and physiographic barriers reinforced this separation, and the realm's major population agglomerations still adhere to the coast, mainly the eastern and northern coasts (Fig. G-9). The viceroyalties existed

FROM THE FIELD NOTES

"Salvador is one of Brazil's oldest and most vibrant cities. Magnificent churches, public buildings, and mansions were built during the time when this was, by many measures, the most important city in the Southern Hemisphere. Long the capital of Brazil, Salvador was the point of entry for tens of thousands of Africans, and the fortune-making plantation economy, augmented by the whaling industry, concentrated enormous wealth here, some of which went into the construction of an opulent city center. But fortunes change, Salvador lost its political as well as economic advantages, and the city fell into disrepair. Walking the streets of the old town in 1982, I noted the state of decay of much of Salvador's architectural heritage and wondered about its survival: weathering in this tropical environment was destroying woodwork, façades, and roofs. But then the United Nations proclaimed Salvador's old town a World Heritage site, and massive restoration began (left). By the late 1990s, much of the district had been revived (right), and tourism's contribution to the local economy was on the rise."

primarily to extract riches and fill Spanish coffers. In Iberia there was little interest in developing the American lands for their own sake. Only after those who had made Spanish and Portuguese America their home and who had a stake there rebelled against Iberian authority did things begin to change, and then very slowly. South America was saddled with the values, economic outlook, and social attitudes of eighteenth-century Iberia—not the best tradition from which to begin the task of forging modern nation-states.

Independence

Certain isolating factors had their effect even during the wars for independence. Spanish military strength was always concentrated at Lima, and those territories that lay farthest from their center of power—Argentina and Chile—were the first to establish their independence from Spain (in 1816 and 1818, respectively). In the north, Simón Bolívar led the burgeoning independence movement, and in 1824 two decisive military defeats there spelled the end of Spanish power in South America.

This joint struggle, however, did not produce unity because no fewer than nine countries emerged from the three former viceroyalties. It is not difficult to understand why this fragmentation took place. With the Andes intervening between Argentina and Chile and the Atacama Desert between Chile and Peru, overland distances seem even greater than they really were, and these obstacles to contact proved quite effective. Hence, from their outset the new countries of South America began to grow apart, and friction and even wars have been frequent. Only within the past ten years have the countries of this realm finally begun to recognize the mutual advantages of increasing cooperation and to make lasting efforts to steer their relationships in this direction.

CULTURAL FRAGMENTATION

When we speak of the "interaction" of South American countries, it is important to keep in mind just who does the interacting. The fragmentation of colonial South America into ten individual republics, and the subsequent postures of each of these states, was the work of a small minority that constituted the landholding, upper-class elite. Thus in every country a vast majority—be they Amerindians in Peru or people of African descent in Brazil—could only watch as their European masters struggled with each other for supremacy.

South America, then, is a continent of **plural societies**, where Amerindians of different cultures, Europeans from Iberia and elsewhere, blacks from western tropical Africa, and Asians from India, Japan, and Indonesia cluster in adjacent areas but do not mix. The result is a cultural kaleidoscope of almost endless variety, whose internal divisions are also reflected in the realm's economic landscape. This is readily visible in the map of South America's dominant livelihood, agriculture (Fig. 5-4). Here **commercial** (for-profit) and **subsistence** (minimum-life-sustaining) farming exist side by side to a greater degree than anywhere else in the world, where one usually dominates the other. The geography of commercial agricultural systems (map categories 1, 2, 3, 5, 6, and 9) is heavily tied to the distribution of landholders of European background, while subsistence farming (categories 4, 7, and 8) is overwhelmingly associated with the spatial patterns of indigenous peoples as well as populations of African and Asian descent.

Certainly, calling this complex human spatial mosaic "Latin" America (as is so often the case) is not very useful. Is there a more meaningful approach to a regional generalization that would better represent and differentiate the continent's cultural and economic spheres? John Augelli, who also developed the Mainland-Rimland concept for Middle America, made such an attempt. His map (Fig. 5-5) shows that five **South American culture spheres**—internal cultural regions—blanket the realm.

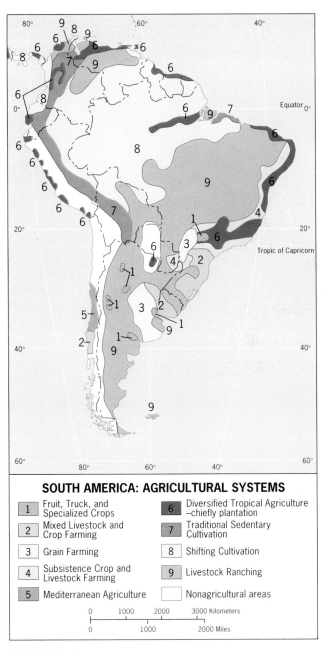

SOUTH AMERICA: AGRICULTURAL SYSTEMS

1 Fruit, Truck, and Specialized Crops	6 Diversified Tropical Agriculture –chiefly plantation
2 Mixed Livestock and Crop Farming	7 Traditional Sedentary Cultivation
3 Grain Farming	8 Shifting Cultivation
4 Subsistence Crop and Livestock Farming	9 Livestock Ranching
5 Mediterranean Agriculture	Nonagricultural areas

0   1000   2000   3000 Kilometers

0   1000   2000 Miles

FIGURE 5-4

- **Tropical-plantation sphere:** Reminiscent of Middle America's Rimland, consists of five Atlantic- or Caribbean-facing coastal strips in northern South America. Location and tropical environmental conditions favored plantation crops, especially sugar. Millions of African slave laborers were brought here and still strongly influence local cultures. Failure of the plantation economy reduced people to the poverty and subsistence that now dominate these five areas.

- **European-commercial sphere:** Covers most of southern, mid-latitude South America, including the core areas of Argentina, Brazil, Chile, and Uruguay. Dominated by populations of European descent with a strong Hispanic cultural imprint. Productive commercial agriculture is the major reason this region is economically more advanced than the rest of the continent.

- **Amerind-subsistence sphere:** Elongated zone lying astride the central Andes from southern Colombia to northern Chile/northwestern Argentina. The feudal socioeconomic structure established by the Spanish conquerors still prevails, with the Amerindian population forming a huge, landless peonage. Much of the

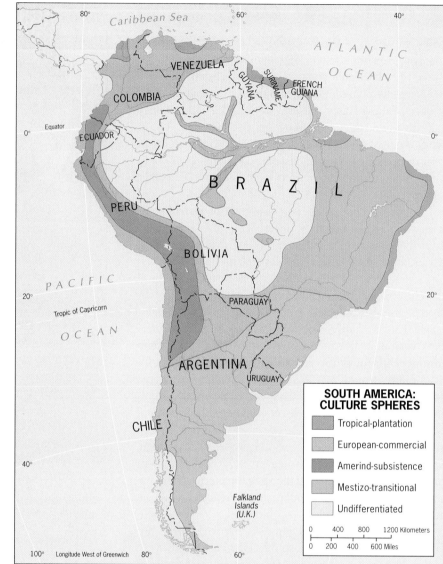

FIGURE 5-5

SOUTH AMERICA:
CULTURE SPHERES

- Tropical-plantation
- European-commercial
- Amerind-subsistence
- Mestizo-transitional
- Undifferentiated

population lives in high-altitude environments that are marginal for farming. The region includes some of the realm's poorest areas.

- **Mestizo-transitional sphere:** Surrounds Amerind-subsistence region and covers much of central and northern South America. Zone of mixture between Europeans and Amerindians—or Africans—around coastal strips of the tropical-plantation culture sphere. Transitional economy with respect to extremes of European-commercial and Amerind-subsistence regions.

- **Undifferentiated sphere:** The remaining, least accessible areas of the realm that lie in and around the Amazon Basin and in southernmost Chile. Sparsely populated and exhibit only very limited economic development.

ECONOMIC INTEGRATION

As noted above, the separatism that has so long characterized international relations is giving way as South American countries discover the benefits of forging new partnerships with one another. With mutually advantageous trade the catalyst, a new continentwide spirit of cooperation is blossoming at every level. Periodic flare-ups of boundary disputes now rarely escalate into open conflict. Cross-border rail, road, and pipeline projects, stalled for years, are multiplying steadily. In southern South America, five formerly contentious nations are developing the *Hidrovia*, a system of river locks that is opening most of the Paraná-Paraguay Basin to barge transport. Investments today flow freely from one country to another, particularly in the agricultural sector.

Recognizing that free trade can solve many of the realm's economic-geographic problems, governments are pursuing several avenues of economic supranationalism. In 2003, South America's republics were affiliating with the following major trading blocs:

Mercosur Launched in 1995, this Southern Cone Common Market established a free-trade zone and customs union linking Brazil, Argentina, Uruguay, and Paraguay.

Andean Community Formed as the Andean Pact in 1969 but restarted in 1995 as a customs union with common tariffs for imports, its members are Venezuela, Colombia, Peru, Ecuador, and Bolivia.

Group of Three (G-3) Launched in 1995, this free-trade agreement involves Mexico, Venezuela, and Colombia, and aims to eliminate all internal tariffs by 2005.

North American Free Trade Agreement (NAFTA) Launched by the United States, Canada, and Mexico in 1994, NAFTA is seeking to expand into South America to include Chile.

These organizations represent only an intermediate step toward a much grander goal: the creation of the **Free Trade Area of the Americas (FTAA)**, a single market of nearly 900 million consumers that would extend from the Arctic shores of Alaska and Canada to Chile's southernmost Cape Horn. This hemisphere-spanning free-trade area is to be formed by combining 24 existing multi-national agreements (including the four listed above) into one. Despite formidable obstacles to assembling this gigantic trading bloc, its architects remain optimistic that FTAA can meet its target completion date of 2005.

URBANIZATION

As in most other less advantaged realms, South Americans are leaving the land and moving to the cities. This **urbanization** process intensified throughout the twentieth century, and it persists so strongly that South America's urban population percentage (79 percent, or 289 million of the continent's 365 million inhabitants) now exceeds both those of Europe and the United States. And nowhere is South America's population increasing faster than in the towns and cities. We usually assume that the populations of rural areas grow more rapidly than urban areas because farm families traditionally have more children than city dwellers. Yet, overall, the urban population of South America has grown annually by about 5 percent since 1950, while the increase in rural areas was less than 2 percent. These numbers underscore not only the dimensions but also the durability of the **rural-to-urban migration** from the countryside to the cities.

The generalized spatial pattern of South America's urban transformation is displayed in Figure 5-6, which shows a *cartogram* of the continent's population.* Here we see not only the realm's 13 countries in population-space relative to each other, but also the proportionate sizes of individual large cities within their total national populations.

Regionally, the realm's highest urban percentages are found in southern South America. Today in Argentina,

---

*A cartogram is a specially transformed map in which countries and cities are represented in proportion to their populations. Those containing large numbers are "blown up" in population-space, while those containing lesser numbers are "shrunk" in size accordingly. This technique is used to map the countries of the world in Figure G-10.

FROM THE FIELD NOTES

"Two unusual perspectives of Rio de Janeiro form a reminder that here the wealthy live near the water in luxury highrises, such as these overlooking Ipanema Beach, while the poor have million-dollar views from their hillslope *favelas*, such as Rocinho."

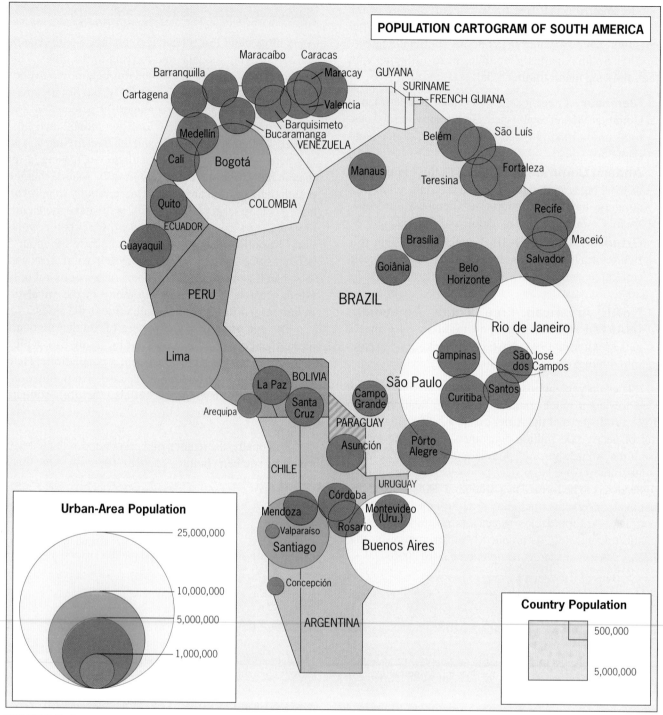

POPULATION CARTOGRAM OF SOUTH AMERICA

FIGURE 5-6

Chile, and Uruguay, at least 86 percent of the people reside in cities and towns (statistically on a par with Europe's most urbanized countries). Ranking next among the most heavily urbanized populations is Brazil at 81 percent. The next highest group of countries, averaging 78 percent urban, border the Caribbean in the north. Not surprisingly, the Amerind-subsistence-dominated Andean countries constitute the realm's least urbanized zone. (Peru, whose population is 72 percent urbanized, is the region's leader.)

Figure 5-6 also tells us a great deal about the relative positions of major metropolises in their countries. Three of them—Brazil's São Paulo and Rio de Janeiro, and Argentina's Buenos Aires—rank among the world's **megacities** (whose populations exceed 10 million).

In South America, as in Middle America, Africa, and Asia, people are attracted to the cities and driven from the poverty of the rural areas. Both *pull* and *push factors* are at work. Rural land reform has been slow in coming, and every year tens of thousands of farmers simply give up and leave, seeing little or no possibility for economic advancement. The cities lure them because they are perceived to provide opportunity—the chance to earn a regular wage. Visions of education for their children, better medical care, upward social mobility, and the excitement of life in a big city draw hordes to places such as São Paulo and Caracas.

But the actual move can be traumatic. Cities in developing countries are surrounded and often invaded by squalid slums, and this is where the urban immigrant most often finds a first—and frequently permanent—abode in a makeshift shack without even the most basic amenities and sanitary facilities. And unemployment is persistently high, often exceeding 25 percent of the available labor force. But still the people come, the overcrowding in the shantytowns worsens, and the threat of epidemic-scale disease (and other disasters) rises.

The "Latin" American City Model

Although the urban experience in the South and Middle American realms has varied because of diverse historical, cultural, and economic influences, there are many common threads that have prompted geographers to search for useful generalizations. One is the model of the intraurban spatial structure of the **"Latin" American city** proposed by Ernst Griffin and Larry Ford (Fig. 5-7), which may have wider application to cities of developing countries in other geographic realms as well.

The basic spatial framework of city structure, which blends traditional elements of South and Middle American culture with modernization forces now reshaping the urban scene, is a composite of radial sectors and concentric zones. Anchoring the model is the CBD, which is the primary business, employment, and entertainment focus of the surrounding metropolitan agglomeration. The landscape of the CBD contains many modern highrise buildings but also mirrors its colonial beginnings. As shown in Figure 4-3, when the Spanish colonizers laid out their cities they created a central square, or *plaza*, dominated by a church and government buildings. Santiago's *Plaza de Armas*, Bogotá's *Plaza Bolívar*, and Buenos Aires' *Plaza de Mayo* (see photo, p. 234) are classic examples. The plaza was the hub of the city, which later outgrew its old center as new commercial districts formed nearby; but to this day the plaza remains an important ceremonial link with the past.

Emanating outward from the urban core along the city's most prestigious axis is the commercial *spine*, which is surrounded by the *elite residential sector* (shown in green in Fig. 5-7). This widening corridor is essentially an extension of the CBD, featuring offices, shopping, and housing for the upper and upper-middle classes.

The three remaining concentric zones are home to the less fortunate residents of the city, with income level

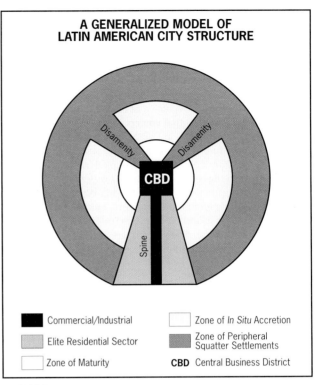

A GENERALIZED MODEL OF LATIN AMERICAN CITY STRUCTURE

Legend:
- Commercial/Industrial
- Elite Residential Sector
- Zone of Maturity
- Zone of *In Situ* Accretion
- Zone of Peripheral Squatter Settlements
- **CBD** Central Business District

FIGURE 5-7

and housing quality decreasing as distance from the CBD increases. The *zone of maturity* in the inner city contains housing for the middle class, who invest sufficiently to keep their aging dwellings from deteriorating. The adjacent *zone of in situ accretion* is one of much more modest housing interspersed with unkempt areas, representing a transition from inner-ring affluence to outer-ring poverty.

The outermost *zone of peripheral squatter settlements* is home to the impoverished and unskilled hordes that have recently migrated to the city from rural areas. Despite the shantytown living conditions, many newcomers achieve their first economic success by becoming part of the **informal sector**. This primitive form of capitalism is common in many lower-income countries, taking place beyond the control of the government. Participants are unlicensed sellers of homemade goods and services, and their willingness to engage in this hard work has transformed many a shantytown into a beehive of activity that can propel resourceful residents toward a middle-class existence.

A final structural element of many South American cities is the *disamenity sector* that contains relatively unchanging slums, known as **barrios** or **favelas**. The worst of these poverty-stricken areas often include sizeable numbers of people who are so poor that they are forced literally to live in the streets. Thus the realm's cities present stupendous contrasts between poverty and affluence, squalor and comfort—a harsh juxtaposition all too frequently observed in the cityscape.

"At the heart of Buenos Aires lies the Plaza de Mayo, flanked by impressive buildings but, unusual for such squares in Iberian America, carpeted with extensive lawns shaded by century-old trees. Getting a perspective of the plaza was difficult until I realized that you could get to the top of the "English Tower" across the avenue you see in the foreground. From there, one can observe the prominent location occupied by the monument to the approximately 700 Argentinian military casualties of the Falklands War of 1982 (center), where an eternal flame behind a brass map of the islands symbolizes Argentina's undiminished determination to wrest the islands from British control."

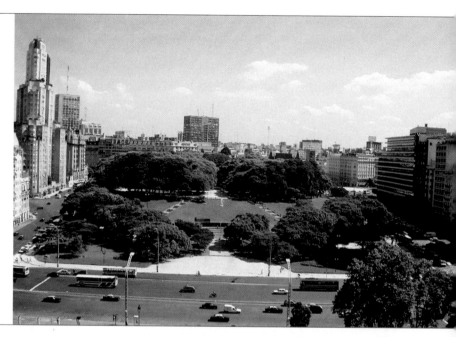

REGIONS OF THE REALM

SOUTH AMERICA DIVIDES geographically into four rather clearly defined regions (see Fig. 5-3). In the north, five entities display a combination of Caribbean and South American properties: Colombia, Venezuela, and the three small Guianas arising from colonial outposts of Britain (Guyana), the Netherlands (Suriname), and France (where Paris still retains control). In the west, four republics share a strong Amerindian cultural heritage and major influences from their Andean physiography: Ecuador, Peru, Bolivia, and, transitionally, Paraguay. In the south lie three countries that actually conform to the much-misused regional term "Latin" America: Argentina, Chile, and Uruguay, where European imprints are strong and Amerindian and the non-Iberian stamp relatively weak.

The fourth South American region also is the largest and in many ways the most important: Brazil, which occupies about half of the continent and contains nearly half of its total population. In Brazil, unlike the other regions, the Iberian influence is Portuguese, not Spanish; also in Brazil, and again unlike its neighbors, Africans, not Amerindians, form a significant element in the population's makeup. African cultures strongly influenced Brazilian society, making Brazil part of what the Portuguese like to call the Lusitanian (as opposed to the Hispanic) world.

THE NORTH: FACING THE CARIBBEAN

As another look at Figure 5-5 confirms, the countries of South America's northern tier have something in common besides their coastal location: each has a tropical-plantation zone signifying early European plantation development, the in-migration of black laborers, and the absorption of this African element into the population

MAJOR CITIES OF THE REALM	
City	**Population*** **(in millions)**
Asunción, Paraguay	1.5
Belo Horizonte, Brazil	4.8
Bogotá, Colombia	7.6
Buenos Aires, Argentina	12.4
Caracas, Venezuela	3.3
Guayaquil, Ecuador	2.4
La Paz, Bolivia	1.7
Lima, Peru	8.2
Manaus, Brazil	1.8
Montevideo, Uruguay	1.4
Quito, Ecuador	1.8
Rio de Janeiro, Brazil	11.2
Santiago, Chile	5.9
São Paulo, Brazil	24.8

*Based on 2004 estimates.

The Guianas

*O*n the north coast of South America, three small countries lie east of Venezuela and adjacent to northern-most Brazil (Fig. 5-8). In "Latin" South America, these territories are anomalies of a sort: their colonial heritage is Western European, and formerly they were known as British, Dutch, and French Guiana—and called *the Guianas*. But two of the three have long been independent: Guyana, Venezuela's neighbor, and Suriname, the Dutch-influenced country in the middle. The easternmost territory, French Guiana, continues to be governed from Paris, where its representatives sit in the Senate and the National Assembly.

None of the three countries has a population over 1 million. Guyana is the largest with 790,000, Suriname has 440,000 inhabitants, and French Guiana a mere 210,000. Culturally and spatially, patterns here are Caribbean: peoples of South Asian and African descent are in the majority, with whites forming a small minority. In Guyana, Asians make up just under half the population, blacks and mixed Africans 44 percent, and the others, including Europeans, are tiny minorities. In Suriname, the ethnic picture is even more complicated, for the colonists brought not only Indians from South Asia (now 37 percent of the population) and blacks (31 percent) but also Indonesians (15 percent) to the territory in servitude. The remaining 10 percent of the Surinamese population consists of a separate category of black communities, peopled by descendants of African slaves (known as Bush blacks) who escaped from the coastal plantations and fled into the forests of the interior. French Guiana is the most European of the three countries. About three-quarters of its small population speak French, and the ethnic majority (66 percent) is of mixed African, Asian, and European ancestry—the *creoles*.

French Guiana, an overseas *département* of France, is the least developed of the three Guianas. The European Space Agency's launch complex at Kourou accounts for fully half of the country's modest productive activity. Most food, however, must be brought in from abroad.

Suriname progressed more rapidly after independence in 1975, but unremitting political instability soon ensued. More than 100,000 residents emigrated to the Netherlands. Since the mid-1990s, bankruptcy has threatened the stagnating economy. Nonetheless, Suriname not only remains self-sufficient in rice production but also exports surpluses of that staple crop along with bananas and seafood. In most years, the leading Surinamese income earner is bauxite (aluminum ore), mined in a zone across the middle of the country. That output may soon be surpassed by oil production, however, because significant reserves have been discovered just offshore.

British-influenced Guyana became independent in 1966 amid internal conflict that basically pitted people of African origins against people of Asian descent. Such problems continue to plague this country, in which the great majority of the people live in small villages near the coast. Bauxite, gold from mineral-rich western Guyana, sugar, rice, shrimp, and molasses earn most of the country's annual revenues. Oil is also likely to become a major economic activity in the near future because the newly discovered offshore oilfield that adjoins Suriname extends westward into Guyanese waters. The two countries, however, do not agree on the location of their maritime boundary, and that is certain to increase tensions between these two already-uneasy neighbors.

Today a major environmental controversy engulfs all three countries. With the economies of the two largest Guianas starved for funds, timber companies from Asia's Pacific Rim are offering to pay handsomely to exploit the most precious natural resource—the vast tracts of tropical rainforest that blanket the landscape beyond the populated coastal strips. Concessions (land development rights) have already been granted, but the battle has now been joined by conservationists. The latter, who aim to purchase their own concessions to preserve as much of the rainforest as possible, are only too well aware that the loggers they oppose have ravaged similar woodland environments across many parts of Southeast Asia.

---

matrix. Not only did black workers arrive; many thousands of South Asians also came to the realm's northern shores as contract laborers. The pattern is familiar: in the absence of large local labor sources, the colonists turned to slavery and indentured workers to serve their lucrative plantations. Between Spanish Venezuela and Colombia on the one hand, and the non-Hispanic "three Guianas" on the other, there is this difference: in Venezuela and Colombia, the population center of gravity soon moved inland, and the plantation phase was followed by a to-

tally different economy. In the Guianas, coastal settlement and the plantation economy still predominate in the twenty-first century (see box titled "The Guianas").

Figure 5-8 reveals that Venezuela and Colombia have assets that the Guianas lack: much larger territories and populations; more varied natural environments, including a share of the Andes Mountains; and greater economic opportunities, based on oil reserves in each country that rank among the world's major deposits. Yet each of these geographic advantages is constrained by negative qualities

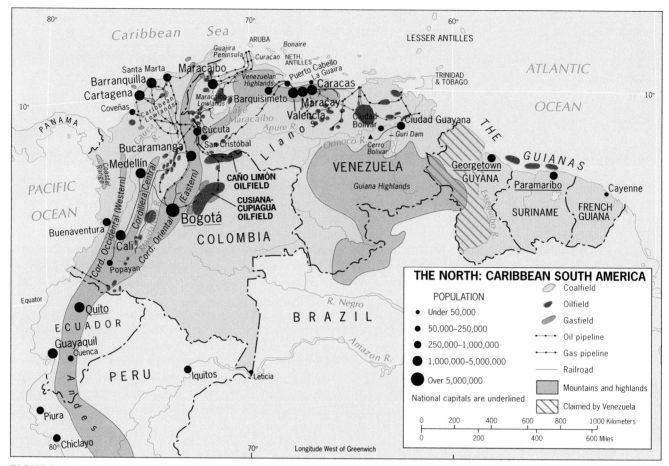

FIGURE 5-8

that continue to prevent the realization of their full potential. Physiographically and demographically, both countries exhibit a pronounced clustering of often-isolated populations in their highland zones and share a huge, near-empty, and relatively low-lying interior. Economically, both depend on a small number of products for the bulk of their export revenues—while the vast majority of their inhabitants practice subsistence agriculture and labor under the social and economic inequalities common to most of Iberian America. And politically, both remain under a dark cloud of instability, particularly violence-riddled Colombia, a state bedeviled by multiple ongoing insurgencies and a society that has all but collapsed.

Colombia

Imagine a country more than twice the size of France, not at all burdened by overpopulation, with a physical geography so varied that it can produce crops ranging from the temperate to the tropical, possessing world-class reserves of oil and natural gas, plus 1200 miles of coastline on Atlantic and Pacific waters. Situated in the northwestern cor-

ner of South America, closer than any of its neighbors to the markets of the north, with a single language and one dominant religion, would such a country not assuredly thrive amidst the burgeoning economic geography of its hemisphere? The answer is no. Colombia today is a country as troubled as any in the world, riven by violence, its economy imperiled, its politics destabilized, its future a gigantic question mark. Colombia's cultural uniformity has not produced social cohesion. The same physical geography that diversifies its environments also divides its population clusters, which are so tenuously interconnected that the country totals less than 250 miles (400 km) of paved four-lane highway. Its proximity to U.S. markets is a curse as well as a blessing: at the root of Colombia's current problems lies its role in the international drug trade.

Colombia's current disorder is not its first. In the past, civil wars between conservatives and liberals (on Roman Catholic religious issues) developed into conflicts pitting rich against poor, elites against workers. In Colombia today, people still refer to the last of these wars as *La Violencia*, a decade of strife beginning in 1948 during which as many as 200,000 people died. When, in the late 1950s, the two sides—the Conservative and Liberal parties—called a truce and agreed on constitutional reform to preclude a recurrence, Colombia at last seemed able to realize its economic potential. Exports of coffee, for which the country was famous, soared. Oil exports from reserves in the northern part of the country increased. Minerals, fruits, and cut flowers flowed to foreign markets, half by value to the United States, a quarter to Europe.

But in the 1970s, disaster struck. In remote parts of the country, groups opposed to the power-sharing monopoly between the political parties began a campaign of terrorism, damaging the developing infrastructure and destroying confidence in the future. Simultaneously, the U.S. market for narcotics expanded rapidly, and many Colombians got involved in the drug trade. Powerful and wealthy drug cartels formed in major cities such as Medellín and Cali, with networks that influenced all facets of Colombian life from the peasantry to the politicians. The fabric of Colombian society unraveled.

As Figure 5-8 shows, western Colombia is mountainous while the east lies in the Orinoco and Amazon Basins. Most of the country's 45.2 million people (the largest population in all of Spanish-speaking South America) live in the western and northern parts of Colombia. Physiography has contributed to the concentration of this population into more than a dozen separate clusters. Some of them are located in the Caribbean Lowlands; others lie in the Magdalena, Cauca, and lesser valleys between the parallel ranges of the Andes (including Medellín and Cali); and still others are found in basins within the mountains themselves, the most important being the cluster around the capital city of Bogotá (elevation: 8500 feet [2700 m]). Colombia's huge coffee crop—second only to Brazil's on the continent and in the world—comes from the numerous *templada* zones on the Andean slopes. But the more sparsely populated east and south have become the production centers for the coca crop that sustains the country's illicit narcotics industry (see box titled "The Geography of Cocaine").

Colombia's belt of oil and natural gas reserves flanks the border with Venezuela from Lake Maracaibo south to the Caño Limón oilfield, then continues onward along the Andean front through the country's largest petroleum deposits (the Cusiana-Cupiagua oilfield), and terminates at the newly discovered Guando oilfield in the upper Magdalena Valley southwest of Bogotá (Fig. 5-8). Pipelines connect the oilfields to the Caribbean ports of Cartagena, Barranquilla, Santa Marta, and Coveñas. Beneath that coastal lowland, Colombia also has substantial reserves of coal, a product that now ranks high on the list of exports. With its fuel, mineral, and agricultural productivity, Colombia's economy might have been one of South America's most prosperous. But the rise of the narcotics industry, coupled with the legacy of *La Violencia*, has dealt the country a devastating blow.

Today, Colombia is in combat. In the cities, narco-terrorists have committed acts of violence ranging from an invasion of the Supreme Court to the downing of a passenger-filled airliner. In the countryside, drug-financed armies of the political "left" fight equally vicious paramilitary forces of the "right." Rebels sabotage pipelines; oil companies hire and arm personnel to protect them. Civilians by the thousands die in the crossfire of all these belligerent groups, and countless more are dislocated.

Parts of Colombia are beyond the control of its government and its national armed forces, and are essentially ruled by rebels who "protect" the local citizenry against Bogotá's efforts to reestablish authority there. In effect, insurgents in several parts of Colombia have created their own domains, and some of their rulers have even negotiated their status with the legitimate government in the capital. What is happening here is a process long studied by political scientists. It involves three stages: (1) *contention*, the period of initial rebellion of the kind that marked the decade of *La Violencia*; (2) *equilibrium*, when the rebels gain effective control over some of the national territory, as has happened in parts of Colombia; and (3) *counteroffensive*, in which the government and the rebels, usually following the breakdown of negotiations, engage in armed conflict that will decide the future of the country.

The geographer Robert McColl translated this sequence into spatial terms, suggesting that the equilibrium stage represents the emergence of an **insurgent** **14** **state**. The formation of an insurgent state is the key stage in the process because it creates an informally bounded territory, usually with a distinct core area centered on a captured town, establishes a parallel government, installs a form of administration, and provides some social services such as schools and medical clinics that substitute for those the formal state may have provided. McColl pointed to Cuba and South Vietnam as countries within which insurgent states had developed before the final stage of decisive counteroffensive occurred.

Importantly, not all insurgent states succeed. They did in Cuba and Vietnam, but not in Peru or Nigeria. In some cases, the process never advances beyond the initial stage of contention. In Chapter 8 we examine the case of South Asia's Sri Lanka, where an insurgent state has existed for two decades and the counteroffensive has been in progress for years—producing stalemate, not decision.

At the beginning of the twenty-first century, Colombia contained not just one, but several enclaves that were

The Geography of Cocaine

Any geographical discussion of northwestern South America today must take note of one of its most widespread activities: the production of illegal narcotics. Of the enormous flow of illicit drugs that enter the United States each year, the most widely used substance undoubtedly is cocaine—all of which comes from South America, mainly Bolivia, Peru, and Colombia (the sources of more than 75 percent of the total world supply). Within these three countries, cocaine annually brings in billions of (U.S.) dollars and "employs" tens of thousands of workers, constituting an industry that functions as a powerful economic force. Those who operate the industry have accumulated considerable power through bribery of politicians, threats of terrorism, and alliances with guerrilla groups in outlying zones beyond governmental control. The cocaine industry itself is structured within a tightly organized network of territories that encompass the various stages of this drug's production.

The first stage of cocaine production is the extraction of coca paste from the coca plant, a raw-material-oriented activity that is located near the areas where the plant is grown. The coca plant was domesticated in the Andes by the Incas centuries ago; millions of their descendants still chew coca leaves for stimulation and brew them into coca tea, the leading beverage of highland South America. The main zone of coca-plant cultivation is along the eastern slopes of the Andes and in adjacent tropical lowlands in Bolivia, Peru, and Colombia.

Today, five areas dominate in the growing of coca leaves for narcotic production (Fig. 5-10): Bolivia's Chaparé district in the marginal Amazon lowlands northeast of the city of Cochabamba; the Yungas Highlands north of the Bolivian capital, La Paz; north-central Peru's Huallaga Valley; south-central Peru's Apurimac Valley, east of Huancayo; and the green-colored zones around the FARC-controlled territory of southern Colombia (Fig. 5-9). These areas, which produce as many as nine crops per year, are especially conducive to high leaf yields thanks to optimal local climatic and soil environments that also allow plants to develop immunity to many diseases and insect ravages. Despite harassment by (U.S.-supported) government efforts to aerially spray herbicides on them, all of these areas continue to thrive as coca cultivation has become a full-fledged cash crop. Operations in guerrilla-controlled south-central Colombia have now reached the scale of plantations, inducing thousands of nearby subsistence farmers (sometimes at gunpoint) to join the more lucrative ranks of the field workforce. In Peru and Bolivia, the specialized coca-cultivation zones have lured an even larger peasant-farmer population to abandon the less profitable production of food crops, thereby further reducing the capability of these nutrition-poor countries to feed themselves. The coca leaves harvested in these eastern Andean source areas make their way to nearby centers—located at the convergence of rivers and trails where coca paste is extracted and prepared.

The second stage of production involves refining that coca paste (about 40 percent pure cocaine) into cocaine hydrochloride (more than 90 percent pure), a lethal concentrate that is diluted with substances such as sugar or flour before being sold on the streets to consumers. Cocaine refining requires sophisticated chemicals, carefully controlled processes, and a labor force skilled in their supervision, and here Colombia has predominated. Most of this activity takes place in ultramodern processing centers located in the rebel-held territory of the lowland central-south and east, well beyond the reach of the Bogotá government. Interior Colombia also possesses the geographic advantage of intermediate location, lying between the source areas to the south (as well as locally) and the U.S. market to the north, a short hop for smugglers by air or sea across the Caribbean.

The final stage of production entails the distribution of cocaine to the U.S. marketplace, which depends on an efficient (but clandestine) transportation network. Most common is the use of private planes that operate directly out of remote airstrips near the refineries in areas beyond effective government control. The transport of cocaine to the United States is often a two-step process, making use of such intermediate Middle American transshipment points as the Bahamas, Dominican Republic, Haiti, Puerto Rico, Panama, and overland through Mexico.

taking on the properties of insurgent states (Fig. 5-9). This is because Colombia confronts not just one but several different guerrilla forces, each operating in particular areas of the country. Thus the M-19 (April 19) Movement focuses on the Urabá area in the north, near the Panamanian border. The Popular Liberation Army is active in the Cauca Valley. The National Liberation Army (ELN) operates in the Arauca area on the Venezuelan border, where it attacks oil facilities, and in the middle Magdalena Valley. And the Revolutionary Armed Forces of Colombia (FARC), by far the most powerful insurgent group, has most recently concentrated on the central-south and east.

In 1999, as an incentive to negotiate, FARC compelled the government to demilitarize a large zone south of Bogotá (now popularly called *Farclandia*), in effect leaving the insurgents there in control of an area the size of Switzerland (red-striped zone in Fig. 5-9). Also im-

FIGURE 5-9

portant was FARC's simultaneous initiative to take over the far eastern town of Mitú and its environs, an area whose remoteness made it particularly difficult for the Colombian army to regain control. And although the insurgents claimed they were suppressing drug production in the zones under their control, the evidence indicates otherwise. According to the latest official estimates and aerial imagery, coca production has skyrocketed in the FARC-held insurgent state.

Colombia's twin scourges, narcotics and guerrillas, have set the country back immeasurably and pose a colossal threat to its future as a coherent state. In Chapter 1 we discussed the centrifugal forces that affect all countries to some degree, and the centripetal forces that must override them for the state to function. In Colombia, the divisive forces grow ever stronger as the future of the state becomes more and more uncertain. Guerrilla movements that literally and/or virtually control large parts of Colombia no longer base their actions on political philosophies: despite a superficial veneer of socialist ideology, they have evolved from "leftist," "Marxist," or radical religious groups into criminal gangs that specialize in terror, kidnapping, extortion, and racketeering. They obstruct foreign investment and destroy facilities,

organize and tax coca farmers, operate airstrips and charge takeoff and landing fees, run sophisticated communications networks, and control hundreds of towns and even entire provinces. In the process, they have driven well over 1 million Colombian farmers off their land and into the cities as refugees, and convinced an additional million (mostly of middle-class status) to flee to neighboring countries and the United States. On the confiscated real estate the rebels establish their own ranches, drug laboratories, and command posts. As the cash from the narcotics trade and from ransoms flows in (estimates for FARC range as high as U.S. $500 million a year), they buy sophisticated weapons that allow them to stalemate the national army.

The election in 2002 of President Alvaro Uribe, who in his campaign took a hard line against the rebels, coupled with large increases in U.S. financial and advisory support for Colombia's armed forces, may change the situation over the long term. At present, however, large parts of the country remain beyond the reach of law and order. And Colombia's problems are spilling over into neighboring states—not only Panama (as noted earlier) but also Ecuador, where a sizeable Colombian refugee population is forming across the border), Peru, which is engaged in its own struggle against cocaine and faces collusion between Peruvian and Colombian operators, and Venezuela, whose actions in giving refuge to rebels being pursued by Colombian authorities have soured diplomatic relations. Colombia is a cornerstone of South America, an important nation in a crucial location where Central America, the Caribbean mainland, and Andean America converge. This country's destabilizing problems are of international, not just domestic, concern.

Venezuela

A long and tortuous boundary separates Colombia from Venezuela, its neighbor to the east. Much of what is important in Venezuela is concentrated in the northern and western parts of the country, where the Venezuelan Highlands form the eastern spur of the north end of the Andes system. Most of Venezuela's 26.1 million people are concentrated in these uplands, which include the capital of Caracas, its early rival Valencia, and the commercial/industrial centers of Barquisimeto and Maracay.

The Venezuelan Highlands are flanked by the Maracaibo Lowlands and Lake Maracaibo to the northwest and by a vast plainland of savanna country, known as the *Llanos*, in the Orinoco Basin to the south and east (Fig. 5-8). The Maracaibo Lowlands, once a disease-infested, sparsely peopled coastland, today constitute one of the world's leading oil-producing areas. Much of the oil is drawn from reserves that lie beneath the shallow waters of the lake itself. Actually, Lake Maracaibo is a misnomer, for the "lake" is open to the ocean and is in fact a gulf with a very narrow entry. The country's third-largest city, Maracaibo, is the focus of the petroleum industry that transformed the Venezuelan economy in the 1970s; however, as we shall see, since then oil has been more of a curse than a blessing.

The *Llanos* on the southern side of the Venezuelan Highlands and the Guiana Highlands in the country's southeast, like much of Brazil's Interior, are in an early stage of development. The 200- to 400-mile (300- to 650-km)-long *Llanos* slope gently from the base of the Andean spur to the Orinoco River. Its mixture of savanna grasses and scrub woodland support cattle grazing on higher ground, but widespread wet-season flooding of the more fertile lower-lying areas has thus far inhibited the *Llanos'* commercial agricultural potential (much of the development of the *Llanos* to date has been limited to the exploitation of its substantial oil reserves). Crop-raising conditions are more favorable in the *tierra templada* areas (see discussion of the concept of altitudinal zonation on p. 205) of the Guiana Highlands. Economic integration of this more remote interior zone with the rest of Venezuela has been spearheaded by the discovery of rich iron ores on the northern flanks of the Guiana Highlands southwest of Ciudad Guayana. Local railroads now connect with the Orinoco, and from there ores are shipped directly to foreign markets.

Despite these opportunities, Venezuela since 1998 has been in upheaval as longstanding economic and social problems finally intensified to the point where the electorate decided to push the country into a new era of radical political change. We noted above that, since the

The oil reserves beneath the Lake Maracaibo Basin have transformed the economic geography of this shallow inland sea. Drilling platforms stud the lake; pipelines link the wells to shoreside refineries and containers. In one way or another, the lives of virtually the entire population of the town of Cabimas (in the background) are dominated or affected by this industry. Shown here are the storage and export facilities near Cabimas, and a tanker being loaded; Venezuela is one of the United States' leading foreign suppliers of petroleum, and oil is the mainstay of the Venezuelan economy.

euphoric 1970s, oil had not bettered the lives of most Venezuelans. A major reason was that the government unwisely acquired the habit of living off oil profits, forcing the country to suffer the consequences of the long global oil depression that began in the early 1980s. Now Venezuela found itself heavily burdened by a huge foreign debt it had incurred in borrowing against future oil revenues that were not materializing fast enough. By the mid-1990s, the government was required to sharply devalue the currency, and a political crisis ensued that resulted in a severe recession and widespread social unrest. With more and more Venezuelans enraged at the way their oil-rich country was approaching bankruptcy without making progress toward the more equitable distribution of the national wealth, voters resoundingly turned in an extreme direction in the 1998 presidential election. Expressing their disgust with Venezuela's ruling elite of both political parties, they elected Hugo Chavez, a former colonel who in 1992 had led a failed military coup. Reaffirming their decision on July 30, 2000, Venezuelans gave nearly 60 percent of their votes to Chavez, apparently giving him a mandate to act as strongman on behalf of the urban poor and the increasingly penurious middle class.

Chavez's election had significant geopolitical consequences. His overtures toward OPEC to resist Western pressures to increase oil production led to an informal agreement with Arab oil exporters and to an invitation to visit Baghdad. His opposition to American involvement in the battle against the Colombian drug cartels soured relations with Bogotá. His revival of a century-old claim on territory in Guyana ruined relations with Venezuela's other neighbor on the Caribbean coast. His hosting of Cuba's Fidel Castro, amid expressions of admiration for Cuba's communist system, led to an agreement whereby Venezuela would sell Cuba oil at a deep discount. His vocal opposition to "superpower domination" made him an icon among millions of the disadvantaged beyond Venezuela's borders. At home, he stirred up racial divisions by actively promoting mestizos (67 percent of the population) over those of European background (20 percent), and he quarreled with unions and workers' federations. In 2002, following weeks of street demonstrations and social unrest, an attempt was made to remove Chavez by arresting and imprisoning him; but with the help of his supporters he survived and resumed his strongman tactics.

Venezuela is a key country in the Americas, and not only because its oil reserves rank among the world's top ten. Mismanagement of its economy led to popular dissatisfaction and the rise of a leader whose earlier actions evinced a disdain for democracy. This was a serious setback for a continent just emerging from an era of disastrous dictatorial rule.

THE WEST: THE ANDEAN, AMERINDIAN REGION

The second regional grouping of South American states (Fig. 5-10) combines countries where the Andes Mountains dominate the topography and where the Amerindian presence remains strong. This most obviously encompasses Peru, Ecuador, and Bolivia, but we also discuss some aspects of the geography of Paraguay here; other references to transitional Paraguay appear elsewhere.

Peru

Peru, which straddles the Andean spine for more than 1000 miles (1600 km), is the largest of the region's three republics in terms of territory as well as population (27.7 million). Its half-million square miles (1.3 million sq km) divide both physiographically and culturally into three subregions: (1) the desert coast, the European-mestizo region; (2) the Andean highlands or *Sierra*, the Amerindian region; and (3) the eastern slopes and adjoining *montaña*, the sparsely populated Amerindian-mestizo interior (Fig. 5-10). It is symptomatic of the cultural division still prevailing in Peru that Lima, the capital whose population constitutes more than one-quarter of the national total, is not located centrally in a basin of the Andes but in the peripheral coastal zone.

At Lima in the heart of the desert coastal strip, the Spaniards avoided the greatest of the Amerindian empires, choosing a site a few miles inland from a suitable anchorage that became the modern outport of Callao. From an economic point of view, the Spaniards' choice of a headquarters on the Pacific coast proved to be sound, for the coastal subregion has become commercially the most productive part of the country. A thriving fishing industry based on the cool productive waters of the Peru (Humboldt) Current offshore contributes significantly to the export trade. Irrigated agriculture in some 40 oases distributed all along the arid coast produces cotton, sugar, rice, vegetables, fruits, and wheat, much of which is exported as well.

The Sierra (Andean) subregion occupies about one-third of the country and is the ancestral home of the largest component in the total population, the Quechua-speakers who had been subjugated by the Inca rulers when the Spanish conquerors arrived. Their survival during the harsh colonial regime was made possible by their adaptation to the high-altitude environments they inhabited, but their social fabric was ripped apart by communalization, forced cropping, religious persecution, and outward migration to towns and haciendas where many became serfs. Another Amerindian people, the Aymara,

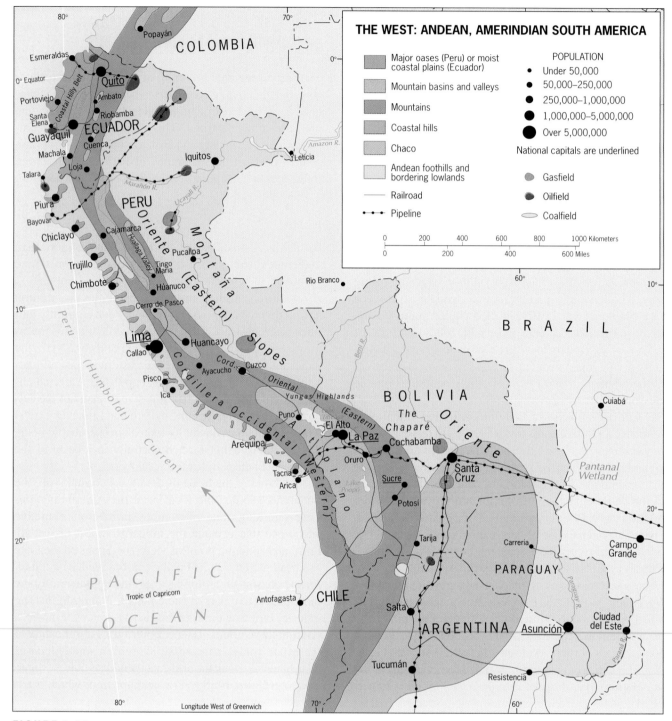

FIGURE 5-10

live in the area of Lake Titicaca, making up about 6 percent of Peru's population but larger numbers in neighboring Bolivia.

Although these Amerindian peoples make up about half the population of Peru, their political influence remains slight—despite the recent election of a president, Alejandro Toledo, whose mestizo ancestry made him popular in the Andean districts. Nor is the Andean sub-

region a major factor in Peru's commercial economy—except, of course, for its mines. In the high valleys and intermontane basins, the Amerindian population is clustered either in isolated villages, around which people practice a precarious subsistence agriculture, or in the more favorably located and fertile areas where they are tenants, peons on white- or mestizo-owned haciendas. Most of these people never receive an adequate daily

AMONG THE REALM'S GREAT CITIES . . . **LIMA**

*E*ven in a realm marked by an abundance of primate cities, Lima (8.2 million) stands out. Here reside 30 percent of the Peruvian population—who produce over 70 percent of the country's gross domestic product, 90 percent of its collected taxes, and 98 percent of its private investments. Economically, at least, Lima *is* Peru.

Lima began as a modest oasis in a narrow coastal desert squeezed between the cold waters of the Pacific and the soaring heights of the nearby Andes. Near here the Spanish conquistadors discovered one of the best natural harbors on the western shoreline of South America, where they founded the port of Callao; but they built their city on a site 7 miles (11 km) inland where soils and water supplies proved more beneficial. This city was named Lima and quickly became the Spaniards' headquarters for all their South American territories.

Since independence, Lima has continued to dominate Peruvian national life. The city's population growth was manageable through the end of the 1970s, but the past quarter-century has witnessed a disastrous doubling in its size. As usual, the worst problems are localized in the peripheral squatter shantytowns that now house close to one-half of the

metropolitan population. Lima's CBD, however, is worlds removed from this squalor, its historic landscape dotted with five-star hotels, new office towers, and prestigious shops as foreign investments have poured in since 1990.

---

caloric intake or balanced diet of any sort. The wheat produced around Huancayo, for example, is mainly exported and would in any case be too expensive for the Amerindians themselves to buy. Potatoes, barley, and corn are among the subsistence crops grown here in the *tierra fría* zone, and in the *tierra helada* of the higher basins the Amerindians graze their llamas, alpacas, cattle, and sheep (these environmental zones are discussed and illustrated on pp. 205–206). The major mineral products from the Sierra are copper, zinc, lead, and several other metallic minerals, with the largest mining complex centered on Cerro de Pasco.

Of Peru's three subregions, the *Oriente*, or East— the inland slopes of the Andes and the Amazon-drained, rainforest-covered *montaña*—is the most isolated. The focus of the eastern subregion, in fact, is Iquitos, a city that looks east rather than west and can be reached by oceangoing vessels sailing 2300 miles (3700 km) up the Amazon River across northern Brazil. Iquitos grew rapidly during the Amazon wild-rubber boom of a century ago and then declined; now it is finally growing again and reflects Peruvian plans to open up the eastern interior. Petroleum was discovered west of Iquitos in the 1970s, and since 1977 oil has flowed through a pipeline

built across the Andes to the Pacific port of Bayóvar. As Figure 5-10 shows, the eastern interior also contains major natural gas deposits, and their exploitation is a leading priority in Peru's current development plans.

Peru today appears to have put behind it the lengthy period of instability during which its government was threatened by well-organized guerrilla movements. The most serious threat, which for a time during the 1980s seemed on the verge of achieving an insurgent state in its Andean base, was the so-called *Sendero Luminoso* (Shining Path) which, at its height, even threatened Lima. But forceful action under the later-disgraced president of Japanese ancestry, Alberto Fujimori, decimated the rebel movement and returned Peru to political stability and economic growth. In the process, the government also tackled the problem of cocaine production in the Huallaga Valley and made progress against that scourge as well.

In 2000 Peru had to cross a political minefield when President Fujimori attempted to hold on to power and the issue led to the exposure of political fraud, evidence of bribery, gunrunning, and worse. Having succeeded in getting himself elected for an illegal third term, Fujimori was confronted by large-scale riots in the capital. He

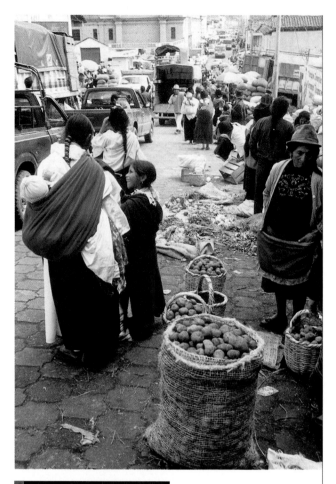

"I walked through the street market in Otavalo, Ecuador, on the Andean road north from Quito to the Colombian border. This is Inca country, and at these elevations (Mount Cayambe at nearly 19,000 feet [5800 m] lies nearby) the staple is the potato. Beets and turnips are also for sale, and many other crops—common and unusual—are cultivated in this area. The people here, both sellers and buyers, seem quite prosperous. They were well dressed against the cool mountain weather and business was brisk. Comparing this scene to what I had seen in the Ecuadorian port city of Guayaquil helped me understand this small country's intense regionalism better. Mestizo coast and Amerindian interior are worlds apart."

finally left for Japan, announced that he would not return to Peru, and thus cleared the way for the installation of Toledo as president.

The episode underscored the weaknesses of Peru's democratic institutions, and while investors and tourists returned to the country, its underlying disabilities were painfully exposed. Even as the economic statistics turned more favorable, another measure of the condition of the country looked worse: the percentage of the people living in poverty increased from 46 in 1995 to more than 50

in 2003. The challenges facing Peru's leaders are daunting indeed.

Ecuador

On the map, Ecuador, smallest of the three Andean West republics, appears to be just a corner of Peru. But that would be a misrepresentation because Ecuador possesses a full range of regional contrasts (Fig. 5-10). It has a coastal belt; an Andean zone that may be narrow (under 150 miles or 250 km) but by no means of lower elevation than elsewhere; and an *Oriente*—an eastern subregion that is as sparsely populated and as economically marginalized as that of Peru. As in Peru, nearly half of the people of Ecuador are concentrated in the Andean intermontane basins and valleys, and the most productive region is the coastal strip. Here, however, the similarities end.

Ecuador's Pacific coastal zone consists of a belt of hills interrupted by lowlands, of which the most important lies in the south between the hills and the Andes, drained by the Guayas River. Guayaquil—the country's largest city, main port, and leading commercial center—forms the focus of this subregion. Unlike Peru's coastal strip, Ecuador's is not a desert: it consists of fertile tropical plains not afflicted by excessive rainfall. Seafood (especially shrimp) is a leading product, and these lowlands support a thriving commercial agricultural economy built around bananas, cacao, cattle-raising, and coffee on the hillsides. Moreover, Ecuador's western subregion is also far less Europeanized than Peru's because its white component of the national population of 13.6 million is only about 10 percent.

A greater proportion of whites are engaged in administration and hacienda ownership in the central Andean zone, where most of the 25 percent of the Ecuadorians who are Amerindian also reside—and, not surprisingly, where land-tenure reform is an explosive issue. The differing interests of the Guayaquil-dominated coastal lowland and the Andean-highland subregion focused on the capital (Quito) have long fostered a deep regional cleavage between the two. This schism has intensified in recent years, and autonomy and other devolutionary remedies are now being openly discussed in the coastlands.

In the jungles of the *Oriente* subregion, petroleum production has reached substantial levels, and oil tops the country's export list. A trans-Andean pipeline was opened in 1972 to the port of Esmeraldas, but poor management contributed to a continuing pollution nightmare of repeated oil spills and toxic-waste dumping along the route; despite recent cleanup efforts, this remains one of the continent's worst environmental disasters. Stimulating the wider development of the interior remains a leading priority of the Ecuadorian government. That policy was given a major boost by the 1998 settlement of the long-running southeastern border dispute with Peru, which resulted in the Ecuadorians gaining navigation rights on the Amazon and its northern Peruvian tributaries.

These and other growth opportunities notwithstanding, most development is now on hold because Ecuador is in the vanguard of a regionwide social movement, led by indigenous peoples, which is threatening to destabilize the Andean West. This movement, grounded in the persistent exclusion of Amerindians within each of the region's three national societies, has spawned a number of uprisings as frustrations can no longer be contained over repeatedly failed promises by governments that have raised expectations for social and economic improvement. Ongoing geographic change further fuels this unrest because the poorest people increasingly urbanize and come to witness close-up how their second-class standing and lack of education bar them from making progress. In Ecuador, the widely supported Confederation of Indigenous Nationalities of Ecuador (CONAIE) spearheads the protest movement, demanding a multiethnic state in which the Amerindians are guaranteed greater political involvement, better access to landownership, and an end to government neglect, which ensures that indigenous communities end up with inadequate educational systems and infrastructural linkages.

Bolivia

From Ecuador southward through Peru, the Andes broaden until in Bolivia they reach a width of some 450 miles (720 km). Here, between the eastern and western Andean cordilleras (ranges), lies the *Altiplano* proper (Fig. 5-10; as noted earlier, an altiplano is an elongated, high-altitude basin). On the boundary between Peru and Bolivia, freshwater Lake Titicaca lies at 12,507 feet (3700 m) above sea level and helps make the adjacent *Altiplano* livable by ameliorating the coldness in its vicinity, where the snow line lies just above the plateau surface. On the surrounding cultivable land, grains have been raised for centuries dating back to pre-Inca times, and the Titicaca Basin still supports a major cluster of Aymara subsistence farmers. This portion of the *Altiplano* is the heart of modern Bolivia and also contains the capital city, La Paz.

Modern Bolivia is the product of the European impact, and the country's Amerindians (who still make up 55 percent of the national population of 9.2 million) no more escaped the loss of their land than did their Peruvian or Ecuadorian counterparts. What made the richest Europeans in Bolivia wealthy, however, was not land but minerals. The town of Potosí in the eastern cordillera became a legend for the immense deposits of silver in its vicinity; tin, zinc, copper, and several ferroalloys were also discovered there. Over most of the past century, Bolivia's tin deposits yielded much of the country's annual export income, but since 1980 declining tin reserves and falling world prices have forced much of the industry to shut down. Today zinc has replaced tin as the leading metallic mineral export, and natural gas and oil are accounting for a growing share of foreign revenues. Bolivia now exports large quantities of gas to Brazil and Argentina, and in return Brazil is assisting in the development of the southeastern lowlands, where commercial agriculture—particularly the production of soybeans—is steadily expanding in the fertile savanna around Santa Cruz.

Bolivia has had a turbulent history. Apart from endless internal struggles for power, the country first lost its corridor to the Pacific coast during the 1880s in a disastrous conflict with Chile; then in 1903 it lost its northern territory of Acre to Brazil in a dispute involving the rubber boom in the Amazon Basin; and finally it lost a huge chunk of southeastern Gran Chaco territory to Paraguay in the 1930s. The most critical loss by far was that of its outlet to the sea: although Bolivia has acquired rail connections to the Chilean ports of Arica and Antofagasta, it remains severely disadvantaged by its landlocked situation. Because the western Andean cordillera and barren *Altiplano* form the country's inhospitable western margins, one might suppose that Bolivia would look eastward and that its *Oriente* would be somewhat better developed than that of Peru or Ecuador. As we noted, this interior subregion is now slowly awakening, but for decades to come the densest settlement clusters will remain in the valleys and basins of the eastern cordillera, where agricultural opportunities are more favorable than on the southern *Altiplano* to the west.

Regional development here in South America's second-poorest country is also constrained because fully 70 percent of all Bolivians still live in poverty. The vast majority of the poor are the long-neglected Amerindian peoples, who finally achieved some recognition in Bolivian society during the 1990s through such limited reforms as bilingual education programs (less than half of the country's citizens speak Spanish as a first language). The insufficiency of the government's response is evident in Bolivia's vigorous participation in the social movements that now bedevil the Andean region. After a number of local confrontations, the bitter feelings of the indigenous peoples burst into the open all across the *Altiplano* with a week-long protest in 2000 that blockaded roads and brought the country to a standstill. This uprising was triggered by a riot in defiance of increased water rates in the city of Cochabamba, and within a few days it forced the La Paz government to capitulate. But nothing was done to address the social inequalities that cause the discontent experienced by so many Bolivians, which can only be mitigated by the integration of the impoverished masses into the mainstream of national life.

Paraguay

As Figure 5-3 indicates, Paraguay is one of those countries (Kazakhstan is another) that lies across a regional transition zone. In terms of economic development, Paraguay's per-capita GNP is more typical of the Andean West than the Southern Cone. Ethnically, too, the pattern is more characteristic of the West: about 95 percent of Paraguay's 6.3 million people are mestizo, but with so pervasive an Amerindian influence that any white ancestry is almost totally submerged. As for languages, Amerindian Guaraní is so widely spoken alongside Spanish that the country is surely one of the world's most completely bilingual. The physiography, however, is decidedly non-Andean because all of Paraguay lies to the east of the mountains in the center of the Paraná-Paraguay Basin; from here, the outlet to the sea has always been to the south (Fig. 5-11). As the Basin's waterways are improved, that regional orientation is steadily strengthening.

Paraguay's landlocked position has much to do with its modest economic development to date, but political instability contributes greatly to the problem. Among South American states, only Guyana is less urbanized than Paraguay, and the vast majority of the people living on the land are mired in an inescapable cycle of poverty. In 2003, about 60 percent of the people, in the slums around the capital, Asunción, and on the land from the moist plains in the east to the dry, scrub-forested Chaco in the west, lived at or below the poverty level.

As Figure 5-11 shows, Paraguay is flanked to the east by a narrow corridor called Entre Rios ("between the rivers") belonging to Argentina, as well as Brazil's southern State of Paraná. In recent years, this complicated border area has taken on regional and even global significance because it lies remote from the capitals and core areas of all three of the countries that converge here, and has become a haven not only for smugglers but also for people with terrorist ties. A substantial number of Arab immigrants has settled in what representatives of national governments refer to as this *Triple Frontier*, where law enforcement is lax, money is made illegally, and dangerous causes are given financial support. The great majority of these Arab immigrants, who live on all three sides of the two borders but concentrate especially in Foz do Iguaçu, Brazil, and Ciudad del Este, Paraguay, are not involved in terrorist activities and attend their mosques (there are dozens in the two towns) for religious, not political, purposes. But Argentinian authorities have long been concerned over the small minority that has links with the terrorist organization Hizbollah in Lebanon, and over the suspected roles this organization may have played in the two deadly attacks on Jewish targets in the Argentine capital, Buenos Aires, in 1992 (the Israeli embassy) and 1994 (the Jewish Information Center, in which 87 persons perished). When it was reported that maps of the Triple Frontier had been found in an al-Qaeda hideout in Kabul, Afghanistan, following the fall of the Taliban regime there, this concern led to an investigation that exposed financial ties involving millions of dollars between certain Muslim businesspeople in the Triple Frontier and Hizbollah in Lebanon. What once may have been a flow of funds from the Middle East to the Triple Frontier now appears to have a reverse direction, suggesting to investigators that the prevailing commercial chaos in the area has become a boon to supporters of what Western governments designate as terrorist organizations and cells, not only in the Middle East but also in other locales in South America as well as North America. A political geographer in Argentina described landlocked and lawless eastern Paraguay as an Afghanistan in South America, dangerously neglected in the War on Terror. Whether Paraguay's fragile political system will be able to deal with this emerging challenge is a key question for the future.

▶ THE SOUTH: MID-LATITUDE SOUTH AMERICA

South America's three southern countries—Argentina, Chile, and Uruguay—constitute a region sometimes referred to as the *Southern Cone* because of its pointed, ice-cream-cone-like shape (Fig. 5-11). As noted earlier, Paraguay has strong links to this region and in some ways forms part of it, although social contrasts between Paraguay and those in the Southern Cone remain sharp.

Since 1995, the countries of this region have been drawing closer together in an economic union named Mercosur, the hemisphere's second-largest trading bloc after NAFTA. Despite setbacks and disputes, Mercosur has expanded and today encompasses Argentina, Uruguay, Paraguay, and Brazil, with Chile and Bolivia participating as associate members.

The Southern Cone represents South America's European-commercial culture sphere (Fig. 5-5). Mercosur has had the effect of extending these cultural and economic norms not only into Brazil but also into Paraguay, which lies in the vast basin of the Paraguay-Paraná river system that flows into the Rio de la Plata (Fig. 5-11). Two of the world's largest dams, the Yacyretá and the Itaipu, were constructed on the Paraná River along Paraguay's eastern border, and one of Mercosur's biggest infrastructure programs envisages the completion of the 2150-mile (3450-km) *Hidrovia* waterway that would open most of

FIGURE 5-11

the Paraguay-Paraná river basin to barge transport. Other Mercosur initiatives, including roads through the Andes, bridges, highways, and railroads, are intended to strengthen intraregional ties.

Argentina

The largest Southern Cone country by far is Argentina, whose territorial size ranks second only to Brazil in this geographic realm; its population of 37.3 million ranks third after Brazil and Colombia. Argentina exhibits a great deal of physical-environmental variety within its boundaries, and the vast majority of the Argentines are concentrated in the physiographic subregion known as the *Pampa* (a word meaning "plain"). Figure G-9 underscores the degree of clustering of Argentina's inhabitants on the land and in the cities of the Pampa. It also shows the relative emptiness of the other six subregions (mapped in Fig. 5-11): the scrub-forest *Chaco* in the northwest; the mountainous *Andes* in the west, along whose crestline lies the boundary with Chile; the arid plateaus of *Patagonia* south of the Rio Colorado;

and the undulating transitional terrain of intermediate *Cuyo*, *Entre Rios* (also known as "Mesopotamia" because it lies between the Paraná and Uruguay rivers), and the *North*.

The Argentine Pampa is the product of the past 150 years. During the second half of the nineteenth century, when the great grasslands of the world were being opened up (including those of the interior United States, Russia, and Australia), the economy of the long-dormant Pampa began to emerge. The food needs of industrializing Europe grew by leaps and bounds, and the advances of the Industrial Revolution—railroads, more efficient ocean transport, refrigerated ships, and agricultural machinery—helped make large-scale commercial meat and grain production in the Pampa not only feasible but also highly profitable. Large haciendas were laid out and farmed by tenant workers; railroads radiated ever farther outward from the booming capital of Buenos Aires and brought the entire Pampa into production.

Over the decades, within the framework of the **von Thünen model** (see pp. 45–46), several specialized agricultural areas appeared on the Pampa. As we would ex-

---

AMONG THE REALM'S GREAT CITIES . . . **BUENOS AIRES**

*T*ts name means "fair winds," which first attracted European mariners to the site of Buenos Aires alongside the broad estuary of the muddy Rio de la Plata. The shipping function has remained paramount, and to this day the city's residents are known throughout Argentina as the *porteños* (the "port dwellers"). Modern Buenos Aires was built on the back of the nearby Pampa's grain and beef industry. It is often likened to Chicago and the Corn Belt in the United States because both cities have thrived as interfaces between their immensely productive agricultural hinterlands and the rest of the world.

Buenos Aires (12.4 million) is yet another classic South American primate metropolis, housing more than one-third of all Argentines, serving as the capital since 1880, and functioning as the country's economic core. Moreover, Buenos Aires is a cultural center of global standing, a monument-studded city that contains the world's widest boulevard (*Avenida 9 de Julio*).

During the half-century between 1890 and 1940, the city was known as the "Paris of the South" for its architecture, fashion leadership, book publishing, and performing arts activities (it still has the world's biggest opera house, the *Teatro Colón*). With the recent restoration of democracy, Buenos Aires is now trying to recapture its golden years. Besides reviving these cultural functions, the city

has added a new one: the leading base of the hemisphere's motion picture and television industry for Spanish-speaking audiences.

"Arriving in Argentina during the political and economic up-heavals that had begun in 2001, I saw signs of dislocation and trouble everywhere. Beggars pursued pedestrians on the once-fashionable Avenida Florida. Banks had installed protective shutters against angry crowds demanding return of their frozen and devalued deposits. A bus trip on the Patagonian Highway turned into an adventure when masked protesters carrying rocks and burning rags stopped vehicles and threatened their occupants. Newspapers carried reports of starvation in Tucumán Province—in a country capable of producing seven times the food its population needs. Twenty years ago, Argentina ended its most recent period of disastrous military rule, but its democratically-elected politicians have not managed the country well."

pect, a zone of fruit and vegetable production became established near the huge Buenos Aires conurbation located beside the estuary of the Rio de la Plata. To the southeast is the predominantly pastoral district, where beef cattle and sheep are raised. In the drier west, northwest, and southwest, wheat becomes the important commercial grain crop, but half the land remains devoted to grazing. Among the exports, cereals have usually led by value, followed by meat products, animal feed, and vegetable oils. Yet even with all this output the Pampa has hardly begun to fulfill its productive potential, which could easily double with more intensive agricultural practices.

Argentina, as we noted in our introductory chapter, once was one of the richest countries in the world. Its historic affluence is still reflected in its architecturally splendid cities whose plazas and avenues are flanked by ornate public buildings and private mansions. This is true not only of the capital, Buenos Aires, at the head of the Rio de la Plata estuary; it also applies to interior cities such as Mendoza and Córdoba. The cultural imprint is dominantly Spanish, but the cultural landscape was diversified by a massive influx of Italians and smaller but influential numbers of British, French, and German immigrants. A sizeable immigration from Lebanon resulted in the diffusion of Arab ancestry to more than 12 percent of the Argentinian population among whom the most prominent is a recent president, Carlos Menem. Argentina has long been one of the realm's most urbanized countries: 90 percent of its population is concentrated in cities and towns, a higher percentage even than Western Europe or the United States. One-third of all Argentinians live in metropolitan Buenos Aires, also by far the leading industrial complex where processing Pampa

products dominates. Córdoba has become the second-ranking industrial center and was chosen by foreign automobile manufacturers as the car-assembly center for the expanding Mercosur market. One in three Argentinian wage earners is engaged in manufacturing, another indication of the country's economic progress. But what concentrates the urban populations is the processing of products from the vast, sparsely peopled interior: Tucumán (sugar), Mendoza (wines), Santa Fe (forest products), and Salta (livestock). Argentina's product range is enormous. There is even an oil reserve near Comodoro Rivadavia on the coast of Patagonia.

Despite all these riches, Argentina's economic history is one of boom and bust. With fewer than 40 million inhabitants, a vast territory with diverse natural resources, adequate infrastructure, and good international linkages, Argentina should still be one of the world's wealthiest countries, as it once was. But political infighting and economic mismanagement have combined to ruin a vibrant and varied economy. What began as a severe recession toward the end of the 1990s became an economic collapse in the first years of the new century.

To understand how Argentina finds itself in this situation, a map of its administrative structure is useful. On paper, Argentina is a federal state consisting of the Buenos Aires Federal District and 23 provinces (Fig. 5-12). As would be expected from what we have just learned, the urbanized provinces are populous while the mainly rural ones have smaller populations—but the gap between Buenos Aires Province (whose capital is La Plata), with nearly 15 million, and Tierra del Fuego (capital: Ushuaia), with barely over 100,000, is wide indeed. Several other provinces have under 500,000 inhabitants, so that the

PROVINCES OF ARGENTINA

- – – – International boundary
- – – – Provincial boundary
- ● Location of Provincial capital

Provincial population

■	Greater than 5 million
■	3–5 million
■	1–3 million
■	500,000–1 million
□	Less than 500,000

0   200   400   600 Kilometers
0   100   200   300 Miles

FIGURE 5-12

base lay among the poor and the working class, which he rewarded with raises and benefits even while embarking on huge infrastructure projects. Argentina's coffers soon were empty, but Perón used the military to keep control. In a sequence of events worthy of an opera, he was overthrown, exiled, succeeded by incompetent military rulers, invited back to resume the presidency, and finally replaced upon his death by his third wife. She, in turn, was overthrown in 1976 by a military junta that plunged Argentina into its darkest days, the notorious "Dirty War" (1976–1983). More than 10,000 people suspected of subversion disappeared; children were taken from their parents and turned over to military families; torture of alleged opponents was routine. In 1982 this ruthless clique played the Malvinas card, invading the British-administered Falkland Islands (which the Argentines call the Malvinas) in an effort to rouse Argentinian nationalism. With the country's economy a shambles and its military despised, that plan failed even before British forces ousted the Argentinians from those islands (Fig. 5-12) a few months later. In 1983, the junta resigned and permitted a democratic election (see the Issue Box entitled "Who Needs Democracy?").

By the time civilian government returned (again) to Argentina, inflation was soaring and the national debt was staggering. But at least it was now possible to gauge the residual influence of the Peronistas in the country's politics, and in 1989 their candidate, Carlos Menem, was elected president. Menem and his advisers did what his predecessor could not: he tightened Argentina's belt, established policies to counter inflation and pay off the public debt, and gave Argentina hope after the horror of the Dirty War and the ignominy of the Malvinas rout. But by the end of his ten-year (two-term) tenure, Argentina was sliding back into recession. Part of the problem was beyond the government's control: international prices for Argentina's goods stopped rising; Brazil devalued its currency, making Argentinian goods less competitive in its Mercosur neighbor (and leading trade partner); foreign investors shied away, seeing trouble ahead. International lenders were forcing Argentina into privatization policies that worsened rather than eased the situation. And in a throwback to Argentina's familiar problems, Menem tried to buy political support for an unconstitutional third term, causing him to abandon difficult reforms that still were needed.

Nor did the federal system help. Under the 150-year-old rules governing Argentina's fiscal system, the provinces were given an automatic share of the taxes collected by the central government, without conditions on spending. That led to corruption and laxity. But when the late-1990s recession hit, and tax collection (already scandalously inefficient) yielded less to hand out, the poorer provinces of the interior suffered severely.

larger ones in addition to dominant Buenos Aires are also disproportionately influential in domestic politics, especially Córdoba and Santa Fe, both with capitals of the same name.

Argentina for most of the twentieth century was unable to achieve stable democratic government. In this supposedly federal system (as in the case of the former Soviet Union), the provinces held no real power, and civilian governments were ineffective and corrupt. In 1943 a group of military plotters that included a young officer named Juan Perón overthrew the government, and by 1946 Perón, using the police and strong-arm squads to quell any opposition, was elected president. His popular

Who Needs Democracy?

SOUTH AMERICA DOES!

"I'M OLD ENOUGH TO remember the good as well as the bad old days here in Argentina. As a young man I worked for the national railroad company when that guy Perón came on the scene. I guess he learned his craft from the fascists in Italy, because he excelled in election by intimidation. He divided this country as never before, spending money on the workers and the poor while curbing freedoms and even abolishing freedoms guaranteed under our constitution. To tell you the truth, my salary actually went up a bit, and I even voted for him the second time around—largely, I think, because I wanted to do something for his beautiful wife Evita, who was loved by the whole country (except the military).

"Anyway, if you looked around South America in those days, we could have done worse. Wall to wall dictators. The military and police, secret and otherwise, put the fear of God into the populace, helped by the church. At least we had a guy who went through the motions of election. While those others divided the wealth among their cronies, Perón tried to spread it around. Even wealth we didn't have! Yes, he certainly put us into debt. And he had his cronies too.

"When they finally deposed Perón, I thought that we might see democratic elections here. (In retrospect, I don't know why I thought that—Spain and Portugal were ruled by two of the worst, Franco and Salazar; that's our "Latin" heritage). Well, you already know how wrong I was. Instead of more democracy, we got military repression. We lived in terror, not just fear. You've probably heard of those nuns who asked questions about the fate of a baby whose mother had 'disappeared.' They loaded them into a helicopter, in their outfits, and threw them out over the Rio de la Plata. Later I heard some soldiers joke about "the flying nuns." Rumors of what could happen to you were rife. Everyone knew someone who knew someone who was a victim.

"Our cloud of military terror lifted after they made their mistake in the Malvinas, 20 years ago, and since then we've had a taste of real democracy. It hasn't been easy, but let me tell you, it's better than 'strongman' rule so many of us South Americans are familiar with. Most of all, there's openness. You can express your views without fear. Political parties can argue their positions without military intimidation. Corrupt public officials can be found out by reporters, and they can't send machine-gun-toting colonels to kill their pursuers. Yes, there are downsides—that openness applies to the economy too, and as we've discovered, outsiders can interfere in our financial affairs. But I'd rather have democracy and open disorder than dictatorship and isolated order, and I've known them both."

SAY NO TO THE EXCESSES OF DEMOCRACY!

"You can't help noticing that the Great Democratic Revolution that was supposed to be sweeping Middle and South America is not exactly a success. Here in Brazil we've had democratic government of a sort since 1989, when our man Fernando Collor de Mello was allowed by the military to win an election. You'll remember what happened to him. He resigned three years later after being implicated in a corruption and influence-peddling scheme. I'm sure the poor benefited hugely from this return to democracy.

"South American countries have a history of being led by men who captured the imagination of the nation and whose vision forged the character of the state. These men knew that the state must serve the people. The state must ensure that its railroads and bus services are available at low cost to all citizens. The state must provide electricity, fuel, and clean water. The state must staff and maintain the schools. The state and the church are indivisible, and the networks of the church are in the service of the state. If the military is needed to maintain order in the interest of stability, so be it. Call them strongmen if you like, but they personified their nations and did so for centuries.

"Democracy is a luxury for rich countries. The Americans like to teach us about democracy, but as a geography high-school teacher here in Santa Catarina, let me ask you this: how can the U.S. call itself a democracy when, in their Senate, a few hundred thousand people in Wyoming have the same representation as over 30 million in California? Oh, I see. You call yourself a 'representative republic.' Well, then don't lecture us about democracy. Even Juan Perón was 'democratically' elected, you know. In fact, most of those allegedly dictatorial rulers you so harshly criticize would probably win at the ballot box.

"Anyway, look at what democracy has done for Argentina and for Venezuela before the Venezuelans had the good sense to throw the rascals out and elect Chavez. Look what it is doing for Peru, where democratically elected President Toledo was trying to sell off the electric company serving Arequipa before the citizens stopped him. That "openness" you hear pro-democracy advocates talk about just means that outsiders can come in, put your country in debt, demand 'privatization' of public companies, buy up corporations and banks and haciendas, fire thousands of workers to increase profits and share prices back home, and make the poor even poorer. What we need in this part of the world is a strong hand, not only to guide the nation but to protect the state."

Vote your opinion at www.wiley.com/college/deblij

Argentina did not receive the speedy international bailouts provided to Mexico or Brazil, even when its banking system collapsed and help was desperately needed to stave off social disorder. To the Argentines this proved their lowly rank in the priorities of the United States and the lending agencies it controls, but it was nothing new. The U.S. administration had not even bothered to appoint an ambassador to Buenos Aires between December 1996 and July 2000. Argentina learned that the cost of high foreign debt, for certain countries, can be low esteem.

"Arrived in Mendoza (Argentina) yesterday to begin a week's study of the wine industry here. Stayed at an old hotel still showing cracks in the walls from the 1861 earthquake. Was invited to the local opera performing *Carmen,* starting time: 11:30 PM. Mendoza is so hot in the daytime that people here use the night to work as well as play, and I was told that the city is 'dead' from about 1 to 6 PM. This morning's guided excursion produced some remarkable viticultural vistas, but nothing topped the drama of the mountains and the famous 'Christ of the Andes' statue at the foot of what is actually the Andean range called the *Sierra de Los Paramillos.* Beyond the mountains in this westward view lies Chile, and during the movement for independence from Spain, General José de San Martin planned and executed an expedition to liberate Chile in 1817 from his headquarters in Mendoza."

Chile

For 2500 miles (4000 km) between the crestline of the Andes and the coastline of the Pacific lies the narrow strip of land that is the Republic of Chile. On average just 90 miles (150 km) wide (and only rarely over 150 miles or 250 km in width), Chile is the world's quintessential example of what *elongation* means to the functioning of a state. Accentuated by its north-south orientation, this severe territorial attentuation not only results in Chile extending across numerous environmental zones; it has also contributed to the country's external political, internal administrative, and general economic problems. Nonetheless, throughout most of their modern history, the Chileans have made the best of this potentially disastrous centrifugal force: from the beginning, the sea has constituted an avenue of longitudinal communication; the Andes Mountains continue to form a barrier to encroachment from the east; and when confrontations loomed at the far ends of the country, Chile proved to be quite capable of coping with its northern rivals, Bolivia and Peru, as well as Argentina in the extreme south.

As Figures G-8 and 5–11 indicate, Chile is a three-subregion country. About 90 percent of its 16 million people are concentrated in what is called Middle Chile, where Santiago, the capital and largest city, and Valparaíso, the chief port, are located. North of Middle Chile lies the Atacama Desert, which is wider, drier, and colder than the coastal desert of Peru. South of Middle Chile, the coast is broken by a plethora of fjords and is-

lands, the topography is mountainous, and the climate—wet and cool near the Pacific—soon turns drier and colder against the Andean interior. South of the latitude of Chiloé Island there are no permanent overland routes, and there is hardly any settlement. These three subregions are also clearly apparent on the map of culture spheres (Fig. 5-5), which displays a mestizo north, a European-commercial zone in Middle Chile, and an undifferentiated south. In addition, a small Amerind-subsistence zone in northern Chile's Andes is shared with Argentina and Bolivia.

Some intraregional differences exist between northern and southern Middle Chile, the country's core area. Northern Middle Chile, the land of the hacienda and of Mediterranean climate with its dry summer season, is an area of (usually irrigated) crops that include wheat, corn, vegetables, grapes, and other Mediterranean products. Livestock raising and fodder crops also take up much of the productive land but are continuing to give way to the more efficient and profitable cultivation of fruits for export. Southern Middle Chile, into which immigrants from both the north and Europe (especially Germany) have pushed, is a better-watered area where raising cattle has predominated. But here, too, more lucrative fruit, vegetable, grain, and other expanding food crops are changing the area's agricultural specializations.

Prior to the 1990s, the arid Atacama region in the north accounted for more than half of Chile's foreign revenues. The Atacama contains the world's largest exploitable deposits of nitrates. At first these provided the country's economic mainstay, but this mining industry declined after the discovery of methods of synthetic nitrate production early in the twentieth century. Subsequently, copper became the chief export (Chile possesses the world's largest reserves). It is found in several places, but the main concentration lies on the eastern margin of the Atacama Desert near the town of Chuquicamata, not far from the port of Antofagasta.

Chile today is emerging from a development boom that transformed its economic geography during the 1990s, a growth spurt that established its reputation as South America's greatest success story. Following the withdrawal of its repressive military dictatorship in 1990, Chile embarked on a program of free-market economic reform that brought stable growth (of up to 8 percent a year), dramatically lowered inflation and unemployment, reduced the poverty rate by more than 20 percent, and attracted massive foreign investment. The last is of particular significance because these new international connections enabled the export-led Chilean economy to diversify and develop in some badly needed new directions. Copper—whose prices have fallen steadily over the past several years—remains the single leading export, but many other mining ventures (including gold) have been launched. In the agricultural sphere, fruit and vegetable production for export has soared because Chile's harvests coincide with the winter farming lull in the high-income countries of the Northern Hemisphere. Other primary economic activities that are increasing their foreign earnings are seafood and wood products, especially paper. Industrial expansion is occurring as well, though at a more leisurely pace, and new manufactures include a modest array of goods that range from basic chemicals to computer software.

Chile's newly internationalized economy has propelled the country to forge a prominent role for itself on the global trading scene. Even the Japanese expressed interest in building ties, and for a time during the 1990s Japan was its leading trading partner. This foothold in the Pacific Rim arena notwithstanding, Chile's trade linkages are now more heavily oriented toward its own hemisphere. The Chileans quickly realized that regional economic integration offered the best opportunities for sustained progress and eagerly accepted the 1994 invitation of the United States, Canada, and Mexico to join the North American Free Trade Agreement. Although the U.S. Congress continues to delay ratification, Chile remains first in line to join NAFTA. In the meantime, responding to warm overtures received from the countries beyond its Andean border, Chile has affiliated with Mercosur as an associate member. However, despite exhortations from its Mercosur partners, Chile in 2000 decided against seeking full membership, preferring instead to negotiate its own free-trade agreement directly with the United States.

Chile ended the twentieth century on a less encouraging note as its still-dominant copper sector dragged the national economy down to a growth standstill. With this commodity earning more than a third of all foreign revenues, there was little the Chileans could do in 1998 when the Asian economic downturn stifled demand and caused a worldwide collapse in the already low price of copper. This was followed by a period of painful readjustment throughout Chile but little loss of confidence in the economy because the financial system retained its strength. Chile's planners also learned some valuable lessons from this experience. One was that the country's economic diversification away from its reliance on copper had neither been as comprehensive nor as rapid as first believed. Among the

FROM THE FIELD NOTES

"Nowhere in South America is the Pan-American Highway in better shape than in Chile, reflecting Chile's economic progress and political stability. My front-row seat in the bus from Valdivia to Puerto Montt afforded a good view of a traffic artery that plays a major role in bringing southern Chile into the national economy."

other lessons were that Chile remains too heavily dependent on its natural resources, urgently needs to expand the skills of its workforce, and must work harder to develop its manufacturing, services, and quaternary sectors.

Chile's recent transformation is widely touted as the "economic model" for all of Middle and South America to emulate. That is almost certainly an overstatement because few countries in these geographic realms can match the Chilean combination of natural and human resources and development circumstances. Yet all of those countries strongly aspire to follow in Chile's footsteps and for the first time in their modern histories are cooperating toward making that a reality. Their goal is the economic integration of all the Americas within the next few years—an accomplishment that could lead to a better life for every inhabitant of the hemisphere.

Uruguay

Uruguay, unlike Argentina or Chile, is compact, small, and rather densely populated. This buffer state of old became a fairly prosperous agricultural country, in effect a smaller-scale Pampa (though possessing less favorable soils and topography). Figures 5-1 and G-8 show the similarity of physical conditions on the two sides of the Plata estuary. Montevideo, the coastal capital, contains more than 40 percent of the country's population of 3.4 million; from here, railroads and roads radiate outward into the productive agricultural interior (Fig. 5-11). In the immediate vicinity of Montevideo lies Uruguay's major farming area, which produces vegetables and fruits for the metropolis as well as wheat and fodder crops. Most of the rest of the country is used for grazing cattle and sheep, with beef products, wool and textile manufactures, and hides dominating the export trade. Tourism is another major economic activity as Argentines, Brazilians, and other visitors increasingly flock to the Atlantic beaches at Punta del Este and other thriving resort towns.

Earlier we noted Thünian patterns in the Argentine Pampa. Uruguay provides another real-world example of this model (Fig. 5-11, inset map). Market gardens and dairying cluster nearest the urban area of the Montevideo national market, with increasingly extensive agriculture prevailing in land-use zones extending to the Brazilian border.

Uruguay's area of 68,500 square miles (177,000 sq km)—less territory even than Guyana—does not leave much room for population clustering. Nevertheless, a special quality of the land area of Uruguay is that it is rather evenly peopled right up to the boundaries with Brazil and Argentina. And of all the countries in South America, Uruguay is the most truly European, notably lacking the racial minorities found even in Chile and Argentina, but

with a sizeable non-Spanish European component in its population. As for future development prospects as Mercosur matures (Montevideo is the bloc's administrative capital), Uruguay is in an excellent position to capitalize on its location between the Pampa to the south and Brazil's most dynamic areas to the north.

BRAZIL: GIANT OF SOUTH AMERICA

By any measure, Brazil is South America's giant. It is so large that it has common boundaries with all the realm's other countries except Ecuador and Chile (Fig. 5-3). Its tropical and subtropical environments range from the equatorial rainforest of the Amazon Basin to the humid temperate climate of the far south. Territorially, Brazil occupies just under 50 percent of South America and is exceeded on the global stage only by Russia, Canada, the United States, and China. In population size, it also accounts for just below half the continental total and is again the world's fifth-largest country (surpassed only by China, India, the United States, and Indonesia). The Brazilian economy is now the eighth largest on Earth, and its modern industrial base is the seventh largest. Moreover, the country is likely to continue advancing in the international ranks and is poised to become a world force as the twenty-first century unfolds.

Population Patterns

Brazil's population of 178 million is as diverse as that of the United States. In a pattern quite familiar in the Americas, the indigenous inhabitants of the country were decimated following the European invasion (fewer than 200,000 Amerindians now survive deep in the Amazonian interior). Africans came in great numbers, too, and today there are more than 10 million Afro-Brazilians in Brazil. Significantly, however, there was also much racial mixing, and 70 million Brazilians have combined European, African, and minor Amerindian ancestries. The remaining 97 million—now barely in the majority at 54 percent—are mainly of European origin, the descendants of immigrants from Portugal, Italy, Germany, and Eastern Europe. The complexion of the population was further diversified by the arrival of Lebanese and Syrian Muslims following the collapse of the Ottoman Empire nearly a century ago. Estimates of the Muslim component in the Brazilian population suggest that about 10 million can claim to be of Arab descent, of whom about 1 million are nominally Muslims. But practicing Muslims are far fewer in number: in 2003 Brazil had only about 35 functioning mosques.

Another significant though small minority began arriving in Brazil in 1908: the Japanese, who today are concentrated in São Paulo State. The more than 1 million Japanese-Brazilians form the largest ethnic Japanese community outside Japan, and in their multicultural environment they have risen to the top ranks of Brazilian society as business leaders, urban professionals, farmers—and even as politicians in the city of São Paulo. Committed to their Brazilian homeland as they are, the Japanese community also retains its contacts with Japan, resulting in many a trade connection.

Brazilian society, to a greater degree than is true elsewhere in the Americas, has made progress in dealing with its racial divisions. To be sure, blacks are still the least advantaged among the country's major population groups, and community leaders continue to complain about discrimination. But ethnic mixing in Brazil is so pervasive that hardly any group is unaffected, and official census statistics about "blacks" and "Europeans" are meaningless. What the Brazilians do have is a true national culture, expressed in an historic adherence to the Catholic faith (now eroding under Protestant-evangelical and secular pressures), in the universal use of a modified form of Portuguese as the common language, and in a set of lifestyles in which vivid colors, distinctive music, and a growing national consciousness and pride are fundamental ingredients.

Brazil's population grew rapidly during the world's twentieth-century population explosion. But over the past three decades, the rate of natural increase has slowed from nearly 3.0 percent to 1.2 percent, and the average number of children born to a Brazilian woman has been halved from 4.4 in 1980 to 2.2 in 2004. These dramatic reductions occurred in the absence of any active family-planning policies by Brazil's federal or State governments and in direct contradiction to the teachings of the Catholic Church, reflecting religion's changing role in this former bastion of Roman Catholicism. Brazil's rapid urbanization, its prevalent economic uncertainties, and the widespread use of contraceptives are among factors influencing this significant change.

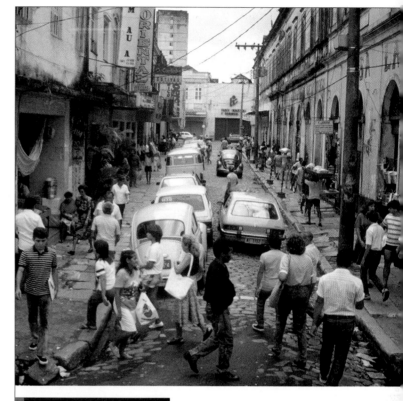

FROM THE FIELD NOTES

"Near the waterfront in Belém lie the now-deteriorating, once-elegant streets of the colonial city built at the mouth of the Amazon. Narrow, cobblestoned, flanked by tiled frontages and arched entrances, this area evinces the time of Dutch and Portuguese hegemony here. Mapping the functions and services here, we recorded the enormous diversity of activities ranging from carpentry shops to storefront restaurants and from bakeries to clothing stores. Dilapidated sidewalks were crowded with shoppers, workers, and people looking for jobs (some newly arrived, attracted by perceived employment opportunities in this growing city of 1.9 million). The diversity of population in this and other tropical South American cities reflects the varied background of the region's peoples and the wide hinterland from which these urban magnets have drawn their inhabitants."

Development Prospects

Brazil is richly endowed with mineral resources, including enormous iron and aluminum ore reserves, extensive tin and manganese deposits, and sizeable oil- and gasfields (Fig. 5-13). Other significant energy developments involve massive new hydroelectric facilities and the successful substitution of sugarcane-based alcohol (gasohol) for gasoline—allowing well over half of Brazil's cars to use this fuel instead of costly imported petroleum. Besides these natural endowments, Brazilian soils sustain a bountiful agricultural output that makes the country a global leader in the production and export of coffee, soybeans, and orange juice concentrate. Commercial agriculture, in fact, is now the fastest growing economic sector, propelled by mechanization and the opening of a major new farming frontier in the fertile grasslands of southwestern Brazil.

As already noted, Brazil ranks seventh among the world's industrial powers. Much of the momentum for this continuing development was unleashed in the early 1990s after the government opened the country's long-protected

FIGURE 5-13

industries to international competition and foreign invest-ment. These new policies are proving to be effective be-cause productivity has risen by over a third since 1990 as Brazilian manufactures attain world-class quality (you may very well fly in a Brazil-made passenger jet the next time you travel by air within the United States). In the mid-1990s the revenues from industrial exports surpassed those from agriculture. Commerce with Argentina made

that member of *Mercosul* (as the Portuguese-speaking Brazilians call Mercosur) one of Brazil's leading trade partners. On the global stage, Brazil became a formidable presence in other ways. The country's enormous and eas-ily accessible iron-ore deposits, the relatively low wages of its workers, and the mechanized efficiency of its steel-makers enable Brazil to produce that commodity at half the cost of steel made in the United States. This causes

American steel producers to demand government protection through tariffs, which goes against purported U.S. principles of free trade.

Even though Brazil has become a major player on the world's economic stage, it has not escaped the cycles of boom and bust that affect transforming economies. Needing loans from international lending agencies to weather the downturns, Brazil's government had to agree to terms that included the privatization of public companies ranging from telephones to utilities. This often meant layoffs, rising prices to consumers, and even civil unrest. Brazil's currency, the *real*, was devalued in 1999, resulting in the outflow of billions of dollars withdrawn from Brazil's banks by panicked depositors. Strikes and marches against increased gas prices, highway tolls, and utility charges paralyzed the country for days. Newspapers reported scandals and corruption involving leading public figures and hundreds of millions of *reals*. In the political campaign that led up to the presidential election of 2002, these issues guaranteed the defeat of the right-of-center government held responsible for it all. In that election a leftist candidate, Luiz Inácio Lula da Silva, won in a landslide victory. He faced the daunting task of adhering to Brazil's existing financial commitments while steering a new course toward more fairness, openness, and honesty in government.

Inequality in Brazil

For all its accomplishments in multiculturalism, Brazil remains a country of stark, appalling social inequalities. Although such inequality is hard to measure precisely, South America is often cited as the geographic realm exhibiting the world's sharpest division between affluence and poverty. And in South America, Brazil is reputed to have the widest gap of all.

Data that underscore this situation include the following. Today the richest 10 percent of the population own two-thirds of all the land and control more than half of Brazil's wealth. The poorest 20 percent of the people live in the most squalid conditions prevailing anywhere on the planet, even including the megacities of Africa and Asia. According to UN estimates, in this age of adequate available (but not everywhere affordable) food, about half the population of Brazil suffers from some form of malnutrition and, in the northeastern States, even hunger. Packs of young orphans and abandoned children roam the cities, sleeping where they can, stealing or robbing when they must. Some of the world's most magnificent central cities are ringed and sectored by the most wretched *favelas* where poverty, misery, and crime converge (see Fig. 5-7 and photo, p. 231).

Most Brazilian voters in the 2002 election believed that these conditions were worsening, that what economic development was taking place benefited those already well off, and that powerful corporations and foreign investors were to blame for the spiral of rising costs and growing unemployment (about 10 percent of the workforce was jobless in 2001). In 2003, President Lula and his new administration faced a formidable set of social as well as economic challenges.

African Heritage

As we noted earlier, Afro-Brazilians still suffer disproportionately from the social ills of this relatively integrated nation. And yet Brazil's culture is infused with African themes, a quality that has marked it from the very beginning. Three centuries ago, the African sculptor and architect affectionately called Aleijadinho was Brazil's most famous artist. The world-famous composer Heitor Villa-Lobos used numerous Afro-Brazilian folk themes in his music. So many Africans were brought in bondage to the city and hinterland of Salvador (Bahia State) from what is today Benin (formerly Dahomey) in West Africa that Bahia has become a veritable outpost of African culture. Candomblé, one of the Macumba religious sects arising from the arrival of African cultures in Brazil, is concentrated in Salvador and Bahia. Its presence there has led to a reconnection with modern Benin as Afro-Brazilians travel to West Africa to search for their roots and West Africans come to Bahia to experience the cultural landscape to which their ancestors gave rise.

In some parts of Bahia, the rural Northeast, and urban areas in Brazil's core you can almost imagine being in Africa. As a nation, Brazil is taking an increased interest in African linkages, not least because the Lusitanian world incorporates not only Portugal and Brazil but also Angola and Moçambique, where the Portuguese colonial experiment failed but where the Lusitanian imprint also survives.

Brazil's Subregions

Brazil is a federal republic consisting of 26 States and the federal district of the capital, Brasília (Fig. 5-13). As in the United States, the smallest States lie in the northeast and the larger ones farther west; their populations range from less than 3 million in the huge, sparsely populated State of Amazonas to 38 million in burgeoning São Paulo State. Although Brazil is about as large as the 48 contiguous United States, it does not exhibit a clear physiographic regionalism. Even the Amazon Basin, which covers almost 60 percent of the country, is not entirely a plain: between the tributaries of the great river lie low but extensive tablelands. Given this physiographic ambiguity, the six

Brazilian subregions discussed next have no absolute or even generally accepted boundaries. In Figure 5-13, those boundaries have been drawn to coincide with the borders of States, making identifications easier.

The **Northeast** was Brazil's source area, its culture hearth. The plantation economy took root here at an early date, attracting Portuguese planters who soon imported the country's largest group of African slaves to work in the sugar fields. But the ample rainfall occurring along the coast soon gives way to lower and more variable patterns in the interior, which is home to about half of the region's 50 million people. This drier inland backcountry—called the *sertão*—is not only seriously overpopulated but also contains some of the worst poverty to be found anywhere in the Americas. The Northeast produces less than one-sixth of Brazil's gross domestic product, but its inhabitants constitute almost one-third of the national population. Given this staggering imbalance, it is not surprising that the region contains half of the country's poor, a literacy rate 20 percent below Brazil's mean, and an infant mortality rate twice the national average.

Much of the Northeast's misery is rooted in its unequal system of land tenure. Farms must be at least 250 acres to be profitable in the hardscrabble *sertão*, a size that only large landowners can afford. Moreover, the Northeast is plagued by a monumental environmental problem: the recurrence of devastating droughts at least partly attributable to **El Niño** (periodic sea-surface-warming events off the continent's northwestern coast). Indeed, extended dry periods occur so regularly that the region has become known as the *Polygon of Drought*.

AMONG THE REALM'S GREAT CITIES . . . **RIO DE JANEIRO**

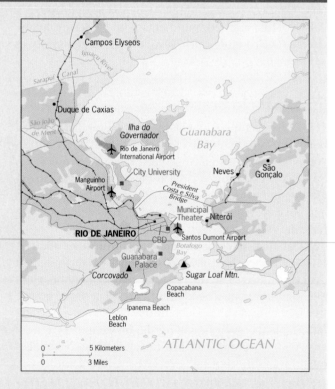

*S*ay "South America" and the first image most people conjure up is Sugar Loaf Mountain, Rio de Janeiro's landmark sentinel that guards the entrance to beautiful Guanabara Bay. Nicknamed the "magnificent city" because of its breathtaking natural setting, *Rio* replaced Salvador as Brazil's capital in 1763 and held that position for almost two centuries until the federal government shifted its headquarters to interior Brasília in 1960. Rio de Janeiro's primacy suffered yet another blow in the late 1950s: São Paulo, its hated urban rival 250 miles (400 km) to the southwest, surpassed Rio to become Brazil's largest city—a gap that has been widening ever since. Although these events triggered economic decline, Rio (11.2 million) remains a major entrepôt, air travel and tourist hub, and center of international business. As a focus of cultural life, however, Rio de Janeiro is unquestionably Brazil's leader, and the *cariocas* (as Rio's residents call themselves) continue to set the national pace with their entertainment industries, universities, museums, and libraries.

On the darker side, this city's reputation is increasingly tarnished by the widening abyss between Rio's affluent and poor populations—symbolizing inequities that rank among the world's most extreme (see photos, p. 231). All great cities experience problems, and Rio de Janeiro is currently bedeviled by the drug use and crime waves emanating from its most desperate hillside *favelas* (slums) that continue to grow explosively.

Nonetheless, Rio's planners recently launched a major project (known as "Rio-City") to improve urban life for all residents. This ambitious scheme is designed to reshape nearly two dozen of the aging city's neighborhoods, introduce an ultramodern crosstown expressway to relieve nightmarish traffic congestion, and—most importantly—bring electrical power, paved streets, and a sewage-disposal network to the beleaguered *favelas*.

Understandably, these conditions propel substantial emigration toward the coastal cities and, increasingly, out of the Northeast to the more prosperous Brazilian subregions that line the Atlantic seaboard to the south. To stem that human tide, Brazil's government has pursued the purchase of underutilized farmland for the settling of landless peasants; although tens of thousands of families have benefited, overall barely a dent has been made in the Northeast's massive rural poverty crisis.

The Northeast today is Brazil's great contradiction. In cities such as Recife and Salvador, hordes of peasants driven from the land constantly arrive to expand the surrounding shantytowns. As yet, few of the generalizations about emerging Brazil apply here, but there are some bright spots. A petrochemical complex has been built near Salvador, creating thousands of jobs and luring foreign investment. Irrigation projects have nurtured a number of productive new commercial agricultural ventures. Tourism is booming along the entire Northeast coast, whose thriving beachside resorts attract thousands of vacationing Europeans. Recife has spawned a budding software industry and a major medical complex. And Fortaleza is the center of new clothing and shoe industries that have already put the city on the global economic map.

The **Southeast** has been modern Brazil's *core area*, with its major cities and leading population clusters. Gold first drew many thousands of settlers, and other mineral finds also contributed to the influx—with Rio de Janeiro itself serving as the terminus of the "Gold Trail." Later, the region's agricultural possibilities ensured its stabilization and growth. The third quarter of the twentieth century brought another mineral age to the Southeast, based on the iron ores around Lafaiete carried to the steelmaking complex at Volta Redonda (Fig. 5-13).

The surrounding State of Minas Gerais (the name means "General Mines") formed the base from which industrial diversification in the Southeast has steadily expanded. The booming metallurgical center of Belo Horizonte paved the way and is now the endpoint of a rapidly developing, ultramodern manufacturing corridor that stretches 300 miles (500 km) southwest to metropolitan São Paulo (gray-striped zone, Fig. 5-13).

São Paulo State is the leading industrial producer and primary focus of ongoing Brazilian development. This economic-geographic powerhouse accounts for nearly half of the country's gross domestic product, with an economy that today matches Argentina's in overall size. Not surprisingly, this region is growing phenomenally (it already contains 21 percent of Brazil's population) as a magnet for migrants, especially from the Northeast.

The wealth of São Paulo State was built on its coffee plantations (known as *fazendas*), and Brazil is still the

Brazil is the leading citrus producer, these days averaging 25 percent of the world's output and well surpassing the 15-percent share of number two, the United States. Contributing mightily to this Brazilian output is the country's thriving orange-juice-concentrate industry, which is based in São Paulo State. This ultramodern processing plant in the heart of the State's citrus belt is located next to Matão, a railroad town about 250 miles (400 km) northwest of Santos (the city of São Paulo's large outport, gateway to the global sea-lane network).

world's leading producer. But coffee today has been eclipsed by other farm commodities. One of them is orange juice concentrate (here, too, Brazil leads the world). São Paulo State now produces more than double the annual output of Florida, thanks to a climate all but devoid of winter freezes, to ultramodern processing plants (see photo above), and to a fleet of specially equipped tankers that ship the concentrate to foreign markets. Another leading pursuit is soybeans, in which Brazil ranks second among the world's producers.

Matching this agricultural prowess is the State's industrial strength. The revenues derived from the coffee plantations provided the necessary investment capital, ores from Minas Gerais supplied the vital raw materials, the nearby outport of Santos facilitated access to the ocean, and immigration from Europe, Japan, and other parts of Brazil contributed the increasingly skilled labor force. As the capacity of the domestic market grew, the advantages of central location and agglomeration secured São Paulo's primacy. This also resulted in metropolitan São Paulo becoming the country's—and South America's—leading industrial complex and megacity (population: 24.8 million).

AMONG THE REALM'S GREAT CITIES . . . SÃO PAULO

São Paulo, which lies on a plateau 30 miles (50 km) inland from its Atlantic outport of Santos, possesses no obvious locational advantages. Yet here on this site we find the third largest metropolis on Earth, whose population has multiplied so uncontrollably that São Paulo has more than doubled in size (from 11 to almost 25 million) over the past three decades.

Founded in 1554 as a Jesuit mission, modern São Paulo was built on the nineteenth-century coffee boom. It has since grown steadily as both an agricultural processing center (soybeans, orange juice concentrate, and sugar besides coffee) and a manufacturing complex (accounting for about half of all of Brazil's industrial jobs). It has also become Brazil's primary focus of commercial and financial activity. Today São Paulo's distinctive, highrise CBD is the very symbol of urban South America and attracts the continent's largest flow of foreign investment as well as the trade-related activities that befit the city's rise as the business capital of Mercosul (Mercosur).

Nonetheless, even for this biggest industrial metropolis of the Southern Hemisphere, the increasingly global tide of postindustrialism is rolling in and São Paulo is being forced to adapt. To avoid becoming a Detroit-style Rustbelt, the aging automobile-dominated manufacturing zone on the central city's southern fringes is today attracting new industries. Internet companies, particularly, have flocked here in such numbers that they already constitute Brazil's largest cluster of e-commerce activity—and prompt many to now call this area *Silicon Village*.

Elsewhere in São Paulo's vast urban constellation—whose suburbs now sprawl outward up to 60 miles (100 km) from the CBD—additional opportunities are being exploited. In the outer northeastern sector, new research facilities as well as computer and telecommunications-equipment factories are springing up. And to the west of central São Paulo, lining the ring road that follows the Pinheiros River, is South America's largest suburban of-

fice complex replete with a skyline of ultramodern highrises.

The huge recent growth of this megacity has also been accompanied by massive problems of overcrowding, pollution, and congestion. Traffic jams here are among the world's worst, with more than twice as many gridlocked motor vehicles on any given day than in Manhattan. Staggering poverty—on a par with Mexico City's—is the most pressing crisis as the ever-expanding belt of shantytowns tightens its grip on much of the metropolis. With its prodigious growth rate expected to persist through the foreseeable future, can anything prevent Greater São Paulo from approaching an unimaginable population of *50 million* barely 20 years from today?

The **South** consists of three States, whose combined population exceeds 25 million: Paraná, Santa Catarina, and Rio Grande do Sul. Southernmost Brazil's excellent agricultural potential has long attracted large numbers of European immigrants. Here the newcomers introduced their advanced farming methods to several areas. Portuguese rice farmers clustered in the valleys of Rio Grande do Sul, where tobacco production has now propelled Brazil to become the world's leading exporter. The Germans, specialists in raising grain and cattle, occupied the somewhat higher areas to the north and in

Santa Catarina. The Italians selected the highest slopes, where they established thriving vineyards. All of these fertile lands proved highly productive, and with growing markets in the large urban areas to the north, this tristate subregion became Brazil's most affluent corner.

With the South firmly rooted in the European-commercial culture sphere (Fig. 5-5), European-style standards of living match the diverse Old World heritage that is reflected in the towns and countryside (where German and Italian are spoken alongside Portuguese). This has led to hostility against non-European Brazilians, and many communities actively discourage poor, job-seeking migrants from the north by offering to pay return bus fares or even blocking their household-goods-laden vehicles. Moreover, extremist groups have arisen to openly espouse the secession of the South from Brazil. The largest, calling itself "The South Is My Country," promotes the creation of a *Republic of the Pampas*. Accusing the separatist movement of subversion and racism, State governments respond with the claim that the issue is an economic one, complaining to Brasília that the South receives only about a 60 percent return on the federal taxes it pays. But with polls showing at least one-third of the population sympathetic to the idea of secession, officialdom may well have a longer-term devolutionary struggle on its hands.

Economic development in the South is not limited to the agricultural sector. Coal from Santa Catarina and Rio Grande do Sul is shipped north to the steel plants of Minas Gerais. Local manufacturing is growing as well, especially in Pôrto Alegre and Tubarão. During the 1990s, a major center of the computer software industry was established in Florianópolis, the island city and State capital just off Santa Catarina's coast. Known as *Tecnópolis*, this budding technopole continues to grow by capitalizing on its seaside amenities, skilled labor force, superior air-travel and global communications linkages, and government and private-sector incentives to support new companies. Only the sparsely populated interior portion of the South has lagged behind the rest of the region, despite the recent completion of massive Itaipu Dam in westernmost Paraná State.

The **Interior** subregion—constituted by the States of Goiás, Mato Grosso, and Mato Grosso do Sul—is also known as the Central-West. This is the region that Brazil's developers have long sought to make a part of the country's productive heartland, and in 1960 the new capital of Brasília was deliberately situated on its margins (Fig. 5-13).

By locating the new capital city in the wilderness 400 miles (650 km) inland from its predecessor, Rio de Janeiro, the nation's leaders dramatically signaled the opening of Brazil's development thrust toward the west. Brasília is also noteworthy in another regard because it represents what political geographers call a **forward capital**. A state will sometimes relocate its capital to a sensitive area, perhaps near a peripheral zone under dispute with an unfriendly neighbor, in part to confirm its determination to sustain its position in that contested zone. Brasília does not lie near a contested area, but Brazil's interior was an internal frontier to be conquered by a growing nation. Spearheading that drive, the new capital occupied a decidedly forward position.

Despite the subsequent growth of Brasília to 2.3 million inhabitants today, it was not until the 1990s that the Interior began its economic integration with the rest of Brazil. The catalyst was the exploitation of the vast *cerrado*—the fertile savannas that blanket the Central-West and make it one of the world's most promising agricultural frontiers (at least two-thirds of its arable land still awaits development). As with the U.S. Great Plains, the flat terrain of the *cerrado* is one of its main advantages because it facilitates the large-scale mechanization of farming with a minimal labor force. Another advantage is rainfall, which is more prevalent than in the Great Plains or Argentine Pampa (Fig. G-7).

The leading crop is soybeans, whose output per acre here exceeds even that of the U.S. Corn Belt. Other grains and cotton are also expanding across the farmscape of the *cerrado*, but the current pace of regional development is inhibited by a serious accessibility problem. Unlike the Great Plains and Pampa, the growth of an efficient transportation network did not accompany the opening of this farming frontier. Thus the Interior's products must travel along poor roads and intermittent railroads to reach the markets and ports of the Atlantic seaboard. Today several projects are finally underway to alleviate these bottlenecks, including the privately financed *Ferronorte* railway that links Santos to the southeastern corner of Mato Grosso State.

The **North** is Brazil's territorially largest and most rapidly developing subregion, which consists of the seven States of the Amazon Basin (Fig. 5-13). This was the scene of the great rubber boom a century ago, when the wild rubber trees in the *selvas* (tropical rainforests) produced huge profits and the central Amazon city of Manaus enjoyed a brief period of wealth and splendor. But the rubber boom ended in 1910, and for most of the seven decades that followed Amazonia was a stagnant hinterland lying remote from the centers of Brazilian settlement. All that changed dramatically during the 1980s as new development began to stir throughout this awakening region, which currently is the scene of the world's largest migration into virgin territory as over 200,000 new settlers arrive each year. Most of this influx is occurring south of the Amazon River, in the tablelands between the major waterways and along the Basin's wide rim.

Two ongoing development schemes are especially worth noting. The first is the *Grande Carajás Project* in southeastern Pará State, a huge multifaceted scheme centered on one of the world's largest known deposits of iron ore in the hills around Carajás (Fig. 5-13). In addition to a vast mining complex, other new construction here includes the Tucuruí Dam on the nearby Tocantins River and a 535-mile (850-km) railroad to the Atlantic port of São Luis. This ambitious development project also emphasizes the exploitation of additional minerals, cattle raising, crop farming, and forestry. What is occurring is an application of the **growth-pole concept**. A

This is what the Amazon's equatorial rainforest looks like from an orbiting satellite after the human onslaught in preparation for settlement. The colors on this Landsat image emphasize the destruction of the trees, with the dark green of the natural forest contrasted against the pale green and pinks of the leveled forest. The linear branching pattern of deforestation here in Rondônia State's Highway BR-364 corridor is explained in the text. But farming here is not likely to succeed for very long, and much of the cleared land is likely to be abandoned. Then the onslaught will resume to clear additional land—as an entire ecosystem comes ever closer to failing forever.

growth pole is a location where a set of activities, given a start, will expand and generate widening ripples of development in the surrounding area. In this case, the stimulated hinterland could one day cover one-sixth of all Amazonia.

Understandably, tens of thousands of settlers have descended on this part of the Amazon Basin. Those seeking business opportunities have been in the vanguard, but they have been followed by masses of lower-income laborers and peasant farmers in search of jobs and land ownership. The initial stage of this colossal enterprise has boosted the fortunes of many towns, particularly the revival of Manaus northwest of Carajás. Here, a thriving industrial complex (specializing in the production of electronic goods) has emerged in the free-trade zone adjoining the city thanks to the outstanding air-freight operations at Manaus's ultramodern airport. But many problems have also arisen as the tide of pioneers rolled across central Amazonia. One of the most tragic involved the primitive Yanomami people, whose homeland in Roraima State was overrun by thousands of claim-stakers (in search of newly discovered gold) who triggered violent confrontations that ravaged the fragile aboriginal way of life.

The second leading development scheme, known as the *Polonoroeste Plan*, is located about 500 miles (800 km) to the southwest of Grande Carajás in the 1500-mile-long Highway BR-364 corridor that parallels the Bolivian border and connects the western Brazilian towns of Cuiabá, Pôrto Velho, and Rio Branco (Fig. 5-13). Although the government had planned for the penetration of western Amazonia to proceed via the east-west Trans-Amazon Highway, the migrants of the 1980s and 1990s preferred to follow BR-364 and settle within the Basin's southwestern rim zone, mostly in Rondônia State. Agriculture has been the dominant activity here, but in the quest for land, bitter conflicts continue to break out between peasants and landholders as the Brazilian government pursues the volatile issue of land reform.

The usual pattern of settlement in this part of the Brazilian North is something like this. As main and branch highways are cut through the wilderness, settlers, enticed by cheap land, follow and move out laterally to clear spaces for farming (see photo above). Crops are planted, but within three years the heavy equatorial-zone rains leach out soil nutrients and accelerate surface erosion. As soil fertility declines, pasture grasses are planted, and the plot of land is soon sold to cattle ranchers. The peasant farmers then move on to newly opened areas, clear more land for planting, and the cycle repeats itself. Unfortunately, this not only establishes low-grade land uses across widespread areas but also requires the burning and clearing of enormous stands of tropical woodland. For most of the past decade, an area of rainforest one-third the size of Tennessee has been disappearing *annually* in Amazonia—accounting for over half of all tropical deforestation and exacerbating an environmental crisis of global proportions (see box titled "Tropical Deforestation," p. 212).

Brazil is the cornerstone of South America, the dominant economic force in Mercosur/Mercosul, the only dimensional counterweight to the United States in the Western Hemisphere, a maturing democracy, and an emerging global giant. The future of the entire South American realm depends on Brazil's stability and social as well as economic progress.

Continued on pages 196-197

FIGURE 6-1

Relief

Meters	Feet
3050	10 000
1525	5000
610	2000
305	1000
0 Sea Level	0
152.5	500 Below
1525	5000 Sea Level
3050	10 000
6100	20 000

Longitude West of Greenwich   Longitude East of Greenwich

0   200   400   600   800   1000 Miles
0   400   800   1200   1600 Kilometers

Scale 1:40 000 000; one inch to 630 miles. Lambert's Azimuthal, Equal Area Projection

Elevations and depressions are given in feet.

A-580000-76 13-18-36
COPYRIGHT BY
RAND McNALLY & COMPANY
MADE IN U.S.A.

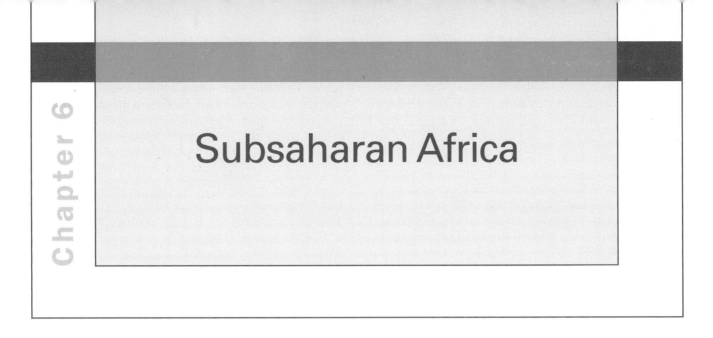

Chapter 6

Subsaharan Africa

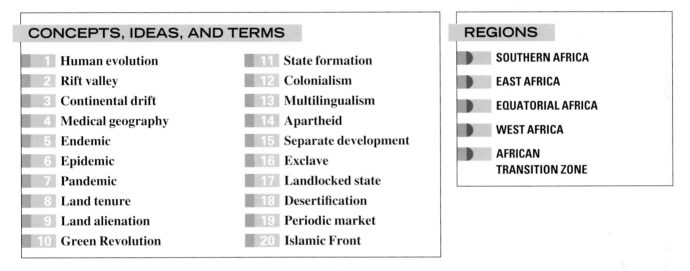

THE AFRICAN CONTINENT occupies a special place in the physical as well as the human world. You need a globe to confirm the former: the Earth has a Land Hemisphere and a Water Hemisphere, and Africa lies at the center of the Land Hemisphere, surrounded in all directions by the other landmasses. As for Africa's human significance, this is where the great saga of **human evolution** began. In Africa we formed our first communities, spoke our first words, created our first art and our first weapons. From Africa our ancestor hominids spread outward into Eurasia more than two million years ago. From Africa our own species emigrated, beginning perhaps 130,000 years ago, northward into present-day Europe and eastward via southern Asia into Australia and, much later, farther afield into the Americas. Disperse our forebears did, but we should remember that, at the source, we are all Africans.

For millions of years, therefore, Africa served as the cradle for humanity's emergence. For tens of thousands of years, Africa was the source of human cultures. For thousands of years, Africa led the world in countless spheres ranging from tool manufacture to plant domestication.

But in this chapter we will encounter an Africa that has been struck by a series of disasters ranging from environmental deterioration to human dislocation on a scale unmatched anywhere in the world. When we assess Africa's misfortunes, though, we should remember that they have lasted hundreds, not thousands, or millions, of years. Africa's catastrophic interlude will end, and Africa's time and turn will come again.

The focus in this chapter will be on Africa south of the Sahara, for which the unsatisfactory but convenient name *Subsaharan Africa* has come into use to signify not physically "under" the great desert but directionally "below" it. The African continent contains two geographic realms: the African, extending from the southern margins of the Sahara to the Cape of Good Hope, and the western flank of the realm dominated by the Muslim faith and Islamic culture whose heartland lies in the Middle East and the Arabian Peninsula. The great desert forms a formidable barrier between the two, but the powerful influences of Islam crossed it centuries before the first Europeans set foot in West Africa. By that time, the African kingdoms in what is known today as the Sahel had been converted, creating an Islamic foothold all along the northern periphery of the African realm (see Fig. G-2, p. 5). As we note later, this cultural and ideological penetration had momentous consequences for Subsaharan Africa.

In the three previous chapters on the Americas, we made frequent reference to the forced migration of Africans to Brazil, the Caribbean region, and the United States. The slave trade was one of those African disasters alluded to above, and it was facilitated in part by what we may call the peril of proximity. The northeastern tip of Brazil, by far the largest single destination for the millions of Africans forced from their homes in bondage, lies about as far from the nearest West African coast as South Carolina lies from Venezuela, a short maritime intercontinental journey indeed (it is more than twice as far from West Africa to South Carolina). That proximity facilitated the forced migration of millions of West Africans to Brazil, which in turn contributed to the emergence of an African cultural diaspora in Brazil that is without equal in the New World. It is therefore logical to focus next on the African realm.

DEFINING THE REALM

THE AFRICAN CONTINENT may be partitioned into two human-geographic realms, but the landmass is indivisible. Before we investigate the human geography of Subsaharan Africa, therefore, we should take note of the entire continent's unique physical geography (Fig. 6-1). We have already noted Africa's situation at the center of the planet's Land Hemisphere; moreover, no other landmass is positioned so squarely astride the equator, reaching almost as far to the south as to the north. This location has much to do with the distribution of Africa's climates, soils, vegetation, agricultural potential, and human population.

MAJOR GEOGRAPHIC QUALITIES OF

Subsaharan Africa

1. Physiographically, Africa is a plateau continent without a linear mountain backbone, with a set of Great Lakes, variable rainfall, generally low-fertility soils, and mainly savanna and steppe vegetation.

2. Dozens of nations, hundreds of ethnic groups, and many smaller entities make up Subsaharan Africa's culturally rich and varied population.

3. Most of Subsaharan Africa's peoples depend on farming for their livelihood.

4. Health and nutritional conditions in Subsaharan Africa need improvement as the incidence of disease remains high and diets are often unbalanced. The AIDS pandemic began in Africa and has become a major health crisis in this realm.

5. Africa's boundary framework is a colonial legacy; many boundaries were drawn without adequate knowledge of or regard for the human and physical geography they divided.

6. The realm is rich in raw materials vital to industrialized countries, but much of Subsaharan Africa's population has little access to the goods and services of the world economy.

7. Patterns of raw-material exploitation and export routes set up in the colonial period still prevail in most of Subsaharan Africa. Interregional and international connections are poor.

8. During the Cold War, great-power competition magnified conflicts in several Subsaharan African countries, with results that will be felt for generations.

9. Severe dislocation affects many Subsaharan African countries, from Liberia to Rwanda. This realm has the largest refugee population in the world today.

10. Government mismanagement and poor leadership afflict the economies of many Subsaharan African countries.

AFRICA'S PHYSIOGRAPHY

Africa accounts for about one-fifth of the Earth's entire land surface. The north coast of Tunisia lies 4800 miles (7700 km) from the south coast of South Africa. Coastal Senegal, on the extreme western *Bulge* of Africa, lies 4500 miles (7200 km) from the tip of the *Horn* in easternmost Somalia. These distances have environmental implications. Much of Africa is far from maritime sources of moisture. In addition, as Figure G-8 shows, large parts of the landmass lie in latitudes where global atmospheric circulation systems produce arid conditions. The Sahara in the north and the Kalahari in the south form part of this globe-girdling desert zone. Water supply is one of Africa's great problems.

Africa's topography reveals several properties that are not replicated on other landmasses. Alone among the continents, Africa does not have an Andes-like linear mountain backbone; neither the northern Atlas nor the southern Cape Ranges are in the same league. Where Africa does have high mountains, as in Ethiopia and South Africa, these are really deeply eroded plateaus or, as in East Africa, high, snowcapped volcanoes. Furthermore, Africa is one of only two continents containing a cluster of Great Lakes, and the only one whose lakes result from powerful tectonic forces in the Earth's crust. These lakes (with the exception of Lake Victoria) lie in deep trenches called **rift valleys**, which form when huge parallel cracks or faults appear in the Earth's crust and the strips of crust between them sink, or are pushed down, to form great, steep-sided, linear valleys. In Figure 6-2 these rift valleys, which stretch more than 6000 miles (9600 km) from the Red Sea to Swaziland, are indicated by red lines.

Africa's rivers, too, are unusual: their upper courses often bear landward, seemingly unrelated to the coast toward which they eventually flow. Several rivers, such as the Nile and the Niger, have inland as well as coastal deltas. Major waterfalls, notably the Victoria Falls on the Zambezi, or lengthy systems of cataracts, separate the upper from the lower courses.

Finally, Africa may be described as the "plateau continent." Except for some comparatively limited coastal plains, almost the entire continent lies above 1000 feet (300 m) in elevation, and fully half of it lies over 2500 feet (800 m) high. As Figure 6-2 shows, the plateau surface has sagged under the weight of accumulating sediments into a half dozen major basins (three of them in the Sahara). The margins of Africa's plateau are marked by escarpments, often steep and step-like. Most notable among these is the Great Escarpment of South Africa, marking the eastern edge of the Drakensberg Mountains.

Continental Drift and Plate Tectonics

Africa's remarkable and unusual physiography was one piece of evidence that geographer Alfred Wegener used to construct his hypothesis of **continental drift**. The present continents, Wegener reasoned, lay assembled as one giant landmass called *Pangaea* not very long ago geologically (220 million years ago). The southern part of this supercontinent was *Gondwana*, of which Africa formed the core (Fig. 6-3). When, about 200 million years ago, tectonic forces began to split Pangaea apart, Africa (and the other landmasses) acquired their present configurations. That process, now known as *plate tectonics*, continues, marked by earthquakes and volcanic eruptions. By the time it started, however, Africa's land surface had begun to acquire some of the features that mark it today—and make it unique. The rift valleys, for example, demarcate the zones where plate movement continues—hence the linear shape of the Red Sea, where the Arabian Plate is separating from the African Plate (Fig. G-4). And yes, the rift valleys of East Africa probably mark the further fragmentation of the African Plate (some geophysicists have already referred to a "Somali Plate" which in the future will separate, Madagascar-like, from the rest of Africa).

So Africa's ring of escarpments, its rifts, its river systems, its interior basins, and its lack of Andes-like mountains, all relate to the continent's central location in Pangaea, all pieces of the puzzle that led to the plate-tectonics solution. Once again, geography was the key.

NATURAL ENVIRONMENTS

Only the southernmost tip of Subsaharan Africa lies outside the tropics. Although African elevations are comparatively high, they are not high enough to ward off the heat that comes with tropical location except in especially favored locales such as the Kenya Highlands and parts of Ethiopia. And, as we have noted, Africa's bulky shape means that much of the continent lies far from maritime moisture sources. Variable weather and frequent droughts are among Africa's environmental problems.

It is useful at this point to refer back to Figure G-8 on page 15. As that map shows, Africa's climatic regions are distributed almost symmetrically about the equator, though more so in the center of the landmass than in the east, where elevation changes the picture. The hot, rainy climate of the Congo Basin merges gradually, both northward and southward, into climates with distinct winter-dry seasons. "Winter," however, is marked more by drought than by cold. In parts of the area mapped *Aw* (savanna), the annual seasonal cycle produces two rainy seasons, often

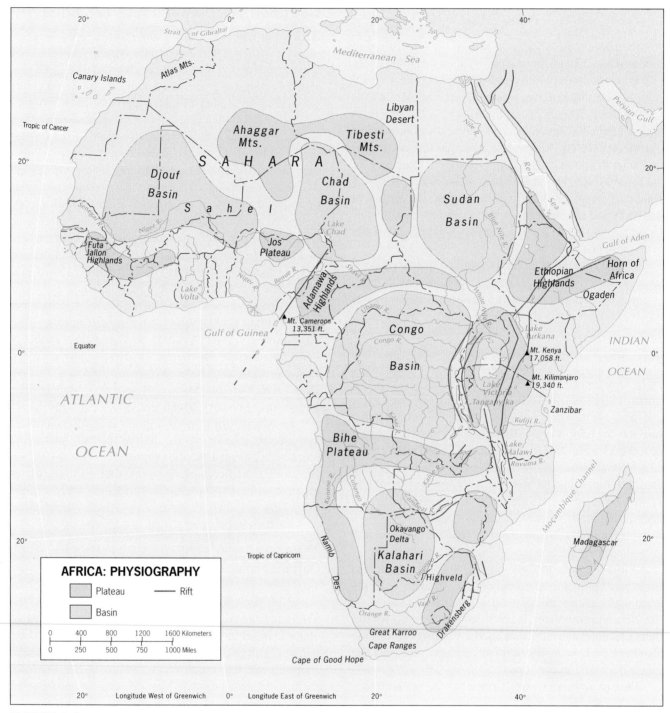

FIGURE 6-2

referred to locally as the "long rains" and the "short rains," separated by two "winter" dry periods. As you go farther north and south, away from the moist Congo Basin, the dry season(s) grow longer and the rainfall diminishes and becomes less and less dependable.

Most Africans make their living by farming, and many grow crops in marginal areas where rainfall variability can have catastrophic consequences. Compare Figures G-7

and G-8 and note the steep decline in annual rainfall from more than 80 inches (200 cm) around the equator in the Congo Basin to a mere 4 inches (10 cm) in parts of Chad to the north and Namibia to the south. Millions of farmers till the often unproductive soils of the *savanna*, and many mix livestock herding with crop-raising to reduce their risk in this difficult environment. But the savanna's wildlife carries diseases that also infect livestock, which makes herd-

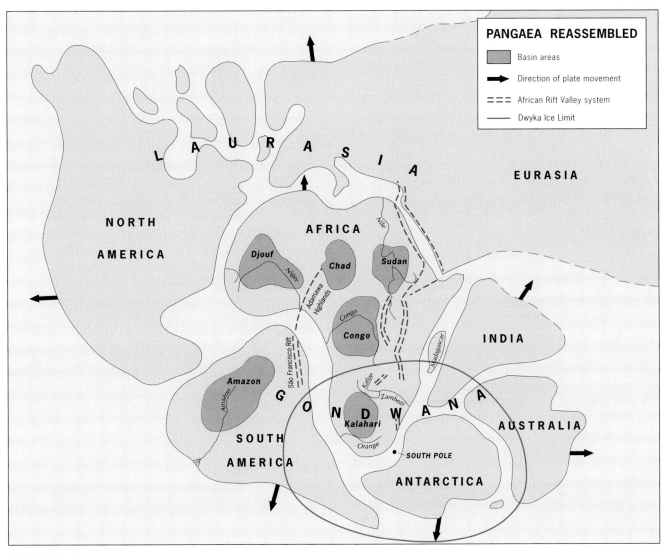

FIGURE 6-3

ing a risky proposition, too. Even where the savanna gives way to the still-drier *steppe*, human pressure continues to grow, and people as well as animals trample fragile, desert-margin ecologies.

Africa's shrinking rainforests and vast savannas form the world's last refuges for wildlife ranging from primates to wildebeests. Gorillas and chimpanzees survive in dwindling numbers in threatened forest habitats, while millions of herbivores range in great herds across the savanna plains where people compete with them for space. European colonizers, who introduced hunting as a "sport" (a practice that was not part of African cultural traditions) and who brought their capacities for mass destruction to animals as well as people in Africa, helped clear vast areas of wildlife and push species to near extinction. Later they laid out game reserves and other types of conservation areas, but these were not sufficiently large or well enough connected

to allow herd animals to follow their seasonal and annual migration routes. The same climatic variability that affects farmers also affects wildlife, and when the rangelands wither, the animals seek better pastures. When the fences of a game reserve wall them off, they cannot survive. When there are no fences, the wildlife invades neighboring farmlands and destroys crops, and the farmers retaliate. After thousands of years of equilibrium, the competition between humans and animals in Africa has taken a new turn. It is the end of an era.

And yet it would seem that there could be room for humans as well as wildlife in Subsaharan Africa. As Table G-1 confirms, all the countries of this realm *combined* have a population only half that of China alone. Indeed, Africa does have some areas of good soils, ample water, and high productivity: the volcanic soils of Mount Kilimanjaro and those of the highlands around the Western Rift Valley, the soils in the Ethiopian Highlands, and those in the moister areas of higher-latitude South Africa and parts of West Africa yield good crops when social conditions do not disrupt the farming communities. But such areas are small in the vastness of Africa. This realm has nothing to compare to the huge alluvial basins of India and China, not even a lower Nile Valley and Delta, where comparatively small areas of fertile, irrigated soils can support tens of millions of people. Except for small (in some cases experimental) patches of rice and wheat, Africa is the land of corn (maize), millet, and root crops, far less able to provide high per-acre yields. Africa's natural environment poses a formidable challenge to the millions who depend directly on it.

ENVIRONMENT AND HEALTH

Africa's natural environment creates other difficulties as well. High temperatures and humidity, where these prevail, create breeding grounds for organisms that carry disease. Mosquitoes, flies, fleas, worms, even snails can transmit such diseases. From birth, Africans (especially the people living in A climates) are exposed to a wide range of dangerous diseases. Over the centuries they have suffered from repeated outbreaks of killer plagues, some of them related to the wildlife that shares their habitat.

Medical geography studies people's health in spatial context. Medical geographers employ methods of spatial analysis to track disease outbreaks, predict how they will spread, identify their sources, detect their carriers, and prevent their repetition. Today GIS (geographic information systems) and other state-of-the-art methods are employed to achieve these goals, and the alliance between doctors and geographers has already yielded significant results. Doctors know how a disease ravages the body; geographers know how climatic conditions such as wind direction or variations in river flow can affect the distribution and effectiveness of disease carriers. This collaboration helps protect vulnerable populations.

Tropical Africa, the source of many serious illnesses, is the focus of much of medical geography's work. Not only the carriers (*vectors*) of infectious diseases but also cultural traditions that facilitate transmission, such as sexual practices, food selection and preparation, and personal hygiene, play their role—and all can be mapped. Comparing medical, environmental, and cultural maps can lead to crucial evidence that helps combat the scourge.

In Africa today, hundreds of millions of people carry one or more maladies, often without knowing exactly what ails them. A disease that infects many people (the *hosts*) in a kind of equilibrium, without causing rapid and widespread deaths, is said to be **endemic** to the population. People affected may not die suddenly or dramatically, but their health deteriorates, energy levels fall, and the quality of life declines. In tropical Africa hepatitis, venereal diseases, and hookworm are among the public health threats in this category.

When a disease outbreak has local or regional dimensions, it is called **epidemic**. It may claim thousands, even tens of thousands of lives, but it remains confined to a certain area, perhaps one defined by the range of its vector. In tropical Africa, trypanosomiasis, the disease known as sleeping sickness and vectored by the tsetse fly, has regional proportions. The great herds of savanna wildlife form the *reservoir* of this disease, and the tsetse fly transmits it to livestock and people. It is endemic to the wildlife, but it kills cattle, so Africa's herders try to keep their animals in tsetse-free zones. African sleeping sickness appears to have originated in a West African source area during the fifteenth century, and it spread throughout much of tropical Africa (Fig. 6-4). Its epidemic range was limited by that of the tsetse fly: where there are no tsetse flies, there is no sleeping sickness. More than anything else, the tsetse fly has kept Subsaharan Africa's savanna-lands largely free of livestock and open to wildlife. Should a remedy be found, livestock would replace the great herds on the grasslands.

When a disease spreads worldwide, it is described as **pandemic**. Africa's and the world's most deadly vectored disease is malaria, transmitted by a mosquito and

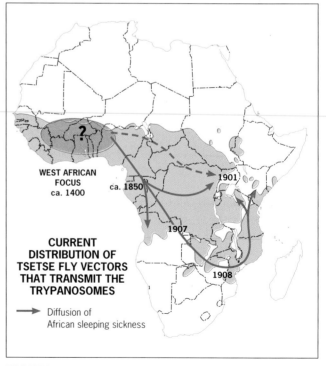

WEST AFRICAN FOCUS ca. 1400

ca. 1850

1901

1907

1908

CURRENT DISTRIBUTION OF TSETSE FLY VECTORS THAT TRANSMIT THE TRYPANOSOMES

→ Diffusion of African sleeping sickness

FIGURE 6-4

killer of as many as 1 million children each year. Whether malaria has an African origin is not known, but it is an ancient affliction. Hippocrates, the Greek physician of the fifth century BC, mentions it in his writings. Apes, monkeys, and several other species also suffer from it. Fever attacks, anemia, and enlargement of the spleen are its symptoms. Malaria has spread around the world and prevails not only in tropical but also in temperate areas. Eradication campaigns against the mosquito vector have had some success, but always the carrier has come back with renewed vigor. At present as many as 300 million people annually are recorded as having malaria, but the actual number probably is much greater because in much of tropical Africa malaria is, simply, a way of life affecting entire populations. The short life expectancies for tropical Africa reported in Table G-1 reflect infant and child mortality from malarial infection.

Another pandemic disease with African origins is yellow fever, also vectored by a mosquito. African monkeys form the reservoir for this scourge, which once spread worldwide but is now largely confined to Africa and South America. From its source area in West Africa, yellow fever not only reached South America but expanded into Central America and even the United States; its defeat in Panama nearly a century ago made possible the construction of the Panama Canal. In Africa it remains a threat: Senegal had an outbreak during the 1960s that claimed more than 20,000 lives.

During the 1970s, Africa was afflicted by yet another malady that appears to have started in the equatorial forest, became an epidemic, and grew into a pandemic: AIDS (see box titled "AIDS in Subsaharan Africa"). Monkeys may have activated this disease in humans, but whatever the source, Africa has been the most severely affected by it. In 2003, there were an estimated 38 million AIDS cases in the world; more than 31 million were in Africa. Again, these official calculations probably conceal a much worse crisis, and the population growth rates for many African countries are declining as a result. During the early part of the twenty-first century, some countries will experience actual population decreases. Life expectancies are dropping to levels not seen for centuries.

The great incubator of vectored and nonvectored diseases in humid equatorial Africa continues to threaten local populations. Sudden outbreaks of Ebola fever in Sudan during the 1970s, in The Congo* in the 1990s, and in Uganda in 2000–2001 projected the risks that many Africans live with onto the world's television screens. But Africa's woes are soon forgotten. Only the fear that an

African epidemic might evolve into a global pandemic mobilizes world attention. Africans cope with the planet's most difficult environments, but they are least capable of combating its hazards because their resources are so limited.

Even when international efforts are made to assist Africa, the unintended results may be catastrophic. Among Africa's endemic diseases is schistosomiasis, also called bilharzia. The vector is a snail, and the transmitted parasites enter via body openings when people swim or wash in slow-moving water where the snails thrive (fast-moving streams are relatively safe). When development projects designed to help African farmers dammed rivers and streams and sent water into irrigation ditches, farm production rose—but so did the incidence of schistosomiasis, which causes internal bleeding and fatigue, although it is rarely fatal. Medical geographers' maps showed the spread of schistosomiasis around the dam projects, and soon the cause was clear: by slowing down the water flow, the engineers had created an ideal environment for the snail vector.

Poverty is a powerful barrier to improving health, and major interventions in the natural environment carry costs as well as benefits. Africa needs improved medical services, more effective child-immunization programs, mobile clinics for remote areas, and similar remedies. While 70 percent of the realm's people continue to live in rural areas, the natural environment will weigh heavily on a vulnerable population.

LAND AND FARMING

In their penetrating book, *Geography of Sub-Saharan Africa*, editor Samuel Aryeetey-Attoh and his colleagues focus on the issue of land tenure, crucial in Africa because most Africans are farmers. **Land tenure** refers to the way people own, occupy, and use land. African traditions of land tenure are different from those of Europe or the Americas. In most of Subsaharan Africa, communities, not individuals, customarily hold land. Occupants of the land have temporary, custodial rights to it and cannot sell it. Land may be held by large (extended) families, by a village community, or even by a traditional chief who holds the land in trust for the people. His subjects may house themselves on it and farm it, but in return they must follow his rules.

When the European colonizers took control of much of Subsaharan Africa, their land ownership practices clashed head-on with those of Africa. Africans believed that their land belonged to their ancestors, the living, and the yet-unborn; Europeans saw unclaimed space and felt justified in claiming it. What Africans called

---

*Two countries in Africa have the same short-form name: Congo. In this book, we use *The Congo* for the larger Democratic Republic of the Congo, and *Congo* for the smaller, neighboring Republic of Congo.

AIDS in Subsaharan Africa

*A*IDS—Acquired Immune Deficiency Syndrome—spread in Africa during the 1980s and became a pandemic during the 1990s. Its impact on Africa is devastating. In 2002, more than 2.5 million people died of AIDS in Subsaharan Africa, 75 percent of the global toll.

Persons infected by HIV (human immunodeficiency virus) do not immediately or even soon display symptoms of AIDS. In the early stages, only a blood test will reveal infection, and then only by indicating that the body is mobilizing antibodies to fight HIV. People can carry the virus for years without being aware of it; during that period they can unwittingly transmit it to others. For that reason, official statistics of AIDS cases lag far behind the reservoir of those infected, especially in countries where populations cannot be thoroughly tested. Since AIDS was identified in the early 1980s, nearly 25 million people have died of it.

By 2003, more than 37 million people were known to be infected by AIDS worldwide, and of these about 31 million lived in 34 tropical African countries according to United Nations sources. In the early 1990s, the worst-hit countries were in Equatorial Africa in what was called the AIDS Belt from (then) Zaïre to Kenya. But a decade later, the most severely afflicted countries lay in Southern Africa. In Zimbabwe, Botswana, and Namibia, more than 25 percent of all persons 15–49 were infected with HIV; in Zambia, nearly 20 percent; and in South Africa, the region's most populous country, about 13 percent (5.8 million). These are the official numbers; medical geographers estimate that 20 to 25 percent of the *entire* population of several tropical African countries is infected.

As the number of deaths from AIDS rises, African countries' vital statistics are showing the demographic impact. In Botswana, life expectancy in 1994 was 60 years; by 2003 it had declined to 39. In South Africa it declined from 66 to 51 in just ten years. In Zimbabwe, population growth was 3.3 percent per year in the early 1980s; now it is 0.9 percent. Meanwhile, AIDS is killing parents at such a rate that, by 2010, it is projected by the UN, it will be responsible for 20 million African orphans.

Geographer Peter Gould, in his 1993 book *The Slow Plague*, called Africa a "continent in catastrophe," pointing out that the official data on which these depressing reports are based understate the magnitude of the calamity. African governments, Gould wrote, conceal the real dimensions of the tragedy for various reasons: a sense of shame, concern that the real figures will drive investors out and keep tourists away, and the arrogance of military rulers used to denying reality. Many doctors in Africa, he added, say that AIDS cases are 80 to 90 percent underreported.

No part of tropical Africa is spared. West African countries, too, are reporting growing numbers of AIDS cases.

Ivory Coast, reporting that more than 10 percent of adults are infected there, may provide the most accurate data. In Nigeria, where the full onslaught may not yet be known, AIDS incidence among adults is more than 4 percent (which, in that populous country, means that more than 5 million people are ill with AIDS).

The misery the AIDS pandemic has created is beyond measure, but its economic impact is beyond doubt. Workers are sick on the job; but jobs are scarce and medical help is unavailable or unaffordable for many, so the workers stay on the job until they collapse. In a realm where unemployment is high, new workers immediately fill the vacancies, but they, too, are unwell. Corporations report skyrocketing AIDS-related costs. In Botswana, whose economy has been prospering due to increased diamond production, highly trained diamond sorters are being lost to AIDS, and training replacement workers is depressing profits. Medical expenses, multiplying death benefits, funeral payments, and recruiting and training costs are reducing corporate incomes from Kenya's sugar plantations to South Africa's gold mines. Thus the impact of AIDS is recorded in the continent's economic statistics. The pattern shown in Figure G-11, where Africa is concerned, is due in substantial part to the AIDS pandemic.

In countries of the global core, progress has been made in containing AIDS and prolonging the lives of those afflicted. This has taken AIDS off the medical center stage, but such a turnaround is not in prospect for Africa. The cost of the medications that combat AIDS is too high, and private drug companies will not dispense their products at prices affordable to African patients. Thus the best short-term strategy is for governments to attempt to alter public attitudes and behavior, notably by making condoms available as widely and freely as possible. Uganda, once one of tropical Africa's worst-afflicted countries, managed to slow the rate of AIDS dispersal by these methods—and through a vigorous advertising campaign on billboards, in newspapers, and on radio and television.

But what Africa needs is a coordinated attack on AIDS through the provision of medicines. In March 2003 President George W. Bush announced an initiative to provide Subsaharan Africa with the financial means to combat the disease: a (U.S.) $10 billion fund to facilitate treatment as well as prevention. The urgency of the matter is underscored by estimates that, in the absence of a massive campaign, some of Africa's countries may see population declines of as much as 10 to 20 percent, a disaster comparable only to Europe's fourteenth-century bubonic plague and the collapse of Native American populations after the introduction of smallpox by the European invaders in the sixteenth. Once again environmental, cultural, and historical factors combine to afflict Africa's peoples.

Shifting agriculture in Africa's tropical forest zones still forms the mainstay for millions. Here in the Ituri Forest in the northeastern corner of The Congo, trees have been felled, the land cleared, and between stumps and rotting trunks the people farm cassava, yams, and bananas. The soil will sustain these crops for some years, but then yields will diminish and the land will be abandoned, so that another plot must be cleared. Women do the traditional farming in Africa; here three women from the village of Ngodi are weeding the field while two youngsters look on.

9 **land alienation**—the expropriation of (often the best) land by Europeans—changed the pattern of land tenure in Africa. By the time the Europeans withdrew, private land ownership was widely dispersed and could not be reversed. Postcolonial African states tried to deal with this legacy by nationalizing all the land and doing away with private ownership, reverting in theory to the role of the traditional chief. But this policy has not worked well. In the rural areas, the government in the capital is a remote authority often seen as unsympathetic to the plight of farmers. Governments keep the price of farm products low on urban markets, pleasing their supporters but frustrating farmers. Large landholdings once owned by Europeans seem now to be occupied by officials or those the government favors. The colonial period's approach to land tenure left Africa with a huge social problem.

Rapid population growth makes the problem worse. Traditional systems of land tenure, which involve subsistence farming in various forms ranging from shifting cultivation to pastoralism, work best when population is fairly stable. Land must be left fallow to recover from cultivation and pastures kept free of livestock so the grasses can revive. A population explosion of the kind Africa experienced during the mid-twentieth century destroys this equilibrium. Soils cannot rest, and pastures are overgrazed. As land becomes degraded, yields decline.

This is the situation in much of Subsaharan Africa, where fully 70 percent of the people depend on farming for their livelihood. That percentage is slowly declining, but it remains the highest for any realm in the global periphery. Subsaharan Africa is sometimes thought of as a storehouse of minerals and fuels and great underground riches, but farming is the key to its economy. And not just subsistence farming: agricultural exports also produce foreign exchange revenues needed to pay for essential imports. The European colonizers identified suitable environments and laid out plantations to grow cocoa, coffee, tea, and other luxury crops for the markets of the global core. Irrigation schemes enabled the export of cotton, groundnuts (peanuts), and sesame seeds. Today, national governments and some private owners continue to operate these projects.

But most African farmers are subsistence farmers who grow grain crops (corn, millet, sorghum) in the drier areas and root crops (yams, cassava, sweet potatoes) where moisture is ample. Shifting cultivation still occurs in remote parts of The Congo and neighboring locales in Equatorial Africa, but this practice is dwindling as the forest shrinks. Others are pastoralists, driving their cattle and goats along migratory routes to find water and pasture, sometimes clashing with sedentary farmers whose fields they invade. Whatever their principal mode of farming, however, all African farmers today feel the effects of population pressure and must cope with dwindling space and damaged ecologies. By *intercropping*, planting several types of crops in one cleared field, some of which resemble those in the forest, shifting cultivators extend the life of their plots. In *compound farming*, usually near a market center, farmers combine the use of compound (village) and household waste as fertilizer with intensive care of the crops to produce vegetables, fruits, and root crops as well as eggs and chickens for urban consumers. Other subsistence farmers use *diversification* to cope with such problems; they combine their cultivation of subsistence crops with nonfarm work, including full-time jobs in nearby towns.

Notwithstanding all these adaptations, African farm production is declining for many reasons, including government mismanagement and corruption. When Nigeria became independent in 1960, its government embarked on costly industrialization, and agriculture was given a lower priority. In that year, Nigeria was the world's largest producer and leading exporter of palm oil; today, Nigeria must import palm oil to meet its domestic demand. Inadequate rural infrastructures create another obstacle. Roads from the rural areas to the market centers are in bad and worsening shape, often impassable during the rainy season and rutted when dry. Ineffective links between producers and consumers create economic losses. Storage facilities in the villages tend to be inadequate, so that untold quantities of grains, roots, and vegetables are lost to rot, pests, and theft. Electricity supply in rural areas also requires upgrading: without power, pumps cannot be run for water supply, and the processing of produce cannot be mechanized. Rural farmers have inadequate access to credit: banks would rather lend to urban homeowners than to distant villagers. This is especially true for women who, according to current esti-

---

A Green Revolution for Africa?

 The **Green Revolution**—the development of more productive, higher-yielding types of grains—has narrowed the gap between world population and food production. Where people depend mainly on rice and wheat for their staples, the Green Revolution pushed back the specter of hunger. The Green Revolution has had less impact in Subsaharan Africa, however. In part, this relates to the realm's high rate of population growth (which is substantially higher than that of India or China). Other reasons have to do with Africa's staples: rice and wheat support only a small part of the realm's population. Corn (maize) supports many more, along with sorghum, millet, and other grains. In moister areas, root crops, such as the yam and cassava, as well as the plantain (similar to the banana) supply most calories. These crops were not priorities in Green Revolution research.

Lately there have been a few signs of hope. Even as terrible famines struck several parts of the continent and people went hungry in places as widely scattered as Moçambique and Mali, scientists worked toward two goals: first, to develop strains of corn and other crops that would be more resistant to Africa's virulent crop diseases, and second, to increase the productivity of those strains. But these efforts faced serious problems. An av-

erage African acre planted with corn, for example, yields only about half a ton of corn, whereas the world average is 1.3 tons. A virus-resistant corn variety has now been developed that also yields more than the old variety, and it has been introduced throughout Subsaharan Africa. In Nigeria, where it was first distributed, yields rose significantly, contributing to a near doubling of farm production in that country since 1980 according to UN data. Hardier types of root crops also raised yields significantly.

But Africa needs even more than this to reverse the cycle of food deficiency. (Nor is the Green Revolution an unqualified remedy: the poorest farmers, who need the help most, can least afford the more expensive higher-yielding seeds and also cannot pay for the pesticides that may be required.) It has been estimated that, for the realm as a whole, food production has fallen about 1 percent annually even as population grew by more than 2 percent. A lack of capital, inefficient farming methods, inadequate equipment, soil exhaustion, male dominance, apathy, and devastating droughts contributed to this decline. The seemingly endless series of civil conflicts (Uganda, The Congo, Liberia, Angola, Rwanda) also reduced farm output. The Green Revolution may narrow the gap between production and need, but the battle for food sufficiency in Africa is far from won.

mates, produce 75 percent of the food in Subsaharan Africa.

Along the banks of Africa's rivers and on the shores of the Great Lakes fishing augments local food supplies, but Africa's fishing industries are still underdeveloped and contribute only modestly to overall nutrition. During the postcolonial period, Subsaharan Africa's population has more than doubled; food production has declined. Agricultural exports have also decreased, but food imports have mushroomed. The situation is especially clear in terms of per-capita output: in 1999 it was 10 percent below that in 1989. Even advances in biotechnology that closed the global gap between food production and demand largely passed Africa by (see box titled "A Green Revolution for Africa?")

AFRICA'S HISTORICAL GEOGRAPHY

Africa is the cradle of humanity. Archeological research has chronicled 7 million years of transition from Australopithecenes to Hominids to *Homo sapiens*. It is therefore ironic that we know comparatively little about Subsaharan Africa from 5000 to 500 years ago—that is, before the onset of European colonialism. This is partly due to the colonial period itself, during which African history was neglected, many African traditions and artifacts were destroyed, and many misconceptions about African cultures and institutions became entrenched. It is also a result of the absence of a written history over most of Africa south of the Sahara until the sixteenth century—and over a large part of it until much later than that. The best records are those of the savanna belt immediately south of the Sahara, where contact with North African peoples was greatest and where Islam achieved a major penetration.

The absence of a written record does not mean, as some scholars have suggested, that Africa does not have a history as such prior to the coming of Islam and Christianity. Nor does it mean that there were no rules of social behavior, no codes of law, no organized economies. Modern historians, encouraged by the intense interest shown by Africans generally, are now trying to reconstruct the African past, not only from the meager written record but also from folklore, poetry, art objects, buildings, and other such sources. Much has been lost forever, though. Almost nothing is known of the farming peoples who built well-laid terraces on the hillsides of northeastern Nigeria and East Africa or of the communities that laid irrigation canals and constructed stone-lined wells in Kenya; and very little is known about the people who, perhaps a thousand years ago, built the great walls of Zimbabwe (photo, p. 276).

Porcelain and coins from China, beads from India, and other goods from distant sources have been found in Zimbabwe and other points in East and Southern Africa, but the trade routes within Africa itself—let alone the products that circulated on them and the people who handled them—still remain the subject of guesswork.

African Genesis

Africa on the eve of the colonial period was a continent in transition. For several centuries, the habitat in and near one of the continent's most culturally and economically productive areas—West Africa—had been changing. For 2000 years, probably more, Africa had been innovating as well as adopting ideas from outside. In West Africa, cities were developing on an impressive scale; in central and Southern Africa, peoples were moving, readjusting, sometimes struggling with each other for territorial supremacy. The Romans had penetrated to southern Sudan, North African peoples were trading with West Africans, and Arab *dhows* were sailing the waters along the eastern coasts, bringing Asian goods in exchange for gold, copper, and a comparatively small number of slaves.

Consider the environmental situation in West Africa as it relates to the past. As Figures G-7 and G-8 indicate, the environmental regions in this part of the continent exhibit a decidedly east-west orientation. The isohyets (lines of equal rainfall totals) run parallel to the southern coast (Fig. G-7); the climatic regions, now positioned somewhat differently from where they were two millennia ago, still trend strongly east-west (Fig. G-8); the vegetation pattern also reflects this situation, with a coastal forest belt yielding to savanna (tall grass with scattered trees in the south, shorter grass in the north) that gives way, in turn, to steppe and desert.

We know that African cultures had been established in all these environmental settings for thousands of years. One of these, the Nok culture, endured for over eight centuries on the Benue Plateau (north of the Niger-Benue confluence in modern Nigeria) from about 500 B.C. to the third century A.D. The Nok people made stone as well as iron tools, and they left behind a treasure of art in the form of clay figurines representing humans and animals. But we have no evidence that they traded with distant peoples. The opportunities created by environments and technologies still lay ahead.

Early Trade

West Africa, over a north-south span of a few hundred miles, displayed an enormous contrast in environments, economic opportunities, modes of life, and products. The peoples of the tropical forest produced and needed goods

FROM THE FIELD NOTES

"Got up before dawn this date in the hope of being the first visitor to the Great Zimbabwe ruins, a place I have wanted to see ever since I learned about it in a historical geography class. Climbed the hill and watched the sun rise over the great elliptical 'temple' below, then explored the maze of structures of the so-called fortification on the hilltop above. Next, for an incredible 90 minutes I was alone in the interior of the great oval walls of the temple. What history this site has seen—it was settled for perhaps six centuries before the first stones were hewn and laid here around A.D. 750, and from the 11th to the 15th centuries this may have been a ceremonial or religious center of a vast empire whose citizens smelted gold and copper and traded them via Sofala and other ports on the Indian Ocean coast for goods from South Asia and even China. No cement was used: these walls are built with stones cut to fit closely together, and in the valley they rise over 30 feet (nearly 10 m) high. And what secrets Great Zimbabwe still conceals—where are the plans, the tools, the quarries? The remnants of an elaborate water-supply system suggest a practical function for this center, but a decorated conical tower and adjacent platform imply a religious role. Whatever the answers, history hangs heavily, almost tangibly, over this African site."

that were different from the products and requirements of the peoples of the dry, distant north. For example, salt is a prized commodity in the forest, where the humidity precludes its formation, but it is plentiful in the desert and steppe. This enabled the desert peoples to sell salt to the forest peoples in exchange for ivory, spices, and dried foods. Thus there evolved a degree of *regional complementarity* between the peoples of the forest and those of the drylands. And the savanna peoples—those located in between—found themselves in a position to channel and handle the trade (which is always economically profitable).

The markets in which these goods were exchanged prospered and grew, and cities arose in the savanna belt of West Africa. One of these old cities, now an epitome of isolation, was once a thriving center of commerce and learning and one of the leading urban places in the world—Timbuktu. Others, predecessors as well as successors of Timbuktu, have declined, some of them into

oblivion. Still other savanna cities, such as Kano in the northern part of Nigeria, are still important.

Early States

Strong and durable states arose in the West African culture hearth (see Fig. 7–3). The oldest state we know anything about is Ghana. Ancient Ghana was located to the northwest of the modern country of Ghana. It covered parts of present-day Mali, Mauritania, and adjacent territory. Ghana lay astride the upper Niger River and included gold-rich streams flowing off the Futa Jallon Highlands, where the Niger has its origins. For a thousand years, perhaps longer, old Ghana managed to weld various groups of people into a stable state. The country had a large capital city complete with markets, suburbs for foreign merchants, religious shrines, and, some miles from the city center, a fortified royal retreat. Taxes were collected from the citizens, and tribute was extracted from subjugated peoples on Ghana's periphery; tolls were levied on goods entering Ghana, and an army maintained control. Muslims from the northern drylands invaded Ghana in about A.D. 1062, when it may already have been in decline. Even so, the capital was protected for 14 years. However, the invaders had ruined the farmlands and destroyed the trade links with the north. Ghana could not survive. It finally broke into smaller units.

In the centuries that followed, the focus of politico-territorial organization in the West African culture hearth shifted almost continuously eastward—first to ancient Ghana's successor state of Mali, which was centered on Timbuktu and the middle Niger River Valley, and then to the state of Songhai, whose focus was Gao, a city on the Niger that still exists. This eastward movement may have been the result of the growing influence and power of Islam. Traditional religions prevailed in ancient Ghana,

but Mali and its successor states sent huge, gold-laden pilgrimages to Mecca along the savanna corridor south of the Sahara, passing through present-day Khartoum and Cairo. Of the tens of thousands who participated in these pilgrimages, some remained behind. Today, many Sudanese trace their ancestry to the West Africa savanna kingdoms.

West Africa's savanna region undoubtedly witnessed momentous cultural, technological, and economic developments, but other parts of Africa also progressed. Early states emerged in present-day Sudan, Eritrea, and Ethiopia. Influenced by innovations from the Egyptian culture hearth, these kingdoms were stable and durable: the oldest, Kush, lasted 23 centuries (Fig. 6-5). The Kushites built elaborate irrigation systems, forged iron tools, and built impressive structures as the ruins of their long-term capital

FIGURE 6-5

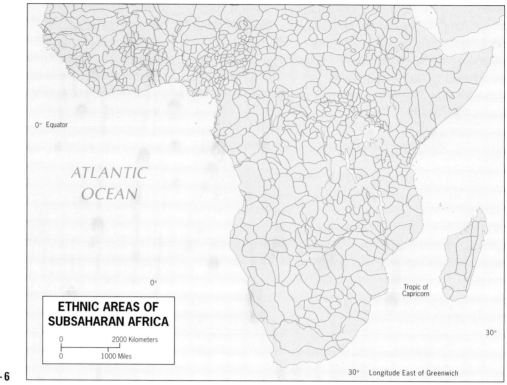

FIGURE 6-6

ETHNIC AREAS OF SUBSAHARAN AFRICA

0   2000 Kilometers

0   1000 Miles

and industrial center, Meroe, reveal. Nubia, to the southeast of Kush, was Christianized until the Muslim wave overtook it in the eighth century. And Axum was the richest market in northeastern Africa, a powerful kingdom that controlled Red Sea trade and endured for six centuries. Axum, too, was a Christian state that confronted Islam, but Axum's rulers deflected the Muslim advance and gave rise to the Christian dynasty that eventually shaped modern Ethiopia.

The process of **state formation** spread throughout Africa and was still in progress when the first European contacts occurred in the late fifteenth century. Large and effectively organized states developed on the equatorial west coast (notably Kongo) and on the southern plateau from the southern part of The Congo to Zimbabwe. East Africa had several city-states, including Mogadishu, Kilwa, Mombasa, and Sofala.

A crucial event affected virtually all of Equatorial, West, and Southern Africa: the great Bantu migration from present-day Nigeria-Cameroon southward and eastward across the continent. This migration appears to have occurred in waves starting as long as 5000 years ago, populating the Great Lakes area and penetrating South Africa, where it resulted in the formation of the powerful Zulu Empire in the nineteenth century (Fig. 6-5).

All this reminds us that, before European colonization, Africa was a realm of rich and varied cultures, diverse lifestyles, technological progress, and external trade (Fig. 6-6). But Europe's intervention would forever change its evolving political map.

The Colonial Transformation

European involvement in Subsaharan Africa began in the fifteenth century. It would interrupt the path of indigenous African development and irreversibly alter the entire cultural, economic, political, and social makeup of the continent. It started quietly in the late fifteenth century, with Portuguese ships groping their way along the west coast and rounding the Cape of Good Hope. Their goal was to find a sea route to the spices and riches of the Orient. Soon other European countries were sending their vessels to African waters, and a string of coastal stations and forts sprang up. In West Africa, the nearest part of the continent to European spheres in Middle and South America, the initial impact was strongest. At their coastal control points, the Europeans traded with African middlemen for the slaves who were needed to work New World plantations, for the gold that had been flowing northward across the desert, and for ivory and spices.

Suddenly, the centers of activity lay not with the cities of the savanna but in the foreign stations on the Atlantic coast. As the interior declined, the coastal peoples thrived. Small forest states gained unprecedented wealth, transferring and selling slaves captured in the interior to the European traders on the coast. Dahomey (now called Benin) and Benin (now part of neighboring Nigeria) were states built on the slave trade. When slavery eventually came under attack in Europe, those who had inherited the power and riches it had brought opposed abolition vigorously in both continents.

Although slavery was not new to West Africa, the *kind* of slave raiding and trading the Europeans introduced certainly was. In the savanna, kings, chiefs, and prominent families traditionally took a few slaves, but those slaves' status was unlike anything that lay in store for those who were shipped across the Atlantic. In fact, large-scale slave trading had been introduced in East Africa long before the Europeans brought it to West Africa. African middlemen from the coast raided the interior for able-bodied men and women and marched them in chains to the Arab markets on the coast (Zanzibar was a notorious market). There, packed in specially built *dhows*, they were carried off to Arabia, Persia, and India. When the European slave trade took hold in West Africa, however, its volume was far greater. Europeans, Arabs, and collaborating Africans ravaged the continent, forcing perhaps as many as 30 million persons away from their homelands in bondage (Fig. 6-7). Families were

destroyed, as were whole villages and cultures; those who survived their exile suffered unfathomable misery.

The European presence on the West African coast completely reoriented its trade routes, for it initiated the decline of the interior savanna states and strengthened the coastal forest states. Moreover, the Europeans' insatiable demand for slaves ravaged the population of the interior. But it did not lead to any major European thrust toward the interior or produce colonies overnight. The African middlemen were well organized and strong, and they held off their European competitors, not just for decades but for centuries. Although the Europeans first appeared in the fifteenth century, they did not carve West Africa up until nearly 400 years later, and in many areas not until after 1900.

In all of Subsaharan Africa, the only area where European penetration was both early and substantial was at its southernmost end, at the Cape of Good Hope. There the Dutch founded Cape Town, a waystation to their developing empire in the East Indies. They settled in the town's hinterland and migrated deeper into the interior. They brought thousands of slaves from Southeast Asia to the Cape, and intermarriage was common. Today, people of mixed ancestry form the largest part of Cape Town's population. Later the British took control not only of the

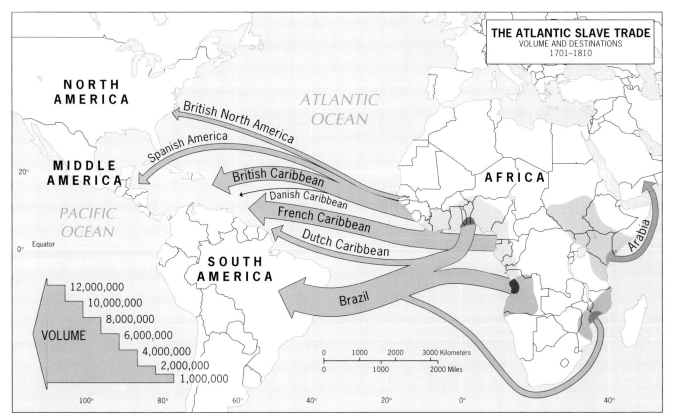

FIGURE 6-7

Cape but also of the expanding frontier and brought in tens of thousands of South Asians to work on their plantations. Multiracial South Africa was in the making.

Elsewhere in the realm, the European presence remained confined almost entirely to the coastal trading stations, whose economic influence was strong. No real frontiers of penetration developed. Individual travelers, missionaries, explorers, and traders went into the interior, but nowhere else in Africa south of the Sahara was there an invasion of white settlers comparable to southernmost Africa's.

Colonization

In the second half of the nineteenth century, after more than four centuries of contact, the European powers finally laid claim to virtually all of Africa. Parts of the continent had been "explored," but now representatives of European governments and rulers arrived to create or expand African spheres of influence for their patrons. Competition was intense. Spheres of influence began to crowd each other. It was time for negotiation, and in late 1884 a conference was convened in Berlin to sort things out. This conference laid the groundwork for the now familiar politico-geographical map of Africa (see box titled "The Berlin Conference").

Figure 6-8 shows the result. The French dominated most of West Africa, and the British East and Southern Africa. The Belgians acquired the vast territory that became The Congo. The Germans held four colonies, one in each of the realm's regions. The Portuguese held a small colony in West Africa and two large ones in Southern Africa (see the map dated 1910).

The European colonial powers shared one objective in their African colonies: exploitation. But in the way they governed their dependencies, they reflected their differences. Some colonial powers were themselves democracies (the United Kingdom and France); others were dictatorships (Portugal, Spain). The British established a system of indirect rule over much of their domain, leaving indigenous power structures in place and making local rulers representatives of the British Crown. This was unthinkable in the Portuguese colonies, where harsh, direct control was the rule. The French sought to create culturally assimilated elites that would represent French ideals in the colonies. In the Belgian Congo, however, King Leopold II, who had financed the expeditions that staked Belgium's claim in Berlin, embarked on a campaign of ruthless exploitation. His enforcers mobilized almost the entire Congolese population to gather rubber, kill elephants for their ivory, and build public works to improve

The Berlin Conference

*I*n November 1884, the imperial chancellor and architect of the German Empire, Otto von Bismarck, convened a conference of 14 states (including the United States) to settle the political partitioning of Africa. Bismarck wanted not only to expand German spheres of influence in Africa but also to play off Germany's colonial rivals against one another to the Germans' advantage. The major colonial contestants in Africa were: (1) the British, who held beachheads along the West, South, and East African coasts; (2) the French, whose main sphere of activity was in the area of the Senegal River and north of the Congo Basin; (3) the Portuguese, who now desired to extend their coastal stations in Angola and Moçambique deep into the interior; (4) King Leopold II of Belgium, who was amassing a personal domain in the Congo Basin; and (5) Germany itself, active in areas where the designs of other colonial powers might be obstructed, as in Togo (between British holdings), Cameroon (a wedge into French spheres), South West Africa (taken from under British noses in a swift strategic move), and East Africa (where German Tanganyika broke the British design for a solid block of territory from the Cape north to Cairo).

When the conference convened in Berlin, more than 80 percent of Africa was still under traditional African rule. Nonetheless, the colonial powers' representatives drew their boundary lines across the entire map. These lines were drawn through known as well as unknown regions, pieces of territory were haggled over, boundaries were erased and redrawn, and African real estate was exchanged among European governments. In the process, African peoples were divided, unified regions were ripped apart, hostile societies were thrown together, hinterlands were disrupted, and migration routes were closed off. Not all of this was felt immediately, of course, but these were some of the effects when the colonial powers began to consolidate their holdings and the boundaries on paper became barriers on the African landscape (Fig. 6-8).

The Berlin Conference was Africa's undoing in more ways than one. The colonial powers superimposed their domains on the African continent. By the time Africa regained its independence after the late 1950s, the realm had acquired a legacy of political fragmentation that could neither be eliminated nor made to operate satisfactorily. The African politico-geographical map is thus a permanent liability that resulted from three months of ignorant, greedy acquisitiveness during a period when Europe's search for minerals and markets had become insatiable.

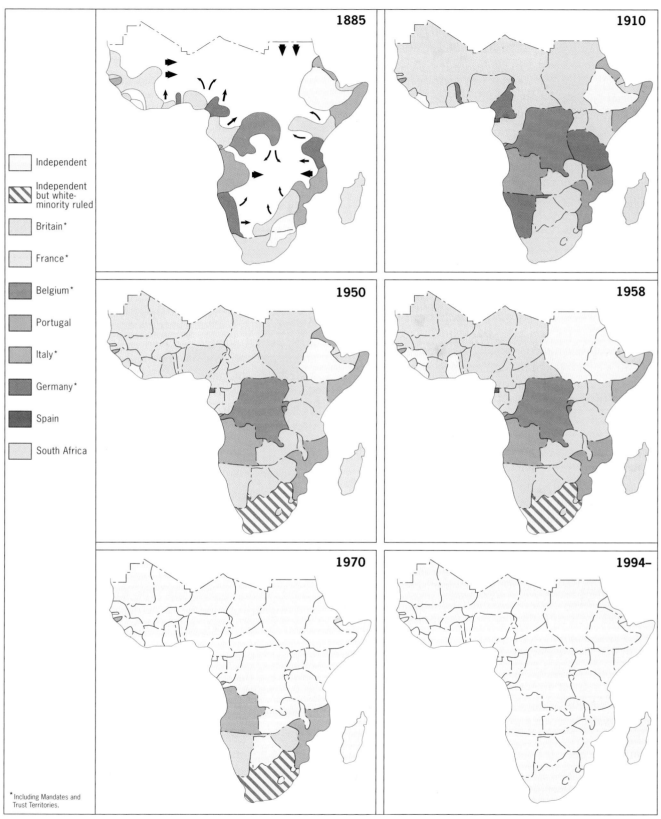

Independent

Independent but white-minority ruled

Britain*

France*

Belgium*

Portugal

Italy*

Germany*

Spain

South Africa

*Including Mandates and Trust Territories.

1885

1910

1950

1958

1970

1994–

COLONIZATION AND DECOLONIZATION, 1885–2004

FIGURE 6-8

export routes. For failing to meet production quotas, entire communities were massacred. Killing and maiming became routine in a colony in which horror was the only common denominator. After the impact of the slave trade, King Leopold's reign of terror was Africa's most severe demographic disaster. By the time it ended, after a growing outcry around the world, as many as 10 million Congolese had been murdered. In 1908 the Belgian government took over, and slowly its Congo began to mirror Belgium's own internal divisions: corporations, government administrators, and the Roman Catholic Church each pursued their sometimes competing interests. But no one thought to change the name of the colonial capital: it was Leopoldville until the Belgian Congo achieved independence in 1960.

12   **Colonialism** transformed Africa, but in its post-Berlin form it lasted less than a century. In Ghana, for example, the Ashanti (Asante) Kingdom still was fighting the British in the early years of the twentieth century; by 1957, Ghana was independent again. In a few years, much of Subsaharan Africa will have been independent for half a century, and the colonial period is becoming an interlude rather than a paramount chapter in modern African history.

Legacies

Nevertheless, some colonial legacies will remain on the map for centuries to come. The boundary framework has been modified in a few places, but it is a politico-geographical axiom that boundaries, once established, tend to become entrenched and immovable. The transport systems were laid out to facilitate the movement of raw materials from interior sources to coastal outlets. Internal circulation was only a secondary objective, and most African countries still are not well connected to each other. Moreover, the colonizers founded many of Subsaharan Africa's cities or built them on the sites of small towns or villages.

In many countries, the elites that gained advantage and prominence during the colonial period also have retained their preeminence. This has led to authoritarianism in some postcolonial African states (for details, see the regional discussion to come), and to violence and civil conflict in others. Military takeovers of national governments have been a byproduct of decolonization, and hopes for democracy in the struggle against colonialism have too often been dashed. Yet a growing number of African countries, including Ghana, Tanzania, Botswana, South Africa, and, most recently, Nigeria, have overcome the odds and achieved representative government under dauntingly difficult circumstances.

Although not strictly a colonial legacy, the impact of the Cold War (1945–1990) on African states should be noted. In three countries—Ethiopia, Somalia, and Angola—great-power competition, weapons, military advisers, and, in Angola, foreign armed forces magnified civil wars that would perhaps have been inevitable in any case. In other countries, ideological adherence to foreign dogmas led to political and social experiments (such as one-party Marxist regimes and costly farm collectivization) that cost Africa dearly. Today, debt-ridden Africa is again being told what to do, this time by foreign financial institutions. The cycle of poverty that followed colonialism exacts a high price from African societies. As for future development prospects, the disadvantages of peripheral location vis-à-vis the world's core areas continue to handicap Subsaharan Africa.

■ CULTURAL PATTERNS

We may tend to think of Africa in terms of its prominent countries and famous cities, its development problems and political dilemmas, but Africans themselves have another perspective. The colonial period created states and capitals, introduced foreign languages to serve as the *lingua franca*, and brought railroads and roads. The colonizers stimulated labor movements to the mines they opened, and they disrupted other migrations that had been part of African life for many centuries. But they did not change the ways of life of most of the people. No less than 70 percent of the realm's population still live in, and work near, Africa's hundreds of thousands of villages. They speak one of more than a thousand languages in use in the realm. The villagers' concerns are local; they focus on subsistence, health, and safety. They worry that the conflicts over regional power or political ideology will engulf them, as has happened to millions in Liberia, Sierra Leone, Ethiopia, Rwanda, The Congo, Moçambique, and Angola since the 1970s. Africa's largest peoples are major nations, such as the Yoruba of Nigeria and the Zulu of South Africa. Africa's smallest peoples number just a few thousand. As a geographic realm, Subsaharan Africa has the most complex cultural mosaic on Earth (see Fig. 6-6).

African Languages

Africa's linguistic geography is a key component of that cultural intricacy. Most of Subsaharan Africa's more than 1000 languages do not have a written tradition, making classification and mapping difficult. Scholars have attempted to delimit an African language map, and Figure 6-9 is a composite of their efforts. One feature is common to all language maps of Africa: the geographic realm begins approximately where the Afro-Asiatic language family (mapped in yellow in Fig. 6-9) ends, al-

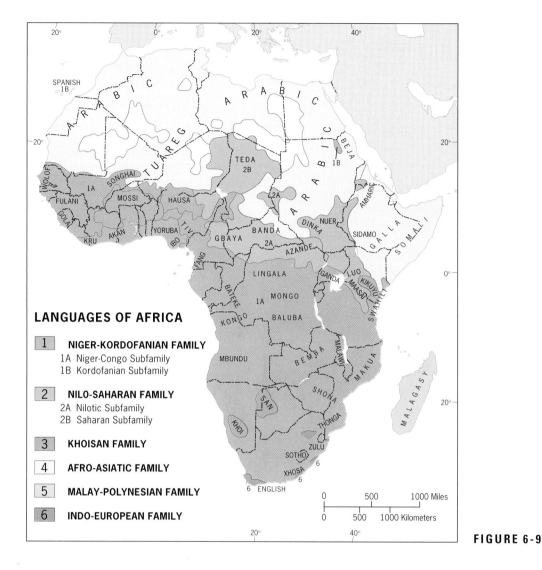

LANGUAGES OF AFRICA

1 **NIGER-KORDOFANIAN FAMILY**
1A Niger-Congo Subfamily
1B Kordofanian Subfamily

2 **NILO-SAHARAN FAMILY**
2A Nilotic Subfamily
2B Saharan Subfamily

3 **KHOISAN FAMILY**

4 **AFRO-ASIATIC FAMILY**

5 **MALAY-POLYNESIAN FAMILY**

6 **INDO-EUROPEAN FAMILY**

FIGURE 6-9

though the correlation is sharper in West Africa than to the east.

In Subsaharan Africa, the dominant language family is the Niger-Kordofanian family. It consists of two subfamilies, the tiny Kordofanian concentration in northeastern Sudan and the pervasive Niger-Congo subfamily that extends across most of the realm from West Africa to East and Southern Africa. The Bantu language forms the largest branch in this subfamily, but Niger-Congo languages in West Africa, such as Yoruba and Akan, also have millions of speakers. Another important language family is the Nilo-Saharan family, extending from Maasai in Kenya northwest to Teda in Chad. No other language families are of similar extent or importance: the Khoisan family, of ancient origins, now survives among the dwindling Khoi and San peoples of the Kalahari; the small white minority in South Africa speak Indo-European languages; and Malay-Polynesian languages prevail in Madagascar, which was peopled from Southeast Asia before Africans reached it.

About 40 African languages are spoken by 1 million people or more, and a half-dozen by about 10 million or more: Hausa (50 million), Yoruba (23 million), Ibo, Swahili, Lingala, and Zulu. Although English and French have become important *linguae francae* in multilingual countries such as Nigeria and Côte d'Ivoire (where officials even insist on spelling the name of their country—Ivory Coast— in the Francophone way), African languages also serve this purpose. Hausa is a common language across the West African savanna; Swahili is widely used in East Africa. And pidgin languages, mixtures of African and European tongues, are spreading along West Africa's coast. Millions of Pidgin English (called *Wes Kos*) speakers use this medium in Nigeria and Ghana.

Multilingualism can be a powerful centrifugal force 13 in society, and African governments have tried with varying success to establish "national" alongside local languages. Nigeria, for example, made English its official language because none of its 250 languages, not even Hausa, had sufficient internal interregional use. But using

The faithful kneel during Friday prayers at a mosque in Kano, northern Nigeria. The survival of Nigeria as a unified state is an African success story; the Nigerians have overcome strong centrifugal forces in a multi-ethnic country that is dominantly Muslim in the north, Christian in the south. In the 1990s, some Muslim clerics began calling for an Islamic Republic in Nigeria, and after the death of the dictator Abacha and the election of a non-Muslim president, the Islamic drive intensified. A number of Nigeria's northern States adopted Sharia (strict Islamic) law, which led to destructive riots between the majority Muslims and minority Christians who felt threatened by this turn of events. The northern city of Kaduna saw a near-civil war in which hundreds were killed, churches and mosques burned, and whole neighborhoods devastated. Late in 2000 the Kaduna State government reversed itself and suspended the imposition of Sharia law, but other States persevered despite social unrest. Can Nigeria avoid the fate of Sudan (see p. 344)?

a European, colonial language as an official medium invites criticism, and Nigeria remains divided on the issue. On the other hand, making a dominant local language official would invite negative reactions from ethnic minorities. Language remains a potent force in Africa's cultural life.

Religion in Africa

Africans had their own religious belief systems long before Christians and Muslims arrived to convert them. And for all of Subsaharan Africa's cultural diversity, Africans had a consistent view of their place in nature. Spiritual forces, according to African tradition, are manifest everywhere in the natural environment, not in a supreme deity that exists in some remote place. Thus gods and spirits affect people's daily lives, witnessing every move, rewarding the virtuous, and punishing (through injury or crop failure, for example) those who misbehave. Ancestral spirits can inflict misfortune on the living. They are everywhere: in the forest, rivers, mountains.

As with land tenure, the religious views of Africans clashed fundamentally with those of outsiders. Monotheistic Christianity first touched Africa in the northeast when Nubia and Axum were converted, and Ethiopia has been a Coptic Christian stronghold since the fourth century A.D. But the Christian churches' real invasion did not begin until the onset of colonialism after the turn of the sixteenth century. Christianity's various denominations made inroads in different areas: Roman Catholicism in much of Equatorial Africa mainly at the behest of the Belgians, the Anglican Church in British colonies, and Presbyterians and others elsewhere. But almost everywhere, Christianity's penetration led to a blending of traditional and Christian beliefs, so that much of Subsaharan Africa is nominally, though not exclusively, Christian. Go to a church in Gabon or Uganda or Zambia, and you may hear drums instead of church bells, sing African music rather than hymns, and see African carvings alongside the usual statuary.

Islam had a different arrival and impact. Long before the colonial invasion, Islam advanced out of Arabia, across the desert, and down the east coast. Muslim clerics converted the rulers of African states and commanded them to convert their subjects. They Islamized the savanna states and penetrated into present-day northern Nigeria, Ghana, and Ivory Coast. They encircled and isolated Ethiopia's Coptic Christians and Islamized the Somali people in Africa's Horn. They established beachheads on the Kenya coast and took over Zanzibar. On the map, the African Transition Zone defines the Islamic Front (see Fig. 6-10). In the field, Arabizing Islam and European Christianity competed for African minds, and Islam proved to be a far more pervasive force. From

Senegal to Somalia, the population is virtually 100 percent Muslim, and Islam's rules dominate everyday life. The Sunni *mullahs* would never allow the kind of marriage between traditional and Christian beliefs seen in much of formerly colonial Africa. This fundamental contradiction between Islamic dogma and Christian accommodation creates a potential for conflict in countries where both religions have adherents.

MODERN MAP AND TRADITIONAL SOCIETY

The political map of Subsaharan Africa has 45 states but no nation-states (apart from some microstates and ministates in the islands and in the south). Centrifugal forces are powerful, and outside interventions during the Cold War, when communist and anticommunist foreigners took sides in local civil wars, worsened conflict within African states. Colonialism's economic legacy was not much better. In tropical Africa, core areas, capitals, port cities, and transport systems were laid out to maximize profit and facilitate exploitation of minerals and soils; the colonial mosaic inhibited interregional communications except where cooperation enhanced efficiency. Colonial Zambia and Zimbabwe, for example (then called Northern and Southern Rhodesia), were landlocked and needed outlets, so railroads were built to Portuguese-owned ports. But such routes did little to create intra-African linkages. The modern map reveals the results: in West Africa you can travel from the coast into the interior of all the coastal states along railways or adequate roads. But no high-standard roadway was ever built to link these coastal neighbors to each other.

To overcome such disadvantages, African states must cooperate internationally, continentwide as well as regionally. The Organization of African Unity (OAU) was established for this purpose in 1963 and in 2001 was superseded by the African Union. In 1975 the Economic Community of West African States (ECOWAS) was founded by 15 countries to promote trade, transportation, industry, and social affairs in the region. And in the early 1990s another important step was taken when 12 countries joined in the Southern African Development Community (SADC), organized to facilitate regional commerce, intercountry transport networks, and political interaction.

Population and Urbanization

Subsaharan Africa remains the least urbanized world realm, but it is urbanizing at a fast pace. By the time you read this, the percentage of urban dwellers will have passed 30. This means that more than 200 million people now live in cities and towns, many of which were founded and developed by the colonial powers.

African cities became centers of embryonic national core areas, and of course they served as government headquarters. This *formal sector* of the city used to be the dominant one, with government control and regulations affecting civil service, business, industry, and workers. Today, however, African cities look different. From a distance, the skyline still resembles that of a modern center. But in the streets, on the sidewalks right below the shopwindows, there are hawkers, basket weavers, jewelry sellers, garment makers, wood carvers—a second economy, most of it beyond government control. This *informal sector* now dominates many African cities. It is peopled by the rural immigrants, who also work as servants, apprentices, construction workers, and in countless other menial jobs.

Millions of urban immigrants, however, cannot find work, at least not for months or even years at a time. They live in squalid circumstances, in desperate poverty, and governments cannot assist them. As a result, the squatter rings around (and also within) many of Africa's cities are unsafe—uncomfortable, unhealthy slums without adequate shelter, water supply, or basic sanitation. Garbage-strewn (no solid-waste removal here), muddy and insect-infested during the rainy season, and stifling and smelly during the dry period, they are incubators of disease. Yet very few residents return to their villages. Every new day brings hope.

In our regional discussion we focus on some of Subsaharan Africa's cities, all of which, to varying degrees, are stressed by the rate of population influx. Despite the plight of the urban poor and the poverty of Africa's rural areas, some of Africa's capitals remain the strongholds of privileged elites who, dominant in governments, fail to address the needs of other ethnic groups. Discriminatory policies and artificially low food prices disadvantage farmers and create even greater urban-rural disparities than the colonial period saw. But today the prospect of democracy brings hope that Africa's rural majorities will be heard and heeded in the capitals.

REGIONS OF THE REALM

ON THE FACE of it, Africa seems to be so massive, compact, and unbroken that any attempt to justify a contemporary regional breakdown is doomed to fail. No deeply penetrating bays or seas create peninsular fragments as in Europe. No major islands (other than Madagascar) provide the broad regional contrasts we see in Middle America.

Nor does Africa really taper southward to the peninsular proportions of South America. And Africa is not cut by an Andean or a Himalayan mountain barrier. Given Africa's colonial fragmentation and cultural mosaic, is regionalization possible? Indeed it is.

Maps of environmental distributions, ethnic patterns, cultural landscapes, historic culture hearths, and other spatial data yield a four-region structure complicated by a fifth, overlapping zone as shown in Figure 6-10. Beginning in the south, we identify the following regions:

1. *Southern Africa*, extending from the southern tip of the continent to the northern borders of Angola, Zambia, Malawi, and Moçambique. Ten countries constitute this region, which extends beyond the tropics and whose giant is South Africa.

2. *East Africa*, where natural (equatorial) environments are moderated by elevation and where plateaus, lakes, and mountains, some carrying permanent snow, define the countryside. Six countries, including the highland part of Ethiopia, comprise this region.

3. *Equatorial Africa*, much of it defined by the basin of the Congo River, where elevations are lower than in East Africa, temperatures are higher and moisture more ample, and where most of Africa's surviving rainforests remain. Among the eight countries that form this region, The Congo dominates territorially and demographically, but others are much better off economically and politically.

4. *West Africa*, which includes the countries of the western coast and those on the margins of the Sahara in the

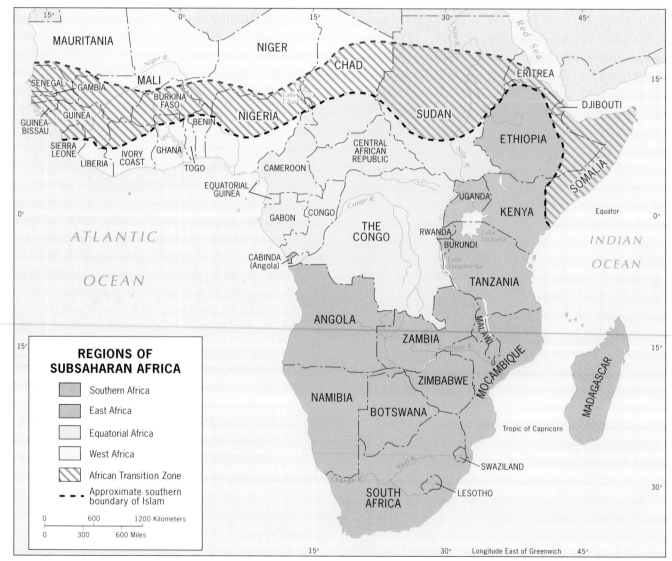

FIGURE 6-10

interior, a populous region anchored in the southeast by Africa's demographic giant, Nigeria. Fifteen countries form this crucial African region.

5. *The African Transition Zone*, the complicating factor on the regional map of Africa. In Figure 6-10, note that this zone of increasing Islamic influence completely dominates some countries (e.g., Somalia in the east and Senegal in the west) while cutting across others, thereby creating Islamized northern areas and non-Islamic southern zones (Nigeria, Chad, Sudan).

Before proceeding, we should take note of one African country not included in any of the regions enumerated above: *Madagascar*, the island state off the southeastern African coast. As we will discover, Madagascar is neither East nor Southern African; it has historic ties to Southeast Asia and cultural links to both Asia and Africa.

► SOUTHERN AFRICA

Southern Africa, as a geographic region, consists of all the countries and territories lying south of Equatorial Africa's The Congo and East Africa's Tanzania (Fig. 6-11). Thus defined, the region extends from Angola and Moçambique (on the Atlantic and Indian Ocean coasts, respectively) to South Africa and includes a half-dozen landlocked states. Also marking the northern limit of the region are Zambia and Malawi. Zambia is nearly cut in half by a long land extension from The Congo, and Malawi penetrates deeply into Moçambique. The colonial boundary framework, here as elsewhere, produced many liabilities.

Southern Africa constitutes a geographic region in both physiographic and human terms. Its northern zone marks the southern limit of the Congo Basin in a broad upland that stretches across Angola and into Zambia (the tan corridor extending eastward from the Bihe Plateau in Fig. 6-2). Lake Malawi is the southernmost of the East African rift-valley lakes; Southern Africa has none of East Africa's volcanic and earthquake activity. Most of the region is plateau country, and the Great Escarpment is much in evidence here. There are two pivotal river systems: the Zambezi (which forms the border between Zambia and Zimbabwe) and the Orange-Vaal (South African rivers that combine to demarcate southern Namibia from South Africa).

The social, economic, and political geographies of Southern Africa also confirm the regional definition. Landlocked Zambia, whose economic core area is the Copperbelt northwest of Ndola, always looked southward for its outlets to the sea, electrical power, and fuels. Malawi's core area and outlets lie in its south. Moçambique, with its lengthy Indian Ocean coastline, has served as an exit not only for Malawi but also for Zimbabwe (through the port of Beira) and South Africa (through Maputo). Offshore Madagascar, however, remains a separate entity (see box titled "Distinctive Madagascar").

Southern Africa is the continent's richest region materially. A great zone of mineral deposits extends through the heart of the region from Zambia's Copperbelt through Zimbabwe's Great Dyke and South Africa's Bushveld Basin and Witwatersrand to the goldfields and diamond mines of the Orange Free State and northern Cape Province in the heart of South Africa. Ever since these minerals began to be exploited in colonial times, many migrant laborers have come to work in the mines. The range and volume of minerals mined in this belt are enormous, from the copper of Zambia and the chrome and asbestos of Zimbabwe to the gold, chromium, diamonds, platinum, coal, and iron ore of South Africa. However, not all of Southern Africa's mineral deposits lie in this central backbone. There is coal in western Zimbabwe at Hwange and in central Moçambique near Tete. In Angola, petroleum from oilfields along the north coast heads the export list; diamonds are mined in the northeast, manganese and iron on the central plateau. Namibia produces copper, lead, and zinc from a major mining complex anchored to the town of Tsumeb in the north, and diamond deposits lie in the beaches along the Atlantic Ocean in the south.

Southern Africa's agricultural diversity matches its mineral wealth. Vineyards drape the slopes of South Africa's Cape Ranges; tea plantations hug the eastern

MAJOR CITIES OF THE REALM	
City	Population* (in millions)
Abidjan, Ivory Coast	4.5
Accra, Ghana	2.2
Adis Abeba, Ethiopia	3.2
Cape Town, South Africa	3.2
Dakar, Senegal	2.5
Dar es Salaam, Tanzania	2.8
Durban, South Africa	2.7
Harare, Zimbabwe	2.2
Ibadan, Nigeria	1.8
Johannesburg, South Africa	3.3
Kinshasa, The Congo	6.2
Lagos, Nigeria	13.9
Lusaka, Zambia	2.0
Mombasa, Kenya	1.1
Nairobi, Kenya	2.8

*Based on 2004 estimates.

FIGURE 6-11

escarpment slopes of Zimbabwe. Before civil war destroyed its economy, Angola was one of the world's leading coffee producers. South Africa's relatively high latitudes and its range of altitudes create environments for apple orchards, citrus groves, banana plantations, pineapple farms, and many other crops. Farther north, tobacco has long been one

of Zimbabwe's leading products. And while corn (maize) is the staple grain for most of Southern Africa's farmers, they also grow wheat and other cereals. Even the pastoral industry is varied, with large herds of beef cattle on the grassy highveld (upland), tens of millions of sheep on the southeast's pastures, and dairy farms around the big cities.

Distinctive Madagascar

*M*aps showing the geographic regions of the Subsaharan Africa realm often omit Madagascar. And for good reason: Madagascar differs strongly from Southern Africa, the region to which it is nearest—just 250 miles (400 km) away. It also differs from East Africa, from which it has received some of its cultural infusions. Madagascar is the world's fourth largest island, a huge block of Africa that separated from the main landmass 160 million years ago. About 2000 years ago, the first settlers arrived—not from Africa (although perhaps via Africa) but from Southeast Asia. Malay communities flourished in the interior highlands of the island, which resembles Africa in having a prominent eastern escarpment and a central plateau (Fig. 6-12). Here was formed a powerful kingdom, the empire of the Merina. Its language, Malagasy, of Malay-Polynesian origin, became the indigenous tongue of the entire island (Fig. 6-9).

The Malay and Indonesian immigrants brought Africans to the island as wives and slaves, and from this forced immigration evolved the African component in Madagascar's population of 17.6 million. In all, nearly 20 discrete ethnic groups coexist in Madagascar, among which the Merina (4 million) and Betsimisaraka (2 million) are the most numerous. Like mainland Africa, Madagascar experienced colonial invasion and competition. Portuguese, British, and French colonists appeared after 1500, but the Merina were well organized and resisted the colonial conquest. Eventually Madagascar became part of France's empire, and French became the *lingua franca* of the educated elite.

Because of its Southeast Asian imprint, Madagascar's staple food is rice, not corn. It has some minerals, including chromite, iron ore, and bauxite, but the economy is weak, damaged by long-term political turmoil and burdened by rapid population growth. The infrastructure crumbled; in 1990 the "main road" from the capital to the nearest port (Fig. 6-12) was a potholed 150 miles that took 10 hours for a truck to navigate.

Meanwhile, Madagascar's unique flora and fauna retreated before the human onslaught. Madagascar's long-term isolation kept evolution here so distinct that the island is a discrete zoogeographic realm. Primates living on the island are found nowhere else; 33 varieties of lemurs are unique to Madagascar. Many species of birds, amphibians, and reptiles are also exclusive to this island. Their home, the rainforest, covered 65,000 square miles (168,000 sq km) in 1950, but today only about one-third of it is left. Logging, introduced by the colonists, damaged it; slash-and-burn agriculture is destroying it; and severe droughts in the 1980s and 1990s intensified the impact. Obviously, Madagascar should be a global conservation priority, but funds are limited and the needs are enor-

FIGURE 6-12

mous. Malnutrition and poverty are powerful forces when survival is at stake for villages and families.

Madagascar's cultural landscape retains its Southeast Asian imprints, in the towns as well as the paddies. The capital, Antananarivo, is the country's primate city, its architecture and atmosphere combining traces of Asia and Africa. Poverty dominates the townscape here, too, and there is little to attract inmigrants (Madagascar is only 20 percent urbanized). But perhaps the most ominous statistic is Madagascar's population doubling time of just 35 years.

FROM THE FIELD NOTES

"My second visit to the Kariba Dam on the Zambezi River between Northern and Southern Rhodesia (now renamed Zambia and Zimbabwe) showed the huge dam wall complete and the lake filling up. The four-lane highway across the top of the dam was not yet open for traffic, and construction was still in progress on the turbine housing and other buildings. This was the time when megadams were thought to hold the solution to development problems: the Kariba Dam would provide electricity to places as far away as Salisbury (now Harare) and the Copperbelt, would support a fishing industry, would enable farming along its shores, would stimulate the tourist industry. It was therefore worth dislocating tens of thousands of people living in the valley and eliminating wildlife refuges (a massive effort was made to rescue wildlife caught on islands as the water rose). But, like other megaprojects of this kind in Africa and elsewhere in the less advantaged world, the Kariba Dam failed to live up to expectations and generated unanticipated problems, medical and otherwise. A hint of the latter can be seen in the extreme left of this photograph: the floating vegetation at the turbine-tunnel entrance. A Brazilian water hyacinth, reportedly introduced by two Brazilian monks who built a small church where the new shoreline would be, clogged the upper reaches of Lake Kariba and spread into Equatorial Africa's rivers."

Despite this considerable wealth and potential, the countries of Southern Africa have not prospered. As Figure G-11 shows, most remain mired in the low-income category; the only exceptions are lower-middle-income Namibia and upper-middle-income South Africa and Botswana (the latter desert state being the realm's second most sparsely populated country). A period of rapid population growth, followed by the devastating onslaught of AIDS, civil conflict, political instability, incompetent government, corruption, and environmental problems have inhibited economic growth. In 2003, millions in this region, notably in misruled Zimbabwe and misguided Zambia as well as Malawi, suffered from malnutrition or hunger and needed foreign assistance to survive.

Despite all this, Southern Africa is better off, on average, than any other African region. Leaders realize that cross-border cooperation helps national economies, but they have also noted that regionwide associations such as the Southern African Development Community (SADC), designed to enhance intraregional commerce, coordinate infrastructure programs, and help member governments facing rebellions, have not been effective. In 2002, five

Southern African states (South Africa, Namibia, Botswana, Swaziland, and Lesotho) tried another tack. Rather than pursuing grandiose objectives, they formed a Southern African Customs Union (SACU), a free-trade area incorporating more than 50 million relatively well-off inhabitants with the South African rand as the organization's international currency. Jointly lowering their tariffs on imports from the European Union and making deals with EU marketers, SACU hopes to stimulate foreign investment for the benefit of all its members.

South Africa

The Republic of South Africa is the giant of Southern Africa, an African country at the center of world attention, a bright ray of hope not only for Africa but for all humankind.

Long in the grip of one of the world's most notorious racial policies (**apartheid**, or "apartness," and its derivative, **separate development**), South Africa today is shedding its past and is building a new future. That virtually all parties to the earlier debacle are now working cooperatively to restructure the country under a new flag, a new national anthem, and a new leadership was one of the great events of the twentieth century. Now, with a new century opened, South Africa is poised to take its long-awaited role as the economic engine for the region—and perhaps beyond.

South Africa stretches from the warm subtropics in the north to Antarctic-chilled waters in the south. With a land area in excess of 470,00 square miles (1.2 million sq km) and a heterogeneous population of 44.5 million, South Africa is the dominant state in Southern Africa. It contains the bulk of the region's minerals, most of its good farmlands, its largest cities, best ports, most productive factories, and most developed transport networks. Mineral exports from Zambia and Zimbabwe move through South African ports. Workers from as far away as Malawi and as nearby as Lesotho work in South Africa's mines, factories, and fields.

Historical Geography

South Africa's location has much to do with its human geography. On the African continent, peoples migrated southward—first the Khoisan-speakers and then the Bantu peoples—into the South African cul-de-sac. On the oceans, the Europeans arrived to claim the southernmost Cape as one of the most strategic places on Earth, the gateway from the Atlantic to the Indian Ocean, a waystation on the route to Asia's riches. The Dutch East India Company founded Cape Town as early as 1652, and soon the Hollanders began to bring Southeast Asians to the Cape to serve as domestics and laborers. When the British took

over about 150 years later, Cape Town had a substantial population of mixed ancestry, the source of today's so-called Coloured sector of the country's citizenry.

The British also altered the demographic mosaic by bringing tens of thousands of indentured laborers from their South Asian domain to work on the sugar plantations of east-coast Natal. Most of these laborers stayed after their period of indenture was over, and today South Africa counts about 1 million Indians among its people. Most are still concentrated in Natal and prominently so in metropolitan Durban.

As we noted earlier, South Africa had been occupied by Europeans long before the colonial "scramble for Africa" gained momentum. The Dutch, after the British took control of the Cape, trekked into the South African interior and, on the high plateau they called the *highveld*, founded their own republics. When diamonds and gold were discovered there, the British challenged the Boers (descendants of the Dutch settlers) for these prizes. In 1899–1902, the British and the Boers fought the Boer War. The British won, and British capitalists took control of South Africa's economic and political life. But the Boers negotiated a power-sharing arrangement and eventually won hegemony. Having long since shed their European links, they now called themselves *Afrikaners*, their word for Africans, and proceeded to erect the system known as apartheid.

These foreign immigrations and struggles took place on lands that Africans had already entered and fought over. When the Europeans reached the Cape, Bantu nations were driving the weaker Khoi and San peoples into less hospitable territory or forcing them to work in bondage. One great contest was taking place in the east and southeast, below the Great Escarpment. The Xhosa nation was moving toward the Cape along this natural corridor. Behind them, in Natal (Fig. 6-11), the Zulu Empire became the region's most powerful entity in the nineteenth century. On the highveld, the North and South Sotho, the Tswana, and other peoples could not stem the tide of European aggrandizement, but their numbers ensured survival (unlike many of Australia's Aboriginal groups, which disappeared).

In the process, South Africa became Africa's most pluralistic and heterogeneous society. People had converged on the country from Western Europe, Southeast Asia, South Asia, and other parts of Africa itself. In the national population matrix today, Africans outnumber non-Africans by about 4 to 1 (Table 6-1).

Social Geography

Heterogeneity also marks the spatial demography of South Africa. Despite centuries of migration and (at the Cape) intermarriage, labor movement (to the mines, farms, and factories), and massive urbanization, regionalism pervades

Table 6-1
DEMOGRAPHIC DATA FOR SOUTH AFRICA

Population Groups	Estimates 2004 (in millions)
African nations	35.9
Zulu	8.9
Xhosa	7.3
Sotho (N and S)	6.6
Tswana	3.4
Others (6)	9.7
Whites	3.9
Afrikaners	2.6
English-speakers	1.2
Others	0.1
Mixed (Coloureds)	3.5
African/White	3.3
Malayan	0.2
South Asian	1.2
Hindus	0.8
Muslims	0.4
TOTAL	44.5

the human mosaic. The Zulu nation still is largely concentrated in the province the Europeans called Natal. The Xhosa still cluster in the Eastern Cape, from the city of East London to the Natal border and below the Great Escarpment. The Tswana still occupy ancestral lands along the Botswana border. Cape Town still is the core area of the Coloured population; Durban still has the strongest Indian imprint. Travel through South Africa, and you will recognize the diversity of rural cultural landscapes as they change from Swazi to Ndebele to Venda.

This persistent regionalism induced the Afrikaner-led white government of the apartheid period, after 1949, to extend its policies and lock the country in a grand design called separate development. By creating "homelands" that would later become "independent" republics, the white minority could assign all Africans to their national source area, making them foreigners in white South Africa while they worked there. This was a form of what is now called *ethnic cleansing*, and between 1960 and 1980 as many as 3.5 million black Africans were forced from their homes and relocated in their assigned "homelands." The social cost of this program was enormous and will be felt well into the twenty-first century. Under the provisions of separate development, nearly 80 percent of South Africa was designated as white-owned, and barely more than 15 percent was to accommodate the African homelands. There, overcrowding led to excessive

pressure on the land, accelerating soil erosion and general environmental deterioration. Crowding also produced conflict among homeland residents, but more serious strife pitted those who were willing to cooperate with the white regime (for example, by serving in government-approved homeland administrations) against those who used all possible means to oppose apartheid and separate development.

For all the misery and dislocation it caused, separate development could not stem the tide of urbanization in South Africa. Millions of workers, jobseekers, and illegal migrants created vast shantytown rings around the cities. The existing "townships" or "locations," as the older African settlements were called, that once were the poorest parts of the cities, now seemed well-off by comparison. Black townships such as Soweto (for South Western Townships) in metropolitan Johannesburg were approved under apartheid laws, even acquiring some services. In the shantytowns, no legalities protected the residents, and the government never stopped trying to eradicate them. Bulldozers would sweep away whole settlements overnight, their residents loaded on trucks and sent to the homelands.

Still the number of both township residents and squatters grew, and eventually the townships became the hubs of opposition to apartheid. Violent uprisings and crippling strikes proved the growing power of the majority and the inability of the state to control it. By the mid-1980s, international attention was focused on South Africa, economic sanctions had been imposed, and the end of the apartheid era was in sight. The South African president, P.W. Botha, began to negotiate with the country's most famous political prisoner, Nelson Mandela. The long-banned African National Congress (ANC) was permitted to engage in political activity again, and exiled leaders began to return home. When, in February 1990, Mandela walked out of prison after 28 years behind bars, he and the last Afrikaner president, F.W. de Klerk, began a process of negotiation and accommodation that led to the transfer of power from the white regime to a government elected by all the voters. In April 1994, a new era began in South Africa. Nelson Mandela, of distinguished Xhosa ancestry, had become president of an ANC-dominated government at Cape Town.

Political Reorganization

The route from prison gate to presidency was not an easy one either for Mandela or for the South Africa he hoped to lead; and again geography played a big role. De Klerk led the white minority, and Mandela the black majority, but there was a third force: the largest single nation in South Africa, the Zulu, whose leader, Chief Mangosuthu Buthelezi, saw a conspiracy between

Mandela and de Klerk that would damage Zulu interests. Buthelezi's political movement, Inkatha, demanded equity in the new South Africa and promised to scuttle the conferences that were negotiating the compromise South Africa needed. Inkatha carried out its threats. In the townships of the major cities and in the hostels of the mining companies, Inkatha "warriors" killed thousands, violence that threatened to become a civil war. Buthelezi and his followers talked of secession and independence for Natal if the new and democratic South Africa was not to their liking.

In the end, geography came to South Africa's rescue. What the country needed was an election that would give the ANC a majority, but not one large enough to overwhelm minority interests. To gain Inkatha's cooperation, Inkatha needed to win in Natal. To give substance to the promise of democracy in South Africa, the ANC would also have to lose in one other province.

To accomplish this (not to say to rig the election), the conference that prepared for the power transfer redrew the map, and thus the electoral map, of South Africa. Since 1910, South Africa had had four provinces: the Cape, Natal, the Orange Free State (north of the great Orange River), and the Transvaal ("across the Vaal," the major Orange tributary). These four provinces essentially represented the European occupation of the coun-

try: the Cape and Natal were British strongholds, and the Orange Free State and Transvaal were Boer republics until their defeat in the Boer War (Fig. 6-13). The new map (Fig. 6-14) created nine provinces, leaving intact Natal (but calling it Kwazulu-Natal) and the Orange Free State. Three new provinces and part of a fourth were carved out of the Cape: the Western Cape, essentially Cape Town and its hinterland, where the electorate is mainly Coloured and white; the Eastern Cape, dominated by the Xhosa nation; the Northern Cape, by far the least populous, mainly rural province; and the Northwest, part Cape and part Transvaal, where the Tswana nation is concentrated. The remainder of the Transvaal also became three provinces: Gauteng, the heart of the country's core area centered on Johannesburg; Northern Province, home to many Afrikaners who opposed the new South Africa; and Mpumalanga, a mix of wealthy white farmers and densely peopled Swazi and other African areas. (A decade later, the country's transforming map remains very much a work in progress: see box titled "Changing Place Names in South Africa.")

In the momentous 1994 elections, the ANC won seven of the nine provinces. It lost in Kwazulu-Natal, where an alliance of Zulu, Asian, and white voters prevailed over the ANC, and in the Western Cape, where white and Coloured voters combined to defeat ANC

FIGURE 6-13

FIGURE 6-14

Changing Place Names in South Africa

The postcolonial map of Africa displays a mixture of the old, the colonial, and the new. Ancient cities such as Ibadan and Kaduna in Nigeria and Harer in Ethiopia kept their names through the colonial period, but other African places were either renamed by the colonists or newly founded and named after European figures such as Belgium's King Leopold (Leopoldville), Britain's Lord Salisbury (Salisbury), and Portugal's Captain Lourenço Marques (Lourenço Marques). When the colonial powers withdrew, these cities were quickly given African names: respectively, Kinshasa, Harare, and Maputo.

In a plural society such as South Africa, such renaming can cause resentment and friction, and the ruling African National Congress (ANC) was careful and indeed diplomatic in not immediately erasing from the map many place names that hark back to the apartheid years, to British-led times, and to Boer domination. When the four provinces became nine (Fig. 6-14), the only new provincial names that appeared on the map were Gauteng (for the Johannesburg-Pretoria Province in the core), Mpumalanga (for the former eastern Transvaal), and Kwazulu, which was added to Natal.

While the ANC proceeded cautiously, city and town councils, now democratically elected and dominated by local majorities, began to agitate for change. A few Boer names disappeared from the map, but South Africa's

government was slow to jump on this bandwagon. For some years there was pressure to change the name of the Northern Province to Limpopo Province, an unobjectionable change (the Limpopo River forms South Africa's northern border), but by late 2003 no authorization had been given. The central government could not stop local town councils from renaming their localities, but that did not mean that the towns they administered were automatically renamed as well.

This produced a curious situation in which Pretoria remains Pretoria for the foreseeable future, but its local council is called the Tshwane Metropolitan Council (Tshwane is the name many locals prefer). Johannesburg remains Johannesburg, but there is pressure to change the city's name to Egoli (city of gold). In the case of Port Elizabeth, however, the weight of the new name carried the day: it will be known as Nelson Mandela. Farther up the coast, East London will become . . . Buffalo City, a neutral appellation.

The transformation of South Africa's map will take considerable time. There are hundreds of names revered by some whites and despised by most blacks, from Pietermaritzburg to Ventersdorp and from Louis Trichardt to Potgietersrus. The national government clearly prefers gradual evolution to precipitous change, and this will keep cartographers busy for years to come.

candidates. Nor did the ANC gain the 70 percent majority it would have needed to govern without opposition support. What all parties in South Africa accomplished after more than four decades of oppression and mismanagement was an example to a world quick to use force to reach political goals.

South Africa's difficult transition to post-apartheid normality and stability, however, is not over. President Mandela provided the leadership the country needed at a crucial time. But now South Africa has entered a post-Mandela era. On June 2, 1999, ANC leader Thabo Mbeki, who had served as President Mandela's deputy, became the country's second popularly elected president. In other African states, the succession from heroic founder-of-the-nation to political inheritor-of-the-presidency has not gone well, but South Africans prepared themselves by negotiating a constitution that contains a structure for presidential succession. That constitution took effect in stages from 1996 to 1999 and may safeguard the country against the problems that have befallen so many other states in the realm.

Soon after assuming the presidency, Mbeki was buffeted by two crises. In 2000, a world conference on AIDS

in Durban focused attention on South Africa's own health emergency: estimates of those infected with HIV in the country in the 15-to-49 age group ranged from 13 to 20 percent. The president, in controversial statements before and during the conference, left observers wondering whether he accepted the scientific evidence involved. And in the same year, the farm invasions in Zimbabwe (see page 297) raised fears among white farmers, especially in South Africa's Northern Province adjacent to Zimbabwe. White representatives in the South African Parliament called on the president to rebuke Zimbabwe's leaders, but some black representatives objected to this demand and even expressed support for the Mugabe regime. President Mbeki, in a difficult political bind, did not speak out.

Economic Geography

Undoubtedly the most serious immediate problems for the new South Africa are economic, and this is one of the president's strengths. Ever since diamonds were discovered at Kimberley in the 1860s, South Africa has been synonymous with minerals. The Kimberley finds, made in a remote corner of what was then the Orange Free

AMONG THE REALM'S GREAT CITIES . . . JOHANNESBURG

Subsaharan Africa displays only one incipient conurbation, and Johannesburg, South Africa, lies at the heart of it. Little more than a century ago, Johannesburg was a small (though rapidly growing) mining town based on the newly discovered gold reserves of the Witwatersrand.

Today Johannesburg forms the focus of a conurbation of more than 6 million, extending from Pretoria to the north to Vereeniging to the south, and from Springs in the east to Krugersdorp in the west. The population of metropolitan Johannesburg itself passed the 3.3 million mark in 2003, thereby overtaking Cape Town to become South Africa's largest metropolis.

Johannesburg's skyline is the most impressive in all of Africa, a forest of skyscrapers reflecting the wealth generated here over the past hundred years. Look southward from a high vantage point, and you see the huge mounds of yellowish-white slag from the mines of the "Rand," the so-called mine dumps, partly overgrown today, interspersed with suburbs and townships. In a general way, Johannesburg developed as a white city in the north and a black city in the south. Soweto, the black township, lies to the south; Houghton and other spacious, upper-class suburbs, formerly exclusively white residential areas, lie to the north.

Johannesburg has neither the scenery of Cape Town nor the climate and beaches of Durban. The city is a mile above sea level, and its thin air often is polluted from smog created by automobiles, factories, mine-dump dust, and countless cooking fires in the townships and shantytowns that ring the metropolis.

In the past century, the Johannesburg area produced nearly one-half of all the world's gold by value. But today,

Johannesburg lies at the heart of an industrial, commercial, and financial complex whose name on the new map of South Africa is *Gauteng*.

State (the British soon annexed it to the Cape), set into motion a new economic geography. Rail lines were laid from the coast to the "diamond capital" even as fortune seekers, capitalists, and tens of thousands of African workers, many from as far afield as Lesotho, streamed to the site. One of the capitalists was Cecil Rhodes, of Rhodes Scholarship fame, who used his fortune to help Britain dominate Southern Africa.

Just 25 years after the diamond discoveries, prospectors found what was long to be the world's greatest goldfield on a ridge called the Witwatersrand (Fig. 6-11). This time the site lay in the so-called South African Republic (the Transvaal), and again the Boers were unable to hold the prize. Johannesburg became the gold capital of the world, and a new and even larger stream of foreigners arrived, along with a huge influx of African workers. Cheap

labor enlarged the profits. Johannesburg grew explosively, satellite towns developed, and black townships mushroomed. The Boer War was only an interlude here on the mineral-rich Witwatersrand.

During the twentieth century, South Africa proved to be even richer than had been foreseen. Additional goldfields were discovered in the Orange Free State. Coal and iron ore were found in abundance, which gave rise to a major iron and steel industry. Other metallic minerals, including chromium and platinum, yielded large revenues on world markets. Asbestos, manganese, copper, nickel, antimony, and tin were mined and sold; a thriving metallurgical industry developed in South Africa itself. Capital flowed into the country, white immigration grew, farms and ranches were laid out, and markets multiplied.

"Looking down on this enormous railroad complex, we were reminded of the fact that almost an entire continent was turned into a wellspring of raw materials carried from interior to coast and shipped to Europe and other parts of the world. This complex lies near Witbank in the eastern Rand, a huge inventory of freight trains ready to transport ores from the plateau to Durban and Maputo. But at least South Africa acquired a true transport network in the process, ensuring regional interconnections; in most African countries, railroads serve almost entirely to link resources to coastal outlets."

South Africa's cities grew apace. Johannesburg was no longer just a mining town; it became an industrial complex and a financial center as well. The old Boer capital, Pretoria, just 30 miles (50 km) north of the Witwatersrand, became the country's administrative center during apartheid's days. In the Orange Free State, major industrial growth (including oil-from-coal technology) matched the expansion of mining. While the core area developed megalopolitan characteristics, coastal cities expanded as well. Durban's port served not only the Witwatersrand but a wider regional hinterland as well. Cape Town was becoming South Africa's largest city; its port, industries, and productive agricultural hinterland gave it primacy over a wide area.

Apartheid ruined these prospects and exposed the economy's weaknesses. The Afrikaner government made huge investments in its separate development policy, which contradicted fundamental principles of economic geography. International sanctions against apartheid hurt the economy, labor unrest weakened it further, and lower prices for those products that did get sold hurt even more. In the black townships, the school systems fell apart under the cry of "liberation before education," creating a huge undereducated (later unemployable) mass of young people.

For South Africa later is now, and the problems these policies created have arrived. As the twenty-first century dawned, South Africa's economy still depended heavily on the export of metals and minerals, but mineworkers demanding higher wages cut into the profits. Manufactured goods figured hardly at all in the export picture. Unemployment stood at high levels even as employed workers sought to raise their wages. In the cities, the whites fled to the suburbs, transforming the downtowns into commuter workplaces and the streets into bazaars. Violent robberies replaced the political crime of the pre-transfer period. Most ominous is the revolution of rising expectations, a phenomenon many African countries experienced after decolonization. Land pressure and housing needs cannot be accommodated overnight, but the newly empowered majority will expect its government to act expeditiously. Joblessness, housing shortages, and land pressure can form a potent mix to destabilize a society, and the South African government faces a daunting challenge.

On maps of development indices, South Africa is portrayed as an upper-middle-income economy, but averages mean little in this country of strong internal core-periphery contrasts. In its great cities, industrial complexes, mechanized farms, and huge ranches, South Africa resembles a high-income economy, much like Australia or Canada. But outside the primary core area (centered on Johannesburg) and beyond the secondary cores and their linking corridors lies a different South Africa, where conditions are more like those of rural Zambia or Zimbabwe. In terms of such indices as life expectancy, infant and child mortality, overall health, nutrition, education, and many others, a wide range marks the population sectors listed in Table 6-1. South Africa has been described as a microcosm of the world, exhibiting in a single state not only a diversity of cultures but also a wide range of human conditions. If South Africa can keep on course, the one-time pariah of apartheid will become a guidepost to a better world.

The Middle Tier

Between South Africa and the region's northernmost countries lies a group of states all of which have a common border with South Africa: large but sparsely-peopled Botswana and Namibia, the ministates of Swaziland and Lesotho, and troubled Zimbabwe. As Figure 6-11 shows, four of these five are landlocked. Diamond-exporting **Botswana** occupies the heart of the Kalahari Desert and surrounding steppe; only 1.6 million people, most of them subsistence farmers, inhabit this Texas-sized country. In 2003, no country in Africa was more severely afflicted by the AIDS epidemic than Botswana. South Africa encircles **Lesotho** (2.3 million) and surrounds most of **Swaziland** (1.1 million), ancestral home of the Swazi nation. Botswana, Lesotho, and Swaziland depend heavily on the income that workers who labor in South African mines, factories, and fields send home.

Apart from South Africa, the region's giant, the most important southern state undoubtedly is **Zimbabwe** (12.5 million), landlocked but well endowed with mineral and agricultural resources. Zimbabwe (the country is named after stone ruins in its interior) is mostly an elevated plateau between the Zambezi and Limpopo rivers, with the desert to the west and the Great Escarpment to the east. Its core area is defined by the mineral-rich Great Dyke and its environs, extending southwest from the vicinity of the capital, Harare, to the country's second city, Bulawayo. Copper, asbestos, and chromium (of which Zimbabwe is one of the world's leading sources) are among its major mineral exports, but Zimbabwe is not just an ore-exporting country. Farms produce tobacco, tea, sugar, cotton, and other crops (corn is the staple). A bitter civil war preceded independence, but many whites, encouraged by the government, stayed on their farms after the black African majority took power. Despite disputes over land policy, Zimbabwe's political transition did not derail the economy as it did in so many other countries.

Two nations form most of Zimbabwe's population: the Shona (71 percent) and the Ndebele (16 percent), the latter concentrated in the southwest of the country. Shona domination produced a government that, by the turn of the century, had been in power for 20 years. The long-term president, Robert Mugabe, commanded a regime marked by deepening corruption, including legislative obstruction of attempts by private Zimbabweans to compete with businesses in which government members had personal financial interest. As attempts to modernize the infrastructure were hampered and the once-promising mixed economy faltered, the government responded to pressure by curbing human rights.

Zimbabwe's problems mounted after 1980, when droughts caused African subsistence farmers to abandon their land. As thousands of livestock died, the Mugabe government in 1998 announced that it would take land from white farmers and distribute it to black African families. Social and political tensions rose, especially when farmers charged the government with assigning newly acquired land to cronies rather than those in need. In 2000, the situation worsened when thousands of so-called war veterans targeted and invaded white farms, drove out and in several cases killed their owners, destroyed the farm workers' dwellings, and occupied the land. The Mugabe regime tolerated these actions with an eye on upcoming elections, and the tactic worked: Mugabe was reelected in 2002 in an election marked by killings and intimidation of voters. In the process, the Zimbabwean agricultural economy was devastated. By early 2003, many citizens faced starvation, with foreign emergency assistance their only hope.

The western tip of Zimbabwe just barely touches the northeastern "Caprivi Finger" of **Namibia**, the region's youngest independent state. Once a German colony named South West Africa, the country was endowed with that narrow strip of land to connect it to the Zambezi River by the mapmakers in Berlin. When South Africa took control of the colony after World War I, the Caprivi Strip became strategically important; in the 1980s, there were rumors of missile (even nuclear) tests in the area. But Namibia achieved independence in 1990, and South Africa abandoned its nuclear program (nonetheless, in 1999, the Lozi people in the Strip launched an ill-fated secession campaign). In 1994, the last territorial vestige of South Africa was erased when the port of Walvis Bay, a South African exclave, was transferred to Namibia.

Namibia is appropriately named after a desert (the coast-paralleling Namib). Only 1.9 million people, most concentrated in the moister northern border areas, inhabit this huge country, which is about as large as Texas and Oklahoma combined. The capital, Windhoek, is centrally situated and therefore distant from this northern cluster. That is because the major economic activities—mining in the Tsumeb area and ranching across the steppe country of the south—happen elsewhere, not in the subsistence-dominated north.

But that is where the voters are, and Namibia will undoubtedly undergo structural readjustment. Much of its productive capacity remains in foreign hands, and Namibia must import food to supplement what little it produces. For Namibia, the long haul toward true sovereignty has just begun.

The Northern Tier

In the four countries that extend across the northern part of the region—Angola, Zambia, Malawi, and Moçambique—problems abound. **Angola** (13.4 million), formerly a Portuguese dependency, is one of Africa's richest countries in terms of raw materials and agricultural potential. It had a thriving economy at independence in 1975 but was engulfed by civil war that outside involvement made worse. The government, centered in the north and supported by northern peoples, chose a communist course. When a rebel movement emerged in the south with support by the (then) South African regime, Angola's leaders called on the Soviet Union for help. The United States also supported the rebel movement. Thousands of Cuban troops arrived in the capital, Luanda, escalating the war but securing the state.

This unrest devastated the Angolan economy. Farms lay abandoned, fields were land-mined, railroads destroyed, ports damaged. The railroad from Katanga to the port of Benguela, once a major income source, lay idle. Civilian casualties were high. But the government could pursue the war for a reason shown in Figure 6-11. Note that a small part of Angola lies separated from the main body of the country across the Congo River in the extreme northwest. This is the **exclave** of Cabinda, which contains oil reserves whose revenues supported the Luanda government during its war.

In the late 1990s, the United Nations brokered an agreement between the government and the rebel movement in the interior that awarded the insurgents a role in the administration. But this arrangement broke down, and as much as half of the country remained under rebel control or in contention. Soon, the rebels were back to their murderous ways, killing 250 people on a train near the capital, Luanda, in August 2001 and taking 50 lives on a bus two weeks later. Oil money and diamond money were funding a civil war seemingly without end.

In early 2002, however, the long-entrenched rebel leader, Jonas Savimbi, was killed by government forces, and the movement collapsed almost immediately. With international help, Angola's government began the task of reintegrating a country that had for so many years been ripped apart by internal as well as external forces. An insurgent state had failed, and now the legitimate administration faced the task of reconstructing a country that could become one of Africa's more prosperous economies.

On the opposite coast, **Moçambique** (population 20.4 million), like Angola a former Portuguese dependency that chose a Marxist course for its economic development, experienced an equally costly struggle, but with far fewer resources than Angola. Before independence, its chief sources of income came from its cashew and coconut plantations and from its relative location. As the map shows, Moçambique's port of Beira is ideally situated to handle the external trade of Zimbabwe and southern Malawi, and its capital, Maputo, is the closest port to the great mining and industrial complex centered on the Witwatersrand in South Africa. In better times, goods streamed through these ports, and when a hydroelectric plant was built on the Zambezi River at Cabora Bassa, prospects seemed fair. Then, however, the economy collapsed under bad management, a rebel movement aided for a time by South African interests destroyed the social order, famines broke out, and Moçambique descended into chaos. More than a million refugees streamed into Malawi, the hydroelectric project was damaged, the port facilities at Beira ceased functioning, and Maputo's transit role ended.

In the late 1990s, some semblance of stability had returned to Moçambique. The government abandoned its Marxist principles, the rebel movement laid down its arms, refugees returned, and the new South Africa was offering agricultural and technical help rather than arms to insurgents. In 1996, the two countries launched the Maputo Development Corridor, a scheme to revitalize the links (rail, road, and air) between Johannesburg and its nearest port, to stimulate trade and investment, and to boost mining, farming, and tourism along the route (Fig. 6-11).

In 2001 Moçambique's modest political and economic progress was imperiled, first by a flareup of antigovernment rebel activity in the far north and then by massive flooding in the lower basin of the Zambezi River, creating a humanitarian crisis of major dimensions. It was a reminder to all of Africa and the entire world how things had changed when the South African Air Force performed heroic helicopter rescues of countless flood victims hanging on to tree branches and rooftops, but it was also clear that Moçambique would soon be off the front pages again, its fragile economy damaged and its people still further impoverished.

Zambia (10.4 million) and Malawi (11.4 million), both former British dependencies, constitute the remainder of the Northern Tier of Southern Africa. Zambia and Malawi are less developed than Zimbabwe, although **Zambia** contains the mineral-rich Copperbelt. The decline of world prices for its minerals, as well as the problems and costs associated with their long-distance transportation, have hurt the economy of Zambia. **Malawi's** economic geography is almost completely agricultural, with a variety of crops including tea, cotton, tobacco, and peanuts. This emphasis on farming has helped cushion the economy against market swings and has meant that more of its labor force has been able to find productive work than in many other African countries. Still, more workers toil in the mines of other countries than are gainfully employed within Malawi itself.

◢ EAST AFRICA

East Africa lies to the east of the row of African Great Lakes that extends from Lake Albert in the north to Lake Malawi in the south (Fig. 6-15). Here the land rises from the low-lying Congo Basin to the west to the East African Plateau which, at various levels of elevation and in varying degrees of dissection, stretches from the spectacular canyons of Ethiopia to the rolling uplands of southern Tanzania. There, the valley of the Rovuma River marks the limit of the region. Hills and valleys, fertile soils, and copious rains mark the transition in Rwanda and Burundi. Eastward the rainforest disappears and open savanna cloaks the countryside. Great volcanoes rise above a rift-valley-dissected highland. At the heart of the region lies Lake Victoria. In the north the surface rises above 10,000 feet (3300 m), and so deep are the trenches cut by faults and rivers there that the land was called, appropriately, Abyssinia (now Ethiopia).

Five countries, in addition to the highland part of Ethiopia, form this East African region: Tanzania, Kenya, Uganda, Rwanda, and Burundi. Here the Bantu peoples that make up most of the population met Nilotic peoples from the north, including the Maasai. In the hills of Rwanda and Burundi, as we noted, a stratified society developed in which the minority Tutsi, in their cattle-owning kingdoms, dominated the Hutu peasantry. The coast was the scene of many historic events: the arrival of Islam, the visit of Ming Dynasty Chinese fleets in the 1400s, the quest for power by the Turks, the Arab slave trade, the European colonial competition. Here developed the East African *lingua franca*, Swahili.

Kenya

Kenya is neither the largest nor the most populous country in East Africa, but over the past half-century it has been the dominant state in the region. Its skyscrapered capital at the heart of its core area, Nairobi, is the region's largest city; its port, Mombasa, is the region's busiest. During the 1950s, the Kikuyu nation led a vicious rebellion that hastened the departure of the British from the region.

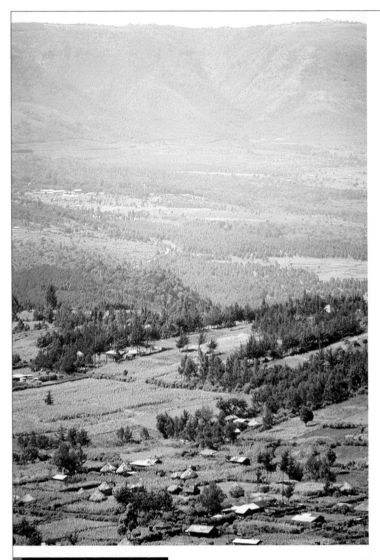

FROM THE FIELD NOTES

"Visiting the village in the Kenya Highlands where a graduate student was doing fieldwork on land reform, I took the long way and drove along the top of the eastern wall of the Eastern Rift Valley. Often the valley wall is not sheer but terraced, and soils on those terraces are quite fertile; also, the west-facing slopes tend to be well-watered. Here African farmers built villages and laid out communal plots, farming these lands in a well-organized way long before the European intrusion."

After independence, Kenya chose a capitalist path of development, aligning itself with Western interests. Without major known mineral deposits, Kenya depended on coffee and tea exports and on a tourist industry based on its magnificent national parks (Fig. 6-15). Tourism became its largest single earner of foreign exchange, and Kenya prospered, apparently proving the wisdom of its capitalist course.

FIGURE 6-15

AMONG THE REALM'S GREAT CITIES . . . **NAIROBI**

*N*airobi is the quintessential colonial legacy: there was no African settlement on this site when, in 1899, the railroad the British were building from the port of Mombasa to the shores of Lake Victoria reached it. However, it had something even more important: water. The fresh stream that crossed the railway line was known to the Maasai cattle herders as Enkare Nairobi (Cold Water).

The railroad was extended into the interior, but Nairobi grew. Indian traders set up shop. The British established their administrative headquarters here. When Kenya became independent in 1963, Nairobi naturally was the national capital.

Nairobi owes its primacy to its governmental functions, which ensured its priority through the colonial and independence periods, and to its favorable situation. To the north and northwest lie the Kenya Highlands, the country's leading agricultural area and the historic base of the dominant nation in Kenya, the Kikuyu. Beyond the rift valley to the west lie the productive lands of the Luo in the Lake Victoria Basin. To the east, elevations drop rapidly from Nairobi's 5000 feet (1660 m), so that highland environs make a swift transition to tropical savanna that, in turn, yields to semiarid steppe.

A moderate climate, a modern city center, several major visitor attractions (including Nairobi National Park, on the

city's doorstep), and a state-of-the-art airport have boosted Nairobi's fortunes as a major tourist destination, though wildlife destruction, security concerns, and political conditions have damaged the industry in recent years.

Nairobi is Kenya's principal commercial, industrial, and educational center. But its growth (to 2.8 million today) has come at a price: its modern central business district stands in stark contrast to the squalor in the shantytowns that house the countless migrants its apparent opportunities attract.

But serious problems arose. Kenya during the 1980s had the highest rate of population growth in the world, and population pressure on farmlands and on the fringes of the wildlife reserves mounted. Poaching became worrisome, and tourism declined. During the late 1990s, violent weather buffeted Kenya, causing landslides and washing away large segments of the crucial Nairobi-Mombasa Highway. This was followed by a severe drought lasting several years, bringing famine to the interior. Meanwhile, government corruption siphoned off funds that should have been invested. Democratic principles were violated, and relationships with Western allies were strained. The AIDS epidemic brought another setback to a country that, in the early 1970s, had appeared headed for an economic takeoff.

Today, Kenya's prospects are uncertain. Geography, history, and politics have placed the Kikuyu (22 percent of the population of 32.4 million) in a position of power. But there are other major peoples (see Fig. 6-15) and several smaller ones. The Luhya, Luo, Kalenjin, and Kamba together constitute about 50 percent of the population,

and on the territorial margins of the country there are peoples such as the Maasai, Turkana, Boran, and Galla. Creating and sustaining a political system that ensures democracy and represents the interests of these disparate peoples is Kenya's unmet challenge.

Tanzania

Tanzania (a name derived from Tanganyika plus Zanzibar) is the biggest and most populous East African country (39.2 million). Its total area exceeds that of the other four countries combined. Tanzania has been described as a country without a core because its clusters of population and zones of productive capacity lie dispersed—mostly on its margins on the east coast (where the capital, Dar es Salaam, is located), near the shores of Lake Victoria in the northwest, near Lake Tanganyika in the far west, and near Lake Malawi in the interior south. This is in sharp contrast to Kenya, which has a well-defined core area in the Kenya Highlands (centered on Nairobi in the heart of the country). Moreover, Tanzania is a country of many

"It was 108 in the shade, but the narrow alleys of Zanzibar's Stone City felt even hotter than that. I spent some time here when I was working on my monograph on Dar es Salaam in the 1960s and had not been back. In those days, African socialism and *uhuru* were the watchwords; since then Tanzania has not done well economically. But here were signs of a new era: a People's Bank in a former government building, and on the old fort's tower a poster saying 'Think Digital Go Tritel.' Nearby, on the sandy beach where I relaxed 35 years ago, was evidence that Zanzibar has not escaped the ravages of AIDS. Now the sand served as a refuge for the sick, who were resting there. 'It's better, bwana, than the corridor of the clinic,' said a young man who could walk only a few steps at a time with the help of a cane, and who breathed with difficulty as he spoke. Here as everywhere in Subsaharan Africa, AIDS has severely strained already-limited medical facilities."

peoples, none numerous enough to dominate the state. About 100 ethnic groups coexist; one-third of the population, mainly those on the coast, are Muslims.

In contrast to Kenya, Tanzania after independence embarked on a socialist course toward development, including a massive farm collectivization program that was imposed without adequate planning. Communist China helped Tanzania construct a railroad, the Tan-Zam Railway, from Dar es Salaam to Zambia; but the project failed. So did an effort to move the capital from colonial Dar es Salaam to Dodoma in the interior (Fig. 6-15). The country's limited tourist infrastructure was allowed to degenerate, giving Kenya virtually the entire tourist market.

But Tanzania did achieve what several other East African countries could not: political stability and a degree of democracy. Thus Tanzania changed economic direction in the late 1980s without social turmoil and embarked on a market-oriented "recovery" program. While Tanzanian coffee, cotton, tobacco, and other farm products sold in increasing amounts on world markets, most Tanzanians remained subsistence farmers. Today, Tanzania still has one of the world's poorest economies (in this realm, only Sierra Leone has a lower per-capita GNP). In addition, demands for greater autonomy, even independence, were rising on the island of Zanzibar, a devolutionary challenge Tanzania may have difficulty overcoming.

Uganda

Uganda contained the most important African political entity in this region when the British arrived in the 1890s. This was the Kingdom of Buganda (shown in dark brown in Fig. 6-15), which faced the north shore of Lake Victoria, had an impressive capital at Kampala, and was stable—as well as ideally suited for indirect rule over a large hinterland. The British established their headquarters at nearby Entebbe on the lake (thus adding to the status of the kingdom) and proceeded to organize their Uganda protectorate in accordance with the principles of indirect rule. The Baganda (the people of Buganda) became the dominant people in Uganda, and when the British departed, they bequeathed a complicated federal system to perpetuate Baganda supremacy.

Although a **landlocked state** dependent on Kenya for an outlet to the ocean, Uganda at independence (1962) had better economic prospects than many other African countries. It was the largest producer of coffee in the British Commonwealth. It also exported cotton, tea, sugar, and other farm products. Copper was mined in the southwest, and an Asian immigrant population of about 75,000 dominated the country's commerce. Nevertheless, political disaster struck. Resentment over Baganda overlordship fueled revolutionary change, and a brutal dictator, Idi Amin, took control in 1971. He ousted the Asians, exterminated his opponents, and destroyed the economy. Eventually, in 1979, an invasion supported by neighboring Tanzania drove Amin from power, but by then Uganda lay in ruins. Recovery has been slow, complicated by the AIDS epidemic, which struck Uganda with particular severity. In the 1990s, Uganda was embroiled in conflict with its northern neighbor, Sudan. Each accused the other of supporting rebellion against itself. Meanwhile, even as economic and political reforms were in progress in Uganda, its government also supported the Tutsi minorities in Rwanda and The Congo. Situated at one of Africa's most volatile crossroads, landlocked Uganda (population: 26.2 million) faces a difficult future.

Rwanda and Burundi

Rwanda and Burundi would seem to occupy Tanzania's northwest corner, and indeed they were part of the German colonial domain conquered before World War I. But during that war Belgian forces attacked the Germans from their Congo bases to the west and were awarded these territories when the conflict was over in 1918. The Belgians used them as labor sources for their mines.

Rwanda (7.7 million) and Burundi (7.0 million) are physiographically part of East Africa, but their cultural geography is linked to the north and west. Here, as we noted earlier, Tutsi pastoralists from the north subjugated Hutu farmers (who had themselves made serfs of the local Twa [pygmy] population), setting up a conflict that was originally ethnic but became cultural. Certain Hutu were able to advance in the Tutsi-dominated society, becoming to some extent converted to Tutsi ways, leaving subsistence farming behind, and rising in the social hierarchy. These so-called moderate Hutu were— and are—often targeted by other Hutus, who resent their position in society, so that today conflict between radical Hutu, on the one hand, and Tutsi and moderate Hutu, on the other, has become a cultural rather than strictly an ethnic one.

This conflict has now had repercussions not only in Rwanda and Burundi but also in The Congo, Uganda, Tanzania, and even Angola and Congo. At the opening of this century, the minority Tutsi and Hutu moderates were in control of the government in both Rwanda and Burundi, where the Tutsi-dominated military had staged a coup. The Hutu majority's leaders, of course, call for democracy, which would put them in power. The situation is one of Africa's most intractable.

Highland Ethiopia

As Figure 6-15 shows, the East African region also encompasses the highland zone of Ethiopia, including the capital, Adis Abeba, the source of the Blue Nile, Lake Tana, and the Amharic core area that was the base of the empire that lost its independence only from 1935 to 1941. Ethiopia, mountain fortress of the Coptic Christians who held their own here, eventually became a colonizer itself. Its forces came down the slopes of the highlands and conquered much of the Islamic part of Africa's Horn, including present-day Eritrea and the Ogaden area, a Somali territory. (Geographically, these are parts of what we have mapped as the African Transition Zone

[Fig. 6-10], which will be discussed in the final section of this chapter.)

Physiographically and culturally, highland Ethiopia is part of the East African region. But because Ethiopia was not colonized and because its natural outlets are to the Red Sea, not southward to Mombasa, effective interconnections between former British East Africa and highland Ethiopia never developed. But the Amhara and Oromo peoples of Ethiopia are Africans, not Arabs; nor have they been Arabized or Islamized as in northern Sudan and Somalia. The independence and secession of Eritrea in 1993 effectively landlocked Ethiopia, but for a few years there was cooperation and Ethiopia used Eritrean Red Sea ports. In 1998, however, a boundary dispute led to a bitter and costly war, and Ethiopia was forced to turn to Djibouti for a maritime outlet. With adversaries on three sides, Ethiopia is likely to turn increasingly toward East Africa, however tenuous the surface links are today.

▶ EQUATORIAL AFRICA

The term *equatorial* is not just locational but also environmental. The equator bisects Africa, but only the western part of central Africa features the conditions associated with the low-elevation tropics: intense heat, high rainfall and extreme humidity, little seasonal variation, rainforest and monsoon-forest vegetation, enormous biodiversity. To the east, beyond the Western Rift Valley, elevations rise, and cooler, more seasonal climatic regimes prevail. As a result, we recognize two regions in these lowest latitudes: East Africa to the east, which we have just discussed, and Equatorial Africa to the west.

Equatorial Africa is physiographically dominated by the giant Congo Basin. The Adamawa Highlands separate this region from West Africa; rising elevations and climatic change mark its southern limits (see the *Cwa* boundary in Fig. G-8). Its political geography consists of eight states, of which The Congo (formerly Zaïre) is by far the largest in both territory and population (Fig. 6-16).

Five of the other seven states—Gabon, Cameroon, São Tomé and Príncipe, Congo, and Equatorial Guinea—all have coastlines on the Atlantic Ocean. The Central African Republic and Chad, the south of which is part of this region, are landlocked. In certain respects, the physical and human characteristics of Equatorial Africa extend even into southern Sudan. This vast and complex region is in many ways the most troubled region in the entire Subsaharan African realm.

During the colonial period, the name *Congo* became attached to two countries on opposite sides of the region's major river: the Belgian Congo and the French Congo. After independence, both countries initially retained this

FIGURE 6-16

identity, but in 1971 the ruler of the larger (former Belgian) Congo changed its name to Zaïre. In 1997, following a prolonged revolution that started in eastern Zaïre, a new regime took power and renamed the country The Democratic Republic of the Congo. As we noted at the beginning of this chapter, we refer to this country as *The* Congo, and to its smaller neighbor simply as Congo.

The Congo

As the map shows, The Congo has but a tiny window (23 miles; 37 km) on the Atlantic Ocean, just enough to accommodate the mouth of the Congo River. Oceangoing ships can reach the port of Matadi, inland from which falls and rapids make it necessary to move goods by road

Substantial areas of The Congo remain beyond the control of the capital and are the scene of costly fighting among rebel armies with devastating impact on local populations. These heavily armed troops of an outfit called the Union of Congolese Patriots are shown having taken over the northeastern town of Bunia in advance of still another round of talks to end more than four years of civil war (the talks failed, and strife continues).

or rail to the capital, Kinshasa. This is not the only place where the Congo River fails as a transport route. Follow it upstream on Figure 6-16, and you note that other trans-shipments are necessary between Kisangani and Ubundu, and at Kindu. Follow the railroad south from Kindu, and you reach another narrow corridor of The Congo's territory at the city of Lubumbashi. That vital part of The Congo contains most of its major mineral resources, including copper and cobalt. As the map shows, the most efficient export route from this area, called Katanga, lies by rail across Angola to the port of Benguela. Civil war in Angola, however, disrupted this route.

With a territory not much smaller than the United States east of the Mississippi, a population of 58.7 million, a rich and varied mineral base, and much good agricultural land, The Congo would seem to have all the ingredients needed to lead this region and, indeed, Africa. But strong centrifugal forces, arising from its physiography and cultural geography, pull The Congo apart. The immense forested heart of the basin-shaped country creates communication barriers between east and west, north and south. Many of The Congo's productive areas lie along its periphery, separated by enormous distances. These areas tend to look across the border, to one or more of The Congo's nine neighbors, for outlets, markets, and often ethnic kinship as well.

The Congo's civil wars of the 1990s started in one such neighbor, Rwanda, and spilled over into what was then still Zaïre. Rwanda, Africa's most densely populated country, has for centuries been the scene of conflict that began as a contest between Hutu farmers and invading, conquering Tutsi pastoralists. Colonial intervention and cultural evolution changed the ground rules. The

colonists' boundaries hemmed in a core of Hutu and Tutsi in hilly, fertile Rwanda and Burundi, but left many Tutsi-related peoples in the neighboring Belgian Congo. The Tutsi, always the minority and always better off than the Hutu, dominated political life in the postcolonial period. A seemingly endless cycle of Hutu revenge and Tutsi retribution cost hundreds of thousands of lives. In the mid-1990s, another such outbreak generated one of the largest refugee streams in human history from Rwanda into Zaïre (and into Tanzania as well). Hutu guerrillas, who had killed hundreds of thousands of Tutsi in Rwanda, now terrorized the refugee camps. Their plan was to take control of then eastern Zaïre, creating a greater Hutu domain.

Eastern Zaïre, however, was peopled by Tutsi and Tutsi-related groups, and soon the Hutu aggressors found themselves outgunned. But they had started a larger war. Soon a Tutsi-dominated force, aided not only by Rwanda's Tutsi-led government but also by a sympathetic Uganda, conquered eastern Zaïre and moved westward. Their leader was an experienced rebel commander from Katanga (named Shaba during the Zaïrian period), Laurent Kabila. In May 1997 Kabila's forces, having overrun Zaïre at astonishing speed, swept into the capital, Kinshasa, and took power. Days later Kabila announced that Zaïre would henceforth be known as The Democratic Republic of the Congo.

Ousting a corrupt, collapsing regime in Kinshasa (long supported by the United States) was one thing; resurrecting The Congo was another. Soon, the Kabila regime was at odds with its former allies. On the one hand, the Kabila regime was accused of failing to institute reforms; on the other, rebel forces that had helped

The failure of The Congolese state and the rebel strife in its eastern and northeastern areas have unleashed other, local conflicts with deadly consequences. In the northeastern Congo, the Hema and Lendu peoples have long been adversaries, requiring effective local and national government to keep the peace and work toward resolution. In April, 2003 the village of Drodro and nearby Hema villages were the scene of indescribable carnage when, it was alleged, a band of Lendu attackers raided the area and massacred nearly 1,000 people. Seen here is a group of traumatized Hema children seeking shelter at a Catholic mission the following morning. For more than a century, The Congo has suffered unimaginably from European tyrants and domestic despots (the most recent, Mobutu, supported by American and other Western interests). Now the country is in the grip of foreign African and local Congolese squads, and the "international community" does virtually nothing to help restore order.

him now saw opportunities to achieve their own territorial aims. As The Congo fell into renewed disarray, its breakdown affected other countries within and beyond its region. In the east, Uganda, Rwanda, and Burundi supported an anti-Kabila campaign even as other groups, also opposed to the new regime, fought among themselves. In the west, Kabila had the support of Zimbabwe, Angola, and Namibia, all of whom sent armed forces. But such support came at a price; Zimbabwean generals secured costly business privileges in The Congo's mineral and agricultural areas.

Eventually, a series of developments brought hope that The Congo's crisis might wane. Following the assassination of President Laurent Kabila in 2001, his son Joseph took control and implemented a United Nations peace plan his father had rejected. Some of the foreign forces in The Congo, notably those of Angola and Uganda, withdrew. In the embattled east of this vast country, the Hutu continued their rampage and the Tutsi-dominated government of Rwanda did little to stop them, but even there the level of violence declined. The death toll by early 2003 was estimated to exceed 3 million.

More than 200 Congolese ethnic groups are encircled by a colonial-era boundary framework that needs realignment. That realignment will undoubtedly cause additional misery for a country that has suffered terribly for more than a century (see the Issue Box entitled "The Impact of Colonialism on Subsaharan Africa").

Across the River

To the west and north of the Congo and Ubangi rivers lie Equatorial Africa's other seven countries (Fig. 6-16). Two of these are landlocked. **Chad**, straddling the African Transition Zone as well as the regional boundary with West Africa, is one of Africa's most remote countries. But several major oil discoveries in the south and a projected pipeline to Kribi in Cameroon will link Chad more effectively with this region. The **Central African Republic**, chronically unstable and poverty-stricken, never was able to convert its agricultural potential and mineral resources (diamonds, uranium) into real progress. And one country consists of two small, densely forested volcanic islands: **São Tomé and Príncipe**, a ministate with a population of only 220,000 and a few exports derived from its cocoa plantations and coconut trees.

The four coastal states present a different picture. All four possess oil reserves and share the Congo Basin's equatorial forests; oil and timber, therefore, rank prominently among their exports. In **Gabon**, this combination has produced Equatorial Africa's only upper-middle-income economy. Of the four coastal states, Gabon also has the largest proven mineral resources, including manganese, uranium, and iron ore. Its capital, Libreville (the only coastal capital in the region), reflects all this in its highrise downtown, bustling port, and fast-growing squatter settlements.

Cameroon, less well endowed with oil or other raw materials, has the region's strongest agricultural sector by virtue of its higher-latitude location and high-relief topography. Western Cameroon is one of the more developed parts of Equatorial Africa and includes the capital, Yaoundé, and the port of Douala. In 1997, Cameroon granted China the right to exploit a bauxite (aluminum) reserve, but relations with neighboring Nigeria remain difficult. Cameroon and Nigeria have fought pitched battles over their joint border on the oil-rich Bakassi Peninsula, an issue that was submitted to the World Court in The Hague in 1994. In 2002 the court ruled in favor of

The Impact of Colonialism on Subsaharan Africa

COLONIALISM IS THE CULPRIT!

"IF YOU WANT to see and feel the way colonialism ruined Africa, you don't have to go beyond the borders of my own country, the so-called Democratic Republic of the Congo, for the evidence. I'm a school teacher here in Kikwit, east of Kinshasa, and my whole life I've lived under the dictatorship left us by the Belgians and in the chaos before and after it. All you have to do is look at the map. Who can say that the political map of Africa isn't a terrible burden? Imagine some outside power, say the Chinese, coming into Europe and throwing the Germans and the French together in one country. There would never be peace! Well, that's what they did here. All over Africa they put old enemies together and parceled up our great nations. And now they tell us to get over our animosities and 'tribalism' and live in democratic peace.

"And let me tell you something about tribalism. Before the conquest, we had ethnic groups and clans like all other peoples have, from the Scots to the Siamese. And we certainly fought with each other. But mostly we got along, and there was lots of gray area between and among us. You weren't immediately labeled a Hutu or a Luba or a Bemba. The colonial powers changed all that. Suddenly we were tribalized. Whether you liked it or not, whether you filled the 'profile' or not, you were designated a Tutsi or a Hutu, a Ganda or a Toro. This enabled the European rulers to use the strong to dominate the weak, to use the rich to tax the poor, to police migrants and assemble labor. The tribalization of Africa was a colonial invention. It worsened our divisions and destroyed our common ground. And now we just have to get over it?

"Of course the colonialists exploited our mineral resources—I tell my pupils how The Congo enriched Europeans and other 'civilized' peoples outside of Africa. But they also grabbed our best land, turning food-producing areas into commercial plantations. Are you surprised that in Zimbabwe Mugabe wants to chase those white farmers off 'their' land? I'm not, and I'm with him all the way. If the colonialists hadn't ruined traditional agriculture, there wouldn't be any food shortages today.

"Look at my country. The war between the Hutu and the Tutsi that started during colonial times has now spread deep into The Congo. We've got so used to taking sides in 'tribal' conflicts that the whole east is out of control. And who are the losers? The poorest of the poor, who get robbed and killed by the Mai Mai, the Interahamwe, and the Tutsis and their friends. What are the ex-colonial powers doing now to help us with the mess they left behind? Nothing! So don't complain to me about poor African leadership and our failure to carry out economic reforms dictated by some foreign bank. The poisonous legacy of colonialism will hold Africa back for generations to come."

COLONIALISM IS A SCAPEGOAT!

"We Africans have many factors to blame for our troubles, but colonialism isn't one of them. Many African countries have been independent for nearly two generations; most Africans were born long after the colonial era ended. Are we better off today than we were at the end of the colonial period? Are those military dictatorships that ran so many countries into the ground any better than the colonizers who preceded them? Are the ghosts of colonialists past rigging our elections, stifling the media? Is our endemic corruption the fault of lingering colonialism? Did the colonial governors set up Swiss bank accounts to which to divert development funds? Can extinct colonial regimes be blamed for the environmental problems we face? Did the bandits of interior Sierra Leone or eastern areas of The Congo learn their mutilation practices from colonialism? As a nurse in a Uganda hospital for AIDS victims, I say that it's time to stop blaming ancient history for our current failures.

"It's not that we Africans aren't at a disadvantage in this unforgiving world. Our postcolonial population explosion was followed by an HIV-induced implosion. Apartheid South Africa spread its malign influence far beyond its borders. The Cold War pitted regimes against rebels and, to use a colonial term, tribe against tribe. Droughts and other plagues of biblical proportions ravaged our continent when we were more vulnerable than ever. Prices for our raw materials on international markets fell when we most needed a boost. Foreign governments protected their farmers against competition from African producers. Those are legitimate complaints. But colonialism?

"Forget about blaming boundaries, white farmers, tribalism, religion, and other residues of colonialism for our present troubles. Look at what happened here in Uganda. A decade ago we were at the heart of the 'AIDS Belt' in tropical Africa, with one of the worst infection rates on the continent. Then came President Museveni, whose government launched a vigorous self-help campaign, educating people to the risks, urging restraint, distributing condoms, encouraging women to resist unwanted sex. Now Uganda's AIDS incidence is dropping and we are held up as an example for all of Africa to follow. While South Africa's President Mbeki reveals his confusion about the causes of AIDS, Zimbabwe's Mugabe is too busy chasing white farmers, and Kenya's Moi is trying to rig his succession, we in Uganda have leadership in an area where we need it desperately. That certainly beats blaming colonialism for everything.

"We have an African Union. If those colonial boundaries are so bad, let's change them, or do what the European Union is doing and begin to erase them. But first, let's isolate tyrants, nurture open democracy, and combat corruption—and stop blaming colonialism."

Cameroon, a decision that would substantially improve Cameroon's position among West and Equatorial African oil producers. Even as Nigeria and Cameroon sought to accept the ruling, local leaders on the peninsula argued that their subjects were ethnically related to Nigeria's Efik people and threatened to declare independence.

With five neighbors, **Congo** could be a major transit hub for this region, especially for The Congo if it recovers from civil war. Its capital, Brazzaville, lies across the Congo River from Kinshasa and is linked to the port of Pointe Noire by road and rail. But devastating power struggles have negated Congo's geographic advantages.

As Figure 6-16 shows, **Equatorial Guinea** consists of a rectangle of mainland territory and the island of Bioko, where the capital of Malabo is located. A former Spanish colony that remained one of Africa's least developed territories, Equatorial Guinea, too, has been affected by the oil business in this area. Petroleum products now dominate its exports, but, as in so many other oil-rich countries, this bounty has not significantly raised incomes for most of the people.

One other territory would seem to be a part of Equatorial Africa: Cabinda, wedged between the two Congos just to the north of the Congo River's mouth. But Cabinda is one of those colonial legacies on the African map—it belonged to the Portuguese and was administered as part of Angola. As we saw earlier, it is an exclave of independent Angola, and a valuable one: it contains major oil reserves.

◗ WEST AFRICA

West Africa occupies most of Africa's Bulge, extending south from the margins of the Sahara to the Gulf of Guinea coast and from Lake Chad west to Senegal (Fig. 6-17). Politically, the broadest definition of this region includes all those states that lie to the south of Western Sahara, Algeria, and Libya and to the west of Chad (itself sometimes included) and Cameroon. Within West Africa, a rough division is sometimes made between the large, mostly steppe and desert states that extend across the southern Sahara (Chad could also be included here) and the smaller, better-watered coastal states.

Apart from once-Portuguese Guinea-Bissau and long-independent Liberia, West Africa comprises four former British and nine former French dependencies. The British-influenced countries (Nigeria, Ghana, Sierra Leone, and Gambia) lie separated from one another, whereas Francophone West Africa is contiguous. As Figure 6-17 shows, political boundaries extend from the coast into the interior, so that from Mauritania to Nigeria, the West African habitat is parceled out among parallel, coast-oriented states. Across these boundaries, especially across those between former British and former French territories, there is only limited interaction. For example, in terms of value, Nigeria's trade with Britain is about 100 times as great as its trade with nearby Ghana. The countries of West Africa are not interdependent economically, and their incomes are largely derived from the sale of their products on the non-African international market.

Given these cross-currents of subdivision within West Africa, why are we justified in speaking of a single West African region? First, this part of the realm has remarkable cultural and historical momentum. The colonial interlude failed to extinguish West African vitality, expressed not only by the old states and empires of the savanna and the cities of the forest, but also by the vigor and entrepreneurship, the achievements in sculpture, music, and dance, of peoples from Senegal to Nigeria's southeastern Iboland. Second, West Africa contains a set of parallel east-west ecological belts, clearly reflected in Figures G-7 and G-8, whose role in the development of the region is pervasive. As the transport-route pattern on the map of West Africa indicates, overland connections within each of these belts, from country to country, are poor; no coastal or interior railroad ever connected this tier of countries. Yet spatial interaction is stronger across these belts, and some north-south economic exchange does take place, notably in the coastal consumption of meat from cattle raised in the northern savannas. And third, West Africa received an early and crucial imprint from European colonialism, which—with its maritime commerce and slave trade—transformed the region from one end to the other. This impact reached into the heart of the Sahara, and it set the stage for the reorientation of the whole area, from which emerged the present patchwork of states.

Despite the effects of the slave trade, West Africa today is Subsaharan Africa's most populous region (Fig. G-9). In these terms, Nigeria (whose census results are in doubt, but with an estimated population of 129 million) is Africa's largest state; Ghana (21.1 million) ranks high as well. As Figure G-9 shows, West Africa also constitutes one of Africa's major population clusters. The southern half of the region, understandably, is home to most of the people. Mauritania, Mali, and Niger include too much of the unproductive Sahel's steppe and the arid Sahara to sustain populations comparable to those of Nigeria, Ghana, or Ivory Coast.

The peoples along the coast reflect the modern era that the colonial powers introduced: they prospered in their newfound roles as middlemen in the coastward trade. Later, they experienced the changes of the colonial period; in education, religion, urbanization, agriculture, politics, health, and many other endeavors, they adopted new ways.

FIGURE 6-17

In contrast, the peoples of the interior retained their ties with a different era in African history. Distant and aloof from the main theater of European colonial activity and often drawn into the Islamic orbit, they experienced a significantly different kind of change. But the map reminds us that Africa's boundaries were not drawn to accommodate such contrasts. Both Nigeria and Ghana possess population clusters representing the interior as well as the coastal peoples, and in both countries the wide cultural gap between north and south has produced political problems.

Nigeria: West Africa's Cornerstone

When Nigeria achieved full independence from Britain in 1960, its government faced the task of administering a European political creation containing three major nations and nearly 250 other peoples ranging from several million to a few thousand in number. The country had been endowed with a federal political system consisting of three regions, each focused on one of the three leading nations. In the southwest lay the Western Region, the traditional home of the Yoruba, a people with long urban traditions and a complex and highly developed culture. In the southeast lay the Eastern Region, historic home of the Ibo people, less affected than the Yoruba by the colonial impact, less urbanized, but widely dispersed to other parts of Nigeria and even beyond. And in the north, the Northern Region was the largest and most populous, the domain of the Hausa-Fulani cluster, a region of ancient cities, Islam, and Sahelian environments.

The map of Nigeria shows a compact territory crossed by the Niger River and its tributary, the Benue, together forming a Y-shaped system (Fig. 6-18). The lower Niger and its delta separated the Yoruba and the Ibo and their

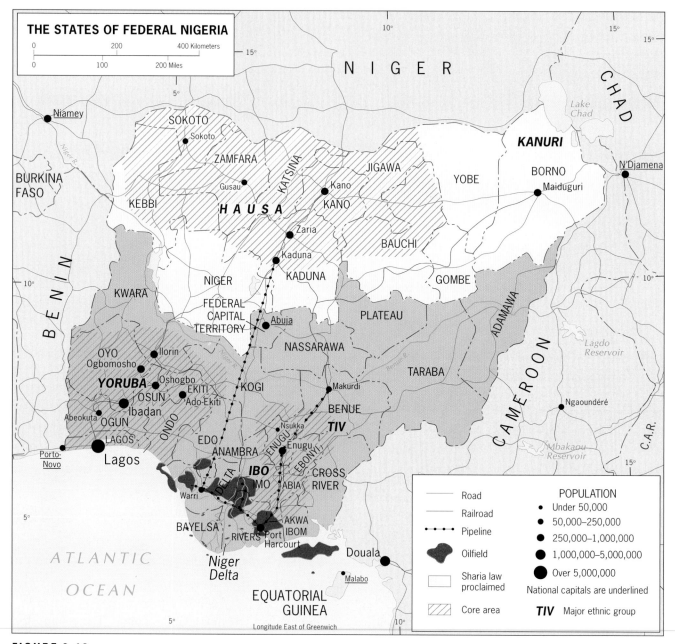

THE STATES OF FEDERAL NIGERIA

FIGURE 6-18

Western and Eastern Regions; the core of the Northern Region was situated between the upper Niger and the Benue. The departing colonial power, recognizing the new state's regional diversity, believed that a federal framework would help Nigeria's government to cope with the resulting stresses. Some progress had been made in the construction of road and railroad links across the "Middle Zone" between north and south, English had been introduced as the common tongue for the educated elite (it is still one of the four official languages today along with Yoruba, Ibo, and Hausa), a considerable civil service had been built up,

and in terms of education Nigeria was in a better position than most other African countries or colonies.

For reasons obvious from the map, Britain's colonial imprint always was stronger in the two southern regions than in the north. Christianity became the dominant faith in the south, and southerners, especially Yoruba, took a lead role in the transition from colony to independent state. The choice of Lagos, the port of the Western Region, as the federal capital (and not one of the cities in the more populous north) reflected British hopes for the country's future.

Nigeria is a large country territorially (about the size of Texas plus Oklahoma), and as Figure 6-17 shows, it extends farther northward than any of the other coastal West African states as far west as Guinea. This has the effect of incorporating much larger populations within its northern borders than its neighbors have. It was an objective of Nigeria's original three-region federal system that the two smaller southern regions would balance the single, larger one in the Islamic north.

But Nigeria's three-region federation did not last long. Initially, it was not the Northern Region but the Eastern Region that mounted a campaign to secede from the federation, although the north did play a role. Many Ibo had moved to northern cities to take jobs northerners were unwilling to do; some became successful and prominent entrepreneurs. These energetic Christians aroused animosities among the majority Muslims, and in September and October 1966 a series of massacres occurred. Surviving

Ibos fled from the north in huge numbers, and in May 1967 the government of the Eastern Region declared its independence as the Republic of Biafra. This led to a civil war that lasted three years and cost an estimated 1 million lives. To ensure that such a secession attempt would not happen again, successive Nigerian governments and regimes repeatedly modified the federal map. Today Nigeria has 36 States as well as a new Federal Capital Territory containing the more centrally located capital, Abuja (Fig. 6-18). But during most of its existence as an independent state, Nigeria has not functioned as a federation. Civilian government was repeatedly abrogated by military dictatorships that centralized power, enriched the rulers, and impoverished the country.

Large oilfields were discovered under the Niger Delta during the 1950s, when Nigeria's agricultural sector produced most of its exports (peanuts, palm oil, cocoa, cotton) and farming still had priority in national and State

AMONG THE REALM'S GREAT CITIES . . . **LAGOS**

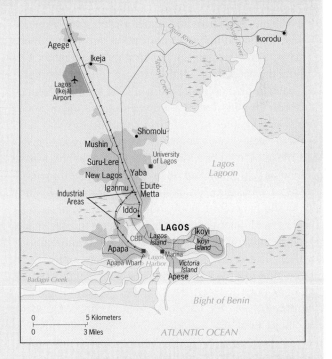

*I*n a realm that is only 30 percent urbanized, Lagos, former capital of federal Nigeria, is the exception: a teeming metropolis of 13.9 million sometimes called the Calcutta of Africa.

Lagos evolved over the past three centuries from a Yoruba fishing village, Portuguese slaving center, and British colonial headquarters into Nigeria's largest city, major port, leading industrial center, and first capital. Situated on the country's southwestern coast, it consists of a group of low-lying islands and sand spits between the swampy shoreline and Lagos Lagoon. The center of the city still lies on Lagos Island, where the highrises adjoining the Marina overlook Lagos Harbor and, across the water, Apapa Wharf and the Apapa industrial area. The city expanded southward onto Ikoyi Island and Victoria Island, but after the 1970s most urban sprawl took place to the north, on the western side of Lagos Lagoon.

Lagos's cityscape is a mixture of modern highrises, dilapidated residential areas, and squalid slums. From the top of a highrise one sees a seemingly endless vista of rusting corrugated roofs, the houses built of cement or mud in irregular blocks separated by narrow alleys. On the outskirts lie the shantytowns of the less fortunate, where shelters are made of plywood and cardboard and lack even the most basic facilities.

By world standards, Lagos ranks among the most severely polluted, congested, and disorderly cities. Misman-

agement and official corruption are endemic. Laws, rules, and regulations, from zoning to traffic, are flouted. The international airport is notorious for its inadequate security and for extortion by immigration and customs officers. In many ways, Lagos is a city out of control.

development plans. Soon, revenues from oil production dwarfed all other sources, bringing the country a brief period of prosperity and promise. But before long Nigeria's oil wealth brought more bust than boom. Misguided development plans now focused on grand, ill-founded industrial schemes and costly luxuries such as a national airline; the continuing mainstay of the vast majority of Nigerians, agriculture, fell into neglect. Worse, poor management, corruption, outright theft of oil revenues during military misrule, and excessive borrowing against future oil income led to economic disaster. The country's infrastructure collapsed. In the cities, basic services broke down. In the rural areas, clinics, schools, water supplies, and roads to markets crumbled. In the Niger Delta area, local people beneath whose land the oil was being exploited demanded a share of the revenues and reparations for ecological damage; the military regime under General Abacha responded by arresting and executing nine of their leaders. On global indices of national well-being, Nigeria sank to the lowest rungs even as its production ranked it as high as the world's tenth-largest oil producer, with the United States its chief customer.

In 1999, Nigeria's hopes were raised when, for the first time since 1983, a democratically elected president was sworn into office. But even as President Obasanjo confronted the country's tough realities—idle, dated factories; abandoned farms and plantations; countless unemployed; ingrained habits of corruption and nonpayment of taxes; a fast-growing population that now nears 130 million; the deepening AIDS crisis—a cultural imbroglio with catastrophic potential loomed. The return to democracy brought with it a loss of power among the northern elite that had long prevailed in the military, and now northern States, predominantly Muslim, began flexing their muscles in other ways. The State of Zamfara, whose politicians had fallen out of favor in the government of President Obasanjo, proclaimed that *Sharia* (Muslim) law would henceforth apply not only to matters such as marriage and inheritance, but to all aspects of life. Sharia criminal punishments are severe and include amputations for theft. But it also affects daily life: men and women cannot ride in taxis together, schools are segregated by sex, many jobs are reserved for males. Muslim leaders in Zamfara insisted that Sharia law would affect only Muslims, not Christians and others in the State, but this was not so. Christian women, for example, had to find public taxis driven by Christians, but most taxis are driven by Muslims who no longer pick up either Muslim or Christian women.

Several other northern States followed Zamfara in proclaiming Sharia law, despite the federal government's appeal to slow their initiatives. When Kaduna State imposed Sharia, riots between Muslims and Christians devastated the old capital city of Kaduna. There and elsewhere, the imposition of Sharia led to the departure of thousands of Christians, intensifying the cultural fault line that threatens the Nigerian federation. By the beginning of 2003, twelve northern States territorially comprising nearly half the country had proclaimed Sharia law (Fig. 6-18). Before long, the horrific punishments meted out to offenders were chronicled in the world's media, including the case of a young woman who had become pregnant and who was condemned to death by stoning following the birth of her child.

All this raises the prospect that Nigeria, West Africa's cornerstone and one of Africa's most important states, may succumb to devolutionary forces arising from its location in the African Transition Zone. This would be a calamitous development. A stable, well-governed, and economically growing Nigeria would be a beacon to region and realm; its collapse could infect Africa's political geography far and wide.

The Coast and Interior

Nigeria is one of 17 states (counting Chad and offshore Cape Verde [not shown in Fig. 6-17]) that constitute the region of West Africa. Four of these countries, comprising a huge territory on the Sahara's margins but containing small populations, are landlocked: Mali, Burkina Faso, Niger, and Chad. Figure G-8 shows clearly how steppe and desert conditions dominate the natural environments of these four interior states. Figure G-9 reveals the concentration of population in the steppe zone and along the ribbon of water the Niger River provides. Scattered oases form the remaining settlements and anchor regional trade.

But even the coastal states do not escape the dominance of the desert over West Africa. Mauritania's environment is almost entirely desert. Senegal, as Figure 6-17 shows, is a Sahel country; and not only northern Nigeria but also northern Benin, Togo, and Ghana have interior steppe zones. The loss of pastures to **desertification** (human-induced desert expansion) is a constant worry for the livestock herders there.

West Africa's states share the effects of the environmental zonation depicted in Figures G-7 and G-8, but they also have distinct regional geographies. **Benin**, Nigeria's neighbor, has a growing cultural and economic link with the Brazilian State of Bahía, where many of its people were taken in bondage and where elements of West African culture have survived. **Ghana**, once known as the Gold Coast, was the first West African state to achieve independence (1957), with a sound economy based on cocoa exports. Two grandiose postindependence

Ivory Coast's (*Côte d'Ivoire's*) first and longtime leader, Félix Houphouët-Boigny, always told his people, especially those who had moved to the capital or overseas, to remember their home villages and support them. He set an example by funding public facilities and modernizing infrastructure in his own village, Yamoussoukro. He invested heavily in the modern highway linking Yamoussoukro to Abidjan. But his most extravagant spending created this elaborate replica of St. Peter's in Rome. Even larger than the original and thus the largest church in Christendom, this basilica (built to the finest detail by Italian craftsmen using imported goods ranging from marble to faucets) was consecrated by Pope John Paul II in 1989. Critics argued that *Côte d'Ivoire's* francs should have been spent on other ventures, and Muslim northerners complained about this "cultural excess." Meanwhile the basilica towers over Yamoussoukro today, a nonfunctional memorial to the man who put his village on the map.

schemes can be seen on its map: the port of Tema, which was to serve a vast West African hinterland, and Lake Volta, which resulted from the region's largest dam project. When neither fulfilled expectations, Ghana's economy collapsed. In the 1990s, stable and democratic government (following a military regime) produced some recovery.

Ivory Coast (which, as we have stated, officially still goes by its French name, *Côte d'Ivoire*) translated three decades of autocratic but stable rule into economic progress that gave it lower-middle-income status based mainly on cocoa and coffee sales. Continued French involvement in the country's affairs contributed to this prosperity; the capital, Abidjan, reflected its comparative well-being. But in a familiar pattern, Ivory Coast's president-for-life first engineered the transfer of the capital to his home village, Yamoussoukro, and then spent tens of millions of dollars building a Roman Catholic basilica there to rival that of St. Peter's in Rome (photo above). It was dedicated just as the country's economy was slowing and its social conditions worsened.

By the late 1990s, Ivory Coast's economy had reverted to the low-income category (Fig. G-11). Worse, regional strife intensified during the run-up to a presidential election in 2000, when southern politicans objected to the ethnic origins and alleged Muslim sympaties of a northern candidate. In September 2002, a series of coordinated attacks by northern rebels failed to take the largest city and former capital, Abidjan, but succeeded in capturing major

northern centers. The attacks cost hundreds of lives and confirmed the worst fears of citizens in this long-stable and comparatively prosperous country, where immigrants have long contributed importantly to economy and society. At the turn of the century, the population of over 16 million included about 2 million Burkinabes (immigrants from Burkina Faso to the north), as many as 2 million Nigerians, about 1 million Malians, 500,000 Senegalese, and smaller groups of Ghanaians, Guineans, and Liberians. This cultural mix had been kept stable by strong central government and a growing economy, but when both these mainstays failed, disorder followed. This is a troubling development in a region where many countries confront similar stresses.

All of West Africa has been affected by what has happened in Liberia and Sierra Leone. **Liberia**, a country founded in 1822 by freed slaves who returned to Africa with the help of American colonization societies, was ruled by their descendants for more than six generations. Rubber plantations and iron mines made life comfortable for the "Americo-Liberians," but among the local peoples, resentment simmered. A military coup in 1980, in which the president was killed, was followed in 1989 by full-scale civil war that pitted ethnic groups against each other and drove hundreds of thousands of refugees into neighboring countries, including Ivory Coast, Guinea, and Sierra Leone. Monrovia, the capital (named after U.S. president James Monroe) was devastated; an estimated 230,000 people, almost 10 percent of the population, perished. In

1997, one of the rebel leaders, Charles Taylor, became president, but his presidency was tainted by allegations of brutality, and by mid-2003 the country had again disintegrated into warfare.

One of the countries affected by these events was **Sierra Leone**, Liberia's coastal neighbor, also founded as a haven for freed slaves, in this case by the British in 1787. Independent since 1961, Sierra Leone went the all-too-familiar route from self-governing Commonwealth member to republic to one-party state to military dictatorship. But in the 1990s a civil war brought untold horror to this small country even as refugees from Liberia's war were arriving. Rebels enabled by diamond sales fought supporters of the legitimate government in a struggle that devastated town and countryside alike and killed and mutilated tens of thousands. Eventually a combination of West African, United Nations, and British forces intervened in the conflict and resurrected a semblance of representative government, but not before Sierra Leone had sunk to dead last on the world's list of national well-being.

Senegal had the advantage of lying separated from Sierra Leone by Guinea, another Francophone country along the West African coast. Its Sahelian environmental problems notwithstanding, Senegal managed to convert its colonial advantage (its capital, Dakar, was the headquarters for France's West African empire) into lasting progress. By no means a rich country, Senegal depends for foreign income on farm exports (chiefly peanuts), fishing (the leading source), phosphate sales, and iron ore production. But its most valuable asset has been its democratic tradition, which has now lasted more than four decades. This has enabled Senegal to weather some storms, including a failed union with its English-speaking enclave, Gambia, and a secessionist movement in the southwestern Casamance District (Fig. 6-17).

Senegal is over 90 percent Muslim, but its population of 10.4 million is dominated by the Wolof, the largest ethnic group constituting about 37 percent of the total, the Fulani, and the Serer (17 percent each). The Wolof are concentrated in and around the capital and wield most of the power. Senegal's leaders have maintained close relationships with France, which is still Senegal's leading trading partner and financial supporter in times of need. Without oil, diamonds, or other lucrative income sources and with an overwhelmingly subsistence-farming population, Senegal nevertheless managed to achieve GNP levels that ranked among the region's highest (see Table G-1). Here is proof that reasonably representative government and stability are greater assets than gold, liquid or otherwise.

Populous, multicultural, divided West Africa remains a region of farmers and herders (the latter range over the savannas and steppes of the north) along a tier of fast-changing environments between ocean and desert. Local village markets drive the traditional economy, and **periodic markets**—not open every day but operating every three, four, or more days—ensure that all villages participate in the exchange network. Traditions of this kind endure here, even as the cities beckon the farmers and burst at the seams. The region's great challenges are economic survival and nation-building, constrained by a boundary framework that is as burdensome as any in Africa.

THE AFRICAN TRANSITION ZONE

The African Transition Zone is Subsaharan Africa's threshold to the world of Islam. As Figures 6-19 and 6-10 show, this region is unlike the four discussed previously. It encompasses some entire countries that are also part of the formal regions on the map (such as Senegal and Burkina Faso), it divides others into Muslim and non-Muslim sectors (such as Chad and Sudan), and, in Africa's Horn, it comprises countries that do not form parts of other regions (Eritrea, Djibouti, Somalia). As elsewhere in the world where geographic realms meet or overlap, complications mark the African Transition Zone. In some areas, the transition from Muslim to non-Muslim society is gradual, as it remains in the northern parts of Ivory Coast and Nigeria despite the recent cultural problems erupting in these important states. Elsewhere, the break is sharp, as it is in eastern Ethiopia where a traditional border between Christian and animist-African cultures and Muslim-African communities is sharp and can be seen in Figures 6-9 and 7-2. The southern border of the African Transition Zone, the religious frontier sometimes referred to as Africa's **Islamic Front**, is therefore neither static nor everywhere the same.

As we noted earlier in this chapter, Islam arrived across the Sahara by caravan and up the Nile by boat, thus reaching and converting the peoples of the interior steppes and savannas on the southern side of the desert. (This wide stretch of semiarid Africa is known today as the *Sahel*, an Arabic word meaning "border" or "margin.") Then came Europe's colonial powers, and the modern politico-geographical map of Subsaharan Africa took shape. Muslims and non-Muslims were thrown together in countries not of their making. Vigorous Christian proselytism slowed the march of Islam, resulting in the distribution shown in Figure 7-2. The north-to-south transition is especially clear in the heart of the continent: Libya is 97 percent Muslim, Chad 54, the Central African Republic 15, and The Congo 1.

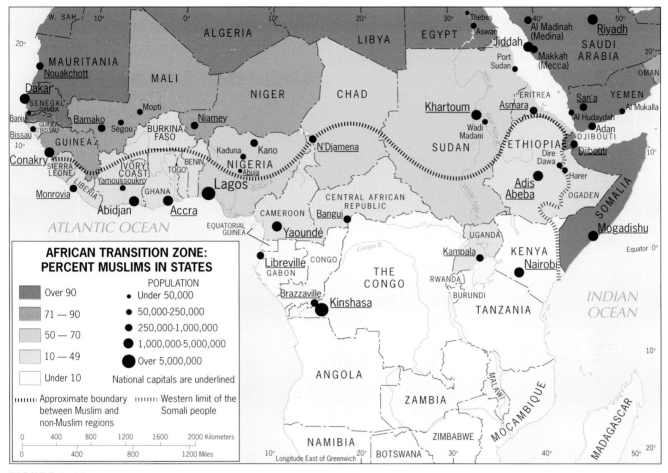

FIGURE 6-19

Conflict marks much of the Islamic Front in Africa: long-term in Sudan, where a 30-year war for independence by non-Islamic African rebels in the south against the Arabized, Muslim north has cost millions of lives; intermittent in Nigeria, where Islamic revivalism is currently clouding the country's prospects; and recent in Ivory Coast, where political rivalries spurred religious strife. Perhaps the most conflict-prone part of the African Transition Zone, however, lies in the east, involving not only Sudan but also the historic Christian state of Ethiopia, nearly encircled by dominantly Muslim societies.

The Horn of Africa (Fig. 6-20), therefore, is an especially volatile part of the African Transition Zone. Landlocked **Ethiopia**, with a Muslim population within its borders constituting 31 percent of the total of 71.1 million, in 1998 got involved with dominantly Muslim, coastal neighbor Eritrea in a costly and senseless border war over a few villages and a small strip of territory. Foreign mediation ended the conflict in 2000, but not before Ethiopia had spent nearly U.S. $3 billion on it even while it was de-

manding cancellation of a U.S. $430 million debt to international lending agencies. By 2003, their agriculture dislocated and their coffers bare, both **Eritrea** and Ethiopia needed massive food assistance to stave off starvation: 2.3 million of Eritrea's 4.8 million inhabitants were malnourished and 14 million Ethiopians faced hunger as well.

In 2002, a ministate in this part of the African Transition Zone, **Djibouti**, came to international attention as a result of the so-called War on Terror declared by the United States. Formerly a French colony and still host to a French military base, coastal Djibouti, ethnically divided between Afars and Issas but politically stable, proved to have a favorable relative location between troubled Somalia, vulnerable Eritrea, and crucial Ethiopia—and directly across the narrow Bab el Mandeb Strait (which separates the Red Sea and the Gulf of Adan) from Yemen, the ancestral home of Usama bin Laden.

But the country whose name is etched most vividly in American and Canadian minds is **Somalia**, where more than 8 million people, virtually all Muslim, live pastoral

FIGURE 6-20

and farming lives at the mercy of a desert-dominated climate. As Figures 6-19 and 6-9 show, Somali ethnic clans also inhabit most of adjoining eastern Ethiopia, so that the Somali "nation" exceeds 10 million. But "nation" is not a term that applies to Somalia, where five major ethnic groups and hundreds of clans are engaged in an unending contest for power as well as survival. In the early 1990s, media coverage of the latest of Somalia's death-dealing droughts led to U.S.-led United Nations intervention with the aim of food provision as well as nation-building, but the donors were immediately caught up in Somalia's relentless strife and a number of U.S. and other forces were killed, their bodies dragged through the streets of the capital, Mogadishu. Clan warfare destroyed what little infrastructure Somalia possessed, and the state essentially ceased to exist.

The fragmentation of Somalia, a merger of former British and Italian dependencies along the Horn's coast soon controlled by a harsh military dictatorship, actually began well before this debacle. In 1991, the former British sector declared its unilateral independence as the Republic of Somaliland (Fig. 6-20). In sharp contrast to the rest of the war-torn country, Somaliland has thrived; its capital, Hargeysa, bustles with commerce; and its political system, based on participation by elders of all clans, has a modicum of democracy. But the international community has not recognized the state. In 1998, another part of Somalia, named Puntland, also proclaimed its autonomy, although it left open the possibility of future participation in a federal Somalia. Meanwhile, the warring factions continue their struggles in the more populous south, vying for control of Mogadishu and the southern port of Kismaayo. And in the interior, the Gedo area has become a hotbed of Islamic extremism, a feature of the African Transition Zone that will be felt from the Bulge to the Horn.

Chapter 7

North Africa/Southwest Asia

FIGURE 7-1

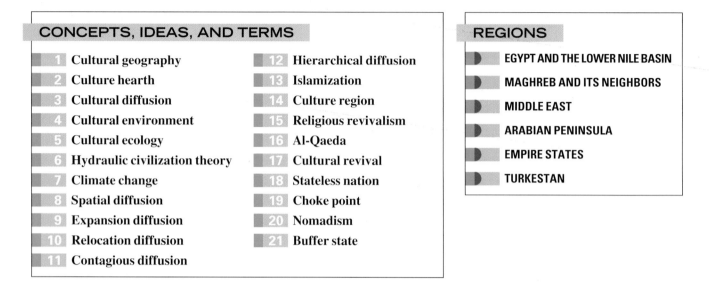

CONCEPTS, IDEAS, AND TERMS

1 Cultural geography
2 Culture hearth
3 Cultural diffusion
4 Cultural environment
5 Cultural ecology
6 Hydraulic civilization theory
7 Climate change
8 Spatial diffusion
9 Expansion diffusion
10 Relocation diffusion
11 Contagious diffusion
12 Hierarchical diffusion
13 Islamization
14 Culture region
15 Religious revivalism
16 Al-Qaeda
17 Cultural revival
18 Stateless nation
19 Choke point
20 Nomadism
21 Buffer state

REGIONS

▶ EGYPT AND THE LOWER NILE BASIN
▶ MAGHREB AND ITS NEIGHBORS
▶ MIDDLE EAST
▶ ARABIAN PENINSULA
▶ EMPIRE STATES
▶ TURKESTAN

$\mathcal{F}$ROM MOROCCO ON the shores of the Atlantic to the mountains of Afghanistan and from the Horn of Africa to the steppes of inner Asia lies a vast geographic realm of enormous cultural complexity. It stands at the crossroads where Europe, Asia, and Africa meet, and it is part of all three (Fig. 7-1). Throughout history, its influences have radiated to these continents and to practically every other part of the world as well. This is one of humankind's primary source areas. On the Mesopotamian Plain between the Tigris and Euphrates rivers (in modern-day Iraq) and on the banks of the Egyptian Nile arose several of the world's earliest civilizations. In its soils, plants were domesticated that are now grown from the Americas to Australia. Along its paths walked prophets whose religious teachings are still followed by hundreds of millions. And at the opening of the twenty-first century, the heart of this realm is beset by some of the most bitter and dangerous conflicts on Earth.

DEFINING THE REALM

IT IS TEMPTING to characterize this geographic realm in a few words and to stress one or more of its dominant features. It is, for instance, often called the "Dry World," containing as it does the vast Sahara as well as the Arabian Desert. But most of the realm's people live where there is water—in the Nile Delta, along the Mediterranean coastal strip (or *tell*) of northwesternmost Africa, along the Asian eastern and northeastern shores of the Mediterranean Sea, in the Tigris-Euphrates Basin, in far-flung desert oases, and along the lower mountain slopes of Iran south of the Caspian Sea and of Turkestan to the northeast. We know this world region as one where water is almost always at a premium, where peasants often struggle to make soil and moisture yield a small harvest, where nomadic peoples and their animals circulate across dust-blown flatlands, where oases are islands of sedentary farming and trade in a sea of aridity. But it also is the land of the Nile, the lifeline of Egypt, the crop-covered tell of northwestern Africa, the verdant coasts of Turkey, the meltwater-fed valleys of Central Asia.

Before we investigate this realm further, we need to look again at Figure G-8, the map of the world climates. Note the high degree of coincidence between the extent of the *B* (desert and steppe) climates and the limits of both the North African and Southwest Asian segments. In North Africa, the border between this realm and Sub-saharan Africa comes close to matching the southern limits of the *BSh* zone. In Southwest Asia, deserts, steppes, and mountain climates dominate. Except for the Mediterranean coasts, this is a realm of low and highly variable annual precipitation, of searing daytime heat and chilling nighttime cold, of strong winds and dust-laden air. Soils are thin; mountain slopes carry little vegetation. To all this, water brings exception—not only along coasts and rivers but also in oases and *qanats*, tunnels dug into water-bearing rock strata at an angle so that the water drains to

North Africa/Southwest Asia

1. North Africa and Southwest Asia were the scene of several of the world's great ancient civilizations, based in its river valleys and basins.

2. From this realm's culture hearths diffused ideas, innovations, and technologies that changed the world.

3. The North Africa/Southwest Asia realm is the source of three world religions: Judaism, Christianity, and Islam.

4. Islam, the last of the major religions to arise in this realm, transformed, unified, and energized a vast domain extending from Europe to Southeast Asia and from Russia to East Africa.

5. Drought and unreliable precipitation dominate natural environments in this realm. Population clusters exist where water supply is adequate to marginal.

6. Certain countries of this realm have enormous reserves of oil and natural gas, creating great wealth for some but doing little to raise the living standards of the majority.

7. The boundaries of the North Africa/Southwest Asia realm consist of volatile transition zones in several places in Africa and Asia.

8. Conflict over water sources and supplies is a constant threat in this realm, where population growth rates are high by world standards.

9. The Middle East, as a region, lies at the heart of this realm; and Israel lies at the center of the Middle East conflict.

10. Religious, ethnic, and cultural discord frequently cause instability and strife in this realm.

the surface. The population map (Fig. G-9) reminds us how scattered and isolated the population clusters are—as a matter of necessity.

An "Arab World"?

North Africa/Southwest Asia is also often referred to as the Arab World. This term implies a uniformity that does not actually exist. First, the name *Arab* is applied loosely to the peoples of this area who speak Arabic and related languages, but ethnologists normally restrict it to certain occupants of the Arabian Peninsula—the Arab "source." In any case, the Turks are not Arabs, and neither are most Iranians or Israelis. Moreover, although the Arabic language prevails from Mauritania in the west across all of North Africa to the Arabian Peninsula, Syria, and Iraq in the east, it is not spoken in other parts of this realm. In Turkey, for example, Turkish is the major language, and it has Ural-Altaic rather than Arabic's Semitic or Hamitic roots. The Iranian language belongs to the Indo-European linguistic family. Other "Arab World" languages that have separate ethnological identities are spoken by the Jews of Israel, the Tuareg people of the Sahara, the Berbers of northwestern Africa, and the peoples of the transition zone between North Africa and Subsaharan Africa to the south.

An "Islamic World"?

Yet another name given to this realm is the World of Islam. The prophet Muhammad (Mohammed) was born in Arabia in A.D. 571, and in the centuries after his death in 632, Islam spread into Africa, Asia, and Europe. This was the age of Arab conquest and expansion. Their armies penetrated southern Europe, their caravans crossed the deserts, and their ships plied the coasts of Asia and Africa. Along these routes they carried the Muslim (Islamic) faith, converting the ruling classes of the states of the West African savanna, threatening the Christian stronghold in the highlands of Ethiopia, penetrating the deserts of inner Asia, and pushing into India and even the island extremities of Southeast Asia. Islam was the religion of the marketplace, the bazaar, the caravan. Where necessary, it was imposed by the sword, and its protagonists aimed directly at the political leadership of the communities they entered. Today, the Islamic faith with its more than 1.3 billion followers extends well beyond the limits of the realm discussed here (Fig. 7-2). It is the major religion in northern Nigeria, in Pakistan, and in Indonesia; it is influential in East Africa; and it still prevails in its old strongholds of Albania, Kosovo, and Bosnia in Eastern Europe. Moreover, in South Asia approximately 150 million Muslims live in Hindu-dominated India, the world's largest religious minority.

On the other hand, the World of Islam is not entirely Muslim either. In Israel, Judaism is the prevailing faith; Christianity remains strong in Lebanon; and ancient Coptic Christian churches still exist in Egypt. Thus the connotation of World of Islam when applied to North Africa/Southwest Asia is far from satisfactory: the religion prevails far beyond these areas, and within the realm there is more religious diversity than it implies.

"Middle East"?

Finally, this realm is frequently called the Middle East. That must sound odd to someone in, say, India, who might think of a Middle West rather than a Middle East!

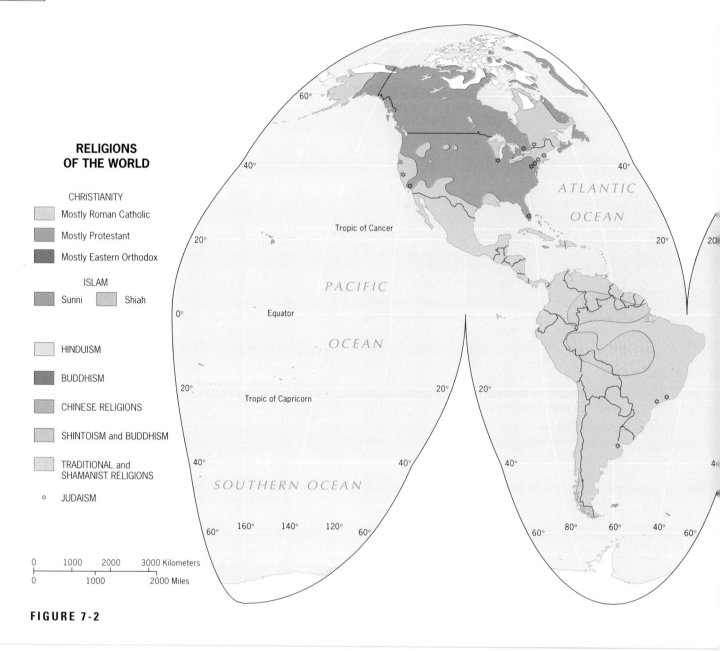

FIGURE 7-2

The name, of course, reflects the biases of its source: the "Western" world, which saw a "Near" East in Turkey, a "Middle" East in Egypt, Arabia, and Iraq, and a "Far" East in China and Japan. Still, the term has taken hold, and it can be seen and heard in everyday usage by scholars, journalists, and members of the United Nations. In view of the complexity of this realm, its transitional margins, and its far-flung areal components, the name *Middle East* need be faulted only for being imprecise—it does not make a single-factor region of North Africa/ Southwest Asia, as do the terms *Dry World*, *Arab World*, and *World of Islam*. In this chapter we do use this name—but only for one of its regions, not for the realm as a whole.

HEARTHS OF CULTURE

This geographic realm occupies a pivotal part of the world: here Eurasia, crucible of human cultures, meets Africa, source of humanity itself. A million years ago, the ancestors of our species walked from East Africa into North Africa and Arabia and spread from one end of Asia to the other. More than one hundred thousand years ago, *Homo sapiens* crossed these lands on the way to Europe, Australia, and, eventually, the Americas. Ten thousand years ago, human communities in what we now call the Middle East began to domesticate plants and animals, learned to irrigate their fields, enlarged their settlements into towns, and formed the earliest states. One thousand

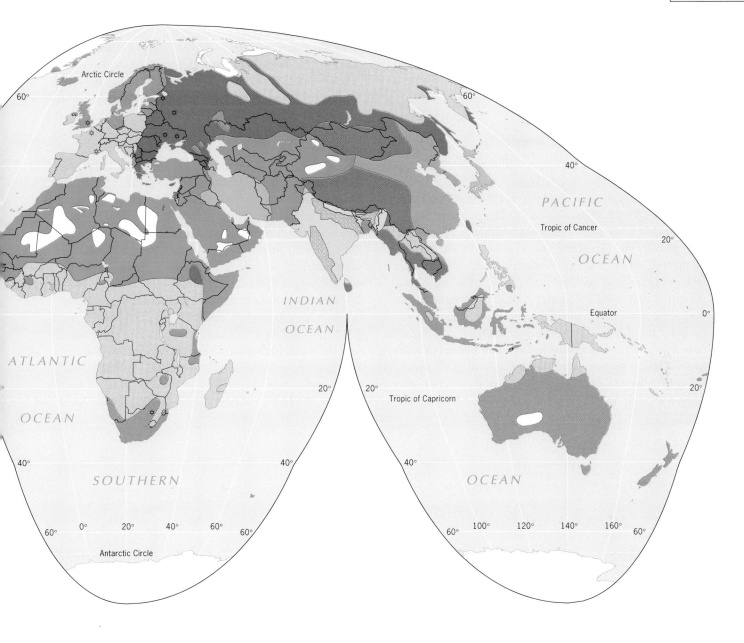

years ago, the heart of the realm was stirred and mobilized by the teachings of Muhammad and the Quran (Koran), and Islam was on the march from North Africa to India. Today this realm is a cauldron of religious and political activity, weakened by conflict but empowered by oil, plagued by poverty but fired by a wave of religious fundamentalism (Muslims prefer the term religious *revivalism*).

In the Introduction (pp. 14–16) we discussed the concept of culture and its regional expression in the cultural landscape. **Cultural geography**, we noted, is a wide-ranging and comprehensive field that studies spatial aspects of human cultures, focusing not only on cultural landscapes but also on **culture hearths**, the crucibles of civilization, sources of ideas, innovations, and ideologies

that changed regions and realms. Those ideas and innovations spread far and wide through a set of processes that we study under the rubric of **cultural diffusion**. Because we understand these processes better today, we can reconstruct ancient routes by which the knowledge and achievements of culture hearths spread (that is, diffused) to other areas. Another topic of cultural geography, also relevant in the context of the North African/Southwest Asian realm, is the **cultural environment** that a dominant culture creates. Human cultures exist in long-term accommodation with (and adaptation to) their natural environments, exploiting opportunities that these environments present and coping with the extremes they can impose. The study of the relationship between human societies and natural environments has become a separate branch of cultural geography called **cultural ecology**. As we will see, the North African/Southwest Asian realm presents many opportunities to investigate cultural geography in regional settings.

Mesopotamia and the Nile

In the basins of the major rivers of this realm (the Tigris and Euphrates of modern-day Turkey, Syria, and Iraq, and the Nile of Egypt) lay two of the world's earliest culture hearths (Fig. 7-3). Mesopotamia, "land amidst the rivers," had fertile alluvial soils, abundant sunshine, ample water, and animals and plants that could be domesticated. Here, in the Tigris-Euphrates lowland between the Persian Gulf and the uplands of present-day Turkey, arose one of humanity's first culture hearths, a cluster of communities that grew into larger societies and, eventually, into the world's first states. (Early state development probably was going on simultaneously in East Asia's river basins as well.) Mesopotamians were innovative farmers who knew when to sow and harvest crops, water their fields, and store their surplus. Their knowledge diffused to villages

near and far, and a *Fertile Crescent* evolved extending from Mesopotamia across southern Turkey into Syria and the Mediterranean coast beyond (Fig. 7-3).

Irrigation was the key to prosperity and power in Mesopotamia, and urbanization was its reward. Among many settlements in the Fertile Crescent, some thrived, grew, enlarged their hinterlands, and diversified socially and occupationally; others failed. What determined success? One theory, the **hydraulic civilization theory**, holds that cities that could control irrigated farming over large hinterlands held power over others, used food as a weapon, and thrived. One such city, Babylon on the Euphrates River, endured for nearly 4000 years (from 4100 B.C.). A busy port, its walled and fortified center endowed with temples, towers, and palaces, Babylon for a time was the world's largest city.

Egypt's cultural evolution may have started even earlier than Mesopotamia's, and its focus lay upstream from (south of) the Nile Delta and downstream from (north of) the first of the Nile's series of rapids, or cataracts (Fig. 7-3). This part of the Nile Valley is surrounded by inhospitable desert, and unlike Mesopotamia (which lay open to all comers), the Nile provided a natural fortress here. The an-

FIGURE 7-3

The ancient Romans transformed the North African countryside by building cities, laying out irrigation systems, and farming fields that had been pastures for the herds of nomads. Their structures, from colonnaded avenues to tiered amphitheaters and from cobbled roadways to covered aqueducts, were built so well that parts of them endure to this day. These are the surviving remnants of a covered aqueduct that carried water for many miles across the undulating landscape of northeastern Tunisia. A cross-section of the "tube" atop the columned structure can be seen at the upper left; the extent of the system can be gauged at the right. Roman engineers sloped the aqueduct in such a way that water could be drawn off to various fields along its course. Not only was the system functional; the structure itself was designed to convey beauty as well as strength and effectiveness. Arched openings between the columns allowed strong Mediterranean winds to pass through, reducing the danger of weakening and collapse. It worked for centuries.

cient Egyptians converted their security into progress. The Nile was their highway of trade and interaction; it also supported agriculture through irrigation. The Nile's cyclical ebb and flow was much more predictable than that of the Tigris-Euphrates river system. By the time Egypt finally fell victim to outside invaders (about 1700 B.C.), a full-scale urban civilization had emerged. Ancient Egypt's artist-engineers left a magnificent legacy of massive stone monuments, some of them containing treasure-filled crypts of god-kings called Pharaohs. These tombs have enabled archeologists to reconstruct the ancient history of this culture hearth.

To the east, separated from Mesopotamia by more than 1200 miles (1900 km) of mountain and desert, lies the Indus Valley (Fig. 7-3). By modern criteria, this eastern hearth lies outside the realm under discussion here; but in ancient times it had cultural and commercial ties with the Tigris-Euphrates region. Mesopotamian innovations reached the Indus region early, and eventually the cities of the Indus became power centers of a civilization that extended far into present-day northern India.

Today, the world continues to benefit from the accomplishments of the ancient Mesopotamians and Egyptians. They domesticated cereals (wheat, rye, barley), vegetables (peas, beans), fruits (grapes, apples, peaches), and many animals (horses, pigs, sheep). They also advanced the study of the calendar, mathematics, astronomy, government, engineering, metallurgy, and a host of other skills and technologies. In time, many of their innovations were adopted and then modified by other cultures in the Old World and eventually in the New World as well. Europe was the greatest beneficiary of these legacies of Mesopotamia and ancient Egypt, whose achievements constituted the foundations of "Western" civilization.

Decline and Decay

Many of the early cities of this realm's culture hearths are archeological curiosities today. In some places, new cities have been built on the sites of the old, but the great cultural traditions of this realm eventually went into deep decline after many centuries of continuity. Several of these ancient urban centers are located in what is now desert. If we presumed that they were not built in the middle of these drylands, we could postulate a hypothesis that also could form an alternative to the hydraulic civilization theory: that of climate change.

Climate change, not a monopoly over irrigation techniques, may have given certain cities in the ancient Fertile Crescent an advantage over others; and climate change, associated with shifting environmental zones after the last Pleistocene glacial retreat, may have destroyed the last of the old civilizations. Perhaps overpopulation and human destruction of the natural vegetation contributed to the process. Indeed, some cultural geographers suggest that the momentous innovations in agricultural planning and irrigation technology were not "taught" by the seasonal flooding of the rivers but were forced on the inhabitants as they tried to survive changing environmental conditions.

The scenario is not difficult to imagine. As outlying areas began to fall dry and farmlands were destroyed, people congregated in the already crowded river valleys—and made every effort to increase the productivity of the land that could still be watered. Eventually overpopulation, destruction of the watershed, and perhaps reduced rainfall in the rivers' headwater areas dealt the final blow. Towns were abandoned to the encroaching desert; irrigation canals filled with drifting sand; croplands dried up. Those who could migrated to areas that were still reputed to be productive. Others stayed, their numbers dwindling, increasingly reduced to subsistence.

As old societies disintegrated, power emerged elsewhere. First the Persians, then the Greeks, and later the Romans imposed their imperial designs on the tenuous lands and disconnected peoples of North Africa/Southwest Asia. Roman technicians converted North Africa's farmlands into irrigated plantations whose products went by the boatload to Roman Mediterranean shores (see photo, p. 325). Thousands of people were carried as slaves to the cities of the new conquerors. Egypt was quickly colonized, as was the area we now call the Middle East. One region that lay distant, and therefore remote from these invasions, was the Arabian Peninsula, where no major culture hearth or large cities had emerged and where the turmoil had not affected Arab settlements and nomadic routes.

STAGE FOR ISLAM

Islam today dominates the cultural geography of this realm, but long before Islam arose, other religions had emerged: Zoroastrianism in what is today Iran and Judaism (and later Christianity) in the area of modern Israel. However, the teachings of Zoroaster remained confined to Persia; Judaism was devastated by the Babylonians and later by the Romans; and Christianity became an alien faith as its center moved to Rome. Until the early seventh century A.D., none of the religions that had their origins in this realm became dominant within it.

Muhammad the Prophet

In a remote place on the Arabian Peninsula, where the foreign invasions of the Middle East had had little effect on the Arab communities, an event occurred early in the seventh century that was to change history and affect the destinies of people in many parts of the world. In a town called Mecca (Makkah), about 45 miles (70 km) from the Red Sea coast in the Jabal Mountains, a man named Muhammad in the year A.D. 611 began to receive revelations from Allah (God). Muhammad (571–632) was then in his early forties and had barely 20 years to live. Convinced after some initial self-doubt that he was indeed chosen to be a prophet, Muhammad committed his life to fulfilling the divine commands he believed he had received. Arab society was in social and cultural disarray, but Muhammad forcefully taught Allah's lessons and began to transform his culture. His personal power soon attracted enemies, and in 622 he fled from Mecca to the safer haven of Medina (Al Madinah), where he continued his work. This moment, the *hejira* ("migration"), marks the starting date of the Muslim era, Year 1 on Islam's calendar. Mecca, of course, later became Islam's holiest place.

The precepts of Islam in many ways constituted a revision and embellishment of Judaic and Christian beliefs and traditions. All of these faiths have but one god, who occasionally communicates with humankind through prophets; Islam acknowledges that Moses and Jesus were such prophets but considers Muhammad to be the final and greatest prophet. What is Earthly and worldly is profane; only Allah is pure. Allah's will is absolute; Allah is omnipotent and omniscient. All humans live in a world that Allah created for their use but only to await a final judgment day.

Islam brought to the Arab World not only the unifying religious faith it had lacked but also a new set of values, a new way of life, a new individual and collective dignity. Islam dictated observance of the Five Pillars: (1) repeated expressions of the basic creed, (2) the daily prayer, (3) a month each year of daytime fasting (Ramadan), (4) the giving of alms, and (5) at least one pilgrimage in each Muslim's lifetime to Mecca. And Islam prescribed and proscribed in other spheres of life as well. It forbade alcohol, smoking, and gambling. It tolerated polygamy, although it acknowledged the virtues of monogamy. Mosques appeared in Arab settlements, not only for the (Friday) sabbath prayer, but also as social gathering places to knit communities closer together. Mecca became the spiritual center for a divided, widely dispersed people for whom a collective focus was something new.

The Arab-Islamic Empire

Muhammad provided such a powerful stimulus that Arab society was mobilized almost overnight. The prophet died in 632, but his faith and fame spread like wildfire. Arab armies carrying the banner of Islam formed, invaded, conquered, and converted wherever they went. As Figure 7-4 shows, by A.D. 700 Islam had reached far into North Africa, into Transcaucasia, and into most of Southwest Asia. In the centuries that followed, it penetrated southern and Eastern Europe, Central Asia's Turkestan, West Africa, East Africa, and South and Southeast Asia, even reaching China by A.D. 1000.

At the heart of this religious realm lay an Arab Empire. Its original capital was at Medina in Arabia, but as it expanded, its headquarters were moved first to Damascus in Syria and later to Baghdad on the Tigris River. Meanwhile, Islam was carried ever farther outward by camel caravan and pilgrim, by sailor, scholar, and sultan. Here was a manifestation of the process of *spatial diffusion* (see box titled "Diffusion Processes"). And while Islam expanded, the Arab Empire that lay at its source matured and prospered. In architecture, mathematics,

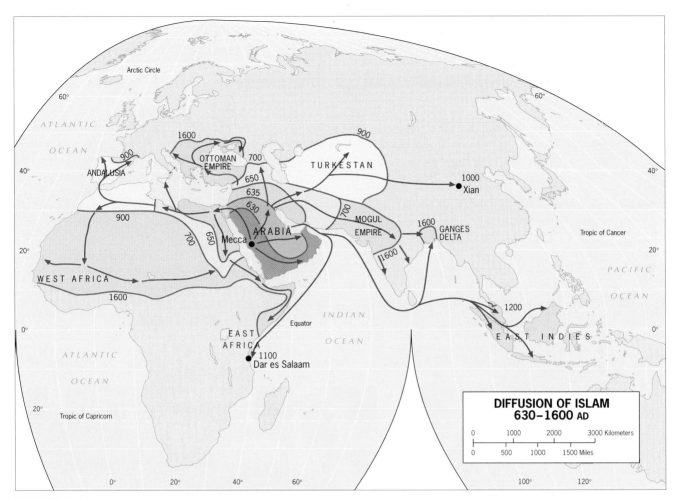

FIGURE 7-4

Diffusion Processes

*O*ne of the most interesting fields in geography studies how ideas, inventions, practices, and other phenomena spread through a population across space and time. A new musical style, a word or phrase, a new way of doing things, new fashions in clothing disseminate by a process called **spatial diffusion**. So do technological innovations, diseases, religious practices, and political ideas.

Understanding diffusion processes allows us to reconstruct the dispersal of cultural and technological ideas in the past. It also helps us predict such things as the impact of epidemics or the probability that an invention will find a market to justify the cost of producing it. Spatial diffusion has many practical applications.

In 1952, the Swedish geographer Torsten Hägerstrand published a fundamental study of spatial diffusion entitled *The Propagation of Diffusion Waves*. Diffusion takes place in two types of processes: **expansion diffusion**, when propagation waves originate in a strong and durable source area and spread outward, affecting an ever larger area and population; and **relocation diffusion**, in which an idea, or innovation (or, as with AIDS, a virus) is carried—usually by migrants—to a distant location and diffuses from there.

Islam, as Figure 7-4 shows, initially spread by rapid expansion diffusion from its western Arabian source. It affected virtually every village, town, and nomadic community through a form of expansion diffusion shaped by local proximity; geographers call this process **contagious diffusion**. But Islam was also disseminated through the conversion of kings, chiefs, and other high officials, who in turn propagated it through their bureaucracies. This is another form of expansion diffusion, called **hierarchical diffusion**, which has many variations; its main spatial expression is the downward filtering of whatever is being diffused from larger to smaller places within a national- or continental-scale urban hierarchy.

The map also shows that Islam spread far and wide by relocation diffusion, notably to the Ganges Delta and to the East Indies, where seafarers established remote new source areas. In East Africa Islam arrived through both expansion and relocation diffusion. Today the diffusion of Islam continues in many areas, including North America, where it arrived by relocation diffusion and now spreads by expansion diffusion.

The Flowering of Islamic Culture

The conversion of a vast realm to Islam was not only a religious conquest: it was also accompanied by a glorious explosion of Arab culture. In science, the arts, architecture, and other fields, Arab society far outshone European society. While the Western European remnants of the Roman Empire languished, Arab energies soared.

When the wave of Islamic diffusion reached the Maghreb (western North Africa), the Arabs saw, on the other side of the narrow Strait of Gibraltar, an Iberia ripe for conquest and ready for renewal. An Arab-Berber alliance, called the Moors, invaded Spain in 711 and controlled all but northern Castile and Catalonia before the end of the eighth century.

It took seven centuries for Catholic armies to recapture all of the Arabs' Iberian holdings, but by then the Muslims had made an indelible imprint on the Spanish and Portuguese cultural landscape. The Arabs brought unity and imposed the rule of Baghdad, and their works soon overshadowed what the Romans had wrought. *Al-Andalus*, as this westernmost outpost of Islam was called, was endowed with thousands of magnificent mosques, castles, schools, gardens, and public buildings. The ultimately victorious Christians destroyed most of the less durable art (pottery, textiles, furniture, sculpture) and burned the contents of Islamic libraries, but the great Islamic structures survived, including the Alhambra in Granada, the Giralda in Seville, and the Great Mosque of Córdoba, three of the world's greatest architectural achievements. While Spanish and Portuguese culture became Hispanic-Islamic culture, the Muslims were transforming their cities from Turkestan to the Maghreb in the image of Baghdad. The lost greatness of a past era still graces those townscapes today.

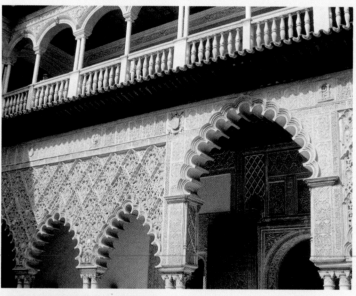

FROM THE FIELD NOTES

"Seville was a Muslim capital of Iberia, and walking the streets of the old city center is an adventure in cultural and historical geography. Seville fell to the Muslim invaders in 711, and the Muslims built a large mosque in the heart of town. Later the Christians, who ousted the Muslims in 1248, destroyed most of the mosque and built a huge cathedral on the site, but they saved the ornate minaret, made it the bell tower, and called it the *Giralda* (left). This minaret had been built by the Muslims to match the 220-foot (67-m) minaret of the Koutoubia Mosque in Marrakech (see page 346). But the finest Islamic legacy in Seville surely is the Alcazar Palace (above), begun in the late twelfth century and embellished and finished by the Christians after their victory. Its arched façades and intricately carved walls remain a monument to the Muslim architects and artists who designed and created them."

and science, the Arabs overshadowed their European contemporaries. The Arabs established institutions of higher learning in many cities including Baghdad, Cairo, and Toledo (Spain), and their distinctive cultural landscapes united their vast domain (see box titled "The Flowering of Islamic Culture"). Non-Arab societies in the path of the Muslim drive were not only Islamized, but also Arabized, adopting other Arab traditions as well. Islam had spawned a culture; it still lies at the heart of that culture today.

As we noted, Islam's expansion eventually was checked in Europe, Russia, and elsewhere. But a map showing the total area under Muslim sway in Eurasia and Africa reveals the enormous dimensions of the domain affected by **Islamization** at one time or another (Fig. 7-5). Islam continues to expand, now mainly by relocation diffusion. There are Islamic communities in cities as widely scattered as Vienna, Singapore, and Cape Town, Islam is also growing rapidly in the United States. With more than

1.3 billion adherents today, Islam is a vigorous and burgeoning cultural force around the world.

ISLAM DIVIDED

For all its vigor and success, Islam still fragmented into sects. The earliest and most consequential division arose after Muhammad's death. Who should be his legitimate successor? Some believed that only a blood relative should follow the prophet as leader of Islam. Others, a majority, felt that any devout follower of Muhammad was qualified. The first chosen successor was the father of Muhammad's wife (and thus not a blood relative). But this did not satisfy those who wanted to see a man named Ali, a cousin of Muhammad, made *caliph* (successor). When Ali's turn came, his followers, the Shi'ites, proclaimed that Muhammad finally had a legitimate successor. This offended the Sunnis, those who did not see a blood relationship as

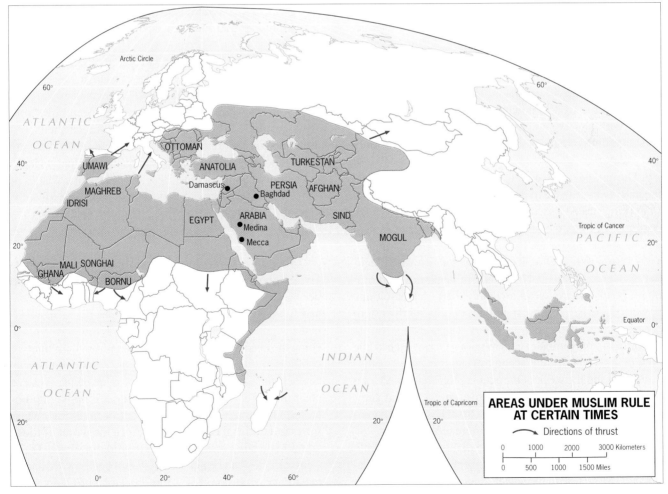

FIGURE 7-5

necessary for the succession. From the beginning of this disagreement, the numbers of Muslims who took the Sunni side far exceeded those who regarded themselves as Shiah (followers) of Ali. The great expansion of Islam was largely propelled by Sunnis; the Shi'ites survived as small minorities scattered throughout the realm. Today, about 85 percent of all Muslims are Sunnis.

But the Shi'ites vigorously promoted their version of the faith. In the early sixteenth century their work paid off: the royal house of Persia (modern-day Iran) made Shi'ism the only legal religion throughout its vast empire. That domain extended from Persia into lower Mesopotamia (modern Iraq), into Azerbaijan, and into western Afghanistan and Pakistan. As the map of religions (Fig. 7-2) shows, this created for Shi'ism a large **culture region** and gave the faith unprecedented strength. Iran remains the bastion of Shi'ism in the realm today, and the appeal of Shi'ism continues to radiate into neighboring countries and even farther afield.

The differences between Sunni and Shiah versions of Islam have profoundly affected the realm. Sunni Muslims believe in the effectiveness of family and community for solving life's problems, whereas Shi'ites consider their infallible *imams* (mosque officials who lead worshipers in prayer) the sole source of true knowledge. Sunnis tend to be comparatively reserved; Shi'ites often are passionate and emotional. For example, the death of Ali's son, Husayn, who would have been caliph and was killed by Sunnis, is commemorated annually with intense processions during which the marchers beat themselves with chains and cut themselves with sharp metal instruments. To many Sunnis, this behavior is unseemly and excessive.

During the late twentieth century, Shi'ism gained unprecedented influence in the realm. In its heartland, Iran, a *shah* (king) tried to secularize the country and to limit the power of the imams; he provoked a revolution that cost him the throne and made Iran an Islamic republic—in fact, a Shi'ite republic. Before long, Iran was at war with neighboring, Sunni-dominated Iraq, and Shi'ite parties and communities elsewhere were invigorated by the newfound power of Shi'ism. From Arabia to Africa's northwestern corner, Sunni-ruled countries warily watched their Shi'ite minorities, newly imbued with religious fervor. Mecca, the holy place for both Sunnis and Shi'ites, became a battleground during the week of the annual pilgrimage, and for a time the (Sunni) Saudi Arabian government denied entry to Shi'ite pilgrims. Recently, that schism has healed somewhat, but intra-Islamic sectarian differences run deep.

Smaller Islamic sects further diversify this realm's religious landscape. Some of these sects play a disproportionately large role in the societies and countries of which they are a part, as we will see in our regional discussion.

Religious Revivalism in the Realm

Another cause of intra-Islamic conflict lies in the resurgence of religious fundamentalism, or as Muslims refer to it, **religious revivalism**. In Shi'ite Iran, the imams wanted to reverse the shah's moves toward liberalization and secularization: they wanted to (and did) recast society in traditional, revivalist Islamic molds. An *ayatollah* (leader under Allah) replaced the shah in 1979; Islamic rules and punishments were instituted. Urban women, many of whom had been considerably liberated and educated during the shah's regime, resumed more traditional Islamic roles. Vestiges of Westernization, encouraged by the shah, disappeared. The war against Iraq (1980–1990) began as a conflict over territory but became a holy war that cost more than a million lives.

Islamic fundamentalism did not rise in Iran alone, nor was it confined to Shi'ite communities. Many Muslims—Sunnis as well as Shi'ites—in all parts of the realm disapproved of the erosion of traditional Islamic values, the corruption of society by European colonialists and later by Western modernizers, and the declining power of the faith in the secular state. As long as economic times were good, such dissatisfaction remained submerged. But when jobs were lost and incomes declined, a return to fundamental Islamic ways became more appealing.

This set Muslim against Muslim in all the regions of the realm (see the Issue Box entitled "Islam in the Twenty-First Century: Revive or Reform?"). Revivalists fired the faith with a new militancy, challenging the status quo from Afghanistan to Algeria. The militants forced their governments to ban "blasphemous" books, to resegregate the sexes in schools, to enforce traditional dress codes, to legitimize religious-political parties, and to heed the wishes of the *mullahs* (teachers of Islamic ways). Militant Muslims proclaimed that democracy inherited from colonialists and adopted by Arab nationalists was incompatible with the rules of the Quran.

This rift between moderate Muslims and militant revivalists began to spill over into other parts of the world decades ago, from Pakistan to the African Transition Zone and from the Caucasus to the Philippines. Even while Islamic countries such as Iran and Algeria struggled to overcome the instability and damage arising from their internal conflicts, militant activists planned and carried out attacks against the allies of the moderates: oil-guzzling Western countries guilty of military intervention, propping up unrepresentative regimes, mistreatment of immigrant Muslims, and cultural corruption. Before the worst of these attacks (through mid-2003) destroyed the World Trade Center in New York and severely damaged the Pentagon outside Washington, D.C., the French had foiled a similar assault on the Eiffel Tower in Paris, the Russians had

Islam in the Twenty-First Century: Revive or Reform?

ISLAMIC REVIVAL IS THE ONLY WAY

"WE MUSLIMS MUST return to, and protect, the fundamental tenets of our faith. Too many of us have strayed, lured and seduced by the vulgarities of Western and other foreign ways. Consider our history. The Prophet set us on course, the Quran was and is the source of all truth, and the message took root from Mecca to Morocco to Malaya. Don't forget: Allah's revelations came many centuries after other religions—Hinduism, Buddhism, Christianity—had captured the minds of millions. But many saw that Islam is the only way, and today practicing Muslims (which means every believer) outnumber Christians, millions of whom never see the inside of a church and do not live pious lives.

Regional ISSUE

"As a devout Muslim living in Egypt and having suffered the consequences of my devotion, let me say this: we Muslims have tried to live harmoniously with other faiths. We not only brought science and enlightenment to the world we converted, but we lived in peace with non-Arabs. In Islamic Andalusia, Jewish communities were safe and secure. It was the Christian Catholics who invented the horrors of the Inquisition. From West Africa to Central Asia, we proved our willingness to live in peace with infidels. So what did we get in return? The "three c's"—the crusades, colonialism, and commercial exploitation! From my American friends I understand that there's an expression, "three strikes and you're out." Well, you non-Muslims are out. And so are those Muslims who still cooperate with foreigners. Including the president of my own country and the entire Saudi royal family. I agree with Sheikh Hamoud: whoever supports the infidel against Muslims is himself an infidel. When you sell your soul for oil dollars or foreign aid, you're subject to a *fatwa*.

"Yes, I'm a revivalist, or what you people call a fundamentalist. Our salvation lies in a return to the strictest rules of Islam. Every Muslim must adhere to every tenet of the faith. Youngsters should learn the Quran by heart: it is the source of all truth beyond which nothing else matters. Women should know and keep their place as it is accorded to them by the Quran. Sharia law should be imposed by all Islamic countries, so that the punishment of transgressors will serve as an example to others. We should dress conservatively, resist the degenerate temptations of other cultures, and follow the teachings of our religious leaders. Let me ask you this: are we better off today than when Sheikh al-Wahhab and Ibn Saud were forging the Islamic state of Arabia? I would say not. We have to go back to our roots to undo the damage that has been done to us.

"Ayatollah Khomeini's Islamic revolution failed in Iran, and our Algerian brothers were prevented from making Algeria an Islamic state, but these were just harbingers of things to come. Our goal is Saudi Arabia, where it all started fourteen centuries ago—and where we will make the new start that will transform the world."

ISLAM NEEDS A REFORMATION

"How is it that we Muslims, with our rich cultural and scientific heritage, wealth of oil, history of interaction with other societies, and central location in the world are held back today by autocratic, corrupt, and incompetent governments, interfering and medieval clerics, stultifying traditions, and internal discord? As a practicing Muslim who is also a realist, I can tell you from my perspective as a Tunisian working in the tourist industry that we need Islamic reform, not the revivalism about which I keep hearing.

"Let's be objective about this. Look at the world's social and economic statistics, and you'll see that countries where Islam is the dominant religion tend to rank near the bottom—not always, but far too often for this to be a coincidence. The world has been swept by a wave of democratization, but not here. The globe is in the grip of a revolution in information and communication technology, but not here. The planet has a mosaic of religions, each professing to be the real path to salvation, and each bedeviled by a fundamentalist fringe, but not here. Here the fundamentalists are a constituency, not a fringe. And they are using tactics ranging from intimidation to terrorism to maintain and increase their clout.

"It all boils down to this: freedom. We need the freedom to debate the issues that confront us without fear of *fatwas* or imprisonment. In no other religious society can you be sentenced to death for expressing an opinion about God or scriptures. It shouldn't happen here. We need freedom of the press and other media, which are more restricted here than anywhere else on Earth. We need freedom of association without fear of persecution. But most of all, we need the right to debate the role of religion in our societies. There is nothing incompatible between Islamic societies and others around the world, but you cannot designate every non-Muslim as an infidel and then expect to have cordial and trusting relations.

"Islam needs a reformation, a reinterpretation of its inspired writings in light of modern times. And the mullahs need to get out of the business of controlling government, which is their constant aim. The separation of mosque and state is indispensable to our progress.

"Our religion-suffused educational system produces graduates who are unable to use what they learned to enter the globalizing job market. I can visualize those kids in the Taliban's *madrassas* doing only one thing, day after day—"studying" to recite the Quran by heart. Even where schools don't go to that extreme, they tend to produce unemployable youngsters whose joblessness and frustration cause them to turn to the mullahs—who are all too happy to direct their anger toward the West.

"And our reformation will have to address the plight of women. As a woman, it infuriates me that half of all women in the Arab world can neither read nor write, that they have unequal citizenship, and that they play virtually no role in government. Nowhere in the world is the domination of men more absolute, and I refuse to believe that the Prophet would approve of this. We need a reformation, not a so-called revival."

suffered hundreds of casualties in an apartment-building bombing in Moscow, and U.S. targets were hit in Lebanon, East Africa, Yemen, and elsewhere.

Although it has been argued that these attacks do not represent Islam as a religion in conflict with the West, they do have a religious context. During the 1990s, following the Cold War defeat of Soviet forces in Afghanistan and the subsequent abandonment of that country by the United States and its allies, an organization named **al-Qaeda** emerged, dedicated to punishing the perceived enemies of Islamic peoples by the most effective means at its disposal: terrorist attacks. However, the leader and chief financier of this movement, Usama bin Laden, had a bigger agenda. His ultimate goal was to overthrow the regime of rich princes ruling his home country, Saudi Arabia, accusing them of collusion with the infidel United States, of apostasy (faithlessness), and worse. Bin Laden wanted to make Saudi Arabia, birthplace of Muhammad and guarantor of holy Mecca, a true Islamic state.

Bin Laden was not the first Saudi to wish to return his country to its Islamic roots. During the eighteenth century an Islamic theologian named Muhammad ibn Abd al-Wahhab (1703–1792) founded a movement to bring Islamic society on the peninsula back to the most fundamental, traditional, and puritanical form of the faith. So strict were his teachings that he was expelled from his home town in 1744—a fateful exile because he moved to a place that happened to be the headquarters of Ibn Saud, ruler over a sizeable swath of the Arabian Peninsula. Ibn Saud formed an alliance with al-Wahhab, who had inherited a fortune when his wife died; between them, they set out on a campaign of conquest that made the Saudi dynasty the rulers of the region and *Wahhabism* the dominant form of Islam. Even the relatively liberal Ottoman sultans could not suppress either the Saudis or the Wahhabis, and in 1932, when the modern Kingdom of Saudi Arabia was established, Wahhabism became its official and dominant form of Islam.

Despite its reference to the name of the founder, Wahhabism today is a term used mostly by non-Muslims and outsiders. Followers of Wahhabi doctrines call themselves *Muwahhidun* or "Unitarians" to signify the strict and fundamental nature of their beliefs. These are the doctrines Usama bin Laden accuses the Saudi princes of having violated, and these are the harsh teachings so often cited in the Western media as emanating from the pulpits of Saudi Arabia's revivalist mosques.

As we will note in the regional section of this chapter, Wahhabism has established footholds in former So-

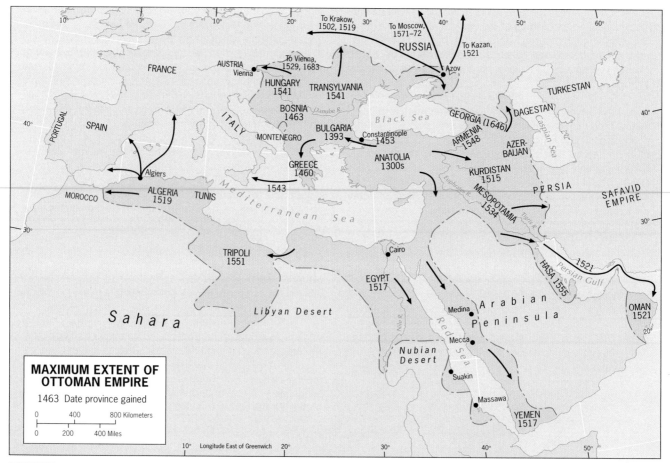

FIGURE 7-6

viet Central Asia despite its setback when Afghanistan's Taliban regime was ousted in 2001 as part of the U.S. response to the 9/11 attacks. As for Usama bin Laden who, like al-Wahhab, derived his wealth from family and who also sought to transform his country in a new political and religious image, his organization continues to threaten moderate governments within the Islamic world and to terrorize others beyond.

Islam and Other Religions

Two other major faiths had their sources in the *Levant* (the area extending from Greece eastward around the Mediterranean coast to northern Egypt), and both were older than Islam. Islam's rise submerged many smaller Jewish communities, but the Christians, not the Jews, waged centuries of holy war against the Muslims, seeking, through the Crusades, not only to drive Islam back but to reestablish Christian communities where they had dominated before the Islamic expansion. The aftermath of that campaign still marks the realm's cultural landscape today. A substantial Christian minority (about one-fourth of the population) remains in Lebanon, and Christian minorities also survive in Israel, Syria, Egypt, and Jordan. Strained relations between the long-dominant

Christian minority in Lebanon and the Muslim majority (itself divided into five sects) contributed to the disastrous armed struggle that engulfed this country in the 1970s and 1980s.

But the most intense conflict in modern times has pitted the Jewish state, Israel, against its Islamic neighbors near and far. Israel's United Nations-sponsored creation in 1948 precipitated more than a half-century of intermittent strife and attempts at mediation; it also caused friction among Islamic states in the region. Jerusalem—holy city for Judaism, Christianity, and Islam—lies in the crucible of this confrontation.

The Ottoman Aftermath

It is a geographic twist of fate that Islam's last great advance into Europe resulted in the European occupation of Islam's very heartland. The Ottomans (named after their leader, Osman I), based in what is today Turkey, conquered Constantinople (now Istanbul) in 1453 and pushed into Eastern Europe. Soon Ottoman forces were on the doorstep of Vienna; they also invaded Persia, Mesopotamia, and North Africa (Fig. 7-6). The Ottoman Empire under Suleyman the Magnificent, who ruled from 1522 to 1560, was the most powerful state in western

FIGURE 7-7

FIGURE 7-8

Eurasia. As Figure 7-6 shows, its armies even advanced toward Moscow, Kazan, and Krakow from their base at Azov (near present-day Rostov in Russia), and the Turks also launched marine attacks on Sicily, Spain, and France.

The Ottoman Empire survived for more than four centuries, but it lost territory as time went on, first to the Hungarians, then to the Russians, and later to the Greeks and Serbs until, after World War I, the European powers took over its provinces and made them colonies—colonies we now know by the names of Syria, Iraq, Lebanon, and Yemen (Fig. 7-7). As the map shows, the French and the British took large possessions; even the Italians annexed part of the Ottoman domain.

The boundary framework that the colonial powers created to delimit their holdings was not satisfactory. As Figure G-9 reminds us, this realm's population of more than 500 million is clustered, fragmented, and strung out in river valleys, coastal zones, and crowded oases. The colonial powers laid out long stretches of boundary as ruler-straight lines across uninhabited territory; they saw no need to adjust these boundaries to cultural or physical features in the landscape. Other boundaries, even some in desert zones, were poorly defined and never marked on the ground. One product of this process was Iraq, delimited by (then) British Colonial Secretary Winston Churchill in 1921 as a kingdom with Baghdad as its capital; another was Jordan, centered on Amman. Later, when the colonies had become independent states, such boundaries led to quarrels, even armed conflicts, among neighboring Muslim states.

THE POWER AND PERIL OF OIL

Travel through the cities and towns of North Africa and Southwest Asia, talk to students, shopkeepers, taxi drivers, and migrants, and you will hear the same refrain: "Leave us in peace, let us do things our traditional way. Our problems—with each other, with the world—seem always to result from outside interference. The stronger countries of the world exploit our weaknesses and magnify our quarrels. We want to be left alone."

That wish might be closer to fulfillment were it not for two relatively recent events: the creation of the state of Israel and the discovery of some of the world's largest oil reserves. We will discuss the founding of Israel in the

regional section of this chapter. Here we focus on the realm's most valuable export product: oil.

It is sometimes said that oil is the realm's most valuable resource. That is not accurate, however. Most of the people here remain farmers, so that the realm's most important resources are actually water and tillable soil. Oil is indeed in demand throughout the world. But there was a Muslim world long before the first barrel of oil was taken from the ground, and there are tens of millions of people whose lives, to this day, are only indirectly affected by oil revenues. For example, Syria and Jordan must import oil, Morocco has produced almost none, and Turkey also depends heavily on oil imports. And it would be a mistake to assume that the lives of all the citizens in the oil-rich countries have been transformed. There, too, the majority still live traditional lives as farmers, traders, and workers in other industries. Nearly everyone, of course, is affected by the fluctuating price of oil, which impacts national economies and influences incomes. The GDP of Saudi Arabia, for example, by 2002 had declined to only about one-third of what it was in just a decade ago, creating major economic and social problems for a Saudi regime already facing political and religious pressures.

Location and Dimensions of Known Reserves

About 30 of the world's countries have significant oil reserves. But five of these countries, all located in the North Africa/Southwest Asia realm, in combination possess larger reserves (ca. 77 percent of the world total) than the rest combined. The estimated reserves of this Big Five in 2003, averaged from several sources including the governments of the countries themselves and reported in billions of barrels, were as follows: (1) Saudi Arabia, 264; (2) Iraq, 117; (3) United Arab Emirates, 99; (4) Kuwait, 97; and (5) Iran, 90. The next-ranking country, Venezuela, had known reserves below 80 billion barrels, Russia below 60, the United States below 40, and all other countries combined below 30.

In general terms, oil (and associated natural gas) exists in this realm in three discontinuous zones (Fig. 7-8). The most productive of these zones extends from the southern and southeastern part of the Arabian Peninsula northwestward around the rim of the Persian Gulf, reaching into Iran and continuing northward into Iraq, Syria, and southeastern Turkey, where it peters out. The second zone lies across North Africa and extends from north-central Algeria eastward across northern Libya to Egypt's Sinai Peninsula, where it ends. The third zone begins on the margins of the realm in eastern Azerbaijan, continues eastward under the Caspian Sea into Turkmenistan and Kazakhstan, and also reaches into Uzbekistan, Tajikistan, Kyrgyzstan, and Afghanistan.

Saudi Arabia is the world's largest oil exporter, but in recent years Russia has risen to second place. As we noted in Chapter 2, oil and natural gas are by far Russia's most valuable commodities, and the Russian state direly needs the revenues they produce. Thus Russia, with modest (though expanding) known reserves, is vigorously exporting to international markets and is now a leading factor in the global energy picture. In combination, though, production by the countries of the North Africa/Southwest Asia realm far exceeds that from all other sources. The United States, not surprisingly, is the world's leading importer even while it consumes almost all of its own production.

As Figure G-11 indicates, the production and export of oil and gas has elevated several of this realm's countries into the higher-income categories. But petroleum wealth also has enmeshed these Islamic societies and their governments in global strategic affairs. When regional conflicts create instability in producing countries that have the potential to disrupt supply lines, powerful consumers are tempted to intervene—and have done so.

When the colonial powers laid down the boundaries that partitioned this realm among themselves, no one knew about the riches that lay beneath the ground. A few wells had been drilled, and production in Iran had begun as early as 1908 and in Egypt's Sinai Peninsula in 1913. But the major discoveries came later, in some cases after the colonial powers had already withdrawn. Some of the newly independent countries, such as Libya, Iraq, and Kuwait, found themselves with wealth undreamed of when the Turkish Ottoman Empire collapsed. As Figure 7-8 shows, however, others were less fortunate. A few countries had (and still have) potential. The smaller, weaker emirates and sheikdoms on the Arabian Peninsula always feared that powerful neighbors would try to annex them (Kuwait faced this prospect in 1990 when Iraq invaded it). The unevenly distributed oil wealth, therefore, created another source of division and distrust among Islamic neighbors.

A Foreign Invasion

The oil-rich countries of the realm found themselves with a coveted energy source, but without the skills, capital, or equipment to exploit it. These had to come from the Western world and entailed what many tradition-bound Muslims feared most: a strong foreign presence on Islamic soil, foreign intervention in political as well as economic affairs, and penetration of Islamic societies by the vulgarities of Western ways. Moreover, the advent of the oil industry created a veneer of modernity in the cultural landscape (where gleaming skyscrapers towered over ornate and historic minarets) as well as among the elites who benefited most directly from the energy riches. This set up tensions and even clashes between the modern and the traditional that continue to plague oil-rich Islamic countries today. It also created contrasts between countries that took different paths as their cultures changed. In the United Arab Emirates, it is no surprise to see a woman emerge from a supermarket, get into her car, and drive home with the groceries. That would be unthinkable in neighboring Saudi Arabia, where women are prohibited from driving cars.

In geographic terms, the impact of oil, its production and sale, in the exporting countries of this realm can be summarized as follows:

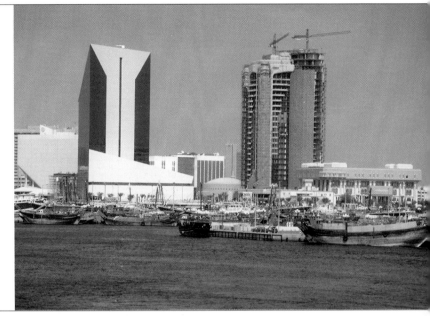

"The impact of oil wealth is nowhere better illustrated than it is along the waterfront of the famous Creek of Dubai (adjoining the city of the same name, largest in the United Arab Emirates). We stopped on the bridge that links the two sectors of Dubai to observe the old *dhows* (most of these wooden boats now with motors as well as triangular sail) still at their centuries-old moorings, overlooked not by traditional Arab buildings but by modern skyscrapers. Oil may have transformed the local economy, but don't count the role of the dhows out just yet. They still carry trade and contraband. We went to the dock and asked what was being transported. 'Jeans and cassettes to Iran,' we were told, 'and caviar and carpets from Iran. It's a good business.' So the traditional role of the dhows goes on, even in the age of oil."

1. *High Incomes.* When oil prices on international markets were high, several countries in this realm ranked as the highest-income societies in the world. Even when oil prices declined, virtually all the petroleum-exporting states remained in the upper-middle-income category (Fig. G-11).

2. *Modernization.* Notably in Saudi Arabia but also in Kuwait and the UAE, huge oil revenues have been used to modernize infrastructure. Futuristic city skyscrapers are only one manifestation of this; modern ports, airports, and superhighways are others.

3. *Industrialization.* Farsighted governments among those with oil wealth, realizing that petroleum reserves will not last forever, have invested some of their income in industrial plants that will outlast the era of oil. Petrochemical industries, plastics fabrication, and desalinization plants are among these facilities.

4. *Intra-Realm Migration.* The oil wealth has attracted millions of workers from less favored parts of the realm to work in the oilfields, in the ports, and in many other, mainly menial capacities. This has brought many Shi'ites to the countries of eastern Arabia; hundreds of thousands of Palestinians also work there as laborers. Saudi Arabia, with a population of 25 million, has more than 5 million foreign workers.

5. *Inter-Realm Migration.* The willingness of workers from such countries as Pakistan, India, and Sri Lanka to work for wages even lower than those the oil industry pays has attracted a substantial flow of temporary immigrants from outside the realm. These workers serve mostly as domestics, gardeners, refuse collectors, and the like.

6. *Regional Disparities.* Oil wealth and its manifestations in the cultural landscape create strong contrasts with areas not directly affected. The ultramodern east coast of Saudi Arabia is a world apart from large areas of its interior, where it becomes a land of desert, oasis, and camel, of vast distances, slow change, and isolated settlements. To some degree, this phenomenon affects all oil-rich countries.

7. *Foreign Investment.* Governments and Arab businesspeople have invested oil-generated wealth in foreign countries. These investments have created a network of international involvement that links many of this realm's countries not only to the economies of foreign states, but also to growing Arab (and thus Islamic) communities in those states.

The map (Fig. 7-8) contains a warning that came home to the oil-rich states (not only in this realm but elsewhere as well) in the 1980s when, after a period of high oil prices and huge revenues, oil prices fell and incomes plummeted. Even the power of an 11-member cartel, OPEC (Organization of Petroleum Exporting Countries), could not recover the lost advantage. Figure 7-8 shows a system of oil and gas pipelines that strongly resembles the exploitative interior-to-coast railroad lines (such as we noted for much of Subsaharan Africa) in a mineral-rich colony of the past. Such a pattern spells disadvantage for the exporter, whether colony or independent country.

Markets, not raw-material exporters, dominate international trade. Oil brought this realm into contact with the outside world in ways unforeseen just a century ago. Oil has strengthened and empowered some of its peoples; it has dislocated and imperiled others. It has truly been a double-edged sword.

REGIONS OF THE REALM

IDENTIFYING AND DELIMITING regions in this vast geographic realm is quite a challenge. Population clusters tend to be widely scattered in some areas, highly concentrated in others. Cultural transitions and landscapes—internal as well as peripheral—make it difficult to discern a regional framework. This, furthermore, is a highly changeable realm and always has been. Several centuries ago it extended into Eastern Europe; now it reaches into Central Asia, where an Islamic **cultural revival** (the regeneration of a long-dormant culture through internal renewal and external infusion) is under way.

The following are the regional components of this far-flung realm today (Fig. 7-9):

1. **Egypt and the Lower Nile Basin**. This region in many ways constitutes the heart of the realm as a whole. Egypt (together with Iran and Turkey) is one of the realm's three most populous countries. It is the historic focus of this part of the world and a major political and cultural force. It shares with its southern neighbor, Sudan, the waters of the lower Nile River.

2. **The Maghreb and Its Neighbors**. Western North Africa (the Maghreb) and the areas that border it also form a region, consisting of Algeria, Tunisia, and Morocco at the center and Libya, Chad, Niger, Mali, and Mauritania along the broad periphery. The last four of these countries also lie astride or adjacent to the broad transition zone where the Arab-Islamic realm of northern Africa merges into Subsaharan Africa (a region we discussed on pp. 314–317).

3. **The Middle East**. This region includes Israel, Jordan, Lebanon, Syria, and Iraq. In effect, it is the crescent-like zone of countries that extends from the eastern Mediterranean coast to the head of the Persian Gulf.

4. **The Arabian Peninsula**. Dominated by the large territory of Saudi Arabia, the Arabian Peninsula also includes the United Arab Emirates, Kuwait, Bahrain, Qatar, Oman, and Yemen. Here lies the source and focus of Islam, the holy city of Mecca; here, too, lie many of the world's greatest oil deposits.

5. **The Empire States**. We refer to this region as the Empire States because two of the realm's giants, both the centers of historic empires, dominate its geography. Turkey, heart of the Ottoman Empire, now is the realm's most secular state and aspires to joining the European Union. Shi'ite Iran, the core of the former Persian Empire, has become an Islamic Republic. To the north, Azerbaijan, once a part of the Persian Empire, lies in the turbulent, Muslim-infused Transcaucasian Transition Zone. To the south, the island of Cyprus is divided between Turkish and Greek spheres, the Turkish sector being another remnant of imperial times.

6. **Turkestan**. Turkish influence ranged far and wide in southwestern Asia, and following the Soviet collapse that influence proved to be durable and strong. In the five former Soviet Central Asian republics the strength and potency of Islam vary, and the new governments deal warily (sometimes forcefully) with Islamic revivalists. Russian influence continues, democratic institutions remain weak, and Western (chiefly American) involvement has increased since the War on Terror began and Afghanistan—which also forms

MAJOR CITIES OF THE REALM

City	Population* (in millions)
Algiers, Algeria	3.3
Almaty, Kazakhstan	1.1
Baghdad, Iraq	5.4
Beirut, Lebanon	2.3
Cairo, Egypt	10.1
Casablanca, Morocco	3.8
Damascus, Syria	2.4
Istanbul, Turkey	10.0
Jerusalem, Israel	0.7
Khartoum, Sudan	3.3
Riyadh, Saudi Arabia	5.6
Tehran, Iran	7.3
Tel Aviv, Israel	2.1
Toshkent, Uzbekistan	2.2
Tunis, Tunisia	2.1

*Based on 2004 estimates.

part of this region—became the major target of that campaign.

EGYPT AND THE LOWER NILE BASIN

Egypt occupies a pivotal location in the heart of a realm that extends over 6000 miles (9600 km) longitudinally and some 4000 miles (6400 km) latitudinally. At the northern end of the Nile and of the Red Sea, at the eastern end of the Mediterranean Sea, in the northeastern corner of Africa across from Turkey to the north and Saudi Arabia to the east, adjacent to Israel, to Islamic Sudan, and to militant Libya, Egypt lies in the crucible of this realm. Because it owns the Sinai Peninsula (recently lost to and regained from Israel), Egypt, alone among states on the African continent, has a foothold in Asia, a foothold that gives it a coast overlooking the strategic Gulf of Aqaba (the northeasternmost arm of the Red Sea). Egypt also controls the Suez Canal, vital link between the Indian and Atlantic oceans and lifeline of Europe. The capital, Cairo (El Qahira) is the realm's largest city and a leading center of Islamic civilization. We hardly need to further justify Egypt's designation (together with northern Sudan) as a discrete region.

Egypt's Nile is the aggregate of two great branches upstream: the White Nile, which originates in the streams that feed Lake Victoria in East Africa, and the Blue Nile, whose source lies in Lake Tana in Ethiopia's highlands. The two Niles converge at Khartoum in modern-day Sudan.

About 95 percent of Egypt's 74.1 million people live within a dozen miles (20 km) of the great river's banks or in its delta (Fig. 7-10). It has always been this way: the Nile rises and falls seasonally, watering and replenishing soils and crops on its banks.

The ancient Egyptians used *basin irrigation*, building fields with earthen ridges and trapping the floodwaters with their fertile silt, to grow their crops. That practice continued for thousands of years until, during the nineteenth century, the construction of permanent dams made it possible to irrigate Egypt's farmlands year round. These dams, with locks for navigation, controlled floods, expanded the country's cultivable area, and allowed the farmers to harvest more than one crop per year on the same field. In a single century, all of Egypt's farmland was brought under *perennial irrigation*.

The greatest of all Nile projects, the Aswan High Dam (which began operating in 1968), creates Lake Nasser, one of the world's largest artificial lakes (Fig. 7-10). As the map shows, the lake extends into Sudan, where 50,000 people had to be resettled to make way for it. The Aswan High Dam increased Egypt's irrigable land by nearly 50 percent and today provides the country with about 40 percent of its electricity. But, as is so often the case with megaprojects of this kind, the dam also produced serious problems. Snail-carried schistosomiasis and mosquito-transmitted malaria thrived in the dam's standing water, afflicting hundreds of thousands of people living nearby. By blocking most of the natural fertilizers in the Nile's annual floodwaters, the dam necessitated the widespread use of artificial fertilizers, very costly to small farmers and damaging to the natural environment. And the now fertilizer- and pesticide-laden Nile no longer supports the fish fauna offshore, reducing the catch and depriving coastal populations of badly needed proteins.

FROM THE FIELD NOTES

"On the eastern edge of central Cairo we saw what looked like a combination of miniature mosques and elaborate memorials. Here lie buried the rich and the prominent of times past in what locals call the 'City of the Dead.' But we found it to be anything but a dead part of the city. Many of the tombs here are so large and spacious that squatters have occupied them. Thus the City of the Dead is now an inhabited graveyard, home to at least one million people. The exact numbers are impossible to determine; indeed, whereas metropolitan Cairo in 2004 has an official population of just over 10 million, many knowledgeable observers believe that 16 million is closer to the mark."

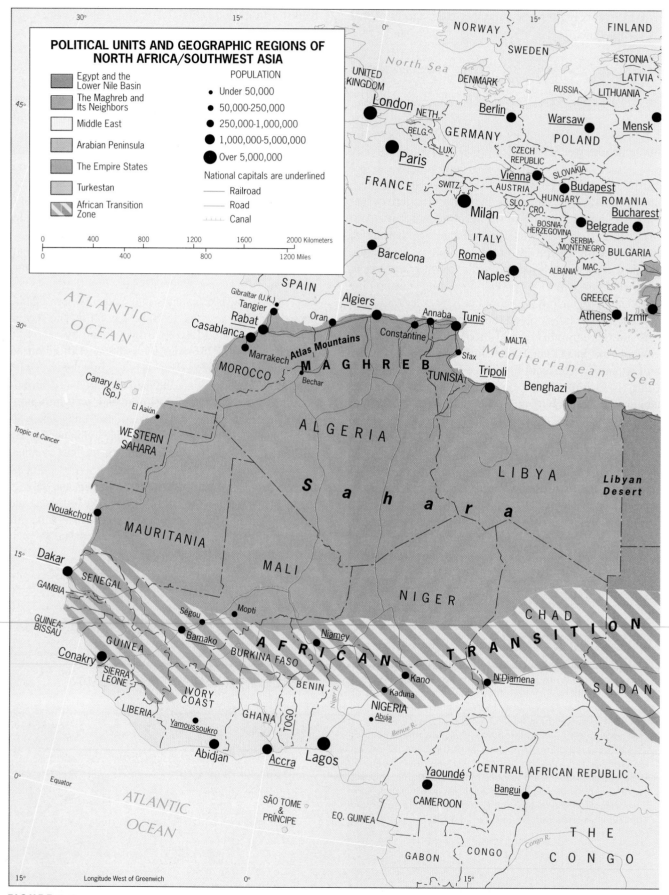

POLITICAL UNITS AND GEOGRAPHIC REGIONS OF NORTH AFRICA/SOUTHWEST ASIA

Egypt and the Lower Nile Basin
The Maghreb and Its Neighbors
Middle East
Arabian Peninsula
The Empire States
Turkestan
African Transition Zone

POPULATION
● Under 50,000
● 50,000-250,000
● 250,000-1,000,000
● 1,000,000-5,000,000
● Over 5,000,000

National capitals are underlined
Railroad
Road
Canal

FIGURE 7-9

FIGURE 7-10

Egypt's elongated oasis along the Nile, just 3 to 15 miles (5 to 25 km) wide, broadens north of Cairo across a delta anchored in the west by the great city of Alexandria and in the east by Port Said, gateway to the Suez Canal. The delta contains extensive farmlands, but it is a troubled area today. The ever more intensive use of the Nile's water and silt upstream is depriving the delta of much needed replenishment. And the low-lying delta is geologically subsiding, raising fears of salt-water invasion from the Mediterranean Sea that would damage soils here.

Egypt's millions of subsistence farmers, the *fellaheen*, still struggle to make their living off the land, as did the peasants of the Egypt of five millennia past. Rural landscapes seem barely to have changed; ancient tools are still used, and dwellings remain rudimentary. Poverty, disease, high infant mortality rates, and low incomes prevail. The Egyptian government is embarked on grandiose plans to expand its irrigated acreage, but without modernization and greater efficiency in the existing farmlands, the future is dim.

Egypt's Regional Geography

Egypt has six subregions, mapped in Figure 7-10. Most Egyptians live and work in Lower (i.e., northern) and Middle Egypt (regions ① and ②), the country's core area anchored by Cairo and flanked by the leading port and second industrial center, Alexandria. The economy has benefited from further oil discoveries in the Sinai (region ⑥) and in the Western Desert (region ④), so that Egypt now is self-sufficient and even exports some petroleum. Cotton and textiles are the other major source of external income, but the important tourist industry has repeatedly been hurt by Islamic extremists. As the population mushrooms, the gap between food supply and demand widens,

AMONG THE REALM'S GREAT CITIES . . . **CAIRO**

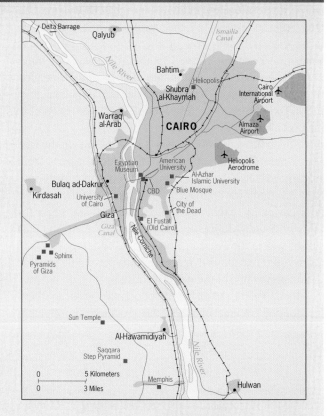

Stand on the roof of one of the highrise hotels in the heart of Cairo, and in the distance you can see the great pyramids, monumental proof of the longevity of human settlement in this area. But the present city was not founded until Muslim Arabs chose the site as the center of their new empire in A.D. 969. Cairo became and remains Egypt's primate city, situated where the Nile River opens into its delta, home today to almost one-seventh of the entire country's population.

Cairo at 10.1 million ranks among the world's 20 largest urban agglomerations, and it shares with other cities of the poorer world the staggering problems of crowding, inadequate sanitation, crumbling infrastructure, and substandard housing. But even among such cities, Cairo is noteworthy for its stunning social contrasts. Along the Nile waterfront, elegant skyscrapers rise above carefully manicured, Parisian-looking surroundings. But look eastward, and the urban landscape extends gray, dusty, almost featureless as far as the eye can see. Not far away, more than a million squatters live in the sprawling cemetery known as the City of the Dead (see photo, p. 339). On the outskirts, millions more survive in overcrowded shantytowns of mud huts and hovels.

And yet Cairo is the dominant city not only of Egypt but for a wider sphere; it is the cultural capital of the Arab World, with centers of higher learning, splendid museums, world-class theater and music, and magnificent mosques and Islamic learning centers. Although Cairo always has been primarily a center of government, administration, and religion, it also is a river port and an industrial complex, a commercial center, and, as it sometimes seems, one giant bazaar. Cairo is the heart of the Arab World, a creation of its geography, and a repository of its history.

and Egypt must import grain. Since the late 1970s, Egypt has been a major recipient of U.S. foreign aid.

Egypt today is at a crossroads in more ways than one. Its planners know that reducing the high birth rate would improve the demographic situation, but revivalist Muslims object to any programs that promote family planning. Its accommodation with Israel helps ensure foreign aid but divides the people. Its government faces a fundamentalist challenge. Egypt's future, in this crucial corner of the realm, is uncertain.

Divided Sudan

As Figure 7-10 shows, Egypt is flanked by two countries that have posed challenges to its leadership: Sudan to the south and Libya to the west. Sudan, more than twice as large as Egypt and with nearly 35 million people, lies centered on the confluence of the White Nile (from Uganda) and the Blue Nile (from Ethiopia). Here the twin capital, Khartoum-Omdurman, anchors a large agricultural area where cotton was planted during colonial times. The British administration combined northern Sudan, which was Arabized and Islamized, with a large area to the south, which was African and where many villagers had been Christianized. After the British left, the regime in Khartoum wanted to impose its Islamic rule on this portion of the African Transition Zone that formed southern Sudan, and a bitter civil war ensued. The cost in human lives and dislocation over the better part of the past three decades is incalculable; in 2003 an estimated 4 million people remained refugees in their own land.

Sudan has a 300-mile (500-km) coastline on the Red Sea, where Port Sudan lies almost directly across from Jiddah and Makkah (Mecca) in Saudi Arabia.* The country's economy for many years was typical of the energy-poor periphery, exchanging sheep, cotton, and sugar for oil, with Saudi Arabia the main trading partner. The pursuit of war in the African Transition Zone impoverished the Islamic regime in Khartoum, and per-capita income in Sudan was one of the world's lowest.

In the 1990s, the discovery of oil reserves in southern Kordofan Province changed the situation significantly

(Fig. 7-10). As the map shows, these reserves lie close to the border with Bahr el Ghazal and Upper Nile Provinces, where rebels have the upper hand. Exploitation of these reserves, involving foreign companies, was accompanied by reports that the Sudanese army drove hundreds of thousands of villagers in the oil-rich areas from their land, killing those who resisted and burning their villages to the ground. Outraged protests in Canada persuaded a Canadian oil company to withdraw from the scene, but otherwise the world paid little heed.

Now oil that once flows in the pipeline from Port Sudan to the capital flows the other way. By 2002 the Khartoum regime was earning U.S. $1 million per day from oil revenues, allowing it to double its military expenditures and giving it renewed advantage over the African rebels in the south. Meanwhile, the civil war and a searing drought combined to put the lives of another three million people at risk. Sudan's drive to Islamize the South and the South's determination not to be Islamized or Arabized is one of Africa's most intractable and underreported problems.

THE MAGHREB AND ITS NEIGHBORS

The countries of northwestern Africa are collectively called the *Maghreb*, but the Arab name for them is more elaborate than that: *Djezira-al-Maghreb*, or "Isle of the West," in recognition of the great Atlas Mountain range rising like a huge island from the Mediterranean Sea to the north and the sandy flatlands of the immense Sahara to the south.

The countries of the Maghreb (sometimes spelled *Maghrib*) are Morocco, last of the North African kingdoms; Algeria, a secular republic beset by the religious-political problems we noted earlier; and Tunisia, smallest and most Westernized of the three (Fig. 7-11). Libya, facing the Mediterranean between the Maghreb and Egypt, is unlike any other North African country: an oil-rich desert state whose population is almost entirely clustered in settlements along the coast.

Whereas Egypt is the gift of the Nile, the Atlas Mountains form the nucleus of the settled Maghreb. These high ranges wrest from the rising air enough orographic rainfall to sustain life in the intervening valleys, where good soils support productive farming. From the vicinity of Algiers eastward along the coast into Tunisia, annual rainfall averages more than 30 inches (75 cm), a total more than three times as high as that recorded for Alexandria in Egypt's delta. Even 150 miles (240 km) inland, the slopes of the Atlas still receive over 10 inches (25 cm) of rainfall. The effect of the topography can be read on the world map of

---

*Although the legal boundary between Egypt and Sudan during the period of British power in this area coincided with the 22nd parallel, the Halaib Triangle, facing the Red Sea north of this line and a part of Egypt, was administered from Khartoum, not from Cairo. After Sudan became independent in 1956, this arrangement continued informally until 2003, when the two governments agreed to recognize the 22nd parallel as their *de facto* as well as their legal boundary. This development occurred too late to be incorporated into the maps for the present edition.

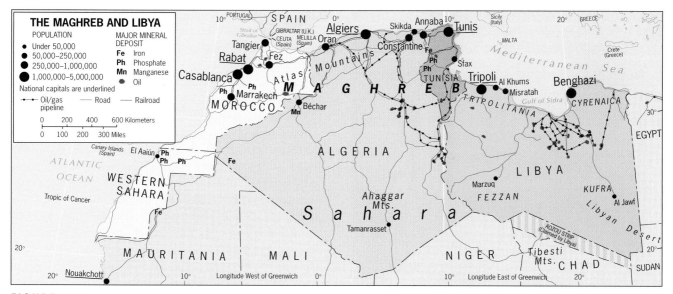

FIGURE 7-11

precipitation (Fig. G-7): where the highlands of the Atlas terminate, desert conditions immediately begin.

The Atlas Mountains are structurally an extension of the Alpine system that forms the orogenic backbone of Europe, of which the Alps and Italy's Appennines are also parts. In northwestern Africa, these mountains trend southwest-northeast and begin in Morocco as the High Atlas, with elevations close to 13,000 feet (4000 m). Eastward, two major ranges dominate the landscapes of Algeria proper: the Tell Atlas to the north, facing the Mediterranean, and the Saharan Atlas to the south, overlooking the great desert. Between these two mountain chains, each consisting of several parallel ranges and foothills, lies a series of intermontane basins (analogous to South America's Andean *altiplanos* but at lower elevations), markedly drier than the northward-facing slopes of the Tell Atlas. In these valleys, the rain shadow effect of the Tell Atlas is reflected not only in the steppe-like natural vegetation but also in land-use patterns: pastoralism replaces cultivation, and stands of short grass and bushes blanket the countryside.

During the colonial era, which began in Algeria in 1830 and lasted until the early 1960s, well over a million Europeans came to settle in North Africa—most of them French, and a large majority bound for Algeria—and these immigrants soon dominated commercial life. They stimulated the renewed growth of the region's towns; Casablanca, Algiers, Oran, and Tunis became the urban foci of the colonized territories. Although the Europeans dominated trade and commerce and integrated the North African countries with France and the European Mediterranean world, they did not confine themselves to the cities and towns. They recognized the agricultural possi-

bilities of the favored parts of the tell (the lower Tell Atlas slopes and narrow coastal plains that face the Mediterranean) and established thriving farms. Not surprisingly, agriculture here is Mediterranean. Algeria soon became known for its vineyards and wines, citrus groves, and dates; Tunisia has long been one of the world's leading exporters of olive oil; and Moroccan oranges went to many European markets.

Between desert and sea, the Maghreb states display considerable geographic diversity. **Morocco**, a conservative kingdom in a revolutionary region, is tradition-bound, weak economically, and has only recently found an oilfield within its borders. Its core area lies in the north, anchored by four major cities; but the Moroccans' attention is focused on the south, where the government is seeking to absorb its neighbor, Western Sahara (a former Spanish dependency with about 300,000 inhabitants, many of them immigrants from Morocco). Even if this campaign is successful, it will do little to improve the lives of most of Morocco's 30.8 million people. Hundreds of thousands have emigrated to Europe, many of them via Spain's two tiny exclaves on the Moroccan coast, Ceuta and Melilla (Fig. 7-11).

Algeria, rich in oil and natural gas, fought a bitter war of liberation against the French colonizers and has been in turmoil virtually ever since, with devastating economic and social consequences. Algiers, Algeria's primate city, is centrally situated along its Mediterranean coast and is home to more than 3 million of the country's 32.5 million citizens—but there are more Algerians in France than in the capital today. Conflict between those wanting to make Algeria an Islamic republic and the military-backed

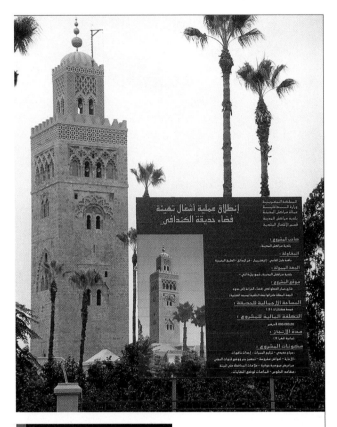

FROM THE FIELD NOTES

"Having seen the *Giralda* in Seville, Spain (photo, p. 328), I was interested to view the tower of the Koutoubia Mosque in Marrakech, Morocco, also built in the twelfth century and closely resembling its Andalusian counterpart. Some 220 feet (67 m) tall, the Koutoubia minaret was built by Spanish slaves and is a monument to the heyday of Islamic culture of the time. I soon discovered that it is not as easy as it looks to get a clear view of the tower, and a boy offered his help, guiding me around the block on which the mosque is situated. He was well informed on the building's history and on the varying fortunes of his home town. It was European mispronunciation of the name 'Marrakech' that led to Morocco being called Morocco, he reported. And Marrakech was Morocco's capital for a very long time, and still should be. 'But back to the mosque,' I said. 'What does the sign say?' He smiled. 'You'll just have to learn some Arabic, just as we know French,' he said. Obviously a teacher in the making, I thought."

regime has cost an estimated 100,000 lives; today, ethnic strife also threatens in the Kabylia region east of Algiers involving Algeria's Berber minority.

As Table G-1 shows, the highest GNP per capita in the Maghreb is earned in **Tunisia**, smallest of the North African countries territorially but with a sizeable population of 10 million. Successive governments have severely repressed Muslim radicals, and comparative stability has allowed the relatively diversified economy, ranging from textiles and farm products to mineral production and tourism, to expand and prosper. Tunis, the historic primate city in the northeastern corner, dominates this most urbanized of the Maghreb states.

Almost rectangular in shape, **Libya** is a country whose four corners matter most (Fig. 7-11). What limited agricultural possibilities exist lie in the northwest in Tripolitania, centered on the capital, Tripoli, and in the northeast in Cyrenaica, where Benghazi is the urban focus. Between these two coastal clusters, which are home to 90 percent of the Libyans, lies the Gulf of Sidra, a deep Mediterranean bay. Libya (5.7 million) has claimed this gulf as its sea, but other countries have not accepted that claim. Libya's two southern corners are the desert Fezzan, a mountainous area near the southwestern border with Algeria and Niger, and the sparsely populated Kufra oasis in the southeast. In recent years, Libya's ruler, Muammar Gadhafi, buoyed by increased oil revenues, has sought to strengthen his country's influence in Africa. This turnabout appeared to run counter to public opinion in his country, where in 2000 Libyans in large numbers attacked African workers, killing hundreds and driving tens of thousands across the border. A few months later, at the 2001 meeting of the Arab League in Amman, Jordan, Gadhafi urged Arabs to "forget about Jerusalem and join Africa." In South Africa, former president Nelson Mandela and incumbent Thabo Mbeki publicly expressed support for the Libyan dictator, whose African initiative is part of a campaign to reposition himself and his country in the international community in the past-9/11 era.

As Figure 7-11 shows, the Maghreb countries and Libya adjoin several desert-dominated states in the African Transition Zone. These are African countries with strong Islamic imprints, but as Figure 6-19 shows, they lie north of the Islamic Front with only one exception, Chad. Vast and sparsely populated **Mauritania** is the most Islamized of these countries, with about half its population of 2.7 million concentrated in and near the capital, Nouakchott, base of a small fishing fleet. Neighboring **Mali** depends on the waters of the upper Niger River, on which its capital of Bamako lies. A multiparty, democratic state, Mali is also multicultural, with a sizeable minority adhering to traditional beliefs and a small Christian minority. To the east of Mali lies **Niger**, which shares boundaries with Algeria and Libya; like Mali, this is one of the world's least urbanized countries (17 percent). Unlike Mali, however, Niger has only a short stretch of the middle course of the Niger River, where the capital, Niamey, is situated. And **Chad**, directly south of Libya, has the lowest percentage of Muslim inhabitants but the strongest division between Islamized north and animist/Christian south. As Figure 6-19 reminds us, the capital, N'Djamena, lies on the southern edge of the African Transition Zone, directly on the Islamic Front.

The regional term *Middle East*, we noted earlier, is not satisfactory, but it is so common and generally used that avoiding it creates more problems than it solves. It originated when Europe was the world's dominant realm and when places were "near," "middle," and "far" from Europe: hence a Near East (Turkey), a Far East (China, Japan, Korea, and other countries of East Asia), and a Middle East (Egypt, Arabia, Iraq). If you check definitions used in the past, you will see that the terms were applied inconsistently: Syria, Lebanon, Palestine, even Jordan sometimes were included in the "Near" East, and Persia and Afghanistan in the "Middle" East.

Today, the geographic designation *Middle East* has a more specific meaning. And at least half of it has merit: this region, more than any other, lies at the middle of the vast Islamic realm (Fig. 7-9). To the north and east of it, respectively, lie Turkey and Iran, with Muslim Turkestan beyond the latter. To the south lies the Arabian Peninsula. And to the west lie the Mediterranean Sea and Egypt, and the rest of North Africa. This, then, is the pivotal region of the realm, the very heart of it.

Five countries form the Middle East (Fig. 7-12): Iraq, largest in population and territorial size, facing the Persian Gulf; Syria, next in both categories and fronting the Mediterranean; Jordan, linked by the narrow Gulf of Aqaba to the Red Sea; Lebanon, whose survival as a unified state has come into question; and Israel, Jewish nation in the crucible of the Muslim world. Because of the extraordinary importance of this region in world affairs, we focus in some detail on issues of cultural, economic, and political geography in the following discussion.

Iraq

In the aftermath of the 2001 terrorist attacks in the United States, the U.S. Government declared war on terrorism. The first regional target of this war was Muslim-fundamentalist, Taliban-ruled Afghanistan, base of al-Qaeda and home to Usama bin Laden, architect of the 9/11 assault. Early in 2002, however, U.S. President George W. Bush, in his State of the Union address before Congress, declared Iraq, Iran, and North Korea to be parts of an "axis of evil," adversaries of the United States and potentially subject to preemptive action.

From a geopolitical perspective, *axis* was not a good term. Apparently referring back to the Nazi-fascist Axis of World War II, it implied an ideological, strategic, and logistical link that did not exist. In the year that followed, North Korea declared itself to be in possession not only of long-range missiles but also of at least one nuclear

weapon; its technical connections were with Pakistan, newfound U.S. ally in the War on Terror, not with Iran or Iraq. Iran, historically supportive of organizations designated as terrorist, in fact cooperated with the United States to some extent during the anti-Taliban campaign in Afghanistan and became embroiled in an internal doctrinal struggle between reformers and revivalists. But it was Iraq that rose to prominence as the first U.S. target on the "axis of evil," and specifically its ruthless ruler, Saddam Hussein.

A combination of factors led to this designation. Part of it lies in unfinished business arising from the American-led United Nations campaign (the 1991 Gulf War) that ousted Saddam's armies from Kuwait following his 1990 invasion. Proceeding to Baghdad to destroy the Iraqi regime was not part of the authorized mission, so Saddam and his ruling clique survived. Finding and destroying weapons of mass destruction still held by Iraq, however, was. Following obfuscation and obstruction, UN weapons inspectors withdrew, leaving the Iraqi regime an opportunity to resurrect its capacity to create and deploy biological and chemical weapons, and to seek ways to build nuclear weapons as well. By the time President Bush described Iraq as part of the "axis of evil," U.S. intelligence agencies were reporting that Iraq was making progress in the production of these weapons of mass destruction (WMD) and constituted a threat to the region and a potential threat to America. During difficult and divisive negotiations at the United Nations, the U.S. Administration was unable to muster support for multinational intervention. In March, 2003 United States military forces, allied with British and Australian troops, attacked and invaded Iraq from the south, and in a matter of weeks toppled the Saddam regime from power. Now the United States faced a massive task of transitional administration (to a newly constituted, representative Iraqi government), reconstruction of the country's heavily damaged infrastructure, and peacekeeping in a volatile and dangerous social environment.

It is therefore important to understand key aspects of the regional geography of California-sized Iraq and its official population of 24.8 million (which may in fact be two million less because of unregistered emigration). The country constitutes nearly 60 percent of the total area of the Middle East as we define it and has more than 40 percent of its population. With its world-class oilfields (revenues from which are to pay for much of Iraq's reconstruction), major gas reserves, and large areas of irrigable farmland, Iraq also is best endowed of all the region's countries with natural resources. Iraq is heir to the early Mesopotamian states and empires that emerged in the Tigris-Euphrates Basin, and the country is studded with matchless archeological sites and museum collections, to

FIGURE 7-12

which disastrous damage was done during and after the 2003 war from combat and looting.

Today, Iraq is bounded by six neighbors and has in recent times had adversarial relations with most of them. To the north lies Turkey, source of both of Iraq's vital rivers. To the east lies Iran, target of a destructive war during the 1980s. Kuwait, at the head of the Persian Gulf, was invaded by Iraq's armies in 1990. To the south, Saudi Arabia long was a friend of Iraq's enemies. And to the west, Iraq is adjoined by Jordan, which gave Iraq some help during the 1991 Gulf War but later turned its back on the Saddam regime, and Syria, functionally most remote from Iraq but also a country ruled by a strongman and his minority clique. As Figure 7-13 shows, pipelines across Syria link Iraq's oilfields with Mediterranean terminals. During the 2003 war and its aftermath, Syria's long desert border became an escape route for Iraqi militants, and for some time it was believed that Saddam Hussein himself had left Iraq for refuge there.

The map also displays Iraq's regional components. The capital and core area, Baghdad and its environs, lie about midway between the Persian Gulf and the Turkish border astride the Tigris River. Three historic domains lie north and south of this heartland. To the north and northwest extends the Sunni stronghold from which Saddam's repressive regime was drawn. In the mountains

United States armed forces, allied with British and Australian troops, invaded Iraq from the south in March, 2003 and in a matter of weeks toppled the Saddam regime. Despite the brevity of the conflict, Iraq sustained major infrastructure damage as well as substantial numbers of civilian and military casualties. Massive postwar looting ransacked hospitals, colleges, schools, libraries, museums, and archeological sites. Reconstructing the country while keeping control over the restive population proved a difficult and dangerous task for military forces trained for combat, not policing. Here American soldiers control a checkpoint in Baghdad, where a dusk-to-dawn curfew failed to end attacks on the U.S. military.

along the Turkish-Iranian border lies the Kurdish zone that forms but a small part of a much larger ethnic Kurdish region (Fig. 7-13; also see box titled "A Future Kurdistan?"). From the eastern suburbs of Baghdad southward stretches the third and largest domain in terms of both territory and population: The Shiah area extending all the way to Basra and the shores of the Persian Gulf. The vast, nearly empty desert bordering Saudi Arabia, Jordan, and Syria has few inhabitants and, as the map shows, no oil to attract settlers.

Population size, however, has not been matched by political power or influence in Iraq. By far the largest group is the Shi'ite sector, constituting a bit more than 60 percent of Iraq's population. Next comes the Sunni cluster, about 22 percent, historically the most powerful but fragmented along numerous clan and factional lines. The Kurds, victims of Saddam's chemical WMD in 1990, make up around 16 percent. There also are numerous smaller minorities, among which is the ancient, mainly Christian, non-Arab Assyrian group living today scattered among the Sunnis and the Kurds in northern Iraq.

Each of Iraq's three subregions has distinguishing geographical properties. The Sunni domain extends in a rough triangle northward from Baghdad along the Tigris River, and it incorporates some of the country's major oilfields and crucial pipelines (Fig. 7-13). On the map, the oil centers of Kirkuk and Mosul are shown to lie within this area, but both cities lie in the transition zone that marks the Sunni-Kurdish border. The heart of Sunni country is the lowland of the Tigris River with its low-rise towns such as Samarra and Tikrit, the latter known as the base of the clan that produced Saddam Hussein. The Kurdish domain has long been split into two parts rep-

resented by two opposing political movements, the northern segment, bordering Turkey and centered on Irbil, and the southern part, adjoining Iran and anchored by As Sulaymaniyah (Fig. 7-13).

The Shiah domain is not only the largest and most populous, but also the most diverse subregion of Iraq, incorporating the eastern sector of Baghdad, several of the holiest sites of the faith (notably An Najaf), the vibrant city of Basra, a number of major oilfields and installations, and the Persian Gulf waterfront. In addition, this cultural subregion extends into the Iranian province of Khuzestan (capital Ahvaz), where several million Arab (not Persian) Shi'ites live (Fig. 7-13). As we will note later, Iran is the Shiah core area, but the great majority of Iranians are not Arabs, and this ethnic difference is more significant than the religious link of Shi'ism. Iraq's Shi'ites, and the Arab Shi'ites in Khuzestan, adhere to a form of Shiah known as Akhbari, which does not have strong political motivation and does not readily generate a political power structure. Iranian Shi'ites, on the other hand, follow Usuli doctrines, in which the link between religion and politics is far stronger. When Saddam's Iraq fought its long and bitter conflict with Iran in the 1980s, Iraqi Shi'ites sided with Iraq and did not sabotage the war effort. Only after the Gulf War of 1991 did anti-Saddam rebellions occur in the Shi'ite south, ruthlessly put down by the Republican Guard. Subsequently, certain Shi'ite religious observances were prohibited and much damage was done to settlements and shrines.

Baghdad, Iraq's ancient capital and historic center of Islam, straddles the meandering Tigris River and has about 5.5 million inhabitants. During the murderous rule of Saddam, the more than 2.5 million Shi'ites in the east-

Sivas

TURKEY

Longitude East of Greenwich

39°   42°

Erzurum

Yerevan   51°
ARMENIA   Xankandi   AZERBAIJAN
        AZER.
            Naxcivan

39°

Elazig

Malatya

Adiyaman

Diyarbakir   Batman

Gaziantep   Sanliurfa

Kilis

Aleppo
36°

Lake
Asad

SYRIA

Homs

Tatvan   Lake
         Van   Van

Hakkari

Zakhu

Al Qamishli

M
e
s
o
p
o
t
a
m
i
a

Tigris River

Dayr az Zawr

Euphrates River   Qadisiyah
                   Lake

Desert

IRAQ

Ar Rutbah

Nukhayb

Mosul

Tikrit

Samarra

Tharthar
Lake

Ar Ramadi

Razzaza
Lake

Karbala

KURDISTAN
DEMOCRATIC
PARTY

Irbil   Kurdish
         Domain

Salmas

Orumiyeh

Kirkuk

As Sulaymaniyah

PATRIOTIC
UNION OF
KURDISTAN

Sunni
Domain

Ba'qubah   Mandali

Baghdad

Al Hillah   Al Kut

Caspian
Sea

Tabriz   Ardabil

Mianeh   Rasht

Zanjan   Elburz Mt.

Qazvin

IRAN

Sanandaj

Hamadan

Kermanshah   Arak

Borujerd

Khorramabad

Zagros Mts.

Dezful

Shiah

JORDAN

Badanah

Al Jawf

Ad Diwaniyah

An Najaf

Al 'Amarah   Domain

An Nasiriyah

Desert

Rafha

Ahvaz

KHUZESTAN
PROVINCE

Basra
Abadan

Shatt al
Arab

Persian

Kuwait   Gulf

Hafar al Batin

KUWAIT

As Subayhiyah

SAUDI
ARABIA

Ha'il

Jubail

IRAQ

POPULATION
· Under 50,000
● 50,000–250,000
● 250,000–1,000,000
● 1,000,000–5,000,000
● Over 5,000,000

National capitals are underlined

▰ Oilfield
— Road
— Railroad
·-·-· Oil pipeline
⊢⊣ Canal

0   40   80   120   160   200 Kilometers
0   20   40   60   80   100 Miles

Longitude East of Greenwich

A Future Kurdistan?

*P*olitical maps of the region do not show it, but where Turkey, Iraq, and Iran meet, the cultural landscape is not Turkish, Iraqi, or Iranian. Here live the Kurds, a fractious and fragmented nation of at least 25 million (their numbers are uncertain). More Kurds live in Turkey than in any other country (perhaps as many as 14 million); possibly as many as 8 million in Iran; about half that number in Iraq; and smaller numbers in Syria, Armenia, and even Azerbaijan (see Fig. 7-12).

The Kurds have occupied this isolated, mountainous, frontier zone for over 3000 years. They are a nation, but they have no state; nor do they enjoy the international attention that peoples of other **stateless nations** (such as the Palestinians) receive. Turkish and Iraqi repression of the Kurds, and Iranian betrayal of their aspirations, briefly make the news but are soon forgotten. Relative location has much to do with this: spatial remoteness and the obstacles created by the ruling regimes inhibit access to their landlocked domain.

Many Kurds dream of a day when their fractured homeland will be a nation-state. Most would agree that the city of Diyarbakir, now in southeastern Turkey, would become the capital. It is the closest any Kurdish town comes to a primate city, although the largest urban concentration of Kurds today is in the shantytowns of Istanbul, where more than 3 million have migrated. In their heartland, meanwhile, the Kurds, their shared goals notwithstanding, are a divided people whose intense disunity has thwarted their objectives; and as in the case of Europe's Basques, a small minority of extremists has tended to use violence in pursuit of their aims.

The Kurds—without oil reserves, without a seacoast, without a powerful patron, without a global public relations machine—are victims of a conspiracy between history and geography. One of the world's largest stateless nations hopes for a deliverance that is unlikely to come.

ern part of the city were tightly controlled and severely punished for even minor infractions. Their sector of Baghdad was called "Saddam City" and its neighborhoods were the poorest and most dilapidated in the entire metropolis. The western sector was the headquarters of the regime, the Ba'ath (Revival) Party that empowered it, and home to the Sunni population that supported it. Here were the key public buildings, several of Saddam's monumental palaces, impressive mosques, statues, and other edifices of the regime. The brief 2003 war inflicted heavy damage on this part of Baghdad, which for months afterward was without reliable power, water, medical facilities, and schools. It was a priority for American planners to get the capital functioning again, but the task was overwhelming in the face of continued opposition, sabotage, and growing resentment. The lengthy delay in the capture of Saddam and his two henchmen sons contributed significantly to the problems the occupiers confronted, because the fear that they might return to punish "collaborators" led many Iraqis to resist involvement in the reconstruction.

In Baghdad's hinterland and beyond, meanwhile, the challenge proved to be no less daunting. The freedoms brought about by the collapse of the Saddam regime enabled Shi'ites to gather at holy sites for the first time in several decades, and local clerics vied for power without fear of retribution from Baghdad. However, Saddam-era protections of secular life and women's rights fell victim to Islamic strictures in the Shi'ite domain, and there was little the occupying forces could do to stem that cultural

tide. While American planners talked of "representative" government, it was evident to all concerned that in a democratic Iraq, proportional representation would mean a Shi'ite-dominated government. Various proposals, from a federal Iraq consisting of autonomous regions to a multiparty system that would preclude Shi'ite domination, had little practical prospect.

Political geographers pointed to the difficulties faced by Russia in its effort to control events in Chechnya, a Massachusetts-sized territory with a population of 1.2 million. How could the United States expect to succeed with barely double the number of troops in a country twenty times as populous and almost twenty times as large? The future of the entire region depends on the fate of this postwar campaign.

Syria

Like Iraq, Syria is also ruled by a minority. Although Syria's population of 18.1 million is about 75 percent Sunni Muslim, the ruling elite comes from a smaller Islamic sect based there, the Alawites. Leaders of this powerful minority have retained control over the country for decades, at times by ruthless suppression of dissent. In 2000, president-for-life Hafez al-Assad died and was succeeded by his son Bashar, signaling a continuation of the political status quo.

Like Lebanon and Israel, Syria has a Mediterranean coastline where crops can be raised without irrigation. Behind this densely populated coastal zone, Syria has a

FIGURE 7-13

much larger interior than its neighbors, but its areas of productive capacity are widely dispersed. Damascus, in the southwest corner of the country, was built on an oasis and is considered to be the world's oldest continuously inhabited city. It is now the capital of Syria, with a population of 2.4 million.

The far northwest is anchored by Aleppo, the focus of cotton- and wheat-growing areas in the shadow of the Turkish border. Here the Orontes River is the chief source of irrigation water, but in the eastern part of the country the Euphrates Valley is the crucial lifeline. It is in Syria's interest to develop its eastern provinces, building upon recent discoveries of oil there (Fig. 7–13).

A lack of economic reform and continued authoritarian political rule combine to deny Syrians their opportunities in this changing world. As we note later, the Golan Heights issue and Syria's continuing armed intervention in Lebanon dominate its foreign affairs.

Jordan

Syria's southern neighbor is Jordan, a classic example (and victim) of changing relative location. This desert kingdom was a product of the Ottoman collapse, but when Israel was created it lost its window on the Mediterranean Sea, the (now Israeli) port of Haifa, previously in the British-administered Mandate of Palestine. Following its independence in 1946 with about 400,000 inhabitants, the creation of Israel bequeathed Jordan some 500,000 West Bank Palestinians and, later, a huge flow of refugees. Today Palestinians outnumber original residents by more than two to one in the population of 5.5 million. It may be said that the 46-year rule by King Hussein, ending in 1999, was the key centripetal force that held the country together.

With an impoverished capital, Amman, without oil reserves, and possessing only a small and remote outlet to the Gulf of Aqaba, Jordan has survived with U.S., British, and other aid. It lost its West Bank territory in the 1967 war with Israel, including its sector of Jerusalem (then the kingdom's second-largest city). No third country has a greater stake in a settlement between Israel and the Palestinians than Jordan.

Lebanon

The map suggests that Lebanon has significant geographic advantages in this region: a lengthy coastline on the Mediterranean Sea; a world-class capital, Beirut, on its shoreline; oil terminals on its shores; and a major capital in its hinterland. The map at the scale of Figure 7-12 cannot reveal still another asset: the fertile, agriculturally productive Bekaa Valley in the eastern interior.

French colonialism in the post-Ottoman era created a territory which in 1930 was about equally Muslim and Christian. Beirut became known as the "Paris of the Middle East." After independence in 1946, Lebanon functioned as a democracy and did well economically as the region's leading banking and commercial center. But the Muslim sector of the population grew much faster than the Christian one, and in the late 1950s the Arabs launched their first rebellion against the established order. Following the first of several waves of influx by Palestinian refugees, Lebanon fell apart in 1975 in a civil war that wrecked Beirut, devastated the economy, and left the country at the mercy of Syrian forces that entered to stabilize the situation.

Today Lebanon (population: 4.4 million) has restabilized and Beirut is being rebuilt, but it will take generations to overcome the losses sustained during a quarter-century of dislocation.

Israel

Israel, the Jewish state, lies at the heart of the Arab world (Fig. 7-9). Since 1948, when Israel was created as a homeland for the Jewish people on the recommendation of a United Nations commission, the Arab-Israeli conflict has overshadowed all else in the region.

Figure 7-14 helps us understand the complex issues involved here. In 1946 the British, who had administered this area in post-Ottoman times, granted independence to what was then called "Transjordan," the kingdom east of the Jordan River. In 1948, the orange-colored area became the UN-sponsored state of Israel—including, of course, land that had long belonged to Arabs in this territory called Palestine.

As soon as Israel proclaimed its independence, neighboring Arab states attacked it. Israel, however, not only held its own but pushed the Arab forces back beyond its borders, gaining the green areas shown in Figure 7-14. Meanwhile, Jordanian armies crossed the Jordan River and annexed the yellow-colored area named the West Bank, including part of the city of Jerusalem. The king called his newly enlarged country Jordan.

More conflict followed. In 1967 a week-long war produced a major Israeli victory: Israel took the Golan Heights from Syria, the West Bank from Jordan, and the Gaza Strip from Egypt, and conquered the whole Sinai Peninsula all the way to the Suez Canal. In later peace agreements, Israel returned the Sinai but not the Gaza Strip.

All this strife produced a huge outflow of Palestinian Arab refugees and displaced persons. The Palestinians are a *stateless nation*; more than one million continue to live as Israeli citizens within the borders of Israel, but 2.3 million are in the West Bank and 1.3 million in the Gaza Strip (the Golan Heights population is comparatively insignificant). An even larger number, however, live in neighboring and nearby countries including Jordan (2.7 million), Syria (423,000), Lebanon (403,000), and Saudi Arabia (296,000); another quarter of a million reside in Iraq, Egypt, Kuwait, and Libya. Many have been assimilated into the local societies, but tens of thousands of others continue to live in refugee camps. In 2003 the scattered Palestinian population was estimated to number 9.3 million.

Israel is about the size of Massachusetts and has a population of 6.8 million (including 1.3 million Arabs), but because of its location amid adversaries and its strong international links, these data do not reflect Israel's importance. Israel has built a powerful military even as its regional Arab neighbors have grown stronger; its policies arouse Arab passions; and Palestinian aspirations for a territorial state have not borne fruit. As a democracy with

FIGURE 7-14

strong Western ties, Israel has been a recipient of massive U.S. financial aid, and U.S. foreign policy has been to seek an accommodation between Jews and Palestinians as well as between Israel and its Arab neighbors.

In effect, a state of war had existed for several years between Palestinians and Israel when, in June 2003, the President of the United States announced a new initiative to seek a permanent solution. In the nearly 40 years since the end of the 1967 War, more than 200,000 Israelis have settled in the West Bank, building prosperous hilltop settlements and modern access routes or clustering in small, vulnerable communities on land deemed holy and God-given by orthodox Jews. Protecting these settlements (which also exist in Gaza) makes Israel an occupying enemy in Palestinian eyes, arousing resentment and retaliation. The Palestinian Arabs responded by attacking Jews in the West Bank and in Gaza and by sending suicide bombers into restaurants and onto buses within Israel itself. Each attack was followed by another round of retaliation, a cycle of destruction that brought Israeli tanks to the courtyard of the Palestinian Authority's (government's) headquarters and Palestinian bombers into hotel lobbies and shopping malls in Tel Aviv.

Every U.S. President in modern history has sought to mitigate the problem, and late in his second term President Clinton came close—only to see the accords his administration engineered scuttled by Chairman Arafat's failure to carry them through. President Bush's proposal invoked a so-called "Road Map" that would ultimately lead to an independent Palestinian state adjoining Israel and consisting of most of the West Bank and Gaza; it would guarantee Israel's right to exist in safety and would oblige Israel to cease expanding its West Bank settlements. Following intensive negotiations, even the Palestinian militant movement Hamas signed on to a three-month truce in July 2003. New leadership in the Palestinian Authority made this progress possible, but the plan remained fragile. Orthodox and right-wing Israelis vehemently object to the removal of "illegal" West Bank settlements, which is part of the "Road Map." Among the Palestinians, many militants object to any suggestion that Israel has a right to exist on land from which Arabs were ousted.

In mid-2003, the "Road Map" still was neither a clear path nor a delimited map. Here are the key geographic obstacles:

1. *The West Bank.* In 1977 only about 5000 Jews lived in the West Bank; by 2003 they numbered more than 200,000, making up nearly 10 percent of the population and creating a seemingly inextricable jigsaw of Jewish and Arab settlements (Fig. 7-15). Palestinians envisage the repatriation of most if not all these set-

tlers; Israelis want to redraw boundaries to incorporate a majority into an expanded Israel.

2. *The Gaza Strip.* Small Jewish communities continue to live under armed protection in a poverty-stricken, barren strip of waterfront land inhabited by 1.3 million Palestinians, many of whom need jobs across the border in Israel to survive. Even now, a half-century later, squalid refugee camps form a constant reminder of the consequences of Israel's creation.

3. *Jerusalem.* The United Nations intended Tel Aviv to be Israel's capital, and Jerusalem an international city. But the Arab attack and the 1948–1949 war allowed Israel to drive toward Jerusalem (see Fig. 7-14, the green wedge into the West Bank). By the time a cease-fire was arranged, Israel held the western part of the city, and Arab forces the eastern sector. But in this eastern sector lay major Jewish historic sites, including the Western Wall. Still, in 1950 Israel declared the western sector of Jerusalem its capital, making this, in effect, a *forward capital*. Figure 7-16 shows the position of the armistice line, leaving most of the Old City in Jordanian hands. But then, in the 1967 war, Israel conquered all of the West Bank, including East Jerusalem, and in 1980 the Jewish state reaffirmed Jerusalem's status as capital, calling on all nations to move their embassies from Tel Aviv. Meanwhile, the government redrew the map of the ancient city, building Jewish settlements in a ring around East Jerusalem that would end the old distinction between Jewish west and Arab east. This enraged Palestinian leaders, who still view Jerusalem as the eventual headquarters of their hoped-for Palestinian state.

4. *The Golan Heights.* This is an issue between Israel and Syria, not primarily involving Palestinians. The Golan Heights were captured by Israel during the 1967 War, and the inset map in Figure 7-14 suggests how complicated the situation is. The strategic "heights" overlook a large area of northern Israel, and they flank the Jordan River and crucial Lake Kinneret (the Sea of Galilee), Israel's chief surface water reservoir. Autocratic Syria demands the return of the Golan Heights, but in democratic Israel political realities may make ceding this war prize impossible.

Israel lies at the center of a fast-moving geopolitical storm; the four issues raised above are only part of the overall problem (others are water rights, compensation for land expropriation, and the Palestinian's "right to return" to pre-refugee abodes). Now, in the age of nuclear, chemical, and biological weapons and longer-range missiles, Israel's search for an accommodation with its Arab neighbors is a race against time.

FIGURE 7-15

THE WEST BANK,
2002

City under total
Palestinian control

Areas of joint control

Areas of total
Israeli control

Aquifer

• Palestinian
town/village

▲ Major Israeli
settlement

National water carrier

Road

Railroad

National capital is underlined

0   5   10   15 Kilometers
0   5   10 Miles

LEBANON

GOLAN
HEIGHTS

SYRIA

Sea of
Galilee
(Lake
Kinneret)

Haifa

Nazareth

Afula

Irbid

Mediterranean Sea

Hadera

Netanya

Tulkarm

Jenin

Nablus

Qalqiya

Jordan R.

Tel Aviv-
Jaffa

WEST

BANK

Ramla

Ramallah

Jericho

'Ammān

JORDAN

Ashdod

Jerusalem

Bethlehem

ISRAEL

Dead
Sea

Hebron

Beersheba

Longitude East of Greenwich

355

FROM THE FIELD NOTES

"Cultural contrast pervades Jerusalem's urban landscape; social tension is palpable. From almost any high vantage point you can see the places of worship and the holy ground that means so much to Jew, Christian, and Muslim: synagogues, churches, and mosques, walled cemeteries, sacred shrines, historic sites, all juxtaposed as in this panorama. In the distance, in another direction, you can see the Jordan Valley; Jerusalem is where the West Bank meets Israel. When the Israeli government decided to make Jerusalem its political headquarters, it was a move designed to proclaim the nation's dominance here: Jerusalem would not become an international city. This makes Jerusalem another example of a forward capital, the vanguard of the state in potentially contested territory."

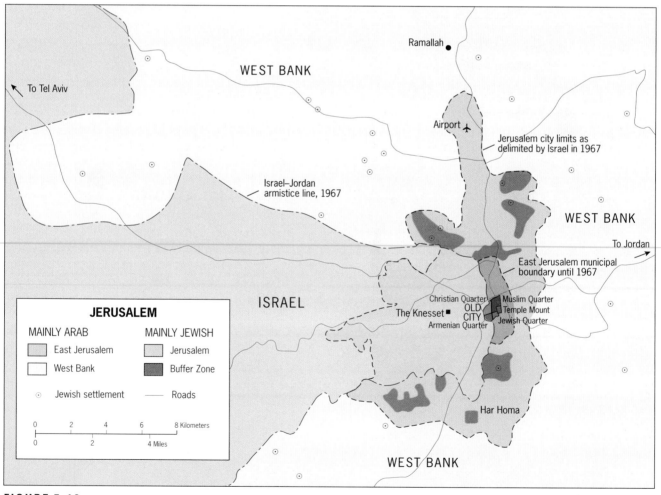

FIGURE 7-16

THE ARABIAN PENINSULA

The regional identity of the Arabian Peninsula is clear: south of Jordan and Iraq, the entire peninsula is encircled by water. This is a region of old-style emirates and sheikdoms made wealthy by oil; it also contains the site where Islam originated.

Saudi Arabia

Saudi Arabia itself has only 25.4 million inhabitants in its vast territory, but we can see the kingdom's importance in Figure 7-8: the Arabian Peninsula contains the Earth's largest concentration of known petroleum reserves. Saudi Arabia occupies most of this area and by some estimates may possess as much as one-quarter of the world's oil deposits. As Figure 7-17 shows, these reserves lie in the eastern part of the country, particularly along the Persian Gulf coast and in the Rub al Khali (Empty Quarter) to the south.

As a region, the Arabian Peninsula is environmentally dominated by a desert habitat and politically dominated by the Kingdom of Saudi Arabia (Fig. 7-17). With 830,000 square miles (2,150,000 sq km), Saudi Arabia is the realm's fourth biggest state; only Kazakhstan, Algeria, and Sudan are larger. On the peninsula, Saudi Arabia's neighbors (moving clockwise from the head of the Persian Gulf) are Kuwait, Bahrain, Qatar, the United Arab Emirates, the Sultanate of Oman, and the Republic of Yemen (created in 1990 through the unification of former North Yemen and South Yemen). Together, these countries on the eastern fringes of the peninsula contain about 30 million inhabitants; the largest by far is Yemen, with 19.8 million. The interior boundaries of several of these states, however, are still inadequately defined here in one of the world's last remaining frontier-dominated zones.

Figure 7-17 reveals that most economic activities in Saudi Arabia are concentrated in a wide belt across the "waist" of the peninsula, from the boomtown of Dhahran on the Persian Gulf through the national capital of Riyadh in the interior to the Mecca-Medina area near the Red Sea. A modern transportation and communications network has recently been completed. But in the more remote zones of the interior, Bedouin nomads still ply their ancient caravan routes across the vast deserts. For decades, Saudi Arabia's royal families were virtually the sole beneficiaries of their country's incredible wealth, and there was hardly any impact on the lives of villagers and nomads. When the oil boom arrived in the 1950s, foreign laborers were brought in (today there are between 3 and 4 million, many of them Shi'ite) to work in the oilfields, ports, factories, and as servants. The east boomed, but the rest of the country lagged behind.

Efforts to reduce Saudi Arabia's regional economic disparities have been impeded by the enormous cost of bringing water from deep-seated sources to the desert surface to stimulate agriculture, by the sheer size of the country, and by the very high rate of population growth (2.9 percent, thereby doubling in just 24 years). Still, housing, health care, and education have seen major improvement, and industrialization also has been stimulated in such places as Jubail on the Persian Gulf and Yanbu, near Medina, on the Red Sea.

Saudi Arabia's prospects have been shaken in the aftermath of the 9/11 attacks. Sixteen of the 19 hijackers who crashed passenger-filled jetliners into the Trade Towers and Pentagon were Saudis representing al-Qaeda and Usama bin Laden, whose objective it was not only to strike the United States but also to damage the rich, conservative monarchy ruling the kingdom. American dependence on Saudi oil and long-term military cooperation with that monarchy muted the investigations that would ordinarily have followed such an event. Eventually, however, rumors began to surface relating to alleged money transfers from members of the Saudi oligarchy to the terrorists' bank accounts, and members of the U.S. Congress expressed views concerning Saudi Arabia that evinced the fraying of the American-Saudi relationship. This would represent a success for Usama bin Laden: the Saudi regime, which exiled him from his home country, is one of his primary targets.

But even without the threat posed by Usama bin Laden and his terrorist movement, the Saudi state would be facing serious challenges. Oil income fueled rapid economic growth and gave Saudi citizens privileges that ranged from free education for all to imported labor to perform work that Saudis disdained. For example, for decades one never saw a Saudi citizen working as a parking attendant, bellhop, or room steward in any of the country's numerous luxury hotels: you expected to see Indians, Pakistanis, Palestinians, or other foreigners employed in such menial jobs. But then even Saudi Arabia experienced a period of declining income, its GDP dropping sharply and some of the citizens' expectations no longer met. This worsened the social contrasts inevitably associated with Saudi Arabia's oil boom, providing the revivalist Muslim clerics, preaching fundamentalism from their pulpits, an ever-larger market. For the growing political opposition in Saudi Arabia there are no adequate alternate channels, so that long-term stability in Islam's birthplace is at risk.

On the Periphery

Five of Saudi Arabia's six neighbors on the Arabian Peninsula face the Persian and Oman Gulfs (Fig. 7-17) and are monarchies in the Islamic tradition. All five also derive

FIGURE 7-17

substantial revenues from oil. Their populations range from 600,000 to 3.6 million in addition to hundreds of thousands of foreign workers; these are not powerful states, although their energy resources give them disproportionate influence. In the new era of the War on Terror, the growing difficulties between the United States and Saudi Arabia, and the campaign involving Iraq, however, the relative location of these peripheral ministates has changed and their importance has risen. They lie on the route from the United

States to Afghanistan, and they are not far from Iraq at a time when Saudi Arabia's regime may not allow its soil to be used to launch attacks on its neighbor to the north. The profiles of countries named Qatar, Bahrain, the United Arab Emirates, Kuwait, and Oman are strengthening.

Proceeding from north to south, let us note the considerable geographic diversity displayed by these states. **Kuwait**, at the head of the Persian Gulf, would undoubtedly be Iraq's nineteenth province today if it did not possess the world's fourth-largest oil reserve, which mobilized the United States and its allies to rescue it following the invasion by Saddam Hussein's armies in 1990. **Bahrain**, an island state with a population of approximately 700,000 (50 percent Shi'ite, 35 percent Sunni) and a labor force that is two-thirds foreign, for decades suffered from chronic violent unrest because the majority resented rule by the Sunni royal dynasty. In 1999, King Hamad set an example for the entire region by establishing a two-chamber legislature, one of them elected by universal suffrage, moving the state toward democracy and, conceivably, constitutional monarchy. Nearby **Qatar** consists of a peninsula jutting out into the Persian Gulf (Fig. 7-17), a featureless, sandy wasteland made habitable by oil, whose volume and revenues are now declining even as those from natural gas are rising. Qatar's rulers have permitted U.S. forces to establish a strategic base of military operations on their territory in pursuit of the War on Terror and the campaign against Iraq's regime. The **United Arab Emirates (UAE)**, a federation of seven emirates, faces the Persian Gulf between Qatar and Oman. The reigning sheik is an absolute monarch in each of the emirates, and the seven sheiks together form the Supreme Council of Rulers. In terms of oil revenues, however, there is no equality: two emirates—Abu Dhabi and Dubai—have most of the reserves.

The eastern corner of the Arabian Peninsula is occupied by the Sultanate of **Oman**, another absolute monarchy, centered on the capital, Muscat. Figure 7-17 shows that Oman consists of two parts: the large eastern corner of the peninsula and a small but critical cape to the north, the Musandam Peninsula, that protrudes into the Persian Gulf to form a narrow *choke point* (see box

titled "Choke Points: Danger on the Sea Lanes"). Here the narrow Hormuz Strait separates the Arabian Peninsula from Iran. Tankers that leave the other Gulf states must negotiate this narrow channel at slow speed, and during politically tense times warships have had to protect them. Iran's claim to several small islands near the Strait that are owned by the UAE is a potential source of dispute. Oman, with only limited oil resources that are likely to run out in two decades, has used its income frugally to improve infrastructure (a good road system and clean water supplies for all outlying areas) and to prepare for an oil-less future. A giant container port and free-trade zone have been built at Salalah, and major investments have been made in luxury tourist facilities, capitalizing on the country's sunshine and beaches. Oman also has gone further than most of the region's countries in attracting foreign investment and privatizing facilities such as its international airport.

At the southern tip of the Arabian Peninsula lies what is in many ways the key state on the region's periphery: **Yemen**. As Figure 7-17 shows, Yemen's boundary with Saudi Arabia has only recently been delimited in a zone where there are known to be oil reserves. Yemen today is the product of a merger between two neighbors, North Yemen and South Yemen, which agreed to join in 1990. San'a, the interior capital in the north, became the new country's official seat of government; Adan (Aden), the port on the southern coast, continues to anchor the south. Given Yemen's complex cultural geography (the population of 19.8 million is 60 percent Sunni, 40 percent Shi'ite), the country has made remarkable progress toward representative government; the minister for human rights

FROM THE FIELD NOTES

"The port of Mutrah, Oman, like the capital of Muscat nearby, lies wedged between water and rock, the former encroaching by erosion, the latter crumbling as a result of tectonic plate movement. From across the bay one can see how limited Mutrah's living space is, and one of the dangers here is the frequent falling and downhill sliding of large pieces of rock. It took about five hours to walk from Mutrah to Muscat; it was extremely hot under the desert sun but the cultural landscape was fascinating. Oil also drives Oman's economy, but here you do not find the total transformation seen in Kuwait or Dubai. Townscapes (as in Mutrah) retain their Arab-Islamic qualities; modern highways, hotels, and residential areas have been built, but not at the cost of the older and the traditional. Oman's authoritarian government is slowly opening the country to the outside world after long-term isolation."

Choke Points: Danger on the Sea Lanes

The Hormuz Strait between the Arabian Peninsula and Iran is only one of dozens of maritime choke points that have taken on added significance during this time of war on terrorism. A **choke point** is defined as the narrowing of an international waterway causing marine traffic congestion, requiring reduced speeds and/or sharp turns, and increasing the risk of collision as well as vulnerability to attack.

For many years, Southeast Asia's Strait of Malacca between the Malay Peninsula and the Indonesian island of Sumatera (Sumatra) has been a notorious choke point because reduced speeds in the Strait give pirates the opportunity to board vessels and plunder them or worse, kill their crews, and take them over. Since the 1980s, hundreds of acts of piracy have made this the world's least safe waterway, especially for smaller ships.

One of the world's busiest choke points is the Strait of Gibraltar between Spain and Morocco, which was in the news in late 2002 when an al-Qaeda document was found that referred to plans for an attack on Western ships slowing down through this 36-mile (58-km) long funnel that narrows to 8 miles (13 km) in width and is subject to high winds and fast currents. Another risky choke point is the Bab al Mandeb Strait between Africa and the Arabian Peninsula, or more specifically, between Djibouti and Yemen, at the southern entrance to the Red Sea. Barim (Perim) Island partially blocks this 18-mile (30-km)-wide strait. Oil tankers, one of which was attacked by al-Qaeda in 2002, and freighters as well as cruise ships approach this strait via all-too-predictable routes.

The English Channel between Britain and France, Turkey's Bosporus between the Black Sea and the Mediterranean Sea, Indonesia's Sunda Strait between the Java Sea and the Indian Ocean, the Strait of Magellan for ships rounding the southern end of South America, and the Hainan Strait between southernmost China's Leizhou Peninsula and Hainan Island are just some of the many other choke points on the world map.

And some choke points are artificial. The Suez Canal between the Red Sea and the Mediterranean Sea and the Panama Canal between Atlantic (Caribbean) and Pacific waters are the busiest and most prominent, but many other route-shortening canals are vulnerable choke points at a time when the risk of hostile action looms larger than the risk of collision.

---

is a woman, Wahiba Fare. In terms of political geography, Yemen lies across the narrow Bab el Mandeb Strait from the Horn of Africa, an extremely strategic location at a choke point on the route between the Indian Ocean and the Red Sea that leads to the Suez Canal. It is also the ancestral home of the terrorist leader Usama bin Laden, whose family moved to Saudi Arabia before he was born; in October 2000, the port of Adan was the scene of a successful terrorist attack on a U.S. warship, the *USS Cole*.

The Arabian Peninsula is a region in difficult and dangerous transition. In the south, monarchical, oil-rich, polarizing Saudi Arabia stands in stark contrast to democratic, agricultural, poorer Yemen. Historic conflicts, cultural disharmony, and latent territorial disputes, now complicated by the threat of terrorism, make this a particularly troubled region.

► THE EMPIRE STATES

Two major states, both with imperial histories, dominate the region that lies immediately to the north of the Middle East (Fig. 7-9), where Arab ethnicity gives way but Islamic culture endures: Turkey and Iran. On its periphery lie two entities that are strongly entwined with these states: Azerbaijan to the north and the northern part of Cyprus to the south (Fig. 7-18).

Turkey

Earlier in this chapter we chronicled the historical geography of the Ottoman Empire, its expansion, cultural domination, and collapse. By the beginning of the twentieth century, the country we now know as Turkey lay at the center of this decaying and corrupt state, ripe for revolution and renewal. This occurred in the 1920s and thrust into prominence a leader who became known as the father of modern Turkey: Mustafa Kemal, known after 1933 as Atatürk, meaning "Father of the Turks."

The ancient capital of Turkey was Constantinople (now Istanbul), located on the Bosporus, part of the strategic straits connecting the Black and Mediterranean seas. But the struggle for Turkey's survival had been waged from the heart of the country, the Anatolian Plateau, and it was here that Atatürk decided to place his seat of government. Ankara, the new capital, possessed certain advantages: it would remind the Turks that they were (as Atatürk always said) Anatolians; it lay nearer the center of the country than Istanbul; and it could therefore act as a stronger unifier. Istanbul lies on the threshold of Europe, with the minarets and mosques of this largest and most varied Turkish city rising above a townscape that resembles Eastern Europe.

Although Atatürk moved the capital eastward and inward, his orientation was westward and outward. To implement his plans for Turkey's modernization, he initiated reforms in almost every sphere of life within the country. Islam, formerly the state religion, lost its official status. The state took over most of the religious schools that had controlled education. The Roman alphabet replaced the Arabic. A modified Western code supplemented Islamic law. Symbols of old—growing beards, wearing the fez— were prohibited. Monogamy was made law, and the emancipation of women was begun. The new government emphasized Turkey's separateness from the Arab World, and it has remained aloof from the affairs that engage other Islamic states.

Turkey is a mountainous country of generally moderate relief and, as Figure G-8 indicates, considerable environmental diversity ranging from steppe to highland. On the dry Anatolian Plateau villages are small, and subsistence farmers grow cereals and raise livestock. Coastal plains are not large, but they are productive and densely populated. Textiles (from home-grown cotton) and farm products dominate the export economy, but Turkey also has substantial mineral reserves, some oil in the southeast, massive dam-building projects on the Tigris and Euphrates rivers, and a small steel industry based on domestic raw

materials. Normally self-sufficient in staples, Turkey's prospects would be even brighter if its government and administration improved.

From before Atatürk's time, Turkey has had a history of mistreating minorities. Soon after the outbreak of World War I, the (pre-Atatürk) regime decided to expel all the Armenians, concentrated in the country's northeast. Nearly 2 million Turkish Armenians were uprooted and brutally forced out; an estimated 600,000 died in a campaign that still arouses anti-Turkish emotions among Armenians today. In modern times, Turkey has been criticized for its treatment of its large and regionally concentrated Kurdish population. About one-fifth of Turkey's population of almost 70 million is Kurdish, and successive Turkish governments have mishandled relationships with this minority nation, even prohibiting the use of Kurdish speech and music in public places during one especially repressive period. The historic Kurdish homeland lies in southeastern Turkey, centered on Diyarbakir, but millions of Kurds have moved to the shantytowns around Istanbul—and to jobs in the countries of the European Union. The Kurdish issue has been a negative factor

AMONG THE REALM'S GREAT CITIES . . . **ISTANBUL**

From both shores of the Sea of Marmara, northward along the narrow Bosporus toward the Black Sea, sprawls the fabulous city of Istanbul, known for centuries as Constantinople, headquarters of the Byzantine Empire, capital of the Ottoman Empire, and, until Atatürk moved the seat of government to Ankara in 1923, capital of the modern Republic of Turkey as well.

Istanbul's site and situation are incomparable. Situated where Europe meets Asia and where the Black Sea joins the Mediterranean, the city was built on the requisite seven hills (as Rome's successor), rising over a deep waterway that enters the Bosporus from the west, the famous Golden Horn. A sequence of empires and religions endowed the city with a host of architectural marvels that give it, when approached from the water, an almost surreal appearance—and what is seen today is a mere remnant of history's accretion, survivals of earthquakes, fires, and combat.

Turkey's political capital may have moved to Ankara, but Istanbul remains its cultural and commercial headquarters. It also is the country's leading urban magnet, luring millions from the poverty-stricken countryside. In the heart of the city known as Stamboul and in the "foreign" area north of the Golden Horn called Beyoglu, modern buildings vie for space and harbor views, blocking the vis-

tas that long made Istanbul's cultural landscape unique. On the outskirts, shantytowns emerge so rapidly that Istanbul's population is doubling every 15 years at current rates, having reached 10 million in 2004. Istanbul's infrastructure is crumbling under this unprecedented influx, its growth is virtually without planning controls, and its legacy of two millennia of cultural landscapes is threatened.

FIGURE 7-18

surrounding Turkey's aspiration to join the EU. During the past five years, the Turkish government has made efforts to improve its image in this arena, having a history of violating the human rights of an entire minority in its effort to suppress a small but violent minority of Kurdish extremists. The death penalty has been abolished, and Kurds have been given freedom to express their culture in ways long forbidden. But some Kurds still resort to violence, and Turkish forces still pursue some of them in the Iraqi Kurdish Security Zone; reports of mistreatment in jail and abrogations of due process continue to surface.

Another cultural issue in modern Turkey involves one of Islam's many sects, in this case the Alevis. Most of Turkey's Muslims are Sunnis, and some are orthodox and strict. The Alevis, who number about 15 million (nearly 20 percent of the population), are Shi'ites—but they practice a rather nonchalant version of the faith. While orthodox Sunnis worship in mosques, live by Islam's daily rules, and fast during the daytime during Ramadan, Alevis do not gather in mosques and do not follow the rules of the Ramadan fast. This has led to serious cultural conflict, including murder and arson. The historic center of Turkey's Alevi sect lies in Sivas, in the east-central part of the country northwest of the Kurdish zone. Like the Kurds, many Alevis have left their cultural base, where they are under pressure from orthodox Sunnis, for the more cosmopolitan cities.

In the autumn of 2002 a significant event occurred: in Turkey's general election, a party with Islamist roots led by a politician with a record of revivalist statements

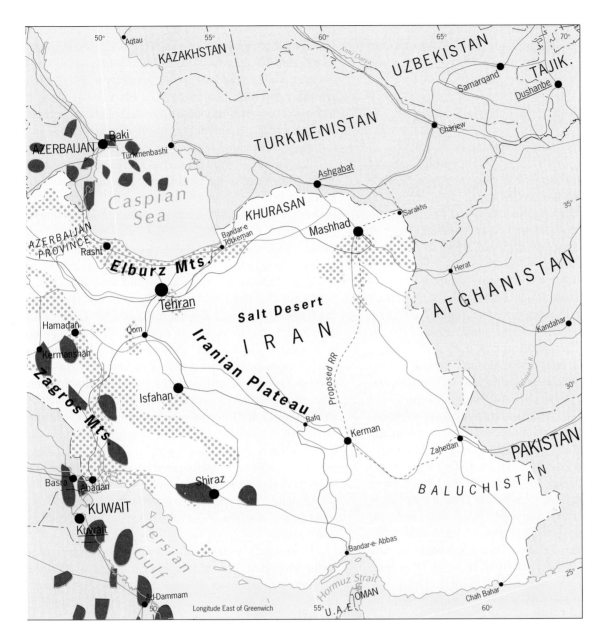

led the voting and became the governing majority. The leader who would be prime minister, Recep Tayyip Erdogan, immediately proclaimed his and his party's intention to retain Turkey's secular status. In previous elections, Islamist parties were either prevented from running or were barred by the army from taking office. The 2002 election was thus a milestone for Turkey and a boost for its EU prospects.

In early 2003, progress appeared to be on the horizon in another context: the ongoing problem of Cyprus (see page 86). A United Nations plan for the island would reunite the two sectors of the Greek-Turkish island in a confederation of two so-called component states, loosely associated under a presidency that would rotate every ten months. The plan might make it possible

for such a "united" Cyprus to join the EU along with the other nine new entrants. It received favorable responses in both Greece and Turkey, but the harder part will be to convince the locals, especially the Cypriot Turks. Until now, the Turks, with about 20 percent of the population, have controlled 36 percent of Cyprus's land area. Under the plan, that would go down to 28.5 percent, still well over one-fifth of the total area. EU membership, on the other hand, would favor the Turkish sector, which would become eligible for the EU subsidies given to poorer members.

Whatever the outcome in Cyprus, the dominant issue for Ankara during the next decade will be EU membership for Turkey itself. The expansion of the European Union into a neighboring geographic realm will be a momentous event. For Turkey, joining the EU will represent a milestone comparable to Atatürk's ascent after the Ottoman collapse.

Iran

Iran, Turkey's neighbor to the east, also has a history of imperial conquest. In 1971, the then-reigning shah and his family celebrated the 2500th anniversary of Persia's first monarchy with unmatched royal splendor. But by 1979, revolution had engulfed Iran, and Shi'ite fundamentalists drove the shah from power. The monarchy was replaced by an Islamic republic, and a frightful wave of retribution followed.

As Figure 7-18 shows, Iran also occupies a critical area in this turbulent realm. It controls the entire corridor between the Caspian Sea and the Persian Gulf. To the west it adjoins Turkey and Iraq, both historic enemies. To the north (west of the Caspian Sea) Iran borders Azerbaijan and Armenia, where once again Muslim confronts Christian. To the east Iran meets Pakistan and Afghanistan, and east of the Caspian Sea lies volatile Turkmenistan.

Iran, as Figure 7-18 demonstrates, is a country of mountains and deserts. The heart of the country is an upland, the Iranian Plateau, that lies surrounded by even higher mountains, including the Zagros in the west, the Elburz in the north along the Caspian Sea coast, and the mountains of Khurasan to the northeast. The Iranian Plateau therefore is actually a huge highland basin marked by salt flats and wide expanses of sand and rock. The highlands wrest some moisture from the air, but elsewhere only oases break the arid monotony—oases that for countless centuries have been stops on the area's caravan routes.

In eastern Iran, neighboring Pakistan, and Afghanistan, people still move with their camels, goats, and other livestock along routes that are almost as old as the human history of this realm. Usually they follow a seasonal and annual cycle, visiting the same pastures year after year, pitching their tents near the same stream. It is a lifestyle especially associated with this realm: **nomadism**. In Iran as elsewhere, nomads are not aimless wanderers across boundless plains. They know their terrain intimately, and they carefully judge how long to linger and when to depart based on many years of experience along the route.

In ancient times, Persepolis in southern Iran (located near the modern city of Shiraz) was the focus of a powerful Persian kingdom, a city dependent on *qanats*, underground tunnels carrying water from moist mountain slopes to dry flatland sites many miles away. Today, Iran's population of 67.1 million is 66 percent urban, and the capital, Tehran, lies far to the north, on the southern slopes of the Elburz Mountains. This mushrooming metropolis of 7.3 million,

lying at the heart of modern Iran's core area, still depends in part on the same kinds of qanats that sustained Persepolis more than 2000 years ago. As such, Tehran symbolizes the internal contradictions of Iran: a country in which modernization has taken hold in the cities, but little has changed in the vast countryside, where the *mullahs* led their peasant followers in a revolution that overthrew a monarchy and installed a theocracy.

As Figure 7-8 shows, Iran has a large share of the oil riches in this part of the world. Petroleum and petroleum products provide about 90 percent of the country's income. The reserves lie in a zone along the southwestern periphery of Iran's territory, and Abadan became its "oil capital" near the head of the Persian Gulf. But Iran is a large and populous country, and the wealth oil generated could not transform it the way the last shah intended, a transformation that had it occurred might have staved off the revolution. Modernization remained but a veneer: in the villages away from Tehran's polluted air, the holy men continued to dominate the lives of ordinary Iranians. As elsewhere in the Muslim world, urbanites, villagers, and nomads remained enmeshed in a web of production and profiteering, serfdom, and indebtedness that has always characterized traditional society here. The revolution swept this system away, but it did not improve the lot of Iran's millions. A devastating war with Iraq (1980–1990), into which Iran ruthlessly poured hundreds of thousands of its young men, sapped both the coffers and energies of the state. When it was over, Iran was left poorer, weaker, and aimless, its revolution spent on unproductive pursuits.

In the early years of the twenty-first century, evidence abounded that the people of Iran remain divided between conservatives determined to protect the power of the mullahs and reformers intent on modernizing and liberalizing Iranian society. In 2002, a conservative court sentenced a university professor to death for publishing a proposal for an Islamic "enlightenment"; thousands of students took to the streets in protest. That such a sentence could be handed down at all is indicative of the gulf between the dogmatic and the rational in this historic and civilized society.

Iran has long been accused of supporting terrorist activities in other countries, and President George W. Bush in 2002 designated the country as part of his "axis of evil." Although Revolutionary Iran disavowed its Persian imperial memories and ambitions, that does not mean its national interests now stop at its borders. Iran remains a major force in this region with cultural links to minorities in Iraq, Azerbaijan, and Afghanistan, and with economic interests in oil reserves and pipelines in the Caspian Sea basin. Iran's revolution ended a monarchy, but it did not extinguish all ties to an imperial past.

TURKESTAN

For centuries Turkish (Turkic) peoples held sway over a vast Central Asian domain that extended from Mongolia and Siberia to the Black Sea. Propelled by population growth and energized by Islam, they penetrated Iran, defeated the Byzantine Empire, and colonized much of Eastern Europe. Eventually, their power declined as Mongols, Chinese, and Russians invaded their strongholds. But these conquerors could not expunge them, as the names on the modern map prove (Fig. 7-19). The

FIGURE 7-19

SOUTH-CENTRAL EURASIA

FIGURE 7-20

MAJOR ETHNOLINGUISTIC GROUPS

ALTAIC
1 Azerbaijani
2 Kalmyk
3 Kazakh
4 Kyrgyz
5 Turk
6 Turkmen
7 Uzbek

CAUCASIAN
8 Chechen-Ingush
9 Dagestani
10 Georgian

AFRO-ASIATIC
11 Arabic

INDO-EUROPEAN
12 Armenian
13 Bakhtiari
14 Baluchi
15 Chahar Aimak
16 Hazara
17 Kurdish
18 Lur
19 Persian
20 Punjabi
21 Pushtun
22 Russian
23 Sindhi
24 Tajik

25 Other
26 Uninhabited or sparsely populated

REGIONAL CONFLICTS
A Abkhazia – Georgia
B South Ossetia – Georgia
C North Osetiya-Alaniya – Ingushetiya
D Chechnya – Russia
E Turkey – Kurdish separatists
F Nagorno-Karabakh – Azerbaijan
G Afghan war

latest conquerors, the Russian czars and their communist successors, created five Soviet Socialist Republics named after the majority peoples within their borders. Thus the Kazakhs, Turkmen, Kyrgyz, and other Turkic peoples retained some geographic identity in what was Soviet Central Asia.

Turkestan is a still-changing region. In some areas, the cultural landscapes of neighboring realms overlap into it, for example, in northern (Russian) Kazakhstan. In other areas, Turkestan extends into adjacent realms, as in Uyghur-influenced western China (Xinjiang). Some areas once penetrated by Turkic peoples are no longer dominated by them, for instance, Afghanistan. And certain peoples now living in Turkestan are not of Turkic ancestry, notably the Tajiks. This is a fractious region in sometimes turbulent transition.

The underlying cultural-geographic reason for this contentiousness—not only in Turkestan but in neighboring regions in Southwest and South Asia as well—is illustrated in Figure 7-20. This detailed map of ethnolinguistic groups actually is a generalization of an even more complex mosaic of peoples and cultures. Every cultural domain on the map also includes minorities that cannot be mapped at this scale, so that people of different faiths, languages, and ways of life rub shoulders everywhere. Often this results in friction, and at times such friction escalates into ethnic conflict.

As we define it, Turkestan includes most or all of six states: (1) Kazakhstan, territorially larger than the other five combined but situated astride an ethnic transition zone; (2) Turkmenistan, with important frontage on the Caspian Sea and bordering Iran and Afghanistan; (3) Uzbekistan, the most populous state and situated at the heart of the region; (4) Kyrgyzstan, wedged between powerful neighbors and chronically unstable; (5) Tajikistan, regionally and culturally divided as well as strife torn; and (6) Afghanistan, engulfed in war almost continuously since it was invaded by Soviet forces in 1979.

The Five States of Former Soviet Central Asia

During their hegemony over Central Asia, the Soviets tried to suppress Islam and install secular regimes (this was their objective when they invaded Afghanistan as well), but today Islam's revival is one of the defining qualities of this region. From Almaty to Samarqand, mosques are being repaired and revived, and Islamic dress again is part of the cultural landscape. In other ways, too, Turkestan reflects the norms of this realm: in its dry-world environments and the clustering of its population, its mountain-fed streams irrigating farms and fields, its sectarian conflicts, its oil-based economies. It also is a region where democratic government remains an elusive goal.

Kazakhstan is the territorial giant of the region and borders two greater giants: Russia and China (Fig. 7-19). During the Soviet period, northern Kazakhstan was heavily Russified and became, in effect, part of Russia's Eastern Frontier (see Fig. 2–10). Rail and road links crossed the area mapped as the Russian Transition Zone, connecting the north to Russia. The Soviets made Almaty, in the heart of the Kazakh domain, the territory's capital. Today the Kazakhs are in control, and they in turn have moved the capital to Astana, right in the heart of the Russian

The Caspian Sea area is becoming an arena for competition—a contest between governments and corporations over the oil and natural gas resources this part of the former Soviet Union contains. Among the countries with Caspian Sea coasts (and therefore with offshore as well as land-based potential), Kazakhstan is in a strong position. One of Kazakhstan's most promising oilfields is the Tengiz Basin (see map, p. 126), and the Kazakh government has issued leases and permits for foreign companies to exploit it. This huge installation, recently built by Chevron, brings American technology to this remote but profitable part of Turkestan.

Transition Zone, where some 5 million Russians still live. Astana is clearly a *forward capital*.

Figure 7-19 reveals Kazakhstan's situation as a corridor between the Caspian Basin's oil reserves and China. Oil and gas pipelines across Kazakhstan could eliminate, or greatly reduce, China's dependence on oil carried by tankers along distant sea lanes.

Uzbekistan , the most populous state in this region (26.3 million), occupies the heart of Turkestan and borders every other state in it (Fig. 7-19). Uzbeks not only make up 75 percent of this population, but also form substantial minorities in several neighboring states. The capital, Toshkent, lies in the far eastern core area of the country, where most of the people live in towns and farm villages, and the crowded Farghona Valley is the focus. In the west lies the shrinking Aral Sea (photo below), whose feeder streams were diverted into cotton fields and croplands during the Soviet occupation; heavy use

of pesticides contaminated the groundwater, and countless thousands of local people suffered severe medical problems as a result.

In the post-Soviet period, Islamic (Sunni) revivalism has become a problem for Uzbekistan. *Wahhabism*, the particularly virulent form of Sunni fundamentalism, took root in eastern Uzbekistan; government efforts to suppress it have only increased its strength in the countryside.

Turkmenistan , a desert republic that extends all the way from the Caspian Sea to the borders of Afghanistan, has a population of just 5.8 million, of which about three-quarters are Turkmen. During Soviet times, the communist planners began work on a massive project: the Garagum (Kara Kum) Canal designed to bring water from Turkestan's eastern mountains into the heart of the desert. Today the canal is 700 miles (1100 km) long, and it has enabled the cultivation of some 3 million acres of cotton, vegetables, and fruits. The plan is to extend the canal all

The Aral Sea is a saltwater lake on the boundary between Kazakhstan and Uzbekistan, both former Soviet Socialist Republics. During the communist period, the lake's two inflowing rivers were diverted for irrigation, resulting in the dramatic shrinkage illustrated by these two Landsat images. The image on the left was taken in the 1960s, when the Aral Sea covered an area of about 26,300 square miles (68,000 sq km). Thirty years later, the lake had been cut in two and the total water area combined was less than 13,000 square miles (33,700 sq km). The surface had dropped about 50 feet (15 m) and a large island in what is now known as the "Greater Sea" (as opposed to the "Lesser Sea" to the north) threatened to become another land bridge. The fate of the Aral Sea is but one chapter in Soviet-era mismanagement of the environment in the headlong rush to increase agricultural productivity. The Kazakh and Uzbek governments, with assistance from international agencies, are now in the process of restoring the flow of the two feeder rivers to arrest the Aral Sea's decline, but it is unlikely that a significant rise in the water level can be achieved.

the way to the Caspian Sea, but meanwhile Turkestan has hopes of greatly expanding its oil and gas output from Caspian Basin reserves. But look again at Figure 7-19: Turkmenistan's relative location is not advantageous for export routes.

Kyrgyzstan's topography and political geography are reminiscent of the Caucasus. Shown in yellow in Figure 7-19, Kyrgyzstan lies intertwined with Uzbekistan and Tajikistan to the point of exclaves and enclaves along its borders. The Kyrgyz, for whom the Soviets established this "republic," constitute barely 60 percent of the population of 5.1 million. Uzbeks and other minorities form a complex cultural geography; mountains and valleys isolate communities and make nation-building difficult. The agricultural economy is weak, consisting of pastoralism in the mountains and farming in the valleys. About 70 percent of the people profess allegiance to Islam, and Wahhabism has gained a strong foothold here. The town of Osh is often referred to as the headquarters of the movement in Turkestan.

Tajikistan's mountainous scenery is even more spectacular than Kyrgyzstan's, and here, too, topography is a barrier to the integration of a multicultural society. The Tajiks, who constitute 62 percent of the population of 6.5 million, are of Persian (Iranian), not Turkic, origin and speak a Persian (and thus Indo-European) language. Most Tajiks, despite their Persian affinities, are Sunni Muslims, not Shi'ites. Small though Tajikistan is, regionalism plagues the state: the government in Dushanbe often is at odds with the barely connected northern area (see Fig. 7-19), a hotbed not only of Islamic revivalism but also of anti-Tajik, Uzbek activism.

Afghanistan

Turkestan's southernmost country, Afghanistan, exists because the British and Russians, competing for hegemony in this area during the nineteenth century, agreed to tolerate it as a cushion—or **buffer state**—between them. This is how Afghanistan acquired the narrow extension leading from the main territory eastward to the Chinese border—the Vakhan Corridor (Fig. 7-21). As the colonialists delimited it, Afghanistan adjoined the domains of the Turkmen, Uzbeks, and Tajiks to the north, Persia (now Iran) to the west, and the western flank of British India (now Pakistan) to the east.

Geography and history seem to have conspired to divide Afghanistan. As Figure 7-21 shows, the towering Hindu Kush range dominates the center of the country, creating three broad environmental zones: the relatively well-watered and fertile northern plains and basins; the rugged, earthquake-prone central highlands; and the desert-dominated southern plateaus. Kabol (Kabul),

the capital, lies on the southeastern slope of the Hindu Kush, linked by narrow passes to the northern plains and by the Khyber Pass to Pakistan.

Across this variegated landscape moved countless peoples: Greeks, Turks, Arabs, Mongols, and others. Some settled here, their descendants today speaking Persian, Turkic, and other languages. Others left archeological remains or no trace at all. The present population of Afghanistan (29.2 million, plus 4 million refugees living outside the country) has no ethnic majority. This is a country of minorities in which the Pushtuns (or Pathans) of the east are the most numerous but make up less than 40 percent of the total (substantially less if those across the borders are subtracted). The second-largest minority are the Tajiks, a world away across the Hindu Kush, concentrated in the zone near Afghanistan's border with Tajikistan. The Hazara of the central highlands and the south, the Uzbeks and Turkmen in the northern border areas, the Baluchi of the southern deserts, and other, smaller groups scattered across this riven country create one of the world's most complex cultural mosaics (Fig. 7-20).

Episodes of conflict have marked the history of Afghanistan, but none was as costly as its involvement in the Cold War. Following the Soviet intervention of 1979, the United States supported the Muslim opposition, the *Mujahideen* ("strugglers"), with modern weapons and money, and the Soviets were forced to withdraw. Soon the factions that had been united during the anti-Soviet campaign were in conflict, delaying the return of some 4 million refugees who had fled to Pakistan and Iran. The situation resembled the pre-Soviet past: a feudal country with a weak and ineffectual government in Kabol.

In 1994, what at first seemed to be just another warring faction appeared on the scene: the so-called Taliban ("students of religion") from religious schools in Pakistan. Their avowed aim was to end Afghanistan's chronic factionalism by instituting strict Islamic law. Popular support in the war-weary country, especially among the Pushtuns, led to a series of successes, and by 1996 the Taliban had taken Kabol. Afghanistan's physical geography and northern opposition then slowed the Taliban's drive, but in 1998 they penetrated the Uzbek and Tajik strongholds and captured Mazar-e-Sharif, a key northern city.

The Taliban's imposition of Islamic law was so strict and severe that Islamic as well as non-Islamic countries objected. Restrictions on the activities of women ended their professional education, employment, and freedom of movement, and had a devastating impact on children as well. Public amputations and stonings enforced the Taliban's code. In the process, Afghanistan became a haven for groups of revolutionaries whose goals went far beyond those of the Taliban: they plotted attacks on Western interests throughout the realm and threatened Arab regimes

FIGURE 7-21

they deemed compliant with Western priorities. Taliban-ruled, cave-riddled, remote, and isolated Afghanistan was an ideal locale for these outlaws. Already in possession of arms and ammunition (Soviet as well as American) left over from the Cold War, they benefited also from Afghanistan's huge, illicit opium trade. In 2000, according to United Nations data, Afghanistan produced between 70 and 79 percent of the world's opium. Much of the revenue found its way into the coffers of the conspirators.

With such resources, the militants were able to organize and launch attacks against several Western targets in the realm and elsewhere, but events took a fateful turn in 1996 that would empower them as never before. During the conflict with the Soviets, the Mujahideen cause had been supported not only by the United States but also by the Saudi extremist referred to earlier in this chapter,

Usama bin Laden. The son of a Saudi construction billionaire who had more than 50 children, Usama graduated from a university in Jiddah in 1979 and headed for Afghanistan with an inherited fortune estimated at about U.S. $300 million to help the anti-Soviet campaign. Well-connected in Saudi Arabia and now a pivotal figure in Afghanistan, bin Laden saw his fame as well as his war chest grow as the Soviet intervention collapsed. Following the withdrawal of the communist forces, he returned to Saudi Arabia, where he denounced his government for allowing U.S. troops on Saudi soil during the Gulf War. The Saudi regime responded by stripping him of his citizenship and expelling him, and bin Laden fled to Sudan. There he set up several legitimate businesses to facilitate his now-global financial transactions, but he also established terrorist training camps. Under international pressure, the Khartoum regime ousted him in 1996, and bin Laden returned to a country he knew would welcome him again—Afghanistan.

Bin Laden used his fortune to support the Taliban's drive to conquer Afghanistan, and in close association with the movement's leader, Mullah Omar, he established the headquarters of the terrorist organization al-Qaeda in the country's mountainous east. From this base he coordinated the global network that had already committed numerous terrorist acts, including the first (1993) attack on New York's World Trade Center. Al-Qaeda's exploits were funded by numerous Muslim sources, including many in Saudi Arabia. This made possible well-coordinated attacks on the U.S. embassies in Nairobi, Kenya, and Dar es Salaam, Tanzania, the bombing of the *USS Cole* in Adan, Yemen, and other strikes. There was no effective response from U.S. or other Western sources. After the embassy bombings, President Clinton ordered an inconsequential missile strike on a suspected terrorist camp in eastern Afghanistan.

Meanwhile, the suicide attacks of September 11, 2001 were being planned as operatives traveled the world and moved hundreds of thousands of dollars through secret channels to enable al-Qaeda's teams to prepare. Three of the four planned attacks succeeded, killing thousands and causing billions of dollars in damage. Several weeks later (having given al-Qaeda leaders time to plan their escape), United States and British forces, with the acquiescence of the president of Pakistan, attacked both the Taliban regime and the al-Qaeda infrastructure in Afghanistan, the first stage in the proclaimed "War on Terror" that would leave virtually no country unaffected.

The allied assault on Afghanistan quickly led to the collapse and destruction of the Taliban regime, many of its soldiers and supporters killed by warlords settling old scores. But neither the leader of the Taliban nor Usama bin Laden was captured, and only a few of al-Qaeda's top 40 lieutenants were apprehended or eliminated. Usama bin Laden, whose cameraman had taped a celebratory dinner in a restaurant in Kandahar after the 9/11 attacks, continued to distribute recordings of polemics and threats.

Meanwhile, Allied forces ended the terror the Taliban had imposed on Afghan society and stabilized the political situation insofar as possible, focusing their efforts on the capital, Kabol, and its environs. Reconstruction of the devastated country started, an elected government and president (Hamid Karzai) were installed, and women were allowed once again to attend schools, serve as teachers and nurses and other professionals, and move about with some freedom. But Afghanistan is a large country, and in outlying cities such as Herat, Mazar-e-Sharif, and Feyzabad, regional governors and warlords made deals to control vast stretches of it. By the summer of 2003, only a fraction of the billions of dollars Afghanistan will need to recover had become available. Meanwhile, remnants of the Taliban and operatives of al-Qaeda lurk in the shadows of this blighted country.

As we noted earlier, several million Pushtuns and other refugees from Afghanistan have for many years lived across the border in Pakistan, where the Taliban movement was founded and where al-Qaeda terrorists have taken refuge. The events of the past several years have greatly altered the regional geography of this part of the world, including the functioning of the Afghanistan-Pakistan boundary, the movement of people and goods, the diffusion of religious doctrines, and the operation of political systems. There was a time when the transition from Southwest Asia to South Asia was evident to anyone emerging from the Khyber Pass onto the plain at Peshawar, but those days are over. Pakistan now lies directly in the path of the winds of change coming from the southern flank of Turkestan.

Within weeks following the al-Qaeda-mounted suicide attacks on New York's World Trade Center and the Pentagon, the United States military was sent into action in Afghanistan, where the terrorist organization was headquartered and given refuge by the repressive Taliban regime. Following a brief campaign, Taliban rule ended, al-Qaeda's operations were disrupted, and a new government was installed in Kabol (Kabul), the capital. But the government's powers are limited, much of the countryside remains beyond Kabol's authority, and promises of massive U.S. financial assistance had not materialized by mid-2003 even as the search for al-Qaeda's leadership continued in Afghanistan as well as neighboring Pakistan. These American soldiers are part of a search operation for militants and weapons in the town of Ahad in Paktia Province, south of Kabol, where an Afghan man coming home from the plot he is farming is told to wait with others for investigation.

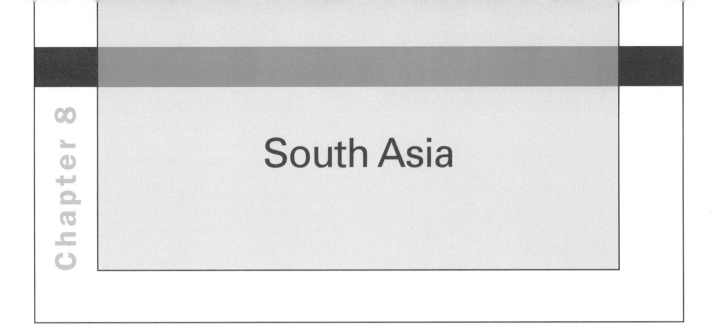

Chapter 8

South Asia

CONCEPTS, IDEAS, AND TERMS

1 Social stratification	7 Physiologic density	12 Forward capital
2 Wet monsoon	8 Rate of natural population increase	13 Irredentism
3 Refugees		14 Caste system
4 Population geography	9 Doubling time	15 Intervening opportunity
5 Population distribution	10 Demographic transition	16 Natural hazards
6 Population density	11 Population explosion	17 Insurgent state

ROM IBERIA TO Arabia and from Malaysia to Korea, Eurasia is a landmass fringed by peninsulas. The largest of all is the great triangle of India that divides the northern Indian Ocean into two seas: the Arabian Sea to the west and the Bay of Bengal to the east (Fig. 8-1). The peninsula of India forms the heart of South Asia, a vast, varied, volatile geographic realm.

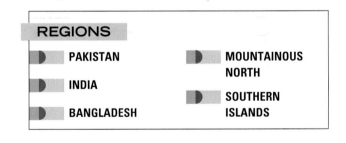

REGIONS

PAKISTAN	MOUNTAINOUS NORTH
INDIA	SOUTHERN ISLANDS
BANGLADESH	

DEFINING THE REALM

MOUNTAINS, DESERTS, AND coastlines combine to make South Asia one of the world's most vividly defined physiographic realms. To the north, the Himalaya Mountains create a natural wall between South Asia and China. To the east, mountain ranges and dense forests mark the boundary between South and Southeast Asia. To the west, rugged highlands and expansive deserts separate South Asia from its neighbors. Within these confines lies a geographic realm that is more densely populated than any other. If current population trends continue, it will soon be the most populous realm on Earth as well.

South Asia consists of five regions (Fig. 8-2). Its keystone is India, whose population passed the 1 billion mark in 1999. In the west lies Pakistan. South Asia's eastern flank is centered on Bangladesh. The northern region consists of the mountainous lands of Kashmir, Nepal, and Bhutan. And the southern region includes the islands of Sri Lanka and the Maldives. As the map shows, India divides into several subregions.

South Asia's physiographic boundaries are formidable barriers, but they have not prevented conquerors or proselytizers from penetrating it. As a result, today this

FIGURE 8-1

South Asia

1. South Asia is clearly defined physiographically and is bounded by mountains, deserts, and ocean; the Indian peninsula is Eurasia's largest.

2. South Asia is the world's most poverty-filled realm, with low average incomes, low levels of education, poorly balanced diets, and poor overall health.

3. With only 3 percent of the world's land area but 22 percent of its population, more than half of it engaged in subsistence farming, South Asia's economic prospects are bleak.

4. Population growth rates in South Asian countries remain high; India's population surpassed the 1 billion mark in 1999.

5. The North Indian Plain, the lower basin of the Ganges River, contains the heart of the world's second largest population cluster.

6. Despite encircling mountain barriers, invaders from ancient Greeks to later Muslims penetrated South Asia and complicated its cultural mosaic.

7. British colonialism unified South Asia under a single flag, but the empire fragmented into several countries along cultural lines after Britain's withdrawal.

8. Pakistan, South Asia's western region, lies on the flanks of two realms: largely Muslim North Africa/Southwest Asia and dominantly Hindu South Asia.

9. India is the world's largest federation and most populous democracy, but its political achievements have not been matched by enlightened economic policies.

10. Religion remains a powerful force in South Asia. Hinduism in India, Islam in Pakistan, and Buddhism in Sri Lanka all show tendencies toward fundamentalism and nationalism.

11. Active and potential boundary problems involve internal areas (notably between India and Pakistan in Kashmir) as well as external locales (between India and China in the northern mountains).

realm is a patchwork of religions, languages, traditions, and cultural landscapes. So complex is this mosaic that, remarkably, its political geography numbers just seven states (and only five on the mainland).

Among the cultural infusions was Islam. Today, Pakistan is an Islamic republic, and Islam provides the cement for the state. Pakistan's eastern border with India is a cultural divide in more ways than one: dominantly Hindu India is a secular, not a theocratic, state. Why, then, do we include Pakistan in the South Asian rather than the Southwest Asian/North African realm? One criterion is ethnic continuity, which links Pakistan to India rather than to Afghanistan or Iran. Another is historical geography. Pakistan was part of Britain's South Asian Empire, and it originated from the partition of that domain between Muslim and Hindu majorities. Although Urdu is the official national language of Pakistan, English is the *lingua franca*, as it is in India. Furthermore, the border between India and Pakistan does not signify the eastern frontier of Islam in Asia. About 150 million of India's more than 1 billion citizens are Muslims, and in South Asia's eastern region, Bangladesh (population: 140 million) is more than 85 percent Muslim. Finally, Pakistan and India are locked in a struggle to control a vital mountainous area in the far north, where the British withdrawal in 1947 left the boundary between them unresolved. Even as Indian and Pakistani cricket teams play each other on sun-baked pitches, their armies face off in a deadly conflict—a conflict that has the potential to unleash a nuclear war.

South Asia and the War on Terror

The western boundary of South Asia has long had the characteristics of a frontier—a zone of transition and contention not all of which is fully integrated into the political systems of the adjoining states. The fateful Soviet invasion and occupation of Afghanistan (1979–1989), when Moscow tried to stabilize that fractious country and establish a secular regime in Kabol (Kabul), caused a bitter conflict in which Afghans fought each other even as they resisted their joint enemy. That war drove several million Pushtun and other Afghan refugees across the Pakistani border. Already along this same border, inside Pakistan but effectively not under governmental control, lay a mountainous, forbidding, inaccessible strip of land called the *Tribal Areas* (see the map on page 370)—where ethnicity, traditions, clan systems, and cultural landscapes were more representative of remote parts of Turkestan than of mainstream Pakistan. Here no one, not even a Pakistani official, goes without the consent of chief and mullah.

In the proxy-war context of the Cold War, the United States saw an opportunity to damage Soviet interests and so dispatched an enormous quantity of weapons to the Afghan resistance fighters via Pakistan. This infusion, in addition to the impact of the ever-growing refugee population, changed the ideological landscape in Pakistan itself. Even as the Soviets were defeated (Usama bin Laden was on the Afghan scene and financed the campaign as well), Islamic revivalism mushroomed in neighboring

FIGURE 8-2

Pakistan. Islamic schools teaching strict Islamic doctrines, approved by Pakistan and supported financially by Saudi Arabians, gave rise to the fundamentalist Taliban movement that swept into power in chaotic Afghanistan during the 1990s.

The Taliban regime brought Afghanistan stability at a dreadful social price; it also created a safe base for the al-Qaeda terrorist movement that was to strike at West-

ern interests the world over, including, ultimately, the United States itself. As noted in Chapter 7, the attacks of September 11, 2001 provoked massive U.S. retaliation in Afghanistan, where the Taliban regime was eliminated and al-Qaeda members were killed or arrested (most of the leadership escaped and went into hiding, possibly in the Tribal Areas). But for Pakistan this campaign had momentous consequences. The leader of its military regime,

General Pervez Musharraf, decided to side with the United States in its war against terrorism—a complete reversal for a country that had spawned the Taliban, recognized the Taliban regime, and sympathized with the movement's Islamist aspirations.

We defer the implications of all this for Pakistan's political geography until our discussion of this realm's regions, but in terms of regional definitions it is clear that things are changing on South Asia's western flank. Pakistan and India continue their deadly South Asian embrace over Kashmir and other issues, but meanwhile Pakistan looks westward, toward its porous western frontier with Afghanistan and Turkestan beyond, as it contemplates its role in the world of Islam in the era of the War on Terror.

A REALM OF POVERTY

At the opening of the twenty-first century, South Asia accounts for more than one-fifth of the world's population and two-thirds of its poorest inhabitants. Its literacy rates are among the lowest in the world. Nearly half of the people in this realm earn less than the equivalent of one U.S. dollar per day. It is estimated that half the children in South Asia are malnourished and underweight, most of them girls. South Asia is often called the most deprived realm in the world.

A combination of geographic factors underlies this tragic picture. With 22 percent of the world's population but just 3 percent of its land area, South Asia lacks the natural resources to raise living standards for its hundreds of millions of subsistence farmers. Governmental policies contribute to the problem: while East and Southeast Asia forged ahead by looking outward, encouraging exports and foreign investment, and spending heavily on literacy and technical education, health care, and land reform, South Asian governments tended to adopt bureaucratic controls and state planning. Cultural traditions also play their role. Resistance to change and reluctance by the privileged to open doors of opportunity to the less advantaged inhibit economic advancement for all.

Here is a grim reality: we will note repeatedly in this chapter that South Asia is self-sufficient in food and that India produces enough grain to not only be able to feed its own population but to fill the needs of its neighbors. In early 2003, official statistics showed that India had a stored wheat surplus of more than 50 million metric tons and a rice surplus of 5 million tons, enough to feed the entire realm's population for more than a year, with another good harvest in sight. But this apparent triumph of the Green Revolution did not end malnourishment in India. Despite growing annual surpluses, India's children remain among the most malnourished in the world:

at the turn of the millennium, 47 percent were reported to be either severely malnourished or moderately malnourished. While food rots in government warehouses in Punjab, children die in villages in the neighboring State of Rajasthan.

How can a government hoard food in huge quantities while letting the poor die of starvation? The answer is economics—and politics. India's government responds to the pressures of powerful farm lobbies, raising the price of grain and then buying it from commercial farmers at higher prices for storage and redistribution, thereby increasing the market cost for poor villagers who cannot afford to buy it. While millions of villagers go hungry, the government is also under pressure from international organizations, which lend India money, to limit food subsidies to "consumers." In addition, India's nationwide subsidized food distribution system, set up four decades ago, suffers from inefficiency, corruption, and massive theft. As a result, poverty and government mismanagement combine to condemn tens of millions of India's rural poor to malnutrition and even starvation at a time when official statistics proclaim that the country produces enough food not only to feed itself but even to export grains to hungry neighbors. We will revisit this issue later in the chapter; let us next assess the realm's physical geography.

PHYSIOGRAPHIC REGIONS OF SOUTH ASIA

Before we look into South Asia's complex and fascinating cultural geography, we need to discuss the physical stage of this populous realm. South Asia is a realm of immense physiographic variety, of snowcapped peaks and forest-clad slopes, of vast deserts and broad river basins, of high plateaus and spectacular shores. The collision of two of the Earth's great tectonic plates created the world's highest mountain ranges, their icy crests yielding meltwaters for great rivers below. The workings of the Earth's atmosphere put South Asia in the path of tropical cyclones and produce reversing seasonal windflows known as *monsoons*. This is a realm of almost infinite variety, a world unto itself.

In general terms, we can recognize three clearly defined physiographic zones in South Asia: the northern mountains, the southern peninsular plateaus, and between them a belt of river lowlands. Superimposed on this configuration is an east-west precipitation gradient from wet (Bangladesh) to dry (western Pakistan) that is clearly visible in Figures G-7 and G-8, broken only by the strip of high moisture along India's southwestern Malabar Coast (Fig. 8-1).

Northern Mountains

The northern mountains extend from the Hindu Kush and Karakoram Ranges in the northwest through the Himalayas in the center (Mount Everest, the world's tallest peak, lies in Nepal) to the ranges of Bhutan and the Indian State of Arunachal Pradesh in the east. Dry and barren in the west on the Afghanistan border, the ranges become green and tree-studded in Kashmir, forested in the lower sections of Nepal, and even more densely vegetated in Arunachal Pradesh. Transitional foothills, with many deeply eroded valleys cut by rushing meltwaters, lead to the river basins below.

River Lowlands

The belt of river lowlands extends eastward from Pakistan's lower Indus Valley (the area known as Sind) through the wide plain of the Ganges Valley of India and on across the great double delta of the Ganges and Brahmaputra in Bangladesh (Fig. 8-1). In the east, this physiographic region often is called the North Indian Plain. To the west lies the lowland of the Indus River, which rises in Tibet, crosses Kashmir, and then bends southward to receive its major tributaries from the Punjab ("Land of Five Rivers").

Southern Plateaus

Peninsular India is mostly plateau country, dominated by the massive Deccan, a tableland built of basalt that poured out when India separated from Africa during the breakup of Gondwana (see Fig. 6-3). The Deccan (meaning "South") tilts to the east, so that its highest areas are in the west and the major rivers flow into the Bay of Bengal. North of the Deccan lie two other plateaus, the Central Indian Plateau to the west and the Chota-Nagpur Plateau to the east (Fig. 8-1). On the map, note the Eastern and Western Ghats ("hills") that descend from Deccan plateau elevations to the narrow coastal plains below. Onshore winds of the annual wet monsoon bring ample rain to the Western Ghats (see box titled "South Asia's Life-Giving Monsoon" and Fig. 8-3). As a result, here lies one of India's most productive farming areas and one of southern India's largest population concentrations.

THE HUMAN SEQUENCE

Great river basins mark the physiography of South Asia; in one of these basins, that of the Indus River in present-day Pakistan, lies evidence of the realm's oldest major civilization. It existed at the same time, and interacted with, ancient Mesopotamia, and it was centered on large, well-organized cities (Fig. 8-4). From here, influences

and innovations diffused into India. In fact, India's very name is believed to derive from the ancient Sanskrit word *sindhu*, used to identify the ancient civilizations in the Indus Valley.

Eventually, such cities as Harappa and Mohenjo-Daro seem to have experienced the same fate as those of Mesopotamia, perhaps also because of environmental change. About 3500 years ago Aryan (Indo-European)-speaking peoples migrated into the region from the northwest, penetrating India and starting the process of welding the Ganges Basin's isolated tribes and villages into an organized system. Having absorbed much of the culture of the Indus Valley, they brought a new order to this region. A new belief system, *Hinduism*, arose and with it a new way of life. A complex **social stratification** developed in which Brahmans, powerful priests, stood at the head of a caste system in which soldiers, merchants, artists, peasants, and all others had their place.

Hinduism was restrictive, especially for those of the lower castes, and in the sixth century BC (more than 2500 years ago) a prince born into one of northern India's kingdoms sought a better way. Prince Siddhartha, better known as Buddha, gave up his royal position to teach religious salvation through meditation, the rejection of Earthly desires, and a reverence for all forms of life. His teachings did not have a major impact on the Hindu-dominated society of his time, but what he taught was not forgotten. Centuries later, the ruler of a powerful Indian state decided to make *Buddhism* the state religion, following which the faith spread far and wide.

But first South Asia was penetrated by other foreign influences: the Persians were followed by the Greeks under Alexander the Great in the late fourth century B.C. This is why, on a world map of languages, Hindi (the major domestic language of India) lies at the eastern end of the region of Indo-European languages, in the same family as Persian, Italian, and English. But the map of India's languages (Fig. 8-5) shows that southern India's languages are *not* Indo-European. Indeed, while northern India was a subregion of cultural infusion and turmoil, the south lay remote and isolated. This part of the peninsula had apparently been settled long before the Indus and Ganges civilizations arose, and southern India developed into a distinctive subregion with its own cultures. The *Dravidian* languages spoken here—Telugu, Tamil, Kanarese (Kannada), and Malayalam—have long literary histories.

Aśoka's Mauryan Empire

When the Greeks withdrew from the Ganges Basin and the Hindu heartland was once again free, a powerful empire arose there—the first true empire in the realm. This,

South Asia's Life-Giving Monsoon

The dynamics of daily summer sea breezes are well known. The hot sun heats up the land, the air over this surface rises, and cool air from over the water blows in across the beach, replacing it. Teeming beaches and high-cost seafront apartments attest to the efficacy of this natural air conditioning.

Under certain circumstances, this kind of circulation can affect whole regions. When an entire landmass heats up, a huge low-pressure system forms over it. This system can draw vast volumes of air from over the oceans onto the land. It is not just a local, daily phenomenon. It takes months to develop, but once it is in place, it also persists for months. When the inflow of moist oceanic air starts over South Asia, the **wet monsoon** has arrived. It may rain for 60 days or more. The countryside turns green, the paddies fill, and another dry season's dust and dirt are washed away. The region is reborn (see photo pair, p. 406).

Not all continents or coasts experience monsoons. A particular combination of topographic and atmospheric circumstances is needed. Figure 8-3 shows how it works in South Asia. It is June. For months, a low-pressure system has been building over northern India. Now, this cell has become so powerful that it draws air from over a large area of the warm, tropical Indian Ocean toward the interior. Some of that moisture-laden air is forced upward against the Western Ghats ①, cooling as it rises and condensing large amounts of rainfall. Other streams of air flow across the Bay of Bengal and get caught up in the convection over northeastern India and Bangladesh ②. Seemingly endless rain now inundates a much larger area, including the whole North Indian Plain. The mountain wall of the Himalayas stops the air from spreading and the rain from dissipating ③. The air moves westward, drying out as it flows toward Pakistan ④.

After persisting for weeks, the system finally breaks down and the wet monsoon gives way to periodic rains and, eventually, another dry season. Then the anxious wait begins for the next year's monsoon, for without it India would face disaster. In recent years, the monsoon has shown signs of irregularity and, as in 1987, partial failure. In India, life hangs by a meteorological thread.

FIGURE 8-3

the Mauryan Empire, extended its influence over India as far west as the Indus Valley (thus incorporating the populous Punjab) and as far east as Bengal (the double delta of the Ganges and Brahmaputra); it reached as far south as the modern city of Bangalore.

This Mauryan Empire was led by a series of capable rulers who achieved stability over a vast domain. Undoubtedly the greatest of these leaders was Aśoka, who reigned for nearly 40 years during the middle of the third century B.C. Aśoka was a believer in Buddhism, and it was he who elevated this religion from obscurity to regional and ultimately global importance.

In accordance with Buddha's teachings, Aśoka reordered his government's priorities from conquest and expansion to a Buddhist-inspired search for stability and peace. He sent missionaries to the outside world to carry

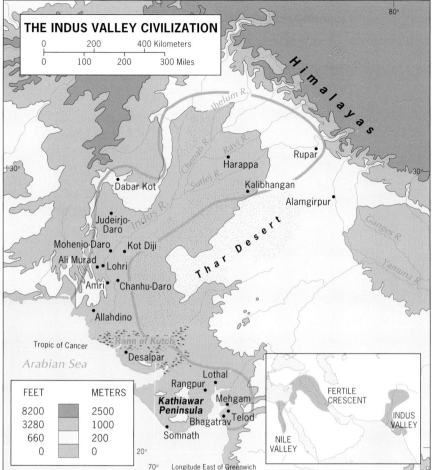

THE INDUS VALLEY CIVILIZATION

FEET	METERS
8200	2500
3280	1000
660	200
0	0

FIGURE 8-4

Buddha's teachings to distant peoples, thereby also contributing to the diffusion of Indian culture. As a result, Buddhism became permanently established as the dominant religion in Sri Lanka (formerly Ceylon), and it established footholds as far afield as Southeast Asia and Mediterranean Europe. Ironically, Buddhism thrived in these remote places even as it declined in India itself. With Aśoka's death, the faith lost its strongest supporter.

The Mauryan Empire represented India's greatest political and cultural achievements in its day, and when the empire collapsed, late in the second century A.D., India fragmented into a patchwork of states. Once again, India lay open to infusions from the west and northwest, and across present-day Pakistan they came: Persians, Afghans, Turks, and others driven from their homelands or attracted by the lands of the Ganges.

The Power of Islam

In the late tenth century, Islam came rolling like a giant tide across the subcontinent, spreading from Persia in the west and Afghanistan in the northwest. Of course, the Indus Valley lay directly in the path of this Islamic advance, and virtually everyone was converted. Next the Muslims penetrated the Punjab, the subregion that lies astride the present Pakistan-India border, and there perhaps as many as two-thirds of the inhabitants became converts. Then Islam crossed the bottleneck where Delhi is situated and diffused east- and southeastward into the Gangetic Plain and the subregion known as Hindustan—India's evolving core area. Here Islam's proselytizers had less success, persuading perhaps one in eight Indians to become Muslims. In the meantime, Islam arrived at the Ganges Delta by boat, and present-day Bangladesh became overwhelmingly Islamic. (To the south of the Ganges heartland, however, Islam's diffusion wave lost its energy: Dravidian India never came under Muslim influence.)

Islam's vigorous, often violent, onslaught changed Indian society. As in West Africa, Islam frequently was superimposed through political control: when the rulers were converted, their subjects followed. By the early fourteenth century, a sultanate centered at Delhi controlled more of the subcontinent than even the Mauryan Empire had earlier. Later, the Islamic Mogul Empire (the similarity to the word "Mongol" is by no means a coincidence) constituted the largest political entity ever to unify the realm in precolonial times. To many Hindus of lower caste,

FIGURE 8-5

Islam represented a welcome alternative to the rigid socio-religious hierarchy in which they were trapped at the bottom. Thus Islam was the faith of the ruling elites and of the disadvantaged, a powerful cultural force in the heartland of Hinduism.

Just as Islam weakened in southern Europe, so its force ultimately became spent in vast and populous India. For all the Muslims' power, they never managed to convert a majority of South Asians. They dominated the northwest corner of the realm (present-day Pakistan), where Lahore became one of Islam's greatest cities. But in all of what is today India, less than 15 percent of the population became and remained Muslim. And throughout the period of Islamic intervention, the struggle for cultural supremacy continued. Placid Hinduism and aggressive Islam did not easily coexist.

The European Intrusion

Into this turbulent complexity of religious, political, and linguistic disunity yet another element began to intrude after 1500: European powers in search of raw materials, markets, and political influence. Because the Europeans profited from the Hindu-Muslim contest, they exploited local rivalries, jealousies, and animosities. British merchants gained control over the trade with Europe in spices, cotton, and silk goods, ousting the French, Dutch, and Portuguese. The British East India Company's ships also took over the intra-Asian sea trade between India and Southeast Asia, which had long been in the hands of Arab, Indonesian, Chinese, and Indian merchants. In effect, the East India Company (EIC) became India's colonial administration.

In time, however, the EIC faced problems that made it increasingly difficult to combine commerce with administration. Eventually, a mutiny among Indian troops in the service of the EIC led to the abolition of the company. The British government took over in 1857 and maintained its rule (*raj*) until 1947.

Colonial Transformation

When the British took power over South Asia, they controlled a realm with considerable industrial development (notably in metal goods and textiles) and an active trade with both Southwest and Southeast Asia. The colonialists saw this as competition, and soon India was exporting raw materials and importing manufactured goods—from Europe, of course. Local industries declined and Indian merchants lost their markets.

Unifying their realm was a tougher task for the British. In 1857 about 750,000 square miles (2 million sq km) of South Asian territory still was beyond British control, including hundreds of entities that had been guaranteed autonomy by the EIC during its administration. These "Native States," ranging in size from a few acres to Hyderabad's 80,000 square miles (200,000 sq km), were assigned British advisors, but in fact India was a near-chaotic amalgam of modern colonial and traditional feudal systems.

Colonialism did produce assets for India. The country was bequeathed one of the best transport networks of the colonial domain, especially the railroad system (although the network focused on interior-seaport linkages rather than fully interconnecting the various parts of the country). British engineers laid out irrigation canals through which millions of acres of land were brought into cultivation. Settlements that had been founded by Britain developed into major cities and bustling ports, led by Bombay (now Mumbai), Calcutta (now Kolkata), and Madras (now Chennai). These three cities are still three of India's largest urban centers, and their cityscapes bear the unmistakable imprint of colonialism (see photo at right). Modern industrialization, too, was brought to India by the British on a limited scale. In education, an effort was made to combine English and Indian traditions; the Westernization of India's elite was supported through the education of numerous Indians in Britain. Modern practices of medicine were also introduced. Moreover, the British administration tried to eliminate features of Indian culture that were deemed undesirable by any standards—such as the burning alive of widows on their husbands' funeral pyres, female infanticide, child marriage, and the caste system. Obviously, the task was far too great to be achieved in barely three generations of colonial rule, but independent India itself has continued these efforts where necessary.

Partition

Even before the British government decided to yield to Indian demands for independence, it was clear that British India would not survive the coming of self-rule as a single political entity. As early as the 1930s, the idea of a separate Pakistan was being promoted by Muslim activists, who circulated pamphlets arguing that British India's Muslims were a nation distinct from the Hindus and that a separate state consisting of Sind, Punjab, Baluchistan, Kashmir, and a portion of Afghanistan should be created from the British South Asian Empire

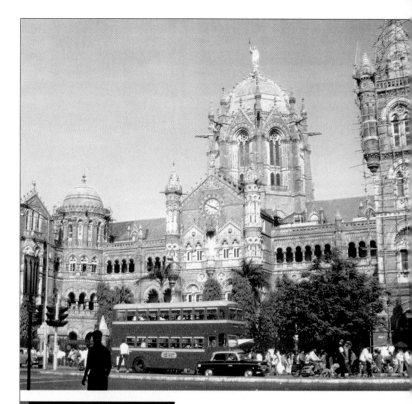

FROM THE FIELD NOTES

"More than a half-century after the end of British rule, the centers of India's great cities continue to be dominated by the Victorian-Gothic buildings the colonists constructed here. Here is evidence of a previous era of globalization, when European imprints transformed urban landscapes. Walking the streets of Mumbai (the British called it Bombay) you can turn a corner and be forgiven for mistaking the scene for London, double-deckered buses and all. One of the British planners' major achievements was the construction of a nationwide railroad system, and railway stations were given great prominence in the urban architecture. I had walked up Naoroji Road, having learned to dodge the wild traffic around the circles in the Fort area, and watched the throngs passing through Victoria Station. Inside, the facility is badly worn, but the trains continue to run, bulging with passengers hanging out of doors and windows."

in this area. The first formal demand for such partitioning was made in 1940, and, as later elections proved, the idea had almost universal support among the realm's Muslims.

As the colony moved toward independence, a political crisis developed: India's majority Congress Party would not even consider partition, and the minority Muslims refused to participate in any future unitary government. But partition would not be a simple matter. True, Muslims were in the majority in the western and eastern sectors of British India, but Islamic clusters were scattered throughout the realm (Fig. 8-6). Any new boundaries between Hindus and Muslims to create an Islamic Pakistan and a Hindu India would have to be drawn right through

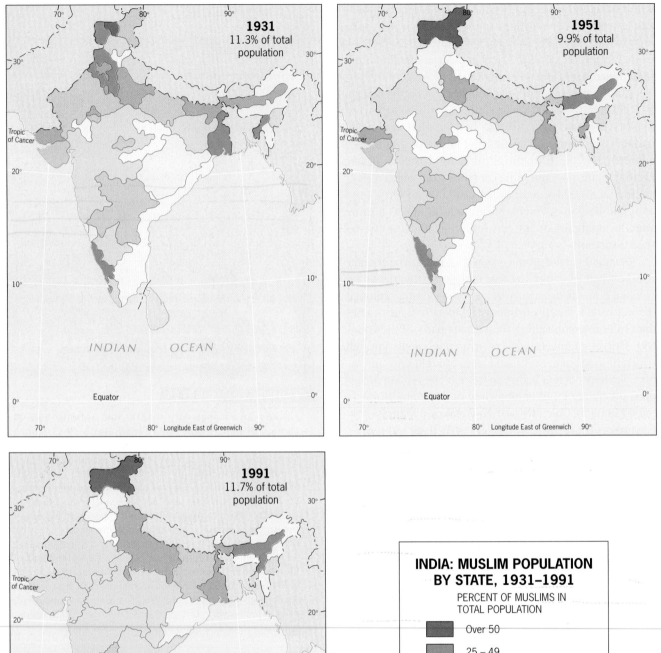

1931
11.3% of total population

1951
9.9% of total population

1991
11.7% of total population

**INDIA: MUSLIM POPULATION
BY STATE, 1931–1991**

PERCENT OF MUSLIMS IN
TOTAL POPULATION

Over 50
25 – 49
15 – 24
10 – 14
5 – 9
2.5 – 4
Under 2.5

FIGURE 8-6

Flight was one response to the 1947 partition of what had been British India, resulting in one of the greatest mass population transfers in human history. Here, two trainloads of eastbound Hindu refugees fleeing (then) West Pakistan arrive at the station in Amritsar, the first city inside India.

areas where both sides coexisted. People by the millions would be displaced.

Nor were Hindus and Muslims the only people affected by partition. The Punjab area, for example, was home to millions of Sikhs, whose leaders were fiercely anti-Muslim. But a Hindu-Muslim border based only on those two groups would leave the Sikhs in Pakistan. Even before independence day, August 15, 1947, Sikh leaders talked of revolt, and there were some riots. But no one could have foreseen the dreadful killings and mass migrations that followed the creation of the boundary and the formation of independent Pakistan and India. The precise number of people who felt compelled to participate in the ensuing migrations will never be known; 15 million is the most common estimate. It was human suffering on an incomprehensible scale.

Even so large a flow of cross-border **refugees**, however, hardly began to "purify" India of Muslims. After the initial mass exchanges, there still were tens of millions of Muslims in India (Fig. 8-6). Today, the Muslim minority in Hindu-dominated India is about as large as the whole Islamic population of Pakistan, having more than tripled since the late 1940s. It is the world's largest cultural minority, far more than a mere remnant of the days when Islam ruled the realm. This force will play a growing role in the India of the future.

Twenty-First-Century Regional Flashpoints

Ethnically complex, multicultural South Asia is a realm of remarkable accommodation as well as persistent conflict. The conflicts have ancient, recent, and current causes. The Islamic invasion of Hindu society is an ancient source of continuing strife; the imperial partition and colonial restructuring of South Asia created enduring causes of conflict; the Cold War produced hotbeds of strife still smoldering; and now the War on Terror is spawning still another basis for discord. No country in South Asia has been left unaffected. India's huge Muslim minority constitutes only one of its social and political challenges. Sri Lanka's Tamil minority, brought to the island under colonial auspices, generated a revolt that has destabilized the country for decades. In Nepal, Maoist rebels struggle to overthrow the government in Cold War fashion.

But the most serious flashpoint in South Asia is Kashmir, where India and Pakistan meet in the mountains of the far north, because this is an international, not a domestic, problem. When the boundary between Hindu India and Muslim Pakistan was hastily delimited in 1947, it stopped short of the northern territory of Jammu and Kashmir, one of the 562 "Native States" recognized by the British colonial administration. Kashmir (the short form of the territory's name) had a Hindu Maharajah but a majority Muslim population, and both Pakistan and India wanted it. When British rule ended, Pakistani and Indian armies were soon at war in Kashmir. After repeated bouts of conflict, Kashmir today is divided along the latest armistice line, but neither side is prepared to yield. When this protracted and costly struggle started, India and Pakistan were armed with conventional weapons. Today they are both nuclear powers, which transforms Kashmir from another of the world's problem frontiers into a potentially catastrophic flashpoint for nuclear war. We focus on the background and implications in the regional section of this chapter.

SOUTH ASIA'S POPULATION DILEMMA

We noted earlier that poverty is pervasive in South Asia and that hunger goes with it. Poverty's persistence also is related to *demographic issues*, that is, the size, distribution, and growth of population in a realm, region, country, or province and factors associated with these conditions. The field of **population geography**, the spatial expressions of demography, focuses on such matters. It is useful to refer back to Figure G-9, which displays the continuing concentration of population in South Asia's major river valleys, underscoring the enduring dependence of the majority of the people on these ribbons of life.

Figure G-9 reveals a crucial facet of population geography: global and regional **population distribution**. Using various symbols (in this case, dots representing quantity), geographers represent the demographic situation spatially so that we can observe from a glance at the map the overall location of densely and sparsely peopled areas. But the general population distribution does not tell us much about **population density**, a more specific measure that reports the number of people who reside in a unit area (e.g., a square mile or square kilometer). Column 4 of Table G-1 lists population density by country, per square mile. That average is often cited in official statistics, but without additional data it is not very meaningful. When millions of people live in a river valley and almost nobody lives in deserts or mountains, what does Pakistan's "arithmetic" average of 487 really mean? Even parts of India, populous as it is, are very sparsely populated. A more insightful measure of density is based on the amount of agriculturally productive land. How many people are there per unit area of land suitable for farming or pastoralism? This **physiologic density** measure, listed in column 5 of Table G-1, shows that Sri Lanka and Bangladesh, not India, are the most densely populated countries in the realm. (The very high figure for the tiny islands of the Maldives reflects the virtual absence of agricultural potential there.)

Critical to any assessment of South Asia's social and economic geography is the realm's **rate of natural population increase**, the number of births minus the number of deaths, usually given as a percentage or per thousand in the population. By world standards these rates, for the realm's individual countries, remain high at around 1.8 percent per year. When populations grow this rapidly on already-large bases, economic advancement is impeded. From the rate of annual population increase, we can calculate a useful measure of a country's demographic condition: its **doubling time**, the number of years it would take, at current rates, for a population to grow twice as large. Some Muslim countries have high growth rates and thus low doubling times, reflecting the youthfulness (and restiveness) of their populations. Check column 9 of Table G-1 and compare South Asia's collective and national doubling times with those of other areas of the world.

Dynamics of Population Growth

South Asia is a realm of massive population clusters. Its five mainland countries today contain about 1.4 billion people—more than one-fifth of all humankind. India alone has 1.086 billion inhabitants, second only to China among countries of the world, and is on course to overtake it. Pakistan, with a population of 150 million, and Bangladesh, with 140 million, also rank among the dozen most populous countries of the world.

Not only is South Asia populous, but its population also is growing rapidly at 1.8 percent per year, yielding a doubling time of only 39 years. Pakistan is growing even faster at 2.1 percent, doubling in just 33 years. As a result, economic gains are being overtaken by growing numbers, and although the Green Revolution narrowed the food deficit, hundreds of millions of children do not get balanced meals or adequate calories.

Three-quarters of South Asia's population live in India, so we should consider this global giant in the context of its population geography and its impact on the realm as a whole. Comparing the map of world economies (Fig. G-11) with the list of population growth rates in Table G-1, we note a clear pattern: the bulk of population growth is occurring in the lower-income economies. In many of the high-income economies, population growth is small, has leveled off, or is even negative. These higher-income economies have gone through the so-called **demographic transition**, a four-stage sequence that took them from high birth rates and high death rates in preindustrial times to very low birth rates and very low death rates today (Fig. 8-7). Stages 2 and 3 in this model constitute the **population explosion**, a hallmark of the twentieth century: death rates in the industrializing and urbanizing countries dropped, but birth rates took longer to decline. In 1900, the world's population was about 1.5 billion; by 2000 it had surpassed 6 billion.

When the British ruled India during the nineteenth century, the country still was in the first stage, with high birth rates and high death rates; the high death rates were caused not only by a high incidence of infant and child mortality but also by famines and epidemics. As Figure 8-7 indicates, the population during Stage 1 does not grow or decline much, but it is not stable. Famines and disease outbreaks kept erasing the gains made during better times. But then India entered the second stage. Birth rates remained high, but death rates declined because medical

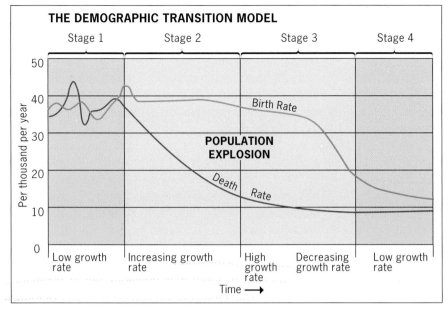

THE DEMOGRAPHIC TRANSITION MODEL

FIGURE 8-7

services improved (soap came into widespread use), food distribution networks became more effective, farm production expanded, and urbanization developed. In the 1920s, India's population still was growing at a rate of only 1.04 percent, but by the 1970s, that rate had shot up to 2.22 percent per year (Fig. 8-8). Note that India gained 28 million people during the 1920s but a staggering 135 million during the 1970s.

Has India entered the third stage, when the death rate begins to level off and birth rates decline substantially, narrowing the gap and slowing the annual increase? The rate of increase suggests it has: from 2.22 percent during the 1970s, it dropped to 2.11 percent in the 1980s and to a projected 1.88 percent during the 1990s (Fig. 8-8). But India has another problem. During its population explosion, its numbers grew so large that even a declining rate of natural increase continues to add ever greater numbers to its total. In Figure 8-8, we see that while the decadal rate of increase dropped from 2.22 to 2.11 between 1970 and 1990, the millions added grew from 135 in the 1970s to 161 during the 1980s. The rate of natural increase for the 1990s is estimated to have been 1.88 percent, adding a still larger number—175 million, taking the total past 1 billion during 1999. If India has indeed entered the third stage of the demographic transition, it will not feel its effects for some time.

FIGURE 8-8

Some population geographers theorize that all countries' populations will eventually stabilize at some level, just as Europe's did. Certain governments, notably China's, have instituted regulations to limit family size, but this policy is more easily implemented by dictatorships than democracies. And even if such stabilization were just one doubling time away in India, the country still would have an astronomical total of more than 2 billion residents.

Geography of Demography

Statistics for a country as large as India tend to lose their usefulness unless they are put in geographic context. In its demographic as in so many of its other aspects, there is not just one India but several regionally different and distinct Indias. Figure 8-9 takes population growth down to the State level and provides a comparison between the census periods of 1981–1991 and 1991–2001 (the names of India's States are shown in Fig. 8-12).

During the period from 1981 to 1991, the highest growth rates were recorded in the States of the northeast, and only eight States had growth rates below 2 percent. But between 1991 and 2001, no fewer than ten States, in-

cluding Goa, had growth rates below 2 percent. Comparing these two maps tells us that little has changed in India's heartland, where the populous States of West Bengal and Madhya Pradesh show slight decreases but Uttar Pradesh and Bihar display similarly slight increases. The most important reductions in rates of natural increase are recorded in parts of the northeast, the east, and the south. In the tip of the peninsula, Kerala and neighboring Tamil Nadu have growth rates lower than that of the region's leading country, Sri Lanka.

One reason for these contrasts lies in India's federal system: individual States pursued population-control policies to reduce population growth, including mass sterilizations. But another, as we note in more detail in the regional section, relates to spatial differences in India's development. As in the world at large, India's economically better-off States have lower rates of population growth.

India is a Hindu-dominated, officially secular country in which population policies can be debated and implemented. Pakistan, on the other hand, is a strictly Islamic state in which no such options exist. Population-control policies are regarded as incompatible with Islamic tenets, and Pakistan remains one of the world's fast-growing nations.

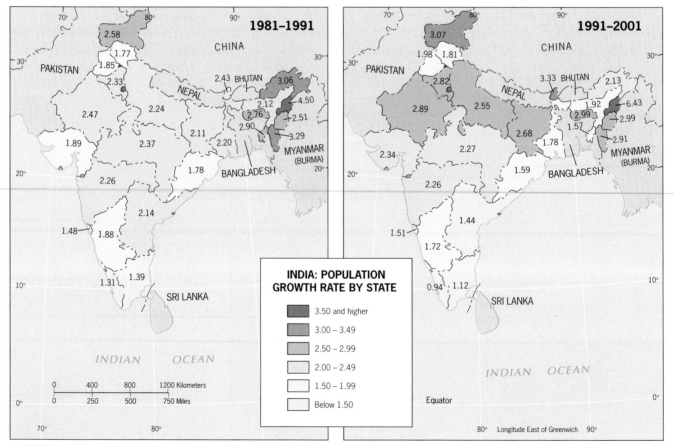

FIGURE 8-9

Among world realms, only Subsaharan Africa suffers more severely from poverty and social dislocation than South Asia—but, as Table G-1 indicates, South Asian conditions are only marginally better than Africa's, and South Asia has more than twice as many people as Subsaharan Africa. We turn next to South Asia's regional components.

REGIONS OF THE REALM

PAKISTAN: ON SOUTH ASIA'S WESTERN FLANK

If India is the dominant entity in South Asia, why focus first on Pakistan? There are several reasons, both historic and geographic. Here lay South Asia's earliest urban civilizations, whose innovations radiated into the great peninsula. Here lies South Asia's Muslim frontier, contiguous to the great Islamic realm to the west and irrevocably linked to the enormous Muslim minority to its east. Pakistan's cultural landscapes bear witness to its transitional location. Teeming, disorderly Karachi is the typical South Asian city; as in India, the largest urban center lies on the coast. Historic, architecturally Islamic Lahore is reminiscent of the scholarly centers of Muslim Southwest Asia. In Pakistan's east, the partition boundary divides a Punjab that is otherwise continuous, a land of villages, wheat fields, and irrigation ditches. In the northwest, Pakistan resembles Afghanistan in its huge migrant populations and its mountainous frontier. And in the far north, Pakistan and India are locked in a deadly conflict over Jammu and Kashmir. The western flank is South Asia's most critical region.

This was perhaps never more true than it is today. Pakistan lies on the eastern periphery of the vast contiguous Islamic realm that extends from Casablanca to Kashmir, in the shadow of Afghanistan and in the fulcrum of the War on Terror. Islamic Pakistan's coherence and stability have become crucial at a time when the global struggle against terrorism is entangling its leaders with Western power and priorities. Pakistan's role in this struggle, dictated by its chief of state, does not have the approval of tens of millions of Pakistanis, who evince their distaste by voting for militant Islamic parties, voicing support for the vanquished Taliban movement, and giving refuge to fugitive al-Qaeda operatives in their homes. The future of Pakistan is once again in the balance, and with it the stability of South Asia.

Gift of the Indus

If, as is so often said, Egypt is the gift of the Nile, then Pakistan is the gift of the Indus. The Indus River and its principal tributary, the Sutlej, nourish the ribbons of life that form the heart of this populous country (Fig. 8-10). Territorially, Pakistan is not large by Asian standards; its area is about the same as that of Texas plus Louisiana. But Pakistan's population of 149.6 million makes it one of the world's ten most populous states. Among Muslim countries (officially, it is known as the Islamic Republic of Pakistan), only Southeast Asia's Indonesia is larger, but Indonesia's Islam has long been less militant than Pakistan's. Pakistan has been in the forefront of support for Islamic revivalism; it lies near the heart of Central Asia; it has the nuclear bomb. Pakistan is a giant in the world of Islam.

Pakistan lies like a giant wedge between Iran and Afghanistan to the west and India to the east. This wedge extends from the Arabian Sea in the south, where it is widest, to the snowcapped mountains of the north. Here in the far north of Pakistan, the boundary framework is complex and jurisdictions are uncertain. India, China, Pakistan,

MAJOR CITIES OF THE REALM	
City	Population* (in millions)
Ahmadabad, India	5.2
Bangalore, India	6.5
Chennai (Madras), India	6.9
Colombo, Sri Lanka	1.0
Delhi-New Delhi, India	15.3
Dhaka, Bangladesh	15.9
Hyderabad, India	6.1
Karachi, Pakistan	11.8
Kathmandu, Nepal	0.8
Kolkata (Calcutta), India	14.3
Lahore, Pakistan	6.4
Mumbai (Bombay), India	18.3
Varanasi, India	1.3

*Based on 2004 estimates.

FIGURE 8-10

and, until recently, the former Soviet Union have all vied for control over parts of this mountainous northern frontier. The Soviets, by virtue of their temporary rule over Afghanistan, controlled the Vakhan Corridor (the long narrow segment of land lying just south of Tajikistan in Fig. 8-10). The Chinese have claimed areas to the east. But the major territorial conflict has been between Pakistan and India over the areas of Jammu and Kashmir. This issue has plagued relations between these two countries for decades (see box titled "The Problem of Kashmir").

When Pakistan became an independent state following the partition of British India in 1947, its capital was Karachi on the south coast, near the western end of the Indus Delta. As the map shows, however, the present

capital is Islamabad, near the larger city of Rawalpindi in the north, not far from Kashmir. By moving the capital from the "safe" coast to the embattled interior and by placing it on the doorstep of contested territory, Pakistan announced its intent to stake a claim to its northern frontiers. And by naming the city Islamabad, Pakistan proclaimed its Muslim foundation, here in the face of the Hindu challenge. This politico-geographical use of a national capital can be assertive, and Islamabad exemplifies the principle of the **forward capital**.

At independence, Pakistan had a bounded national territory, a capital, a cultural core, and a population— but it had few centripetal forces to bind state and nation. The disparate regions of Pakistan shared the Islamic faith and an aversion for Hindu India, but little else. Karachi and the coastal south, the desert of Baluchistan, the city of Lahore and the Punjab, the rugged northwest along Afghanistan's border, and the mountainous far north are worlds apart, and a Pakistani nationalism to match that of India at independence did not exist. Successive Pakistani governments, civilian as well as military, turned to Islam to provide the common bond that history and geography had denied the nation. In the process, Pakistan became one of the world's most theocratic states; its common law, based on the English model, was gradually transformed into a Quranic (Koranic) system with Islamic Sharia courts and associated punishments.

But even Islam itself is not unified in restive Pakistan. About 80 percent of the people are Sunni Muslims, and the Shiah minority numbers about 16 percent. Sunni fanatics intermittently attack Shi'ites, leading to retaliation and creating grounds for subsequent revenge.

Despite the Islamization of Pakistan's plural society, it remains a strongly regionalized country in which Urdu is the official language and English is still the *lingua franca* of the elite. Yet several other major languages prevail in diverse parts, and lifeways vary from nomadism in Baluchistan to irrigation farming in the Punjab to pastoralism in the northern highlands.

Subregions

Punjab

Pakistan's core area is the Punjab, the Muslim heartland across which the postcolonial boundary between Pakistan and India was superimposed. (As a result, India also has a region called Punjab, sometimes spelled Panjab, there.) Pakistan's Punjab is home to nearly 60 percent of the country's population. In the triangle formed by the Indus River and its tributary, the Sutlej, live some 90 million people. Punjabi is the language here, and intensive farming of wheat is the mainstay.

Three cities anchor this core area: Lahore, the outstanding center of Islamic culture in the realm, Faisalabad, and Multan. Lahore, now home to 6.4 million people, lies close to the India-Pakistan border. Founded about 2000 years ago, Lahore was situated favorably to become a great Muslim center during the Mogul period, when the Punjab was a corridor into India. As a center of royalty, Lahore was adorned with numerous magnificent buildings, including a great fort, multiple palaces, and several mosques displaying superb stonework and marble embellishments. The site of an old university and many majestic gardens, Lahore is the cultural focus of Islam. The city received hundreds of thousands of refugees and grew rapidly after partition. Lahore did lose its eastern

hinterland, but its new role in independent Pakistan sustained its growth.

Punjab's relationship with Pakistan's other three provinces is one of the country's weak points. Both the governments and peoples of the other three provinces feel uneasy about the dominance of Punjab, the populous, powerful core of the country.

Sind

South of Punjab lies Sind, centered on the chaotic port city of Karachi on the coast and on Hyderabad upriver. As Figure 8-10 suggests, Sind has three dominant landscapes: arid rocky hills to the west, desert to the east, and between these the fertile, irrigated alluvial valley of the lower Indus River. The British built large irrigation systems here, and improved seed strains have made Sind a Pakistani breadbasket for wheat and rice. Commercially, cotton is king here, supporting major textile industries in the cities and towns.

A flood of immigrants from India moved into Sind during and after partition, and relations between immigrants and locals have not always been good. Karachi has become the focus of conflict between immigrant Muhajirs and their descendants and other groups in this poverty-stricken, crime-ridden megacity of 11.8 million. Karachi grew explosively during and after the refugee influx, and its infrastructure crumbled even as its population increased by millions. The social order broke down, law enforcement deteriorated, and in the mushrooming shantytowns gang battles over turf spilled over into the built-up areas. When Pakistan joined the War on Terror, Karachi's vast, disorderly and ungovernable interior became the scene of al-Qaeda criminality and a haven for its operatives. Multicultural megacities can put severe strain on low-income economies when conditions spin out of control.

Meanwhile, Punjab's domination of the national scene rankles political and business leaders in Sind, and calls for greater autonomy, and even secession, are heard. Certainly, Sind's contributions to Pakistan's economy outweigh its influence in distant Islamabad.

Baluchistan

A trip westward from Sind and Karachi into the rocky and sandy deserts makes one wonder whether this is the same country. Baluchistan's arid landscapes extend far into neighboring Iran, and the sparse population of this remote and forbidding territory still moves across the borders into and out of Iran along ancient caravan routes. Pakistan is a Sunni Islamic country, but about 16 percent of the people adhere to Shi'ite principles. Shi'ism arrived in Pakistan along these desert routes from its Iranian stronghold.

The Problem of Kashmir

*K*ashmir is a territory of high mountains surrounded by Pakistan, India, China, and along several miles in the far north, Afghanistan (Fig. 8-11). Although known simply as Kashmir, the area actually consist of several political divisions, including the State property referred to as Jammu and Kashmir (one of the 562 Indian States at the time of independence) and the administrative areas of Gilgit in the northwest and Ladakh (including Baltistan) in the east. The main conflict between India and Pakistan over the final disposition of this territory has focused on the southwest, where the State of Jammu and Kashmir is located.

When partition took place in 1947, the existing States of British India were asked to decide whether they wanted to be incorporated into India or Pakistan. In most of the States, the local ruler made this decision, but Kashmir was an unusual case. It had about 5 million inhabitants at the time, nearly half of them Muslims concentrated in the basin known as the Vale of Kashmir, where the capital, Srinagar, is located (Fig. 8-11). Another 45 percent of the people, dominantly Hindu, were concentrated in Jammu, which leads down the foothill slopes of the Himalayas to the edge of the Punjab. A small minority of Buddhists made the area called Ladakh their mountainous home.

But in the State of Jammu and Kashmir the ruler was Hindu, not Muslim, although the overall population was more than 75 percent Muslim. Thus the ruler had to make a difficult decision in 1947—to go with Pakistan and thereby exclude the State from Hindu India, or to go with India and thereby incur the wrath of most of the people. Hence the maharajah of Kashmir sought to remain outside both Pakistan and India and to retain autonomous status. This decision was followed, after the partitioning of India and Pakistan, by a Muslim uprising against Hindu rule in Kashmir. The maharajah asked for Indian help, and Pakistan's forces came to the aid of the Muslims. After more than a year's fighting and through the intervention of the United Nations, a cease-fire line left Srinagar, the Vale, and most of Jammu and Kashmir (including nearly four-fifths of the territory's population) in Indian hands (Fig. 8-11). Eventually, this line began to appear on maps

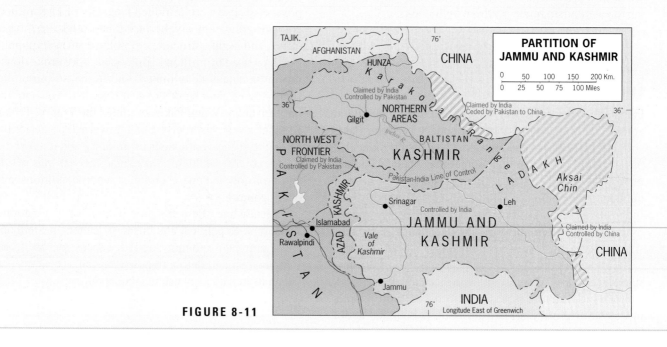

FIGURE 8-11

Baluchistan may figure more importantly in Pakistan's economic future than its present. The geology indicates considerable mineral potential, and oil and natural gas have been found here.

North West Frontier

In November 2002, an alliance of anti-American, pro-Taliban, militant Islamic parties took office in Peshawar, the capital of this northernmost Pakistani province wedged between Afghanistan to the west, Kashmir to the east, and the powerful Punjab to the south. Vowing to end the American military presence in their province and to institute strict Islamic laws, members of the alliance stated that they would lead all of Pakistan in a new direction commanded by the faith. Included in the alliance were representatives from the Tribal Areas, nominally but not functionally a part of the province (Fig. 8-10).

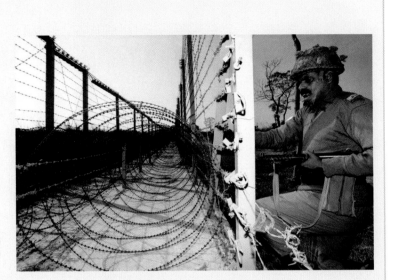

A member of India's Border Security Force checks the barbed-wire fortification that parallels the Line of Control in the Vale of Kashmir.

as the final boundary settlement, and Indian governments have proposed that it be so recognized.

Why should two countries, whose interests would be served by peaceful cooperation, allow a distant mountain-land to trouble their relationship to the point of war? There is no single answer to this question, but there are several areas of concern for both sides. In the first place, Pakistan is wary of any situation whereby India would control vital irrigation waters needed in Pakistan. As the map shows, the Indus River, the country's lifeline, crosses Kashmir. Moreover, other tributary streams of the Indus originate in Kashmir, and it was in the Punjab that Pakistan learned the lessons of dealing with India for water supplies. Second, the situation in Kashmir is analogous to the one that led to the partition of the subcontinent: Muslims are under Hindu domination. The majority of Kashmir's people are Muslims, so Pakistan argues that free choice would deliver Kashmir to the Islamic Republic. A free plebiscite is what Pakistanis have sought—and the Indians have thwarted. Furthermore, Kashmir's connections with Pakistan before partition were much stronger than those between Kashmir and India, although India has invested heavily in improving its links to Jammu and Kashmir since the military stalemate. In recent years, it did seem more likely that the cease-fire line (now called the Line of Control) would indeed become a stable boundary between India and Pakistan. (The incorporation of the State of Jammu and Kashmir into the Indian federal union was accomplished as far back as 1975, when India was able to reach an agreement with the State's chief minister and political leader.)

The drift toward stability was reversed in the late 1980s when a new crisis engulfed Kashmir. Extremist Muslim groups demanding independence escalated a long-running insurgency in 1988. That brought a swift crackdown by the Indian military, whose harsh tactics prompted the citizenry to support the separatists far more strongly. In the early 1990s, Pakistan charged the Indians with widespread human-rights violations, and the Indians accused the Pakistanis of inciting secession and supplying arms. Between 1990 and 1995, nearly 10,000 people were killed in the sporadic fighting that continues to afflict this scenic area. In 1995, Muslim extremists began a campaign that involved capturing and killing foreign visitors, a move that was condemned by representatives of both sides. Nevertheless, the memory of three wars between India and Pakistan defied all attempts at compromise. But all this was overshadowed by the implications of the 1998 test explosion of nuclear bombs by India and Pakistan. This development transformed Kashmir from a problem frontier to a potentially calamitous flashpoint.

The North West Frontier's physical geography is dominated by mountain ranges, mountain-encircled basins, and strategic passes, of which the most famous is the Khyber Pass (see photo, p. 392). The Turks invaded the upper Indus Valley via this route; later, the Mongols (Moguls) streamed through it on their way to India; recently, millions of Pushtun and other refugees used it to escape the carnage in Afghanistan. Almost directly across from the Khyber Pass lies Peshawar in an alluvium-filled, fertile valley where wheat and corn cover the countryside. After Baluchistan, this remains Pakistan's least-urbanized and poorest province.

Peshawar also is home to some of Pakistan's most militant Islamic groups, who forged an uneasy alliance with the several million refugees who fled Afghanistan and settled in United Nations-assisted refugee camps in the province. In pre-Taliban times these refugees were subject to the pressures of **irredentism** from their kin in 13

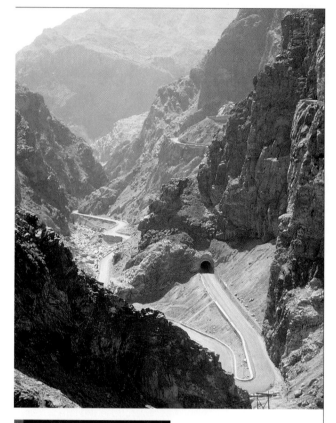

FROM THE FIELD NOTES

"Driving through the Khyber Pass that links Afghanistan and Pakistan was a riveting experience. From the dry and dusty foothills on the Afghanistan side, with numerous ruins marking historic battles, I traversed the rugged Hindu Kush via multiple hairpin turns and tunnels into Pakistan. This is one of the most strategic passes in the world, but for invading armies it was no easy passage: defensive forces had the advantage in these treacherous valleys. But now the roads and tunnels facilitate the movement of refugees, drugs, and arms, and they are used by militant separatist forces from both sides of the border. Pakistan's North West Frontier Province, and especially the city of Peshawar, are hotbeds of these activities."

tinues at the relatively high annual rate of 2.1 percent, implying a doubling time of only 33 years. Subsistence farming of food crops still occupied most of the people. Life expectancy for a child born today is just 63 years. Illiteracy still afflicts about 60 percent of the population over age 15.

Nevertheless, Pakistan has made significant economic progress during its five decades-plus of independence. Irrigation has expanded enormously; land reform has progressed; and the cotton-based textile industry has generated important revenues. Despite a limited mineral resource base, manufacturing has grown, including a steel mill at Port Qasim near Karachi.

To the rest of the world, Pakistan sells textiles, carpets, tapestries, leather goods, and rice. Despite its large and growing population, Pakistan during the 1990s was able to export rice (although it did import some wheat), which is a measure of the effect of the Green Revolution as well as its program of farm expansion. In recent years, Pakistan has also become one of the largest producers of heroin in southern Asia, and its opium and hashish enter the international drug trade despite government efforts to interdict it. Afghanistan is the leading producer of heroin in this part of the world; sales were the biggest source of income for the Taliban and continue to support warlords and regional "commanders" in the post-Taliban era.

Pakistan's legitimate economy has shown substantial growth in recent years, and the country's huge public debt has been partially forgiven in return for its support in the War on Terror. As Table G-1 shows, Pakistan ranks second in per-capita GNP among the major states of mainland South Asia, and a reduction in the rate of population growth would further improve its prospects. As we will see, these prospects are now endangered by a new threat: a nuclear arms race with its giant neighbor.

Afghanistan, who urged them to demand autonomy within Pakistan. The government noted the risks involved and improved infrastructural linkages between the North West Frontier and the Punjab, strengthening the integration of the aptly named frontier province into the national framework. But, as the 2002 election showed, the North West Frontier still goes its own way, resisting national policies and obstructing the War on Terror.

Livelihoods

For all its size and growing regional influence, Pakistan remains a low-income economy. The level of urbanization currently stands at only 33 percent. Population growth con-

Emerging Regional Power

A country's underdevelopment is not necessarily a barrier against its emergence as a power to be reckoned with in international affairs. When Pakistan became independent in 1947, the country was weak, disorganized, and divided. Just five and a half decades later, Pakistan is a major military force and a nuclear power.

To say that Pakistan in 1947 was a divided country is no exaggeration. In fact, upon independence, present-day Pakistan was united with present-day Bangladesh, and the two countries were called West Pakistan and East Pakistan, respectively. The basis for this scheme was Islam: in Bangladesh, too, Islam is the state religion. Between the two Islamic wings of Pakistan lay Hindu India. But there was little else to unify the easterners and westerners, and their union lasted less than 25 years. In 1971, a costly war of secession led to independence for East Pakistan, which took the name Bangladesh; at the same time, West Pakistan became Pakistan.

The loss of Bangladesh was no disaster for Pakistan: in virtually every respect, Bangladesh was (and remains) even more severely impoverished than Pakistan (see Table G-1). Pakistan's challenges lay closer to home—in Kashmir, in the Northwest Frontier, and, most importantly, to the east in India. India had supported independence for Bangladesh and did not resume diplomatic relations with Pakistan until five years later. But the political relationship between India and Pakistan has remained tense. Apart from the conflict in Kashmir (elaborated further in the Issue Box entitled "Who Should Govern Kashmir?"), their other differences are numerous. India remained a democracy while Pakistan became a military dictatorship; India's treatment of its Muslim minorities frequently riled Pakistan; India developed a close relationship with the same Soviet Union seen as a threat by Islamabad. During the Cold War, India tilted toward Moscow, while Pakistan, despite frequent disputes, was favored by Washington in order to curb Soviet influence in Afghanistan.

Pakistan was confronted by new challenges across its eastern border in the 1990s when India exploded several nuclear bombs at its test site. Worldwide consternation at this provocation put intense pressure on Pakistan not to follow suit, but Pakistan had little choice. Days later, it set off nuclear explosions at its own atomic test site. The "Islamic Bomb" was now reality, and the prospect of nuclear war, which had seemed to be receding, once again reasserted itself.

No longer merely a newly decolonized, economically disadvantaged country trying to survive, Pakistan has taken a crucial place in the political geography of two neighboring realms in turbulent transition.

▶ INDIA: GIANT OF THE REALM

Nearly three-quarters of the great land triangle of South Asia is occupied by a single country—India, the world's most populous democracy and, in terms of human numbers, the world's largest federation. Consider this: India has nearly as many inhabitants as live in all the countries of Subsaharan Africa *plus* North Africa/Southwest Asia combined—75 of them.

That India has endured as a unified country is a politico-geographical miracle. India is a cultural mosaic of immense ethnic, religious, linguistic, and economic diversity and contrast; it is a state of many nations. The period of British colonialism gave India the underpinnings of unity: a single capital, an interregional transport network, a *lingua franca*, a civil service. Upon independence in 1947, India adopted a federal system of government, giving regions and peoples some autonomy and identity, and allowing others to aspire to such status. Unlike Africa, where federal systems failed and where military dictatorships replaced them, India remained essentially democratic and retained a federal framework in which States have considerable local authority.

This political, democratic success has been achieved despite the presence of powerful centrifugal forces in this vast, culturally diverse country. Relations between the Hindu majority and the enormous Muslim minority, better in some States than in others, have at times threatened to destabilize the entire federation. Local rebellions, demands by some minorities for their own States, frontier wars, even involvement in a foreign but nearby civil war (in Sri Lanka) have buffeted the system—which has bent but not collapsed. India has succeeded where others in the postcolonial world have failed.

This success has not been matched in the field of economics, however. After more than 50 years of independence India remains a very poor country, and not all of this can be blamed on the colonial period or on population growth, although overpopulation remains a strong impediment to improvement of living standards. Much of it results from poor and inconsistent economic planning, too much state ownership of inefficient industries, excessive government control over economic activities, bureaucratic suppression of initiative, corruption, and restraints against foreign investment. As we shall presently see, a few bright spots in some of India's States contrast sharply to the overwhelming poverty of hundreds of millions.

States and Peoples

The map of India's political geography shows a federation of 28 States, 6 Union Territories (UTs), and 1 National Capital Territory (NCT) (Fig. 8-12). The federal government retains direct authority over the UTs, all of which are small in both territory and population. The NCT, however, includes Delhi and the capital, New Delhi, and has more than 15 million inhabitants.

The political spatial organization shown in Figure 8-12 is mainly the product of India's restructuring following independence from Britain. Its State boundaries reflect the broad outlines of the country's cultural mosaic: as far as possible the system recognizes languages, religions, and cultural traditions. Indians speak 14 major and numerous minor languages, and while Hindi is the official language (and English is the *lingua franca*), it is by no means universal. The map is the product of endless compromise—endless because demands for modifications of it continue to

Who Should Govern Kashmir?

KASHMIR SHOULD BE PART OF PAKISTAN!

"I DON'T KNOW why we're even debating this. Kashmir should and would have been made part of Pakistan in 1947 if that colonial commission hadn't stopped mapping the Pakistan-India boundary before they got to the Chinese border. And the reason they stopped was clear to everybody then and there: instead of carrying on according to their own rules, separating Muslims from Hindus, they reverted to that old colonial habit of recognizing "traditional" States. And what was more traditional than some Hindu potentate and his minority clique ruling over a powerless majority of Muslims? It happened all over India, and when they saw it here in the mountains they couldn't bring themselves to do the right thing. So India gets Jammu and Kashmir and its several million Muslims, and Pakistan loses again. The whole boundary scheme was rigged in favor of the Hindus anyway, so what do you expect?

"Here's the key question the Indians won't answer. Why not have a referendum to test the will of all the people in Kashmir? India claims to be such a democratic example to the world. Doesn't that mean that the will of the majority prevails? But India has never allowed the will of the majority even to be expressed in Kashmir. We all know why. About two-thirds of the voters would favor union with Pakistan. Muslims want to live under an Islamic government. So people like me, a Muslim carpenter here in Srinagar, can vote for a Muslim collaborator in the Kashmir government, but we can't vote against the whole idea of Indian occupation.

"Life isn't easy here in Srinagar. It used to be a peaceful place with boats full of tourists floating on beautiful lakes. But now it's a violent place with shootings and bombings. Of course we Muslims get the blame, but what do you expect when the wishes of a religious majority are ignored? So don't be surprised at the support our cause gets from Pakistan across the border. The Indians call them terrorists and they accuse them of causing the 60,000 deaths this dispute has already cost, but here's a question: why does it take an Indian army of 600,000 to keep control of a territory in which they claim the people prefer Indian rule?

"Now this so-called War on Terror has made things even worse for us. Pakistan has been forced into the American camp, and of course you can't be against 'terrorism' in Afghanistan while supporting it in Kashmir. So our compatriots on the Pakistani side of the Line of Control have to stay quiet and bide their time. But don't underestimate the power of Islam. The people of Pakistan will free themselves of collaborators and infidels, and then they will be back to defend our cause in Kashmir."

KASHMIR BELONGS TO INDIA!

"Let's get something straight. This stuff about that British boundary commission giving up and yielding to a maharajah is nonsense. Kashmir (all of it, the Pakistani as well as the Indian side) had been governed by a maharajah for a century prior to partition. What the maharajah in 1947 wanted was to be ruled by neither India nor Pakistan. He wanted independence, and he might have gotten it if Pakistanis hadn't invaded and forced him to join India in return for military help. As a matter of fact, our Prime Minister Nehru prevailed on the United Nations to call on Pakistan to withdraw its forces, which of course it never did. As to a referendum, let me remind you that a Kashmir-wide referendum was (and still is) contingent on Pakistan's withdrawal from the area of Kashmir it grabbed. And as for Muslim 'collaborators,' in the 1950s the preeminent leader on the Indian side of Kashmir was Sheikh Muhammad Abdullah (get it?), a Muslim who disliked Pakistan's Muslim extremism even more than he disliked the maharajah's rule. What he wanted, and many on the Indian side still do, is autonomy for Kashmir, not incorporation.

"In any case, Muslim states do not do well by their minorities, and we in India generally do. As far as I am concerned, Pakistan is disqualified from ruling Kashmir by the failure of its democracy and the extremism of its Islamic ideology. Let me remind you that Indian Kashmir is not just a population of Hindus and Muslims. There are other minorities, for example the Ladakh Buddhists, who are very satisfied with India's administration but who are terrified at the prospect of incorporation into Islamic Pakistan. You already know what Sunnis do to Shi'ites in Pakistan. You are aware of what happened to ancient Buddhist monuments in Taliban Afghanistan (and let's not forget where the Taliban came from). Can you imagine the takeover of multicultural Kashmir by Islamabad?

"To the Muslim citizens of Indian Kashmir, I, as a civil servant in the Srinagar government, say this: look around you, look at the country of which you are a subject. Muslims in India are more free, have more opportunities to participate in all spheres of life, are better educated, have more political power and influence than Muslims do in Islamic states. Traditional law in India accommodates Muslim needs. Women in Muslim-Indian society are far better off than they are in many Islamic states. Is it worth three wars, 60,000 lives, and a possible nuclear conflict to reject participation in one of the world's greatest democratic experiments?

"Kashmir belongs to India. All inhabitants of Kashmir benefit from Indian governance. What is good for all of India is good for Kashmir."

FIGURE 8-12

this day. As recently as late 2000, the federal government authorized the creation of three new States. In the northeast lie very small States established to protect the local traditions of small populations; minority groups in the larger States ask why they should not receive similar recognition.

With only 28 States for a national population of 1.086 billion, several of India's States contain more peo-

ple than many countries of the world (Table 8-1). As Figure 8-12 shows, the (territorially) largest States lie in the heart of the country and on the great southward-pointing peninsula. Uttar Pradesh (166 million, according to the 2001 Census) and Bihar (about 83 million) constitute much of the Ganges River Basin and are the core area of modern India (see box titled "Solace and Sickness from the Holy Ganges"). Maharashtra (almost 97 million),

"We were docked in the port of Mumbai (Bombay), literally surrounded by a fleet of warships including an aircraft carrier (foreground) and a half-dozen submarines. India is often in the news because of its ongoing conflict with Pakistan, but here was evidence of India's other agenda: its projection of strength as a regional power in the Indian Ocean." Note: after this photograph was taken, India as well as Pakistan successfully tested nuclear bombs, and the potential for devastating conflict grows."

Table 8-1
INDIA: POPULATION BY STATE

State	1991 Census	2001 Census
Andhra Pradesh	66,508,000	75,728,000
Arunachal Pradesh	865,000	1,091,000
Assam	22,414,000	26,638,000
Bihar	86,374,000	82,879,000
Chhattisgarh*		20,796,000
Goa	1,170,000	1,344,000
Gujarat	41,310,000	50,597,000
Haryana	16,464,000	21,083,000
Himachal Pradesh	5,171,000	6,077,000
Jammu and Kashmir	7,719,000	10,070,000
Jharkhand*		26,909,000
Karnataka	44,977,000	52,734,000
Kerala	29,099,000	31,839,000
Madhya Pradesh	66,181,000	60,385,000
Maharashtra	78,937,000	96,752,000
Manipur	1,837,000	2,389,000
Meghalaya	1,775,000	2,306,000
Mizoram	690,000	891,000
Nagaland	1,210,000	1,989,000
Orissa	31,660,000	36,707,000
Punjab	20,282,000	24,289,000
Rajasthan	44,006,000	56,473,000
Sikkim	406,000	541,000
Tamil Nadu	55,859,000	62,111,000
Tripura	2,757,000	3,191,000
Uttar Pradesh	139,112,000	166,053,000
Uttaranchal*		8,480,000
West Bengal	68,078,000	80,221,000
National Capital Territory**		13,783,000
Union Territories	1,998,000	2,670,000

*Established 2000.
**Established 1993.

anchored by the great coastal city of Bombay (renamed Mumbai in 1996), also has a population larger than that of most countries. West Bengal, the State that adjoins Bangladesh, contains more than 80 million residents, 14.3 million of whom live in its urban focus, Calcutta (renamed Kolkata in 2000).

These are staggering numbers, and they do not decline much toward the south. Southern India consists of four States linked by a discrete history and by their distinct Dravidian languages. Facing the Bay of Bengal are Andhra Pradesh (just under 76 million) and Tamil Nadu (62 million), both part of the hinterland of the megacity of Madras (renamed Chennai in 1997) and located on the coast near their joint border. Facing the Arabian Sea are Karnataka (almost 53 million) and Kerala (32 million). Kerala, often at odds with the federal government in New Delhi, has long had the highest literacy rate in India and one of the lowest rates of population growth owing to strong local government and strictly enforced policies. "It's a matter of geography," explained a teacher in the Kerala city of Cochin. "We are here about as far away as you can get from the capital, and we make our own rules."

As Figure 8-12 shows, India's smaller States lie mainly in the northeast, on the far side of Bangladesh, and in the northwest, toward Jammu and Kashmir. North of Delhi, India is flanked by China and Pakistan, and physical as well as cultural landscapes change from the flatlands of the Ganges to the hills and mountains of

Solace and Sickness from the Holy Ganges

*S*tand on the banks of the Ganges River in Varanasi, Hinduism's holiest city, and you will see people bathing in holy water, drinking it, and praying as they stand in it—while the city's sewage flows into it nearby, and the partially cremated corpses of people and animals float past. It is one of the world's most compelling—and disturbing sights.

The Ganges (Ganga, as the Indians call it) is Hinduism's sacred river. Its ceaseless flow and spiritual healing power are earthly manifestations of the Almighty. Therefore, tradition has it, the river's water is immaculate, and no amount of human (or other) waste can pollute it. On the contrary: just touching the water can wash away a believer's sins.

At Varanasi, Prayagraj (Allahabad), and other cities and towns along the Ganges, the river banks are lined with Hindu temples, decaying ornate palaces, and dozens of wide stone staircases called ghats. These stepped platforms lead down to the water, enabling thousands of bathers to enter the river. They come from the city and from afar, many of them pilgrims in need of the healing and spiritual powers of the water. It is estimated that more than a million people enter the river somewhere along its 1600-mile (2600-km) course every day. During religious festivals, the number may be ten times as large.

By any standards, the Ganges is one of the world's most severely polluted streams, and thousands among those who enter it become ill with diarrhea or other diseases; many die. In 1986, then-Prime Minister Rajiv Gandhi launched a major scheme to reduce the level of pollution in the river, a decade-long construction program of sewage treatment plants and other facilities. In the mid-1980s, Gandhi was told, nearly 400 million gallons of sewage and other wastes were being disgorged into the Ganges every day. The plan called for the construction of nearly 40 sewage treatment plants in riverfront cities and towns.

Many Hindus, however, opposed this costly program to clean up the Ganges. To them, the holy river's spiritual purity is all that matters. Getting physically ill is merely incidental to the spiritual healing power contained in a drop of Ganga's water.

The stone steps leading into the Ganges' waters in Varanasi, India. Varanasi is India's holiest city, and millions descend these *ghats* every year.

spurs of the Himalayas. In the State of Himachal Pradesh, forests cover the hillslopes and relief reduces living space; only 6 million people live here, many in small, comparatively isolated clusters. Before independence and political consolidation, the colonial government called this area the "Hill States."

But the map becomes even more complex in the distant northeast, beyond the narrow corridor between Bhutan and Bangladesh. The dominant State here is Assam (26.6 million), famed for its tea plantations and important because its oil and gas production amounts to more than 40 percent of India's total.

In the Brahmaputra Valley, Assam resembles the India of the Ganges. But in almost all directions from Assam, things change. To the north, in sparsely populated Arunachal Pradesh (1.1 million), we are in the

Himalayan offshoots again. To the east, in Nagaland (2 million), Manipur (2.4 million), and Mizoram (0.9 million), lie the forested and terraced hillslopes that separate India from Myanmar (Burma). This is an area of numerous ethnic groups (more than a dozen in Nagaland alone) and of frequent rebellion against Delhi's government. And to the south, the States of Meghalaya (2.3 million) and Tripura (3.2 million), hilly and still wooded, border the teeming floodplains of Bangladesh. Here in the country's northeast, where peoples are always restive and where population growth is still soaring, India faces one of its strongest regional challenges.

India's Changing Map

After independence, the Indian government began phasing out the privileged "Princely States" which the British had protected during the colonial period. Next, the government reorganized the country on the basis of its major regional languages (see Fig. 8-5). Hindi, spoken by more than one-third of the population, was designated the country's official language, but the Indian constitution gave national status to 13 other major languages, including the four Dravidian languages of the south. English, it was anticipated, would become India's common language, its *lingua franca* at government, administrative, and business levels. Indeed, English not only remained the language of national administration but also became the chief medium of commerce in growing urban India. English was the key to better jobs, financial success, and personal advancement, and the language constituted a common ground in higher education.

The newly devised framework based on the major regional languages, however, proved to be unsatisfactory to many communities in India. In the first place, many more languages are in use than the 14 that had been officially recognized. Demands for additional States soon arose. As early as 1960, the State of Bombay was divided into two language-based States, Gujarat and Maharashtra.

This devolutionary pressure has continued throughout India's existence as an independent country. In 2000, the three newest States were recognized: Jharkhand, carved from southern Bihar State on behalf of 18 poverty-stricken districts there; Chhattisgarh, where tribal peoples had been agitating since the 1930s for separation from the State of Madhya Pradesh; and Uttaranchal, which split from India's most populous, Ganges Basin, core-area State of Uttar Pradesh on the basis of its highland character and lifeways (Fig. 8-12).

For many years India has faced quite a different set of cultural-geographic problems in its northeast, where numerous ethnic groups occupy their own niches in a varied, forest-clad topography. The Naga, a cluster of peoples whose domain had been incorporated into Assam State, rebelled soon after India's independence. A protracted war brought federal troops into the area; after a truce and lengthy negotiations, Nagaland was proclaimed a State in 1961. That led the way for other politico-geographical changes in India's problematic northeastern wing.

The Sikhs

A further dilemma involves India's Sikh population. The Sikhs (the word means "disciples") adhere to a religion that was created about five centuries ago to unite warring Hindus and Muslims into a single faith. This faith's principles rejected the negative aspects of Hinduism and Islam, and it gained millions of followers in the Punjab and adjacent areas. During the colonial period, many Sikhs supported British administration of India, and by doing so they won the respect and trust of the British, who employed tens of thousands of Sikhs as soldiers and policemen. By 1947, there was a large Sikh middle class in the Punjab. When independence came, many left their rural homes and moved to the cities to enter urban professions. Today, they still exert a strong influence over Indian affairs, far in excess of the less than 2 percent of the population (about 19 million) they constitute.

After independence, the Sikhs demanded that the original Indian State of Panjab (Punjab) be divided into a Sikh-dominated northwest and a Hindu-majority southeast. The government agreed, so that Punjab as now constituted (Fig. 8-12) is India's Sikh stronghold, whereas neighboring Haryana State is mainly Hindu.

The Muslims

These ethnic, cultural, and regional problems are but a sample of the stresses on India's federal framework. There is no Muslim State in India, but India has about 150 million Muslims within its borders—the largest cultural minority in the world. As Figure 8-6 shows, the percentage of Muslims is highest in remote Jammu and Kashmir, but it also is substantial in such widely dispersed States as Kerala, Assam, and Uttar Pradesh. Moreover, the Muslim population today (approximately 14 percent) constitutes a larger percentage than it did after partition (9.9 percent). This Islamic minority also ranks among the most rapidly growing sectors of India's population and is strongly urbanized as well—nearly one-third of the population of India's largest city, Mumbai (Bombay), is Muslim.

Relations between the Hindu majority and the Muslim minority are complex. What makes the news is conflict—

for example, in 2002 when Muslims attacked a train carrying Hindus to a contested holy site in Ayodhya, killing dozens, which was followed by retaliation in several towns (but not others) in Gujarat. What does not make the news is that a Muslim population surpassing 150 million lives and participates in the kind of democracy that is all but unknown in the Muslim world itself. In the first decade of the twenty-first century, a Hindu nationalist party was the leading force in the federal government, but a Muslim scientist, the man who developed India's first nuclear weapons, was the country's president. On the other hand, many Muslims have been angered by the Hindu-inspired name changes appearing on India's map, most recently the substitution of Prayagraj for the venerable Muslim name of Allahabad in the State of Uttar Pradesh.

Concern over links between local Muslim subversives and foreign terrorists is changing Hindu attitudes, as is reflected by growing cooperation between Israeli and Indian intelligence operations and official statements linking a recent terrorist attack on the Indian Parliament not only to Pakistan (the usual suspect) but also to domestic collaborators. Pakistan's prominent role in the War on Terror is seen as a contradiction in terms by Indians facing continuous terrorism in their sector of Kashmir. Achieving a lasting accommodation of the Muslim minority will be a key challenge for India in the decades ahead.

Centrifugal Forces: From India to Hindustan?

In Chapter 1 we introduced the concept of centrifugal and centripetal forces affecting the fabric of the state. No country in the world exhibits greater cultural diversity than India, and variety in India comes on a scale unmatched anywhere else on Earth. Such diversity spells strong centrifugal forces, although, as we will see, India also has powerful consolidating bonds.

Among the centrifugal forces, Hinduism's stratification of society into castes remains pervasive. Under Hindu dogma, *castes* are fixed layers in society whose ranks are based on ancestries, family ties, and occupations. The **caste system** may have origins in the early social divisions into priests and warriors, merchants and farmers, craftspeople and servants; it may also have a racial basis, for the Sanskrit term for caste is color. Over the centuries, its complexity grew until India had thousands of castes, some with a few hundred members, others containing millions. Thus, in city as well as in village, communities were segregated according to caste, ranging from the highest (priests, princes) to the lowest

(the untouchables). The term *untouchable* has such negative connotations that some scholars object to its use. Alternatives include *dalits* (oppressed), the common term in Maharashtra State but coming into general use; *harijans* (children of God), which was Gandhi's designation, still widely used in the State of Bihar; and *Scheduled Castes*, the official government label.

A person was born into a caste based on his or her actions in a previous existence. Hence, it would not be appropriate to counter such ordained caste assignments by permitting movement (or even contact) from a lower caste to a higher one. Persons of a particular caste could perform only certain jobs, wear only certain clothes, worship only in prescribed ways at particular places. They or their children could not eat, play, or even walk with people of a higher social status. The untouchables occupying the lowest tier were the most debased, wretched members of this rigidly structured social system. Although the British ended the worst excesses of the caste system, and postcolonial Indian leaders—including Mohandas (Mahatma) Gandhi (the great spiritual leader who sparked the independence movement) and Jawaharlal Nehru (the first prime minister)—worked to modify it, a few decades cannot erase centuries of class consciousness. In traditional India, caste provided stability and continuity; in modernizing India, it constitutes an often painful and difficult legacy.

Today we can discern a geography of caste—a degree of spatial variation in its severity. Cultural geographers estimate that about 15 percent of all Indians are of lower caste, about 40 percent of backward caste (one important rank above the lower caste), and some 18 percent of upper caste, at the top of which are the Brahmans, men in the priesthood. (The caste system does not extend to the Muslims, Sikhs, and other non-Hindus in India, which is why these percentages do not total 100.) The colonial government and successive Indian governments have tried to help the lowest castes. This effort has had more effect in the urban than in the rural areas of India. In the isolated villages of the countryside, the untouchables often are made to sit on the floor of their classroom (if they go to school at all); they are not allowed to draw water from the village well because they might pollute it; and they must take off their shoes, if they wear any, when they pass higher-caste houses. But in the cities, untouchables have reserved for them places in the schools, a fixed percentage of State and federal government jobs, and a quota of seats in national and State legislatures. Mohandas Gandhi, who took a special interest in the fate of the untouchables (*harijans*) in Indian society, accomplished much of this reform.

The caste system remains a powerful centrifugal force, not only because it fragments society but also because efforts to weaken it often result in further division. Gandhi himself was killed, only a few months after independence, by a Hindu fanatic who opposed his work for the least fortunate in Indian society. Today, India is being swept by a wave of Hindu fundamentalism that is caused, at least in part, by continuing efforts to help the poorest. Higher castes see themselves as disadvantaged, and they take refuge in a "return" to fundamental Hindu values.

Hindutva

Another growing centrifugal force in India has to do with a concept known as *Hindutva* or Hinduness—a desire to remake India as a society in which Hindu principles prevail. This concept has become the guiding agenda for a political party that has become a powerful component of the federal government, and it is variously expressed as Hindu nationalism, Hindu patriotism, and Hindu heritage. This naturally worries Muslims and other minorities, but it also concerns those who understand that India's secularism, its separation of religion and state, are indispensable to the survival of its democracy. *Hindutva* enthusiasts want to impose a Hindu curriculum on schools, change the flexible family law in ways that would make it unacceptable to Muslims, inhibit the activities of non-Hindu religious proselytizers, and forge an India in which non-Hindus are essentially outsiders. Moderate Hindus and non-Hindus in India oppose such notions, which are as divisive as any India has faced. They nevertheless acknowledge the appeal of the Bharatiya Janata Party (BJP), which swept to power on a Hindu-nationalist platform, and realize that only the constraints of a coalition government kept the party from pursuing its more extreme goals.

To get a sense of what a BJP-governed India might be like, Indians need only look at the sole State where the party has been in full control: Gujarat. That is where the worst riots between Hindus and Muslims regularly take place, where Muslims and Hindus have been self-segregating in ethnic-cultural neighborhoods, where extremist groups waving "Hindu" flags threaten minorities, and where schools use textbooks that contain language that offends non-Hindus. Gujarat is economically one of India's more prosperous States, but it neglects its rural poor and may be India's most culturally polarized State. Is this, critics of *Hindutva* ask, the image of the India of the future?

The radicalization of Hinduism and the infusion of Hindu nationalism into federal politics loom today as twin threats to India's unity. The BJP and other Hindu-nationalist parties are polarizing electorates at State as well as federal levels; the recent spate of name changes on India's map is one manifestation of this polarization. If Indian politics were to fragment along religious lines, the miracle of Indian unity would be at risk.

Centripetal Forces

In the face of all these divisive forces, what bonds have kept India unified for so long? Without question, the dominant binding force in India is the cultural strength of Hinduism, its sacred writings, holy rivers, and influence over Indian life. For most Indians, Hinduism is a

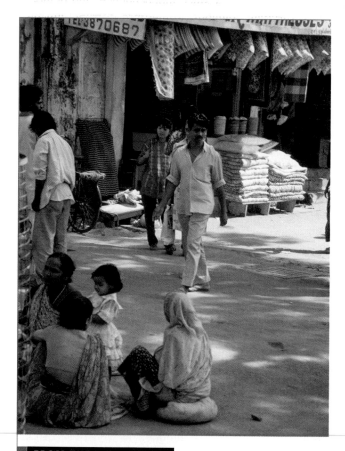

FROM THE FIELD NOTES

"The streets of India's cities often seem to be one continuous market, with people doing business in open storefronts, against building walls, or simply on the sidewalk. Here the formal and informal sectors of India's economy intermingle. I walked this way in Delhi every morning, and the store selling mattresses and pillows was always open. But the women in the foreground, selling handkerchiefs and other small items from a portable iron rack, were sometimes here, and sometimes not; one time I ran into them about a half-mile down the road. As I learned one tumultuous day, every time the government tries to exercise some control over the street hawkers and sidewalk sellers, massive opposition results and chaos can ensue. A few days later, everything is as it was. Change comes very slowly here."

way of life as much as it is a faith, and its diffusion over virtually the entire country (regardless of the Muslim, Sikh, and Christian minorities) brings with it a national coherence that constitutes a powerful antidote to regional divisiveness. Over the long term, however, the key ingredients of this Hinduism have been its gentility and introspection, radical outbursts notwithstanding. Now the specter of Hindu fanaticism threatens this vital bond.

Another centripetal force lies in India's democratic institutions. In a country as culturally diverse and as populous as India, reliance on democratic institutions has been a birthright ever since independence, and democracy's survival—raucous, sometimes corrupt, always free—has been a crucial unifier.

Furthermore, communications are better in much of India than in many other countries in the global periphery, and the continuous circulation of people, ideas, and goods helps bind the disparate state together. Before independence, opposition to British rule was a shared philosophy, a strong centripetal force. After independence, the preservation of the union was a common objective, and national planning made it possible.

India's capacity for accommodating major changes and its flexibility in the face of regional and local demands are also a centripetal force. Boundaries have been shifted; internal political entities have been created, relocated, or otherwise modified; and secessionist demands have been handled with a mixture of federal power and cooperative negotiation. Indians in South Asia have accomplished what Europeans in Yugoslavia could not, and India's history of success is itself a centripetal force.

Finally, no discussion of India's binding forces would be complete without mentioning the country's strong leadership. Gandhi, Nehru, and their successors did much to unify India by the strength of their compelling personalities. For many years, leadership was a family affair: Nehru's daughter, Indira Gandhi, twice took decisive control (in 1966 and 1980) after weak governments, and her son, Rajiv Gandhi (who, in 1991, like his mother seven years earlier, also was assassinated), was prime minister in the late 1980s. Since his death, the crucial question of India's leadership has hung in the balance.

Urbanization

India is famous for its great and teeming cities, but is not yet an urbanized society. Only 28 percent of the population lived in cities and towns in 2004—but in terms of

---

FROM THE FIELD NOTES

"Searing social contrasts abound in India's overcrowded cities. Even in Mumbai (Bombay), India's most prosperous large city, hundreds of thousands of people live like this, in the shadow of modern apartment buildings. Within seconds we were surrounded by a crowd of people asking for help of any kind, their ages ranging from the very young to the very old. Somehow this scene was more troubling here in well-off Mumbai than in Kolkata (Calcutta) or Chennai (Madras), but it typified India's urban problems everywhere."

AMONG THE REALM'S GREAT CITIES . . . MUMBAI (BOMBAY)

*A*nother historic name is disappearing from the map: Bombay. In precolonial times, fishing folk living on the seven small islands at the entrance to this harbor named the place after their local Hindu goddess, *Mumbai*. The Portuguese, first to colonize it, called it Bom Bahia, "Beautiful Bay." The British, who came next, corrupted both to *Bombay*, and so it remained for more than three centuries. Now local politics has taken its turn. Leaders of a Hindu nationalist party that control the government of the State of which Bombay is the capital, Maharashtra, passed legislation to change its name back to Mumbai. In 1996, India's federal government approved this change.

Mumbai's 18.3 million people make this India's largest city. Maharashtra is India's economic powerhouse, the State that leads the country in virtually every respect. Locals dream of an Indian Ocean Rim of which Mumbai will be the anchor.

As such, Mumbai is a microcosm of India, a burgeoning, crowded, chaotic, fast-moving agglomeration of humanity (see photo, p. 401). The gothic-Victorian architecture of the city center is a legacy of the British colonial period (see photo, p. 381). Shrines, mosques, temples, and churches evince the pervasive power of religion in this multicultural society. Street signs come in a bewildering variety of scripts and alphabets. Creaking double-decker buses compete with oxwagons and handcarts on the congested roadways. The throngs on the sidewalks—businesspeople, holy men, sari-clad women, beggars, clerks, homeless wanderers—spill over into the streets.

Mumbai is an urban agglomeration of city-sized neighborhoods, each with its own cultural landscape. The seven islands have been connected by bridges and causeways, and the resulting Fort District still is the center of the city, with many of its monuments and architectural landmarks. Marine Drive leads to the wealthy Malabar Hill area, across Back Bay. Northward lie the large Muslim districts, the Sikh

neighborhoods, and other ethnic and cultural enclaves. Beyond are some of the world's largest and poorest squatter settlements.

It is situation, not site, that gave Bombay primacy in South Asia. The opening of the Suez Canal in 1869 made Bombay the nearest Indian port to Europe. Today, Mumbai is the focus of India's fastest-growing economic zone—not yet a tiger but on the move.

---

sheer numbers, that 28 percent amounts to 304 million people, more than the entire population of the United States.

And India's rate of urbanization is on the upswing. People by the hundreds of thousands are arriving in the already-overcrowded cities, swelling urban India by about 5 percent annually, almost three times as fast as the overall population growth. Not only do the cities attract as they do in every geographic realm; many villagers are being driven off the land by the desperate conditions in the countryside. As villagers manage to establish themselves in Mum-

bai or Kolkata or Chennai, they allow their relatives to join them in squatter settlements that often are populated by newcomers from the same area, bringing their language and customs with them and cushioning the stress of the move.

As a result, India's cities display staggering social contrasts. Squatter shacks without any amenities at all crowd against the walls of modern high-rise apartments and condominiums. Hundreds of thousands of homeless roam the streets and sleep in parks, under bridges, on sidewalks. As crowding intensifies, social stresses multi-

AMONG THE REALM'S GREAT CITIES . . .   KOLKATA (CALCUTTA)

Calcutta is synonymous with all that can go wrong in large cities: poverty, dislocation, disease, pollution, crime, corruption. To call a city the Calcutta of its region is to summarize urban catastrophe. West Bengal's leaders had this in mind when they proposed that the city's local name, *Kolkata*, be restored. The city got its dreadful reputation (as well as its Anglicized name) during colonial times, when plague, malaria, and other diseases claimed countless thousands in legendary epidemics. The British chose the site, 80 miles (130 km) up the Hooghly River from the Bay of Bengal and less than 30 feet (9 m) above sea level, not far from some unhealthy marshes but well placed for commerce and defense. When the British East India Company was granted freedom of trade in the populous hinterland, Calcutta's heyday began; when (in 1772) the British made it their colonial capital, the city prospered. The British sector of the city was drained and raised, and so much wealth accumulated here that Calcutta became known as the "city of palaces." Outside the British town, rich Indian merchants built magnificent mansions. Beyond lay neighborhoods that were often based on occupational caste, whose names are still on the map today (such as Kumartuli, the potters' district). Almost everywhere, on both banks of the Hooghly, lay the huts and hovels of the poorest of the poor. Searing social contrasts characterized Calcutta.

The twentieth century was not kind to Calcutta. In 1912 the British moved their colonial capital to New Delhi. The 1947 partition that created Pakistan also created then-East Pakistan (now Bangladesh), cutting off a large part of Calcutta's hinterland and burdening the city with a flood of refugees. The Indian part of the city had arisen virtually with-

out any urban planning, and the influx created almost unimaginable conditions. Today, Kolkata counts 14.3 million residents, including as many as 500,000 homeless. Beyond the façade of the downtown lies what may be the sickest city of all.

---

ply. Disorder never seems far from the surface; sporadic rioting, often attributable to rootless urban youths unable to find employment, has become commonplace in India's cities.

India's modern urbanization has its roots in the colonial period, when the British selected Calcutta (Kolkata), Bombay (Mumbai), and Madras (Chennai) as regional trading centers and fortified ports. Madras was fortified as early as 1640; Bombay (1664) had the situational advantage of being the closest of all Indian ports to Britain; and Calcutta (1690) lay on the margin of India's largest population cluster and had the most productive hinterland, to which the Ganges Delta's countless channels connected it. This natural transport network made Calcutta an

ideal colonial headquarters, but the population of Bengal was often rebellious. In 1912 the British moved their colonial government from Calcutta to the safer interior city of New Delhi, built adjacent to the old Mogul capital of Delhi.

Figure 8-12 displays the distribution of major urban centers in India. Except for Delhi-New Delhi, the largest cities have coastal locations: Kolkata (Calcutta) dominates the east, Mumbai (Bombay) the west, and Chennai (Madras) the south. But urbanization also has expanded in the interior, notably in the Ganges Basin core area. The surface interconnections among India's cities remain inadequate (notably the road network), but an Indian urban system is emerging.

AMONG THE REALM'S GREAT CITIES . . . DELHI NEW AND OLD

Fly directly over the Delhi–New Delhi conurbation into the regional airport, and you may not see the place at all. A combination of smog and dust creates an atmospheric soup that can limit visibility to a few hundred feet for weeks on end. Relief comes when the rains arrive, but Delhi's climate is mostly dry. The tail-end of the wet monsoon reaches here during late June or July, but altogether the city only gets about 25 inches (60 cm) of rain a year.

When the British colonial government decided to leave Calcutta and build a new capital city adjacent to Delhi, conditions were different. South of the old city lay a hill about 50 feet (15 m) above the surrounding countryside, on the right bank of the southward-flowing Yamuna tributary of the Ganges. Compared to Calcutta's hot, swampy environment, Delhi's was agreeable. In 1912, it was not yet a megacity. Skies were mostly clear. Raisina Hill became the site of a New Delhi.

This was not the first time rulers chose Delhi as the seat of empire. Ruins of numerous palaces mark the passing of powerful kingdoms. But none brought to the Delhi area the transformation the British did. In 1947 the Indian government decided to keep its headquarters here. In 1970 the metropolitan population exceeded 4 million. By 2004 it was 15.3 million.

Delhi is popular as a seat of government for the same reason as its present expansion: the city has a fortuitous relative location. The regional topography creates a narrow

corridor through which all land routes from northwestern India to the North Indian Plain must pass, and Delhi lies in this gateway. Thus the twin cities not only contain the government functions; they also anchor the core area of this populous country.

Old Delhi was once a small, traditional, homogeneous town. Today Old and New Delhi form a multicultural, multifunctional urban giant.

Economic Geography

If India has faced problems in its great effort to achieve political stability and national cohesion, these problems are more than matched by the difficulties that lie in the way of economic growth and development. The large-scale factories and power-driven machinery of the colonial powers wiped out a good part of India's indigenous industrial base. Indian trade routes were taken over. European innovations in health and medicine sent the rate of population growth soaring, without introducing solutions for the many problems this spawned. Surface communications improved and food distribution systems became more efficient, but it would be a mistake to assume that this adequately protected Indian peasants from starvation. Local and regional food shortages occurred as El Niño-generated droughts struck India repeatedly, grain prices rose beyond what villagers could afford, and the colonial (and later sovereign) governments failed to intervene effectively. Today, more than 70 percent of Indi-

ans still live on the land, where, as we noted earlier, their social, economic, and political disadvantages sometimes spell disaster. Although tens of millions of Indians have been lifted above the poverty line over the past decade, an estimated 38 percent of the population still lives below it.

Nevertheless, and despite India's low per-capita GNP, the sheer size of India's economically active population makes this one of the world's largest economies, ranking sixth in the world according to the latest data. International agencies complain about India's resistance to change, obstructions to foreign investment, and endemic corruption, but there are signs that economic reform is accelerating.

One of these signs is India's decision to authorize Special Economic Zones on the East Asian model. The first of these zones, Navi Mumbai (New Bombay), is being built directly across from Mumbai Harbor adjacent to the country's most modern facility, Jawaharlal

FROM THE FIELD NOTES

"Negotiating the traffic in India's chaotic cities is always a challenge as creaking buses, vintage cars, taxis large and small, scooters, and bicycles mingle in a mass of movement that sometimes makes streets look like rivers of humanity. I watched the scene from a vantage point on Nungambakkam Road in Chennai (formerly Madras), impressed that so much high-speed congestion produced no accidents. 'You have to have a sense of humor to be part of all that,' said a man who stopped to chat. 'Just down the street, make a left and then a right, and you'll see that even the authorities do.' He was referring to the sign appealing to drivers to obey the seemingly nonexistent rules. India is in dire need of highway and road improvement; in cities like Chennai, the road system is pretty much the way the British left it in the 1940s when India became independent."

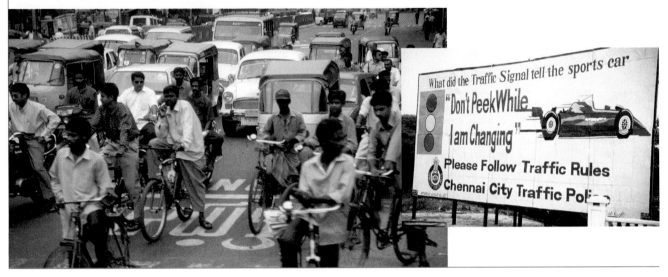

Nehru Port. It will encompass more than 11,000 acres (4300 hectares) and will legally be considered a foreign territory within India. It is designed to attract both exporters from India itself as well as major investors from abroad who want to reach India's huge market. For the foreseeable future, however, most Indians will continue to depend directly on the land for their livelihoods, and it is to this sector of the economy we turn first.

Agriculture

Growing contradictions mark agriculture in India. In Punjab, Haryana, and western Uttar Pradesh, farming remains relatively large-scale and comparatively efficient; from these areas come much of the huge grain surplus India has been accruing over the last seven years. In Gujarat, Chhattisgarh, and Bihar, some children still starve and tens of millions are malnourished as staples do not reach them or are unaffordable when they do.

Traditional farming methods persist as they do in Subsaharan Africa, and yields per acre are low by world standards for virtually every crop grown this way. Transport systems are inadequate, hampering the movement of farm produce. In 2003, only about half of India's 600,000 villages were accessible by motorable road, and in this era of modern transportation animal-drawn carts still outnumber vehicles nationwide.

As the total population grows, the amount of cultivated land per person declines. Today, this physiologic density is 1688 per square mile (652 per sq km). However, this is nowhere near as high as the physiologic density in neighboring Bangladesh, where the figure is more than twice as great (3800 and 1467, respectively). But India's farming is so inefficient that this comparison is deceptive. Fully two-thirds of India's huge working population depends directly on the land for its livelihood, but the great majority of Indian farmers are poor and cannot improve their soils, equipment, or yields. Those areas in which India has substantially modernized its agriculture (as in Punjab's wheat zone) remain islands in a sea of agrarian stagnation.

This stagnation has persisted in large measure because India, after independence, failed to implement a much-needed nationwide land reform program. Roughly one-quarter of India's entire cultivated area is still owned by less than 5 percent of the country's farming families, and after independence little land redistribution took place. Perhaps half of all rural families own either as little as an acre or no land at all. Independent India inherited inequities from the British colonial period, but the individual States of the federation would have had to cooperate in any national land reform program. As always, the large landowners retained considerable political influence, so the program never got off the ground.

"Travel into rural India, and you soon grasp the realities of Indian agriculture. Human and animal labor predominate; farming methods and equipment are antiquated. At this village I watched the women feed scrawny sticks of sugarcane into a rotating press turned slowly by a pair of bullocks prodded by a boy. In this family enterprise, the grandfather is responsible for boiling the liquid sugar in a huge pan over the fire pit (upper right). The dehydrated sugar is then molded into large round cakes to be sold at the local market. As I observed this slow and inefficient process, I understood better why India is among the world's leaders in terms of sugarcane *acreage*—and among the world's last in terms of *yields*."

To make matters worse, much of India's farmland is badly fragmented as a result of local rules of inheritance, thereby inhibiting cooperative farming, mechanization, shared irrigation, and other opportunities for progress. Not surprisingly, land consolidation efforts have had only limited success except in the States of Punjab, Haryana, and parts of Uttar Pradesh, where modernization has gone farthest. Official agricultural development policy, at the federal and State levels, has also contributed to India's agricultural malaise and the uneven distribution of progress. Unclear priorities, poor coordination, inadequate information dissemination, and other failures have been reflected in the country's disappointing output.

It is instructive to compare Figure 8-13, showing the distribution of crop regions and water supply systems in India, with Figure G-7, which shows mean annual precipitation in India and the world. In the comparatively dry northwest, notably in the Punjab and neighboring areas of the upper Ganges, wheat is the leading cereal crop. Here, India has made major gains in annual production through the introduction of high-yielding grain varieties developed under the banner of the Green Revolution, the international research program that played a key role in overcoming the food crises of the 1960s. These "miracle crops" led to the expansion of cultivated areas, the construction of new irrigation systems, and the more intensive use of fertilizer (a mixed blessing, for fertilizers tend to be ex-

The arrival of the annual rains of the wet monsoon transforms the Indian countryside. By the end of May, the paddies lie parched and brown, dust chokes the air and it seems that nothing will revive the land. Then the rains begin, and blankets of dust turn into layers of mud. Soon the first patches of green appear on the soil, and by the time the monsoon ends all is green. The photograph on the left, taken just before the onset of the wet monsoon in the State of Goa, shows the paddies before the rains begin; three months later the countryside looks as on the right.

FIGURE 8-13

pensive and the "miracle" crops are more heavily dependent on them).

Toward the moister east, and especially in the wet-monsoon-drenched areas (Figure 8-13), rice becomes the dominant staple. About one-fourth of India's total farmland lies under rice cultivation, most of it in the States of Assam, West Bengal, Bihar, Orissa, and eastern Uttar Pradesh and along the Malabar-Konkan coastal strip fac-

ing the Arabian Sea. These areas receive over 40 inches (100 cm) of rainfall annually, and irrigation supplements precipitation where necessary.

India devotes more land to rice cultivation than any other country, even China, but despite the Green Revolution and its "miracle rice" strains, yields per acre in India remain among the world's lowest. As Figure 8-13 shows, most of the rice-growing areas lie in the east and south, where water supply makes paddies possible. Although annual surpluses of wheat far surpass those of rice, it is noteworthy that, over the last seven years, India has produced more rice than it needs despite its agricultural inefficiencies. But the risks India's villagers face are not just those of cost and distribution. As the population map (Fig. G-9) shows, there is a considerable degree of geographic covariation between India's rice-producing zones and its most densely populated areas. India is just one poor-harvest year away from another food crisis.

Subsistence remains the fate of tens of millions of Indian villagers who cannot afford fertilizers, cannot cultivate the new and more productive strains of rice or wheat, and cannot escape the cycle of poverty. Perhaps as many as 175 million of these people do not even own a plot of land and must live as tenants, always uncertain of their fate. This is the enduring reality against which optimistic predictions of improved nutrition in India must be weighed. True, rice and wheat yields have increased at slightly more than the rate of population growth since the Green Revolu-

tion. But food security remains elusive, and India continues to face the risks inherent in the ever-growing needs of its burgeoning population.

Industrialization

In 1947 India inherited the mere rudiments of an industrial framework. After more than a century of British control over the economy, only 2 percent of India's workers were engaged in industry, and manufacturing and mining combined produced only about 6 percent of the national income. Textile and food processing were the dominant industries. Although India's first iron-making plant opened in 1911 and the first steel mill began operating in 1921, the initial major stimulus for heavy industrialization did not come until after the outbreak of World War II. Manufacturing was concentrated in the largest cities: Kolkata (Calcutta) led, Mumbai (Bombay) was next, and Chennai (Madras) ranked third.

The geography of manufacturing still reflects those beginnings, and industrialization in India has proceeded slowly, even after independence (Fig. 8-14). Kolkata

FIGURE 8-14

now anchors India's eastern industrial region—the Bihar-Bengal District—where jute manufactures dominate, but cotton, engineering, and chemical industries also operate. On the nearby Chota-Nagpur Plateau to the west, coal-mining and iron and steel manufacturing have developed.

On the opposite side of the subcontinent, two industrial areas dominate the western manufacturing region: one is centered on Mumbai and the other on Ahmadabad. This dual region, lying in Maharashtra and Gujarat States, specializes in cotton and chemicals, with some engineering and food processing. Cotton textiles have long been an industrial mainstay in India, and this was one of the few industries to benefit from the nineteenth-century economic order the British imposed. With the local cotton harvest, the availability of cheap yarn, abundant and inexpensive labor, and the power supply from the Western Ghats' hydroelectric stations, the industry thrived.

The southern industrial region consists chiefly of a set of linear, city-linking corridors focused on Chennai, specializing in textile production and light engineering activities. Today, all of India's manufacturing regions are increasing their output of ready-to-wear garments—another legacy of the early development of cotton textiles. Clothing has become India's second-leading export by value; the production of gems and jewelry, another growing specialization, ranks first.

An important development in the south is occurring in and around Bangalore in the State of Karnataka, India's "Silicon Plateau." Several hundred software companies are based here, one-third of them foreign with names such as IBM, Texas Instruments, and Motorola. What attracts them, and makes them profitable, is the low cost of India's software engineers, who earn about one-fifth of what their foreign colleagues earn. The emigration of technicians is becoming a problem, but replacements are still plentiful. Bangalore is proof of India's potential in the modern world.

Despite some imbalances and inefficiencies, India's industrial resource base is well endowed. Limited high-quality coal deposits are exploited in the Chota-Nagpur area. In combination with large lower-grade coalfields elsewhere, the country's total output is high enough to rank it among the world's ten leading coal producers. With no known major petroleum reserves (some oil comes from Assam, Gujarat, Punjab, and offshore from Mumbai), India must spend heavily on fuel imports every year. Major investments have been made in hydroelectric plants, especially multipurpose dams that provide electricity, enhance irrigation, and facilitate flood control. India's iron ore deposits in Bihar (northwest of Kolkata) and Karnataka (in the heart of the Deccan) may rank among the largest in the world. Jamshedpur, located west of Kolkata in the eastern industrial region, has become India's steel-making and metals-fabrication center. Yet India still exports iron ore as a raw material to the higher-income industrialized countries, mainly Japan. For low-income, revenue-needy countries, entrenched practices are difficult to break.

India East and West

The most commonly cited, and most clearly evident, regional division of India is between north and south. The north is India's heartland, the south its Dravidian appendage; the north speaks Hindi as its *lingua franca*, the south prefers English over Hindi; the north is bustling and testy, the south seems slower and less agitated.

Bangalore, near the meeting point of the States of Karnataka, Andhra Pradesh, and Tamil Nadu, is the center of what has come to be called India's "Silicon Plateau." This is the heart of the country's high-tech development, and from here many skilled workers have made their way to Europe and the United States, where their abilities earn them more than at home. Even so, Bangalore is one of India's highest-income locations, as can be seen from the appearance of this workforce leaving its shift at a branch of International Instruments. Why Bangalore? The city has great centrality with good connections in all directions; it lies more than 3000 feet (900 m) above sea level, making for tolerable summers; it has long been an educational center and has highly skilled residents; it is a cultural meeting (and competing) place of Kannada-, Tamil-, and Telugu speakers, making English a needed common language; and its State government has encouraged the high-tech sector through critical infrastructure improvements.

But there is another, as yet less obvious, but potentially more significant divide across India. In Figure 8-14, draw a line from Lucknow, on the Ganges River, south to Madurai, near the southern tip of the peninsula. To the west of this line, India is showing signs of economic progress, the kind of economic activity that has brought Pacific Rim countries such as Thailand and Indonesia a new life. To the east, India has more in common with less promising countries also facing the Bay of Bengal: Bangladesh and Myanmar (Burma).

As with other regional divides, there are exceptions to our east-west delineation. Indeed, our map seems to suggest that much of India's industrial strength lies in the east. But what the map cannot reveal is the profitability of those industries. True, the east is rich in iron and coal, but the heavy industries built by the state in the 1950s are now outdated, uncompetitive, and in decline. The hinterland of Kolkata now contains India's Rustbelt. The government keeps many industries going but at a high cost. Old industries, such as carpetmaking and cottonweaving, continue to use child labor to remain viable. The State of Bihar represents the stagnation that afflicts much of India east of our line: by several measures it ranks among the poorest of the 28 States.

Compare this to western India. The State of Maharashtra, the hinterland of Mumbai, leads India in many categories, and Mumbai leads Maharashtra. Many smaller, private industries have emerged here, manufacturing goods ranging from umbrellas to satellite dishes and from toys to textiles. Across the Arabian Sea lie the oil-rich economies of the Arabian Peninsula. Hundreds of thousands of workers from western India have found jobs there, sending money back to families from Punjab to Kerala. More importantly, many have used their foreign incomes to establish service industries back home. Outward-looking western India, in contrast to the inward-looking east, has begun to establish other ties to the outside world. Satellite links have enabled Bangalore to become the center of a growing software-producing complex reaching world markets. The beaches of Goa, the small State immediately to the south of Maharashtra, appeal to the tourist markets of Europe. This is, in fact, a classic case of **intervening opportunity** because resorts have sprung up along Goa's coast, and European tourists who once went to the more distant Maldives and Seychelles are coming to Goa. Maharashtra's economic success also has spilled over into Gujarat to the north, and even landlocked Rajasthan (the next State to the north) is experiencing the beginnings of what, by Indian standards, is a boom.

The boom has created political problems, however. Not only is Maharashtra State a rising economic power; it also is the base of a strong Hindu nationalist political movement whose leaders object to foreign intrusions and have blocked major development projects and other enterprises. They halted a huge industrial scheme about halfway through and closed a fast-food operation that they deemed incompatible with local culture. Such clashes between foreign interests and domestic traditions are not unique to India, but the rising tide of Hindu fundamentalism has uncertain prospects and unsettles investors to whom India remains a high-risk calculation.

Nevertheless, India's east-west divide shows a growing contrast that puts the west far ahead. The hope is that Maharashtra's success will spread northward and southward along the Arabian Sea coast and will ultimately diffuse eastward as well. But for this to happen, India will have to bring its population spiral under control.

▶ BANGLADESH: PERSISTENT POVERTY

On the map of South Asia, Bangladesh looks like another State of India: the country occupies the area of the double delta of India's great Ganges and Brahmaputra rivers, and India almost completely surrounds it on its landward side (Fig. 8-15). But Bangladesh is an independent country, born in 1971 after its brief war for independence against Pakistan, with a territory about the size of Wisconsin. Today it is one of the poorest and least developed countries on Earth, with a population of 139.5 million that is growing at an annual rate of 2.2 percent.

Not only is Bangladesh a poor country; it also is highly susceptible to damage from **natural hazards**. During the twentieth century, eight of the ten deadliest natural disasters in the entire world struck this single country. In 1991, a cyclone (as hurricanes are called in this part of the world) killed over 150,000 people.

The reasons for Bangladesh's vulnerability can be deduced from Figures 8-1 and 8-15. Southern Bangladesh is the deltaic plain of the Ganges-Brahmaputra river system, combining fertile alluvial soils that attract farmers with the low elevations that endanger them when water rises. The shape of the Bay of Bengal forms a funnel that sends cyclones and their storm surges barreling into the delta coast. Without money to build seawalls, floodgates, elevated shelters in sufficient numbers, or adequate escape routes, hundreds of thousands of people are at continuous risk, with deadly consequences.

The risks from water are not confined to floods and cyclones. Today Bangladesh is confronting the dreadful reality that much of its drinking water, drawn from

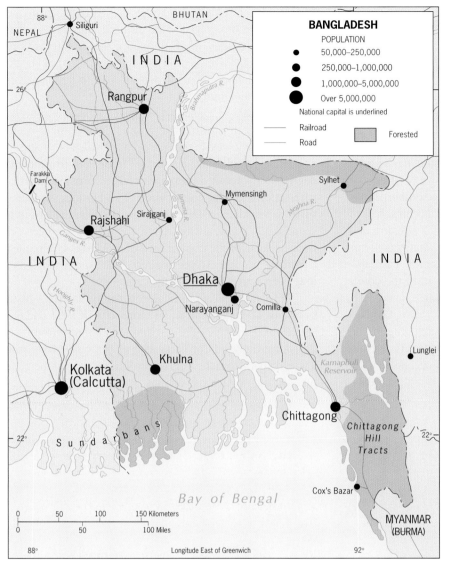

FIGURE 8-15

millions of shallow wells, is poisoned by arsenic. The source of this arsenic is still uncertain; some scientists blame oxidizing pyrite at the top of the underground water table; others conclude that it derives from decomposing peat. What *is* certain is that between 40 and 80 million people are developing symptoms of arsenic ingestion (the World Health Organization has called this the largest mass poisoning in human history), starting with skin lesions and leading to cancer after prolonged exposure. The solution is to drill deeper wells, but only the rich world could make this possible. And Bangladesh has little influence there.

Bangladesh remains a nation of subsistence farmers. The urbanization level stands at only 23 percent, with Dhaka (the megacity capital), the southeastern port of Chittagong, and the western-border-zone cities of Khulna and Rajshahi the only urban centers of consequence.

Moreover, Bangladesh has one of the highest physiologic densities in the world (3800 people per square mile/1467 per sq km), and only higher-yielding varieties of rice and the introduction of wheat in the crop rotation (where climate allows) have improved diets and food security. But diets remain poorly balanced, and overall, the people's nutrition is unsatisfactory.

Bangladesh is a dominantly Muslim society. Its relations with neighboring India have at times been strained over water resources (India's control over the Ganges that is Bangladesh's lifeline), cross-border migration (10 percent of the population is Hindu), and transit between parts of India across Bangladesh's north (refer to Fig. 8-14 to see the reason why). All the disadvantages of the global periphery afflict this populous, powerless country where survival is the leading industry and all else is luxury.

THE MOUNTAINOUS NORTH

As Figure 8-1 shows, a tier of landlocked countries and territories lies across the mountainous zone that walls India off from China. One of them, Kashmir, is in a state of near-war. Another, Sikkim, was absorbed by India in 1975 and made into one of its federal States. But Nepal and Bhutan retain their independence.

Nepal, northeast of India's Hindu core, has a population of just under 25 million and is the size of Illinois. It has three geographic zones (Fig. 8-16): a southern, subtropical, fertile lowland called the Terai; a central belt of Himalayan foothills with swiftly flowing streams and deep valleys; and the spectacular high Himalayas themselves (topped by Mount Everest) in the north. The capital, Kathmandu, lies in the east-central part of the country in an open valley of the central hill zone.

Nepal is materially poor but culturally rich. The Nepalese are a people of many sources, including India, Tibet, and interior Asia. About 90 percent are Hindu, and Hinduism is the country's official religion; but Nepal's Hinduism is a unique blend of Hindu and Buddhist ideals.

Thousands of temples and pagodas ranging from the simple to the ornate grace the cultural landscape, especially in the valley of Kathmandu, the country's core area. Although over a dozen languages are spoken, 90 percent of the people also speak Nepali, a language related to Indian Hindi.

As the data in Table G-1 suggest, Nepal is a troubled country suffering from severe underdevelopment—and its GNP is the lowest in the entire realm. It also faces strong centrifugal social and political forces. Environmental degradation, crowded farmlands and soil erosion, and deforestation scar the countryside. The Himalayan peaks form a world-renowned tourist attraction, but tourist spending in Nepal, always relatively modest, has been cut back because of the current revival of Maoist-communist terrorism, not only in the western hills where they have long been active, but also in the capital itself.

In fact, Nepal's political geography has long been troubled. The end of absolute monarchy in 1991 and the advent of democracy did not end the country's regional divisions: the southern Terai with its tropical lowlands is a world apart from the hills of central Nepal, and the peoples of the west have origins and traditions different

FIGURE 8-16

from those in the east. Moreover, in 2001 the assassination of Nepal's king threatened the disintegration of the state, narrowly averted by the leadership of the Nepal Congress Party. Then in 2002 the country's stability was imperiled again as the Maoist insurgency spread. The tourist industry collapsed and the economy withered in this scenic, divided, endangered state.

Mountainous **Bhutan**, wedged between India and China's Tibet, is the only other buffer between Asia's giants. In landlocked, fortress-like Bhutan, time seems to have stood still. Bhutan is officially a constitutional monarchy, but its king rules the country with virtually absolute power; economic subsistence and political allegiance are the norms of life for most of the population of just under one million. Thimphu, the capital, has about 50,000 inhabitants. The symbols of Buddhism, the state religion, dominate its cultural landscape. Social tensions arise from the large but diminishing Nepalese minority, most of whom are Hindus and some of whom have been persecuted by the dominant Bhutia.

Forestry, hydroelectric power, and tourism all have potential here, and Bhutan has considerable mineral resources. But isolation and inaccessibility preserve traditional ways of life in this mountainous buffer state.

THE SOUTHERN ISLANDS

As Figure 8-1 shows, South Asia's continental landmass is flanked by several sets of islands: Sri Lanka off the southern tip of India, the Maldives in the Indian Ocean, and the Andaman Islands (belonging to India) marking the eastern edge of the Bay of Bengal.

The **Maldives** consists of more than a thousand tiny islands whose combined area is just 115 square miles (less than 300 sq km) and whose highest elevation is barely over 6 feet (2 m) above sea level. Its population of approximately 300,000 from Dravidian and Sri Lankan sources is now 100 percent Muslim, one-quarter of which is concentrated on the capital island named Maale. The Maldives might be unremarkable, except that, as Table G-1 shows, this country has by far the realm's highest GNP. The locals have translated their palm-studded, beach-fringed islands into a tourist mecca that attracts tens of thousands of mainly European visitors annually.

Sri Lanka: South Asian Tragedy

Sri Lanka (known as Ceylon before 1962), the compact, pear-shaped island located just 22 miles (35 km) across the Palk Strait from India, became independent from

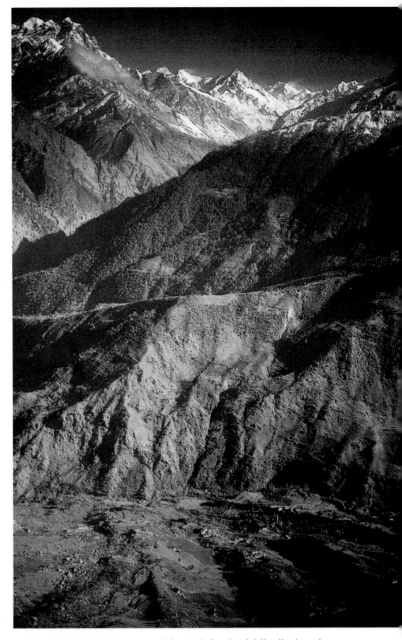

Human-induced environmental degradation is vividly displayed on Nepal's highest inhabitable slopes. Against the spectacular backdrop of the main Himalaya range north of Kathmandu, population pressures in the teeming valleys have forced farming and fuel-wood extraction to ever higher elevations. The inevitable result is deforestation and when the dense vegetation and binding roots are removed, copious summer-monsoon rainwater is free to cascade down the steep, denuded mountainsides. This massive runoff soon produces myriad gullies that further destabilize slopes by triggering mass movements of soil and rock. The lower half of this photo exhibits the evidence of this erosional process, which affects the more level terrain in the foreground as well. Ominously, we can also discern human dwellings creeping upslope, which portends a final onslaught on the single remaining forested ridge.

"Some countries have to take warnings of global warming more seriously than others. As we approached the Maldives, the islands lay like lilypads on the surface of a pond. No part of this country's natural surface lies more than 6 feet (less than 2 meters) above sea level. The upper floors of the buildings in the capital, Maale, form the Maldives's highest points. Almost any rise in sea level would threaten this Indian Ocean outpost of South Asia."

Britain in 1948 (Fig. 8-17). There were good reasons to create a separate sovereignty for Sri Lanka. This is neither a Hindu nor a Muslim country: the majority of its nearly 19 million people—about 70 percent—are Buddhists. Furthermore, unlike India or Pakistan, Sri Lanka is a plantation country, and commercial agriculture is still the mainstay of the agricultural economy.

The great majority of Sri Lanka's people are descended from migrants who came to this island from northwest India beginning about 2500 years ago. Those migrants introduced the advanced culture of their source area, building towns and irrigation systems and bringing Buddhism. Today, their descendants, known as the Sinhalese, speak a language (Sinhala) that belongs to the Indo-European language family of northern India.

The Dravidians who lived on the mainland, just across the Palk Strait, came later and in far smaller numbers—until the British colonialists intervened. During the nineteenth century, the British brought hundreds of thousands of Tamils to work on their tea plantations, and soon a small minority became a substantial segment of Ceylonese society. The Tamils brought their Dravidian tongue to the island and introduced their Hindu faith. At the time of independence, they constituted about 15 percent of the total population, a figure that rose to 18 percent by the mid-1970s, when Sri Lanka had about 14 million citizens. But for reasons we address below, Sri Lanka's Tamil population has been shrinking since. Estimates vary, but the Tamil population today may constitute only 8 to 10 percent

of the total; it is a matter of record that more than half a million Tamils have fled the island. This leaves the Tamil minority only slightly larger than the Muslim sector at less than 2 million.

When Ceylon became independent, it was one of the great hopes of the postcolonial world. The country had a sound economy and a democratic government, and it was renowned for its tropical beauty. Its reputation soared when a massive campaign succeeded in eradicating malaria and when family-planning campaigns reduced population growth when the rest of the realm was experiencing a population explosion. Rivers from the cool, forested interior highlands fed the paddies that provided ample rice; crops from the moist southwest paid the bills; and the capital, Colombo, grew to reflect the optimism that prevailed.

In the midst of this glowing scenario, the seeds of disaster were already being sown. Sri Lanka's Tamil minority soon began proclaiming its sense of exclusion, demanding better treatment from the Sinhalese majority. Although the government recognized Tamil as a "national language" in 1978, sporadic violence marked the Tamil campaign, and in 1983 full-scale civil war began. Now many in the Tamil community demanded a separate Tamil state to encompass the north and east of the country (see Fig. 8-17), and a rebel army called the Tamil Tigers confronted Sri Lanka's national forces.

The sequence of events fits the model of the evolution of the **insurgent state** discussed in Chapter 5. In Sri

FIGURE 8-17

Lanka, the *equilibrium* stage was reached in the early 1990s, when the Tamil Tigers claimed the Jaffna Peninsula and set up their headquarters there. The *counteroffensive* stage of the late 1990s brought defeat to the insurgents on some battlefields, but Sri Lanka reeled under the Tigers' terrorist tactics when bombs exploded in Colombo and its international airport and many civilian aircraft were destroyed in an attack in July 2001. By the time the two sides agreed to seek a peaceful solution, the Tamil-held portion of Sri Lanka had taken on the trappings of an insurgent state. *Eelam*, as it is called, has a flag and even a national anthem; its leaders answer their mail on official-looking stationary with "Eelam" on the letterhead. There is an Eelam police force, and there are banks and even a system for tax collection. Meanwhile, the Tigers had come into conflict with Sri Lanka's Muslims as well as its Sinhalese.

"The long civil war involving Tamil demands for a separate state has severely damaged Sri Lanka's economy. Its impact can be seen in the capital, Colombo, where the tourist industry collapsed and new investment dwindled. In the mid-1990s, with hopes for a resolution of the crisis rising, some new building finally began in Colombo: the twin towers seen rising here above the cityscape were being built by Singaporean investors anticipating a resumption of economic progress in this embattled country. But even the capture of Jaffna by government forces did not lower the tension in the capital. We saw armed military checkpoints everywhere." Note: The towers were completed, but badly damaged later by a massive bomb set off by the Tamil Tigers in the business district. A suicide bomber made an attempt on the life of the Prime Minister; the Temple of the Tooth, Sri Lankan Buddhism's holiest shrine, was damaged by a car bomb that killed many. Sri Lanka's future is threatened by a conflict from which there seems to be no escaping.

Clashes between Tamils and Muslims in the hinterland of Batticaloa underscored the unwillingness of local Muslims to submit to Tamil rule should Sri Lanka be forced to yield territory to the rebels.

In 2002 a Norwegian team of negotiators tried to help find a solution to Sri Lanka's complex crisis, and hostilities were terminated by both sides. It was a tenuous peace interrupted by occasional minor skirmishes, but it was crucial to the overall peace process. In mid-2003 it appeared that Tamil leaders would accept a territorial status short of independence, a major concession given their 20-year insistence on outright sovereignty. But the degree of autonomy a future Eelam would be able to claim remained a difficult issue. In any case, it is clear

"Sri Lanka is a dominantly Buddhist country, the religion of the majority Sinhalese. Most areas of the capital, Colombo, have numerous reminders of this in the form of architecture and statuary: shrines to the Buddha, large and small, abound. But walk into the Tamil parts of town, and the cultural landscape changes drastically. This might as well be a street in Chennai or Madurai: elaborate Hindu shrines vie for space with storefronts and Buddhist symbols are absent. The people here seemed to be less than enthused about the Tamil Tigers' campaign for an independent state. 'This would never be a part of it anyway,' said the fellow walking toward me as I took this photograph. 'We're here for better or worse, and for us the situation up north makes it worse.' But, he added, Sri Lankan governments of the past had helped create the situation by discriminating against Tamils."

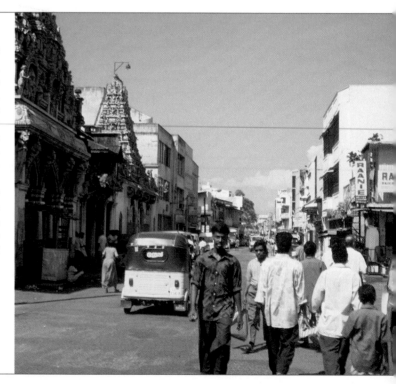

that Sri Lanka will emerge from its struggle a divided country, if indeed its peace holds.

Once-promising Sri Lanka is paying dearly for its politicians' failures. The cost of the civil war is enormous. The economic consequences are incalculable, with the tourist industry having been reduced to a fraction of what it could be. And South Asia has lost a beacon of opportunity and progress.

Continued on pages 198-199

Scale 1:16 000 000; one inch to 250 miles. Polyconic Projection
Elevations and depressions are given in feet

East Asia

Chapter 9

FIGURE 9-1

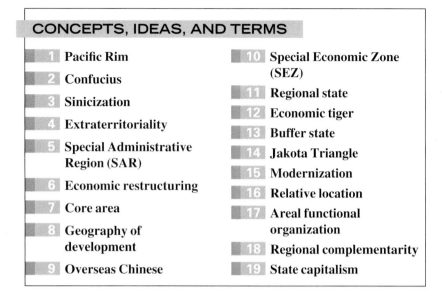

CONCEPTS, IDEAS, AND TERMS

1 Pacific Rim
2 Confucius
3 Sinicization
4 Extraterritoriality
5 Special Administrative Region (SAR)
6 Economic restructuring
7 Core area
8 Geography of development
9 Overseas Chinese
10 Special Economic Zone (SEZ)
11 Regional state
12 Economic tiger
13 Buffer state
14 Jakota Triangle
15 Modernization
16 Relative location
17 Areal functional organization
18 Regional complementarity
19 State capitalism

REGIONS

● CHINA PROPER
● XIZANG (TIBET)
● XINJIANG
● MONGOLIA
● JAKOTA TRIANGLE (JAPAN-KOREA-TAIWAN)

*E*AST ASIA IS a geographic realm like no other. At its heart lies the world's most populous country. On its periphery lies one of the globe's most powerful national economies. Along its coastline, on its peninsulas, and on its islands an economic boom has transformed cities and countrysides. Its interior contains the world's highest mountains and vast deserts. It is a storehouse of raw materials. The basins of its great rivers produce food that can sustain more than a billion people.

DEFINING THE REALM

THE EAST ASIAN geographic realm consists of six political entities: China, Mongolia, North Korea, South Korea, Japan, and Taiwan. Note that we refer here to "political entities" rather than "states." In changing East Asia, the distinction is significant. Taiwan, which its government officially calls the Republic of China, functions as a state but is regarded by mainland China (the People's Republic of China) as a temporarily wayward province. North Korea is not a full member of the United Nations, and the division of the Korean Peninsula may be temporary.

As defined here, East Asia lies between the vast expanses of Russia to the north and the populous countries of South and Southeast Asia to the south. This geographic realm extends from the deserts of Central Asia to the Pacific islands of Japan and Taiwan. Environmental diversity is one of its hallmarks.

East Asia also is the hub of the evolving regional
1 phenomenon called the **Pacific Rim**. From Japan to Taiwan and from South Korea to Hong Kong (Xianggang), the Pacific frontage of East Asia is being transformed (see box titled "Names and Places"). Skyscrapers tower over Haikou, capital of the once-dormant southern Chinese island of Hainan. Luxury automobiles from Europe and America ply the streets of Dalian, long the drab port for China's Northeast. A forest of construction cranes marks the emergence of Pudong, a huge commercial and industrial zone where Shanghai stakes its entry into the Pacific Rim's booming economy. Millions of people are on the move, abandoning their farms and villages and seeking work in such projects.

Japan was the leader in East Asia's Pacific Rim development. Long before this regional term even came into general use, Japan had built a giant economy with global connections and was the only highly developed country in Asia not plagued by stagnation. Although China lay isolated and South Korea struggled in the aftermath of its terrible war against communism (1950–1953), Japan built a society unlike that of any other in eastern Eurasia and merited recognition as a discrete geographic realm. But Pacific Rim developments are diminishing the contrasts between Japan and its East Asian neighbors. South Korea

MAJOR GEOGRAPHIC QUALITIES OF

East Asia

1. East Asia is encircled by snowcapped mountains, vast deserts, cold climates, and Pacific waters.

2. East Asia was one of the world's earliest culture hearths, and China is one of the world's oldest continuous civilizations.

3. East Asia is the world's most populous geographic realm, but its population remains strongly concentrated in its eastern regions.

4. China, the world's largest nation-state demographically, is the current rendition of an empire that has expanded and contracted, fragmented and unified many times during its long existence.

5. China today remains a mainly rural society, and its vast eastern river basins feed hundreds of millions in a historic pattern that continues today.

6. China's sparsely peopled western regions are strategically important to the state, but they lie exposed to minority pressures and Islamic influences.

7. Along China's Pacific frontage an economic transformation is taking place, affecting all the coastal provinces and creating an emerging Pacific Rim region.

8. Increasing regional disparities and fast-changing cultural landscapes are straining East Asian societies.

9. Japan, the economic giant of the East Asian realm, has a history of colonial expansion and wartime conduct that still affects international relations here.

10. East Asia may witness the rise of the world's next superpower as China's economic and military strength and influence grow—and if China avoids the devolutionary forces that fractured the Soviet Union.

11. The political geography of East Asia contains a number of flashpoints that can generate conflict, including Taiwan, North Korea, and several island groups in the realm's seas.

Names and Places

In 1958, the government of the People's Republic of China adopted the so-called *pinyin* system of standard Chinese, which replaced the Wade-Giles system used since colonial times. Pinyin was adopted not to teach foreigners how to spell and pronounce Chinese names and words but to establish a standard form of the Chinese language throughout China. The pinyin system is based on the pronunciation of Chinese characters in Northern Mandarin, the Chinese spoken in the region of the capital and the north in general.

The new linguistic standard caught on slowly outside China, but today it is in general use. The old name of the capital, Peking, has become Beijing. Canton is now Guangzhou. The Yangtze Kiang (River) is now the Chang Jiang/Yangzi. Tientsin, Beijing's port, is now Tianjin. A few of the old names persist, however. The Chinese call their colony Xizang, but many maps still carry the name Tibet.

Pinyin usage also affected personal names. China's long-time ruler, Mao Tse-tung, is now called Mao Zedong. His eventual successor, Teng Hsiao-ping, was more simply Deng Xiaoping. And remember: when the Chinese write their names, they use the last name first. The current president, Hu Jintao, is Jintao to friends and Mr. Hu to others. To the Chinese, it is Bush George and Cheney Dick, not the other way around.

today challenges Japan on international markets, selling goods ranging from electronics to automobiles. In southeastern China, Guangdong Province is an economic juggernaut that includes Shenzhen, recently the world's fastest growing city. Japan's own investments in Pacific Rim economies have helped not only to lessen the contrasts, but also to enhance the linkages between Tokyo and the mainland. In short, Japan has again become part of a functional region within the East Asian realm.

No matter how powerful Japan's economy became during the second half of the twentieth century, China remains the colossus of East Asia. One of the world's oldest continuous civilizations, China was a major culture hearth when Japan was an isolated frontier inhabited by the Ainu. Migrations that originated in China advanced through the Korean Peninsula and reached the islands. Philosophies, religions, and cultural traditions (including, for example, urban design and architecture) diffused from China to Korea and Japan, and to other locales on the margins of the Chinese hearth. Over the past millennium, China repeatedly expanded to acquire empires, only to collapse in chaos when its center failed to maintain control. European colonial powers took such opportunities to partition China among themselves. The Russians and the Japanese pushed their empires deep into China. But always China recovered, and today China may again be poised to expand its sphere of influence and, indeed, to take its place as a world power in the twenty-first century.

● NATURAL ENVIRONMENTS

Figure 9-1 dramatically illustrates the complex physical geography of the East Asian realm. In the southwest lie ice-covered mountains and plateaus, the Earth's crust in this region crumpled up like the folds of an accordion. A gigantic collision of tectonic plates is creating this landscape as the Indian Plate pushes northward into the underbelly of the Eurasian Plate (Fig. G-4). The result is some of the world's most spectacular scenery, but snow, ice, and cold are not the only dangers to human life here. Earthquakes and tremors occur almost continuously, causing landslides and avalanches. As the map shows, the high mountains and plateaus widen from a relatively narrow belt in the Karakoram to form Xizang's (Tibet's) vast plateau, flanked by the Himalayas to the south. Then, east of Tibet, the mountain ranges converge again and bend southward into Southeast Asia, where they lose their high relief.

As Figure G-9 shows, this Asian interior is one of the world's most sparsely populated areas, but it is nevertheless critical to the lives of hundreds of millions of people. In these high mountains, fed by the melting ice and snow, rise the great rivers that flow eastward across China and southward across Southeast and South Asia. Throughout the Holocene, these rivers have been eroding the uplands and depositing their sediments in the lowlands, in effect creating the alluvium-filled basins that now sustain huge populations. Fertile alluvial soils and adequate growing seasons, combined with ample water and millions of hands to sow the wheat and plant the rice, have allowed the emergence of one of the great population concentrations on Earth.

Figure G-8 records this high-elevation zone of East Asia as category *H*, highland climate, which is by far the largest area of its kind in the world. Here one criterion overpowers the rules of climatic classification: altitude. The mountains wrest virtually all moisture from air masses

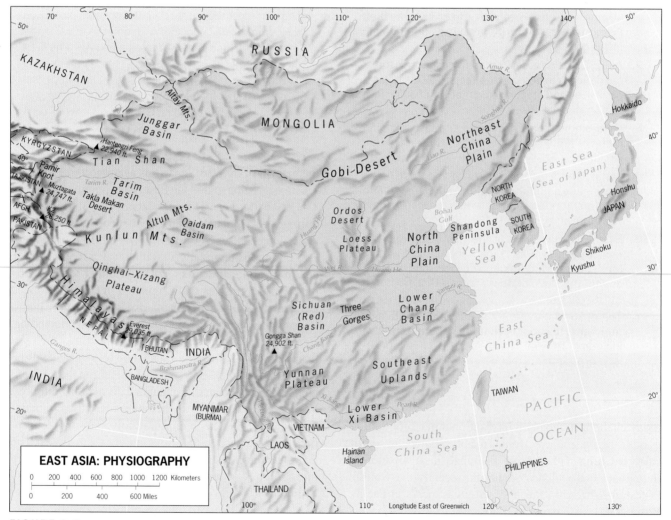

EAST ASIA: PHYSIOGRAPHY

FIGURE 9-2

moving northward and block their path into interior Asia. As Figure G-8 shows, desert conditions prevail in East Asia's northern interior. Figure 9-2 names two of the more famous ones: the Takla Makan of far western China and the Gobi of Mongolia.

Physiography, therefore, has much to do with East Asia's population distribution, but even the more habitable and agriculturally productive east has its limitations. The northeast suffers from severe continentality, with long and bitterly cold winters. High relief encircles the river basins north of the Yellow Sea and dominates much of the northern part of the Korean Peninsula, creating strong local environmental contrasts. South Korea, as Figure G-8 shows, experiences relatively moderate conditions comparable to those of the U.S. Southeast; North Korea has a harsh continental climate like that of North Dakota.

In general, coastal, peninsular, and insular East Asia possess more moderate climates than the interior. Like South Korea, southern Japan and southeastern China have humid temperate climates, and southernmost Taiwan and Hainan Island even have areas of tropical (*A*) conditions. Proximity to the ocean tends to moderate climatic environments, and coastal East Asia proves the point.

Before we focus on East Asia's human geography, it is useful to look at a map of this realm's complex physical stage (Fig. 9-2). From the high-relief interior come three major river systems that have played crucial roles in the human drama. In the north, the Huang He (Yellow River) arises deep in the high mountains, crosses the Ordos Desert and the Loess Plateau, and deposits its fertile sediments in the vast North China Plain, where East Asia's earliest states emerged. In the center, the Chang Jiang (Long River), called the Yangzi downstream, crosses the Sichuan Basin and the Three Gorges, where a huge dam project is under way, and waters extensive ricefields in the Lower Chang Basin. And in the south, the Xi Jiang (West River) originates on the Yunnan Plateau and becomes the Pearl River in its lowest course. Its estuary, flanked by several of China's largest urban-industrial complexes, has become one of the hubs of the evolving Pacific Rim.

Further scrutiny of Figure 9-2 indicates that a fourth river system plays a role in China: the Liao River in the northeast and its basin, the Northeast China Plain. As the map suggests, however, the Liao is not comparable to the great rivers to its south, its course being shorter and its basin, in this higher-latitude area, much smaller. As we will see, China's Northeast was for some time the country's industrial, not agricultural, heartland.

Looking again toward the interior, note the Loess Plateau south of the Ordos Desert, where the Huang He (Yellow River) makes its giant loop. Loess is a fertile,

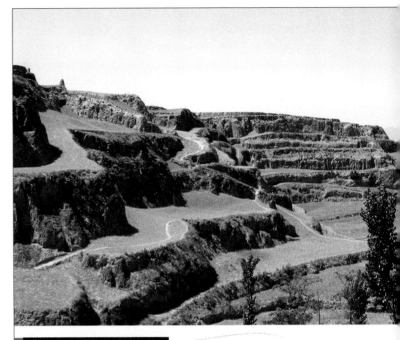

FROM THE FIELD NOTES

"From the train, travelling from Beijing to Xian, we had a memorable view of the Loess Plateau. Loess is a fine-grained dust formed from rocks pulverized by glacial action, blown away by persistent winds and deposited in sometimes well-defined locales. Loess covers much of the North China Plain, which is what makes it so fertile, but there it is not as thick as in the Loess Plateau in the middle basin of the Huang He (Yellow River). Here the loess averages 250 feet (75 m) in thickness and in places reaches as much as 600 feet (180 m). Because loess has some very distinctive physical properties, it tends to create unusual landscapes. Through a complicated physical process following deposition, loess develops the capacity to stand upright in walls and columns, and resists collapse when it is excavated. As a result the landscape looks terraced: streams cut deep and steep-sided valleys. The Loess Plateau is a physiographic region, but it is a cultural region too. Hundreds of thousands of people have literally dug their homes into vertical faces of the kind shown here, creating cave-like, multi-room dwellings with wooden exterior doors."

windblown deposit composed of rock pulverized by glaciers (see photo above). Add water (in this case the middle Huang and its tributaries) and an adequate growing season, and a sizeable population will arise. To the south, deep in the interior, lies the Sichuan (Red) Basin, crossed by the Chang Jiang. This basin has supported human communities for a long time, and, as we noted in the Introduction, you can actually see its current population cluster on the world population map (Fig. G-9). The Sichuan Basin, encircled as it is by mountains, is one of the world's most clearly defined physiographic regions, and the concentration of its approximately 120 million inhabitants reflects that definition.

Still farther to the south lies the Yunnan Plateau, source of the tributaries that feed the Xi River. Much of southeastern China has comparatively high relief; it is hilly and in places mountainous. This high relief has helped limit contacts between China and Southeast Asia.

East Asia's Pacific margin is a jumble of peninsulas and islands. The Korean Peninsula looks like a near-bridge from Asia to Japan, and indeed it has served as such in the past. The Liaodong and Shandong peninsulas protrude into the Yellow Sea, which continues to silt up from the sediments of the Huang and Liao rivers. Off the mainland lie the islands that have played such a crucial role in the modern human geography of Asia and, indeed, the world: Japan, Taiwan, and Hainan. Japan's environmental range is expressed by cold northern Hokkaido and warm southern Kyushu, but Japan's core area lies on its main island, Honshu. As Figure 9-1 shows, myriad smaller islands flank the mainland and dot the East and South China Seas. As we will discover, some of these smaller islands have major significance in the human geography of this realm.

HISTORICAL GEOGRAPHY

Consider this: there is no evidence that *Homo sapiens*, modern humans, reached the Americas any earlier than 40,000 years ago, and many paleoanthropologists argue that the only available evidence indicates that humans crossed from Eurasia to Alaska a mere 14,000 to 13,000 years ago. On the opposite side of the Pacific, however, the story of hominid and human settlement spans hundreds of thousands of years—perhaps 1 million.

What took so long? If hominids like *Homo erectus* left their ancestral African homelands and migrated as far away as eastern Asia more than half a million years ago, and if modern humans followed them and reached Australia some 50,000 years ago, what kept the hominids from the Americas altogether, and what delayed the humans?

No satisfactory answers have yet been found to questions like these. The width of the Pacific Ocean may have doomed maritime migrants, although trans-Pacific migration may eventually have occurred. The frigid temperatures along the northern route from Siberia to Alaska may have stopped hominids and humans alike. But the Pleistocene was punctuated by warm interglaciations; why did it take until the early Holocene for humans to make the Bering Strait crossing?

All this implies that for hundreds of thousands of years East Asia was a cul-de-sac, a dead-end for migrants out of Africa. Many archeological sites in the realm have yielded evidence of *Homo erectus*, including what is perhaps the most famous of all: Peking Man. In a cave near the Chinese capital, Beijing, an archeologist in 1927 found a single tooth he recognized as representing a hominid. Later excavations proved him right as parts of more than three dozen skeletons were unearthed. They proved that Peking Man and his associates made stone and bone tools, had a communal culture, controlled the use of fire, hunted wildlife, and cooked their meat. Later arrivals undoubtedly challenged older communities, and the frontier of settlement must have expanded across the river basins of the east. But when *Homo sapiens*, modern humans, arrived in East Asia, the hominids could not compete. Current anthropological theory holds that the better-equipped humans eliminated the hominids, perhaps between 60,000 and 40,000 years ago, after which the human migration into the Americas could begin.

In the context of East Asia we should mention a minority view. Some anthropologists state that the evidence supports the notion that modern humanity developed *not* from one stock in Africa, but from four stocks in widely separated parts of the world: Africa, western Eurasia (the Caucasoids), Australia (the Australoids), and eastern Eurasia (the Mongoloids). According to this idea, today's East Asians trace their ancestry to Peking Man and beyond.

Early Cultural Geography

Whatever the outcome of the debate over human origins in East Asia, it is clear that humans have inhabited the plains and river basins, foothills, and islands of this realm for a very long time. Hunting sustained both the hominids and the early human communities; fishing drew them to the coasts and onto the islands. The first crossing into Japan may have occurred as long as 10,000 to 12,000 years ago, possibly much longer, when the Jomon people, a Caucasoid population of uncertain geographic origins, entered the islands; their modern descendants, the Ainu, spread throughout the archipelago. Today, only about 20,000 persons living in northernmost Hokkaido trace their ancestry to Ainu sources.

About 2300 years ago the Yayoi people, rice farmers who had settled in Korea, appear to have crossed by boat to Kyushu, Japan's southernmost island, from where they advanced northward. The Ainu, who subsisted by fishing, trapping, and hunting, were driven back, but gene-pool studies show that much mixing of the groups took place; they also show that the Yayoi invasion was followed by other incursions from the Asian mainland. By then, powerful dynastic states had already arisen in what is today China, and early Chinese culture traits thus found their way into Japan through the process we know as relocation diffusion.

On the Asian mainland, plant and animal domestication had begun as early as anywhere on Earth. We in the Western world take it for granted that these momentous processes began in what we now call the Middle East and diffused from the Fertile Crescent to other parts of Eurasia and the rest of the world. But the taming of animals and the selective farming of plants may have begun as early, or earlier, here in East Asia. As in Southwest Asia, the fertile alluvial soils of the great river basins and the ebb and flow of stream water created an environment of opportunity, and millet and rice were being harvested between 7000 and 8000 years ago.

Even during this Neolithic period of increasingly sophisticated stone tools, East Asia was a mosaic of regional cultures. Their differences are revealed by the tools they made and the decorations on their bowls, pots, and other utensils. An especially important discovery of two 8000-year-old pots in the form of a silkworm cocoon, from China's Hebei Province, suggests a very ancient origin for one of the region's leading historic industries.

As noted earlier, plant and animal domestication produced surpluses and food storage, enabling population growth and requiring wider regional organization. Here as elsewhere during the Neolithic, settlements expanded, human communities grew more complex, and power became concentrated in a small group, an *elite*.

This process of *state formation* is known to have occurred in only a half dozen regions of the world, and China was one of these. But evidence about China's earliest states has long been scarce. Today, however, archeologists are focusing on the lower Yi-Luo River Valley in the western part of Henan Province, where the first documented Chinese dynasty, the Xia Dynasty (2200–1770 B.C.) existed. The capital of this ancient state, Erlitou, has been found, and archeologists now refer to the Xia Dynasty as the Erlitou culture. Secondary centers are being discovered, and Erlitou tools and implements in a wider area prove that the Xia Dynasty represents a substantial state.

All early states were ruled by elites, but China's political history is chronicled in *dynasties* because here the succession of rulers came from the same line of descent, sometimes enduring for centuries. In the transfer of power, family ties counted for more than anything else. Dynasties were overthrown, but the victors did not change this system. Dynastic rule lasted into the twentieth century.

The Xia Dynasty may have been the earliest Chinese state, but it lay in the area where, later, more powerful dynastic states arose: the North China Plain. Here the tenets of what was to become Chinese society were implanted early and proved to be extremely durable. From this culture hearth ideas, innovations, and practices diffused far and wide. From agriculture to architecture, poetry to porcelain making, influences radiated southward into Southeast Asia, westward into interior Asia, and eastward into Korea and Japan. In the North China Plain lay the origins of what was to become the Middle Kingdom, which its citizens considered the center of the world.

Table 9-1 summarizes the sequence of dynasties that may have begun with Xia more than four millennia ago and ended in 1911 when the last emperor of the Qing (Manchu) Dynasty, a boy 6 years old, was forced to abdicate the Chinese throne. As the table shows, every dynasty contributed importantly to the development of Chinese society, sometimes progressively through enlightened policies and efficient administration, at other times regressively when capricious cruelty, authoritarian excesses, and xenophobia prevailed. During those 4000 years, China evolved into the world's most populous nation, always rebounding after disastrous famines and floods, its territory covering more than 80 percent of the East Asian realm (Fig. 9-3).

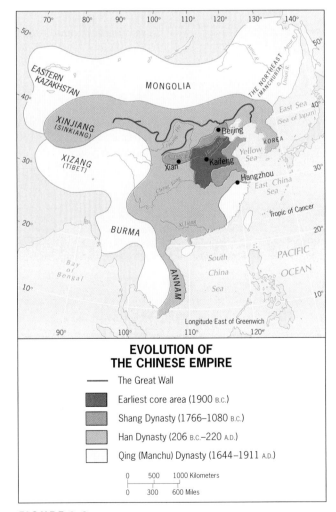

EVOLUTION OF THE CHINESE EMPIRE

—— The Great Wall

■ Earliest core area (1900 B.C.)

■ Shang Dynasty (1766–1080 B.C.)

■ Han Dynasty (206 B.C.–220 A.D.)

□ Qing (Manchu) Dynasty (1644–1911 A.D.)

0 500 1000 Kilometers
0 300 600 Miles

FIGURE 9-3

	Table 9-1			
	THE CHINESE DYNASTIES, 2200 BC to AD 1911			
Dynasty Name(s)	**Date(s)**	**Areas Governed**	**Major Features**	**Geographical Impact**
Xia	c2200–c1770 B.C.	Small part of Huang River Basin, centered in Yi-Luo tributary.	First dynastic state (?); Neolithic technology, sophisticated stone tools.	Beginnings of stream diversion, irrigation.
Shang (latter part Yin)	c1766–c1080 B.C.	Huang-Wei River confluence across Shandong and Henan Provinces. Anyang capital.	Neolithic to Bronze Age transition. Timber houses with thatched roofs, beginnings of Chinese writing, moon-cycle-based calendar. Superb bronze vases and other implements.	Unifies large region in lower Huang River Basin.
Zhou	c1027–221 B.C. Two periods: Spring and Autumn c1027–481 B.C.; Warring States 475–221 B.C.	A people centered in Shaanxi Province to the west of the Shang area overthrow the Shang rulers and expand their domain eastward. Warring States is a period of feudalism.	Formative period for China. Irrigation systems, iron smelting, horses, expanded farm production. Major towns develop. Writing system is established. Chopsticks come into use. Daoism and Confucianism arise; Daoism centers on mysticism, individualism, the personal "way," and Confucianism on duties, community standards, respect for government.	Taoism diffuses widely throughout East Asia, reaches South Asia, and influences Buddhism. Confucianism becomes China's guiding philosophy for more than 2000 years. Building of *Great Wall* begins.
Qin (Ch'in)	221–206 B.C.	Most of North China Plain, south into Lower Chang Basin; first consolidation of large Chinese state.	Time of bureaucratic dictatorship after ferment of Zhou: Confucian Classics burned, Taoism combated. Much of Great Wall is built at huge cost in lives. Cruel despotism, excesses. Emperor's tomb discovered at Xian in 1976 with 6000 terracotta life-sized men and horses.	Ruler Ch'in (Qin)'s name immortalized as *China*.
Han	206 B.C.–A.D. 220	Major territorial expansion adds Xinjiang in interior Asia, Vietnam in Southeast Asia.	Second formative dynasty for China; restoration of Confucian principles and a flowering of culture. First use of eunuchs by authoritarian government, planting a long-term weakness in administration. Xian becomes one of the greatest cities of the ancient world. Chinese refer to themselves as the *People of Han*.	About the same time, same size as the Roman Empire. *Silk Road* carries goods from China across inner Asia to Syria and on to Rome.
Period from AD 220–580 witnessed disunion and division. Important developments included the arrival of Buddhism in East Asia and the diffusion of Chinese influence and ideas into Korea and across Korea into Japan.				
Sui	A.D. 581–618	North and South China reunified, Xian rebuilt and greatly expanded. Conflict with Turks in Central Asia and with Koreans in east.	Brief but important dynasty with revival of Confucian rituals and practices in education. Brutal but modernizing regime conducts a census, establishes a penal code, and engages in massive public works.	Construction of key segments of Grand Canal links Huang and Yangzi Rivers and thus connects northern and southern economies.
Tang	A.D. 618–907	Defeat of Turks in Central Asia. Campaign to put down rebellions in the south succeeds, though northern frontier remains unstable.	A golden age for China and a third formative dynasty. Enormous and efficient bureaucracy dictates virtually all aspects of life. Xian now the cultural capital and largest city in the world. *Buddhism* thrives, pagodas arise everywhere. The arts, Chinese and foreign, flourish.	Arab and Persian seafarers visit Chinese ports, the Silk Route is loaded with trade, and China seems set for an international era. It is not to be: *Islam* arrives in Central Asia and Tang armies are defeated by Arabs in 751. The Silk Route breaks down. In the east, however, Chinese influences permeate Korea and penetrate Japan.

Table 9-1
(Continued)

Dynasty Name(s)	Date(s)	Areas Governed	Major Features	Geographical Impact

The 907–960 period is known as the Five Dynasties because five leaders tried to establish dynasties in northern China in quick succession, each failing in turn. In southern China it is known as the time of the Ten Kingdoms, for ten regimes ruled various parts of the region. Cultural continuity prevails despite political instability.

Dynasty Name(s)	Date(s)	Areas Governed	Major Features	Geographical Impact
Song (Northern (Southern	A.D. 960–1279 A.D. 960–1127) A.D. 1127–1279)	Rulers consolidate the Five (northern) Dynasties, then begin capturing the Ten (southern) Kingdoms. They are unable to control the Liao state north of the Great Wall. Mongols invade, drive Song rulers southward, and end dynasty in 1279.	Preoccupied with organization and administration, Song rulers create a competent, efficient *civil service*. A defensive rather than an expansionist period. An era of rich cultural and intellectual achievement in mathematics, astronomy, mapmaking. Paper and movable type are invented, as is *gunpowder*. Large ships are built. The arts, from porcelain making to poetry, continue to thrive. Practice of footbinding begins. Improved rice varieties feed estimated 100 million inhabitants.	Capital is moved from northern Kaifeng on Huang River to Hangzhou in the south as Song rulers lose control over the north. Mongol invasion is followed by Kublai Khan's choice of Beijing as his capital (1272). Mongols use Chinese authoritarian and bureaucratic traditions to forge their own dynastic rule. Several cities have populations exceeding 1 million.
Yuan (Mongol)	A.D. 1264–1368	Controls eastern Asia from Siberia in the north to the Vietnam border in the south, including present-day Russian Far East and most of Korea.	Dynasty begins in north before Song ends in south. Repressive regime, social chasms between Mongol rulers and Chinese ruled. Deep and enduring hostility between Chinese and Mongols. Cultural *isolationism* results; Mongols are acculturated to Chinese norms. Marco Polo visits East Asia.	Most of present-day Mongolia is part of the Chinese sphere for the first time. Mongol effort to enter Japan fails. Mongol power over much of Eurasia stimulates trade and flow of information. Beijing grows into major city.
Ming	A.D. 1368–1644	Rules over all eastern China from Amur River in north to Red River (Vietnam) in south, as far west as Inner Mongolia and Yunnan. At various times incorporates North Korea, most of Mongolia, even Myanmar (Burma).	Chinese rule again after Mongol Yuan dynasty. Stable but autocratic and inward-looking regime. Major advances in science and technology. In first half of fifteenth century, Chinese oceangoing vessels, larger than any in Europe, explore Pacific and Indian Ocean waters and reach East Africa. Farming expands, silk and cotton industries improve, printing plants multiply. Defensive walls are built in the north and around cities, including the Forbidden City in Beijing, the Ming capital. But mismanagement, infighting, eunuch influence in the palaces, and anticommercialism weaken Ming rule.	Early Ming *ships* and fleets reach Southeast Asian and Indian Ocean shores long before smaller European vessels do, but the Chinese advantage is lost when Ming rulers turn isolationist and order maritime ventures halted. Population growth causes serious problems of food supply and political control.
Qing (Manchu)	A.D. 1644–1911	*Largest* China-centered empire ever incorporates Mongolia, much of Turkestan, Xizang (Tibet), Myanmar, Indochina, Korea, Taiwan.	Foreign rule again: Manchu, a people with Tatar links living in present-day Northeast China, seize an opportunity and take control of Beijing. Improbably, a group numbering about 1 million rule a nation of several hundred million. It is done by keeping Ming systems of administration and retaining (and rewarding) Ming officials. Expansion creates an empire. Population pressure and the concentration of land ownership in fewer hands create problems worsened by floods and famines. European powers and Japan force concessions on weakening Qing rulers; defeats in war lead to revolution and collapse.	Manchu (Qing) rulers create the map that today forms the justification for China's claims to Xizang (Tibet), Taiwan, and the central Eurasian interior; at the height of its power, the Qing empire forces the Russians to recognize Manchu authority over China's Northeast as far north as the Argun River. Latent claims to Russia's Far East are based on Manchu predominance there.

Dominant as China is in its regional sphere, there is more to East Asia than the Chinese giant. To the north, the landlocked state of Mongolia is what remains of the time when Mongol armies conquered much of Eurasia (as Table 9-1 shows, even China fell under the Mongol sway). To the east, the Korean Peninsula escaped incorporation into China. Offshore, Japan first surpassed China as a military power, colonizing Manchuria (as it was then called) and penetrating deep into China's heartland. Later, Japan outclassed China as an economic power, achieving world status while China languished under Maoist communism. And as the twenty-first century dawned, Taiwan remained a separate political entity, its detachment from China resulting from the struggle that brought the communists to power in Beijing.

REGIONS OF THE REALM

EAST ASIA PRESENTS us with an opportunity to illustrate the changeable nature of regional geography. Our regional delimitation is based on current circumstances, and it predicts ways the framework may change. It is anything but static. At the beginning of the twenty-first century, we can identify five geographic regions in the East Asian realm (Fig. 9-4). These are:

1. *China Proper.* Almost any map of China's human geography—population distribution, urban centers, sur-

FIGURE 9-4

MAJOR CITIES OF THE REALM

City	Population* (in millions)
Beijing (Shi), China	10.9
Busan, South Korea	3.8
Chongqing (Shi), China	30.0**
Guangzhou, China	3.9
Hong Kong (Xianggang) SAR, China	7.3
Macau SAR, China	0.5
Nanjing, China	2.8
Osaka, Japan	11.0
Seoul, South Korea	9.9
Shanghai (Shi), China	12.7
Shenyang, China	4.9
Shenzhen, China	3.8
Taipei, Taiwan	2.5
Tianjin (Shi), China	9.3
Tokyo, Japan	26.7
Wuhan, China	6.0
Xian, China	3.3

*Based on 2004 estimates.
**See discussion of this estimate on pp. 438–439.

face communications, agriculture, industry—emphasizes the strong concentration of Chinese activity in the country's eastern sector. This is the "real" China, where its great cities, populous farmlands, and historic sources are located. Long ago, scholars called this *China Proper*, and it is a good regional designation. But China is a large and complex country, and a number of subregions are nested within China Proper. Some of these, such as the North China Plain and the Sichuan Basin, are old and well-established geographic units. One in particular is new: China's Pacific Rim, still growing, yet poorly defined, and shown as "formative" in Figure 9-4.

2. *Xizang (Tibet)*. The high mountains and plateaus of Xizang, ruled by China but still widely known by its older name of Tibet, form a stark contrast to teeming China Proper. Here, next to one of the world's largest and most populous regions, lies one of the emptiest and, in terms of inhabited space, smallest regions.

3. *Xinjiang*. The vast desert basins and encircling mountains of Xinjiang form a third East Asian region. Again, physical as well as human geographic criteria come into play: here China meets Islamic Central Asia.

4. *Mongolia*. The desert state of Mongolia forms East Asia's fourth region. Like Tibet, landlocked Mongolia, vast but sparsely peopled, stands in stark contrast to populous China Proper.

5. *Jakota Triangle*. East Asia's fifth region is defined by its economic geography. *Ja*pan, South *Ko*rea, and *Ta*iwan (the name "Jakota" derives from the first two letters of each) were transformed by Pacific Rim economic developments during the second half of the twentieth century. Its recent emergence foreshadows further changes in the decades ahead as Korean unification becomes a possibility and contrasts between the Jakota Triangle and Pacific Rim China diminish.

◗ CHINA PROPER

When we in the Western world chronicle the rise of civilization, we tend to focus on the historical geography of Southwest Asia, the Mediterranean, and Western Europe. Ancient Greece and Rome were the crucibles of culture; Mediterranean and Atlantic waters were the avenues of its diffusion. China lay remote, so we believe, barely connected to this Western realm of achievement and progress. When an Italian adventurer named Marco Polo visited China during the thirteenth century and described the marvels he saw there, his work did little to change European minds. Europe was and would always be the center of civilization.

The Chinese, naturally, take a different view. Events on the western edge of the great Eurasian landmass were deemed irrelevant to theirs, the most advanced and refined culture on Earth. Roman emperors were rumored to be powerful, and Rome was a great city, but nothing could match the omnipotence of China's rulers. Certainly the Chinese city of Xian far eclipsed Rome as a center of sophistication. Chinese civilization existed long before ancient Greece and Rome emerged, and it was still there long after they collapsed. China, the Chinese teach themselves, is eternal. It was, and always will be, the center of the civilized world.

We should remember this notion when we study China's regional geography because 4000 years of Chinese culture and perception will not change overnight—not even in a generation. Time and again, China overcame the invasions and depredations of foreign intruders, and afterward the Chinese would close off their vast country against the outside world. Just 30 years ago, in the early 1970s, there were just a few *dozen* foreigners in the entire country with its (then) nearly 1 billion inhabitants. The institutionalization of communism required this insularity, and even the Soviet advisors had been thrown out. But by the early 1970s, China's rulers decided that an opening to the Western world would be advantageous, and so U.S. President Richard Nixon was invited to visit Beijing. That historic occasion, in 1972, ended this latest period of isolation—as always, on China's terms. Since then, China has been open to tourists and businesspeople, teachers, and investors. Tens of thousands of Chinese students have been sent to study at American and other Western institutions. Long-suppressed ideas flowed into China, and the pro-democracy movement arose and climaxed in 1989. China's rulers knew that their violent repression of this movement would anger the world, but that did not matter because they deemed foreign condemnation irrelevant. Foreigners in China had done much worse. Moreover, Westerners had no business interfering in China's domestic affairs.

RELATIVE LOCATION

Throughout their nation's history, the Chinese have at times decided to close their country to any and all foreign influences; the most recent episode of exclusion occurred just a few decades ago. Exclusion is one of China's recurrent traditions, made possible by China's relative location and Asia's physiography. In other words, China's "splendid isolation" was made possible by geography.

Earlier we noted the role of relief and desert in encircling the culture hearth of East Asia, but equally telling is the factor of distance. Until recently, China lay far from the modern source areas of innovation and change. True, China—as the Chinese emphasize—was itself such a hearth, but China's contributions to the outside world remained limited, essentially, to finely made arts and crafts. China did interact with Korea, Japan, Taiwan, and parts of Southeast Asia, and eventually millions of Chinese emigrated to neighboring countries. But compare these regional links to those of the Arabs, who ranged worldwide and who brought their knowledge, religion, and political influence to areas from Mediterranean Europe to Bangladesh and from West Africa to Indonesia. Later, when Europe became the center of intellectual and material innovation, China found itself farther removed, by land or sea, than almost any other part of the world.

Today, modern communications notwithstanding, China still is distant from almost anywhere else on Earth. Going by rail from Beijing to Moscow, the capital of China's Eurasian neighbor, involves a tedious journey that takes the better part of a week. Direct surface connections with India are practically nonexistent. Overland linkages with Southeast Asian countries, though improving, remain tenuous.

But for the first time in its history, China now lies near a world-class hearth of technological innovation and financial power: Japan. This proximity to an industrial and financial giant is critical for the momentous economic developments taking place in China's coastal provinces. Japanese investments and business partnerships have transformed the economic landscape of Pacific coast China. American and European trade links also are important, but Japan's role has been crucial. Japan's economic success set the Pacific Rim engine in motion, and Japan's best financial years happened to coincide with China's reopening to foreigners in the 1970s. Geographic and economic circumstances combined to transform the map and made "Pacific Rim" a household word around the world.

EXTENT AND ENVIRONMENT

China's total area is slightly smaller than that of the United States including Alaska: each country has about 3.7 million square miles (9.6 million sq km). As Figure 9-5 reveals, the longitudinal extent of China and the 48 contiguous U.S. States also is similar. Latitudinally, however, China is considerably wider. Miami, near the southern limit of the United States, lies halfway between Shanghai and Guangzhou. Thus China's lower-latitude southern region takes on characteristics of tropical Asia. In the Northeast, too, China incorporates much of what in North America would be Quebec and Ontario. Westward, China's land area becomes narrower and physiographic similarities increase. But, of course, China has no west coast.

Now compare the climate maps of China and the United States in Figure 9-6 (which are enlargements of the appropriate portions of the world climate map in Fig. G-8). Note that both have a large southeastern climatic region marked *Cfa* (that is, humid, temperate, warm-summer), flanked in China by a zone of *Cwa* (where winters become drier). Westward in both countries, the *C* climates yield to colder, drier climes. In the United States, moderate *C* climates develop again along the Pacific coast. China, however, stays dry and cold as well as high in elevation at equivalent longitudes.

FIGURE 9-5

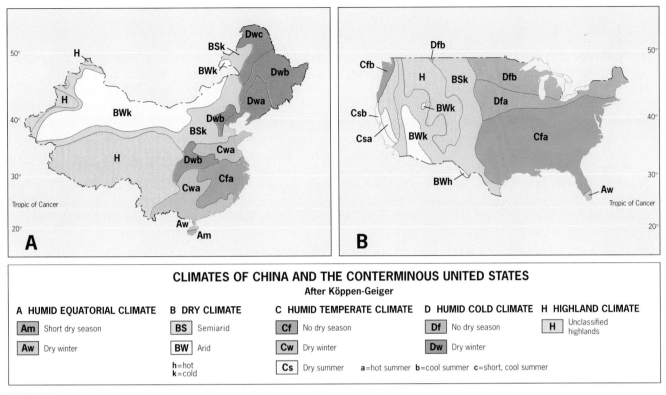

CLIMATES OF CHINA AND THE CONTERMINOUS UNITED STATES
After Köppen-Geiger

A HUMID EQUATORIAL CLIMATE	B DRY CLIMATE	C HUMID TEMPERATE CLIMATE	D HUMID COLD CLIMATE	H HIGHLAND CLIMATE
Am Short dry season	**BS** Semiarid	**Cf** No dry season	**Df** No dry season	**H** Unclassified highlands
Aw Dry winter	**BW** Arid	**Cw** Dry winter	**Dw** Dry winter	
	h=hot k=cold	**Cs** Dry summer a=hot summer b=cool summer c=short, cool summer		

FIGURE 9-6

Note especially the comparative location of the U.S. and Chinese *Cfa* areas in Figure 9-6. China's lies much farther to the south. In the United States, the *Cfa* climate extends beyond 40° North latitude, but in China, cold and generally winter-dry *D* climates take over at the latitude of Virginia. Beijing has a warm summer but a bitterly cold and long winter. Northeast China, in the general latitudinal range of Canada's lower Quebec and Newfoundland, is much more severe than its North American equivalent. Harsh environments prevail over vast regions of China, but, as we will see, nature compensates in spectacular fashion. From the climatic zone marked *H* (for highlands) in the west come the great life-giving rivers whose wide basins contain enormous expanses of fertile soils. Without these waters, China would not have a population more than four times that of the United States.

EVOLVING CHINA

Even if China is not the world's longest continuous civilization (Egypt may claim this distinction), no other state on Earth can trace its cultural heritage as far back as China can. China's fortunes rose and fell, but over more than 40 centuries its people created a society with strong traditions, values, and philosophies. Kongfuzi (Confu-

cius), whose teachings and writing still influence China, lived during the Zhou Dynasty, 2500 years ago (see box titled "Kongfuzi [Confucius]"). The Chinese still refer to themselves as the "People of Han," the dynasty that marks the breakdown of the old feudal order, the rise of military power, the unification of a large empire, the institution of property rights, the flourishing of architecture, the arts, and the sciences, and the development of trade along the Silk Route (Table 9-1). The Han Dynasty reigned 2000 years ago. The walled city of Xian, then called Ch'angan, was the early Han capital and one of the greatest cities of the ancient world.

As we try to gain a better understanding of China's present political, economic, and social geography, we should not lose sight of the China that endured for thousands of years under a sequence of imperial dynasties as a single entity. Larger and more populous than Europe, China had its divisive feudal periods, but always it came together again, ruled—mostly dictatorially—from a strong center as a unitary state. China was, and remains, a predominantly rural society, a powerless and subservient population controlled by an often ruthless bureaucracy. Retributions, famines, epidemics, and occasional uprisings took heavy tolls in the countryside. But in the cities, Chinese culture flourished, subsidized by the taxes and tribute extracted from the hinterlands. As China grew, it incorporated

Kongfuzi (Confucius)

*C*onfucius (*Kongfuzi* or *Kongzi* in pinyin) was China's most influential philosopher and teacher. His ideas dominated Chinese life and thought for over 20 centuries.

Kongfuzi was born in 551 BC and died in 479 BC. Appalled at the suffering of ordinary people during the Zhou Dynasty, he urged the poor to assert themselves and demand explanations for their harsh treatment by the feudal lords. He tutored the indigent as well as the privileged, giving the poor an education that had hitherto been denied them and ending the aristocracy's exclusive access to the knowledge that constituted power.

Kongfuzi's revolutionary ideas extended to the rulers as well as the ruled. He abhorred supernatural mysticism and cast doubt on the divine ancestries of China's aristocratic rulers. Human virtues, not godly connections, should determine a person's place in society, he taught. Accordingly, he proposed that the dynastic rulers turn over the reins of state to ministers chosen for their competence and merit. This was another Kongzi heresy, but in time this idea came to be accepted and practiced.

His Earthly philosophies notwithstanding, Kongfuzi took on the mantle of a spiritual leader after his death. His thoughts, distilled from the mass of philosophical writing (including Daoism) that poured forth during his lifetime, became the guiding principles of the formative Han Dynasty. The state, he said, should not exist just for the power and pleasure of the elite; it should be a cooperative system for the well-being and happiness of the people.

With time, a mass of writings evolved, much of which Kongfuzi never wrote. At the heart of this body of literature lay the Confucian Classics, 13 texts that became the basis for education in China for 2000 years. From government to morality and from law to religion, the Classics were Chinese civilization's guide. The entire national system of education (including the state examinations through which everyone, poor or privileged, could enter the civil service and achieve political power) was based on the Classics. Kongfuzi championed the family as the foundation of Chinese culture, and the Classics prescribe a respect for the aged that was a hallmark of Chinese society.

But Kongfuzi's philosophies also were conservative and rigid, and when the colonial powers penetrated China, Kongzi's Classics came face to face with practical Western education. For the first time, some Chinese leaders began to call for reform and modernization, especially of teaching. Kongzi principles, they said, could guide an isolated China, but not China in the new age of competition. But the Manchu rulers resisted this call, and the Nationalists tried to combine Kongzi and Western knowledge into a neo-Kongzi philosophy, which was ultimately an unworkable plan.

The communists who took power in 1949 attacked Kongzi thought on all fronts. The Classics were abandoned, indoctrination pervaded education, and, for a time, even the family was viewed as an institution of the past. Here the communists miscalculated. It proved impossible to eradicate two millennia of cultural conditioning in a few decades. When China entered its post-Mao period, public interest in Kongfuzi surged, and the shelves of bookstores again sag under the weight of the *Analects* and other Confucian and post-Kongfuzi writings. The spirit of Kongfuzi will pervade physical and mental landscapes in China for generations to come.

---

minorities ranging from Koreans and Mongols to Uyghurs and Tibetans; in the far south, it now includes several peoples with Southeast Asian affinities. Like the Chinese citizenry itself, these minorities have experienced both benevolent government and brutal subjugation. But it has always been China's wish to **Sinicize** them: to endow them with the elements of Chinese culture.

A Century of Convulsion

When the European colonialists appeared in East Asia, China long withstood them with a self-assured superiority based on the strength of its culture and the reassuring continuity of the state. There was no market for the British East India Company's rough textiles in a country long used to finely fabricated silks and cottons. There was little interest in the toys and trinkets the Europeans produced in the hope of barter for Chinese tea and porcelain. Even key European inventions, such as the mechanical clock, though considered amusing and entertaining, were ignored and even deprecated as irrelevant to Chinese culture.

The self-confident Ming emperors even beat the Europeans at their own game. In the fourteenth century, great oceangoing vessels sailed the South China Sea; Chinese fleets carrying as many as 20,000 men reached Southeast and South Asia and even East Africa. Several times as large and far more sophisticated than anything built in Europe, China's ships, together with their sailors and the goods they carried, demonstrated the potential of the Middle Kingdom. But suddenly China faced a crisis at home: the impact of the onset of the Little Ice Age that also

AMONG THE REALM'S GREAT CITIES . . . **XIAN**

The city known today as Xian is the site of one of the world's oldest urban centers. It may have been a settlement during the Shang-Yin Dynasty more than 3000 years ago; it was a town during the Zhou Dynasty, and the Qin emperor was buried here along with 6000 life-sized terracotta soldiers and horses, reflecting the city's importance. During the Han Dynasty the city, then called Ch'angan, was one of the greatest centers of the ancient world, the Rome of ancient China. Ch'angan formed the eastern terminus of the Silk Route, a storehouse of enormous wealth. Its architecture was unrivalled, from its ornamental defensive wall with elaborately sculpted gates to the magnificent public buildings and gardens at its center.

Situated on the fertile loess plain of the upper Wei River, Ch'angan was the focus of ancient China during crucial formative periods. After two centuries of Han rule, political strife led to a period of decline, but the Sui emperors rebuilt and expanded Ch'angan when they made it their capital. During the Tang Dynasty, Ch'angan again became a magnificent city with three districts: the ornate Palace City; the impressive Imperial City, which housed the national administration; and the busy Outer City containing the homes and markets of artisans and merchants.

After its Tang heyday the city again declined, although it remained a bustling trade center. During the Ming Dynasty it was endowed with some of its architectural landmarks, including the Great Mosque marking the arrival of Islam; the older Big Wild Goose Pagoda dates from the influx of Buddhism. After the Ming period, Ch'angan's name was changed to Xian (meaning "Western Peace"), then to Siking, and in 1943 back to Xian again.

Having been a gateway for Buddhism and Islam, Xian in the 1920s became a center of Soviet communist ideology. The Nationalists, during the struggle against the Japanese, moved industries from the vulnerable east to Xian, and when the communists took power, they enlarged Xian's industrial base still further. The present city (population: 3.3 million) lies southwest of the famed tombs, its cultural landscape now dominated by a large industrial complex that includes a steel mill, textile factories, chemical plants, and machine-making facilities. Little remains (other than some prominent historic landmarks) of the splendor of times past, but Xian's location on the railroad to the vast western frontier of China sustains its long-term role as one of the country's key gateways.

FROM THE FIELD NOTES

"To pass through the gate from Tiananmen Square and enter the Manchu Emperors' Forbidden City was a riveting experience. Look back now toward the gate, and you see throngs of Chinese visiting what less than a century ago was the exclusive domain of the Qing Dynasty's absolute rulers—so absolute that unauthorized entry was punished by instant execution. The vast, walled complex at the heart of Beijing not only served as the imperial palace; it was a repository of an immense collection of Chinese technological and artistic achievements. Some of this is still here, but most of all you are awed by the lingering atmosphere, the sense of what happened here in these exquisitely constructed buildings that epitomized what high Chinese culture could accomplish. Inevitably you think of the misery of millions whose taxes and tribute paid for what stands here."

affected Europe. Cold and drought decimated the wheat harvest on the North China Plain, and boats were needed by the thousands to carry rice from the Chang Basin to the hungry north. That spelled the end of China's long-range maritime explorations; it may have changed the historical geography of the world.

Even when Europe's sailing ships made way for steam-driven vessels and newer and better European products (including weapons) were offered in trade for China's tea and silk, China continued to reject European imports and resisted commerce in general. (When the Manchus took control of Beijing in 1644, their bows and arrows proved superior to Chinese-manufactured muskets which were so heavy and hard to load that they were almost useless.) The Chinese kept the Europeans confined to small peninsular outposts, such as Macau, and minimized interaction with them. Long after India had succumbed to mercantilism and economic imperialism, China maintained its established order. This was no surprise to the Chinese. After all, they had held a position of undisputed superiority in their Celestial Kingdom as long as could be remembered, and they had dealt with foreign invaders before.

A (Lost) War on Drugs

All this confidence was shattered during the Manchu (Qing) Dynasty, China's last imperial regime. The Manchus, who had taken control in Beijing as a small minority and who had grafted their culture onto China's, had the misfortune of reigning when the standoff with Europeans ended and the balance of power shifted in favor of the colonialists.

On two fronts in particular, the economic and the political, the European powers destroyed China's invincibility. Economically, they succeeded in lowering the cost and improving the quality of manufactured goods, especially textiles, and the handicraft industries of China began to collapse in the face of unbeatable competition. Politically, the demands of the British merchants and the growing English presence in China led to conflicts. In the early part of the nineteenth century, the central issue was the importation into China from British India of opium, a dangerous and addictive intoxicant. Opium was destroying the very fabric of Chinese culture, weakening the society, and rendering China easy prey for colonial profiteers. As the Manchu government moved to stamp out the opium trade in 1839, armed hostilities broke out, and soon the Chinese found themselves losing a war on their own territory. The First Opium War (1839–1842) ended in disaster: China's rulers were forced to yield to British demands, and the breakdown of Chinese sovereignty was under way.

British forces penetrated up the Chang Jiang and controlled several areas south of it (Fig. 9-7); Beijing hurriedly sought a peace treaty by which it granted leases and concessions to foreign merchants. In addition, China ceded Hong Kong Island to the British and opened five ports, including Guangzhou (Canton) and Shanghai, to foreign commerce. No longer did the British have to accept a status that was inferior to the Chinese in order to do business; henceforth, negotiations would be pursued on equal terms. Opium now flooded into China, and its impact on Chinese society became even more devastating. Fifteen years after the First Opium War, the Chinese again tried to stem the disastrous narcotic tide, and the foreigners who had attached themselves to their country again defeated them. Now the government legalized cultivation of the opium poppy in China itself. Chinese society was disintegrating; the scourge of this drug abuse was not defeated until after the revival of Chinese power in the twentieth century.

But before China could reassert itself, much of what remained of its independence steadily eroded (see box titled "Extraterritoriality"). In 1898 the Germans obtained a lease on Qingdao on the Shandong Peninsula, and the French acquired a sphere of influence in the far south at Zhanjiang (Fig. 9-7). The Portuguese took Macau; the Russians obtained a lease on Liaodong in the Northeast as well as railway concessions there; even Japan got into the act by annexing the Ryukyu Islands and, more importantly, Formosa (Taiwan) in 1895.

After four millennia of recurrent cultural cohesion, economic security, and political continuity, the Chinese world lay open to the aggressions of foreigners whose innovative capacities China had negated to the end. Now the ships flying European flags lay in the ports of China's coasts and rivers, but China had not learned to manufacture the cannons to blast them out of the water. The smokestacks of foreign factories rose over the townscapes of its great cities, and no Chinese force could dislodge them. The Japanese, seeing China's weakness, invaded Korea. The Russians entered Manchuria, the home base of the Manchus. The foreign invaders even took to fighting among themselves, as Japan and Russia did in Manchuria in 1904. (Today this part of China is called the *Northeast*, the name Manchuria having been—understandably—rejected.)

A New China Rises

In the meantime, organized opposition to the foreign presence in China was gathering strength, and the twentieth century opened with a large-scale revolt against all outside elements. Bands of revolutionaries roamed both cities and countryside, attacking not only the hated foreigners but also Chinese who had adopted Western cultural traits. Known as the Boxer Rebellion (after a loose translation of

FIGURE 9-7

the Chinese name for these revolutionary groups), this 1900 uprising was put down with much bloodshed by an international force consisting of British, Russian, French, Italian, German, Japanese, and American soldiers. Simultaneously, another revolutionary movement was gaining support, aimed against the Manchu leadership itself. In 1911, the emperor's garrisons were attacked all over China,

and in a few months the 267-year-old Qing Dynasty was overthrown. Indirectly, it too was a casualty of the foreign intrusion, and it left China divided and disorganized.

The fall of the Manchus and the proclamation of a republican government in China did little to improve the country's overall position. The Japanese captured Germany's holdings on the Shandong Peninsula, including

Extraterritoriality

*D*uring the nineteenth century, as China weakened and European colonial invaders entered China's coastal cities and sailed up its rivers, the Europeans forced China to accept a European doctrine of international law—**extraterritoriality**. Under this doctrine, foreign states and their representatives are immune from the jurisdiction of the country in which they are based. Today, this applies to embassies and diplomatic personnel. But in Qing (Manchu) China, it went far beyond that.

The European, Russian, and Japanese invaders established as many as 90 *treaty ports*—extraterritorial enclaves in China's cities under unequal treaties enforced by gunboat diplomacy. In their "concessions," diplomats and traders were exempt from Chinese law. Not only port areas but also the best residential suburbs of large cities were declared to be "extraterritorial" and made inaccessible to Chinese citizens. In the city of Guangzhou (Canton in colonial times), Sha Mian Island in the Pearl River was a favorite extraterritorial enclave. A sign at the only bridge to the island stated, in English and Cantonese, "No Dogs or Chinese."

Christian missionaries fanned out into China, their residences and churches fortified with extraterritorial security. In many places, Chinese found themselves unable to enter parks and buildings without permission from foreigners. This involved a loss of face that contributed to bitter opposition to the presence of foreigners—a resentment that exploded in the Boxer Rebellion of 1900.

After the collapse of the Qing Dynasty in 1911, the Chinese Nationalists negotiated an end to all commercial extraterritoriality in China Proper; the Russians, however, would not yield in then-Manchuria. Only Hong Kong and Macau retained their status as colonies.

When China's government in 1980 embarked on a new economic policy that gave major privileges and exemptions to foreign firms in certain coastal areas and cities, opponents argued that this policy revived the practice of extraterritoriality in a new guise. This issue remains a sensitive one in a China that has not forgotten the indignities of the colonial era.

the city of Qingdao, during World War I. When the victorious European powers met at Versailles in 1919 to divide the territorial spoils, they affirmed Japan's rights in the area. This led to yet another Chinese effort to counter the foreign scourge. Nationwide protests and boycotts of Japanese goods were organized in what became known as the May Fourth Movement. One participant in these demonstrations was a charismatic young man named Mao Zedong.

Nationalists and Communists

During the chaotic 1920s, the Nationalists and the Communist Party at first cooperated, with the remaining foreign presence their joint target. After Sun Yat-sen's death in 1925, Chiang Kai-shek became the Nationalists' leader, and by 1927 the foreigners were on the run, escaping by boat and train or falling victim to rampaging Nationalist forces. But soon the Nationalists began purging communists even as they pursued foreigners, and in 1928, when Chiang established his Nationalist capital in the city of Nanjing, it appeared that the Nationalists would emerge victorious from their campaigns. They had driven the communists ever deeper into the interior, and by 1933 the Nationalist armies were on the verge of encircling the last communist stronghold in the area of Ruijin in Jiangxi Province.

This led to a momentous event in Chinese history: the *Long March*. Nearly 100,000 people—soldiers, peas-

ants, leaders—marched westward from Ruijin in 1934, a communist column that included Mao Zedong and Zhou Enlai. The Nationalist forces rained attack after attack on the marchers, and of the original 100,000, about three-quarters were killed. But new sympathizers joined along the way (see the route marked on Fig. 9-7), and the 20,000 survivors found a refuge in the mountainous interior of Shaanxi Province, 2000 miles (3200 km) away. There, they prepared for a renewed campaign that would bring them to power.

Japan in China

Although many foreigners fled China during the 1920s and 1930s, others seized the opportunity presented by the contest between the Nationalists and communists. The Japanese took control over the Northeast, and when the Nationalists proved unable to dislodge them, they set up a puppet state there, appointed a Manchu ruler to represent them, and called their possession Manchukuo.

The inevitable full-scale war between the Chinese and the Japanese broke out in 1937, with the Nationalists bearing the brunt of it (which gave the communists an opportunity to regroup). The gray boundary on Figure 9-7 shows how much of China the Japanese conquered. The Nationalists moved their capital to Chongqing, and the communists controlled the area centered on Yanan. China had been broken into three pieces.

The Japanese committed unspeakable atrocities in their campaign in China. Millions of Chinese citizens were shot, burned, drowned, subjected to gruesome chemical and biological experiments, and otherwise wantonly victimized. Years later, when China's economic reforms of the 1980s and 1990s led to a renewed Japanese presence in China, the Chinese public and its leaders called for Japan to acknowledge and apologize for these wartime abuses. In Japan, this pitted apologists against strident nationalists, causing a political crisis. In 1992, Emperor Akihito visited China and referred to the war but stopped short of a formal apology. The book is not yet closed on this most sensitive issue.

Communist China Arises

After the U.S.-led Western powers defeated Japan in 1945, the civil war in China quickly resumed. The United States, hoping for a stable and friendly government in China, sought to mediate the conflict but at the same time recognized the Nationalists as the legitimate government. The United States also aided the Nationalists militarily, destroying any chance of genuine and impartial mediation. By 1948, it was clear that Mao Zedong's well-organized militias would defeat Chiang Kai-shek. Chiang kept moving his capital—back to Guangzhou, seat of Sun Yat-sen's first Nationalist government, then back to Chongqing. Late in 1949, after a series of disastrous defeats in which hundreds of thousands of Nationalist forces were killed, the remnants of Chiang's faction gathered Chinese treasures and valuables and fled to the island of Taiwan. There, they took control of the government and proclaimed their own Republic of China.

Meanwhile, on October 1, 1949, standing in front of the assembled masses at the Gate of Heavenly Peace on Beijing's Tiananmen Square, Mao Zedong proclaimed the birth of the People's Republic of China.

■ CHINA'S HUMAN GEOGRAPHY

After more than a half-century of communist rule, China is a society transformed. It has been said that the year 1949 actually marked the beginning of a new dynasty not so different from the old, an autocratic system that dictated from the top. In that view, Mao Zedong simply bore the mantle of his dynastic predecessors. Only the family lineage had fallen away; now communist "comrades" would succeed each other.

And certainly some of China's old traditions continued during the communist era, but in many other ways Chinese society was totally overhauled. Benevolent or otherwise, the dynastic rulers of old China headed a country in which—for all its splendor, strength, and cultural richness—the fate of landless people and of serfs often was undescribably miserable; in which floods, famines, and diseases could decimate the populations of entire regions without any help from the state; in which local lords could (and often did) repress the people with impunity; in which children were sold and brides were bought. The European intrusion made things even worse, bringing slums, starvation, and deprivation to millions who had moved to the cities.

The communist regime, dictatorial though it was, attacked China's weaknesses on many fronts, mobilizing virtually every able-bodied citizen in the process. Land was taken from the wealthy; farms were collectivized; dams and levees were built with the hands of thousands; the threat of hunger for millions receded; health conditions improved; child labor was reduced. But China's communist planners also made terrible mistakes. The *Great Leap Forward*, requiring the reorganization of the peasantry into communal brigade teams to speed industrialization and make farming more productive, had the opposite effect and so disrupted agriculture that between 20 and 30 million people died of starvation between 1958, when the program was implemented, and 1962, when it was abandoned.

Mao ruled China from 1949 to 1976, long enough to leave lasting marks on the state. Another of his communist dictums had to do with population. Like the Soviets (and influenced by a horde of Soviet advisors and planners), Mao refused to impose or even recommend any population policy, arguing that such a policy would represent a capitalist plot to constrain China's human resources. As a result, China's population grew explosively during his rule.

Yet another costly episode of Mao's rule was the so-called *Great Proletarian Cultural Revolution*, launched by Mao Zedong during his last decade in power (1966–1976). Fearful that Maoist communism was being contaminated by Soviet "deviationism" and worried about his own stature as its revolutionary architect, Mao unleashed a campaign against what he viewed as emerging elitism in society. He mobilized young people living in cities and towns into cadres known as Red Guards and ordered them to attack "bourgeois" elements throughout China, criticize Communist Party officials, and root out "opponents" of the system. He shut down all of China's schools, persecuted untrustworthy intellectuals, and encouraged the Red Guards to engage in what he called a renewed "revolutionary experience." The results were disastrous: Red Guard factions took to fighting among themselves, and anarchy, terror, and economic paralysis followed. Thousands of China's leading intellectuals died, moderate leaders were purged, and teachers, elderly citizens, and older revolutionaries were tortured to make them confess to "crimes" they did not commit. As the economy suffered, food and industrial production declined. Violence and famine killed as many as 30 million people as the Cultural Revolution spun out of control. One of those who survived was a Communist Party leader who had himself been purged and then been reinstated—Deng Xiaoping. Deng was destined to lead the country in the post-Mao period of economic transformation.

Political and Administrative Divisions

Before we investigate the emerging human geography of contemporary China, we should acquaint ourselves with the country's political and administrative framework (Fig. 9-8). For administrative purposes, China is divided into the following units:

4 Central-Government-Controlled Municipalities (Shi's)

5 Autonomous Regions

22 Provinces

2 Special Administrative Regions

The four central-government-controlled municipalities are the capital, Beijing; its nearby port city, Tianjin; China's largest metropolis, Shanghai; and the Chang river port of Chongqing, in the interior. These *Shi's* form the cores of China's most populous and important subregions, and direct control over them from the capital entrenches the central government's power.

We should note that the administrative map of China continues to change—and to pose problems for geographers. The city of Chongqing was made a *shi* in 1996, and its "municipal" area was enlarged to incorporate not only

FIGURE 9-8

the central urban area but a huge hinterland covering all of eastern Sichuan Province. As a result, the "urban" population of Chongqing is officially 30 million, making this the world's largest metropolis—but in truth, the central urban area has no more than about 6 million inhabitants. And because Chongqing's population is officially *not* part of the province that borders it to the west (Fig. 9-8), the official population of Sichuan declined by 30 million when the *Chongqing Shi* was created. Despite what Table 9-2 suggests, therefore, the population within Sichuan Province's outer borders remains the largest of any of China's administrative divisions.

The five Autonomous Regions were established to recognize the non-Han minorities living there. Some laws

Table 9-2
CHINA: POPULATION BY MAJOR ADMINISTRATIVE DIVISIONS*
(in Millions)

Provinces	1990 Population	2001 Population
Anhui	56.2	63.3
Fujian	30.0	34.4
Gansu	22.4	25.8
Guangdong	62.8	77.8
Guizhou	32.4	38.0
Hainan	6.6	8.0
Hebei	61.1	67.0
Heilongjiang	35.2	38.1
Henan	85.5	95.6
Hubei	54.0	59.8
Hunan	60.7	66.0
Jiangsu	67.1	73.6
Jiangxi	37.7	41.7
Jilin	24.7	26.9
Liaoning	39.5	41.9
Qinghai	4.5	5.2
Shaanxi	32.9	36.6
Shandong	84.4	90.4
Shanxi	28.8	32.7
Sichuan	107.2	86.4
Yunnan	37.0	42.9
Zhejiang	41.5	46.1
Autonomous Regions		
Guangxi Zhuang	42.2	47.9
Nei Mongol	21.5	23.8
Ningxia Hui	4.7	5.6
Xinjiang Uyghur	15.2	18.8
Xizang	2.2	2.6

*Population data for the four centrally administered municipalities and the Hong Kong (Xianggang) and Macau Special Administrative Regions are shown in the table on page 429.

that apply to Han Chinese do not apply to certain minorities. As we saw in the case of the former Soviet Union, however, demographic changes and population movements affect such regions, and the policies of the 1940s may not work in the twenty-first century. Han Chinese immigrants now outnumber several minorities in their own regions. The five Autonomous Regions (A.R.'s) are: (1) Nei Mongol A.R. (Inner Mongolia); (2) Ningxia Hui A.R. (adjacent to Inner Mongolia); (3) Xinjiang Uyghur A.R. (China's northwest corner); (4) Guangxi Zhuang A.R. (far south, bordering Vietnam); and (5) Xizang A.R. (Tibet).

China's 22 Provinces, like U.S. States, tend to be smallest in the east and largest toward the west. The territorially smallest are the three easternmost provinces on China's coastal bulge: Zhejiang, Jiangsu, and Fujian. The two largest are Qinghai, flanked by Tibet, and Sichuan, China's Midwest.

As with all large countries, some provinces are more important than others. The Province of Hebei nearly surrounds Beijing and occupies much of the core of the country. The Province of Shaanxi is centered on the great ancient city of Xian. In the southeast, momentous economic developments are occurring in the Province of Guangdong, whose urban focus is Guangzhou. When, in the pages that follow, we refer to a particular province or region, Figure 9-8 is a useful locational guide.

In 1997, the British dependency of Hong Kong (Xianggang) was taken over by China and became the country's first **Special Administrative Region (SAR)**. In 1999, Portugal similarly transferred Macau, opposite Hong Kong on the Pearl River Estuary, to Chinese control, creating the second SAR under Beijing's administration.

Population Issues

Table 9-2 underscores the enormity of China's population: many Chinese provinces have more inhabitants than do most of the world's countries. With a 2004 population of 1.306 billion, China remains the largest nation on Earth. We should view these huge numbers in the context of China's population growth rate. High rates of population growth dim national development prospects in many countries in the periphery, but China has a special problem. Even a modest growth rate—such as 0.7 percent annually—still adds more than 9 million to the total every year.

Mao Zedong, as we have noted, refused to institute policies that would reduce the rate of population increase because he believed (encouraged by Soviet advisors) that this would play into the hands of his capitalist enemies. Fast-growing populations, he said, constituted the only asset many poor, communist, or socialist countries had. If the capitalist world wanted to reduce growth rates, it must be an anticommunist plot.

But other Chinese leaders, realizing that rapid population growth would stymie their country's progress, tried to stem the tide through local propaganda and education. Some of these leaders lost their positions as "revisionists," but their point was not lost on China's future leaders. After Mao's death, China embarked on a vigorous population-control program. In the early 1970s, the annual rate of natural increase was about 3 percent; by the mid-1980s, it was down to 1.2 percent. Families were ordered to have one child only, and those who violated the policy were penalized by losing tax advantages, educational opportunities, and even housing privileges. Today, China's census bureau officially reports a moderate growth rate of 0.7 percent.

Although the one-child policy gained general acceptance, since it was vigorously policed throughout China by Communist Party members, it also had serious negative consequences. Abortions soared, sometimes in the third trimester and coerced by Party enforcers. Female infanticide also rose because families wanted their only child to be a boy. Families tried to hide their second and third children by sending them to family members in other villages; upon discovery by Party members, some parents had their houses burned down. When reports of such dreadful byproducts of the population-control campaign leaked to the outside world, the policy was relaxed during the late 1980s. Certain farming and fishing families were allowed to have a second child to fulfill future labor needs, and rural families could request permission to have a second child if their first baby was a girl. Ethnic minority groups were mostly exempted.

Partly because of disobedience and partly because of some relaxation of the one-child policy, China's growth rate may be inching upward again. The objective of China's demographic planners was to reach 1.2 billion no earlier than the year 2000, but that goal was missed, perhaps by as many as 70 million. This means that an additional 70 million people must be housed, educated, treated medically, and, eventually, employed—a population almost the size of the largest country in Europe, Germany. Against this background, China's desire for rapid economic progress will not easily be fulfilled.

The Minorities

We asserted earlier that China remains an empire; its government controls territories and peoples that are in effect colonized. The ethnolinguistic map (Fig. 9-9) apparently confirms this proposition. This map should be seen in context, however. It does not show local-area majority populations but instead reveals where minorities are concentrated. For example, the Mongolian population in China is shown to be clustered along the southeastern border of Mongolia in the Autonomous Region called Nei Mongol (see Fig. 9-8). But even in that A.R., the Han Chinese, not the Mongols, are now in the majority.

Nevertheless, the map gives definition to the term *China Proper* as the home of the People of Han, the ethnic (Mandarin-speaking) Chinese depicted in tan and light orange in Figure 9-9. From the upper Northeast to the border with Vietnam and from the Pacific coast to the margins of Xinjiang, this Chinese majority dominates. When you compare this map to that of population distribution (to be discussed shortly), it will be clear that the minorities constitute only a small percentage of the country's total. The Han Chinese form the largest and densest clusters.

In any case, China controls non-Chinese areas that are vast, if not populous. The Tibetan group numbers under 3 million, but it extends over all of settled Xizang. Turkic peoples inhabit large areas of Xinjiang. Thai, Vietnamese, and Korean minorities also occupy areas on the margins of Han China. As we will see later, the Southeast Asian minorities in China have participated strongly in the Pacific Rim developments on their doorsteps. Hundreds of thousands have migrated from their Autonomous Regions to the economic opportunities along the coast.

Numerically, Chinese dominate in China to a far greater degree than Russians dominated their Soviet Empire. But territorially, China's minorities extend over a proportionately larger area. The Ming and Manchu rulers bequeathed the People of Han an empire.

PEOPLE AND PLACES OF CHINA PROPER

A map of China's population distribution (Fig. 9-10) reveals the continuing relationship between the physical environment and its human occupants. In technologically advanced countries, we have noted, people shake off their dependence on what the land can provide; they cluster in cities and in other areas of economic opportunity. This depopulates rural areas that may once have been densely inhabited. In China, that stage has not yet been reached. Although China has large cities, as in India a sizeable majority of the people (62 percent) still live on—and from—the land. Thus the map of population distribution reflects the livability and productivity of China's basins, lowlands, and plains. Compare Figures 9-6A and 9-10, and China's continuing dependence on soil, water, and warmth will be evident.

FIGURE 9-9

The population map also suggests that in certain areas environmental limitations are being overcome. Industrialization in the Northeast, irrigation in the Inner Mongolia Autonomous Region, and oil-well drilling in Xinjiang enabled millions of Chinese to migrate from China Proper into these frontier zones, where they now outnumber the indigenous minorities.

Nevertheless, physiography and demography remain closely related in China. To grasp this relationship, it is useful to compare Figures 9-2 (physiography) and 9-10 (population) as our discussion proceeds. On the population map, the darker the color, the denser the population: in places we can follow the courses of major rivers through this scheme. Look, for example, at China's Northeast. A ribbon of population follows the Liao and the Songhua rivers. Also note the huge, nearly circular population con-

centration on the western edge of the red zone; this is the Sichuan Basin, in the upper course of the Chang Jiang. Four major river basins contain more than three-quarters of China's 1.3 billion people:

1. The Liao-Songhua Basin or the Northeast China Plain
2. The Lower Huang He (Yellow River) Basin, known as the North China Plain
3. The Upper and Lower Basins of the Chang Jiang (Yangzi River)
4. The Basins of the Xi (West) and Pearl River

Not all the people living in these river plains are farmers, of course. China's great cities also have developed in these populous areas, from the industrial centers of the

CHINA: POPULATION
DENSITY PER SQUARE KILOMETER

0 1 25 50 100 200 400 600 800 1000

0   200   400   600   800   1000   1200 Kilometers

0   200   400   600 Miles

FIGURE 9-10

Northeast to the Pacific Rim upstarts of the South. Both Beijing and Shanghai lie in major river basins. And, as the map shows, the hilly areas between the river basins are not exactly sparsely populated either. Note that the hill country south of the Chang Basin (opposite Taiwan) still has a density of 130 to 260 people per square mile (50 to 100 per sq km).

Northeast China Plain

The Northeast China Plain is the heartland of China's Northeast, ancestral home of the Manchus who founded China's last dynasty, battleground between Japanese and Russian invaders, Japanese colony, heartland of communist industrial development, and now losing ground to the industries of the Pacific Rim. This used to be called Manchuria. As the administrative map (Fig. 9-8) shows, there are three provinces here: Liaoning in the south, fac-

ing the Yellow Sea; Jilin in the center; and Heilongjiang, by far the largest, in the north.

When you are familiar with the vast farmlands of the North China Plain, the southernmost part of the Northeast China Plain looks so similar that it seems to be a continuation. Near the coast, farms look as they do near the mouth of the great Huang He; to the east, the Liaodong Peninsula looks like the Shandong Peninsula across the water. That impression soon disappears, however. Travel northward, and the Northeast China Plain reveals the effects of cold climate and thin soils. Farmlands are patchy; smokestacks seem to rise everywhere. This is industrial, not primarily farming, country. But the landscape looks like a Rustbelt: equipment often is old and outdated, transport systems are inadequate, buildings are in disrepair. Stream, land, and air pollution is dreadful. The Northeast in some ways resembles industrial zones in former communist Eastern Europe.

The Northeast has experienced many ups and downs. Although Japanese colonialism was ruthless and exploitive, Japan did build railroads, roads, bridges, factories, and other components of the regional infrastructure. (At one time, half of the entire railroad mileage in China was in the Northeast.) After the Japanese were ousted, the Soviets looted the area of machinery, equipment, and other goods. During the late 1940s, the Northeast was a ravaged frontier. But then the communists took power and made the industrial development of the Northeast a priority. From the 1950s until the 1970s, the Northeast led the nation in manufacturing growth. Its population, just a few million in the 1940s, mushroomed to more than 100 million. Towns and cities grew exponentially.

All this growth was based on the Northeast's considerable mineral wealth (Fig. 9-11). Iron ore deposits and

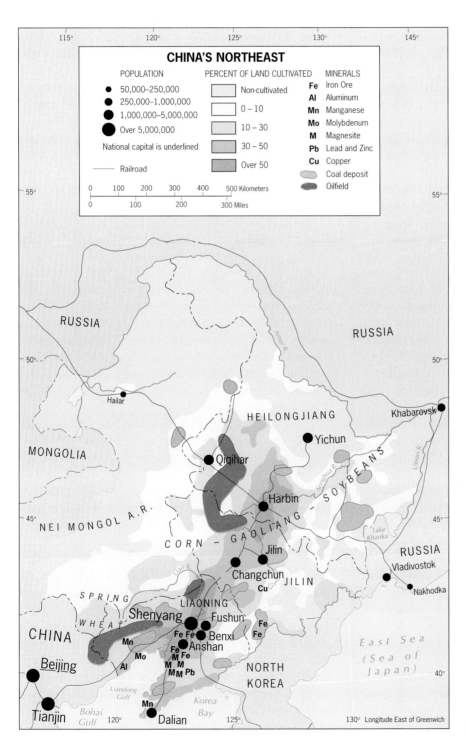

FIGURE 9-11

coalfields lie concentrated in the Liao Basin; aluminum ore, ferroalloys, lead, zinc, and other metals are plentiful, too. As the map shows, the region also is well endowed with oil reserves; the largest lie in the Daqing Reserve between Harbin and Qiqihar. The communist planners made Changchun capital of Jilin Province, the Northeast's automobile manufacturing center. Shenyang, capital of Liaoning Province, became the leading steelmaking complex. Fushun and Anshan were assigned other functions, and the communists encouraged Han Chinese to migrate to the Northeast to find employment there. To attract them, the government built apartments, schools, hospitals, recreational facilities, even old-age homes, all to speed the region's industrial development.

For a time, it worked. During the 1970s, the Northeast was responsible for fully one-quarter of the entire country's industrial output. Stimulated by the needs of the growing cities and towns, farms expanded and crops diversified. The Songhua River Basin in the hinterland of Harbin, capital of Heilongjiang Province, became a productive agricultural zone despite the environmental limitations here. The Northeast was a shining example of what communist planning could accomplish.

Then came the **economic restructuring** of the post-Mao era. Under the new, market-driven economic order, the state companies that had led China's industrial resurgence were unable to adapt. Suddenly they were aging, inefficient, obsolete. Official policy encouraged restructuring and privatization, but the huge state factories could not (and much of their labor force would not) comply. Today, the Northeast's contribution to China's industrial output is only about 10 percent as the production of factories in the Southeast, from Shanghai to Guangdong, rises annually.

Economic geographers, however, will not write off the prospects of this northeastern corner of China; this area, as we noted, has had its downturns before. The Northeast still contains a storehouse of resources, including extensive stands of oak and other hardwood forests on the mountain slopes that encircle the Northeast China Plain. The locational advantages of the major centers are undiminished. Harbin, for example, lies at the convergence of five railroads and is situated at the head of navigation on the Songhua River, providing a link to the northeast corner of the region and, beyond the Amur River, to Russia's Khabarovsk. Moreover, when the Koreas reunite, producing what may become the western Pacific Rim's second Japan, and when the Russian Far East takes off (Vladivostok and Nakhodka are the natural ports for the upper Northeast), this northern frontier of China may yet reverse its decline and enter still another era of economic success.

North China Plain

Take the new superhighway between the port of Tianjin and the capital, Beijing, or the train from Beijing southward to Anyang, and you see a physical landscape almost entirely devoid of relief and a cultural landscape that repeats itself endlessly. Villages, many showing evidence of the collectivization program of the early communist era, pass at such regular intervals that some kind of geometry seems to control the cultural landscape (see photo, p. 445). Carefully diked canals stretch in all directions. Rows of trees mark fields and farms. Depending on the season, the air is clear, as it is during the moist summer when the wheat crop covers the countryside like a carpet, or choked with dust, which can happen almost any other time of the year. The soil here is a fertile mixture of alluvium and loess (deposits of fine, windborne silt or dust of glacial origin), and the slightest breeze stirs it up. In the early spring the dust hangs like a fog over the landscape, covering everything with a fine powder. Add this to urban pollution and the cities become as unhealthy as any in the world.

For all we see, there are geographic explanations. The geometry of the settlement pattern results from the flatness of the uniformly fertile surface and the distance villagers could walk to neighboring villages and back during daylight. The canals form part of a vast system of irrigation channels designed to control the waters of the Huang He (Yellow River). The rows of trees serve as windbreaks. And while wheat dominates the crops of the North China Plain, you will also see millet, sorghum, soybeans, cotton for the textile industry, tobacco (the Chinese are heavy smokers), and fruits and vegetables for the urban markets.

The North China Plain is one of the world's most heavily populated agricultural areas. Figure 9-10 shows that most of it has a density of more than 1000 people per square mile (400 per sq km), and in some parts the density is twice as high. Here, the ultimate hope of the Beijing government lay less in land redistribution than in raising yields through improved fertilization, expanded irrigation facilities, and the more intensive use of labor. A series of dams on the Huang River, including the Xiaolangdi Dam upstream in Henan Province, now reduce the flood danger, but outside the irrigated areas the ever-present problem of rainfall variability and drought persists. The North China Plain has not produced any substantial food surplus even under normal circumstances; thus when the weather turns unfavorable, the situation soon becomes precarious. The specter of famine may have receded, but the food situation is still uncertain in this very critical part of China Proper.

AMONG THE REALM'S GREAT CITIES . . . **BEIJING**

*B*eijing (10.9 million), capital of China, lies at the northern apex of the roughly triangular North China Plain, just over 100 miles (160 km) from its port, Tianjin, on the Bohai Gulf. Urban sprawl has reached the hills and mountains that bound the Plain to the north, a defensive barrier fortified by the builders of the Great Wall. You can reach the Great Wall from central Beijing by road in about an hour.

Although settlement on the site of Beijing began thousands of years ago, the city's rise to national prominence began during the Mongol (Yuan) Dynasty, more than seven centuries ago. The Mongols, preferring a capital close to their historic heartland, endowed the city with walls and palaces. Following the Mongols, later rulers at times moved their capital southward, but the government always returned to Beijing. From the third Ming emperor onward, Beijing was China's focus; it was ideally situated for the Manchus when they took over in 1644. During the twentieth century, China's Nationalists again chose a southern capital, but when the communists prevailed in 1949 they reestablished Beijing as their (and China's) headquarters.

Ruthless destruction of historic monuments carried out from the time of the Mongols to the communists (and now during China's modernization) has diminished, but not destroyed, Beijing's heritage. From the Forbidden City of the Manchu emperors to the fifteenth-century Temple of Heaven, the capital remains an open-air museum and the cultural focus of China. Successive emperors and aristocrats, moreover, bequeathed the city numerous parks and additional recreational spaces that other cities lack. Today Beijing is being transformed as a result of China's new economic policies. A forest of highrises towers above the retreating traditional cityscape, avenues have been widened, and a beltway has been built. A new era dawns in this historic capital.

"We flew over the North China Plain for an hour, and one could not miss the remarkably regular spacing of the villages on this seemingly table-flat surface. Many of these villages became communes during the communist reorganization of China's social order. The elongated buildings (many added since the 1950s) each were designed to contain one or more extended families. After the demise of Mao and his clique the collectivization effort was reversed and the emphasis returned to the small nuclear family. In this village and others we overflew, the long sides of the buildings almost always lay east-west, providing maximum sun exposure to windows and doors, and enhancing summer ventilation from longitudinal breezes. The more distant view looks hazy, because the North China Plain during the spring is wafted by loess-carrying breezes from the northwest. On the ground, a yellow-gray hue shrouds the atmosphere until a passing weather front temporarily clears the air."

The North China Plain is an excellent example of the concept of a national **core area**. Not only is it a densely populated, highly productive agricultural zone, but it also is the site of the capital and other major cities, a substantial industrial complex, and several ports, among which Tianjin ranks as one of China's largest (Fig. 9-12). Tianjin, on the Bohai Gulf, is linked by rail and highway (less than a two-hour drive) to Beijing. Like many of China's harbors, that of Tianjin's river port is not particularly good; but Tianjin is well situated to serve not only the northern sector of the North China Plain and the capital, but also the Upper Huang Basin and Inner Mongolia beyond (see Fig. 9-1). Tianjin, again like several other Chinese ports, had its modern start as a treaty port, but the city's major growth awaited communist rule. For decades it was a center for light industry and a flood-prone harbor,

FIGURE 9-12

"Beijing seems to be one gigantic construction project, with six-lane ringroads and arteries connecting clusters of high-rises still sprouting from the ground. But there is another side to this. Old, historic Beijing's neighborhoods are being swept away, often without any consideration either for the heritage being lost or for the families being displaced. Here, near the old city wall in the southwest, part of a neighborhood already has been bulldozed, and locals salvage what they can (including reusable bricks) from the rubble. Laundry still hangs from lines in front of the still-occupied houses in the background, but any day they, too, will be gone."

but after 1949 the communists constructed a new artificial port and flood canals. They also chose Tianjin as a site for major industrial development and made large investments in the chemical industry (in which Tianjin still leads China), iron and steel production, heavy machine manufacturing, and textiles. Today, with a population of 9.3 million, Tianjin is China's fourth largest metropolis and the center of one of its leading industrial complexes.

Northeast of Tianjin lies the smaller city of Tangshan. In 1976, a devastating earthquake, with its epicenter between Tangshan and Tianjin, killed an estimated 750,000 people, the deadliest natural disaster of the twentieth century. So inept was the government's relief effort in nearby Beijing that the widespread resentment it provoked may have helped end the Maoist period of communist rule in China.

Beijing, unlike Tianjin, is China's political, cultural, and educational center. Its industrial development has not matched Tianjin's. The communist administration did, however, greatly expand the municipal area of Beijing, which (as we noted) is not controlled by the Province of Hebei but is directly under the central government's authority. In one direction, Beijing was enlarged all the way to the Great Wall—30 miles (50 km) to the north—so that the "urban" area includes hundreds of thousands of farmers. Not surprisingly, this has circumscribed an enormous total population, enough to rank Beijing, with its 10.9 million inhabitants, among the world's megacities.

Inner Mongolia

Northwest of the core area as defined by the North China Plain, along the border with the state of Mongolia, lies Inner Mongolia, administratively defined as the Nei Mongol Autonomous Region (see Fig. 9-8, p. 438). Originally

established to protect the rights of the approximately 5 million Mongols who live outside the Mongolian state, Inner Mongolia has been the scene of massive immigration by Han Chinese. Today Chinese outnumber Mongols here by nearly four to one. Near the Mongolian border, Mongols still traverse the steppes with their tents and herds, but elsewhere irrigation and industry have created an essentially Chinese landscape. The A.R.'s capital, Hohhot, has been eclipsed by Baotou on the Huang He, which supports a corridor of farm settlements as it crosses the dry land here on the margins of the Gobi and Ordos deserts. With about 24 million inhabitants, Inner Mongolia still cannot compare to the huge numbers that crowd the North China Plain, but recent mineral discoveries have boosted industry in Baotou and livestock herding is expanding. Nei Mongol may retain its special administrative status, but it functions as part of Han China in all but name.

Basins of the Chang/Yangzi

In contrast to the contiguous, flat agricultural-urban-industrial North China Plain defined by the sediment-laden Huang River, the basins and valleys of the Chang Jiang (Long River) display variation in elevation and relief. The Chang River, whose lower-course name becomes the Yangzi, is an artery like no other in China. Near its mouth lies the country's largest city, Shanghai. Part of its middle course is now being transformed by a gigantic engineering project. Farther upstream, the Chang crosses the populous, productive Sichuan Basin. And unlike the Huang, the Chang Jiang is navigable to oceangoing ships for over 600 miles (1000 km) from the coast all the way to Wuhan (Fig. 9-12). Smaller ships can reach Chongqing, even after dam construction. Several of the Chang River's

AMONG THE REALM'S GREAT CITIES . . . SHANGHAI

*S*ail into the mouth of the great Chang/Yangzi River, and you see little to prepare you for your encounter with China's largest city. For that, you turn left into the narrow Huangpu River, and for the next several hours you will be spellbound. To starboard lies a fleet of Chinese warships. On the port side you pass oil refineries, factories, and neighborhoods. Soon, rusty tankers and freighters line both sides of the stream. Highrise tenements tower behind the cluttered, chaotic waterfront where cranes, sheds, boatyards, piles of rusting scrap iron, mounds of coal, and stacks of cargo vie for space. In the river, your boat competes with ferries, barge trains, and cargo ships. Large vessels, anchored in midstream, are being offloaded by dozens of lighters tied up to them in rows. The air is acrid with pollution. The noise—bells, horns, whistles—is deafening.

What strikes you is the vastness and sameness of Shanghai's cityscape, until you pass beneath the first of two gigantic suspension bridges. Suddenly, everything changes. To the left, or east, the space-age Oriental Pearl Television Tower looking like a rocket on its launchpad rises above a row of modern, glass-and-chrome skyscrapers that make the Huangpu look like Hong Kong's Victoria Harbor. To the right stand the remnants of Victorian buildings, monuments to the British colonialists who made Shanghai a treaty port and started the city on its way to greatness. Everywhere, construction cranes rise above the cityscape. Shanghai looks like one vast construction zone—complete with a 1510-foot office tower (called the World Financial Center) that will become one of the world's tallest buildings when it opens.

Shanghai is an Open Coastal City, where China invites investment by offering privileges to foreign companies. The government is spending heavily to make Shanghai a tiger on the Pacific Rim, and the scene on the east side of the Huangpu may be unmatched in Asia (photo, p. 463). There, an immense business and industrial complex, the

Special Economic Zone (SEZ) named *Pudong*, has arisen since 1990. Its growth rivals that of Shenzhen. Its future, say locals, will eclipse that of Xianggang (Hong Kong).

Meanwhile, Shanghai's past is being obliterated. Whole neighborhoods in the central city are falling to the wreckers' bulldozers, their occupants sent far way to those tenements we saw along the river. Even as citizens are ousted, jobseekers from the hinterland jam the streets. Nearly 3 million converged on Shanghai when word of the Pudong project spread. About 1 million found work. With 12.7 million residents today, Shanghai remains China's biggest city and also ranks among the world's dozen largest. But Shanghai may have traded its soul for a stock market.

tributaries also are navigable, so that 18,500 miles (30,000 km) of water transport routes serve its drainage basin. Thus the Chang Jiang constitutes one of China's leading transit corridors. With its tributaries it handles the trade of a vast area, including nearly all of middle China and sizeable parts of the north and south. The North China Plain may be the core area of China, but in many ways the Lower Chang Basin is its heart.

Early in Chinese history, when the Chang's basin was being opened up and rice and wheat cultivation began, a canal was built to link this granary to the northern core of old China. Over 1000 miles (1600 km) long, this was the longest artificial waterway in the world, but during the nineteenth century it fell into disrepair. Known as the Grand Canal, it was dredged and rebuilt when the Nationalists controlled eastern China. After 1949, the communist regime continued this restoration effort, and much of the canal is now again open to barge traffic, supplementing the huge fleet of vessels that hauls domestic interregional trade along the east coast (Fig. 9-12).

As Figure 9-13 shows, the Lower Chang Basin is an area of both rice and wheat farming, offering further proof of its pivotal situation between south and north in the heart of China Proper. Shanghai lies at the coastal gateway to this productive region on a small tributary of the Chang (now Yangzi) River, the Huangpu. The city has an immediate hinterland of some 20,000 square miles (50,000 sq km)—smaller than West Virginia—containing more than 50 million people. About two-thirds of this population are farmers who produce food, silk filaments, and cotton for the city's industries.

Travel upriver along the Yangzi/Chang Jiang, and you meet an unending stream of vessels large and small, including numerous "barge trains"—as many as six or more barges pulled by a single tug in an effort to save fuel (they are slow, but time is not the primary concern). The traffic between Shanghai and Wuhan—Wuhan is short for Wuchang, Hanyang, and Hankou, a coalescing conurbation of three cities—makes the Yangzi one of the world's busiest waterways. Smoke-belching boat engines create a more or less permanent plume of pollution here, worsening the regional smog created by the factories of Nanjing and others perched on the waterfront. Here China's Industrial Revolution is in full gear, with all its environmental consequences.

Above Wuhan the river traffic dwindles because the depth of the Chang reduces the size of vessels that can reach Yichang. River boats carry coal, rice, building materials, barrels of fuel, and many other items of trade. But the middle course of the Chang River is becoming more than a trade route. In the northward bend of the great river between Yichang and Chongqing, where the Chang has cut deep troughs on its way to the coast, one of the world's largest engineering projects is in progress. It has several names: the Sanxia Project, the Chang Jiang Water Transfer Project, the Three Gorges Dam, and the New China Dam. It is most commonly called the Three Gorges Dam because it is here that the great river flows

CHINA: AGRICULTURAL REGIONS

FIGURE 9-13

FROM THE FIELD NOTES

"China's new consumer culture is perhaps nowhere more evident than in burgeoning, modernizing, globalizing Shanghai. I walked today down Nanjing Road, exactly 20 years after I first saw it in 1982, but only four years after my last excursion here. The rate of transformation is increasing. Thousands of locals have taken up the fast-food habit and are eating at Kentucky Fried Chicken (the colonel's face is everywhere) or McDonald's (I counted five of the latter along my route). A section of Nanjing Road has been made into a pedestrian mall, and here too is evidence of a new habit: private automobile transportation. Car sales lined the street, and potential buyers thronged the displays. Given the existing congestion on local roads, where will these cars fit?"

through a 150-mile (240-km) series of steep-walled valleys less than 360 feet (110 m) wide. Near the lower end of this natural trough, the dam will rise to a height of over 600 feet (180 m) above the valley floor to a width of 1.3 miles (2.1 km), creating a reservoir that will inundate the Three Gorges and extend more than 380 miles (over 600 km) upstream. China's project engineers say that the dam will end the river's rampaging flood cycle, enhance navigation, stimulate development along the new lake perimeter, and provide at least one-tenth, and perhaps as much as one-eighth, of China's electrical power supply. As such it will transform the heart of China, which is why many Chinese like to compare this gigantic project to the Great Wall and the Grand Canal, and call it the New China Dam.

The Chinese government approved the massive project in 1992. Work began in 1993, and the Chang Jiang's course was diverted late in 1997, a year ahead of schedule. In 1998, the engineers forecast completion of the scheme on or before the target date, which is mid-2009. All this is happening despite the paucity of foreign investment. Citing its growing doubts about megaprojects of this kind and its concern over the environmental damage, the World Bank declined to provide funds to support the Three Gorges Dam. Other lenders also withheld their money, but China, barring an economic collapse,

has the financial resources to keep the project on track. As construction proceeds, the Chinese will be able to offer foreign companies attractive contracts for the many related projects (power generation and transmission, roads, bridges, high-tech facilities) that will arise from it. Many will participate, whatever their governments' official view of the New China Dam.

It is a measure of the post-Mao changes in China's still-communist society that the Three Gorges project can be criticized—not only in foreign circles, but in China itself. Most Chinese citizens seem to support the initiative, but some scholars as well as nonprofessionals have spoken out against it. In a country where such dissent still tends to be regarded as "anticommunist propaganda," this takes courage: the risks are considerable, and some opponents have been subjected to harsh treatment. Those within China who oppose the project include poets who bemoan the disappearance of one of China's most fabled and inspiring slivers of scenery, archeologists who decry the loss of prehistoric sites, biologists who abhor the ecological damage the dam will cause, and social workers who fear that the nearly 1.5 million people who will be displaced by the rising water will not be adequately protected. Some engineers fear that the silt-laden Chang Jiang will soon clog up both the reservoir and the river channel above it, possibly blocking the very port of Chongqing which the project is supposed to benefit. And there are those who remember the fate of countless thousands who had the misfortune to live below dams constructed by Mao-era builders. They worry about insufficient quality control on the Three Gorges project and fear that an unprecedented catastrophe may lie ahead.

One city that will be (and has already been) strongly affected by the Three Gorges Dam project is Chongqing, upstream from the reservoir. Boats of up to 1000 tons could reach this river port before dam building began, and the city's planners always hoped that this link to the coast would yield economic benefits arising from China's Pacific Rim development. Dam construction has disrupted and delayed river traffic, but the diversion channel around the Three Gorges Dam will restore Chongqing's port functions. Still, its relative location would suggest that its greater economic opportunities lie westward, in the fast-developing Sichuan Basin, rather than eastward through its tenuous link with the coast. Having recently been made a *Shi* by the regime in Beijing, its municipal territory enlarged to incorporate some 30 million people, Chongqing now finds itself near the upper end of China's largest reservoir. The last Nationalist capital may yet become the number one growth pole in China's interior.

Sichuan, its basin crossed by the Chang after the river emerges from the mountains of the west, frequently was a problem province for both China's dynastic rulers and the communist regime. Its thriving capital, Chengdu, was one of China's most active centers of dissent during the 1989 pro-democracy movement. With nearly 120 million inhabitants (including Chongqing) today, Sichuan Province would be one of the world's ten largest states, but it does not even contain 10 percent of China's population. Sichuan's fertile, well-watered soils yield a huge variety of crops: grains (rice, wheat, corn), soybeans, tea, sugarcane, and many fruits and vegetables.

In many ways Sichuan is a country unto itself, isolated by topography and distance from the government in Beijing. The people here are Han Chinese, but cultural landscapes reveal the area's proximity to Southeast Asia. Along the great riverine axis of the Chang/Yangzi, China changes character time and again.

Basins of the Xi (West) and Pearl River

As Figure 9-12 shows, the Xi River and its basins are no match for the Chang or Huang, not even for the Liao. This southernmost river even seems to have the wrong name: Xi means West! It reaches the coast in a complex area where it forms a delta immediately adjacent to the estuary of the Pearl River. In this subtropical part of China, local relief is higher than in the lowlands and basins of the center and north, so that farmlands are more

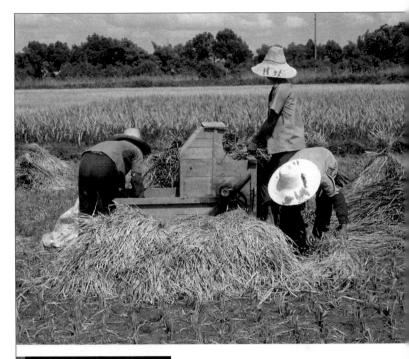

FROM THE FIELD NOTES

"It may not look like much, but this home-made, powered thresher is doing the work of several people in this paddy near Guilin in the Guangxi Zhuang Autonomous Region of southern China. Three people do the bunching, feeding, and bagging! This rural scene led me to wonder: what will happen when China's agriculture becomes more mechanized? How will millions of redundant farmers be employed?"

confined. Especially in the interior areas, water supply is a recurrent problem. On the other hand, its warm climate permits the double-cropping of rice (see photo above). Regional food production, however, has never approached that of the North China Plain or the Lower Chang Basin. And, as Figure 9-10 shows, the population in this southern part of China is smaller than that in the more northerly heartlands.

Again unlike the Huang and Chang rivers, which rise in snow-covered interior mountains, the Xi is a shorter stream whose source is on the Yunnan Plateau (see Fig. 9-2). Just up the coast from its delta, however, lies one of the most important areas of modern China. As we will note later, the mouth of the Pearl River is flanked by some of China's fastest-growing economic complexes, including Guangzhou, capital of Guangdong Province, Hong Kong, the former British dependency, and adjoining Shenzhen, one of the world's fastest-growing cities.

This is all the more remarkable because, as the maps emphasize, South China is not especially well endowed with geographic advantages, locational or otherwise. The

CHINA: ENERGY RESOURCES AND SURFACE COMMUNICATIONS

▲ Gasfield	—— Major road
Oilfield	●—◆—● Oil pipeline
Manufacturing region	—— Railroad
Coalfield	Major coal mine

0   300   600   900 Kilometers
0 100 200 300 400 500 Miles

FIGURE 9-14

Pearl River is navigable for larger ships only to Guangzhou, and the Xi for small river boats no farther than Wuzhou. Guangdong Province, the key province of southern China, lies far from China's heartlands. Nor is South China blessed with the resources that propelled the industrial development of the Northeast and other parts of China. Indeed, one of the principal objectives of the Three Gorges Dam project is to provide South China with ample electricity. As Figure 9-14 shows, northern, central, and western China have most of the coal, oil, and natural gas resources, not the South.

China Proper, with its populous and powerful subregions, is in many ways the dominant region in the East Asian geographic realm. But China Proper is only one part of China. To the west lie two vast regions with relatively sparse populations: Xizang (Tibet) and Xinjiang.

And here, in this remote western periphery adjoining Turkestan, China meets Muslim Central Asia (Fig. 9-15).

XIZANG (TIBET)

As the climate and physiographic maps demonstrate, harsh physical environments dominate this colony of China. It is a vast, sparsely peopled region designated as an Autonomous Region, but in fact Xizang (Tibet) is an occupied society. In terms of its human geography, we should take note of two subregions.

The first is the core area of Tibetan culture, between the Himalayas to the south and the Trans-Himalayas lying not far to the north. In this area, some valleys lie below 7000 feet (2130 m); the climate is comparatively

CHINA'S WESTERN FLANK

- Under 50,000
- 50,000–250,000
- 250,000–1,000,000
- 1,000,000–5,000,000
- Over 5,000,000

Oil/gas reserve
Existing pipeline
Proposed pipeline

National capitals are underlined

FIGURE 9-15

mild, and some cultivation is possible. Here lies Tibet's main population cluster, including the crossroads capital of Lhasa. The Chinese government has made investments to develop these valleys, which contain excellent sites for hydroelectric power projects (some have been used for a few light industries) and promising mineral deposits. The second area of interest is the Qaidam Basin in the north (Fig. 9-15). This basin lies thousands of feet below the surrounding Kunlun and Altun Mountains and has always contained a concentration of nomadic pastoralists. Recently,

however, exploration has revealed the presence of oilfields and coal reserves below the surface of the Qaidam Basin, and these resources are now being developed.

As Figure 9-3 reminds us, Tibet came under Chinese domination during the Manchu (Qing) Dynasty in 1720, but it regained its separate status in the late nineteenth century. China's communist regime took control after the invasion of 1950; in 1959, China crushed an uprising after Tibetan villagers tried to resist the Chinese presence. Tibetan society had been organized around the fortress-like monasteries of Buddhist monks who paid allegiance to their supreme leader, the Dalai Lama. The Chinese wanted to modernize this feudal system, but the Tibetans clung to their traditions. In 1959, they proved no match for the Chinese armed forces; the Dalai Lama was ousted, and the monasteries were emptied. The Chinese destroyed much of Tibet's cultural heritage, looting its religious treasures and works of art. Their harsh rule devastated Tibetan society, but after Mao's death in 1976 the Chinese relaxed their tight control. Although amends were made (religious treasures were returned to Xizang, monastery reconstruction was permitted, and Buddhist religious life resumed), pro-independence rioting has been frequent since 1987. As a result, the Chinese have again tightened their grip. Since its formal annexation in 1965, Xizang has been administered as an Autonomous Region. Although it is large in area, its population still totals less than 3 million.

Visiting Chinese-occupied Tibet is a depressing experience. Tibetan domestic architecture has beauty and expressiveness, but the Chinese have built ugly, gray structures to house the occupiers—often directly in front of Tibetan monasteries and shrines. This juxtaposition of Chinese banality and Tibetan civility seems calculated to perpetuate hostility, an anachronism in this supposedly postcolonial world. During the rule of Mao Zedong, the Chinese officially regarded Tibet as a territorial buffer against India (then allied with the Soviet Union). The breakup of the Soviet Empire has not softened Beijing's stand.

XINJIANG

China's northwestern corner consists of the giant Xinjiang-Uyghur Autonomous Region, constituting over one-sixth of the country's land area and situated in a strategic part of Central Asia (Fig. 9-15). Linked to China Proper via the vital Hexi Corridor in Gansu Province to the east, Xinjiang is China's largest single administrative area.

But even today, after more than a half-century of vigorous Sinicization, Chinese remain a minority here. About 40 percent of the population of 19 million is Chinese (up from 5 percent in 1949), and the Chinese control virtually all aspects of life. The majority are Muslim Uyghurs, who make up approximately half the total, and Kazakhs, Kyrgyz, and others with cultural affinities across the western border to all five republics of former Soviet Central Asia.

The physical geography of Xinjiang is dominated by high mountain ranges and vast basins (Fig. 9-2). The southern margin is defined by the Kunlun Shan (Mountains) beyond which lies Xizang (Tibet). Across the north-center of the region lies the mountain wall of the Tian Shan. Between the Kunlun and Tian Shan extends the vast, dry Tarim Basin within which lies the Takla Makan Desert, and north of the Tian Shan lies another depression, the Junggar Basin. Rivers rising on the mountain slopes disappear beneath the sediments of the basins, sustaining a ring of oases where their waters come near the surface or where wells and *qanats* make irrigation possible (Fig. 9-16).

This would not seem to be the most favorable geography, but Xinjiang has other assets that China exploits. The Qing (Manchu) armies that fought the nomadic Muslim warlords here in the 1870s saw the agricultural possibilities and planted the first crops. The communist rulers after 1949 exiled political opponents as well as criminals to notorious prison camps in this remote frontier. Then Xinjiang turned out to have extensive reserves of oil and natural gas (Figs. 9-15, 9-16). Here, too, China could ex-

Here is a cultural landscape not readily associated with China: a mosque and memorial ground near the city of Kashi (Kashgar) in western Xinjiang (see Figure 9-16). The mainly Muslim Uyghur people constitute just over half the population in remote but important Xinjiang, and small minorities of Kazakhs, Kyrgyz, and other Muslim groups with historic ties to Turkestan also continue to practice their faith here. Relations between the Muslim peoples of Xinjiang and the Chinese authorities are often difficult, and Beijing is actively promoting the settlement of Han Chinese in this Autonomous Region while suppressing any Uyghur demands for independence.

XINJIANG UYGHUR AUTONOMOUS REGION

- Under 50,000
- 50,000–250,000
- 250,000–1,000,000
- 1,000,000–5,000,000
- Over 5,000,000

National capitals are underlined

Irrigated fields
Oilfields
Existing pipeline
Road
Railroad

FIGURE 9-16

periment with space technology, rocketry, and atomic weaponry far from densely populated areas (and shielded from foreign spies). And now, as Figure 9-15 reminds us, China's westernmost borders lie relatively close to the great energy reservoirs of the Caspian Sea Basin—and only one neighbor lies in the way. Pipeline construction across Kazakhstan to Xinjiang is in progress.

Xinjiang's evolving human geography is revealed by Figure 9-16. Most of the Han Chinese are concentrated in the north-central cluster of cities and towns between the capital, Ürümqi, and the oil center of Karamay. Of special

interest here is the city of Shihezi, founded by the Chinese army and now China's model for the region. Over 90 percent of its half-million population are ethnic Chinese; Shihezi's comparative modernity and orderly life stand in sharp contrast to the poorer, often squalid towns and settlements outside the Chinese heartland. Also centered on the Ürümqi-Karamay area is Xinjiang's largest agricultural zone, where cotton and fruits yield export revenues. As the map shows, a ring of oases and clusters of irrigated agriculture encircles both the Tarim and Junggar Basins.

Beijing has made extensive efforts to integrate Xinjiang into the Chinese state. A railroad and highway link Ürümqi and Lanzhou via the Hexi Corridor. Westward from Ürümqi, a railroad crosses the border with Kazakhstan and connects to the Öskemen-Almaty line; another railroad, opened in 2000, connects the capital to Kashi (Kashgar) in the distant southwest (Fig. 9-16). But vast spaces in Xinjiang remain distant and inaccessible. As the map shows, just one road, recently upgraded, loops around the vast southern and eastern flank of the Tarim Basin from Kashi via Hotan to Korla.

China's western frontier region, therefore, is a place of challenges as well as opportunities. Here a regime that from time to time displays intolerance of "foreign" belief systems meets a majority Muslim population that resists incorporation and whose leaders sometimes call for secession. Here China is having its brush with the War on Terror, accusing some of these Muslims of al-Qaeda sympathies, if not activism. Here the gap between comparatively well-off Chinese and poorer minorities is large—and spatially vivid. Here, too, China has unresolved boundary disputes with neighbors, including Tajikistan and India. On the other hand, Xinjiang's energy resources (also including some coal), agricultural and industrial potential, and relative location make this a promising frontier, a symbol for the rest of Central Asia of what Chinese administration can accomplish.

CHINA'S PACIFIC RIM: EMERGING REGION

From Dalian on the doorstep of the Northeast to Hainan Island off the country's southernmost point, China is booming. In the cities, old neighborhoods are being bulldozed to make way for modern skyscrapers. In the countryside of the coastal provinces, thousands of factories employing millions of workers who used to farm the land are changing life in the towns and villages. New roads, airports, power plants, and dams are being built. It is an uneven, sometimes chaotic process that is affecting certain parts of China's coast more than others. It is penetrating parts of China's interior, but meanwhile social gaps are widening: between advantaged and disadvantaged, well-off and poor, urbanite and villager. It is creating regional economic disparities that China has never seen before. And it is producing what may become a separate geographic region along China's Pacific littoral, a region that has more in common with Japan, South Korea, and Taiwan than with Yunnan or Gansu.

In the Introduction, we discussed issues relating to the geography of development, emphasizing that, in this complex, globalizing world, the concept of developed versus underdeveloped countries (DCs and UDCs) is no longer tenable. By almost any measure, China remains what used to be called a UDC. As Figure G-11 shows, it barely ranks as a lower-middle-income economy. But China now has its own core-periphery duality: in its Pacific coast provinces, per-capita incomes today are considerably above the national average of U.S. $3920, while hundreds of millions of villagers in China's remote interior earn far less. In its burgeoning Pacific Rim, China resembles a modernizing economy with "developed" characteristics. In much of its interior, it does not. This contrast is one reason a discrete Pacific Rim region appears to be forming in China Proper.

Geography of Development

Many yardsticks are used to gauge a country's development, including, as we noted, GNP per capita; percentage of workers in farms, factories, or other kinds of employment; amount of energy consumed; efficiency of transport and communications; use of manufactured metals like aluminum and steel in the economy; productivity of the labor force; and social measures such as literacy, nutrition, medical services, and savings rates. All these data are calculated per person, yielding an overall measure that ranks the country among all those providing statistics. These data do not, however, tell us *why* some countries, or parts of countries, exhibit the level of development they do.

The **geography of development** leads us into the study of raw-material distributions, environmental conditions, cultural traditions, historic factors (such as the lingering effects of colonialism), and forces of location. In the emergence of China's Pacific Rim, relative location played a major role, as we will see.

Geographers tend to view the development process spatially, whereas other scholars focus on structural aspects. In the 1960s, economist Walt Rostow formulated a global model of the development process that is still being discussed today. This model suggested that all developing countries follow an essentially similar path through five interrelated growth stages. In the earliest of these stages, a *traditional society* engages mainly in subsistence farming, is locked in a rigid social structure, and resists technological change. When it reaches the second stage, *preconditions for takeoff*, progressive leaders move the country toward greater flexibility, openness, and diversification. Old ways are abandoned, workers move from farming into manufacturing, and transport improves. This leads to the third stage, *takeoff*, when the country experiences a type of industrial revolution. Sustained growth takes hold, industrial urbanization proceeds, and technological and mass-production breakthroughs occur. If the economy continues to expand, the fourth stage, *drive to maturity*, brings sophisticated industrial specialization and increasing international trade. Some countries reach a still more advanced fifth stage of *high mass consumption* marked by high incomes, widespread production of consumer goods and services, and most workers employed in the tertiary and quaternary economic sectors.

This model takes little account of core-periphery contrasts within individual countries. In China, takeoff conditions (stage 3) exist in much of the Pacific Rim, but other major areas of this vast country remain in stage 1. China has passed through its transition from tradition-bound to progressive leadership, but the effect of that leadership's reformist policies has not been the same in all parts of the country. The most fertile ground lies along the Pacific, where China has a history of contact with foreign enterprises, where the country is most open to the world, and from where many Chinese departed for Southeast Asian (and other) countries. These **Overseas Chinese** were positioned to play a crucial role in China's Pacific Rim development: many had left kin and community there, and those links facilitated the investment of hundreds of millions of dollars when the opportunity arose. Many of the Pacific Rim's factories were built by means of this repatriated money.

Transforming the Economic Map

China's communist era has witnessed the metamorphosis of the world's most populous nation. Despite the dislocations of collectivization, communization, the Cultural Revolution, power struggles, policy reversals, and civil conflicts, China as a communist state in the aftermath of Mao managed to stave off famine and hunger, control its population growth rate, strengthen its military capacity, improve its infrastructure, and enhance its world position. It took over Hong Kong from the British and Macau from the Portuguese, closing the chapter on colonialism. When economic turbulence struck its Pacific Rim and Southeast Asian neighbors, China withstood the onslaught. Its trade with neighbors near and far is growing. Foreign investment in China has mushroomed. And still its government espouses communist dogma.

How could the government manipulate the coexistence of communist politics and market economics? That was the key question confronting Deng Xiaoping and his comrades when they took power in 1979. In terms of ideology, this objective would seem unattainable; an economic "open-door" policy would surely lead to rising pressures for political democracy.

But Deng thought otherwise. If China's economic experiments could be spatially separated from the bulk of the country, the political impact would be kept at bay.

At the outset, the new economic policies would apply mainly to China's bridgehead on the Pacific Rim, leaving most of the vast country comparatively unaffected. Accordingly, the government introduced a complicated system of **Special Economic Zones (SEZs)**, so-called *Open Cities*, and *Open Coastal Areas*, which would attract technologies and investments from abroad and transform the economic geography of eastern China (Fig. 9-17).

In these economic zones, investors are offered many incentives. Taxes are low. Import and export regulations are eased. Land leases are simplified. The hiring of labor under contract is allowed. Products made in the economic zones may be sold on foreign markets and, under some restrictions, in China as well. Even Taiwanese enterprises may operate here. And profits made may be sent back to the investors' home countries.

When Deng's government made the decisions that would reorient China's economic geography, location was a prime consideration. Beijing wanted China to participate in the global market economy, but it also wanted to cause as little impact on interior China as possible—at least in the first stages. The obvious answer was to position the Special Economic Zones along the coast. Initially, the government established four SEZs, all with particular locational properties (Fig. 9-17):

1. *Shenzhen*, adjacent to then-booming British Hong Kong on the Pearl River Estuary in Guangdong Province
2. *Zhuhai*, across from then still-Portuguese Macau, also on the Pearl River Estuary in Guangdong Province
3. *Shantou*, opposite southern Taiwan, a colonial treaty port, also in Guangdong Province, source of many Chinese now living in Thailand
4. *Xiamen*, on the Taiwan Strait, also a colonial treaty port (then known as Amoy in the local dialect), in Fujian Province, source of many Chinese now based in Singapore, Indonesia, and Malaysia

In 1988 and 1990, respectively, two additional SEZs were proclaimed:

5. *Hainan Island*, declared an SEZ in its entirety, its potential success linked to its location near Southeast Asia
6. *Pudong*, across the river from Shanghai, China's largest city, different from other SEZs because it was a giant state-financed project designed to attract large multinational companies

China also opened 14 other coastal cities for preferential treatment of foreign investors (Fig. 9-17). Again, most of these cities had been treaty ports in the colonial

FIGURE 9-17

extraterritoriality period. They were chosen for their size, overseas trading history, links to emigrated Chinese, level of industrialization, and pool of local talent and labor. These 14 Open Cities are, from north to south:

Dalian (Liaoning Province)	Shanghai Shi
Qinhuangdao (Hebei)	Ningbo (Zhejiang)
Tianjin Shi	Wenzhou (Zhejiang)
Yantai (Shandong)	Fuzhou (Fujian)
Qingdao (Shandong)	Guangzhou (Guangdong)
Lianyungang (Jiangsu)	Zhanjiang (Guangdong)
Nantong (Jiangsu)	Beihai (Guangxi Zhuang A.R.)

When all these Pacific Rim initiatives were implemented, China's market-conscious communist leaders had reason to expect an economic boom in the coastal provinces. (On the map, note that the new economic policy affected every coastal province, as well as the single coastal A.R.) Japan already was a highly developed economy (at the final stage of the Rostow model). South Korea and Taiwan were well beyond takeoff. Hong Kong proved what could be done on this side of the Pacific. Singapore, Southeast Asia's most successful state, is 77 percent Chinese. In truth, China's pragmatic planners were unsure of just what forces they might unleash, which is why foreign investment rules were stricter in the Open Cities and Open Coastal Areas than in the freewheeling SEZs.

CHINA'S PACIFIC RIM TODAY

When China's planners laid out the SEZs, they held the highest hopes for Shenzhen, immediately adjacent to the thriving British dependency of Hong Kong. They knew that the kinds of attractions the SEZs offered to foreign investors would entice many factory owners paying Hong Kong wages to move across the border to Shenzhen. Successful Chinese living in Southeast Asia could use Hong

The story of Shenzhen has no parallel in the modern world. Across Hong Kong's once-sealed, now-porous border with China a fishing village was overrun by urban and industrial development so that its population went from under 20,000 to over 3 million in just 15 years. Investment from Hong Kong, Japan, and China, as well as indirectly from Taiwan, and cheap labor from Guangdong and Fujian Provinces as well as the Guangxi-Zhuang Autonomous Region, in a few years produced an industrial and service center that became the prototype of China's Special Economic Zones, where business and industry were given incentives and advantages not available elsewhere. When you visit Shenzhen today, it looks so established and functional that it is easy to forget that this city has, in effect, no history as yet other than its phenomenal growth . . . Macau, on the other hand, which reverted to China at the end of 1999 and became the second Special Administrative Region (SAR), has a centuries-long history of Portuguese colonialism. Long a gambling mecca and rife with organized crime, Macau has a population just one-eighth of Shenzhen and its recent acquisition presents China's leaders with quite a different prospect. A favorable relative location near the Zhuhai SEZ and major infrastructure improvements, including an international airport and huge island-linking bridges, enhance Macau's potential as a participant in the economic growth of the Pearl River Estuary.

Kong as a base for investment in Shenzhen; Taiwanese, even during the political standoff between Beijing and Taipei, could channel investments into this favored SEZ.

Those hopes were not disappointed. In the late 1970s, Shenzhen was just a sleepy fishing and duck-farming village of about 20,000. By 2004, its population neared 4 million—the fastest growth of any urban center on Earth in human history. Thousands of factories, large and small, moved into the SEZ from Hong Kong. Workers by the hundreds of thousands came from Guangdong Province, Guangxi Zhuang A.R., and beyond to look for jobs. High-rise buildings housing banks, corporate offices, hotels, and other service facilities made Shenzhen look like a new Hong Kong (see photo above).

Shenzhen's economic impact was far-reaching. Persuading Hong Kong industries to cross the border precipitated a sharp modification of Hong Kong's own economy; attracting investments from Overseas Chinese, mainly in Southeast Asia, opened financial coffers long closed to China; and enabling Taiwanese investors to participate in China's Pacific Rim boom created links with the "wayward province" that have economic as well as political

benefit. By its very success (in its 1990s heyday, it produced 15 percent of all of China's exports by value), Shenzhen signaled the way for the other SEZs and Open Cities.

Guangdong Province and the Pearl River Estuary

As Figure 9-17 shows, eight provinces (including Hainan Island) and one Autonomous Region face the Pacific and form the basis for recognizing an emerging Pacific Rim region in eastern China. But among these provinces, one stands out by making the largest contribution by far to the national economy: Guangdong Province. Although Hong Kong and Macau, both SARs, are not administratively a part of Guangdong Province, they are functionally crucial parts of the vast economic complex that fringes the estuary of the Pearl River (Fig. 9-18). Indeed, without Hong Kong, there would not have been a Shenzhen of the kind described above.

The economic core of Guangdong consists of Guangzhou (the capital), Shenzhen SEZ, Zhuhai SEZ, as well as Hong Kong (Xianggang) and Macau. In a few

FIGURE 9-18

years this Pearl River hub will have three cities with more than 5 million inhabitants, five with more than 1 million, and six with more than a half-million. Per-capita incomes here are estimated to be ten times the national average.

The Pearl River Hub fits the model of the **regional state** as defined by the Japanese scholar Kenichi Ohmae. These "natural economic zones," as he called them, defy old political borders and are shaped by the global economy of which they are parts; their representatives deal directly with foreign partners and negotiate the best terms they can with the national governments under which they operate (Europe, we noted, has several examples of this kind). The Pearl River Megalopolis has moved so far ahead of China as a whole that it is a political as well as an economic force in the country.

Hong Kong's Role

The process started with the economic success of Hong Kong while it was still a British dependency, empowered by a unique set of geographic circumstances. Here, on the left bank of the Pearl River Estuary, more than 7 million people (97 percent of Chinese descent) crowded onto 400 square miles (1000 sq km) of fragmented, hilly territory under the tropical sun and created an economy larger than that of a hundred countries.

Hong Kong (the name means "Fragrant Harbor") was ceded by the Chinese to Britain during the colonial period in three parts: the islands in 1841, the Kowloon Peninsula in 1860, and the New Territories (a lease) in 1898 (Fig. 9-19). Until the 1950s, Hong Kong was just another trading colony, but then the Korean War and the United Nations embargo on communist China cut Hong Kong off from its hinterland. With a huge labor force and no domestic resources, the colony needed an alternative—and it came in the form of a textile industry based on imported raw materials. Soon Hong Kong textiles were selling on worldwide markets; capital was invested in light industries; and the industrial base diversified into electrical equipment, appliances, and countless other consumer goods. Meanwhile, Hong Kong quietly served as a back door to China, which rewarded the colony by providing fresh water and staple foods. The colony had become an **economic tiger**.

Hong Kong also became one of the world's leading financial centers. When China changed course and embarked on its "Open Door" policy in the 1980s, Hong

Pedestrians on a downtown Hong Kong street in May 2003 wear surgical masks to protect against SARS (severe acute respiratory syndrome). The SARS virus originated in adjacent southern China and had first been detected in Hong Kong during mid-February. Within days SARS reached pandemic proportions as international air travelers unknowingly dispersed the disease across much of the world (Toronto, Canada was particularly hard hit). During the 100 days that Hong Kong grappled with its local outbreak, 1755 confirmed cases of SARS were recorded here, resulting in 298 deaths. Health authorities responded by imposing stringent quarantine measures to isolate Hong Kong's disease victims. These soon proved to be effective, and on June 23 the UN's World Health Organization (WHO) sounded the all-clear. Two weeks later, WHO pronounced that SARS had been contained globally—but with the caveat that the virus could return to trigger another outbreak in the winter of 2003–2004.

FIGURE 9-19

Kong was ready. Hong Kong factory owners moved their plants to Shenzhen, where wages were even lower; Hong Kong banks financed Shenzhen industries; and Hong Kong's economy adjusted once again to a new era. Hong Kong's port, however, continued to transfer Shenzhen's products. During the 1990s, about 25 percent of all of China's trade passed through Hong Kong harbor.

When the British lease on Hong Kong expired in 1997 and the British agreed to yield authority to China, the Chinese promised to allow Hong Kong's freewheeling economy to continue, in effect creating "one country, two systems." At midnight on June 30, Hong Kong became the Xianggang Special Administrative Region (SAR). True to the letter (if not always the spirit) of this commitment, the name "Hong Kong" has continued in general use. And on December 20, 1999 China took control over the last remaining European colony—Portuguese Macau—and made it a SAR as well, guaranteeing (as in Hong Kong) the continuation of its existing social and economic system for 50 years.

The Other SEZs

Not only did China grant economic advantages to corporations doing business in the coastal SEZs; China made enormous infrastructure investments to enhance their attractiveness and efficiency. In Shanghai, a massive development project transformed the right bank of the Huangpu River (across from the old downtown) into an ultramodern complex whose skyline rivals that of Hong Kong (see photo next page). *Pudong*, as this area is called, is designed to offer competition to the great Pearl River node as part of the Beijing government's effort to reduce the imbalance between north and south along its Pacific Rim.

None of the other SEZs has come close to what has happened in Shenzhen or Pudong. Xiamen was established especially to attract Taiwanese investment; Shantou has benefited greatly from investment by Overseas Chinese in Southeast Asia (notably Thailand); Zhuhai is benefiting from the fact that its wages are even lower than Shenzhen's, causing the transfer of many industries from Shenzhen; and Hainan lags because of relative location and poor administration. The SEZs are no workers' paradise: thousands of factories producing goods ranging from toys to textiles create a pollution-choked landscape, working conditions often are dreadful, and wages are marginal. Unfortunately, China's new industrial revolution has some of the features of the old one.

Toward the Interior

Economic progress on the Pacific Rim has created growing regional disparities in the new China. This is of great concern to China's leaders because such disparities can lead to political unrest and instability. Thus the government has begun to grant SEZ-like privileges to a part of Yunnan Province, which borders Southeast Asian countries; the notion of a Pacific Rim-linked corridor extending westward from Shanghai along the Yangzi/Chang River to the Sichuan Basin also is gaining ground. These, however, are mere band-aids for a massive problem that will play a major role in China during the twenty-first century.

CHINA: GLOBAL SUPERPOWER?

China in the early twenty-first century appears likely to become more than an economic force of world proportions: China also appears on course to achieve global superpower stature. As long as it retains its autocratic form of government (which allowed it to impose draconian population policies and comprehensive economic experiments without having to consult the electorate), China will be able to practice the kind of state capitalism that—in another guise—made South Korea an economic power. And unlike Japan or South Korea, China has no constraints on its military power. Its People's Liberation Army served as security forces during the pro-democracy turbulence of 1989; China's military is the largest standing army in the world, with nearly 3 million soldiers and some 1.2 million reserves. Its military equipment is being updated, and the government made a major investment in its military by raising its armed forces budget enormously in 2001. Already, China is a nuclear power and has a growing arsenal of medium-range and intercontinental ballistic missiles.

During the twentieth century, the United States and the Soviet Union were locked in a 45-year Cold War that repeatedly risked nuclear conflict. That fatal exchange never happened, in part because it was a struggle between superpowers who understood each other comparatively well. While the politicians and military strategists were plotting, the cultural doors never closed: American audiences listened to Prokofiev and Shostakovich, watched Russian ballet, and read Tolstoy and Pasternak even as the Soviets cheered Van Cliburn, read Hemingway, and lionized American dissidents. In short, it was an intracultural Cold War, which reduced the threat of mutual destruction.

The twenty-first century may witness a far more dangerous geopolitical struggle in which the adversaries may well be the United States and China. U.S. power and influence still prevail in the western Pacific, but it is easy to discern areas where Chinese and American interests will diverge (Taiwan is only one example). American bases in Japan, thousands of American troops in South Korea, and American warships in the East and South

Across the Huangpu River from the colonial-era Bund, the Lujiazui Development Zone in the heart of Shanghai's spectacular Pudong District reflects China's new business and commercial era. The futuristic Oriental Pearl Television Tower (left) and the 88-story Mao Observation Tower, containing a hotel with the world's highest atrium (right), will soon be outranked by an even taller structure. There was a time when American cities' skyscrapers symbolized the New World's power and modernity; now it is Asia's turn to show the world that a new era is in the making.

China Seas are potential grounds for dispute. All this might generate the world's first *inter*cultural Cold War, in which the risk of fatal misunderstanding is incalculably greater than it was during the last.

How can such a Cold War be averted? Trade, scientific and educational links, and cultural exchanges are obvious remedies: the stronger our interconnections, the less likely is a deepening conflict. We Americans should learn as much about China as we can, to understand it better, to appreciate its cultural characteristics, and to recognize the historico-geographical factors underlying China's views of the West.

For more than 40 centuries, China has known authoritarianism of both the brutal and the benevolent kind, has been fractured by regionalism only to unify again, and has depended on communalism to survive environmental and despotic depredations. For nearly 25 centuries Confucianism has guided it. Time and again, China has been opened to, and ruled by, foreigners, and time and again it has retreated into isolationism when things went wrong. Such a reversal may no longer be possible, given what we have seen in this chapter. But now a renewed force is rising in modernizing China: nationalism. A spate of recent books published in China, including one by Song Qiang et al. titled *China Can Say No*, reflects the frustrations of many Chinese over what they view as American arrogance and insensitivity to China's traditions and interests. The "inadvertent" bombing of the Chinese embassy in Belgrade by American aircraft operating under NATO command in 1999, and the mid-air collision between what the Chinese

regarded as a U.S. "spy plane" and one of its aircraft near the island of Hainan in 2001, are the kinds of incidents that contribute to this view and that energize Chinese nationalism. American scorn for China's human-rights failings are another source of irritation.

China today is on the world stage, globally engaged and internationally involved. Constraining the forces that tend to lead to world-power competition between China and the United States will be the geopolitical challenge of the twenty-first century.

▶ MONGOLIA

Between China's Inner Mongolia and Xinjiang to the south and Russia's Eastern Frontier region to the north lies a vast, landlocked, isolated country called Mongolia that constitutes a discrete region of East Asia (Fig. 9-4). With only 2.5 million inhabitants in an area larger than Alaska, Mongolia is a steppe- and desert-dominated vacuum between two of the world's most powerful countries.

From the late 1600s until the revolutionary days of 1911, Mongolia was part of the Chinese Empire. With Soviet support the Mongols held off Chinese attempts to regain control, and in the 1920s the country became a People's Republic on the Soviet model. Free elections in 1990 ushered in a new political era, but Mongolia's economy, still largely based on animal products (chiefly cashmere wool), is troubled by mismanagement and environmental problems.

The map reflects Mongolia's difficulties. Despite its historic associations, ethnic affinities, and cultural involvement with China, Mongolia's incipient core area, including the capital of Ulaanbaatar, adjoins Russia's Eastern Frontier. The country's 800,000 herders and their millions of sheep follow nomadic tracks along the fringes of the vast Gobi Desert. When Siberian cold descends on the steppes, as happened in the winters of 2000 and 2001, human and livestock losses are severe and the economy breaks down.

This is a weak country in a vulnerable part of inner Asia, a **buffer state** wedged, Tibet-like, between populous and powerful neighbors. Will it escape the fate of other Asian buffers?

THE JAKOTA TRIANGLE REGION

We turn now to a region that exemplifies the future of East Asia. Along the Pacific Rim, from Japan (through South Korea, Taiwan, China's Guangdong) to Singapore,

rapid economic development has transformed community and society. In China, as we saw, a Pacific Rim region is in the making, still discontinuous today but likely to extend all along the coast in the future. In Japan, South Korea, and Taiwan we can observe that future—today. That is why we recognize an East Asian region we call the **Jakota Triangle**, consisting of Japan, Korea (South Korea at present but probably a reunited Korea later), and Taiwan (Fig. 9-20). This is a region of great cities, huge consumption of raw materials from all over the world, voluminous exports, and global financial linkages. It also is a region of social problems, political uncertainties, and economic vulnerabilities.

JAPAN

When we assess China's prospects of becoming a superpower, we should remember what happened in Japan in the nineteenth century. In 1868, a group of reform-minded modernizers seized power from an old guard,

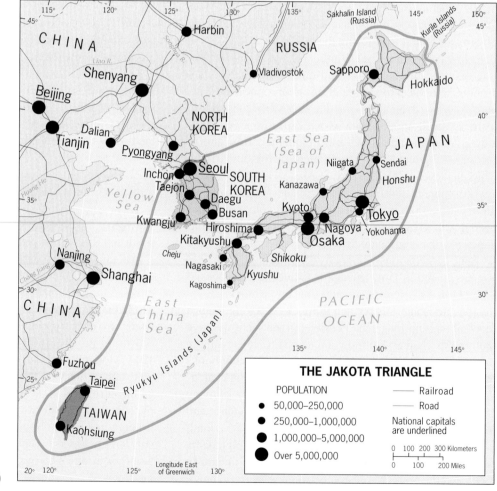

FIGURE 9-20

and by the end of the century Japan was a military and economic force. From the factories in and around Tokyo and from urban-industrial complexes elsewhere poured forth a stream of weapons and equipment the Japanese used to embark on colonial expansion. By the mid-1930s, Japan lay at the center of an empire that included all of the Korean Peninsula, the whole of China's Northeast (which the Japanese called Manchukuo), the Ryukyu Islands, and Taiwan, as well as the southern half of Sakhalin Island (called Karafuto). Not even a disastrous earthquake, which destroyed much of Tokyo in 1923 and killed 143,000 people, could slow the Japanese drive.

World War II saw Japan expand its domain farther than the architects of the 1868 "modernization" could have anticipated. By early December 1941, Japan had conquered large parts of China Proper, all of French Indochina to the south, and most of the small islands in the western Pacific. Then, on December 7, 1941, Japanese-built aircraft carriers moved Tokyo's warplanes within striking range of Hawai'i, and the surprise attack on Pearl Harbor underscored Japan's confidence in its war machine. Soon the Japanese overran the Philippines, the (then) Netherlands East Indies, Thailand, and British Burma and Malaya, and drove a wide corridor through the heart of China to the border with Vietnam.

A few years later, Japan's expansionist era was over. Its armies had been driven from virtually all its possessions, and when American nuclear bombs devastated two Japanese cities in 1945, the country lay in ruins. But once again, Japan, aided this time by an enlightened U.S. postwar administration, surmounted disaster.

Japan's economic recovery and its rise to the status of world economic superpower was the success story of the second half of the twentieth century. Japan lost the war and its empire, but it scored many economic victories in a new global arena. Japan became an industrial giant, a technological pacesetter, a fully urbanized society, a political power, and an affluent nation. No city in the world today is without Japanese cars in its streets; few photography stores lack Japanese cameras and film; laboratories the world over use Japanese optical equipment. From microwave ovens to VCRs, from oceangoing ships to camcorders, Japanese goods flood the world's markets. Even today, while experiencing a severe downturn, the Japanese economy remains the second largest national economy in the world. As we will see later in this chapter, Japan faces several challenges, economic, political, and demographic. But even while in recession, the Japanese economy remains formidable.

Japan's brief colonial adventure helped lay the groundwork for other economic successes along the western Pacific Rim. The Japanese ruthlessly exploited Korean and Formosan (Taiwanese) natural and human resources, but they also installed a new economic order there. After World War II, this infrastructure facilitated an economic transition—and soon made both Taiwan and South Korea competitors on world markets.

Spatial Limitations

In discussing the economic geography of Europe, we noted how the comparatively small island of Britain became the crucible of the Industrial Revolution, which gave it a huge head start and advantage over the continent, across which the Industrial Revolution spread decades later. Britain was not large, but its insular location, its local raw materials, the skills of its engineers, its large labor force, and its social organization combined to endow it not only with industrial strength but also with a vast overseas empire.

After the modernizers took control of Japan in 1868—an event known as the *Meiji Restoration* (the return of "enlightened rule" centered on the Emperor Meiji)—they turned to Britain for guidance in reforming their nation and its economy. In the decades that followed, the British advised the Japanese on the layout of cities and the construction of a railroad network, on the location of industrial plants, and on the organization of education. The British influence still is visible in the Japanese cultural landscape today: the Japanese, like the British, drive on the left side of the road. Consider how this affects the effort to open the Japanese market to U.S. automobiles!

The Japanese reformers of the late nineteenth century undoubtedly saw many geographic similarities between Britain and Japan. At that time, most of what mattered in Japan was concentrated on the country's largest island, Honshu (literally, "mainland"). The ancient capital, Kyoto, lay in the interior, but the modernizers wanted a coastal, outward-looking headquarters. So they chose the town of Edo, on a large bay where Honshu's eastern coastline turns sharply (Fig. 9-21). They renamed the place *Tokyo* ("eastern capital"), and little more than a century later it was the largest urban agglomeration in the world. Honshu's coasts were near mainland Asia, where raw materials and potential markets for Japanese products lay. The notion of a greater Japanese empire followed naturally from the British example.

But in other ways, the British and Japanese archipelagoes, at opposite ends of the Eurasian landmass, differed considerably. In total area, Japan is larger. In addition to Honshu, Japan has three other large islands—Hokkaido to the north and Shikoku and Kyushu to the south—as well as numerous small islands and islets, for a total land area of about 146,000 square miles (377,000 sq km). Much of this area is mountainous and steep-sloped, geologically young, earthquake-prone, and studded with volcanoes.

FIGURE 9-21

Britain has lower relief, is older geologically, does not suffer from severe earthquakes, and has no active volcanoes. And in terms of the raw materials for industry, Britain was much better endowed than Japan. Self-sufficiency in iron ore and high-quality coal gave Britain the head start that lasted a century.

Japan's high-relief topography has been an ever-present challenge. All of Japan's major cities, except the ancient capital of Kyoto, are perched along the coast, and virtually all lie partly on artificial land claimed from the sea. Sail into Kobe harbor, and you will pass artificial islands designed for high-volume shipping and connected to the mainland by automatic space-age trains. Enter Tokyo Bay, and the refineries and factories to your east and west stand on huge expanses of landfill that have pushed the bay's shoreline outward. With 127.4 million people, the vast majority (78 percent) living in towns and cities, Japan uses its habitable living space intensively—and expands it wherever possible.

As Figure 9-22 shows, farmland in Japan is both limited and regionally fragmented. Urban sprawl has invaded much cultivable land. In the hinterland of Tokyo lies the Kanto Plain; around Nagoya, the Nobi Plain; and surrounding Osaka, the Kansai District—each a major farming zone under relentless urban pressure. All three of these plains lie within Japan's fragmented but well-defined core area (delimited by the red line on the map), the heart of Japan's prodigious manufacturing complex.

FIGURE 9-22

From Realm to Region

For nearly a half-century, Japan's seemingly insurmountable economic lead, its almost unique (for so large a population) ethnic homogeneity, its insular situation, and its cultural uniformity justified its designation as a discrete geographic realm, not a region of a greater East Asia. But rapid change on the rim of the western Pacific—both within Japan and beyond it—has eroded the bases for this distinction. In the mid-1990s, Japan's economic engine faltered while those of its Pacific Rim neighbors accelerated; its social fabric was torn by unprecedented acts of violence, and its foreign relations were soured by disputes over wartime atrocities in China and Korea. Japan stood alone in the vanguard of Pacific Rim developments, but it is no longer discrete.

Early Directions

On the map, the Korean Peninsula seems to extend like an unfinished bridge from the Asian mainland toward southern Japan. Two Japanese-owned islands, Tsushima and Iki, in the Korea Strait almost complete the connection (Fig. 9-21). In fact, Korea was linked to Japan in prehistoric times, when sea levels were lower. Peoples and cultural infusions reached Japan from Asia time and again, even after rising waters inundated the land bridge between them.

The Japanese islands were inhabited even before the ancestors of the modern Japanese arrived from Asia. As we noted earlier, the Ainu, a people of Caucasian ancestry, had established themselves on all four of the major

islands thousands of years before. But the Ainu could not hold their ground against the better organized and armed invaders, and they were steadily driven northward until they retained a mere foothold on Hokkaido. Today, the last vestiges of Ainu ancestry and culture are disappearing.

By the sixteenth century, Japan had evolved a highly individualistic culture, characterized by social practices and personal motivations based on the Shinto belief system. People worshiped many gods, venerated ancestors, and glorified their emperor as a divine figure, infallible and omnipotent. The emperor's military ruler, or *shogun*, led the campaign against the Ainu and governed the nation. The capital, Kyoto, became an assemblage of hundreds of magnificent temples, shrines, and gardens. In their local architecture, modes of dress, festivals, theater, music, and other cultural expressions, the Japanese forged a unique society. Handicraft industries, based on local raw materials, abounded.

Nonetheless, in the mid-nineteenth century, Japan hardly seemed destined to become Asia's leading power. For 300 years the country had been closed to outside influences, and Japanese society was stagnant and tradition-bound. When the European colonizers first appeared on Asia's shores, Japan tolerated, even welcomed, their merchants and missionaries. But as the European presence in East Asia grew stronger, the Japanese began to shut their doors. Toward the end of the sixteenth century, the emperor decreed that all foreign traders and missionaries should be expelled. Christianity (especially Roman Catholicism) had gained a considerable foothold in Japan, but now this was feared as being a prelude to colonial conquest. Early in the seventeenth century, the shogun

FROM THE FIELD NOTES

"The city of Kyoto, chronologically Japan's second capital (after Nara; before Tokyo), is the country's principal center of culture and religion, education and the arts. Tree-lined streets lead past hundreds of Buddhist temples; tranquil gardens provide solace from the bustle of the city. I rode the bullet train from Tokyo and spent my first day following a walking route recommended by a colleague, but got only part of the way because I felt compelled to enter so many of the temple grounds and gardens. And not only Buddhism, but also Shinto makes its mark on the cultural landscape. I passed under a *torii*—a gateway usually formed by two wooden posts topped by two horizontal beams turned up at their ends—which signals that you have left the secular and entered the sacred, and found this beautiful Shinto shrine with its orange trim and olive-green glazed tiles."

launched a massive and bloody campaign that practically stamped out Christianity.

Determined not to suffer the same fate as the Philippines, which by then had fallen to Spain, the Japanese henceforth allowed only minimal contact with Europeans. A few Dutch traders, confined to a little island near the city of Nagasaki, were for many decades the sole representatives of Europe in Japan. Thus Japan entered a long period of isolation, lasting past the middle of the nineteenth century.

Japan could maintain its aloofness, while other areas were being enveloped by the colonial tide, because of its strong central government and well-organized military, the natural protection its islands provided, and its remoteness along East Asia's difficult northern coast. Also, Japan's isolation was far less splendid than that of China, whose exquisite silks, prized teas, and skillfully made wares attracted traders and usurpers alike.

When Japan finally came face to face with the new weaponry of its old adversaries, it had no answers. In the 1850s, the steel-hulled "black ships" of the American fleet sailed into Japanese harbors, and the Americans extracted one-sided trade agreements. Soon the British, French, and Dutch were also on the scene, seeking similar treaties. When there was local resistance to their show of strength, the Americans quickly demonstrated their superiority by shelling parts of the Japanese coast. By the late 1860s, even as Japan's modernizers were about to overturn the old order, no doubt remained that Japan's protracted isolation had come to an end.

A Japanese Colonial Empire

When the architects of the Meiji Restoration confronted the challenge to build a Japan capable of competing against powerful adversaries in a changing world, they took stock of the country's assets and liabilities. Material assets, they found, were limited. To achieve industrialization, coal and iron ore were needed. In terms of coal, there was enough to support initial industrialization. Coalfields in Hokkaido and Kyushu were located near the coast; since the new industries also were on the coast, cheap water transportation was possible. As a result, the shores of the (Seto) Inland Sea (Fig. 9-21) became the sites of many factories. But there was little iron ore, certainly nothing like what was available domestically in Britain, and not enough to sustain the massive industrialization the reformers had in mind. This commodity would have to be purchased overseas and imported.

On the positive side, manufacturing—light manufacturing of the handicraft type—already was widespread in Japan. In cottage industries and in community workshops, the Japanese produced textiles, porcelain, wood products,

and metal goods. The small ore deposits in the country were enough to supply these local industries. Power came from human arms and legs and from wheels driven by water; the chief source of fuel was charcoal. Importantly, Japan did have an industrial tradition and an experienced labor force that possessed appropriate manufacturing skills. All this was not enough to lead directly to industrial modernization, but it did hold promise for capital formation. The community and home workshops were integrated into larger units, hydroelectric plants were built, and some thermal (coal-fired) power stations were constructed in critical areas.

For the first time, Japanese goods (albeit of the light manufactured variety) began to compete with Western products on international markets. The Japanese planners resisted any infusion of Western capital, however. Although Western support would have accelerated industrialization, it would have cost Japan its economic autonomy. Instead, farmers were more heavily taxed, and the money thus earned was poured into the industrialization effort.

Another advantage for Japan's planners lay in the country's military tradition. Although the shogun's forces had been unable to repel the invasions of the 1850s, this was the result of outdated equipment, not lack of manpower or discipline. So while its economic transformation gathered momentum, the military forces, too, were modernized. Barely more than a decade after the Meiji Restoration, Japan laid claim to its first Pacific prize: the Ryukyu Islands (1879). A Japanese colonial empire was in the making.

When Japan gained control over its first major East Asian colonies, Taiwan and Korea, its domestic raw-material problems were essentially solved, and a huge labor pool fell under its sway as well. High-grade coal, high-quality iron ore, and other resources were shipped to the factories of Japan; and in Taiwan as well as in Korea, the Japanese built additional manufacturing plants to augment production. This, in turn, provided the equipment to sustain the subsequent drive into China and Southeast Asia.

Modernization

The reformers who set Japan on a new course in 1868 probably did not anticipate that three generations later their country would lie at the heart of a major empire sustained by massive military might. They set into motion a process of **modernization**, but they managed to build on, not replace, Japanese cultural traditions. We in the Western world tend to equate modernization with Westernization: urbanization, the spread of transport and communications facilities, the establishment of a market (money) economy, the breakdown of local traditional communities, the proliferation of formal schooling, the acceptance and adoption of foreign innovations. In the non-Western world, the process often is viewed differently. There, "modernization" is seen as an outgrowth of colonialism, the perpetuation of a system of wealth accumulation introduced by foreigners driven by greed. In this view, the local elites who replaced the colonizers in the newly independent states only continue the disruption of traditional societies, not their true modernization. Traditional societies, they argue, can be modernized without being Westernized.

In this context, Japan's modernization is unique. Having long resisted foreign intrusion, the Japanese did not achieve the transformation of their society by importing a Trojan horse; it was done by Japanese planners, building on the existing Japanese infrastructure, to fulfill Japanese objectives. Certainly Japan imported foreign technologies and adopted innovations from the British and others, but the Japan that was built, a unique combination of modern and traditional elements, was basically an indigenous achievement.

Relative Location

Japan's changing fortunes over the past century reveal the influence of **relative location** in the country's development. When the Meiji Restoration took place, Britain, on the other side of the Eurasian landmass, lay at the center of a global empire. The colonization and Europeanization of the world were in full swing. The United States was still a developing country, and the Pacific Ocean was an avenue for European imperial competition. Japan, even while it was conquering and consolidating its first East Asian colonies (the Ryukyus, Taiwan, Korea), lay remote from the mainstream of global change.

Then Japan became embroiled in World War II and dealt severe blows to the European colonial armies in Asia. The Europeans never recovered: the French lost Indochina, and the Dutch were forced to abandon their East Indies (now Indonesia). When the war ended, Japan was defeated and devastated, but at the same time the Japanese had done much to diminish the European presence in the Pacific Basin. Moreover, the global situation had changed dramatically. The United States, Japan's trans-Pacific neighbor, had become the world's most powerful and wealthiest country, whereas Britain and its global empire were fading. Suddenly Japan was no longer remote from the mainstream of global action: now the Pacific was becoming the avenue to the world's richest markets. Japan's relative location—its situation relative to the economic and political foci of the world—had changed. Therein lay much of the opportunity the Japanese seized after the postwar rebuilding of their country.

FROM THE FIELD NOTES

"Visiting the Peace Memorial Park in Hiroshima is a difficult experience. Over this site on August 6, 1945 began the era of nuclear weapons use, and the horror arising from that moment in history, displayed searingly in the museum, is an object lesson in this time of nuclear proliferation and enhanced risk. In the museum is a model of the city immediately after the explosion (the red ball marks where the detonation occurred), showing the total annihilation of the entire area at an immediate loss of more than 80,000 people and the death from radiation of many more subsequently. In the park outside, the Atomic Bomb Memorial Dome, the only building to partially survive the blast, has become the symbol of Hiroshima's devastation and of the dread of nuclear war."

Japan's Spatial Organization

Imagine this: 128 million people crowded into a territory the size of Montana (population: 950,000), most of it mountainous, subject to frequent earthquakes and volcanism, with no domestic oilfields, little coal, few raw materials for industry, and not much level land for farming. If Japan today were an underdeveloped country in need of food relief and foreign aid, explanations would abound: overpopulation, inefficient farming, energy shortages.

True, only an estimated 18 percent of Japan's national territory is designated as habitable. And Japan's large population is crowded into some very big cities. Moreover, Japan's agriculture is not especially efficient. But Japan defeated the odds by calling on old Japanese virtues: organizational efficacy, massive productivity, dedication to quality, and adherence to common goals. Even before the Meiji Restoration, Japan was a tightly organized country of some 30 million citizens. Historians suggest that Edo, even before it became the capital and was renamed Tokyo, may have been the world's largest urban center. The Japanese were no strangers to urban life, they knew manufacturing, and they prized social and economic order.

Areal Functional Organization

All this proved invaluable to the modernizers when they set Japan on its new course. The new industrial growth of the country could be based on the urban and manufacturing development that was already taking place. As we noted, Japan does not possess major domestic raw-material sources, so no substantial internal reorganization was necessary. However, some cities were better sited and enjoyed better situations relative to those limited local resources and, more importantly, to external sources of raw materials than others. As Japan's regional organization took shape, a hierarchy of cities developed; Tokyo took and kept the lead, but other cities grew rapidly into industrial centers.

This process was governed by a geographic principle that Allen Philbrick called **areal functional organization**, a set of five interrelated tenets that help explain the evolution of regional organization, not only in Japan but throughout the world. Human activity, Philbrick reasoned, has spatial focus. It is concentrated in some locale, whether a farm or factory or store. Every one of these establishments occupies a particular location; no two of them can occupy exactly the same spot on the Earth's surface (even in highrises, there is a vertical form of absolute location). Nor can any human activity proceed in total isolation, so that interconnections develop among these various establishments. This system of interconnections grows more complex as human capacities and demands expand. Each system (for example, farmers sending crops to market and buying equipment at service centers) forms a unit of areal functional organization. In the Introduction, we referred to functional regions as systems of spatial organization; we can map Philbrick's units of areal functional organization as regions. These regions evolve because of what he called "creative imagination" as people apply their total cultural experience and their technological know-how when they organize and rearrange their living space. Finally, Philbrick suggests, we can recognize levels of development in areal functional organization, a ranking of places and regions based on the type, extent, and intensity of exchange.

In the broadest sense, we can divide regions of human organization into three categories: subsistence, transitional, and exchange. Japan's areal organization reflects the exchange category. Within each of these categories, we can also rank individual places on the basis of the number and kinds of activities they generate. Even in regions where subsistence activities dominate, some villages have more interconnections than others. The map of Japan (Fig. 9-22) showing its resources, urban settlements, and surface communications can tell us much about Japan's economy. It looks just like maps of other parts of the world where an exchange type of areal organization has developed—a hierarchy of urban centers ranging from the largest cities to the tiniest hamlets, a dense network of railways and roads connecting these places, and productive agricultural areas near and between the urban centers.

For Japan, however, the map shows us something else: Japan's external orientation, its dependence on foreign trade. All primary and secondary regions lie on the coast. Of all the cities in the million-size class, only Kyoto lies in the interior (Fig. 9-22). If we deduced that Kyoto does not match Tokyo–Yokohama, Osaka–Kobe, or Nagoya in terms of industrial development, that would be correct—the old capital remains a center of small-scale light manufacturing. Actually, Kyoto's ancient character has been deliberately preserved, and large-scale industries have been discouraged. With its old temples and shrines, its magnificent gardens, and its many workshop and cottage industries, Kyoto remains a link with Japan's premodern past.

Leading Economic Regions

As Figure 9-22 shows, Japan's dominant region of urbanization and industry (along with productive agriculture) is the *Kanto Plain*, which contains about one-third of the Japanese population and is focused on the Tokyo–Yokohama–Kawasaki metropolitan area. This gigantic cluster of cities and suburbs (the world's second

AMONG THE REALM'S GREAT CITIES . . . **TOKYO**

*M*any urban agglomerations are named after the city that lies at their heart, and so it is with the second largest of all: Tokyo (26.7 million). Even its longer name—Tokyo–Yokohama–Kawasaki—does not begin to describe the congregation of cities and towns that form this crowded metropolis that encircles the head of Tokyo Bay and continues to grow, outward and upward.

Near the waterfront, some of Tokyo's neighborhoods are laid out in a grid pattern. But the urban area has sprawled over hills and valleys, and much of it is a maze of narrow, winding streets and alleys. Circulation is slow, and traffic jams are legendary. The train and subway systems, however, are models of efficiency—although during rush hours you must get used to the *shirioshi* pushing you into the cars to get the doors closed.

At the heart of Tokyo lies the Imperial Palace with its moats and private parks. Across the street, buildings retain a respectful low profile, but farther away Tokyo's skyscrapers seem to ignore the peril of earthquakes. Nearby lies one of the world's most famous avenues, the Ginza, lined by department stores and luxury shops. In the distance you can see an edifice that looks like the Eiffel Tower, only taller: this is the Tokyo Tower (see photo, p. 475), a multipurpose structure designed to test lighter Japanese steel, transmit television and radio signals, detect Earth tremors, monitor air pollution, and attract tourists.

Tokyo is the epitome of modernization, but Buddhist temples, Shinto shrines, historic bridges, and serene gardens still grace this burgeoning urban complex, a cultural landscape that reflects Japan's successful marriage of the modern and the traditional.

largest urban agglomeration), interspersed with intensively cultivated farmlands, forms the eastern anchor of the country's elongated and fragmented core area. Besides its flatness, the Kanto Plain possesses other advantages: its fine natural harbor at Yokohama, its relatively mild and moist climate, and its central location with respect to the country as a whole. (The region's only disadvantage is its vulnerability to earthquakes—see box titled "When the Big One Strikes.") It has also benefited enormously from Tokyo's designation as the modern capital, which coincided with Japan's embarkation on its planned course of economic development. Many industries and businesses chose Tokyo as their headquarters in view of the advantages of proximity to the government's decision-makers.

The Tokyo–Yokohama–Kawasaki conurbation has become Japan's leading manufacturing complex, producing more than 20 percent of its annual output. The raw materials for all this industry, however, come from far away. For example, the Tokyo area is among the chief steel producers in Japan, using iron ores from the Philippines, Malaysia, Australia, India, and even Africa; most of the coal is imported from Australia and North America, and the petroleum comes from Southwest Asia and Indonesia. The Kanto Plain cannot produce nearly enough food for its massive resident population. Imports must come from Canada, the United States, and Australia as well as other areas in Japan. Thus Tokyo depends completely on its external trading ties for food, raw materials, and markets for its products, which run the gamut from children's toys to high-precision optical equipment to the world's largest oceangoing ships.

The second-ranking economic region in Japan's core area is the Osaka–Kobe–Kyoto triangle—also known as the *Kansai District*—located at the eastern end of the Inland Sea. The Osaka–Kobe conurbation had an advantageous position with respect to Manchuria (Manchukuo) during the height of Japan's colonial empire. Situated at

When the Big One Strikes

*T*he Tokyo–Yokohama–Kawasaki urban area is the second biggest metropolis on Earth (only Mexico City is larger). But Tokyo is more than a large city: it constitutes the most densely concentrated financial and industrial complex in the world. Two-thirds of Japan's businesses worth more than U.S. $50 million are clustered here, many with vast overseas holdings. More than half of Japan's huge industrial profits (averaging about U.S. $100 billion annually between the late 1980s and the downturn of the late 1990s) are generated in the factories of this gigantic metropolitan agglomeration.

But Tokyo has a worrisome environmental history because three active tectonic plates are converging here (Fig. 9-23). All Japanese know about the "70-year" rule: over the past three-and-a-half centuries, the Tokyo area has been struck by major earthquakes roughly every 70 years—in 1633, 1703, 1782, 1853, and 1923. The Great Kanto Earthquake of 1923 set off a firestorm that swept over the city and killed an estimated 143,000 people. Tokyo Bay virtually emptied of water; then a *tsunami* (seismic sea wave) roared back in, sweeping everything before it. That Japan could overcome this disaster was evidence of the strength of its economy.

Today, Tokyo is more than a national capital. It is a global financial and manufacturing center in which so much of the world's wealth and productive capacity are concentrated that an earthquake comparable to the one of 1923 would have a calamitous effect worldwide. Ominously, Tokyo at the outset of our new century is a much more vulnerable place than the Tokyo of the 1920s. True, building regulations are stricter, and civilian preparedness is better. But whole expanses of industries have been

built on landfill that will liquefy; the city is honeycombed by underground gas lines that will rupture and stoke countless fires; congestion in the area's maze of narrow streets will hamper rescue operations; and many older highrise buildings do not have the structural integrity that has lately emboldened builders to erect skyscrapers of 50 stories and more. Add to this the burgeoning population of the Kanto Plain—approaching 30 million on what may well be the most dangerous 4 percent of Japan's territory—and we realize that the next big earthquake in the Tokyo area will not be a remote, local news story.

FIGURE 9-23

---

the head of the Inland Sea, Osaka was the major Japanese base for the China trade and for the exploitation of Manchuria, but it suffered when the empire was destroyed and it lost its trade connections with China after World War II. Kobe (like Yokohama, which is Japan's chief shipbuilding center) has remained one of the country's busiest ports, handling both Inland Sea traffic and extensive overseas linkages. Kyoto, as we noted, remains much as it was before Japan's great leap forward, a city of small workshop industries. The Kansai District is also an important farming area. Rice, of course, is the most intensively grown crop in the warm, moist lowlands, but this agricultural zone is smaller than the one on the Kanto Plain. Here is another huge concentration of people that must import food.

The Kanto Plain and Kansai District, as Figure 9-22 indicates, are the two leading primary regions within

Japan's core. Between them lies the *Nobi Plain* (also called the Chubu District), focused on the industrial metropolis of Nagoya, Japan's leading textile producer. The map indicates some of the Nagoya area's advantages and liabilities. The Nobi Plain is larger than the lowlands of the Kansai District; thus its agricultural productivity is greater, though not as great as that of the Kanto Plain. But Nagoya has neither Tokyo's centrality nor Osaka's position on the Inland Sea: its connections to Tokyo are via the tenuous Sen-en Coastal Strip. Its westward connections are better, and the Nagoya area may be coalescing with the Osaka–Kobe conurbation. Still, the quality of Nagoya's port is not nearly as good as Tokyo's Yokohama or Osaka's Kobe, and it has had silting problems.

Westward from the three regions just discussed—which together constitute what is often called the Tokaido megalopolis—extends the Inland Sea, along whose shores the remainder of Japan's core area is continuing to develop. The most impressive growth has occurred around the western entry to the Inland Sea (the Strait of Shimonoseki), where *Kitakyushu*—a conurbation of five cities on northern Kyushu—constitutes the fourth Japanese manufacturing complex and primary economic region. Road and railway tunnels connect Honshu and Kyushu, but the northern Kyushu area does not have an urban-industrial equivalent on the Honshu side of the strait. The Kitakyushu conurbation includes Yawata, site of northwest Kyushu's (rapidly declining) coal mines. The first steel plant in Japan was built there on the basis of this coal; for many years it was the country's largest. The advantages of transportation here at the western end of the Inland Sea are obvious: no place in Japan is better situated to do business with Korea and China. As relations with mainland Asia expand, this area will reap many benefits. Elsewhere on the Inland Sea coast, the Hiroshima–Kure urban area has a manufacturing base that includes heavy industry. And on the coast of the Korea Strait, Fukuoka and Nagasaki are the principal centers—respectively, an industrial city and a center of large shipyards.

Only one major Japanese manufacturing complex lies outside the belt extending from Tokyo in the east to Kitakyushu in the west—the secondary region centered on Toyama on the Sea of Japan (East Sea). The advantage here is cheap power from nearby hydroelectric stations, and the cluster of industries reflects it: paper manufacturing, chemical factories, and textile plants. Figure 9-22 also gives an inadequate picture of the variety and range of industries that exist throughout Japan, many of them oriented to local (and not insignificant) markets. Thousands of manufacturing plants operate in cities and towns other than those shown on this map, even on the cold northern island of Hokkaido, which is connected to Honshu by the Seikan rail tunnel, the world's longest, beneath the treacherous Tsugaru Strait.

The map of Japan's areal organization shows that four primary regions dominate its core area. Each of them is primary because each duplicates to some degree the contents of the others. Each contains iron and steel plants, is served by one major port, and lies in or near a large, productive farming area. What the map does not show is that each also has its own external connections for the overseas acquisition of raw materials and sale of finished products. These linkages may even be stronger than those among the four internal regions of the core area. Only for Kyushu and its coal have domestic raw materials affected the nature and location of heavy manufacturing, and these resources are almost depleted. In the structuring of the country's areal functional organization, therefore, more than just the contents of Japan itself is involved. In this respect Japan is not unique: all countries that have exchange-type organization must adjust their spatial forms and functions to the external interconnections required for progress. But for few countries is this truer than Japan.

East and West, North and South

The previous discussion shows that Japan has its own Pacific Rim: almost all of Honshu's economic expansion is occurring on the east coast. Japanese geographers often call this eastern zone *Omote Nippon* ("Front Japan"), the Pacific-facing, megalopolitan, high-tech, dynamic sector that embodies the modern economic superpower—as opposed to *Ura Nippon* ("Back Japan"), the comparatively underdeveloped periphery of unkempt farms and poor villages bordering the Sea of Japan (East Sea) to the west. In a country that needs food as much as Japan does, the derelict state of farmlands in *Ura Nippon* is of much concern to the country's planners.

The congestion on Honshu is another concern. Efforts to link Hokkaido and southern Shikoku to the "mainland" have produced world-record tunnels and bridges such as the Akashi Kaikyo Bridge, designed to diffuse Honshu's economic primacy to more remote areas. After the Kobe earthquake of 1995, planning began to relocate Tokyo's capital functions to a site between 40 and 200 miles (65 and 320 km) west of Tokyo, not only for safety reasons but primarily to remove governmental offices from the overcrowded and high-cost Tokyo urban area. In 1999, with the Japanese economy slowed by recession, this project was delayed, but Japanese geographers report that the relocation of the national capital remains on the agenda.

Food and Population

Japan's economic modernization so occupies center stage that we can easily forget the country's considerable achievements in agriculture. Japan's planners, who are as interested in closing the food gap as in expanding industries, have created extensive networks of experiment stations to promote mechanization, optimal seed selection and fertilizer use, and information services to distribute knowledge about enhancing crop yields to farmers as rapidly as possible. Although this program has succeeded, Japan faces the unalterable reality of its stubborn topography: it simply lacks sufficient land to farm. Populous Japan has one of the highest physiologic population densities—7965 per square mile (3075 per sq km)—on Earth. (We discuss this concept on pages 384 and 405.)

The Japanese go to extraordinary lengths to maximize their limited farming opportunities, making huge investments in research on higher-yielding rice varieties, mechanization, irrigation, terracing, fertilization, and other practices. More than 90 percent of farmland is assigned to food

Tokyo, at the center of one of the largest metropolises in the world, continues to change. Land-filling and bridge-building in the bay continue; skyscrapers sprout amid low-rise neighborhoods in this earthquake-prone area; traffic congestion worsens. The red-painted Tokyo Tower, a beacon in this part of the city, was modeled on the Eiffel Tower in Paris but, as a billboard at its base announces, is an improvement over the original: lighter steel, greater strength, less weight. Tokyo Bay, part of which can be seen from this vantage point, was the scene of one of history's most costly environmental disasters during the earthquake of 1923. A giant *tsunami* (seismic sea wave) swept up the bay from the epicenter even as landfills liquefied and buildings sank into the mud. The death toll exceeded 140,000; a much larger population than lived in Tokyo in 1923 is now at risk of a repeat.

crops. Hilly terrain is given over to cash crops, such as tea and grapes (Japan has a thriving wine industry). Vegetable gardens ring all the cities.

But all this costs money, and Japan's food prices are high. Millions of Japanese have moved from the farms to the cities, depopulating the countryside which needs farmers. Japan's political system gives underpopulated rural areas disproportionate influence, and produce prices are kept artificially high, all to induce farmers to stay on the land. It has not worked. Between 1920 and 2000, full-time farmers declined from 50 percent to under 4 percent of the labor force. Yet the policies persist. In the early 1990s, American farmers proved that they could provide rice to Japanese consumers at one-sixth the price the Japanese were paying for home-grown rice. But the government resisted pressure to open its markets to imports.

With their national diet so rich in starchy foods like rice, wheat, barley, and potatoes, the Japanese need protein to balance it. Fortunately, they can secure enough of it, not by buying it abroad but by harvesting it from rich fishing grounds near the Japanese islands. With their customary thoroughness, the Japanese have developed a fishing industry that is larger than that of the United States or any of the long-time fishing nations of northwestern Europe, and it now supplies the domestic market with a second staple after rice. Although mention of the Japanese fishing industry brings to mind a fleet of ships scouring the oceans and seas far from Japan, most of this huge catch (about one-

seventh of the world's annual total) comes from waters within a few dozen miles of Japan itself. Where the warm Kuroshio and Tsushima currents meet colder water off Japan's coasts, a rich fishing ground yields sardines, herring, tuna, and mackerel in the warmer waters and cod, halibut, and salmon in the seas to the north. Japan's coasts have about 4000 fishing villages, and tens of thousands of small boats ply the waters offshore to bring home catches that are distributed to local and city markets. The Japanese also practice *aquaculture*—the "farming" of freshwater fish in artificial ponds and flooded paddy (rice) fields, of seaweeds in aquariums, and of oysters, prawns, and shrimp in shallow bays. They are even experimenting with cultivating algae for their food potential.

When you walk the streets of Japan's cities, you will notice something that Japanese vital statistics are reporting: the Japanese are getting taller and heavier. Coupled with this change is evidence that heart disease and cancer are rising as causes of mortality. The reason seems to lie in the changing diets of many Japanese, especially the younger people. When rice and fish were the staples for virtually all, and the calories available were limited but adequate, the Japanese as a nation were among the world's healthiest and longest-lived. But as fast-food establishments diffused throughout Japan, and Western (especially American) tastes for red meat and fried food spread among children and young adults, their combined impact on the population was soon evident. Coupled with the heavy cigarette smoking that prevails in all the Asian countries of the Pacific Rim, this is changing the region's medical geography.

Japan's Pacific Rim Prospects

As the twenty-first century opens, Japan remains the economic giant on the western Pacific Rim—but a giant with an uncertain future. Japanese products still dominate world markets and Japanese investments span the globe, but in the late 1990s Japan's economy faltered. Its growth rate declined. Many companies scaled back their operations, and some went bankrupt. The banking system proved to be vulnerable. The first Pacific Rim economic tiger was ailing.

Japan's problems stemmed from a combination of circumstances at home and abroad. Overseas, the collapse of Southeast Asian currencies and economies hurt Japanese investments there. From automobile plants to toy factories, long-profitable operations turned into liabilities. In Japan itself, financial mismanagement, long tolerable because of the booming economy, now took its toll. In addition, competition from still-healthy economies, notably from Taiwan but also from struggling South Korea, undercut Japanese goods on world markets.

In the long run, all this may prove to have been a temporary setback, but Japan is unlikely to regain its former economic dominance on the Pacific Rim. Japan's dependence on foreign oil is a potential weakness that could further depress its economy should international supply lines be disrupted. Western competition for Japan's markets is growing. The countless investments Japan made around the world during its time of plenty, from skyscrapers and movie studios to golf courses and hotels, have lost much of their value and must be sold at huge losses. While Japan reorganizes, American and European businesses buy its assets and take advantage of its problems.

Were you to make a field trip to Japan to study its economic geography, none of this might be immediately evident. The cities still bustle with activity, raw materials continue to arrive at the ports, container traffic still stirs the docks. Even in a downturn, this mighty economy outproduces all others on the western Pacific Rim. And, indeed, that observation is relevant to the future. Japan's infrastructure has the capacity to propel an economic rebound. But it needs new opportunities.

One of these opportunities lies in the Russian Far East and beyond, in Siberia. As we noted in Chapter 2, the Russian Far East lies directly across from northern Honshu and Hokkaido and is a storehouse of raw materials awaiting exploitation. For this, the Japanese are ideally located—but political relations with the former Soviet Union, and now with Russia, have stymied all initiatives (see box titled "The Wages of War"). Japan should be one of the major beneficiaries of any substantial oil reserves near Sakhalin (see map, p. 132), but again the necessary connections may not be possible. Japan also has a major potential opportunity in Korea, especially if a unification process takes hold there. But relations with Korea continue to be eroded by Korean memories of Japanese behavior during World War II and by Japan's refusal to acknowledge its misdeeds.

All this must be seen against the backdrop of Japan's changing society. The population of 127 million is aging rapidly, and population geographers expect it to rise only slightly until 2007, when it will begin to decline. This decline will increase quite rapidly, to about 100 million in 2050 and only 67 million in 2100. Already, for some time, Japan has faced a labor shortage (although the recent economic downturn produced rising unemployment, a novelty in the postwar period). Over the long term, Japan will need millions of immigrants to keep its economy going, but for culturally homogeneous, ethnically conscious Nippon, which has historically resisted immigration, this will be a wrenching reality. Still, Japanese leaders will have no choice. The country's "graying" will strain welfare systems and tax burdens.

Modernization, meanwhile, is further stressing Japan's society. The Japanese have a strong tradition of family cohesion and veneration of their oldest relatives, but this tradition is breaking down. Younger people are less willing than their parents and grandparents to accept overcrowded housing, limited comforts, and little privacy. In Japan, for all its material wealth, family homes tend to be small, cramped, often flimsily built, and sometimes lacking the basic amenities other high-income countries consider indispensable. With their disposable income, the Japanese during the boom economy came to rank among the world's most-traveled tourists, and, having seen how Americans and Europeans of similar income levels live, they returned home dissatisfied.

As we just noted, Japan remains one of the most uniform, culturally homogeneous societies in today's interactive world. The Chinese closed their doors to foreigners intermittently; the Japanese have done so far more effectively. Other than the dwindling Ainu, there are no indigenous minorities. Koreans were brought to Japan during World War II to serve as forced laborers, but otherwise Japanese residency and citizenship imply an almost total ethnic sameness. During periods of labor shortage, the notion to invite foreign workers would not be entertained. When the United States, after the Vietnam War, sought new homes for hundreds of thousands of Vietnamese refugees, Japan accepted a few hundred

The Wages of War

Japan and the former Soviet Union never signed a peace treaty to end their World War II conflict. Why? There are four reasons, and all of them are on the map just to the northeast of Japan's northernmost large island, Hokkaido (Fig. 9-21). Their names are Habomai, Shikotan, Kunashiri, and Etorofu. The Japanese call these rocky specks of the Kurile Island chain their "Northern Territories." The Soviets occupied them late in the war and never gave them back to Japan. Now they are part of Russia, and the Russians have not given them back either.

The islands themselves are no great prize. During World War II, the Japanese brought 40,000 forced laborers, most of them Koreans, to mine the minerals there. When the Red Army overran them in 1945, the Japanese were ordered out, and most of the Koreans fled. Today the population of about 50,000 is mostly Russian, many of them members of the military based on the islands and their families. At their closest point the islands are only 3 miles (5 km) from Japanese soil, a constant and visible reminder of Japan's defeat and loss of land. Moreover, territorial waters bring Russia even closer, so that the islands' geostrategic importance far exceeds their economic potential.

Attempts to settle the issue have failed. In 1956, Moscow offered to return the tiniest two, Shikotan and Habomai, but the Japanese declined, demanding all four islands back. In 1989, then-Soviet President Mikhail Gorbachev visited Tokyo in the hope of securing an agreement. The Japanese, it was widely reported, offered an aid-and-development package worth U.S. $26 billion to develop Russia's eastern zone—its Pacific Rim and the vast resources of the eastern Siberian interior. This would have begun the transformation of Russia's Far East, stimulated the ports of Nakhodka and Vladivostok, and made Russia a participant in the spectacular growth of the western Pacific Rim.

But it was not to be. Subsequently, Russian Presidents Yeltsin and Putin also were unable to come to terms with Japan on this issue, facing opposition from the islands' inhabitants and from their own governments in Moscow. And so World War II, almost 60 years after its conclusion, continues to cast a shadow over this northernmost segment of the western Pacific Rim.

---

after much negotiation. Such jealously guarded ethnic-cultural uniformity is unlikely to be viable in the globalizing world of the twenty-first century.

Japan's future in the geopolitical framework of the western Pacific also is uncertain. More than a half-century after the end of World War II, American forces are still based on Japanese territory, guaranteeing Japanese security. China and Korea accept this arrangement, which constrains Japan's rearming, in view of what happened during World War II. But in Japan, where nationalism is rising, domestic military preparedness is a growing political issue. The situation is fraught with potential problems.

Japan is therefore a pivotal country in the western Pacific Rim that stands at economic, social, and geopolitical crossroads. Its future will profoundly influence not only East Asia and the Pacific, but the world as a whole.

KOREA

On the Asian mainland, directly across the Sea of Japan (East Sea), lies the peninsula of Korea (Fig. 9-24), a territory about the size of the State of Idaho, much of it mountainous and rugged, and containing a population of 73 million. Unlike Japan, however, Korea has long been a divided country, and the Koreans a divided nation. For uncounted centuries Korea has been a pawn in the struggles of more powerful neighbors. It has been a dependency of China and a colony of Japan. When it was freed from Japan's oppressive rule at the end of World War II (1945), the victorious Allied powers divided Korea for administrative purposes. That division gave North Korea (north of the 38th parallel) to the forces of the Soviet Union and South Korea to those of the United States. In effect, Korea traded one master for two new ones. The country was not reunited for the rest of the century because North Korea immediately fell under the communist ideological sphere and became a dictatorship in the familiar (but in this case extreme) pattern. South Korea, with massive American aid, became part of East Asia's capitalist perimeter. Once again, it was the will of external powers that prevailed over the desires of the Korean people.

In 1950, North Korea sought to reunite the country by force and invaded South Korea across the 38th parallel. This attack drew a United Nations military response led by the United States. Thus began the devastating Korean War (1950–1953) in which North Korea's forces pushed far to the south only to be driven back across its own half of Korea almost to the Chinese border. Then China's Red Army entered the war and drove the UN troops southward again. A cease-fire was arranged in 1953, but not before the

FIGURE 9-24

The map legend:

KOREA
POPULATION
- Under 50,000
- 50,000–250,000
- 250,000–1,000,000
- 1,000,000–5,000,000
- Over 5,000,000

National capitals are underlined

— Railroad
Rugged mountain area
Industrial area
Main rice producing area
Secondary rice producing area
Free Trade Zone

0 50 100 Kilometers
0 25 50 Miles

people and the land had been ravaged in a way that was unprecedented even in Korea's violent past. The cease-fire line shown in Figure 9-24 became a heavily fortified *de facto* boundary. For more than five decades virtually no contact of any kind has occurred across this border, which divides economies and families alike.

As a result, the Jakota Triangle, as a region of East Asia, must include South Korea but exclude North Korea (see box titled "Obsolescent North Korea"). Times may change, as they did in formerly divided Germany, but at the turn of the century South Korea, with 45 percent of Korea's land area but with two-thirds of the Korean population, has emerged as one of the economic tigers on the Pacific Rim. As an undivided country, Korea might have achieved even more because North and South Korea exhibit what geographers call **regional complementarity**.

Obsolescent North Korea

The Korean Peninsula is physiographically well-defined, and its people—the Korean nation—have occupied it for thousands of years. But internal political division, foreign cultural intrusion, and colonial subjugation have created a turbulent history. Now Koreans are divided again, this time by ideology. Even as the capitalist South became one of the Pacific Rim's economic tigers, North Korea remained an isolated, stagnant state on a Stalinist-communist model abandoned long ago even in the former Soviet Union itself.

For over a half-century, North Korea has been ruled by two men: Kim Il Sung, the "Great Leader," and his son Kim Jong Il, the "Dear Leader." Total allegiance to the dictators is ensured by an omnipotent police and military. Self-imposed isolation is North Korea's hallmark. The state controls all aspects of life in North Korea.

Economic policies under the communist regime have been disastrous. Per-capita incomes in the North are less than one-tenth of those in the South; poverty is rampant. Infant mortality is nearly four times higher than in South Korea. Industrial equipment is 25 years out of date. Exports are negligible. When a combination of failed agricultural policies and environmental extremes created famine in the late 1990s and early 2000s, the regime resisted relief efforts, even from South Korea. As during the Maoist period in China, the world will never know the number of casualties.

North Korea did progress on one front: nuclear capability and associated weaponry. In 1999 it fired a test missile across northern Japan, heightening international concern. The North Koreans used their nuclear threat to negotiate a multibillion-dollar technical aid program to build reactors for peaceful purposes, with Japan and South Korea among the major contributors. But in late 2002, a North Korean source admitted that his country had indeed pursued a nuclear weapons program and was now in possession of "at least one" nuclear bomb. The North Koreans refused to allow visitors to inspect its nuclear facilities at Yongbyon, 60 miles (100 km) due north of the capital, Pyongyang.

President George Bush, in his 2002 State of the Union address, called North Korea part of the "axis of evil" to be dealt with as part of the War on Terror. But while Iraq was forced to admit UN weapons inspectors, North Korea, with 37,000 U.S. troops along its border with the South, remained exempt. Meanwhile, North Korea acknowledged having kidnapped Japanese youngsters from public places in Japan and having forced them to teach Japanese language and culture to North Korean spies. In 2002 this admission led to authorized visits to Japan by some of the survivors and a furor among the Japanese people.

Even as the "Dear Leader" was able to reject U.S. and Japanese demands that nuclear operations at Yongbyon be reviewed, he approved some small gestures toward his southern neighbor. Beginning in 2000, a few hundred members of families separated by the 1953 cease-fire line were allowed a few hours to meet each other. He also allowed the preliminary design of a railroad that would link North and South across the demilitarized zone (DMZ). After authorizing the creation of a barbed-wire-enclosed "free-trade zone" in the far northeast at Rajin-Sonbong, where South Korean cruise ships could dock, he approved another one in the northwest, at Sinuiji, in 2002. But when the Chinese billionaire who was to take charge of this free-trade zone was arrested for corruption in China, that plan fell apart.

Desperate North Koreans flee their country across the Chinese border and on the seas; in foreign countries they invade embassy compounds and plead for asylum. In 2003 the reunification of the two Koreas seemed far off; but so did German reunification in 1985, and then it happened a few years later. But the reunification of the two Koreas will entail far greater difficulties than that of the Germanies. North Korea is far worse off even than former communist East Germany; the cost will be staggering. South Koreans will face far greater sacrifices than West Germans did, with far fewer resources. Still, the boundary superimposed on the Korean nation is a leftover of the Cold War, and its elimination and the reintegration of North Korea into the international community is a laudable goal.

This condition arises when two adjacent regions complement each other in economic-geographic terms. In this case, North Korea has raw materials that the industries of South Korea need; South Korea produces food the North needs; North Korea produces chemical fertilizers farms in the South need.

In the past half-century, however, the two Koreas were cut off from each other, and they developed in opposite directions. North Korea, whose large coal and iron ore deposits attracted the Japanese and whose hydroelectric plants produce electricity that the South could use, carried on what limited trade it generated with the Chinese and the Soviets. South Korea's external trade links were with the United States, Japan, and Western Europe.

In the early postwar years, there were few indications that South Korea would emerge as a major economic force on the Pacific Rim and, indeed, on the world stage. Over 70 percent of all workers were farmers; agriculture was inefficient; the country was stagnant. But huge infusions of aid, first from the United States and then Japan, coupled with the reorganization of farming and stimulation of industries (even if they lost money), produced a dramatic turnaround. Large feudal estates were parceled out to farm families in 3-hectare (7.5-acre) plots, and a program of massive fertilizer importation was begun. Production rose to meet domestic needs, and some years yielded surpluses.

The industrialization program was based on modest local raw materials, plentiful and capable labor, ready overseas (especially American) markets, and continuing foreign assistance. Despite corrupt dictatorial rule, political instability, and social unrest, South Korean regimes managed to sustain a rapid rate of economic growth that placed the country, by the late 1980s, among the world's top ten trading powers. To get the job done, the government borrowed heavily from overseas, and it controlled banks and large industries. South Korea's growth resulted from state capitalism rather than free-enterprise capitalism, but it had impressive results. South Korea be-

came the world's leading shipbuilding nation; its automobile industry grew rapidly; and its iron and steel and chemical industries began to thrive. It has placed less emphasis, however, on smaller, high-technology industries, which are the strengths of other Pacific Rim economic tigers. So the future remains uncertain.

Today, South Korea is prospering, and we can see its economic prowess on the map (Fig. 9-24). The capital, Seoul, with 10 million inhabitants, now ranks among the world's megacities and is the anchor of a huge industrial complex facing the Yellow Sea at the waist of the Korean Peninsula. Hundreds of thousands of farm families migrated to the Seoul area after the end of the Korean War (today only about 20 percent of South Koreans remain on the land). They also moved to Busan, the nucleus of the country's second largest manufacturing zone, located on the Korea Strait opposite the western tip of Honshu. And the government-supported, urban-industrial drive here continues. Just 30 years ago, Ulsan City, 40 miles (60 km) north of Busan along the coast, was a fishing center with perhaps 50,000 inhabitants; today its population exceeds

AMONG THE REALM'S GREAT CITIES . . . **SEOUL**

*S*eoul (9.9 million), located on the Han River, is ideally situated to be the capital of all of Korea, North and South (its name means capital in the Korean language). Indeed, it served as such from the late fourteenth century until the early twentieth, but events in that century changed its role. Today, the city lies in the northwest corner of South Korea, for which it serves as capital; not far to the north lies the tense demilitarized zone (DMZ) that contains the cease-fire line with North Korea. That line cuts across the mouth of the Han, depriving Seoul of its river traffic. Its ocean port, Inchon, has emerged as a result.

Seoul's undisciplined growth, attended by a series of recent accidents including the failure of a major bridge over the Han and the collapse of a six-story department store, reflects the unbridled expansion of the South Korean economy as a whole, as well as the political struggles that carried the country from autocracy to democracy. Central Seoul lies in a basin surrounded by hills to an elevation of about 1000 feet (330 m), and the city has sprawled outward in all directions, even toward the DMZ. An urban plan designed in the early 1960s was overwhelmed by the immigrant flow.

During the period of Japanese colonial control, Seoul's surface links to other parts of the Korean Peninsula were improved, and this infrastructure played a role in the city's

later success. Seoul is not only the capital but also the leading industrial center of South Korea, exporting sizeable quantities of textiles, clothing, footwear, and (increasingly) electronic goods. South Korea is still an economic tiger on the Pacific Rim, and Seoul is its heart.

1.6 million, nearly half of them the families of workers in the Hyundai automobile factories and the local shipyards and docks.

The third industrial area shown in Figure 9-24, anchored by the city of Kwangju, has advantages of relative location that will spur its development, but it faces domestic problems. Kwangju lies at the center of the Cholla region, the southwest corner of South Korea; the southeast is called Kyongsang. For more than three decades beginning in the 1960s, the army generals who ruled South Korea all came from Kyongsang, and through state control of banks and other monopolies they bestowed favors on Kyongsang companies while denying access to potential competitors in Cholla. The regime in Seoul also thwarted infrastructure development in Cholla, even to the point of impeding plans for improved railroad links between Kwangju and the capital. This regional friction led to protests in Cholla and violent repression by the military rulers, the aftermath of which is fading only slowly in this part of the now-democratic country. People in Cholla often say that they feel discriminated against in their own country and that it is prejudice, not ability, that has held their region back. They have reasons: the Samsung group of companies, for example, long had a stated policy not to employ any graduates from Cholla's universities.

Thus South Korea's economic growth has not gone smoothly; dictatorial rule and police-state tactics to keep control were part of the transition. Today, South Korea is a more democratic society, but the Asian economic downturn of the late 1990s exposed the weaknesses of its financial and business structure, inherited from the earlier period. Giant corporate conglomerates called *chaebol* conspired to suppress competition; they controlled the banking system and dictated government economic policy. It was perhaps the ultimate version of what economic geographers call **state capitalism**, and it proved to have weaknesses similar to communist planned economies. When South Korea felt the impact of events in other parts of Asia, the system collapsed and a large infusion of funds from international banks was necessary. As the new century began, South Korea's economy had stabilized and was recovering, but circumstances within as well as outside of the country have combined to slow its climb up the ladder already topped by Japan.

▌ TAIWAN

The third component of the region we have called the Jakota Triangle is another, and more durable, success story on the Pacific Rim: Taiwan. Along with the Penghu (Makung) Islands in the Taiwan Strait and the islands

FROM THE FIELD NOTES

"South Korea was one of the four 'economic tigers' marking the upsurge of the Western Pacific Rim in the 1980s and 1990s, its products ranging from computers to cars selling on world markets and its GDP rising rapidly. Walking through the prosperous central business district of globalizing Seoul you see every luxury name brand on display amid plentiful evidence of wealth and well-being. But you also see evidence of the survival of traditions long lost elsewhere, in the huge number of bookstores (South Koreans still read voraciously), in the stores selling traditional as well as modern musical instruments, in the school uniforms. And in other ways: when I got on a bus with all seats taken, a boy got up and offered me his place."

of Jinmen Dao (Quemoy) near Xiamen and Matsu Tao (Matsu) near Fuzhou at the coast of mainland China, Taiwan is regarded by Beijing's communist regime as a Chinese province temporarily "wayward," that is, disobedient. After the reunification of Portuguese Macau with China in 1999, Taiwan remained the only major unresolved territorial issue in China's postcolonial, postwar sphere.

Beijing's claim to Taiwan has strengths as well as weaknesses. Han people settled on Taiwan, but the island had been settled thousands of years earlier by Malay-Polynesian groups who had split into mountain (east) and plains (west) inhabitants. When the Chinese emigrated from famine-stricken Fujian Province to Taiwan in the seventeenth century, they soon displaced and assimilated the plains people, but the mountain aborigines held out until the Japanese colonial invasion. Formal Chinese control began in 1683 when the expansionist Qing (Manchu) Dynasty made Taiwan part of Fujian Province. Rice and sugar became important exports to the mainland, and by the 1840s Taiwan's population was an estimated 2.5 million. In 1886 the island was declared a province of China, but by then the British had established two treaty ports there, the French had blockaded it, and the Japanese had sent expeditions to it. Nine years later, in 1895, China ceded Taiwan to Japan in the treaty settlement of the Sino-Japanese War, and it became a Japanese colony.

When China's aging Manchu Dynasty was overthrown following the rebellion of 1911, a Nationalist government in 1912 proclaimed the Republic of China (ROC), led by Sun Yat-sen. Naturally, this government wanted to oust all colonialists, not only Europeans but also Japanese, and not just from the mainland, but from Taiwan as well. On Taiwan, however, the Japanese held on, using the island as a source of food and raw materials as well as a market for Japanese products. To ensure all this, the Japanese launched a prodigious development program involving road and railroad construction, irrigation projects, hydroelectric schemes, mines (mainly for coal), and factories. Farmlands were expanded and farming methods improved. While mainland China was engulfed in conflict, Taiwan remained under Japanese control. Not until 1945, when Japan was defeated in World War II, did Taiwan become an administrative part of China again. But only briefly. As we noted earlier, the Nationalists, under the leadership of Chiang Kai-shek, now faced the rising tide of communism on the mainland, and the ROC government fought a losing war. In 1949, the flag of the Republic of China was taken to a last stronghold: offshore Taiwan. There, Chiang and his followers, with their military might, weapons, and wealth taken from the mainland, established what they (and the world, led by the United States) proclaimed as the legitimate government of all China. On the island, the Nationalists, helped by the United States, began a massive reconstruction of the war-damaged infrastructure. In the international arena, the ROC represented the country; in the still-young United Nations, the Nationalists of Taiwan occupied China's seat.

On the huge mainland, however, it was the communists in Beijing who ruled and called their country the People's Republic of China (PRC). Thus was born the two-China dilemma which still confronts the international community. When the ROC was ousted from the United Nations in 1971 to make way for the PRC, legitimacy was also transferred.

All this might have remained an internal dispute but for two major developments: Taiwan's rise as an economic tiger on the Pacific Rim, coupled with remarkable strides toward democracy made in the ROC; and the emergence on the world political and economic stage of

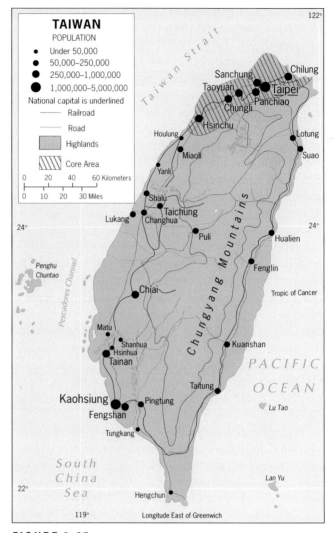

FIGURE 9-25

communist-ruled China following decades of isolation. Today, the power of the PRC is growing rapidly, and Taiwan's heirs to the ROC face an uncertain future; but it is the ROC, not the PRC, that has managed to combine economic success with democratization.

Taiwan, as Figure 9-25 shows, is not a large island. It is smaller than Switzerland but has a population much larger (22.7 million), most of it concentrated in an arc lining the western and northern coasts. The Chungyang Mountains, an area of high elevations (some over 10,000 feet [3000 m]), steep slopes, and dense forests, dominate the eastern half of the island. Westward, these mountains yield to a zone of hilly topography and, facing the Taiwan Strait, a substantial coastal plain. Streams from the mountains irrigate the paddyfields, and farm production has more than doubled since 1950 even as hundreds of thousands of farmers left the fields for work in Taiwan's expanding industries.

Today, the lowland urban-industrial corridor of western Taiwan is anchored by the capital, Taipei (Taibei), at the island's northern end and rapidly growing Kaohsiung (Gaoxiong) in the far south.* The Japanese developed Chilung (Jilong), Taipei's outport, to export nearby coal, but now the raw materials flow the other way. Taiwan imports raw cotton for its textile industry, bauxite (for aluminum) from Indonesia, oil from Brunei, and iron ore from Africa. Taiwan has a developing iron and steel industry, nuclear power plants, shipyards, a large chemical industry, and modern transport networks. Increasingly, however, Taiwan is exporting products of its high-technology industries: personal computers, telecommunications equipment, and precision electronic instruments. Taiwan has enormous brainpower, and many foreign firms join in the research and development carried on in such places as Hsinchu (Xinzhu) in the north, where the government has helped establish a technopole centered on the microelectronics and personal computer industries. In the south, the "science city" of Tainan specializes in microsystems and information technology.

The establishment of China's Special Economic Zones has gone a long way toward solving the economic problems arising from the political discord between China and Taiwan. In the almost-anything-goes environment of the SEZs, Taiwanese business owners are permitted to build or buy factories just as "real" foreigners are, and as we noted earlier, the Xiamen SEZ was situated directly across from Taiwan for just this purpose. As a result, thousands of Taiwanese companies now operate in China even as political tensions continue.

In some ways, Taiwan's emergence as an economic tiger on the Pacific Rim is even more spectacular than Japan's. True, Taiwan received much Western assistance, but it got out of debt faster than anyone expected. Today, annual per-capita income exceeds U.S. $15,000, which is higher than that in many European countries. Taiwan's trade surplus has yielded tens of billions of dollars in reserves, which is being used to further its own development and to invest overseas. Taiwan, for example, is the largest foreign investor in the economy of Vietnam. But although Taiwan weathered the regional economic crisis of the late 1990s rather well, the new century has been problematic. The U.S. and Japanese markets for high-tech exports, especially computer chips, shrank markedly. Thousands of firms long profitable in Taiwan had to move their opera-

tions to China's SEZs, where labor costs were much lower, to stay in business. This reduced economic growth from 6.3 percent in 2000 to 1.1 percent in 2001 and into negative territory in 2002, while unemployment, basically unheard of in Taiwan, shot up to nearly 5 percent. All this happened while mainland China's economy continued to grow at an annual rate of about 7 percent. It was a shock to Taiwan's confidence as an East Asian economic "tiger."

And Taiwan's political future remains clouded. The Nationalists who made the island their base in 1949 established a Republic of China that was authoritarian, but the ROC has evolved into a democratic state. This was not an easy process, but when Taiwan's own democracy movement culminated in 1991, it produced not a massacre, but free elections for a new parliament. Taiwan's leaders now proclaim that the direct and popular election of their president in 1996 was the first such vote not just in Taiwan or the ROC, but in the four-thousand-year history of the Chinese state. Taiwan thus presents itself as an alternative model to the communist-ruled PRC, proving to the world that democracy is compatible with Chinese culture and tradition. Nevertheless, the international community does not recognize Taiwan as an independent state, and a substantial segment of the Taiwanese electorate itself opposes any move toward outright sovereignty. Beijing dictates that Taiwan is off limits to U.S. cabinet members. In 1995, when the United States permitted the ROC's president to visit the American university of which he is a graduate, so that he might attend a class reunion, relations between Washington and Beijing plummeted. In 2000, when Taiwan's voters elected a president who, during the campaign, had proclaimed Taiwan's "equality" with mainland China, Beijing threatened force and massed weaponry across the Taiwan Strait.

The political geography of Taiwan, therefore, is at odds with its economic geography. The ROC stood firm against communism in the 1960s and 1970s; but its long-term ally, the United States, was required to publicly commit itself to a one-China policy, the one being the PRC, the communist giant. Even as goods and capital flow between Taiwan and the mainland, Beijing engages in threatening acts (such as nearby missile tests) when democracy in the ROC takes another step forward.

Today, Taiwan's political options are as limited as its economic horizons are endless. The communist Chinese have made it clear that a move toward independence, based on the likely outcome of any referendum, would lead to intervention. Though militarily well prepared, Taiwan could not long hold off the PRC's enormous armed forces. In any case, independence for Taiwan has philosophical disadvantages for the leaders of the ROC, as it would diminish their claim to legitimacy

---

*The Taiwanese have retained the old Wade-Giles spelling of place names; China now uses the *pinyin* system. Names in parentheses are written according to the pinyin system.

Chinese Rule or Independence for Taiwan?

SHOULD CHINA RULE TAIWAN?

"I AM A graduate student in political geography here at East China Normal University, and I've been listening to your lecture about Chinese-American relations and the disagreement we have over Taiwan. You talk about this as though you have a right to stand between two parts of China, and with the power you have right now, I guess you can. You said that you hoped that mainland Chinese would become more 'reasonable' in their attitude toward the island province, but let me (as my professor would say) erect a hypothetical.

"Imagine that America wakes up one morning and discovers that a band of thugs have flown several 747s, loaded with weapons of all kinds, including those of mass destruction ready for use, into the San Juan, Puerto Rico airport. After taking the airport amid many casualties, they move into the city and enter administration buildings, threatening mass annihilation unless the local government resigns. By evening, Puerto Rico is under their control, American and Puerto Rican flags have been lowered and replaced by the banner of an 'independent' Puerto Rico.

"In Washington, your Congress meets and, realizing that the entire population of the island is at risk, makes the only move it can under the circumstances: it orders your armed forces to assemble a fleet to head for Puerto Rico, to blockade the island so as to create a basis for negotiation and a platform for possible military action. The fleet assembles near Miami and heads eastward through the channel between the Bahamas and Cuba.

"A day before reaching Puerto Rican waters, the path of the hastily assembled American fleet is blocked by a larger fleet from China, which has sailed from its base at Colón near the northern end of the Panama Canal. The commander of the American fleet communicates with his Chinese counterpart. 'We cannot let you pass,' states the Chinese admiral. 'We have not yet determined whether the invaders of Puerto Rico were invited by the people there to liberate them. We may have common ground with them. Please avoid an unnecessary conflict by going back to your bases.'

"How, sir, do you suppose the American people would feel toward China if a version of this scenario took place? They would be outraged at China's interference in a domestic matter. They would order your navy to attack. Well, we in China have had something like this happen with Taiwan. Thugs took it over. American aircraft carriers appeared in the Taiwan Strait. We've 'avoided conflict' for many decades, but that won't last forever. One way or another, Taiwan will submit to Beijing's rule. The Chinese people demand it."

Regional ISSUE

TAIWAN SHOULD BE INDEPENDENT

"Instead of harping constantly on the 'reunification' issue, why don't Beijing's rulers get their own house in order first? Their political system is straight out of the discarded past, they're apparently allergic to democracy, and they can't seem to accept or accommodate any disobedience or dissent. I represent a Taiwanese high-tech corporation and I fly frequently between Taipei and Hong Kong and visit the so-called People's Republic via that route—and I can tell you, the Chinese could learn a lot about democracy, openness, freedom, transparency, and honesty from the Taiwanese. Politically, that 'One Country Two Systems' promise in Hong Kong is eroding as we speak. Economically, the Chinese ought to acknowledge that a lot of the 'Chinese miracle' on the Pacific Rim is actually a 'Taiwanese miracle.' Without our money, skills, discipline, and rigor there wouldn't be much of a miracle, especially if you also subtract what Southeast Asia's Chinese have contributed.

"Of course, Taiwan is a great rallying point for Beijing's autocrats. Gives them a chance to rail against the Western world, against America, against social excesses (we even have freedom of religion in Taiwan, imagine that!), against international conspiracies. But the truth is that the voters of Taiwan long ago ousted the 'thugs' who took us over, forged a thriving economy and a vibrant democracy, and made this island an example of what might have been across the Taiwan Strait.

"Could I say a few words about the 'historic' justification for China's claim to Taiwan? First of all, this island was peopled by Southeast Asian groups long before any Chinese arrived here, so if it's a matter of who got here first, we're last on the list. Beijing didn't integrate Taiwan politically until the 1680s, after groups of hungry citizens started to cross the Strait. That's pretty late in China's vaunted 5000-year history. Taiwan didn't become a province until 1886, which lasted exactly nine years. From 1895 to 1945 Taiwan belonged to Japan as part of the war settlement, and then Beijing took it back only to lose it again four years later to the 'thugs' under Chiang Kai-shek. That makes 13 years as part of the mainland system and 103 years of something else.

"For a country that has expanded and contracted for millennia, has gained and lost territory from Tibet to Taiwan, you would think that China could live with an ethnically compatible, economically successful, politically stable, democratic and independent neighbor. Those stiff bureaucrats in Beijing should have better things to do than threaten force to 'reincorporate' their exemplary neighbor."

as the alternative model for a reunified China. In practical terms, conflict on the Pacific Rim would be disastrous for both sides, whatever course it were to take.

And so Taiwan's future hangs in the balance, a wayward province with national aspirations, a quasi-state with global connections, a test of Chinese capacity for compromise and accommodation (see the Issue Box entitled "Chinese Rule or Independence for Taiwan?"). In this era of devolutionary forces, the fate of Taiwan will be a harbinger of the world order of the twenty-first century.

Scale 1:16 000 000; one inch to 250 miles. Polyconic Projection
Elevations and depressions are given in feet

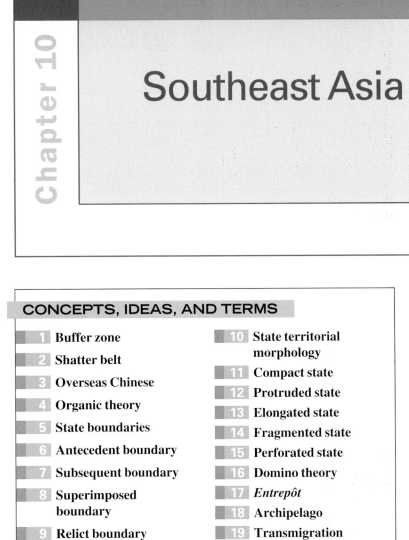

Continued on pages 204-205

Chapter 10

Southeast Asia

CONCEPTS, IDEAS, AND TERMS

1. Buffer zone
2. Shatter belt
3. Overseas Chinese
4. Organic theory
5. State boundaries
6. Antecedent boundary
7. Subsequent boundary
8. Superimposed boundary
9. Relict boundary
10. State territorial morphology
11. Compact state
12. Protruded state
13. Elongated state
14. Fragmented state
15. Perforated state
16. Domino theory
17. *Entrepôt*
18. Archipelago
19. Transmigration

REGIONS

▶ MAINLAND SOUTHEAST ASIA
▶ INSULAR SOUTHEAST ASIA

*S*OUTHEAST ASIA . . . . THE very name roils American emotions. Here the United States owned its only major colony. Here American forces triumphed over Japanese enemies. Here the United States fought the only war it ever lost. Here Washington's worst Cold War fears failed to materialize. Here American companies invested heavily when the Pacific Rim's economic growth transformed dormant economies into potential Pacific tigers. Here American troops and intelligence agents pursue the War on Terror. Southeast Asia, once remote and stagnant, has taken center stage in our globalizing world.

FIGURE 10-1

DEFINING THE REALM

SOUTHEAST ASIA IS a realm of peninsulas and islands, a corner of Asia bounded by India on the northwest and China on the northeast (Fig. 10-1). Its western coasts are washed by the Indian Ocean, and to the east stretches the vast Pacific. From all these directions, Southeast Asia has been penetrated by outside forces. From India came traders; from China, settlers; from across the Indian Ocean, Arabs to engage in commerce and Europeans to build empires; and from across the Pacific, the Americans. Southeast Asia has been the scene of countless contests for power and primacy—the competitors have come from near and far.

Southeast Asia's geography in some ways resembles that of Eastern Europe. It is a mosaic of smaller countries on the periphery of one of the world's largest states. It has been a **buffer zone** between powerful adversaries. It is a **shatter belt** in which stresses and pressures from without and within have fractured the political geography. Like Eastern Europe, Southeast Asia exhibits great cultural diversity. It is a realm of hundreds of cultures, numerous languages and dialects, and several major religions.

MAJOR GEOGRAPHIC QUALITIES OF

Southeast Asia

1. Southeast Asia extends from the peninsular mainland to the archipelagos offshore. Because Indonesia controls part of New Guinea, its functional region reaches into the neighboring Pacific geographic realm.

2. Southeast Asia, like Eastern Europe, has been a shatter belt between powerful adversaries and has a fractured cultural and political geography shaped by foreign intervention.

3. Southeast Asia's physiography is dominated by high relief, crustal instability marked by volcanic activity and earthquakes, and tropical climates.

4. A majority of Southeast Asia's more than half-billion people live on the islands of just two countries: Indonesia, with the world's fourth-largest population, and the Philippines. The rate of population increase in the insular region of Southeast Asia exceeds that of the mainland.

5. Although the overwhelming majority of Southeast Asians have the same ancestry, cultural divisions and local traditions abound, which the realm's divisive physiography sustains.

6. The legacies of powerful foreign influences, Asian as well as non-Asian, continue to affect the cultural landscapes of Southeast Asia.

7. Southeast Asia's political geography exhibits a variety of boundary types and several categories of state territorial morphology.

8. The Mekong River, Southeast Asia's Danube, has its source in China and borders or crosses five Southeast Asian countries, sustaining tens of millions of farmers, fishing people, and boat owners.

9. The realm's giant in terms of territory as well as population, Indonesia, has not asserted itself as the dominant state because of mismanagement and corruption; but Indonesia has enormous potential.

LAND AND SEA BORDERS

The Southeast Asian realm borders South Asia and East Asia on land and the Pacific and Austral realms at sea (Fig. 10-2). Note that one political entity that forms part of it, Indonesia, extends beyond the realm's borders. The easternmost province of Indonesia is the western half of the Pacific island of New Guinea, where indigenous cultures are Papuan, not Indonesian. Today Indonesia rules what is in effect a Pacific island colony called West Papua. We discuss this territory as part of our assessment of Indonesia, but we will also consider all of New Guinea under the Pacific Realm in Chapter 12.

Because the politico-geographical map (Fig. 10-2) is so complicated, it should be studied attentively. One good way to strengthen your mental map of this realm is to follow the mainland coastline from west to east. The westernmost state in the realm is Myanmar (called Burma before 1989 and still referred to by that name), the only country in Southeast Asia that borders both India and China. Myanmar shares the "neck" of the Malay Peninsula with Thailand, heart of the *mainland* region. The south of the peninsula is part of Malaysia—except for Singapore, at the very tip of it. Facing the Gulf of Thailand is Cambodia. Still moving generally eastward, we reach Vietnam, a strip of land that extends all the way to the Chinese border. And surrounded by its neighbors is landlocked Laos, remote and isolated. This leaves the islands that constitute *insular* Southeast Asia: the Philippines in the north and Indonesia in the south, and between them the offshore portion of Malaysia, situated on the largely Indonesian island of Borneo. Also on Borneo lies the ministate of Brunei, small but, as we will see, important in the regional picture. And finally, a new state is just appearing on the map in the realm's southeastern corner: East Timor, a former Portuguese colony annexed by Indonesia in 1976 and released to United Nations supervision in 1999. After a transition period, East Timor became an independent state in 2002.

FIGURE 10-2

No dominant state has arisen among the 11 countries of Southeast Asia. The largest country in terms of population as well as territory, Indonesia, has as-yet unrealized potential to overshadow the rest, but not in the foreseeable future. Neither did any single, dominant core of indigenous culture develop here as it did in East Asia. In the river basins and on the plains of the mainland, as well as on the islands offshore, a flowering of cultures produced a diversity of societies whose languages, religions, arts, music, foods, and other achievements formed an almost infinitely varied mosaic—but none of those cultures rose to imperial power. The European colonizers forged empires here, often by playing one state off against another; the Europeans divided and ruled. Out of this foreign intervention came the modern map of Southeast

Asia; only Thailand (formerly Siam) survived the colonial era as an independent entity. Thailand was useful to two competing powers, the French to the east and the British to the west. It was a convenient buffer, and while the colonists carved pieces off Thailand's domain, the kingdom endured.

Indeed, the Europeans accomplished what local powers could not: the formation of comparatively large, multicultural states that encompassed diverse peoples and societies and welded them together. Were it not for the colonial intervention, it is unlikely that the 17,000 islands of far-flung Indonesia would today constitute the world's fourth largest country in terms of population. Nor would the nine sultanates of Malaysia have been united, let alone with the peoples of northern Borneo across the South China Sea.

For good or ill, the colonial intrusion consolidated a realm of few culture cores and numerous ministates into less than a dozen countries.

PHYSICAL GEOGRAPHY

As Figure 10-1 shows, Southeast Asia is a realm in which high relief dominates the physiography. From the Arakan Mountains in western Myanmar (Burma) to the glaciers (yes, glaciers!) of the Indonesian part of New Guinea, elevations rise above 10,000 feet (3300 m) in many places. In Myanmar, Mount Victoria is such a peak; in northern Laos, Phu Bia comes close; on the Indonesian island of Sumatera (Sumatra under the old spelling) the tallest volcano is Kerinci; on Jawa (Java) Mounts Slamet and Semeru tower over the countryside; and on Borneo Mount Kinabalu, not a volcano but an erosional remnant, is higher than any other mountain in this realm. In the Philippines, Mount Pinatubo at 4874 feet (1477 m) is not especially high, but this volcano's eruptions have had major impact on the country and, in 1991, on global climate.

The relief map reminds us that this is not only the Pacific Rim but also the Pacific Ring of Fire, where the crust is unstable, earthquakes are common, and volcanoes are active. Among the islands, Borneo is the sole exception. Borneo is a slab of ancient crust, pushed high above sea level by tectonic forces and eroded into its present mountainous topography.

As Figure 10-1 underscores, rivers rise in the highland backbones of the islands and peninsulas, and deposit their sediments as they wind their way toward the coast; the physiography of Sumatera demonstrates this unmistakably. The volcanic hills, plateaus, and better-drained lowlands are fertile and, in the warmth of tropical climates, can yield multiple crops of rice.

On the peninsular mainland we see a pattern that is already familiar: rivers rising in the Asian interior that create alluvial plains and deltas. The Mekong River is the Chang/Yangzi of Southeast Asia: you can trace it all the way from China via Laos, Thailand, and Cambodia into southern Vietnam, where it forms a massive and populous delta. In the west, Myanmar's key river is the Irrawaddy; Thailand's is the Chao Phraya. In the north, the Red River Basin is the breadbasket of northern Vietnam.

No survey of the physical geography of Southeast Asia would be complete without reference to the realm's seas, gulfs, straits, and bays. Irregular and indented coastlines such as these, with thousands of islands near and far, create difficult problems when it comes to drawing maritime boundaries (the islands also form havens for rebels and criminals). As we note later, Southeast Asia has one of the most complex maritime boundary frameworks in the world, and it is rife with potential for conflict.

POPULATION GEOGRAPHY

Compared to the huge population numbers and densities in the habitable regions of South Asia and China, demographic totals for the countries of Southeast Asia, with the exception of Indonesia, seem modest. Again, comparisons with Europe come to mind. Three countries—Thailand, the Philippines, and Vietnam—have populations between 60 and 85 million. Laos, quite a large country territorially (comparable to the United Kingdom), had just 5.8 million inhabitants in 2004. Cambodia, substantially larger than Greece, had 12.7 million.

As Table G-1 shows, arithmetic population densities are not especially high in Southeast Asia except in the urbanized ministate of Singapore. Similarly, physiologic densities are lower than in eastern China or neighboring South Asian countries. When the political situation is stable, several Southeast Asian countries are able to export large quantities of rice. Thailand and, in recent years, Vietnam have ranked among the world's leading exporters in this crucial staple of Asian diets.

Viewed in spatial perspective, one can discern both similarities and differences in the population patterns of Southeast Asia and its giant neighbors. As in East and South Asia, population clusters in the large basins and deltas of major rivers, where densities are high and the countryside is parceled into paddies as far as the eye can see. Numerous small, space-conserving villages dot the landscape in such lowlands as the Red and Mekong basins of Vietnam, a scene reminiscent of South China. But between these alluvial lowlands lies higher ground where the landscape takes on savanna characteristics with reddish, leached, less productive soils, scattered villages, and comparatively sparse population. And in the islands of Indonesia and the Philippines, Southeast Asia has environments that are not present at all in India or China—well-watered volcanic soils supporting luxuriant natural vegetation and capable of sustaining intensive agriculture. Whole countrysides have been meticulously terraced to create paddyfields and irrigate them.

People and Land

Of Southeast Asia's 552 million inhabitants, well over half (308 million, 56 percent) live on the islands of Indonesia and the Philippines, leaving the realm's mainland countries with just 44 percent of the population. With such high population pressure in adjacent realms, why has Southeast Asia not been flooded by waves of immigrants?

In fact, Southeast Asia has received its share of immigrants, but overland invasions have been comparatively limited. Several factors have hindered travel along overland routes into Southeast Asia. First, physical obstacles inhibit such movement. In Chapter 8, we noted the barrier effect of the densely forested hills and mountains along the border between northeastern India and northwestern Myanmar (Burma). North of Myanmar lies forbidding Xizang (Tibet), and northeast of Myanmar and north of Laos is the high, rugged Yunnan Plateau. Transit is easier in the east, between southeastern China and northern Vietnam, and this has indeed been an avenue for contact and migration. As the ethnolinguistic map of China (p. 441) shows, migration from Southeast Asia into China also has occurred in this area.

Second—as reflected in the population distribution map (Fig. G-9)—the highly clustered agricultural opportunities in Southeast Asia lie separated by large stretches of less productive land, and these in turn are crossed by high-relief zones that hinder movement. The map of Southeast Asia reveals several concentrated areas of productive capacity but few corridors that would facilitate migration.

As we noted earlier, immigrants to Southeast Asia over the past millennium have come mainly by sea, not by land. That is true not only for Arabs and Europeans but also for neighboring Chinese. Mostly the Chinese immigrants came not as farmers seeking land but as traders and workers looking for opportunities in the cities and towns. "Chinatowns" form a vital part of most Southeast Asian cities today, especially the coastal ones.

The Ethnic Mosaic

Southeast Asia's peoples come from a common stock just as (Caucasian) Europeans do, but this has not prevented the emergence of regionally or locally discrete ethnic or cultural groups. Figure 10-3 displays the broad distribution of ethnolinguistic groups in the realm, but be aware that this is a generalization. At the scale of this map, numerous small groups cannot be depicted.

The map shows the rough spatial coincidence, on the mainland, between major ethnic group and modern political state. The Burman dominate in the country once known as Burma (now Myanmar); the Thai occupy the state once known as Siam (Thailand); the Khmer form the nation of Cambodia and extend northward into Laos; and the Vietnamese inhabit the long strip of territory facing the South China Sea.

Territorially, by far the largest population is classified in Figure 10-3 as Indonesian, the inhabitants of the great archipelago that extends from Sumatera* west of the Malay Peninsula to the Malukus (Moluccas) in the east and from the lesser Sunda Islands in the south to the Philippines in the north. Collectively, all these peoples—

---

*As in Africa, names and spellings have changed with independence. In this chapter, we will use the contemporary spellings, except when we refer to the colonial period. Thus Indonesia's four major islands are Jawa, Sumatera, Kalimantan (the Indonesian part of Borneo), and Sulawesi. The Dutch called them Java, Sumatra, Dutch Borneo, and Celebes, respectively.

FROM THE FIELD NOTES

"It was a relatively cool day in Hanoi, which was just as well because the drive from Haiphong had taken nearly four hours, much of the time waiting behind oxcarts, and the countryside was hot and humid. On my way to an office along Hang Bai Street I ran into this group of youngsters and their teacher on their way to the Ho Chi Minh memorial. He right away realized the level of my broken French, and explained patiently that he taught history, and took his class on a field trip several times a year. 'They get all dressed up and it is a big moment for them,' he said, 'and there is a lot of history to be visited in Hanoi.' The kids had seen me take this picture and now they crowded around. I told them that I noticed that rouge seemed to be a favorite color in their outfits. 'But that should not surprise you, monsieur,' said the teacher, smiling. 'This is Vietnam, communist Vietnam, and red is the color that rallies us!' In the city at the opposite end of the country named after Ho Chi Minh (founder of his country's Communist Party, president from 1945 to 1969, victor over the French colonizers, and leader in the war against the United States), my experiences with teachers and students had been quite different, to the point that most called their city Saigon. But here in Hanoi the ideological flame still burned brightly."

the Filipinos, Malays, and Indonesians—shown in Figure 10-3 are known as Indonesians, but they have been divided by history and politics. Note, on the map, that the Indonesians in Indonesia itself include Javanese, Madurans, Sundanese, Balinese, and other large groups; hundreds of smaller ones are not shown. In the Philippines, too, island isolation and contrasting ways of life are reflected in

ETHNIC MOSAIC OF SOUTHEAST ASIA

INDO–ARYAN
Hindustani

CHINESE

THAI
Siamese, Lao,
Shan, Chuan,
Upper Thai, Nung

MIAO–YAO

TIBETO–BURMAN
Lisu, Burman, Chin,
Kachin, Karen

VIETNAMESE

MON–KHMER
Khmer, Mon

INDONESIAN
Malay, Indonesian,
Filipino

PAPUAN

FIGURE 10-3

the cultural mosaic. Also part of this Indonesian ethnic-cultural complex are the Malays, whose heartland lies on the Malay Peninsula but who form minorities in other areas as well. Like most Indonesians, the Malays are Muslims, but Islam is a more powerful force in Malay society than, in general, in Indonesian culture.

Figure 10-3 also reminds us that (again like Eastern Europe) Southeast Asia has many ethnic minorities. On the Malay Peninsula, note the South Asian (Hindustani) cluster; Hindu communities with Indian ancestries exist in many parts of the peninsula, but here in the southwest they form the majority in a small area. Nearby lies the largest single exception to the rule that Chinese minorities in Southeast Asia are urban-based: along the west coast of the Malay Peninsula, Chinese are farmers and villagers as well as city dwellers. In the north of the realm, minorities share the lands in which the Burman, Thai, and Vietnamese dominate. Those minorities, as a comparison between Figures 10-2 and 10-3 proves, tend to occupy zones peripheral to the regional core areas, where forests often are dense and difficult to penetrate, where isolation from the power cores prevails, where remoteness promotes detachment from the national state, and where, time and again, ethnic conflicts have raged as the core-based majority tried to control the frontier. During the war in Vietnam, peoples in the highlands (the Montagnards) tended to be sympathetic to the American campaign because of their longstanding hostility to the dominant (lowland) Vietnamese. Today, the Shan and the Karen in Myanmar are in a state of conflict with the regime based in the core area.

HOW THE POLITICAL MAP EVOLVED

The leading colonial competitors in Southeast Asia were the Dutch, French, British, and Spanish (with the last replaced by the Americans in their stronghold, the Philippines). The Japanese had colonial objectives here as well, but these came and went during the course of World War II.

The Dutch acquired the greatest prize: control over the vast archipelago now called Indonesia (formerly the Netherlands East Indies). France established itself on the eastern flank of the mainland, controlling all territory east of Thailand and south of China. The British conquered the Malay Peninsula, gained power over the northern part of the island of Borneo, and established themselves in Burma as well. Other colonial powers also gained footholds but not for long. The exception was Portugal, which held on to its eastern half of the island of Timor (Indonesia) until after the Dutch had been ousted from their East Indies.

Figure 10-4 shows the colonial framework in the late nineteenth century, before the United States assumed control over the Philippines. Note that while Thailand survived as an independent state, it lost territory to the British in Malaya and Burma and to the French in Cambodia and Laos.

The Colonial Imprint

The colonial powers divided their possessions into administrative units as they did in Africa and elsewhere. Some of these political entities became independent states when the colonial powers withdrew. France, one of the mainland's leading colonial powers, divided its Southeast Asian empire into five units. Three of these units lay along the east coast: Tonkin in the north next to China, centered on the basin of the Red River; Cochin China in the south, with the Mekong Delta as its focus; and between these two, Annam. The other two French territories were Cambodia, facing the Gulf of Thailand, and Laos, landlocked in the interior. Out of these five French dependencies there emerged the three states of Indochina. The three east coast territories ultimately became one state, Vietnam; the other two (Cambodia and Laos) each achieved separate independence.

The British ruled two major entities in Southeast Asia (Burma and Malaya) in addition to a large part of northern Borneo and many small islands in the South China Sea. Burma was attached to Britain's Indian empire; from 1886 until 1937 it was governed from distant New Delhi. But when British India became independent in 1947 and split into several countries, Burma was not part of the grand design that created West and East Pakistan (the latter now Bangladesh), Ceylon (now Sri Lanka), and India. Instead, in 1948 Burma (now Myanmar) was given the status of a sovereign republic.

In Malaya, the British developed a complicated system of colonies and protectorates that eventually gave rise to the equally complex, far-flung Malaysian Federation. Included were the former Straits Settlements (Singapore was one of these colonies), the nine protectorates on the Malay Peninsula (former sultanates of the Muslim era), the British dependencies of Sarawak and Sabah on the island of Borneo, and numerous islands in the Strait of Malacca and the South China Sea. The original Federation of Malaysia was created in 1963 by the political unification of recently independent mainland Malaya, Singapore, and the former British dependencies on the largely Indonesian island of Borneo. Singapore, however, left the Federation in 1965 to become a sovereign city-state, and the remaining units were later restructured into peninsular Malaysia and, on Borneo, Sarawak and Sabah. Thus the term *Malaya* properly refers to the geographic area of the Malay Peninsula, including Singapore and other nearby islands; the term *Malaysia* identifies the politico-geographical entity of which Kuala Lumpur is the capital city.

The Hollanders took control of the "spice islands" through their Dutch East India Company, and the wealth that they extracted from what is today Indonesia brought the Netherlands its Golden Age. From the mid-seventeenth to the late-eighteenth century, the Dutch could develop their East Indies sphere of influence almost without challenge, for the British and French were preoccupied with the Indian subcontinent. By playing the princes of Indonesia's states against one another in the search for economic concessions and political influence, by placing the Chinese in positions of responsibility, and by imposing systems of

FIGURE 10-4

forced labor in areas directly under its control, the Company had a ruinous effect on the Indonesian societies it subjugated. Java (Jawa), the most populous and productive island, became the focus of Dutch administration; from its capital at Batavia (now Jakarta), the Company extended its sphere of influence into Sumatra (Sumatera), Dutch

Borneo (Kalimantan), Celebes (Sulawesi), and the smaller islands of the East Indies. This was not accomplished overnight, and the struggle for territorial control was carried on long after the Dutch East India Company had yielded its administration to the Netherlands government. Dutch colonialism thus threw a girdle around Indonesia's more than 17,000 islands, paving the way for the creation of the realm's largest and most populous nation-state (224 million today).

In the colonial tutelage of Southeast Asia, the Philippines, long under Spanish domination, had a unique experience. As early as 1571, the islands north of Indonesia were under Spain's control (they were named for Spain's King Philip II). Spanish rule began when Islam was reaching the southern Philippines via northern Borneo. The Spaniards spread their Roman Catholic faith with great zeal, and between them the soldiers and priests consolidated Hispanic dominance over the mostly Malay population. Manila, founded in 1571, became a profitable waystation on the route between southern China and western Mexico (Acapulco usually was the trans-Pacific destination for the galleons leaving Manila's port). There was much profit to be made, but the indigenous people shared little in it. Great landholdings were awarded to loyal Spanish civil servants and to men of the church. Oppression eventually yielded revolution, and Spain was confronted with a major uprising when the Spanish-American War broke out elsewhere in 1898.

As part of the settlement of that war, the United States replaced Spain in Manila. That was not the end of the revolution, however. The Filipinos now took up arms against their new foreign ruler, and not until 1905, after terrible losses of life, did American forces manage to "pacify" their new dominion. Subsequently, U.S. administration in the Philippines was more progressive than Spain's had been. In 1934, Congress passed the Philippine Independence Law, providing for a ten-year transition to sovereignty. But before independence could be arranged, World War II intervened. In 1941, Japan conquered the islands, temporarily ousting the Americans; U.S. forces returned in 1944 and, with strong Filipino support, defeated the Japanese in 1945. The agenda for independence was resumed, and in 1946 the sovereign Republic of the Philippines was proclaimed.

Today, all of Southeast Asia's states are independent, but centuries of colonial rule have left strong cultural imprints. In their urban landscapes, their education systems, their civil service, and countless other ways, this realm still carries the marks of its colonial past.

Cultural-Geographic Legacies

The French, who ruled and exploited a crucial quadrant of Southeast Asia, had a name for their empire: *Indochina*. That name would be appropriate for the rest of the realm as well because it suggests its two leading Asian influences for the past 2000 years. Periodic expansions of the Chinese Empire, notably along the eastern periphery, as well as the arrival of large numbers of Chinese settlers (see box titled "Overseas Chinese"), infused the region with cultural norms from the north. The Indians came from the west by way of the sea, as their trading ships plied the coasts, and settlers from India founded colonies on Southeast Asian shores in the Malay Peninsula, on the lower Mekong Plain, on Jawa and Bali, and on Borneo.

With the migrants from the Indian subcontinent came their faiths: first Hinduism and Buddhism, later Islam. The Muslim religion, promoted by the growing number of Arab traders who appeared on the scene, became the dominant religion in Indonesia (where nearly 90 percent of the population adheres to Islam today). But in Myanmar, Thailand, and Cambodia, Buddhism remained supreme, and in all three countries the overwhelming majority of the people are now adherents. In culturally diverse Malaysia, the Malays are Muslims (to be a Malay *is* to be a Muslim), and almost all Chinese are Buddhists; but most Malaysians of Indian ancestry remain Hindus. Although Southeast Asia has generated its own local cultural expressions, most of what remains in tangible form has resulted from the infusion of foreign elements. For instance, the main temple at Angkor Wat, constructed in Cambodia during the twelfth century, remains a monument to the Indian architecture of that time.

The *Indo* part of Indochina, then, refers to the cultural imprints from South Asia: the Hindu presence, the importance of Buddhism (which came to Southeast Asia via Sri Lanka [Ceylon] and its seafaring merchants), the influences of Indian architecture and art (especially sculpture), writing and literature, and social structures and patterns.

The *China* in the name Indochina signifies the role of the Chinese here. Chinese emperors coveted Southeast Asian lands, and China's power reached deep into the realm. Social and political upheavals in China sent millions of Sinicized people southward. Chinese traders, pilgrims, seafarers, fishermen, and others sailed from southeastern China to the coasts of Southeast Asia and established settlements there. Over time, those settlements attracted more Chinese emigrants, and Chinese influence in the realm grew (Fig. 10-5). Not surprisingly, relations between the Chinese settlers and the earlier inhabitants of Southeast Asia have at times been strained, even violent. The Chinese presence in Southeast Asia is long-term, but the invasion has continued into modern times. The economic power of Chinese minorities and their role in the political life of the realm have led to conflicts.

The Chinese initially profited from the arrival of the Europeans, who stimulated the growth of agriculture, trade, and industries; here the Chinese found opportunities they lacked at home. They harvested rubber, found jobs on the docks and in the mines, cleared the bush, and transported goods in their sampans. They brought useful

Overseas Chinese

*S*outheastern China has been the source of Chinese emigrants to Southeast Asia for as long as 2000 years, but the largest movements have occurred over the past six centuries (Fig. 10-5). During the initially outward-looking Ming Dynasty (1368–1644) and early in the following Manchu (Qing) Dynasty, small groups of Chinese emigrants, sometimes pushed by famine at home, sailed southward to establish small communities in Southeast Asian towns. During the early colonial period (1670–1870), growing commercial opportunities formed the pull factors that attracted greater numbers. But the largest exodus came during the late colonial period (1870–1940) when Southeast Asia's expanding job and trade opportunities attracted more than 20 million Chinese, many from Fujian and Guangdong provinces. During the communist period, additional migrants illegally left China, many to or via Hong Kong.

Toward the end of their rule, colonial regimes began to restrict Chinese immigration, as did independent Thailand. The Chinese had become an influential class of middlemen and controlled ever more commerce, and colonial rulers as well as local peoples feared Chinese influence and "expansionism." During World War II and subsequent decolonization, Chinese were persecuted in Southeast Asia, especially in Malaya while under Japanese control, and in Indonesia in the 1960s when Chinese were targeted as communist sympathizers.

Not counting intermarriages, Southeast Asia today may contain as many as 30 million Chinese (more than half the world total of **Overseas Chinese**), and many have become very successful. When China began its new economic policy, Overseas Chinese invested heavily in the SEZs and Open Cities, playing a large role in stimulating the economic growth experienced there.

But success entails risks. Resentment against Chinese prosperity and business practices led to discriminatory legislation in Malaysia. When Indonesia's economy collapsed in 1998, police stood by as gangs looted, burned, murdered, and raped at will in Chinese commercial districts. Wealthier Chinese fled, exacerbating Indonesia's economic crisis.

FIGURE 10-5

"Like most major Southeast Asian cities, Bangkok's urban area includes a large and prosperous Chinese sector. No less than 14 percent of Thailand's population of 63.6 million is of Chinese ancestry, and the great majority of Chinese live in the cities. In Thailand, this large non-Thai population is well integrated into local society, and intermarriage is common. Still, Bangkok's 'Chinatown' is a distinct and discrete part of the great city. There is no mistaking Chinatown's limits: Thai commercial signs change to Chinese, goods offered for sale also change (Chinatown contains a large cluster of shops selling gold, for example), and the urban atmosphere, from street markets to bookshops, is dominantly Chinese. This is a boisterous, noisy, energetic part of multicultural Bangkok, a vivid reminder of the Chinese commercial success in Southeast Asia."

skills with them, and as tailors, shoemakers, blacksmiths, and fishermen, they prospered. The Chinese were also astute in business; soon, they not only dominated the region's retail trade but also held prominent positions in banking, industry, and shipping. Thus they have always been far more important than their modest numbers in Southeast Asia would suggest. The Europeans used them for their own designs but at times found the Chinese to be stubborn competitors—so much so that eventually they tried to impose restrictions on Chinese immigration. When the United States took control of the Philippines, it also sought to stop the influx of Chinese into those islands.

When the European colonial powers withdrew and Southeast Asia's independent states emerged, Chinese population sectors ranged from nearly 50 percent of the total in Malaysia (in 1963) to barely over 1 percent in Myanmar. In Singapore today, Chinese constitute 77 percent of the population of 4.3 million; when Singapore seceded from Malaysia in 1965, the Chinese component in Malaysia was substantially reduced (it is about 25 percent today). In Indonesia, the percentage of Chinese in the total population is not high (no more than 3 percent), but the Indonesian population is so large that even this small percentage indicates a Chinese sector of more than 6 million. In 1998, this tiny minority reportedly controlled 70 percent of Indonesia's economy. In Thailand, on the other hand, many Chinese have married Thais, and the Chinese minority of about 14 percent has become a cornerstone of Thai society, dominant in trade and commerce.

In general, Southeast Asia's Chinese communities remained aloof and formed their own separate societies in the cities and towns. They kept their culture and language alive through social clubs, schools, and even residential suburbs that, in practice if not by law, were Chinese in character. At one time they were caught in the middle between the Europeans and Southeast Asians when the local people were hostile toward both white people and the Chinese. Since the withdrawal of the Europeans, however, the Chinese have become the main target of this antagonism, which remained strong because of the Chinese involvement in money lending, banking, and trade monopolies (see the Issue Box entitled "The Chinese Presence in Southeast Asia"). Moreover, there is the specter of an imagined or real Chinese political imperialism along Southeast Asia's northern flanks.

The *China* in Indochina, therefore, represents a diversity of penetrations. Southern China was the source of most of the old invasions, and Chinese territorial consolidation provided the impetus for successive immigrants. Mongoloid racial features carried southward from East Asia mixed with the preexisting Malay stock to produce a transition from Chinese-like people in the northern mainland to darker-skinned Malay types in the distant Indonesian east. Although Indian cultural influences remained strong, people throughout Southeast Asia adopted Chinese modes of dress, plastic arts, types of houses and boats, and other cultural attributes. During the past century, and especially during the last half-century, renewed Chinese immigration brought skills and energies that propelled these minorities to comparative wealth and influence.

The Chinese Presence in Southeast Asia

THE CHINESE ARE TOO INFLUENTIAL!

"IT'S HARD TO imagine that there was a time when we didn't have Chinese minorities in our midst. I think I understand how Latin Americans (sic) feel about their "giant to the north." We've got one too, but the difference is that there are a lot more Chinese in Southeast Asia than there are Americans in Latin America. The Chinese even run one country because they are the majority there, Singapore. And if you want to go to a place where you can see what the Chinese would have in mind for this whole region, go there. They're knocking down all the old Malay and Hindu quarters, and they've got more rules and laws than we here in Indonesia could even think of. I'm a doctor here in Bandung and I admire their modernity, but I don't like their philosophy.

"We've had our problems with the Chinese here. The Dutch colonists brought them in to work for them, they got privileges that made them rich, they joined the Christian churches the Europeans built. I'm not sure that a single one of them ever joined a mosque. And then shortly after Sukarno led us to independence they even tried to collaborate with Mao's communists to take over the country. They failed and many of them were killed, but look at our towns and villages today. The richest merchants and the money lenders tend to be Chinese. And they stay aloof from the rest of us.

"We're not the only ones who had trouble with the Chinese. Ask them about it in Malaysia. There the Chinese started a full-scale revolution in the 1950s that took the combined efforts of the British and the Malays to put down. Or Vietnam, where the Chinese weren't any help against the enemy when the war happened there. Meanwhile they get richer and richer, but you'll see that they never forget where they came from. The Chinese in China boast about their coastal economy, but it's coastal because that's where our Chinese came from and where they sent their money when the opportunity arose.

"As a matter of fact, I don't think that China itself cares much for or about Southeast Asia. Have you heard what's happening in Yunnan Province? They're building a series of dams on the Mekong River, our major river, just across the border from Laos and Burma. They talk about the benefits and they offer to destroy the gorges downstream to facilitate navigation, but they won't join the Mekong River Commission. When the annual floods cease, what will happen to Tonle Sap, to fish migration, to seasonal mudflat farming? The Chinese do what they want—they're the ones with the power and the money."

Regional **ISSUE**

THE CHINESE ARE INDISPENSABLE!

"Minorities, especially successful minorities, have a hard time these days. The media portray them as exploiters who take advantage of the less fortunate in society, and blame them when things go wrong, even when it's clear that the government representing the majority is at fault. We Chinese arrived here long before the Europeans did and well before some other 'indigenous' groups showed up. True, our ancestors seized the opportunity when the European colonizers introduced their commercial economy, but we weren't the only ones to whom that opportunity was available. We banded together, helped each other, saved and shared our money, and established stable and productive communities. Which, by the way, employed millions of locals. I know: my family has owned this shop in Bangkok for five generations. It started as a shed. Now it's a six-floor department store. My family came from Fujian, and if you added it all up we've probably sent more than a million U.S. dollars back home. We're still in touch with our extended family and we've invested in the Xiamen SEZ.

"Here in Thailand we Chinese have done very well. It is sometimes said that we Chinese remain aloof from local society, but that depends on the nature of that society. Thailand has a distinguished history and a rich culture, and we see the Thais as equals. Forgive me, but you can't expect the same relationship in East Malaysia. Or, for that matter, in Indonesia, where we are resented because the 3 percent Chinese run about 60 percent of commerce and trade. But we're always prepared to accommodate and adjust. Look at Malaysia, where some of our misguided ancestors started a rebellion but where we're now appreciated as developers of high-tech industries, professionals, and ordinary workers. This in a Muslim country where Chinese are Christians or followers of their traditional religions, and where Chinese regularly vote for the Malay-dominated majority party. So when it comes to aloofness, it's not us.

"The truth is that the Chinese have made great contributions here, and that without the Chinese this place would resemble South or Southwest Asia. Certainly there would be no Singapore, the richest and most stable of all Southeast Asian countries, where minorities live in peace and security and where incomes are higher than anywhere else. In fact, mainland Chinese officials come to Singapore to learn how so much was achieved there. Imagine a Southeast Asia without Singapore! It's no coincidence that the most Chinese country in Southeast Asia also has the highest standard of living."

SOUTHEAST ASIA'S POLITICAL GEOGRAPHY

Southeast Asia's diverse and fractured natural landscapes, as well as varied cultural influences, have combined to create a complex political geography. As we noted earlier, political geography focuses on the spatial expressions of political behavior; we were introduced to several aspects of it when we examined Western Europe's supranationalism, Eastern Europe's balkanization, Pakistan's irredentism toward Kashmiri Muslims, and the geopolitics of eastern Eurasia. Many political geographers have studied the fundamental causes of the cyclical rise and decline of states; their answers have ranged from environmental changes to biological forces. Friedrich Ratzel (1844–1904) conceptualized the state as a biological organism whose life, from birth through expansion and maturation to eventual senility and collapse, mirrors that of any living thing. Ratzel's **organic theory** of state development held that nations, being aggregates of human beings, would over the long term live and die as their citizens did.

Other political geographers sought to measure the strength of the forces that bind states (centripetal forces) and that divide them (centrifugal forces), and thus to assess a state's chances to survive separatism of the kind Canada and Sri Lanka are experiencing today, to name just two examples. Even in today's era of modern warfare, a stretch of water still affects the course of events. Taiwan would not be what it is without the 100 miles (160 km) of water that separates it from mainland China. Singapore's secession from (then) Malaya was facilitated because Singapore is an island; had it been attached to the Malay Peninsula, the centrifugal forces of separatism might not have prevailed. The role of physical geography in political events remains powerful.

In this section we pay particular attention to the boundaries and territories of the component states of Southeast Asia because this realm's political geography displays virtually every possible attribute of both. We focus first on the borders and then on the states' territorial shapes.

The Boundaries

Boundaries are sensitive parts of a state's anatomy: just as people are territorial about their individual properties, so nations and states are sensitive about their territories and borders. The saying that "good fences make good neighbors" certainly applies to states, but, as we know, the boundaries between states are not always "good fences."

Boundaries, in effect, are contracts between states. That contract takes the form of a treaty that contains the *definition* of the boundary in the form of elaborate description. Next, cartographers perform the *delimitation* of the treaty language, drawing the boundary on official, large-scale maps. Throughout human history, states have used those maps to build fences, walls, or other barriers in a process called *demarcation*.

Once established, we can classify boundaries geographically. Some are sinuous, conforming to rivers or mountain crests (*physiographic*) or coinciding with breaks or transitions in the cultural landscape (*anthropogeographic*). As any world political map shows, many boundaries are simply straight lines, delimited without reference to physical or cultural features. These *geometric* boundaries can produce problems when the cultural landscape changes where they exist.

In general, the boundaries of Southeast Asia were better defined than those of several other postcolonial areas of the world, notably Africa, the Arabian Peninsula, and Turkestan. The colonial powers that established the original treaties tried to define boundaries to lie in remote and/or sparsely peopled areas, for example, across interior Borneo. Nevertheless, certain Southeast Asian boundaries have produced problems, among them the geometric boundary between West Papua, the portion of New Guinea ruled by Indonesia, and Papua New Guinea, the eastern component of the island.

Even on a small-scale map of the kind we use in this chapter, we can categorize the boundaries of this realm. A comparison between Figures 10-2 and 10-3 reveals that the boundary between Thailand and Myanmar over long segments is anthropogeographic, notably where the name *Karen*, the Myanmar minority, appears on Figure 10-3. Figure 10-1 shows that a large segment of the Vietnam–Laos boundary is physiographic-political, coinciding with the Annamese Cordillera (Mountains).

Boundaries also can be classified genetically, that is, as their evolution relates to the cultural landscapes they traverse. A leading political geographer, Richard Hartshorne (1899–1992), proposed a four-level *genetic boundary classification*. All four of these boundary types can be observed in Southeast Asia.

Certain boundaries, Hartshorne reasoned, were defined and delimited before the present-day human landscape developed. In Figure 10-6 (upper-left map), the boundary between Malaysia and Indonesia on the island of Borneo is an example of the first boundary type, the **antecedent boundary**. Most of this border passes through sparsely inhabited tropical rainforest, and the break in settlement can even be detected on the small-scale world population map (Fig. G-9).

FROM THE FIELD NOTES

"I stood on the Laotian side of the great Mekong River which, during the dry season, did not look so great! On the opposite side was Thailand, and it was rather easy for people to cross here at this time of the year. But, the locals told me, it is quite another story in the wet season. Then the river inundates the rocks and banks you see here, it rushes past, and makes crossing difficult and even dangerous. The buildings where the canoes are docked are built on floats, and rise and fall with the seasons. The physiographic-political boundary between Thailand and Laos lies in the middle of the valley we see here."

GENETIC POLITICAL BOUNDARY TYPES

FIGURE 10-6

A second category of boundaries evolved as the cultural landscape of an area took shape, part of the ongoing process of accommodation. These **subsequent boundaries** are represented in Southeast Asia by the map in the upper right of Figure 10-6, which shows in some detail the border between Vietnam and China. This border is the result of a long process of adjustment and modification, the end of which may not yet have come.

The third category involves boundaries drawn forcibly across a unified or at least homogeneous cultural landscape. The colonial powers did this when they divided the island of New Guinea by delimiting a boundary in a nearly straight line (curved in only one place to accommodate a bend in the Fly River), as shown in the lower-left map of Figure 10-6. The **superimposed boundary** they delimited gave the Netherlands the western half of New Guinea. When Indonesia became independent in 1949, the Dutch did not yield their part of New Guinea, which is peopled mostly by ethnic Papuans, not Indonesians. In 1962 the Indonesians invaded the territory by force of arms, and in 1969 the United Nations recognized its authority there. This made the colonial, superimposed boundary the eastern border of Indonesia and had the effect of extending Indonesia from Southeast Asia into the Pacific Realm. Geographically, all of New Guinea forms part of the Pacific Realm.

The fourth genetic boundary type is the so-called **relict boundary**—a border that has ceased to function but whose imprints (and sometimes influence) are still evident in the cultural landscape. The boundary between the former North and South Vietnam (Fig. 10-6, lower-right map) is a classic example: once demarcated militarily, it has had relict status since 1976 following the reunification of Vietnam in the aftermath of the Indochina War (1964–1975).

Southeast Asia's boundaries have colonial origins, but they have continued to influence the course of events in postcolonial times. Take one instance: the physiographic boundary that separates the main island of Singapore from the rest of the Malay Peninsula, the Johor Strait (see Fig. 10-11). That physiographic-political boundary facilitated, perhaps crucially, Singapore's secession from the state of Malaysia. Without it, Malaysia might have been persuaded to stop the separation process; at the very least, territorial issues would have arisen to slow the sequence of events. As it was, no land boundary needed to be defined. The Johor Strait demarcated Singapore and left no question as to its limits.

State Territorial Morphology

Boundaries define and delimit states; they also create the mosaic of often interlocking territories that give individual countries their shape, also known as their *morphol-*

ogy. The **territorial morphology** of a state affects its condition, even its survival. Vietnam's extreme elongation has influenced its existence since time immemorial. As we will see, Indonesia has tried to redress its fragmentation into thousands of islands by promoting unity through the "transmigration" of Jawanese from the most populous island to many of the others.

Political geographers identify five dominant state territorial configurations, all of which we have encountered in our world regional survey but which we have not categorized until now. All but one of these shapes are represented in Southeast Asia, and Figure 10-7 provides the terminology and examples:

- **Compact states** have territories shaped somewhere between round and rectangular, without major indentations. This encloses a maximum amount of territory within a minimum length of boundary. Southeast Asian example: Cambodia.

- **Protruded states** (sometimes called *extended*) have a substantial, usually compact territory from which extends a peninsular corridor that may be landlocked or coastal. Southeast Asian examples: Thailand, Myanmar.

- **Elongated states** (also called *attenuated*) have territorial dimensions in which the length is at least six times the average width, creating a state that lies astride environmental and/or cultural transitions. Southeast Asian example: Vietnam.

- **Fragmented states** consist of two or more territorial units separated by foreign territory or by water. Subtypes are mainland-mainland, mainland-island, and island-island. Southeast Asian examples: Malaysia, Indonesia, Philippines.

- **Perforated states** completely surround the territory of other states, so that they have a "hole" in them. No Southeast Asian example; the most illustrative current case is South Africa, perforated by Belgium-sized Lesotho.

In the discussion that follows, we will have frequent occasion to refer to this geographic property of Southeast Asia's states. For so comparatively small a realm with so few states, Southeast Asia displays a considerable variety of state morphologies. When we link these features to other geographic aspects (such as relative location), we obtain useful insights into the regional framework.

One point of caution: states' territorial morphologies do not determine their viability, cohesion, unity, or lack thereof; they can, however, influence these qualities. Cambodia's compactness has not ameliorated its divisive political geography, for example. But as we will find in the pages that follow, shape plays a key role in the still-unfolding political and economic geography of Southeast Asia.

CAMBODIA
Phnom Penh ●

COMPACT

THAILAND
Bangkok ●

PROTRUDED

VIETNAM
Hanoi ●

ELONGATED

Pretoria ●

SOUTH AFRICA

LESOTHO

● Cape Town

PERFORATED

Manila ●

PHILIPPINES

Ho Chi Minh City
(Saigon-Cholon) ●

**STATE TERRITORIAL
CONFIGURATIONS**

FRAGMENTED

FIGURE 10-7

REGIONS OF THE REALM

SOUTHEAST ASIA'S first-order regionalization must be based on its mainland-island fragmentation. But as we have noted there are physiographic, historical, and cultural reasons to include the Malaysian (southern) part of the Malay Peninsula in the insular region, as shown in Figure 10-2. Using the political framework as our grid, we see that the regions of Southeast Asia are constituted as follows:

Mainland Region Vietnam, Cambodia, Laos, Thailand, Myanmar (Burma)

Insular Region Malaysia, Singapore, Indonesia, East Timor, Brunei, Philippines

▶ MAINLAND SOUTHEAST ASIA

Five countries form the mainland region of Southeast Asia: two of them protruded, one compact, one elongated, and one landlocked. Two colonial powers, buffered by Thailand, shaped its modern historical geography. One religion, Buddhism, dominates cultural landscapes, but this is a multicultural, multiethnic region. Although one of the least urbanized regions of the world, it contains several major cities. And as Figure 10-2 shows, two countries (Vietnam and Myanmar) possess more than one core area each. We approach the region from the east.

City	Population* (in millions)
Bangkok, Thailand	8.1
Hanoi, Vietnam	4.1
Ho Chi Minh City (Saigon), Vietnam	8.5
Jakarta, Indonesia	13.1
Kuala Lumpur, Malaysia	1.5
Manila, Philippines	10.7
Phnom Penh, Cambodia	1.2
Singapore, Singapore	4.3
Viangchan, Laos	0.7
Yangon, Myanmar (Burma)	5.0

MAJOR CITIES OF THE REALM

*Based on 2004 estimates.

ELONGATED VIETNAM

Consider this: when the Indochina War ended in 1975, Vietnam had less than half the population it has today. In 2004, Vietnam's population exceeded 80 million, about 60 percent of it under 21 years of age. For the great majority of Vietnamese, therefore, the terrible war of the 1960s and 1970s is history, not memory. What concerns most Vietnamese today is the need to overcome two decades of isolation, to reconnect to the world at large, and to join in the economic boom on the Pacific Rim. In 1995, the United States moved to normalize relations with Hanoi, opening an embassy and authorizing interactions long prohibited in the aftermath of the conflict. Five years later, this was followed by a successful state visit by President Clinton.

Travel up the crowded road from the northern port of Haiphong to the capital, Hanoi, or sail up the river to the southern metropolis officially known as Ho Chi Minh City—but called *Saigon* by almost everyone there—and you are quickly reminded of the cultural effects of Vietnam's elongation (Fig. 10-8). The French colonizers delimited Vietnam as a 1200-mile (2000-km) strip of land extending from the Chinese border in the north to the tip of the Mekong Delta in the south. Substantially smaller than California, this coastal belt was the domain of the Vietnamese (Fig. 10-3). The French recognized that Vietnam, whose average width is under 150 miles (240 km), was not a homogeneous colony, so they divided it into three units: (1) Tonkin, land of the Red River Delta and centered on Hanoi in the north; (2) Cochin China, region of the Mekong Delta and centered on Saigon in the south; and (3) Annam, focused on the ancient city of Hué, in the middle (Fig. 10-4). Today, the Vietnamese

prefer to use *Bac Bo*, *Nam Bo*, and *Trung Bo* to designate these areas.

The Vietnamese (or Annamese, also Annamites, after their cultural heartland) speak the same language, although the northerners can easily be distinguished from southerners by their accent. As elsewhere in their colonial empire, the French made their language the *lingua franca* of Indochina, but their tenure was cut short by the Japanese, who invaded Vietnam in 1940. During the Japanese occupation, Vietnamese nationalism became a powerful force, and after the Japanese defeat in 1945, the French could not regain control. In 1954, the French suffered a disastrous final trouncing on the battlefield at Dien Bien Phu in the northwest and were ousted from the country.

But even after its forces routed the colonizers, Vietnam did not become a unified state. Separate regimes took control: a communist one in Hanoi and a noncommunist counterpart in Saigon. Vietnam's pronounced elongation had made things difficult for the French; now it played its role during the postcolonial period. Note, in Figure 10-8, that Vietnam is widest in the north and south, with a narrow "waist" in its middle zone. North and South Vietnam were worlds apart.

Many Americans still remember how the United States became involved in the inevitable conflict between communists and noncommunists in Vietnam during the 1960s and 1970s. At first, American military advisors were sent to Saigon to help the shaky regime there cope with communist insurgents. When the tide turned against the South, the United States committed military forces and equipment to halt the further spread of communism (see box titled "Domino Theory"); at one time, more than 500,000 U.S. soldiers were in Vietnam.

The conduct of the war, its mounting casualties, and its apparent futility created severe social tensions in the United States. If you had been a student during that period, your college experience would have been radically different from what it is today. Protest rallies, "teach-ins," antidraft demonstrations, "sit-ins," marches, strikes, and even hostage-taking disrupted campus life. The Indochina War threatened the stability of American society. It drove an American president (Lyndon Johnson) from office in 1968 and destroyed the electoral chances of his vice president (Hubert Humphrey), who would not disavow the U.S. role in the war.

In 1975, the Saigon government fell, and the United States was ousted—just a little over two decades after the French defeat at Dien Bien Phu. When the last helicopter left from the roof of the U.S. embassy in Saigon amid scenes of desperation and desertion, it marked the end of a sequence of events that closely conformed to the *insurgent state model* we introduced in the context of contemporary Colombia in Chapter 5 (p. 237).

FIGURE 10-8

POPULATION

- • Under 50,000
- • 50,000–250,000
- ● 250,000–1,000,000
- ● 1,000,000–5,000,000
- ● Over 5,000,000

National capitals are underlined

Vietnam Today

In the aftermath of its devastating war, Vietnam found itself in a territorial conflict with China, and even as countless Vietnamese were boarding boats to escape communist rule, they were joined by thousands of local Chinese. The dimensions of this emigration will never be known; of the estimated 2 million refugee "boat people," more than half perished from storms and sinkings, starvation at sea, and pirate attacks. The majority of those who survived were resettled in America. Meanwhile, the United States placed an embargo on Vietnam, isolating the country and stifling its economy. Only its fertile, rice-yielding river basins allowed it to survive.

Domino Theory

*D*uring the Indochina War (1964–1975), it was U.S. policy to contain communist expansion by supporting the efforts of the government of South Vietnam to defeat communist insurgents. Soon, the war engulfed North Vietnam as U.S. bombers attacked targets north of the border between North and South. In the later phases of the war, conflict spilled over into Laos and Cambodia. U.S. warplanes also used bases in Thailand. Like dominoes, one country after another fell to the ravages of the war or was threatened by it.

Some scholars warned that this domino effect could eventually affect not only Thailand but also Malaysia, Indonesia, and Burma (today Myanmar): the whole Southeast Asian realm, they predicted, could be destabilized. But, as we know, that did not happen. The war remained confined to Indochina. And the domino "theory" seemed invalid.

But is the theory totally without merit? Unfortunately, some political geographers to this day mistakenly define this idea in terms of communist activity. Communist insurgency, however, is only one way a country may be destabilized (Peru came close to collapse from it). But right-wing rebellion (Nicaragua's Contras), ethnic conflict (Bosnia), religious extremism (Algeria), and even economic and environmental causes can create havoc in a country. Properly defined, the **domino theory** holds that destabilization from any cause in one country can result in the collapse of order in a neighboring country, triggering a chain of events that can affect a series of contiguous states in turn.

A recent instance of the domino effect comes from Equatorial Africa, where the collapse of order in Rwanda spilled over into The Congo (then Zaïre) and also affected Burundi, Congo, and Angola.

AMONG THE REALM'S GREAT CITIES . . . **SAIGON**

*O*fficially, it is known as Ho Chi Minh City, renamed in 1975 after the ouster of American forces, but only bureaucrats and atlases call it that. To the people here, it is still Saigon, city on the Saigon River, served by, it seems, a thousand companies using the name. Saigon Taxi, Saigon Hotel, Saigon Pharmacy—there is hardly a city block without the name Saigon on some billboard.

The phenomenon is significant. Hanoi's northerners ordained the change, but this still is South Vietnam, and many southerners see the North as peopled by slow, dogmatic communists obstructing the hopes and plans for their vibrant city. In truth, Saigon may be at least twice as large as Hanoi, but like the capital it suffers from decades of deterioration, a broken-down infrastructure, and undependable public services. When, in the early 1990s, businesspeople could find no modern accommodations or working telephones, an Australian firm built an eight-story floating hotel with its own power plant, towed it up the Saigon River, and parked it on the downtown waterfront.

Therein lies Saigon's great advantage: large oceangoing vessels can reach the city in a three-hour sail up the meandering river. The Chinese developed its potential: in the 1960s, Chinese merchants controlled more than half of the South's exports and nearly all of its textile factories and foreign exchange. Saigon-Cholon (Cholon means "great market") was a Chinese success story, but after the war the Hanoi regime nationalized all Chinese-owned firms. Tens of

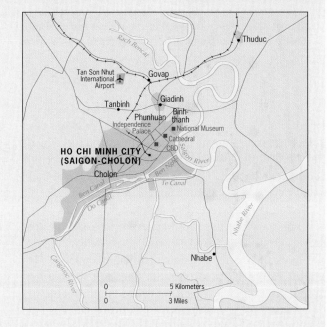

thousands of middle-class Chinese left the city, its economy devastated.

Today, Saigon (population: 8.5 million) is steadily recovering. A modern skyline is rising above the colonial one. A large Special Economic Zone has been created south of the city. Factories are being built. The Pacific Rim era is under way. Will the South rise again?

FROM THE FIELD NOTES

"Although modernization has changed the skylines of Vietnam's major cities (Saigon more so than Hanoi), most of the country's urban areas look and function much as they have for generations. Main arteries throb with commerce and traffic (still mostly bicycles and mopeds); side streets are quieter and more residential. No highrises here, but the problems of dense urban populations are nevertheless evident. Every time I turned into a side street, the need for street cleaning and refuse collection was obvious, and drains were clogged more often than not. 'We haven't allowed our older neighborhoods to be destroyed like they have in China,' said my colleague from Hanoi University, 'but we also don't have an economy growing so fast that we can afford to provide the services these people need.' Certainly the Pacific Rim contrast between burgeoning coastal China and slower-growing Vietnam— a matter of communist-government policy—is evident in their cities. 'As you see, noting much has changed here,' she said. 'Not much evidence of globalization. But we're basically self-sufficient, nobody goes hungry, and the gap between the richest and the poorest here in Vietnam is a fraction of what it is now in China. We decided to keep control.' You would find more agreement on these points here in Hanoi than in Saigon, but the Vietnamese state is stable and progressing."

At a time when the western Pacific Rim was booming, Vietnam was also held back by its dogmatic communist regime and by its ruined infrastructure. In the early 1990s, the 100-km (60-mile) road trip between the dilapidated capital, Hanoi, and the port of Haiphong could take as long as four hours. In the paddies of the northern (Tonkin) core area, anchored by Hanoi, irrigation water was still raised by bucket. On the roads, animal-drawn carts outnumbered vehicles.

Even then, Saigon-Cholon (officially called Ho Chi Minh City) in the south was far ahead of the northern capital. With about 10 percent of the country's population and a far larger share of its best-educated people, Saigon recovered faster and attracted foreign investment sooner. Cholon, the Chinatown that suffered severely in the aftermath of the war, thrived again. When, in 1995, the United States moved to normalize relations, Saigon's economy experienced a surge strong enough to trouble the dogmatic bureaucrats in Hanoi. In the city's Special Economic Zone and in its futuristic New Saigon district, signs of a new age are evident.

But Vietnam's communist government has been slow to open the country to foreign investment and involve-

ment. Instead, it has in some ways confirmed the fears of those who resisted communist rule all along: it has harassed Buddhist and Christian groups and detained their leaders, and it has revived old conflicts between minorities in the mountains and ethnic Vietnamese in control. Researchers and visitors report that land is being taken from the hill peoples in large swaths and planted with coffee and other export crops; in 2002 the government announced that Vietnam had become the world's largest exporter of coffee. That may be premature, but the deforestation of the interior highlands is already having an environmental effect as annually worsening floods afflict the once-stable countryside.

By 2003, Vietnam's economy lagged behind those of other countries with much potential on the Pacific Rim. Foreign investment has been limited, and many plans to privatize state industries have been shelved. Like China, Vietnam remains a communist country, but unlike China it has not had the leadership of a Deng Xiaoping. Hanoi's avowed decision to mix communist ideology with market economics never resulted in far-reaching change: the emphasis remained on communist statism, not on economic reform. As a result, Vietnam is experiencing neither the turmoil seen in China nor the huge regional disparities arising there. During the mid-1990s, when Saigon prospered and Hanoi stagnated, it appeared that North and South were once again growing apart. Today, Saigon's exuberance is restrained by Hanoi's dogmatism—and with it, Vietnam's economic progress.

COMPACT CAMBODIA

Former French Indochina contained two additional entities—Cambodia and Laos. In Cambodia, the French possessed one of the greatest treasures of Hinduism: the city of Angkor, capital of the ancient Khmer Empire, and the temple complex known as Angkor Wat. The Khmer Empire prevailed here from the ninth to the fifteenth centuries, and King Suryavarman II built Angkor Wat to symbolize the universe in accordance with the precepts of Hindu cosmology. When the French took control of this area during the 1860s, Angkor lay in ruins, sacked long before by Vietnamese and Thai armies. The French created a protectorate, began the restoration of the shrines, established Cambodia's permanent boundaries, and restored the monarchy under their supervision.

Geographically, Cambodia enjoys several advantages, most notably its compact shape. Compact states enclose a maximum of territory within a minimum of boundary and are without peninsulas, islands, or other remote extensions of the national spatial framework. Cambodia had the further advantage of strong ethnic and cultural homogeneity: 90 percent of its 12.7 million inhabitants are Khmers, with the remainder equally divided between Vietnamese and Chinese. As Figure 10-8 shows, Southeast Asia's greatest river, the Mekong, enters Cambodia from Laos and crosses it from north to south, creating a great bend before flowing into southern Vietnam (see box titled "The Mighty Mekong"). Phnom Penh, the country's capital, lies on the river. The ancient capital of Angkor lies in the northwest, not far from the Tonle Sap ("Great Lake"), a major lake linked to, and filled by, the waters of the Mekong. Cambodia has a coastline on the Gulf of Thailand, but its core area lies in the interior; the Mekong is more important than the Gulf. Kompong Som, the port at the end of the railroad from Phnom Penh, has only limited facilities.

Cambodia's present social geography continues to suffer from the after-effects of the Indochina War. Neither its compactness nor its isolation could overcome the disadvantage of relative location near a conflict that was bound to spill across its borders (Fig. 10-8). Internal disharmony fueled by external interference led in 1970 to the ouster of the last king by the military; in 1975 that regime was itself overthrown by communist revolutionaries, the so-called Khmer Rouge. These new rulers embarked on a course of terror and destruction in order to

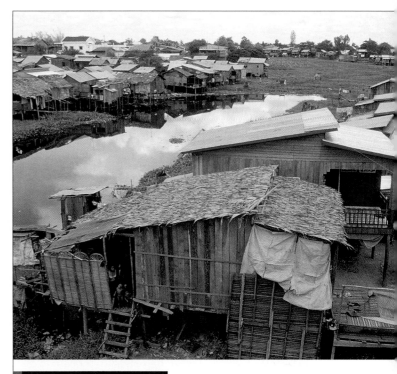

FROM THE FIELD NOTES

"Here near Siem Reap, along a small stream flowing into the Tonle Sap, I had the opportunity to observe one of Cambodia's many Mekong Basin stilt villages. The water rises and falls with the seasons; it is comparatively dry now, and a pool of stagnant water is what is left of the river that will soon form here once again. In the past, stilt houses were built of traditional materials only, but now there are tin roofs and plastic sheeting. Conditions are not sanitary; there was refuse everywhere, waiting to be swept away by the rising waters. Outhouses such as those in the left foreground drained directly into the temporary pond, now mosquito-infested and severely polluted. I wondered how many infants would survive their first year here."

reconstruct Kampuchea (as they called Cambodia) as a rural society. They drove townspeople into the countryside where they had no place to live or work, emptied hospitals and sent the sick and dying into the streets, outlawed religion and family, and in the process killed as many as 2 million Cambodians (out of a population of 8 million). In the late 1970s, Vietnam, having won its own war, invaded Cambodia to drive the Khmer Rouge away. But this action led to new terror, and a stream of refugees crossed the border into Thailand. Eventually, remnants of the Khmer Rouge managed to establish a base in the northwest of Cambodia, where their murderous leader, Pol Pot, who was never brought to account for his actions, committed suicide in 1998.

Cambodia's postwar trauma continues today. Once self-sufficient and able to feed others, it now must import

The Mighty Mekong

From its source among the snowy peaks of China's Qinghai and Xizang, the Mekong River rushes and flows some 2600 miles (4200 km) to its delta in southernmost Vietnam. This "Danube of Southeast Asia" crosses or borders five of the realm's countries (see photo, p. 500), supporting rice farmers and fishing people, forming a transport route where roads are few, and providing electricity from dams upstream. Tens of millions of people depend on the waters of the Mekong, from subsistence farmers in Laos to apartment dwellers in China. The Mekong Delta in southern Vietnam is one of the realm's most densely populated areas and produces enormous harvests of rice.

But problems loom. China is building a series of dams across the Lancang (as the Mekong is called there) to supply Yunnan Province with electricity. Although such hydroelectric dams should not interfere with water flow, countries downstream worry that a severe dry spell in the interior would impel the Chinese to slow the river's flow to keep the reservoirs full. Cambodia is concerned over the future of the Tonle Sap, a large natural lake filled by the Mekong (Fig. 10-8). In Vietnam, farmers worry about salt water invading the paddies should the Mekong's level drop. And the Chinese may not be the only dam builders in the future: Thailand has expressed an interest in building a dam on the Thai-Laos border where it is defined by the Mekong.

In such situations, the upstream states have an advantage over those downstream. Several international organizations have been formed to coordinate development in the Mekong Basin, including the Mekong River Commission (MRC) founded half a century ago. China has offered to sell electricity from its dams to Thailand, Laos, and Myanmar. Coordinated efforts to reduce logging in the Mekong's drainage basin have had some effect. After consultations with the MRC, Australia built a bridge linking Laos and Thailand. There is even a plan to make the Mekong navigable from Yunnan to the coast, creating an alternative outlet for interior China.

Sail the Mekong today, however, and you are struck by the slowness of development along this artery. Wooden boats, thatch-roofed villages, and teeming paddies mark a river still crossed by antiquated ferries and flanked by few towns. Of modern infrastructure, one sees little. And yet the Mekong and its basin form the lifeline of mainland Southeast Asia's dominantly rural societies.

FROM THE FIELD NOTES

"Along a road in Xiangkhoang Province, in north-central Laos, I stopped to watch these people operate their gas-powered rice mill. I had some difficulty communicating with the people, but from the young boy I gathered that some of this equipment had been fashioned from U.S. war materiel, including the drum (I wondered what it might have contained) and the funnel, which was made from bomb casings."

food. Rice and beans are the subsistence crops, but the dislocation in the farmlands set production back severely, and continuing strife in the countryside has disrupted supply routes for years.

Cambodians hope that tourism, focused chiefly on the great temples at Angkor, will become a mainstay of the economy. But political stability is a prerequisite for this industry, and it has not yet returned to this ancient kingdom.

◼ LANDLOCKED LAOS

North of Cambodia lies Southeast Asia's only landlocked country, Laos (Fig. 10-8). Interior and isolated, Laos changed little during 60 years of French colonial administration (1893–1953). Then, along with other French-ruled areas, it became an independent state. Soon this well-entrenched, traditional kingdom fell victim to rivalries between traditionalists and the victorious communists, and the old order collapsed.

Laos has no fewer than five neighbors, one of which is the East Asian giant, China. The Mekong River forms a long stretch of its western boundary, and the important sensitive border with Vietnam to the east lies in mountainous terrain. With 5.8 million people (about half of them ethnic Lao, related to the Thai of Thailand), Laos lies surrounded by comparatively powerful states. The country has no railroads, just a few miles of paved roads, and very little industry; it is only 17 percent urbanized (the capital, Viangchan, lies on the Mekong and has an oil pipeline to Vietnam's coast). Laos remains the region's poorest and most vulnerable entity.

PROTRUDED THAILAND

In virtually every way, Thailand is the leading state of the region under discussion. Its per-capita GNP during the economic boom of the 1990s was higher than that of Vietnam, Cambodia, Laos, and Myanmar combined. Alone among its neighbors, Thailand was a strong participant in the Pacific Rim's economic development. Thailand's capital, Bangkok, is one of the two largest urban centers in the region and one of the world's most prominent primate cities. The country's population, 63.6 million in 2004, is growing at the slowest rate in the entire realm, equivalent to that of fully urbanized Singapore. Over the

past few decades, only political instability and uncertainty have inhibited economic progress. Thailand is a constitutional monarchy; its moves toward stable democracy have been thwarted by graft, corruption, and, sometimes, violent confrontation.

Thailand is the textbook example of a protruded state. From a relatively compact heartland, in which lie the core area, capital, and major areas of productive capacity, a 600-mile (1000-km) corridor of land, in places less than 20 miles (32 km) wide, extends southward to the border with Malaysia (Fig. 10-9). The boundary that defines this protrusion runs down the length of the Malay Peninsula to the Kra Isthmus, where neighboring Myanmar peters out and Thailand fronts the Andaman Sea (an arm of the Indian Ocean) as well as the Gulf of Thailand. In the entire country, no place lies farther from the capital than the southern end of this tenuous protrusion.

Consider this spatial situation in the context of Figure 10-3 (p. 492). Note that the Malay ethnic population group extends from Malaysia more than 300 km (200 mi) into Thai territory. In Thailand's five southernmost

AMONG THE REALM'S GREAT CITIES . . . **BANGKOK**

*B*angkok (8.1 million), mainland Southeast Asia's second largest city, is a massive sprawling metropolis on the banks of the Chao Phraya River, an urban agglomeration without a center, an aggregation of neighborhoods ranging from the immaculate to the impoverished. In this city of great distances and high tropical temperatures, getting around is often difficult because roadways are choked with traffic. A (diminishing) network of waterways affords the easiest way to travel, and life focuses on the busiest waterway of all, the Chao Phraya. Ferries and water taxis carry tens of thousands of commuters and shoppers across and along this bustling artery, flanked by a growing number of highrise office buildings, luxury hotels, and ultramodern condominiums. Many of these modern structures reflect the Thais' fondness for domes, columns, and small-paned windows, creating a skyline unique in Asia.

Gold is Buddhist Thailand's symbol, adorning religious and nonreligious architecture alike. From a high vantage point in the city, you can see hundreds of golden spires, pagodas, and façades rising above the townscape. The Grand Palace, where royal, religious, and public buildings are crowded inside a white, crenellated wall more than a mile long (see p. 512), is a gold-laminated city within a city embellished by ornate gateways, dragons, and statuary. Across the mall in front of the Grand Palace lie government buildings sometimes tar-

geted by rioters in Bangkok's volatile political atmosphere. Not far away are Chinatown and myriad markets; Bangkok has a throbbing commercial life.

Decentralized, dispersed Bangkok has major environmental problems, ranging from sinking ground (resulting in part from excessive water pumping) to some of the world's worst air pollution. Yet industries, port facilities, and residential areas continue to expand as the city's hinterland grows and prospers with Pacific Rim investment. Bangkok is the pulse of Southeast Asia's mainland region.

TERRITORIAL MORPHOLOGY: THAILAND

Core area

Roads — Railroads

0   50   100   150 Kilometers

0   50   100 Miles

POPULATION
- Under 50,000
- 50,000–250,000
- 250,000–1,000,000
- 1,000,000–5,000,000
- Over 5,000,000

National capitals are underlined

FIGURE 10-9

provinces, 85 percent of the inhabitants are Muslims (the figure for Thailand as a whole is less than 5 percent). The political border between Thailand and Malaysia is porous; in fact, you can cross it at will almost anywhere along the Kolok River by canoe, and inland it is a matter of walking a forest trail. For more than a century, the southern provinces have had closer ties with Malaysia across the border than with Bangkok 1000 km (600 mi) away, and the Thai gov-

ernment has not tried to restrict movement or impose oner-ous rules on the Muslim population here. But in the new era of the War on Terror and in view of rising violence in this southern frontier, the south is coming to national atten-tion. The Pattani United Liberation Organization, named after the functional capital of the Muslim south, was active in the l960s and 1970s but has been dormant since. Then a series of bombings at a Pattani hotel, several schools, a

Chinese shrine, and a Buddhist temple in 2001 and 2002 raised fears that Muslim-inspired violence was on the rise again. In April 2002 Bangkok newspapers reported that police had found bomb-making materials in the house of a Muslim religious leader. An al-Qaeda leader allegedly moved through the area about the same time. Thailand's pronounced protrusion, its porous border, its dependence on tourism, and the vulnerable location of its Andaman-coast tourist facilities combine to raise concern over this distant part of its national territory.

As Figure 10-2 shows, Thailand occupies the heart of the mainland region of Southeast Asia. While Thailand has no Red, Mekong, or Irrawaddy Delta, its central lowland is watered by a set of streams that flow off the northern highlands and the Khorat Plateau in the east. One of these streams, the Chao Phraya, is the Rhine of Thailand. From the head of the Gulf of Thailand to Nakhon Sawan, this river is a highway of traffic. Barge trains loaded with rice head for the coast, ferry boats head upstream, and freighters transport tin and tungsten (of which Thailand is among the world's largest producers). Bangkok sprawls on both sides of the lower Chao Phraya, here flanked by skyscrapers, pagodas, factories, boatsheds, ferry landings, luxury hotels, and modest dwellings in crowded confusion. On the right bank of the Chao Phraya, Bangkok's west side, lie the city's remaining *klong* (canal) neighborhoods, where waterways and boats form the transport system. Bangkok still is known as the Venice of Asia, although many *klongs* have been filled in and paved over to serve as roadways.

During the Pacific Rim boom of the 1990s, Thailand's economy grew rapidly. Relative location, natural environment, and social conditions all contributed to this growth. Situated at the head of the Gulf of Thailand, the country's core area opens to the South China Sea and hence the Pacific Ocean. The Gulf itself has long been a rich fishing ground and more recently a source of oil (the major refinery lies in the port area of Laem Chabang, an industrial zone just to the north of the tourist resort of Pattaya). But the key to Thailand's economic explosion was the Thai workforce, laboring under often dreadful conditions for low wages to produce goods sold on foreign markets at cheap prices. This labor force attracted major foreign investment that, in turn, stimulated service industries.

For a time, Thailand became a Pacific Rim economic tiger, producing goods ranging from plastic toys to "Japanese" cars. But while the workers toiled, government and financial mismanagement planted seeds of ruin. In 1997, the country's currency, the *baht*, began a slide in value that undermined the economy and led to numerous bank failures. Exports and worker productivity declined, real estate values plunged, and Thailand be-

came the first casualty in a series of economic setbacks that would eventually affect the entire western Pacific Rim.

Although Thailand and the other Pacific Rim economies suffered severe downturns, we should remember that the boom period had yielded many infrastructural improvements not only in the core area but also in distant areas such as the north, centered on Chiang Mai. Thailand, like other Pacific Rim economies, has improved highways, ports, power systems, airports, and housing. Although a period of poor maintenance followed the events of 1997, Thailand's fundamental assets remained to serve the country well when the world economy revived. By 2003, Thailand's GNP per capita had recovered to exceed U.S. $6000, its balance of trade had been favorable for five consecutive years, and the tourist industry, severely affected by the events of 2001–2002, was recovering.

Under normal circumstances, Thailand's natural environments and cultural landscapes continue to attract millions of visitors annually. The magnificent architecture of Buddhism (the country's old name, Siam, was a Chinese word meaning golden-yellow), swathed in gold and shining in the tropical sun, is without equal, as the photo on the following page underscores. The warm coasts and beaches have long made Thailand a favorite destination for tourists from Germany to Japan. One of the world's most famous resorts is Phuket, on the distant southwest coast near the end of Thailand's Malay Peninsula protrusion.

Tourism is Thailand's leading source of foreign revenues, but there is a dark side to the industry. The Thai people's relaxed attitude toward sex has made Thai tourism in large part a sex industry, candidly advertised in such foreign markets as Japan where "sex tours" attract planeloads of male participants. When the AIDS pandemic reached Thailand in the 1980s, it changed no habits, and by the mid-1990s the country was suffering one of the world's worst outbreaks of the disease. At the turn of the century, more than 1 million of Thailand's population of 60-plus million were infected.

Thailand's territorial morphology creates both problems and opportunities. Problems include the effective integration of the national territory through surface communications, which must cover long distances to reach remote locales; the influx of refugees from neighboring countries dislocated by internal conflict; and control over contraband (opium) drug production and trade in the interior "Golden Triangle" where Myanmar, Thailand, and Laos meet. The government's efforts to speed the development of rural areas, notably the northeast, through megaprojects such as large dams have disrupted local life, damaged ecologies, and deepened the split between city and countryside. On the other hand, Thailand's lengthy Kra Isthmus creates economic opportunities where southernmost Thailand lies

FROM THE FIELD NOTES

"Thailand exhibits one of the world's most distinctive cultural landscapes, both in its urban and its rural areas. Graceful pagodas, stupas, and spirit houses of Buddhism adorn city and countryside alike. Gold-layered spires rise above, and beautify cities and towns that would otherwise be drab and featureless. The architecture of Buddhism has diffused into public architecture, so that many secular buildings, from businesses to skyscrapers, are embellished by something approaching a national style. The grandest expression of the form undoubtedly is a magnificent assemblage of structures within the walls of the Grand Palace in the capital, Bangkok, of which a small sample is shown here. Climb onto the roof of any tall building in the city's center, however, and the urban scene will display hundreds of such graceful structures, interspersed with the modern and the traditional townscape."

close to Malaysia and Indonesia, raising hopes for a three-country development scheme in that area. Given political stability, Thailand's future as a regional leader appears secure.

EXTENDED MYANMAR

Myanmar (formerly Burma) shares with Thailand its protruded morphology but little else. Long languishing under one of the world's most corrupt military regimes,

Myanmar is one of the planet's poorest countries where, it seems, time has stood still for centuries. Whereas air, railroad, and highway routes serve Thailand's peninsular protrusion, there is not even an all-weather track to the southern end of Myanmar's. Whereas busy shipping lanes converge on Thailand's coastal core area, Myanmar's Irrawaddy Delta is more reminiscent of Bangladesh.

Myanmar's territorial morphology is complicated by a shift in the Burmese core area that took place during colonial times. Prior to the colonial period, the focus of embryonic Burma lay in the so-called dry zone between the Arakan Mountains and the Shan Plateau. The urban focus of the state was Mandalay, which had a central situation and relative proximity to the non-Burmese highlands all around (Fig. 10-1). Then the British developed the agricultural potential of the Irrawaddy Delta, and Rangoon (now called Yangon) became the hub of the colony. The Irrawaddy waterway links the old and the new core areas, but the center of gravity has shifted to the south.

The political geography of Myanmar constitutes a particularly good example of the role and effect of state morphology on internal state structure. Not only is Myanmar a protruded state; in addition, its core area is surrounded on the west, north, and east by a horseshoe of great mountains—where many of the country's 11 minority peoples had their homelands before the British occupation. The colonial boundaries had the effect of incorporating these peoples into the domain of the Burman people, who today constitute about 56 percent of the population of 50.3 million. When the British departed and Burma became independent in 1948, the less numerous peoples traded one master for another. In 1976, nine of these indigenous peoples formed a union to demand the right to self-determination in their homelands.

As Figure 10-3 shows, the peripheral peoples of Myanmar occupy a significant part of the state. The Shan of the northeast and far north, who are related to the neighboring Thai, account for about 7 percent of the population, or 3.4 million. An even larger minority, the Karen (just under 10 percent, 4.8 million), live in the neck of Myanmar's protrusion, and have proclaimed that they wish to create an autonomous territory within a federal Myanmar. Although the powerful military have dealt these aspirations a series of setbacks, centrifugal forces continue to bedevil the central regime. Its response has been to exert power by all available means rather than to accommodate these forces, even to the point of stifling political discourse (let alone opposition) among the Burman themselves.

This is all the more tragic because Myanmar has economic potential far superior to that of Bangladesh, its neighbor to the west, and to other countries in its own region. It can feed itself, and it could export significant quantities of rice. Varied soils and diverse environments can yield a range of other crops. Tin and other metals have been found in the highlands. Today, Myanmar is exploiting the forests of the interior at an alarming rate; teak has risen to first place among exports by value—among legitimate exports, that is. In its corner of the "Golden Triangle," Myanmar also is one of the world's largest illicit producers of opium poppies and harvests other illegal crops for the drug trade as well. For a time, the regime in Yangon tried to control the trade, but its constant conflict with peoples outside the core area diverted its attentions and energies from that campaign.

For decades, the repressive policies of a brutal military regime have kept Myanmar in turmoil and all but halted its development. Infrastructure throughout the country is in dismal condition, but this has not stopped the regime from purchasing a nuclear reactor and jet fighters from Russia and conventional weapons from North Korea. Nor has isolated Myanmar escaped the cycle of violence affecting other countries in Asia: religious riots between Muslims and Buddhists in 2001 claimed numerous victims. About 4 percent of Myanmar's population is Muslim.

In the first years of this new century, the military regime made some overtures toward the opposition led by one of the world's most courageous politicians, Daw Aung San Suu Kyi. She has been in various forms of detention for more than a decade but has not yielded to threats or enticements. Progress, however, remained minimal. In any case, a change of heart by the regime will not undo the effects of policies that have pushed impoverished and exhausted Myanmar into the ranks of the world's poorest economies.

▶ INSULAR SOUTHEAST ASIA

On the peninsulas and islands of Southeast Asia's southern and eastern periphery lie six of the realm's 11 states (Fig. 10-2). Few regions in the world contain so diverse a set of countries. Malaysia, the former British colony, consists of two major areas separated by hundreds of miles of South China Sea. The realm's southernmost large state, Indonesia, sprawls across thousands of islands from Sumatera in the west to New Guinea in the east. North of the Indonesian archipelago lies the Philippines, a nation that once was a U.S. colony. These are three of the most severely fragmented states on Earth, and each has faced challenges that such politico-spatial division brings. This insular region of Southeast Asia also contains two small but important sovereign entities: a city-state and a sultanate. The city-state is Singapore, once a part of Malaysia (and one instance in which internal centrifugal forces were too great to be overcome). The sultanate is Brunei, an oil-rich Muslim territory on the island of Borneo that seems

transplanted from the Persian Gulf. In addition, a third small entity, East Timor, achieved independence in May 2002 following three years of United Nations administration. Few parts of the world are more varied or more interesting geographically.

■ MAINLAND-ISLAND MALAYSIA

The state of Malaysia represents one of the three types of fragmented states discussed earlier: the mainland-island type, in which one part of the national territory lies on a continent and the other on an island. Malaysia is a colonial political artifice that combines two quite disparate components into a single state: the southern end of the Malay Peninsula and the northern part of the island of Borneo. These are known, respectively, as West Malaysia and East Malaysia (Fig. 10-2). The name *Malaysia* came into use in 1963, when the original Federation of Malaya, on the Malay Peninsula, was expanded to incorporate the areas of Sarawak and Sabah in Borneo. When the name *Malaya* is used, it refers to the peninsular part of the Federation, whereas Malaysia refers to the total entity.

The Malays of the peninsula, traditionally a rural people, displaced older aboriginal communities there and today make up about 61 percent of the country's population of 25.3 million. They possess a strong cultural identity expressed in adherence to the Muslim faith, a common language, and a sense of territoriality that arises from their perceived Malayan origins and their collective view of Chinese, Indian, European, and other foreign intruders.

The Chinese came to the Malay Peninsula and to northern Borneo in substantial numbers during the colonial period, and today they constitute one-quarter of Malaysia's population (they are the largest single group in Sarawak). During World War II, when the Japanese ruled Malaya, the Chinese were ruthlessly persecuted; many were driven into the forested interior where they founded a communist-inspired resistance movement that long destabilized the area. Today, the commercially active, urbanized Chinese sector still is victimized by discrimination, now by the dominant Malays.

Hindu South Asians were in this area long before the Europeans, and for that matter before the Arabs and Islam arrived on these shores. Today they still form a substantial minority of over 7 percent of the population, clustered, like the Chinese, on the western side of the peninsula (Fig. 10-3).

The populous peninsular part of Malaysia remains the country's dominant sector with 11 of its 13 States and nearly 80 percent of its population. Here the Malay-dominated government has strictly controlled economic and social policies while pushing the country's modernization. During the Asian economic boom of the 1990s, Malaysia's planners embraced the notion of symbols: the capital, Kuala Lumpur, was endowed with the world's tallest building; a space-age airport outpaced Malaysia's needs; a high-tech administrative capital began to be built at Putrajaya, and a nearby development was called Cyberjaya—all part of a so-called *Multimedia Supercorridor* to anchor Malaysia's core area (Fig. 10-10).

The chief architect of this program was Malaysia's autocratic head of state, Mahathir bin Mohamad, leader of the Malay-dominated party that forms the majority in government. Dr. Mahathir, who in 2002 announced his intention to resign, has long had the support not only of a majority of the country's Malays, but also that of the important and influential ethnic Chinese minority. But Malaysia's head-long rush to modernize, coupled with some personal issues

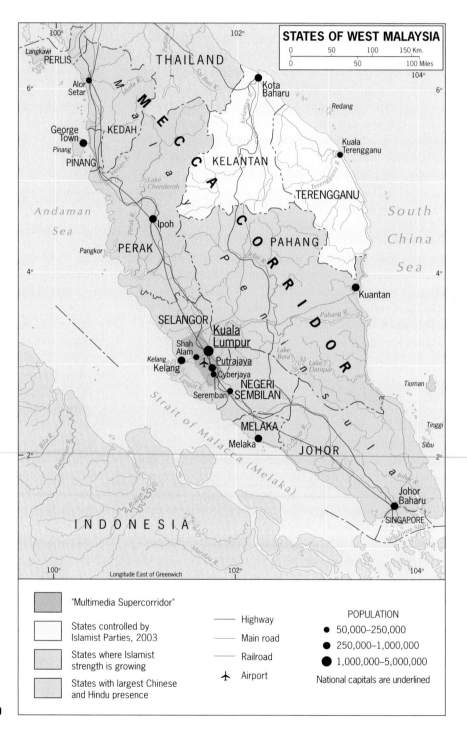

FIGURE 10-10

involving Mahathir and his rivals, caused a backlash among more conservative Muslims in the north and east of the peninsula, where a Muslim fundamentalist party made headway. At the turn of the century, Malaysians talked of two "corridors" marking their country: the Multimedia Supercorridor in the west and the "Mecca Corridor" in the east (Fig. 10-10). By 2001, younger members of the Islamic Party of Malaysia (PAS), which wanted Malaysia to become an Islamic state under Sharia law, had launched a campaign they described as a jihad to pursue this aim. The Mahathir government's response was to arrest and detain them, which elicited strong criticism inside and outside Malaysia. Then came September 11, 2001 and that disapproval evaporated.

The Malay Peninsula's primacy began long ago, during colonial times, when the British created a substantial economy based on rubber plantations, palm-oil extraction, and mining (tin, bauxite, copper, iron). The Strait of Malacca (Melaka) became one of the world's busiest and most strategic waterways, and Singapore, at the southern end of it, a prized possession (Singapore seceded from the Malaysian Federation in 1965).

Greater Malaysia survived decolonization, the loss of Singapore, and ethnic troubles; after 1980, its economy took off and became a major force on the Pacific Rim. Foreign manufacturers, attracted by the skills and modest wages of the Malaysian workforce, established hundreds of factories here (see box titled "Pinang: A Future Singapore?"). Although the old colonial exports still rank high on the list, electronics have made their appearance in a big way. One benefit of the merger with East Malaysia was that sector's production of valuable timber and petroleum. Kuala Lumpur began to resemble Singapore, becoming a forest of highrises. And Malaysia's GNP rose to third place in the realm, after all-urban Singapore and oil-rich Brunei, and well ahead of Thailand.

Territorially, **East Malaysia** makes up 60 percent of the country, but Sabah and Sarawak contain only about 20 percent of Malaysia's population. In terms of development in the East, Sabah, also the more populous, is ahead of remote Sarawak (Fig. 10-6, upper-left map). Kota Kinabalu, Sabah's capital, is the most modern city in all of Borneo (not just the Malaysian part), with a substantial port handling timber, a large tourist-industry infrastructure including a modern airport, and natural gas reserves in production to the north.

The ethnic complexity of Sarawak and Sabah (East Malaysia) defies summation. Ethnic Chinese are the single largest group in Sarawak, but each State counts more than two dozen ethnic identities. As a result, politics in East Malaysia are fractious and often combative, leading to repeated intervention by the federal government. That government, based as it is in West Malaysia, sometimes

FROM THE FIELD NOTES

"Multinational, cosmopolitan Port Kelang, Malaysia made for an interesting field day. One busy side street had a row of restaurants with Chinese, Italian, Malaysian, and Vietnamese menus. Rugby was being played on one field; cricket on another. When I was here in 1995 the area was booming, but in 1999 there were signs of trouble resulting from the general economic downturn of the late 1990s. Having seen several sights of the kind shown in this photograph, I stopped to ask the lone supervisor here why there seemed to be no activity in this large project; building appeared to have come to a stop. 'This company built a whole lot of townhouse estates,' he said. 'People were upwardly-mobile and buying them as fast as we could build them. Then the crisis came and for many the money ran out.' In truth, Malaysia weathered the storm better than other Southeast Asian countries, but even here the impact was noticeable on the landscape."

is accused by Easterners of "colonial" exploitation and infringement. The sprawling country's greatest challenge continues to lie in its ethnic and cultural heterogeneity, heightened by its territorial fragmentation.

SINGAPORE

In 1965, a fateful event occurred in Southeast Asia: Singapore, crown jewel of British colonialism in this realm, seceded from the Malaysian Federation and became a sovereign state, albeit a ministate (Fig. 10-11). With its magnificent relative location, its human resources, and firm government, Singapore then overcame the limitations of space and the absence of raw materials to become one of the economic tigers on the Pacific Rim.

With a mere 240 square miles (600 sq km) of territory, space is at a premium in Singapore, and this is a constant worry for the government. Singapore's only

Pinang: A Future Singapore?

In the northwestern corner of Malaysia's Malay Peninsula, opposite the northern end of Indonesian Sumatera, where the Strait of Malacca (Melaka) opens into the Indian Ocean, lies an island called Pinang (Fig. 10-10). A thriving outpost during the British colonial period, Pinang (capital: George Town), like Singapore, was one of the so-called Straits Settlements.

Today, Singapore no longer is a part of Malaysia, but Pinang's importance is growing rapidly. Now connected to the mainland by one of Asia's longest bridges (9 miles [14 km]), the 100-square-mile island has a population of over 1 million. Chinese outnumber Malays by 60 to 32 percent, and George Town, the old colonial port, has become the focus of an expanding high-technology manufacturing complex for the international computer industry. The 65-story Tun Abdul Razak Complex, housing government offices, businesses, department stores, and recreational facilities, is just one landmark in an ultramodern skyline. From the top you can see modern industrial parks with neon-lit signs proclaiming the presence of such companies as Intel, SONY, Philips, Motorola, and Hitachi.

The rapid development of Pinang is a calculated strategy by the Malaysian government. The emphasis is on building a highly skilled labor force that is locally nurtured in technical schools and training facilities financed by the multinational corporations themselves. Malaysia makes infrastructure investments ranging from the new toll bridge to a major airport. George Town's colonial and multicultural townscape is attracting a growing tourist trade, and new hotels and resorts are enlarging the service industry.

Malaysia hopes that Pinang will become the leading corner of a major growth triangle (a concept that is gaining popularity along the Pacific Rim), together with Indonesian Sumatera to the west and southern Thailand to the north. Meanwhile, Pinang is following in Singapore's economic footsteps. But Kuala Lumpur's politicians surely hope that Chinese-dominated Pinang will not follow Chinese-dominated Singapore's political course.

FIGURE 10-11

local spatial advantage over Hong Kong is that its small territory is less fragmented (there are just a few small islands in addition to the compact main island). With a population of 4.3 million and an expanding economy, Singapore must develop space-conserving, high-tech industries. It has succeeded in this effort, and the urban area's highrise buildings reflect the city-state's prosperity. You will find no slums in Singapore; the dilapidated streets in the old Chinatown were left merely to conserve a fragment of the city's past. Singapore mostly impresses with its newness and its modernity, and with the order of daily life.

As the map shows, Singapore lies where the Strait of Malacca (Melaka), leading westward to India, opens into the South China Sea and the waters of Indonesia (Fig. 10-11 inset). The port developed as part of the British Southeast Asian empire, and when independence came in 1963, it was made a part of the Malaysian Federation. However, two years later Singapore seceded from Malaysia and became a genuine city-state on the Pacific Rim. No reunification, merger, or annexation looms here.

Benefiting from its relative location, the old port of Singapore had become one of the world's busiest (by numbers of ships served) even before independence. It thrived as an *entrepôt* between the Malay Peninsula, Southeast Asia, Japan, and other emerging economic powers on the Pacific Rim and beyond. Crude oil from Southeast Asia still is unloaded and refined at Singapore, then shipped to Asian destinations. Raw rubber from the adjacent Peninsula and from Indonesia's island of Sumatera is shipped to Japan, the United States, China, and other countries. Tim-

ber from Malaysia, rice, spices, and other foodstuffs are processed and forwarded via Singapore. In return, automobiles, machinery, and equipment are imported into Southeast Asia through Singapore.

But that is the old pattern. Singapore's leaders want to redirect the city-state's economy and move toward high-tech industries for the future. In Singapore the government tightly controls business as well as other aspects of life. (Some newspapers and magazines have been banned for criticizing the regime, and there are even fines for such things as eating on the subway and failing to flush a public toilet.) Its overall success after secession has tended to keep the critics quiet: while GNP per capita from 1965 to 2002 multiplied by a factor of more than 15—to U.S. $24,910—that of neighboring Malaysia reached just U.S. $8330. Among other things, Singapore became (and for many years remained) the world's largest producer of disk drives for small computers.

As multinational corporations settled in, Singapore's political stability on the volatile Pacific Rim was another advantage. Nonetheless, a problem emerged: unlike Taiwan, where brainpower is plentiful, Singapore's technical capacities were limited. As recently as 1995, only 8 percent of the population was university educated. So while the government reduced the incentives for low-technology manufacturers to come here and wages rose, the high-tech industries found themselves thinking about moving to nearby Malaysia and Thailand. That slowed Singapore's boom and challenged the government to find ways to stimulate a revival.

FROM THE FIELD NOTES

"My first visit to Singapore was by air, and I got to know the central city, where I was based, rather well. But the second was by freighter from Hong Kong, and this afforded quite a different perspective. Here was one reason for Singapore's success: I have simply never seen a neater, cleaner, better organized, or more modern port facility. Even piles of loose items were carefully stacked. Loading and offloading went quietly and orderly. Singapore is one of the world's leading *entrepôts*, where goods are brought in, stored, and transshipped. Nobody does it better, and Singapore carefully guards its reputation for dependability, on-schedule loading, and lack of corruption."

To accomplish this revival, Singapore has moved in several directions. First, it will focus on three growth areas: information technology, automation, and biotechnology. Second, there are notions of a "Growth Triangle" involving Singapore's developing neighbors, Malaysia and Indonesia; those two countries would supply the raw materials and cheap labor, and Singapore the capital and technical know-how. Third, Singapore opened its doors to capitalists of Chinese ancestry who left Hong Kong when China took it over and who wanted to relocate their enterprises here. Singapore's population is 77 percent Chinese, 14 percent Malay, and 8 percent South Asian. The government is Chinese-dominated, and its policies have served to sustain Chinese control. Indeed, China itself often cites Singapore's combination of authoritarian-

ism and economic success as proof that communism and market economics can coexist.

Singapore continues to build on its success. It has constructed a second bridge across the Johor Strait to Malaysia to accommodate the spiraling truck traffic between the two countries. An active land reclamation program already has added over 4 percent more territory to the crowded island; a recent scheme connected the main island to the Jurong group of islands in the southwest (Fig. 10-11). All seven Jurong Islands, in turn, were glued together by dikes, their intervening waters drained to create space for a giant petrochemical complex. Key to Singapore's progress has been its comparative lack of corruption, attracting many companies to establish themselves here as a base from which to do business in notoriously

FIGURE 10-12

corrupt countries elsewhere in the realm. Still, Singapore is the smallest of the Pacific Rim's economic tigers, and it faces growing competition near and far. Over the long term its relative location, which yielded its early wealth, may turn out to be its most enduring guarantee of survival—if not prosperity.

■ INDONESIA'S ARCHIPELAGO

18 The very term **archipelago** denotes fragmentation, and Indonesia is the world's most expansive archipelagic state. Spread across more than 17,000 islands, Indonesia's 224 million people live separated and clustered— separated by water and clustered on islands large and small.

The map of Indonesia requires some attention (Fig. 10-12). Five large islands dominate the archipelago territorially, one of which, New Guinea in the east, is not part of the Indonesian culture sphere, although its western half is under Indonesian control. The other four major islands are collectively known as the Greater Sunda Islands: *Jawa* (Java), smallest but by far the most populous and important; *Sumatera* (Sumatra) in the west, directly across the Strait of Malacca from Malaysia; *Kalimantan*, the Indonesian sector of large, compact, minicontinent Borneo; and wishbone-shaped, distended *Sulawesi* (Celebes) to the east. Extending eastward from Jawa are the Lesser Sunda Islands, including Bali and, near the eastern end, Timor. Another important island chain within Indonesia is the Maluku (Molucca) Islands, between Sulawesi and New

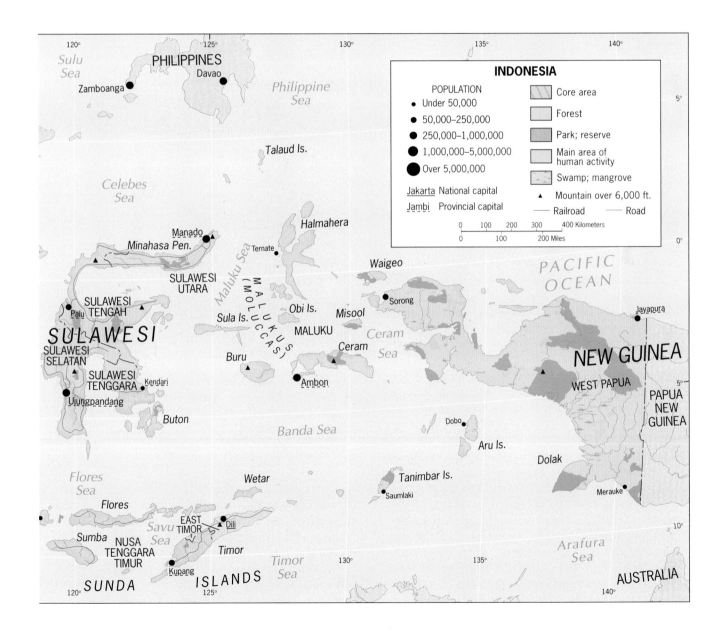

Guinea. The central water body of Indonesia is the Java Sea.

As the map shows, Indonesia does not control all of the archipelago. In addition to East Malaysia's territory on Borneo, there also is Brunei (see box titled "Rich and Broken Brunei"). And on the eastern part of the island of Timor, a long struggle against Indonesian rule led, in 1999, to a referendum on independence. The overwhelming majority of East Timorese voted in favor of sovereignty, and since then East Timor has moved toward statehood under United Nations auspices.

Indonesia is a Dutch colonial creation, and the Dutch chose Jawa as their colonial headquarters, making Batavia (now Jakarta) their capital. Today, **Jawa** remains the core of Indonesia. With about 125 million inhabitants, Jawa is one of the world's most densely peopled places and one of the most agriculturally productive. It also is the most highly urbanized part of a country in which more than 60 percent

of the people still live on the land, and the Pacific Rim boom of the 1990s had strong impact here. The city of Jakarta, on the northwestern coast, became the heart of a larger conurbation now known as *Jabotabek*, consisting of the capital as well as Bogor, Tangerang, and Bekasi. During the 1990s the population of this megalopolis grew from 15 to 20 million, and it is predicted to reach 30 million by 2010. Already, Jabotabek is home to over 10 percent of Indonesia's entire population and 25 percent of its urban population. Thousands of factories, their owners taking advantage of low prevailing wages, were built in this area, straining its infrastructure and overburdening the port of Jakarta. On an average day, hundreds of ships lie at anchor, awaiting docking space to offload raw materials and take on finished products.

Indonesia's experience with rapid economic growth was tempered by the failures of its government. Suharto, the same president who had ruled the country for 25

AMONG THE REALM'S GREAT CITIES . . . JAKARTA

Jakarta, capital of Indonesia, sometimes is called the Kolkata (Calcutta) of Southeast Asia. Stand on the elevated highway linking the port to the city center and see the villages built on top of garbage dumps by scavengers using what they can find in the refuse, and the metaphor seems to fit. There is poverty here unlike that in any other city in the realm.

But there are other sides to Jakarta. Indonesia's economic progress has made its mark here, and the evidence is everywhere. Television antennas and satellite dishes rise like a forest from rusted, corrugated-iron rooftops. Cars (almost all, it seems, late-model), mopeds, and bicycles clog the streets, day and night. A meticulously manicured part of the city center contains a cluster of highrise hotels, office buildings, and apartments. Billboards advertise planned communities on well-located, freshly cleared land.

Jakarta's population is a cross-section of Indonesia's, and the silver domes of Islam rise above the cityscape alongside Christian churches and Hindu temples. The city always was cosmopolitan, beginning as a conglomerate of villages at the mouth of the Ciliwung River under Islamic rule, becoming a Portuguese stronghold and later the capital of the Dutch East Indies (under the name Batavia). Advantageously situated on the northwestern coast of Jawa, Indonesia's most populous island, Jakarta bursts at the seams with growth. Sail into the port, and hundreds of vessels, carrying flags from Russia to Argentina, await berths. Travel to the outskirts, and huge shantytowns are being ex-

panded by a constant stream of new arrivals. So vast is the human agglomeration—nobody knows how many millions have descended on this place (the official figure is 13.1 million)—that the majority live without adequate (or any) amenities.

A seemingly permanent dome of metallic-gray air, heavy with pollutants, rests over the city. The social price of economic progress is high, but Jakartans are willing to pay it.

Rich and Broken Brunei

*B*runei is an anomaly in Southeast Asia—an oil-exporting Islamic sultanate far from the Persian Gulf. Located on the north coast of Borneo, sandwiched between Malaysian Sarawak and Sabah (Fig. 10-6, upper-left map), the Brunei sultanate is a former British-protected remnant of a much larger Islamic kingdom that once controlled all of Borneo and areas beyond. Brunei achieved full independence in 1984. With a mere 2228 square miles (5700 sq km)—slightly larger than Delaware—and only 395,000 people, Brunei is dwarfed by the other political entities of Southeast Asia (Fig. 10-1). But the discovery of oil in 1929 (and natural gas in 1965) heralded a new age for this remote territory.

Today, Brunei is one of the largest oil producers in the British Commonwealth, and recent offshore discoveries suggest that production will increase. As a result, the population is growing rapidly through immigration (64 percent of Brunei's residents are Malay, 20 percent Chinese), and the sultanate enjoys one of the highest standards of living in Southeast Asia. In 1998, however, a financial scandal of gigantic proportions was revealed, involving the disappearance of more than U.S. $15 billion and the collapse of

Brunei's largest private company, which had been under the direction of the Sultan's youngest brother. This financial calamity continued to buffet the country's economy five years later and was reflected by a steep drop in its per-capita GNP.

Most of Brunei's inhabitants live and work near the oil-fields in the western corner of the country and in the capital, Bandar Seri Begawan. Evidence of profligate spending is everywhere, ranging from sumptuous palaces, magnificent mosques, and luxurious hotels. In this respect Brunei offers a stark contrast to neighboring East Malaysia, but even within Brunei there are spatial differences. Brunei's interior remains an area of subsistence agriculture (a small minority of indigenous groups survive here) and rural isolation, its villages a world apart from the modern splendor of the coast.

Brunei's territory is not only small; it is fragmented. A sliver of Malaysian land separates the larger west from the east (where the capital lies), and Brunei Bay provides a water link between the two parts. It has been proposed that Brunei purchase this separating corridor from the Malaysians, but no action has yet been taken.

---

years, and who had enriched himself, his family, and numerous friends and cronies, was reelected in 1997 for another five-year term. His authoritarian rule, which some credited for creating the stability that made Indonesia's economic boom possible, tolerated no free opposition or open election campaigns. But when Indonesia's economy was battered by the slump that had earlier struck Thailand and Malaysia, political turmoil forced Suharto to step down. During the rioting that accompanied this transition, the Jawanese, as they had done in the past, turned on the Chinese in their midst.

Always in Indonesia, Jawa is where the power lies. The Jawanese constitute about 56 percent of the country's population. Sumatera, much larger than Jawa, has only about one-third as many people (45 million). Sulawesi has some 16 million, and Kalimantan, Indonesia's huge territory on Borneo, just 13 million. As for West Papua in the far east, which is larger than California, its population is barely more than 2.5 million.

Diversity in Unity

Indonesia's survival as a unified state is as remarkable as India's and Nigeria's. With more than 300 discrete ethnic clusters, over 250 languages, and just about every religion practiced on Earth (although Islam dominates),

actual and potential centrifugal forces are powerful here. Wide waters and high mountains perpetuate cultural distinctions and differences. Indonesia's national motto is *bhinneka tunggal ika*: diversity in unity.

What Indonesia has achieved is etched against the country's continuing cultural complexity. There are dozens of distinct aboriginal cultures; virtually every coastal community has its own roots and traditions. And the majority, the rice-growing Indonesians, include not only the numerous Jawanese—who share an adherence to Islam but have their own ethnic and cultural identity—but also the Sundanese (who constitute 14 percent of Indonesia's population), the Madurese (8 percent), and others. Perhaps the best impression of the cultural mosaic comes from the string of islands that extends eastward from Jawa to Timor (Fig. 10-12). The rice-growers of Bali adhere to a modified version of Hinduism, giving the island a unique cultural atmosphere; the population of Lombok is mainly Muslim, with some Balinese Hinduism; Sumbawa is a Muslim community; Flores is mostly Roman Catholic. In western Timor, Protestant groups dominate; in now-independent East Timor, where the Portuguese ruled, Roman Catholicism prevails. Nevertheless, Indonesia nominally is the world's largest Muslim country: overall, 88 percent of the people adhere to Islam, and in the cities the silver domes of neighborhood mosques rise above the townscape. For

FROM THE FIELD NOTES

"Getting to Ambon was no easy matter, and my boat trip was enlivened by an undersea earthquake that churned up the waters and shook up all aboard. The next morning our first view of the town of Ambon, provincial capital of Maluku, showed a center dominated by a large mosque with a modern minaret and a large white dome to the right of the main street leading to the port. My host from the local university told me that, as on many of the islands of the Malukus, about half of the people were Christians, not Muslims, and that some large churches were situated in the outskirts. 'But the relationships between Muslims and Christians are worsening,' he said. It had to do with the arrival of Jawanese, some of whom were 'agitators' and Islamic fundamentalists, stirring up religious passions. 'Have you heard of the Taliban?' he asked. 'Well, we have similar so-called religious students here, educated in Islamic schools that teach extremism.' . . . We drove from the town into the countryside to the west, toward Wakisihu and past the old Dutch Fort Rotterdam. 'Let me show you something,' he said as we turned up a dirt road. At its end, in the middle of a field, sat a large single structure called University of Islam, distinguished by a pair of enormous stairways. 'Can you call a single building like this where all they teach is Islam a university?' he asked. 'What they teach here is Islamic fundamentalism and intolerance.' He was prophetic. Weeks later, religious conflict broke out, the mosque in town was burned along with Christian churches; his own university lay in ruins."

centuries, the regional version of Islam was a moderate and tolerant one, and many Indonesians were "nominal" Muslims, practicing their faith irregularly and adhering only loosely to its precepts. In recent years, however, radical Islamic movements have arisen in Indonesia, treated warily by a government unsure of the right response. As the War on Terror reached into Indonesia in the form of pressure on the Indonesian administration to take firmer action and to investigate possible al-Qaeda links, terrorists responded in 2002 by bombing a nightclub in Bali. More than 200 people, mostly vacationing Westerners and about half of them Australians, were killed (see photo next page).

Indonesia's economic downturn and its internal political upheavals had already fostered Islamic revivalism in the sprawling country, and not only in Jawa, before the events of September 2001 changed the world. As we note in some detail later, the government's longstanding population policies unintentionally contributed to rising conflict between Muslims and Christians, and until 2001 Jakarta's biggest problem was to contain this strife and

to restrain Muslim radicals. Now the question is whether this, the world's largest Muslim-dominated country by far, can maintain its historic moderation in the face of provocations near and far.

Transmigration and the Outer Islands

As noted earlier, Indonesia's population of 224 million makes it the world's fourth most populous country, but Jawa, we also noted, contains approximately 56 percent of it. With about 125 million people on an island the size of Louisiana, the population pressure is enormous here. Moreover, Indonesia's annual rate of population growth remains at 1.6 percent, resulting in a doubling time of 44 years. To deal with this problem, and at the same time to strengthen the core's power over outlying areas, the Indonesian government long pursued a policy known as **transmigration**, inducing Jawanese (and Madurese from the adjacent island of Madura) to relocate to other islands. Several million Jawanese have moved to locales as

In the immediate aftermath of the pair of late-Saturday-night bomb blasts that wreaked havoc in the crowded nightclub district of Bali's Kuta Beach on October 13, 2002, foreign tourists carry their baggage through the wreckage as they hastily depart the island. The final death toll came to 202, with many of the victims being young Australian vacationers; in addition, more than 300 people were injured, many of them maimed for life. With the threat of further terrorist attacks by Islamic militants an ever-present one, another casualty has been Bali's tourist boom. Predictions of a swift post-disaster recovery have not been realized, perhaps due in part to the twin scourges of the Iraq War and the SARS pandemic that plagued international air travel during the first half of 2003. According to a World Bank report at the end of that half-year, sales revenues and profits throughout Bali had dropped as sharply as 70 percent from the previous six-month period. With its reputation still badly tarnished, tourists are not likely to return to the island in significant numbers for some time to come.

distant as the Malukus, Sulawesi, and Sumatera; many Madurese were resettled in Kalimantan.

In the government's view, the transmigration policy would serve to counter centrifugal forces arising from Indonesia's geography, spread the official language (Bahasa Indonesia, a modified form of Malay), and bring otherwise unexploited land into production. It has done so, but it also had negative effects. In Kalimantan, the Muslim Madurese found themselves facing armed Dayaks, indigenous people who did not want to give up their land, and thousands were killed in what amounted to a regional civil war. Elsewhere, Jawanese often were blamed for introducing methods and practices of administration and business that ran counter to local traditions. In West Papua, where Indonesia is in effect a colonial power, aboriginal people have attacked Indonesian newcomers.

In fact, the Dutch colonialists started the Transmigration Program, but Sukarno abandoned it and it was not revived until 1974, when Suharto saw economic and political opportunities in it. Over the next 25 years, nearly 8 million people were relocated; the World Bank gave Indonesia nearly U.S. $1 billion to facilitate the program before it pulled out amid reports of forced migration.

The results were mixed. World Bank project reviews suggested that the vast majority of the transmigrants were happy with the land, schools, medical services, and other amenities with which they were provided, but independent surveys painted a different picture. About half of the migrants never managed to rise above a life of subsistence; many never received even the most basic needs for survival in their new environment; land was simply taken away from indigenous inhabitants, notably in Kalimantan; the newcomers did much damage to the traditional cultures they invaded; and the program led to massive forest and wetland destruction. And thousands of migrants died in clashes with local inhabitants.

In March 2001, the Transmigration Program was canceled as one of the final acts of the short-lived government of President Abdurrahman Wahid, but the damage it did will trouble Indonesia for generations. From northern Sumatera to western Kalimantan to central Sulawesi to the islands of the Malukus and West Papua, cultural strife and its polarizing effect will challenge Indonesian governments of the future.

The Major Islands

Indonesia is an important state in a crucial geographic location. Each of its four major territorial components, in addition to Jawa, is geographically distinct.

Sumatera, Indonesia's westernmost major island, lies across the narrow Strait of Malacca from West Malaysia. Tropical rainforest still covers enough of it to sustain refuges for the orangutan population, but logging and human population growth are making ever greater inroads. Sharing Malaya's equatorial environments, Sumatera in colonial times became a base for rubber and palm-oil plantations. Its high relief and fertile volcanic soils make possible the cultivation of a wide range of crops, from vegetables in the highlands, to tea and coffee on the slopes, to subsistence root crops in the lowlands. Sumatera and its neighboring small islands contain large coal reserves, petroleum and natural gas, tin, bauxite (for aluminum), and even gold and silver. As the map suggests, intra-island surface communications are still weak, although the north-south Sumatera Highway was completed in the 1980s. The highest population density is around the northern city of Medan; another cluster lies in the middle zone of the island; and a third, augmented by a large contingent of transmigrated Jawanese, focuses on Palembang and Bandar Lampung in the south.

The northern part of the island has long been the most fractious. While the Batak people (see Fig. 10-3) adjusted to colonization and Westernization, and made Medan one of Indonesia's Pacific Rim boom towns, the Aceh fought the Dutch in a 30-year war that ended only in the twentieth century. In the 1990s, while Indonesia, Malaysia, and Thailand talked of an economic growth triangle including northern Sumatera, the Aceh mounted another rebellion—this time against Jakarta. Having just agreed to allow a referendum on independence to take place in East Timor, the new democratic government under then-president Wahid appeared ready to negotiate a settlement. Then, even as the Aceh problem simmered, the House of Representatives of the central Sumateran Province of Riau approved a "declaration of independence." And the House of East Kalimantan (Kalimantan Timur in Fig. 10-12) proposed that the province be renamed the "Federal State of East Kalimantan." Devolutionary forces are at work throughout Indonesia.

Certainly, the Aceh problem is crucial among these issues. With its population of more than 4 million, its natural resources, its militant history, and its peninsular situation, the province's Free Aceh Movement would seem to have a better chance than most components of the republic to force Jakarta's hand. But in mid-2003 a negotiated settlement, brokered by a Swiss team, appeared in prospect. The Acehnese would shelve their demands for independence, and the Indonesians would agree to consider all other options prior to an election in 2004. Ending the violence and settling the political question would go a long way toward ensuring Indonesia's coherence elsewhere in its far-flung domain.

Kalimantan is the Indonesian part of the minicontinent of Borneo, a slab of the Earth's crystalline crust whose backbone of mountains is of erosional, not volcanic, origin. Compact and massive—at over 290,000 square miles (750,000 sq km) it is larger than Texas—Borneo has a deep, densely forested interior that is a last refuge of some 35,000 orangutans. Elephants, rhinoceroses, and tigers still survive even as the loggers constrict

their habitat. Numerous other species of fauna—insects, reptiles, mammals, birds—inhabit the rich mountain and lowland forests. So much of Borneo's Pleistocene heritage has survived principally because of its comparatively small human population. The Indonesian part, Kalimantan, constitutes 28 percent of the national territory, but, with 13 million people, it contains less than 6 percent of Indonesia's population.

Figure 10-12 shows that the main areas of human activity lie in the west, southeast, and east. All towns of any size, including Pontianak in the west and Balikpapan in the east, are on or near the coast; the rivers still form important routes into the interior. Parks and reserves to protect flora and fauna cover but a small portion of the vast territory. Not only the flora and fauna, but also the human population has ancient roots. Indigenous peoples, including some of the Dayak clans, continue their slash-and-burn subsistence in the interior; others have become sedentary farmers. But Indonesia's transmigration policy has brought about half a million Jawanese and Madurese to Kalimantan, and they are laying out farms and digging drainage canals, bringing drastic change to rural areas. Of Kalimantan's four provinces, only the southernmost exports significant raw materials, including oil from offshore reserves, coal, iron ore, and some gold and diamonds; but in the national picture, Indonesian Borneo remains little developed.

Sulawesi is an island that looks like a set of intersecting mountain ranges rising from the sea. Its northern end is a 500-mile (800-km) chain of extinct, dormant, and active volcanoes known as the Minahasa Peninsula, its future northern extension toward the Philippine Sea already marked by a ribbon of seafloor volcanoes just rising above the surface. So rugged is the relief on this island that you cannot take surface transportation from Palu to Ujungpandang. The ethnic mosaic is quite complex, with seven major groups inhabiting more or less isolated parts of the island as well as a sizeable cluster of Jawanese transmigrants who are concentrated in Sulawesi Tengah (Fig. 10-12). Since 1999, the level of strife between local Christians and immigrant Jawanese Muslims has risen and has destabilized this remote province of Sulawesi—forcing Christians who once lived peacefully among Muslim villagers to flee, and culturally segregating settlements.

The two most populous and developed parts of Sulawesi are the southwestern peninsula centered on Ujungpandang and the eastern end of the Minahasa Peninsula, where Manado is the urban focus. Subsistence farming occupies most of the population of 16 million, but logging and wood products, some mining, and fishing augment the cash economy. Cultural landscapes are varied; in Minahasa, Christian churches remain numerous and form reminders of the colonial period, when relations between this

"I drove from Manado on the Minahasa Peninsula in northeastern Sulawesi to see the ecological crisis at Lake Tondano, where a fast-growing water hyacinth is clogging the water and endangering the local fishing industry. On the way, in the town of Tomolon, I noticed this side street lined with prefabricated stilt houses in various stages of completion. These, I was told, were not primarily for local sale. They were assembled from wood taken from the forests of Sulawesi's northern peninsula, then taken apart again and shipped from Manado to Japan. 'It's a very profitable business for us,' the foreman told me. 'The wood is nearby, the labor is cheap, and the market in Japan is insatiable. We sell as many as we can build, and we haven't even begun to try marketing these houses in Taiwan or China.' At least, I thought, this wood was being converted into a finished product, unlike the mounds of logs and planks I had seen piled up in the ports of Borneo awaiting shipment to East Asia."

province and the Dutch were favorable. During the liberation period, Minahasans sided with the Dutch (as many of their neighbors in the Malukus did) and called their peninsula Holland's "twelfth province." It was an unfulfilled wish, souring relations between Jakarta and Manado for years afterward. But when President Suharto was forced to resign in 1998, his transitional successor, Habibie, was a son of Minahasa, and the past was forgotten.

West Papua, like East Timor, fell to Indonesia well after the Dutch colonial era had ended, but unlike East Timor, the United Nations approved Indonesia's takeover in 1969. Until 2002 this province was known by its Bahasa Indonesian name of *Irian Jaya* (Guinea West), but in that year the government of President Megawati Sukarnoputri bowed to the demands of indigenous leaders of the Papuan communities on the island and changed its name. As the map shows, the eastern half of the island of New Guinea comprises the independent state of Papua New Guinea. The Papuans are divided not only into numerous ethnolinguistic groups but also into two political entities.

West Papua is an Indonesian province, but it lies in the Pacific geographic realm, not in Southeast Asia. As such, it has about 22 percent of the country's land area but just 1 percent of its population, including more than 200,000 ethnic Indonesians, most from Jawa and many of them concentrated in the provincial capital with the non-Papuan name of Jayapura. Forested, high-relief, glacier-peaked West Papua, lying as it does in a different geographic realm, is a world apart from teeming Jawa. It has what is reputedly the world's richest gold mine and second-largest open-pit copper mine; its forests not only harbor the indigenous peoples but also sustain a major logging industry.

Sanctioned as Indonesia's rule in West Papua may be, local opposition has continued for decades. The *Organisasi Papua Merdeka* (Free Papua Movement, or FPM) and occasional rebel attacks remind the Jakarta government of its role here. In 1999, during the turmoil in the other provinces, the FPM held rallies in Jayapura to promote its cause, and in 2000 a congress was held where independence was demanded and a Papuan flag, the Morning Star, displayed. Later that year, Indonesian troops killed demonstrators in Merauke on the province's south coast. All this is deepening Papuan opposition to Jakarta's rule, and the time may come when Indonesia will rue its acquisition of West Papua, potential source of the kinds of centrifugal forces an archipelagic state must constantly confront.

EAST TIMOR

The easternmost of the lesser Sunda Islands is Timor, the eastern part of which was a Portuguese colony, overrun by Indonesia in 1975 and annexed in 1976, which became the scene of a bitter struggle for independence. When the people of East Timor in 1999 were allowed to express their views on independence, this Connecticut-sized territory with fewer than 800,000 inhabitants took the first steps toward nationhood.

Indonesia's armed forces reacted violently, adding hundreds more to the tens of thousands who had already died in the struggle against Jakarta. The country's modest infrastructure was devastated, and its people were dislocated. Foreign intervention, led by Australia (once

Indonesia's sole supporter in its claim to East Timor), established some order. Reconstruction and social reorganization, under UN auspices, have been in progress since.

In Indonesia, many viewed the "loss" of East Timor as the first step in the disintegration of the state, to be followed by secession in Aceh, the Malukus, and West Papua. But in truth, East Timor was a special case, and if devolution were indeed to fragment Indonesia, East Timor was not the prototype. Unlike West Papua, the Malukus, or Aceh, Indonesia's control over East Timor defied UN rules. This was not merely an internal matter.

East Timor became an independent state in May 2002. As the map shows, its sovereignty creates several politico-geographical complications (Fig. 10-13). The political entity of East Timor consists of a main territory,

where the capital named Dili is located, and a small exclave on the north coast of (Indonesian) West Timor called Ocussi (spellings vary, and it is sometimes mapped as Ocussi-Ambeno or Ambeno Province). Although there is a road from the exclave through the center of the island to the main territory, relations between East Timor and Indonesia may not make this link feasible; therefore, the two parts of the fragmented new state may have to be connected by boat traffic.

East Timor's independence also creates a new alignment at sea, and the country's very future may depend on this realignment. As we discuss in detail in Chapter 12, all coastal states have maritime as well as terrestrial boundaries, and these maritime boundaries sometimes endow a country with valuable offshore oil, gas, or other resources.

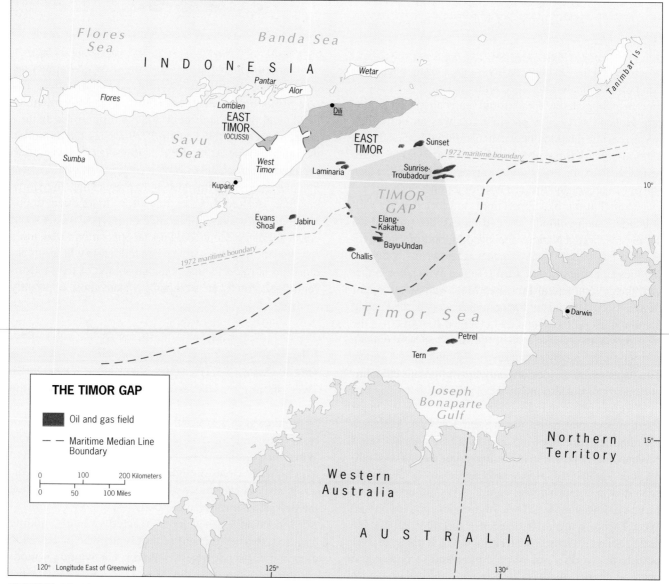

FIGURE 10-13

South of East Timor lies an area called the *Timor Gap* that contains major oil and gas reserves. While Indonesia ruled East Timor, Indonesia and Australia agreed on a maritime boundary that greatly favored the Australians (the 1972 boundary on Figure 10-13). But after East Timor achieved independence, its new government wanted to erase the bilateral 1972 boundary and replace it with one based on United Nations regulations. That would put the boundary midway between Australia and East Timor, and assign to East Timor the bulk of the oil and gas reserves within the Timor Gap as well as newly discovered fields elsewhere under the Timor Sea. But Australia, seeing this coming, withdrew from the UN Convention on the Law of the Sea in 2002 and argued that it is not bound by any UN regulations in this matter. It offered East Timor most of the revenues from one gas field, Bayu-Undan, but nothing from the other reserves. East Timor in 2003 faced a choice between accepting a disadvantageous deal that would net it tens of millions of dollars or challenging Australia over a prize that may amount to as much as U.S. $6 billion. The future of the country depends on the resolution of this dilemma.

FRAGMENTED PHILIPPINES

North of Indonesia, across the South China Sea from Vietnam, and south of Taiwan lies an archipelago of more than 7000 islands (only about 460 of them larger than one square mile in area) inhabited by 83.9 million people. The inhabited islands of the Philippines can be viewed as three groups: (1) Luzon, largest of all, and Mindoro in the north, (2) the Visayan group in the center, and (3) Mindanao, second largest, in the south (Fig. 10-14). Southwest of Mindanao lies a small group of islands, the Sulu Archipelago, nearest to Indonesia, where a Muslim-based insurgency has kept the area in turmoil.

Few of the generalizations we have been able to make for Southeast Asia could apply in the Philippines without qualification. The country's location relative to the mainstream of change in this part of the world has had much to do with this situation. The islands, inhabited by peoples of Malay ancestry with Indonesian strains, shared with much of the rest of Southeast Asia an early period of Hindu cultural influence, which was strongest in the south and southwest and diminished northward. Next came a Chinese invasion, felt more strongly on the largest island of Luzon in the northern part of the Philippine archipelago. Islam's arrival was delayed somewhat by the position of the Philippines well to the east of the mainland and to the north of the Indonesian islands. The few southern Muslim beachheads were soon overwhelmed by the Spanish invasion during the sixteenth century. Today the Philippines, adjacent to

the world's largest Muslim state (Indonesia), is 83 percent Roman Catholic, 7 percent Protestant, and only 5 percent Muslim.

The Muslim population, concentrated on the southeastern flank of the archipelago (Fig 10-14), has long proclaimed its marginalization in this dominantly Christian country. Over the past 30 years, a half dozen Muslim organizations have promoted the Muslim cause with tactics ranging from political pressure to violent insurgency. The campaigns of violence have affected all of the Muslim areas, from southern Mindanao to Palawan, and the strategies of the rebels have ranged from bombings to kidnappings. Densely forested Basilan Island is the headquarters of a group known as Abu Sayyaf, which demanded a separate Muslim state and is believed to have received support through the al-Qaeda network. The War on Terror reached into this part of the Philippines when American troops joined Filipino forces to pursue Abu Sayyaf terrorists. Also active on behalf of Muslim causes in this area are the Moro National Liberation Front and the Moro Islamic Liberation Front; the Philippine government has entered into negotiations with both. Cease-fire agreements, however, have not held. The Muslim challenge is pirating a disproportionate share of Manila's operating budget.

Out of the Philippines melting pot, where Mongoloid-Malay, Arab, Chinese, Japanese, Spanish, and American elements have met and mixed, has emerged the distinctive Filipino culture. It is not a homogeneous or a unified culture, but in Southeast Asia it is in many ways unique. One example of its absorptive qualities is demonstrated by the way the Chinese infusion has been accommodated: although the "pure" Chinese minority numbers less than 2 percent of the population (far lower than in most Southeast Asian countries), a much larger portion of the Philippine population carries a decidedly Chinese ethnic imprint. What has happened is that the Chinese have intermarried, producing a sort of Chinese-mestizo element that constitutes more than 10 percent of the total population. In another cultural sphere, the country's ethnic mixture and variety are paralleled by its great linguistic diversity. Nearly 90 Malay languages, major and minor, are spoken by the 84 million people of the Philippines; only about 1 percent still use Spanish. At independence in 1946, the largest of the Malay languages, Tagalog, or Pilipino, became the country's official language, and the educational system promotes its general use. English is learned as a subsidiary language and remains the chief *lingua franca*; an English-Tagalog hybrid ("Taglish") is increasingly heard today, cutting across all levels of society.

The widespread use of English in the Philippines, of course, results from a half-century of American rule and influence, beginning in 1898 when the islands were ceded to the United States by Spain under the terms of

FIGURE 10-14

the treaty that followed the Spanish-American War. The United States took over a country in open revolt against its former colonial master and proceeded to destroy the Filipino independence struggle, now directed against the new foreign rulers. It is a measure of the subsequent success of U.S. administration in the Philippines that this was the only dependency in Southeast Asia that sided against the Japanese during World War II in favor of the colonial power. The U.S. rule had its good and bad features, but the Americans did initiate reforms that were long overdue, and they were already in the process of negotiating a future independence for the Philippines when the war intervened in 1941.

The Philippines' population, concentrated where the good farmlands lie in the plains, is densest in three general areas (Fig. G-9): (1) the northwestern and south-central part of Luzon, (2) the southeastern extension of Luzon, and (3) the islands of the Visayan Sea between Luzon and Mindanao. Luzon is the site of the capital, Manila-Quezon City (10.7 million, more than one-eighth of the entire national population), a major metropolis facing the South China Sea. Alluvial as well as volcanic soils, together with ample moisture in this tropical environment, produce self-sufficiency in rice and other staples and make the Philippines a net exporter of farm products despite a high population growth rate of 2.4 percent.

The Philippines seems to get little mention in discussions of developments on the Pacific Rim, and yet it would seem to be well positioned to share in the Pacific Rim's economic growth. Governmental mismanagement and political instability have slowed the country's participation, but during the 1990s the situation improved. Despite a series of jarring events—the ouster of U.S. military bases, the damaging eruption of a volcano near the capital, the violence of Muslim insurgents, and a dispute over the nearby Spratly Islands in the South China Sea—the Philippines made substantial economic progress during the decade. Its electronics and textile industries (mostly in the Manila hinterland) expanded continuously, and more foreign investment arrived. The centrally positioned Visayan island of Cebu experienced particularly rapid growth, based on its central location, good port, expanded airport, large and literate labor force, and a cadre of managers experienced in the service industries. A free-trade zone attracted foreign companies from the United States and Japan, and now products from toys to semiconductors flow from Cebu's manufacturers to world markets. But agriculture continues to dominate the Philippines' economy, unemployment remains high, further land reform is badly needed, and social restructuring (reducing the controlling influence over national affairs by a comparatively small group of families) must occur. However, progress is being made. The country now is a lower-middle-income economy, and given a longer period of stability and success in reducing the population growth rate, it will rise to the next level and finally take its place among Pacific Rim growth poles.

In recent years, the population issue has divided this dominantly Roman Catholic society, with the government promoting family planning and the clergy opposing it. But behind this debate lies another of the Philippines' assets: in a realm of mostly undemocratic regimes, the Philippines has come out of its period of authoritarian rule a rejuvenated, if not yet robust, democracy.

AMONG THE REALM'S GREAT CITIES . . . **MANILA**

Manila, capital of the Philippines, was founded by the Spanish invaders of Luzon more than four centuries ago. The colonists made a good choice in terms of site and situation. Manila sprawls at the mouth of the Pasig River where it enters one of Asia's finest natural harbors. To the north, east, and south a crescent of mountains encircles the city, which lies just 600 miles (975 km) across the South China Sea from Hong Kong.

Manila, named after a flowering shrub in the local marshlands, is bisected by the Pasig, which is bridged in numerous places. The old walled city, Intramuros, lies to the south. Despite heavy bombardment during World War II, some of the colonial heritage survives in the form of churches, monasteries, and convents. St. Augustine Church, completed in 1599, is one of the city's landmarks.

The CBD of Manila lies on the north side of the Pasig River. Although Manila has a well-defined commercial center with several avenues of luxury shops and modern buildings, the skyline does not reflect the high level of energy and activity common in Pacific Rim cities on the opposite side of the South China Sea. Neither is Manila a city of notable architectural achievements. Wide, long, and straight avenues flanked by palm, banyan, and acacia trees give it a look not unlike parts of San Juan, Puerto Rico.

In 1948, a newly built city immediately to the northeast of Manila was inaugurated as the *de jure* capital of the Philippines and called Quezon City. The new facilities were eventually to house all government offices, but many functions of the national government never made the move. In the meantime, Manila's growth overtook Quezon City's, so that it became part of the Greater Manila metropolis (which today is home to 10.7 million). Although the proclamation of Quezon City as the Philippines' official capital was never rescinded, Manila remains the *de facto* capital of the country today.

Continued on pages 212-213

FIGURE 11-1

Scale 1:16 000 000; one inch to 250 miles. Lambert's Azimuthal, Equal Area Project

Elevations and depressions are given in feet

Relief

Meters		Feet
3050		10 000
1525		5000
610		2000
305		1000
152.5		500
0	Sea Level	0
152.5		500 Below
1525		5000 Sea Level
3050		10 000
6100		20 000

A-590200-76   6-5-17
COPYRIGHT BY
RAND McNALLY & COMPANY
MADE IN U.S.A.

Longitude 115° East of Greenwich

Chapter 11

The Austral Realm

CONCEPTS, IDEAS, AND TERMS

1. Austral
2. Southern Ocean
3. Subtropical Convergence
4. West Wind Drift
5. Biogeography
6. Wallace's Line
7. Aboriginal population
8. Outback
9. Federation
10. Unitary state
11. Import-substitution industries
12. Aboriginal land issue
13. Immigration policies
14. Environmental degradation
15. Peripheral development

REGIONS

- AUSTRALIA
- CORE AREA
- OUTBACK
- NEW ZEALAND

Cities and Towns

0 to 50,000	500,000 to 1,000,000
50,000 to 500,000	1,000,000 and over

DEFINING THE REALM

THE AUSTRAL REALM is geographically unique. It is the only geographic realm that lies entirely in the Southern Hemisphere. It is also the only realm that has no land link of any kind to a neighboring realm and is thus completely surrounded by ocean and sea. It is second only to the Pacific as the world's least populous realm. Appropriately, its name refers to its location (**Austral** means south)—a location far from the sources of its dominant cultural heritage but close to its newfound economic partners on the western Pacific Rim.

Two countries constitute the Austral Realm: Australia, in every way the dominant one, and New Zealand, physiographically more varied than its giant partner. Between them lies the Tasman Sea. To the west lies the Indian Ocean, to the east the Pacific, and to the south the frigid Southern Ocean (see box titled "Is There a Southern Ocean"?).

This southern realm is at a crossroads. On the doorstep of populous Asia, its Anglo-European legacies are now infused by other cultural strains. Polynesian Maori in New Zealand and Aboriginal communities in Australia are demanding better terms of life. Pacific Rim markets are buying huge quantities of raw materials. Japanese and other Asian tourists fill hotels and resorts. Queensland's tropical Gold Coast resembles Honolulu's Waikiki. The streets of Sydney and Melbourne display a multicultural panorama unimagined just two generations ago. All these changes have stirred political debate. Issues ranging from immigration quotas to indigenous land rights dominate, exposing social fault lines (city versus Outback in Australia, North and South in New Zealand). Aborigines and Maori were here first, and the Europeans came next. Now Asia looms in Australia's doorway.

MAJOR GEOGRAPHIC QUALITIES OF

The Austral Realm

1. Australia and New Zealand constitute a geographic realm by virtue of territorial dimension, relative location, and dominant cultural landscape.

2. Despite their inclusion in a single geographic realm, Australia and New Zealand differ physiographically. Australia has a vast, dry, low-relief interior; New Zealand is mountainous.

3. Australia and New Zealand are marked by peripheral development—Australia because of its aridity, New Zealand because of its topography.

4. The populations of Australia and New Zealand are not only peripherally distributed but also highly clustered in urban centers.

5. The realm's human geography is changing—in Australia because of Aboriginal activism and Asian immigration, and in New Zealand because of Maori activism and Pacific-islander immigration.

6. The economic geography of Australia and New Zealand is dominated by the export of livestock products (and in Australia also by wheat production and mining).

7. Australia and New Zealand are being integrated into the economic framework of the western Pacific Rim, principally as suppliers of raw materials.

LAND AND ENVIRONMENT

Physiographic contrasts between massive, compact Australia and elongated, fragmented New Zealand are related to their locations with respect to the Earth's tectonic plates (consult Fig. G-4). Australia, with some of the geologically most ancient rocks on the planet, lies at the center of its own plate, the Australian Plate. New Zealand, younger and less stable, lies at the convulsive convergence of the Australian and Pacific Plates. Earthquakes are rare in Australia, and volcanic eruptions are unknown; New Zealand has plenty of both. This locational contrast is also reflected by differences in relief (Fig. 11-1). Australia's highest relief occurs in what Australians call the Great Dividing Range, the mountains that line the east coast from the Cape York Peninsula to southern Victoria, with an outlier in Tasmania. The highest point along these old, now eroding mountains is Mount Kosciusko, 7310 feet (2228 m) tall. In New Zealand, entire ranges are higher than this, and Mount Cook reaches 12,315 feet (3754 m).

West of Australia's Great Dividing Range, the physical landscape generally has low relief with some local exceptions such as the Macdonnell Ranges near the center; plateaus and plains dominate (Fig. 11-2). The Great Artesian Basin is a key physiographic region, providing underground water sources in what would otherwise be desert country, and crossed in the south by the continent's major river system, the Murray-Darling. The area mapped as *Shield* in Figure 11-2 contains Australia's oldest rocks and much of its mineral wealth.

Figure G-8 reveals the effects of latitudinal location and interior isolation on Australia's climatology. In this respect, Australia is far more varied than New Zealand, its climates ranging from tropical in the far north, where rainforests flourish, to Mediterranean in parts of the south. The interior is dominated by desert and steppe conditions, the steppes providing the grasslands that sustain tens of millions of livestock. Only in the east does Australia have an area of humid temperate climate, and here lies most of the country's economic core area. New Zealand, by contrast, is totally under the influence of the Southern and Pacific Oceans, creating moderate, moist conditions, temperate in the north and colder in the south.

Biogeography

One of this realm's defining characteristics is its wildlife. Australia is the land of kangaroos and koalas, wallabies and wombats, possums and platypuses. These and numerous other *marsupials* (animals whose young are born very early in their development and then carried in an abdominal pouch) owe their survival to Australia's early isolation during the breakup of Gondwana (see Fig. 6-3). Before more advanced mammals could enter Australia and replace the marsupials, as happened in other parts of the

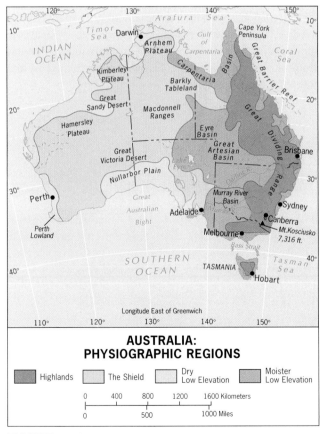

AUSTRALIA: PHYSIOGRAPHIC REGIONS

Highlights   The Shield   Dry Low Elevation   Moister Low Elevation

0   400   800   1200   1600 Kilometers

0   500   1000 Miles

FIGURE 11-2

Is There a Southern Ocean?

*H*ere is an interesting geographic experiment: try to find the Southern Ocean on commonly used maps of the world—commonly used, that is, here in the Northern Hemisphere. You are unlikely to find it, even on maps and globes published by the National Geographic Society, Rand McNally, and other major mapmaking organizations.

Now look in the library for atlases and maps published in Southern Hemisphere countries, for example, New Zealand and Australia. There the Southern Ocean is a prominent geographic feature. And it is not just a matter of opinion: to people whose countries adjoin it, the Southern Ocean is a significant factor in daily life. Its persistent westerly winds influenced the course of history. It is a crucial weather maker. Beneath its vast and frigid surface lie resources yet unknown. Its waters contain biota that are crucial to the global web of marine life.

How is it that the primary issue involving the **Southern Ocean** is whether it exists at all? The all-too-obvious answer is that this ocean, unlike the others on our planet, is not neatly bounded by continental coasts. Of course, it borders Antarctica to the south, but its northern limits are not visible on a relief map. So mapmakers simply extend the Atlantic, Pacific, and Indian Oceans all the way to Antarctic shores, ignoring the very existence of an ocean as large as the Indian.

For us geographers, it is a good exercise to turn the globe upside down now and then. After all, the usual orientation is quite arbitrary. Modern mapmaking started in the Northern Hemisphere, and the cartographers put their hemisphere on top and the other at the bottom. That is now the norm, and it can distort our view of the world. In bookstores in the Southern Hemisphere, you sometimes see tongue-in-cheek maps showing Australia and Argentina at the top, and Europe and Canada at the bottom. But this matter has a serious side. A reverse view of the globe shows us how vast the ocean encircling Antarctica is. The Southern Ocean may be remote, but its existence is real.

Where do the northward limits of the Southern Ocean lie? This ocean is bounded not by land but by a marine transition called the **Subtropical Convergence**. Here the cold, extremely dense waters of the Southern Ocean meet the warmer waters of the Atlantic, Pacific, and Indian Oceans. It is quite sharply defined by changes in temperature, chemistry, salinity, and marine fauna. Flying over it, you can actually observe it in the changing colors of the water: the Antarctic side is a deep gray, the northern side a greenish blue.

Although the Subtropical Convergence moves seasonally, its position does not vary far from latitude 40° South, which also is the approximate northern limit of Antarctic icebergs. Defined this way, the great Southern Ocean is a huge body of water that moves clockwise (from west to east) around Antarctica, which is why we also call it the **West Wind Drift**.

world, the landmass was separated from Antarctica and India, and today it contains the world's largest assemblage of marsupial fauna.

Australia's vegetation also has distinctive qualities, notably the hundreds of species of eucalyptus trees native to this geographic realm. Many other plants form part of Australia's unique flora, some with unusual adaptation to the high temperatures and low humidity that characterize much of the continent.

The study of fauna and flora in spatial perspective combines the disciplines of biology and geography in a field known as **biogeography**, and Australia is a giant laboratory for biogeographers. In the Introduction (pp. 10–14), we noted that several of the world's climatic zones are named after the vegetation that marks them: tropical savanna, steppe, tundra. When climate, soil, vegetation, and animal life reach a long-term, stable adjustment, vegetation forms the most visible element of this ecosystem.

Biogeographers are especially interested in the distribution of plant and animal species, and in the relationships between plant and animal communities and their natural environments. (The study of plant life is called *phytogeography*; the study of animal life is called *zoogeography*.) These scholars seek to explain the distributions the map reveals. In 1876 one of the founders of biogeography, Alfred Russel Wallace, published a book entitled *The Geographical Distribution of Animals* in which he fired the first shot in a long debate: where does the zoogeographic boundary of Australia's fauna lie? Wallace's fieldwork in the area revealed that Australian forms exist not only in Australia itself but also in New Guinea and in some islands to the west. So Wallace proposed that the faunal boundary should lie between Borneo and Sulawesi, and just east of Bali (Fig. 11-3).

Wallace's Line soon was challenged by other researchers, who found species Wallace had missed and who visited islands Wallace had not. There was no question that Australia's zoogeographic realm ended somewhere in the Indonesian archipelago, but where? Western Indonesia was the habitat of nonmarsupial animals such as tigers, rhinoceroses, and elephants, as well as primates; New Guinea clearly was part of the realm of the marsupials. How far had the more advanced mammals progressed eastward along the island stepping stones toward New Guinea? The zoogeographer Max Weber found evidence that led him to postulate his own *Weber's Line*, which, as Figure 11-3 shows, lay very close to New Guinea.

Not all research in zoogeography or phytogeography deals with such large questions. Much of it focuses on the relationships between particular species and their habitats, that is, the environment they normally occupy and of which they form a part. Such environments change, and the changes can spell disaster for the species. In Australia, the arrival of the **Aboriginal population** (between 50,000 and 60,000 years ago) had limited effect on the habitats of the extant fauna. But the invasion of the European colonizers and the introduction of their livestock led to the destruction of habitats and the extinction of many native species. Whether in East Africa or in Western Australia, the key to the conservation of what remains of natural flora and fauna lies in the knowledge embodied by the field of biogeography.

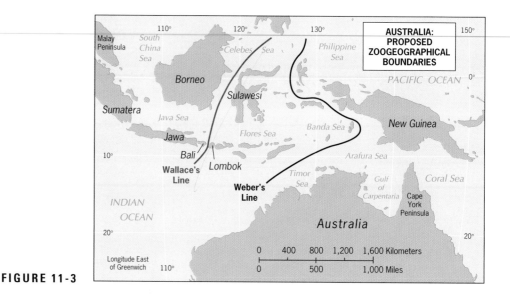

FIGURE 11-3

REGIONS OF THE REALM

AUSTRALIA IS THE dominant component of the Austral Realm, a continent-scale country in a size category that also includes China, Canada, the United States, and Brazil. For two reasons, however, Australia has fewer regional divisions than do the aforementioned countries: Australia's relatively uncomplicated physiography and its diminutive human numbers. Our discussion, therefore, uses the core-periphery concept as a basis for investigating Australia and focuses on New Zealand as a region by itself.

▶ AUSTRALIA

On January 1, 2001, Australia celebrated its hundredth birthday as a state, the Commonwealth of Australia, recognizing (still) the British monarch as the head of state and entering its second century with a strong economy, stable political framework, high standard of living for most of its people, and favorable prospects ahead. Positioned on the Pacific Rim, ten times the size of Texas, well endowed with farmlands and vast pastures, major rivers, ample underground water, minerals, and energy resources, served by good natural harbors, and populated by 20 million mostly-well-educated people, Australia is one of the most fortunate countries on Earth.

Not everyone in Australia shares adequately in all this good fortune, however, and the less advantaged made their voices heard during the celebrations. The country's indigenous (Aboriginal) population, though a small minority today of about 300,000, remains disproportionately disadvantaged in almost every way, from lower life expectancies to higher unemployment than average, from lower

high school graduation rates to higher imprisonment ratios. But the nation is now embarked on a campaign to address these ills, with actions ranging from public demonstrations supporting reconciliation to official expressions of regret for past mistreatment and from enhanced social services to favorable court decisions over Aboriginal land claims.

When Australia was born as a federal state, its per-capita GNP, as reckoned by economic geographers, was the highest in the world. Australia's bounty fueled a huge flow of exports to Europe, and Australians prospered. That golden age could not last forever, and eventually the country's share of world trade declined. Still, Australia today ranks among the top 15 countries in the world in terms of GNP, and for the vast majority of Australians life is comfortable.

This is not to suggest that Australia has escaped every downturn in the global economy of which it is a part or that there are no concerns at all for the future. During the Pacific Rim boom of the 1980s and early 1990s, some locals wryly called Australia an NDC, a Newly Declining Country, a seller of raw materials, not finished ones; a purveyor of livestock, meat, and wheat on variable world markets; a society sinking deeply into debt. When the Pacific Rim economies stalled in the late 1990s, there were dire predictions about the impact on Australia. But Australia's economy proved to be remarkably adaptable. When its Asian markets dried up, Western markets soon took up the slack. The Australian economy weathered the Pacific Rim crisis better than expected.

The statistics bear witness to Australia's current good fortune. In terms of the indicators of development discussed in Chapter 9, Australia is far ahead of all its western Pacific Rim competitors except Japan (which has dropped in the rankings because of its own economic stagnation). As Australians celebrated their first century, they were, on average, earning far more than Thais, Malaysians, or Koreans. In terms of consumption of energy per person, number of automobiles and miles of roads, levels of health, and literacy, Australia had all the properties of a developed country. Australian cities, where 85 percent of all Australians live, are not encircled by crowded shantytowns. Nor is the Australian countryside inhabited by a poverty-stricken peasantry.

Distance

Australians often talk about distance. One of their leading historians, Geoffrey Blainey, labeled it a "tyranny"—an imposed remoteness from without and a divisive part of

MAJOR CITIES OF THE REALM	
City	Population* (in millions)
Adelaide, Australia	1.1
Auckland, New Zealand	1.2
Brisbane, Australia	1.8
Canberra, Australia	0.3
Melbourne, Australia	3.4
Perth, Australia	1.4
Sydney, Australia	4.1
Wellington, New Zealand	0.4

*Based on 2004 estimates.

AUSTRALIA:
POLITICAL DIVISIONS, CAPITALS
AND COMMUNICATIONS

POPULATION

● Under 50,000

● 50,000–250,000

● 250,000–1,000,000

● 1,000,000–5,000,000

Canberra National capital

Sydney State capital

Railroad

Road

Australian core area

0   200   400   600   800   1000 Kilometers

0   250   500 Miles

FIGURE 11-4

life within. Even today, Australia is far from nearly every-where on Earth. A jet flight from Los Angeles to Sydney takes more than 14 hours nonstop and is correspondingly expensive. Freighters carrying products to European markets take ten days to two weeks to get there. Inside Australia, distances also are of continental proportions, and Australians pay the price—literally. Until some upstart

private airlines started a price war, Australians paid more per mile for their domestic flights than air passengers anywhere else in the world.

But distance also was an ally, permitting Australians to ignore the obvious. Australia was a British progeny, a European outpost. Once you had arrived as an immigrant from Britain or Ireland, there were a wide range of environments, magnificent scenery, vast open spaces, and seemingly limitless opportunities. When the Japanese Empire expanded, Australia's remoteness saved the day. When immigration became an issue, Australia in its comfortable isolation could adopt an all-white admission policy that was not officially terminated until 1976. When boat people by the hundreds of thousands fled Vietnam in the aftermath of the Indochina War, almost none reached Australian shores.

Today Australia is changing and rapidly so. Immigration policy now focuses on the would-be immigrants' qualifications, skills, financial status, age, and facility in the English language. With regard to skills, high-technology specialists, financial experts, and medical personnel are especially welcome. Relatives of earlier immigrants, as well as a quota of genuine asylum-seekers, also are readily admitted. In recent years, total immigration has been limited to about 80,000 annually, but the number may have to increase if only to keep the country's population growing. According to recent data, Australia's declining natural rate of increase now hovers around 0.5 percent.

Already, Australia's changed immigration policies have dramatically altered cultural landscapes, especially in the urban areas (see photo, p. 544). The country is fast becoming a truly multicultural society; in Sydney, for example, one in seven residents is of Asian ancestry, a ratio that will rise to one in five by 2010. During the 1990s, when for a time Japan was Australia's leading trade partner, Australian schools began to teach Japanese to tens of thousands of children. The Austral Realm is clearly in transition.

Core and Periphery

Australia is a large landmass, but its population is heavily concentrated in a core area that lies in the east and southeast, most of which faces the Pacific Ocean (here named the Tasman Sea between Australia and New Zealand). As Figure 11-4 shows, this crescent-like Australian heartland extends from north of the city of Brisbane to the vicinity of Adelaide and includes the largest city, Sydney, the capital, Canberra, and the second largest city, Melbourne. A secondary core area has developed in the far southwest, centered on Perth and its outport, Fremantle. Beyond lies the vast periphery, which the Australians call the **Outback**.

To better understand the evolution of this spatial arrangement, it helps to refer again to the map of world climates (Fig. G-8). Environmentally, Australia's most favored strips face the Pacific and Southern Oceans, and they are not large. We can describe the country as a coastal rimland with cities, towns, farms, and forested slopes giving way to the vast, arid, interior Outback. On the western

flanks of the Great Dividing Range lie the extensive grassland pastures that catapulted Australia into its first commercial age—and on which still graze one of the largest sheep herds on Earth (over 160 million sheep, producing more than one-fifth of all the wool sold in the world). Where it is moister, to the north and east, cattle by the millions graze on ranchlands. This is frontier Australia, over which livestock have ranged for almost two centuries.

Aboriginal Australians reached this landmass as long as 50,000 to 60,000 years ago, crossed the Bass Strait into

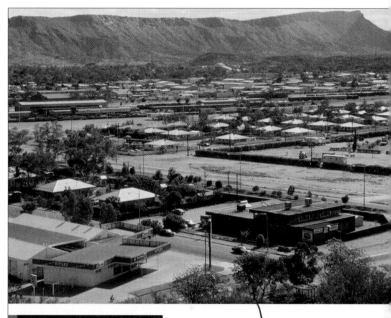

FROM THE FIELD NOTES

"My most vivid memory from my first visit to Alice Springs is spotting vineyards and a winery in this parched, desert environment as the plane approached the airport. I asked a taxi driver to take me there, and got a lesson in economic geography. Drip irrigation from an underground water supply made viticulture possible; the tourist industry made it profitable. None of this, however, is evident from the view seen here: a spur of the Macdonnell Ranges overlooks a town of bare essentials under the hot sun of the Australian desert. What Alice Springs has is centrality: it is the largest settlement in a vast area Australians often call 'the centre.' Not far from the midpoint on the nearly 2000-mile (3200-km) Stuart Highway from Darwin on the Northern Territory's north coast to Adelaide on the Southern Ocean, Alice Springs also is the northern terminus of the Central Australian Railway, seen in the middle distance. The shipping of cattle and minerals is a major industry here. You need a sense of humor to live here, and the locals have it: the town actually lies on a river, the intermittent Todd River. An annual boat race is held, and in the absence of water the racers carry their boats along the dry river bed. No exploration of Alice Springs would be complete without a visit to the base of the Royal Flying Doctor Service, which brings medical help to outlying villages and homesteads."

Tasmania, and had developed a patchwork of indigenous cultures when Captain Arthur Phillip sailed into what is today Sydney Harbor (1788) to establish the beginnings of modern Australia. The Europeanization of Australia doomed the continent's Aboriginal societies. The first to suffer were those situated in the path of British settlement on the coasts, where penal colonies and free towns were founded. Distance protected the Aboriginal communities of the northern interior longer than elsewhere; in Tasmania, the indigenous Australians were exterminated in just decades after having lived there for perhaps 45,000 years.

Eventually, the major coastal settlements became the centers of seven different colonies, each with its own hinterland; by 1861, Australia was delimited by its now-familiar pattern of straight-line boundaries (Fig. 11-4). Sydney was the focus for New South Wales; Melbourne, Sydney's rival, anchored Victoria. Adelaide was the heart of South Australia, and Perth lay at the core of Western Australia. Brisbane was the nucleus of Queensland, and Hobart was the seat of government in Tasmania. The largest clusters of surviving Aboriginal people were in the so-called Northern Territory, with Darwin, on Australia's tropical north coast, its colonial city. Notwithstanding their shared cultural heritage, the Australian colonies were at odds not only with London over colonial policies but also with each other over economic and political issues. The building of an Australian nation during the late nineteenth century was a slow and difficult process.

A Federal State

On January 1, 1901, following years of difficult negotiations, the Australia we know today finally emerged: the Commonwealth of Australia, consisting of six States and two Federal Territories (Table 11-1; Fig. 11-4). The special status of Federal Territory was assigned to the Northern Territory to protect the interests of the Aboriginal population there. A second Federal Territory was delimited in southern New South Wales for the establishment of a new federal capital (Canberra). Both Sydney and Melbourne had fiercely competed for this honor, but it was decided that a completely new seat of government, in a specially designated area, was the best solution. In 1927, the government buildings were ready and the administration moved to Canberra, a city planned from scratch to serve a nation built from a group of quarrelsome colonies.

Today Canberra, Australia's only large city that is not situated on the coast, has a population of 342,000, less than one-tenth of Sydney's. As a growth pole, Canberra has been no great success. But as a symbol of Australian federalism, it has been a triumph. Australia's unification was made possible through the adoption of an idea with ancient Greek and Roman roots, one also familiar to Americans and Canadians: the notion of association. The term to describe it comes from the Latin *foederis*; in practice, it means alliance and coexistence, a union of consensus and common interest—a **federation**. It stands in contrast to the idea that states should be centralized, or unitary. For this, too, the ancient Romans had a term: *unitas*, meaning unity. Most European countries are **unitary states**, including the United Kingdom of Great Britain and Northern Ireland. Although the majority of Australians came from that tradition (a kingdom, no less), they managed to overcome their differences and establish a Commonwealth that was, in effect, a federation of States with different viewpoints, economies, and objectives, separated by vast distances along the rim of an island continent. And yet, the experiment succeeded.

Table 11-1 STATES AND TERRITORIES OF FEDERAL AUSTRALIA, 2004				
State	**Area** **(1000 sq mi)**	**Estimated** **Population** **(millions)**	**Capital**	**Estimated** **Population** **(millions)**
New South Wales	309.5	6.8	Sydney	4.1
Queensland	666.9	3.8	Brisbane	1.8
South Australia	379.9	1.5	Adelaide	1.1
Tasmania	26.2	0.5	Hobart	0.2
Victoria	87.9	4.9	Melbourne	3.4
Western Australia	975.1	1.9	Perth	1.4
Territory				
Australian Capital Territory	0.9	0.3	Canberra	0.3
Northern Territory	519.8	0.2	Darwin	0.1

AMONG THE REALM'S GREAT CITIES . . . **SYDNEY**

*M*ore than two centuries ago, Sydney was founded by Captain Arthur Phillip as a British outpost on one of the world's most magnificent natural harbors. The free town and penal colony that struggled to survive evolved into Australia's largest city. Today metropolitan Sydney (4.1 million) is home to nearly one-fifth of the country's entire population. An early start, the safe harbor, fertile nearby farmlands, and productive pastures in its hinterland combined to propel Sydney's growth. Later, as road and railroad links made Sydney the focus of Australia's growing core area, industrial development and political power augmented its primacy.

With its incomparable setting and mild, sunny climate, its many urban beaches, and its easy reach to the cool Blue Mountains of the Great Dividing Range, Sydney is one of the world's most liveable cities. Good public transportation (including an extensive cross-harbor ferry system from the doorstep of the waterfront CBD), fine cultural facilities headed by the multitheatre Opera House complex, and many public parks and other recreational facilities make Sydney attractive to visitors as well. A healthy tourist trade, much of it from Japan and other Asian countries, bolsters the city's economy. Sydney's hosting of the 2000 Olympic Games was further testimony to its rising visibility.

Increasingly, Sydney also is a multicultural city. Its small Aboriginal sector is being overwhelmed by the arrival of large numbers of Asian immigrants. The Sydney suburb of Cabramatta symbolizes the impact: more than half of its nearly 100,000 residents were born elsewhere, mostly in Vietnam. Unemployment is high, drug use is a problem, and crime and gang violence persist. Yet, despite the deviant behavior of a small minority, tens of thousands of Asian immigrants have established themselves in some profession. As the photo and caption on p. 544 reveal, today things are definitely improving in this dynamic community.

These developments underscore Sydney's coming of age. The end of Australia's isolation has brought Asia across the country's threshold, and again the leading metropolis will show the way.

An Urban Culture

During this century of federal association, the Australians developed an urban culture. Despite those vast open spaces and romantic notions of frontier and Outback, 85 percent of all Australians live in cities and towns. On the map, Australia's areal functional organization is similar to Japan's: large cities lie along the coast, the centers of manufacturing complexes as well as the foci of agricultural areas. Contributing to this situation in Japan was mountainous topography; in Australia, it was an arid interior. There, however, the similarity ends. Australia's territory is 20 times larger than Japan's, and Japan's population is more than six times that of Australia's. Japan's port cities are built to receive raw materials and to export finished products. Australia's cities forward minerals and farm products from the Outback to foreign markets and import manufactures from overseas. Distances in Australia are much greater, and spatial interaction (which tends to decrease with increasing distance) is less. In comparatively small, tightly organized Japan, you can travel from one end of the country to the other along highways, through tunnels, and over bridges with utmost speed and efficiency. In Australia, the overland trip from Sydney to Perth, or from Darwin to Adelaide, is time-consuming and slow. Nothing in Australia compares to Japan's high-speed bullet trains.

For all its vastness and youth, Australia nonetheless developed a remarkable cultural identity, a sameness of urban and rural landscapes that persists from one end of the continent to the other. Sydney, often called the New York of Australia, lies on a spectacular estuarine site, its compact, highrise central business district overlooking a port bustling with ferry and freighter traffic. Sydney is a vast, sprawling metropolis with multiple outlying centers studding its far-flung suburbs; brash modernity and reserved British ways blend here. Melbourne, sometimes regarded as the Boston of Australia, prides itself on its more interesting architecture and more cultured ways. Brisbane, the capital of Queensland, which also anchors Australia's Gold Coast and adjoins the Great Barrier Reef, is the Miami of Australia; unlike Miami, however, its residents can find nearby relief from the summer heat in the mountains of its immediate hinterland (as well as at its beaches). Perth, Australia's San Diego, is one of the world's most isolated cities, separated from its nearest Australian neighbor by two-thirds of a continent and from Southeast Asia and Africa by thousands of miles of ocean.

And yet, each of these cities—as well as the capitals of South Australia (Adelaide), Tasmania (Hobart), and, to a lesser extent, the Northern Territory (Darwin)—exhibits an Australian character of unmistakable quality. Life is orderly and unhurried. Streets are clean, slums are few, graffiti rarely seen. By American and even European standards, violent crime (though rising) is uncommon. Standards of public transportation, city schools, and health-care provision are high. Spacious parks, pleasing waterfronts, and plentiful sunshine make Australia's urban life more acceptable than that almost anywhere else in the world. Critics of Australia's way of life say that this very pleasant state of affairs has persuaded Australians that hard work is not really necessary. Some experience in the commercial centers of the major cities contradicts that assertion: the pace of life is quickening. The country's cultural geography evolved as that of a European outpost, prosperous and secure in its isolation. Now Australia must reinvent itself as a major link in an Australo-Asian chain, a Pacific partner in a transformed regional economic geography.

Economic Geography

From the very beginning, however, goods imported from Britain (and later from the United States) were expensive, largely because of transport costs. This encouraged local entrepreneurs to set up their own industries in and near the developing cities—industries economic geographers call **import-substitution industries**.

When the prices of foreign goods became lower because transportation was more efficient and therefore cheaper, local businesses demanded protection from the colonial governments, and high tariffs were erected against imported goods. Local products now could continue to be made inefficiently because their market was guaranteed. If Japan could not afford this, how could Australia? We can see the answer on the map (Fig. 11-5). Even before federation in 1901, all the colonies could export valuable minerals whose earnings shored up those inefficient, uncompetitive local industries. By the time the colonies unified, pastoral industries were also contributing income. So the miners and the farmers paid for those imports Australians could not produce themselves, plus the products made in the cities. No wonder the cities grew: here were secure manufacturing jobs, jobs in state-run service enterprises, and jobs in the growing government bureaucracy. When we noted earlier that Australians once had the highest per-capita GNP in the world, this was achieved in the mines and on the farms, not in the cities.

But the good times had to come to an end. The prices of farm products fluctuated, and international market competition increased. The cost of mining, transporting, and shipping ores and minerals also rose. Australians like to drive (there are more road miles per person in Australia than in the United States), and expensive petroleum imports were needed. Meanwhile, the government-protected industries had been further fortified by strong labor unions. Not surprisingly, as the economy declined, the national debt rose, inflation grew, and unemployment crept upward. In 1999, only 22 percent by value of Australian exports were the kind of high-tech goods that have made East Asia's economic tigers so successful. That was more than double the 1985 figure but still far below what the country needed to produce.

Agricultural Abundance

And yet, Australia has material assets of which other countries on the Pacific Rim can only dream. In agriculture, sheep-raising was the earliest commercial venture, but it was the technology of refrigeration that brought world markets within reach of Australian beef producers. Wool, meat, and wheat have long been the country's big three income earners; Figure 11-5 displays the vast pastures in the east, north, and west that constitute the ranges of Australia's huge herds. The zone of commercial grain farming forms a broad crescent extending from northeastern New South Wales through Victoria into South Australia, and covers much of the hinterland of Perth. Keep in mind the scale of this map: Australia is only slightly smaller than the 48 contiguous States of the United States! Commercial grain farming in Australia is big business. As the climatic map would suggest, sugarcane grows along most of the warm, humid coastal strip of Queensland, and Mediterranean crops (including grapes for Australia's wines) cluster in the hinterlands of Adelaide and Perth. Mixed horticulture concentrates in the basin of the Murray River system, including rice, grapes, and citrus fruits, all under irrigation. And, as elsewhere in the world, dairying has developed near the large urban areas. With its

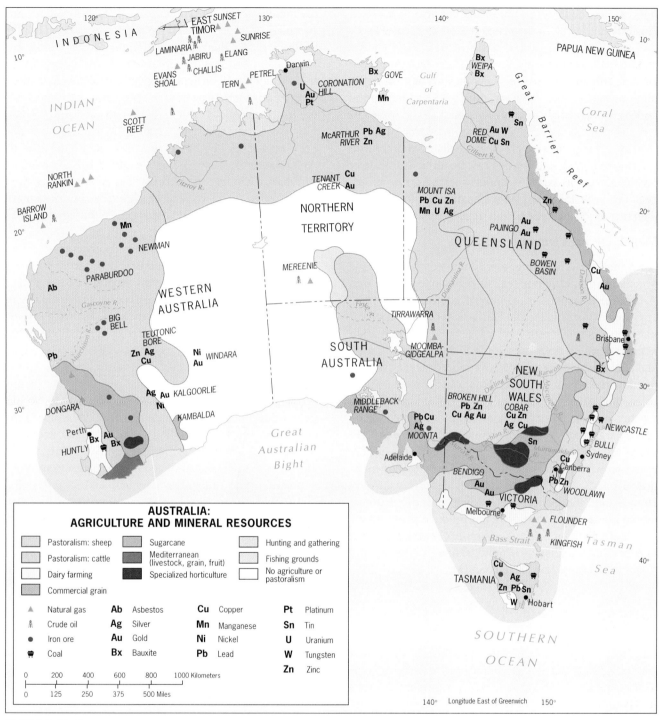

FIGURE 11-5

considerable range of environments, Australia yields a diversity of crops.

Mineral Wealth

Australia's mineral resources, as Figure 11-5 shows, also are diverse. Major gold discoveries in Victoria and New South Wales produced a ten-year gold rush starting in 1851 and ushered in a new economic era. By the mid-

dle of that decade, Australia was producing 40 percent of the world's gold. Subsequently, the search for more gold led to the discoveries of other minerals. New finds are still being made today, and even oil and natural gas have been found both inland and offshore (see the symbols in Fig. 11-5 in the Bass Strait between Tasmania and the mainland, and off the northwestern coast of Western Australia). Coal is mined at many locations, notably in the east near Sydney and Brisbane but also in Western Australia and even in Tasmania; before coal prices fell, this was a valuable export. Major deposits of metallic and nonmetallic minerals abound—from the complex at Broken Hill and the mix of minerals at Mount Isa to the huge nickel deposits at Kalgoorlie and Kambalda, the copper of Tasmania, the tungsten and bauxite of northern Queensland, and the asbestos of Western Australia. A glance at the map reveals the wide distribution of iron ore (the red dots), and for this raw material as for many others, Japan has been Australia's best customer in recent years. In the late 1990s, Japan was buying more than one-third of all Australian mineral exports.

Manufacturing's Limits

Australian manufacturing, as we noted earlier, remains oriented to domestic markets. One cannot expect to find Australian automobiles, electronic equipment, or cameras challenging the Pacific Rim's economic tigers for a place on world markets—not yet, at any rate. Australian manufacturing is diversified, producing some machinery and equipment made of locally produced steel as well as textiles, chemicals, paper, and many other items. These industries cluster in and near the major urban areas where the markets are. The domestic market in Australia is not large, but despite declining real incomes, it remains relatively affluent. This makes it attractive to foreign producers, and Australia's shops are full of high-priced goods from Japan, South Korea, Taiwan, and Hong Kong. Indeed, despite its long-term protectionist practices, Australia still does not produce many goods that could be manufactured at home. Overall, the economy continues to display symptoms of a still-developing rather than a fully developed country.

Australia's Future

A new age is dawning in Australia, a country that not long ago (1988) celebrated the bicentennial of its first European settlement—two centuries as a European outpost facing the Pacific Ocean. Now, with Australia well into its third century, its European bonds are weakening and its Asian ties are strengthening.

On the face of it, Australia and its northern neighbors on the Pacific Rim would seem to exhibit a geographic complementarity: Japan and the economic tigers need Australia's excess food, metals, and minerals, and Australia needs the cheap manufactures Asia produces. But it is not that simple. Australia still has tariff barriers against imported goods, and the Asian countries on the Pacific Rim maintain import barriers against processed foods and minerals, thus discouraging Australia from refining its exports and earning more from them. Asia's

manufacturers, furthermore, are more interested in the potential of large markets (such as Indonesia, with 224 million people) than Australia, and its investors like the low wages of East and Southeast Asia. In any case, Australia would find it difficult to open its economy and lower its protective tariffs without reciprocation from its Asian trading partners.

Australia's allies in the global core also have not been invariably helpful. U.S. government subsidies to American wheat farmers have enabled U.S. farmers to sell their wheat at lower prices than Australian farmers can afford to sell theirs. The U.S. government's attempts to address this issue have failed because of vigorous bargaining by senators representing grain-producing States; this is only part of a giant farm-subsidy program that also hurts Mexican, Caribbean, and Brazilian farmers among others. But as Australia's overall economy grows, it also becomes less dependent on farm exports, so that an America that advocates free trade and then acts against this principle can still be regarded as a satisfactory trading partner and global ally.

Aboriginal Issues

In this first decade of the twenty-first century, Australia faces other, more serious challenges. The Aboriginal population of 300,000 (including many of mixed ancestry) has been gaining influence in the country's affairs. In the 1980s, Aboriginal leaders began a campaign to obstruct exploration and mining on ancestral and sacred lands. Until 1992, Australians had taken it for granted that Aborigines had no rights to land ownership; the continent had been open and undemarcated, and there was no evidence of prior title. In that year, however, the Australian High Court ruled in favor of an Aborigine, Eddie Mabo, and his co-inhabitants of the Murray Islands in the Torres Strait. The court ruled that Mabo and his community owned customary title to their land. The ruling implied that Aborigines elsewhere, too, could claim title to traditional land.

When the federal government passed the law that would codify this far-reaching and astounding ruling, it protected Australia's pastoral leases (the tracts of land leased to ranchers of cattle and sheep by the Commonwealth) by excluding them—but not the holdings of mining companies. This led to an immediate Aboriginal victory at Coronation Hill in the Northern Territory (Fig. 11-5) where the administration prohibited the development of platinum, gold, and palladium deposits because they were deemed to lie on an Aboriginal community's sacred land.

Under the Native Title Bill, passed by the Australian Senate in 1993, Aborigines who had long lived on "vacant" land could petition the courts for title of that land: those who could prove that their land was taken away

FROM THE FIELD NOTES

"While in Darwin I had to visit a government office, and waiting in line with me was this Aboriginal girl, her brother, and her father. They lived about two hours from the city, she said (which I gathered was the length of their bus ride), and they had to come here from time to time for filing forms and visiting the doctor. When I asked whether they enjoyed coming to Darwin, all three shook their heads vigorously. 'It is not a friendly place,' she said. 'But we must come here.' In fact I had been surprised at the comparatively small number of Aboriginal people I had seen in Darwin, far fewer than in Alice Springs. Certainly laid-back Alice Springs is a very different place from larger, busier, and more modern Darwin."

were entitled to compensation. And when a mining company's lease expired, Aborigines could reclaim the land. The result was a crisis of confidence in the mining industry and a sharp drop in investment in exploration, resource development, and infrastructure.

Soon the Aborigines tested the government's move to exclude the pastoral leases, and again they won in the High Court. In 1996, ruling on a case brought before it by the Wik Aboriginal community in Queensland, the court decided that pastoral leases and native title could coexist, thereby voiding the government's exclusion of these leases from the Native Title Bill. This decision implied that vast areas (potentially as much as 78 percent of Australia) could be subject to Aboriginal claims (Fig. 11-6). Now not only miners and ranchers had reason to be concerned: activist Aboriginal groups announced that they would claim ownership to parcels of land in the heart of Sydney and other major cities.

Again the federal government tried to mitigate the impact of the High Court's ruling, this time by passing legislation that would abolish native title and thus preclude claims on pastoral leases where such title would "interfere" with the rights and work of the leaseholding pastoralist. Aborigines would be allowed to enter such land to hold ceremonies and gather food in their traditional ways, but they must not in any way obstruct pastoral activities by taking cultivated crops, moving livestock away, cutting fences, or opening gates. Moreover, all Aboriginal claims on offshore water and marine resources were voided. Aboriginal leaders responded by threatening to inundate the court system with numerous title claims that would cost those affected, and the government, hundreds of millions of dollars.

The **Aboriginal land issue** is essentially (though not exclusively) an Outback issue, notably in Western Australia, South Australia, and Queensland States. It is having a major impact on Australia's political geography; views on the matter in the large cities tend to differ sharply from those in the interior. It has had a polarizing effect on politics and society, strengthening political parties and movements opposed to making further "concessions" to Aborigines. In Queensland, the One Nation Party in 1998 got 25 percent of the vote on such a platform, forcing the ruling conservative party (the Liberals) that controlled the State government to make further concessions

LAND AND SEA AREAS SUBJECT TO ABORIGINAL CLAIMS

- Aboriginal claims
- Northern Territory already Aboriginal land
- Other land

FIGURE 11-6

to its right wing, including a reduction in welfare awards to Aborigines. Polls indicate that most Australians nevertheless oppose the kind of extremism the anti-Aborigine (and anti-immigration) parties represent. But the national government must navigate a difficult course toward reconciliation (see the Issue Box entitled "Indigenous Rights and Wrongs").

It should be noted that only a few Aboriginal communities still pursue their original life of hunting, gathering, and fishing in the remotest corners of the Northern Territory, northern Queensland, and northernmost Western Australia. The remainder are scattered across the continent, subsisting on reserves the government set aside for them (much of the Northern Territory is so designated), working on cattle stations, or performing mostly menial

The name Cabramatta conjures up varied reactions among Australians. During the 1950s and 1960s, many immigrants from southern Europe settled in this western suburb of Sydney attracted by affordable housing. During the 1970s and 1980s, Southeast Asians arrived in large numbers, and during this period Cabramatta, in the eyes of many, became synonymous with gang violence and drug dealing. More recently, however, Cabramatta's ethnic diversity has come to be viewed in a more favorable light, and is seen as the "multicultural capital of Australia," a tourist attraction and proof of Australia's capacity to accommodate non-Europeans. Meanwhile, Cabramatta has been spruced up with Oriental motifs of various kinds. This 'Freedom Gate' in the Vietnamese community is flanked by a Ming horse and a replica of a Forbidden City lion—all reflecting better times for an old transit point for immigrants and refugees.

jobs in cities and towns. In comparatively affluent Australia, the Aboriginal peoples suffer from poverty, disease, inadequate education, and even malnutrition. The Aboriginal rights question will be a major one in Australia's future.

Immigration Issues

The population question has preoccupied Australia as long as Australia has existed—and even longer. Fifty years ago, when Australia had less than half the population it has today, 95 percent of the people were of European ancestry, and more than three-quarters of them came from the British Isles. Eugenic (race-specific) **immigration policies** maintained this situation until the 1970s. Today, the picture is dramatically different: of 19.9 million Australians, only about one-third have British-Irish origins, and Asian immigrants outnumber both European immigrants and the natural increase each year. During the early 1990s, nearly 150,000 legal immigrants arrived in Australia annually, most of them coming from Hong Kong, Vietnam, China, the Philippines, India, and Sri Lanka. Immigration quotas have since been reduced, most recently to 80,000, but Asian immigrants continue to outnumber those from Western sources. Immigration from Britain and Ireland varies from 20,000 to 25,000 annually, and between 5000 and 10,000 New Zealanders move to Australia each year.

The immigration issue roils Australian society. In the mid-1990s, when hundreds of "boat people" landed on Australian shores as illegal immigrants, their arrival and subsequent treatment led to a wave of soul-searching. In 1998, when immigrant workers were among strike breakers at a major dockyard, labor unions denounced their role. Restrictions on foreign ownership of Australian real estate reflect Australian fears of what extremist groups call the "Asianization" of the country. At the same time, the success of many Asian settlers in Australia evinces the opportunities still available in this open and free society.

The Asian immigration issue is essentially an urban one: most Asian immigrants have settled in the cities and towns. Sydney has by far the largest contingent, and many ethnic districts have become part of its urban-cultural mosaic. In the 1970s Sydney, always Australia's most multicultural city, had Italian, Greek, and Yugoslavian neighborhoods. Today it also has Vietnamese, Laotian, Indian, and Chinese districts (see photo at left). Assimilation continues along a bumpy road, and the economy is the barometer: when the economy lags, social problems tend to worsen.

The issue of asylum-seekers suddenly returned to the forefront of Australian concerns in August 2001 when a Norwegian freighter saved more than 400 Afghans from a sinking Indonesian fishing boat and took them toward

Indigenous Rights and Wrongs

THE LEAST WE SHOULD DO IS APOLOGIZE

"I CAN'T BELIEVE this. Here we are more than a dozen years beyond our two-hundredth-year anniversary, the evidence of our mistreatment of the Aborigines is everywhere, and we can't, as a nation, say we're sorry? How could we get ourselves into a public debate over this? What does it take to offer sincere apologies to the original Australians and to tell them (and not only to tell them) that we'll do better?

"We should remind ourselves more often than we do of what happened here. When our ancestors arrived on these shores there were more than a million Aboriginal people in Australia, with numerous clans and cultures. They had many virtues, but one in particular: they didn't appropriate land. Land was assigned to them by the creator (what they call the Dreaming), and their relationship to it was spiritual, not commercial. They didn't build fences or walls. Neither had they adopted some bureaucratic religion. Just think, a world where land was open and free and religion was local and personal. Wouldn't that do something for the Middle East and Northern Ireland! Well, we showed up and started claiming and fencing off the land that, under British rules, was there for the taking. If the Aborigines got in the way, they were pushed out, and if they resisted, they got killed. You don't even want to think about what happened in Tasmania. A campaign of calculated extermination. Between 1800 and 1900 the Aboriginal population dropped from 1 million to around 50,000. And when we became a nation, they weren't even accorded citizenship.

"None of this stopped some white Australian men from getting Aboriginal women pregnant. But from 1910 on church and state managed to make things even worse. They took the young children away from their mothers and put them in institutions, where they would be 'Europeanized' and then married off to white partners, so that they would lose their Aboriginal inheritance. This, if you'll believe it, went on into the 1960s! Think of the scenes, these kids being kidnapped by armed officials from their mothers. And we have to argue about saying we're sorry?

"Well, at least we're discussing it in the open, and the misdeeds of the past are coming out. Aboriginal Australians finally got the vote in 1962, and they're able to get elected to parliament. I work in a State government office that assists Asian immigrants here in Sydney, and I only wish that we'd done for the Aborigines what we've done for the immigrants we admit."

Regional ISSUE

ENOUGH IS ENOUGH

"Prime Minister John Howard was right when he said that apologizing for history should be a private matter. I was born on this Outback sheep station 25 years ago, and we employ a dozen Aboriginal workers, most of whom were born in this area. I had nothing to do with what happened more than a century ago, or even the first 75 years of the past century. What I would have done is irrelevant, and what my great-great-grandparents may have done has nothing to do with me. All over the world people are born into situations not of their making, and they should all go around apologizing?

"And as a matter of fact I believe that this country has bent over backwards, in my time at least, to undo the alleged wrongs of the past. Look, the Aboriginal minority counts about 300,000 or 2 percent of the population. Compare this to the map: they've got the Northern Territory and other parts of the country too, and that's close to 15 percent of Australia. And now we're going to give them more? I could see disaster in the making when that guy Eddie Mabo got the High Court to agree to his so-called 'native title' claim to land with which he and his people hadn't done a thing. Since when do Aboriginal customs take precedence over the law of the land? But even worse was that decision supporting the Wiks up there in Cape York. That was unbelievable: native title on land already held in pastoral lease? That affects all of us. Pretty soon the whole of Australia will be targeted by these Aboriginal claimants and their lawyers. And by the way, in the old days those people moved around all the time. Who's to say what clans owned what land when it comes to claiming native title? These court cases are going to tie us up in legal knots for generations to come.

"And now I gather that some Aboriginal people are calling for a treaty between themselves and the government, for what purpose I'm not sure—I guess it'll be like the one the Kiwis made with the Maori there. Is this really what we need if we're going to put the past behind us? Look, I like the fellows working here, but you've got to realize that no laws or treaties are going to change all of the problems they have. They're getting preference for many kinds of government employment, remedial help in all kinds of areas, but still they wind up in jail, leaving school, giving up their jobs. Yes, I'm sure wrongs were done in the past. But apologizing isn't going to make any difference. They have their opportunity now. In a lot of countries they would have never gotten it: Aussies are pretty decent people. It's up to them to make the best of it."

Australian-administered Christmas Island in the Indian Ocean south of Jawa. Australian prime minister John Howard declared that none of the migrants would be allowed to set foot on Australian soil, and the government sent a military force to transfer them to a troop ship that took about half of them to Nauru and the other half to New Zealand. An Australian federal court objected and international observers protested, but three-quarters of Australians supported this action. Since 2000, there have been riots, breakouts, and other violence at several remote refugee camps; the public's view of illegal immigration had hardened even before the Christmas Island incident, however.

Prosperous Australia's lengthy and open coastline lies at the end of a chain of land and water that starts in South and East Asia and converges on Indonesia. Australia today is open to legal immigration from all parts of the world, but it must constrain what would otherwise become a flood of asylum seekers whose invasion could disrupt the country's steady march toward multiculturalism.

Environmental Degradation

Another growing issue involves the environment and conservation. Australia's wealth was derived not only from its ores and minerals but from its soils and pastures as well. Not surprisingly, its ecology also paid a heavy price.

Great stands of magnificent forest were destroyed. In Western Australia, centuries-old trees were simply "ringed" and left to die, so that the sun could penetrate through their leafless crowns to nurture the grass below. Then the sheep could be driven into these new pastures. In Tasmania, where Australia's native eucalyptus tree reaches its greatest dimensions (comparable to North American redwood stands), tens of thousands of acres of this irreplaceable treasure have been lost to chain saws and pulp mills. Many of Australia's unique marsupial species have been destroyed, and many more are endangered or threatened. "Never have so few people wreaked so much havoc on the ecology of so large an area in so short a time," observed a geographer in Australia recently; today, awareness of this **environmental degradation** is growing. In Tasmania, the "Green" environmentalist political party has become a force in State affairs, and its activism has slowed deforestation, dam-building, and other "development" projects. Still, many Australians fear the environmentalist movement as an obstacle to economic growth at a time when the economy needs stimulation. This, too, is an issue for the future.

Status and Role

Several issues involving Australia's status at home, relations with neighbors, and position in the world are stirring up national debate as well. A domestic question is whether Australia should become a republic, ending the status of the British monarch as the head of state, or continue its status quo in the British Commonwealth. A 1999 referendum proved that a majority of Australians were not prepared to abandon the monarchy—at least not in favor of "a president appointed by a two-thirds majority of the membership of the Commonwealth Parliament" as the ballot put it. Although polls had predicted a republican victory, the language on the ballot probably changed the outcome; most Australians seem to favor a republic, but more of them distrust their politicians. If the ballot had said "in favor of a popularly elected president," chances are that Australia would be on its way to becoming a federal republic. This issue will undoubtedly arise again.

Relations with neighboring Indonesia and East Timor have been complicated. For many years, Australia had what may be called a special relationship with Indonesia, whose help it needs in curbing illegal seaborne immigration. It was also profitable for Australia to counter international (UN) opinion and recognize Indonesia's 1976 annexation of Portuguese East Timor, for in doing so Australia could deal directly with Jakarta for the oil reserves under the Timor Sea. Thus Australia gave neither recognition nor support to the rebel movement that fought for independence in East Timor. But this story has a relatively happy ending. When the East Timorese campaign for independence succeeded in 1999 and Indonesian troops began an orgy of murder and destruction, Australia sent an effective peacekeeping force and spearheaded the United Nations effort to stabilize the situation. Today, a new chapter has opened in Australia's relations with these northern neighbors.

Australia also has a long-term relationship with Papua New Guinea (PNG), as we will note in Chapter 12. This association, too, has gone through difficult times. In recent years, the inhabitants of Papua New Guinea have strongly resisted privatization, World Bank involvement, and globalization generally. Australia assists PNG in several spheres, but its motives are sometimes questioned. In 2001 the construction of a projected gas pipeline from PNG to the Australian State of Queensland precipitated fighting among tribespeople over land rights, resulting in dozens of casualties, and Australian public opinion reflected doubts over the appropriateness of this venture.

Immediately after the September 2001 attack on U.S. targets in New York and Washington, Australian leaders expressed strong support for the American campaign against terrorism, but they made it a point also to assure the Indonesian government that the War on Terror was not a war on Islam. Although the great majority of Australians supported this stance, their will was severely tested in late 2002 when a terrorist attack on a nightclub in Kuta Beach on the island of Bali killed 88 vacationing Australians.

Territorial dimensions, relative location, and raw-material wealth have helped determine Australia's place in the world and, more specifically, on the Pacific Rim. Australia's population, still just 20 million in the first decade of our new century, is smaller than Malaysia's and not much larger than that of the Caribbean island of Hispaniola. But Australia's importance in the international community far exceeds its human numbers.

NEW ZEALAND

Fifteen hundred miles east-southeast of Australia, in the Pacific Ocean across the Tasman Sea, lies New Zealand. In an earlier age, New Zealand would have been part of the Pacific geographic realm because its population was Maori, a people with Polynesian roots. But New Zealand, like Australia, was invaded and occupied by Europeans. Today, its population of 4 million is almost 80 percent European, and the Maori form a minority of less than 400,000, with many of mixed Euro-Polynesian ancestry.

New Zealand consists of two large mountainous islands and many scattered smaller islands (Fig. 11-7). The two large islands, with the South Island somewhat larger than the North Island, look diminutive in the great Pacific Ocean, but together they are larger than Britain. In contrast to Australia, the two main islands are mainly mountainous or hilly, with several peaks rising far higher than any on the Australian landmass. The South Island has a spectacular snowcapped range appropriately called the Southern Alps, with numerous peaks reaching beyond 10,000 feet (3300 m). The smaller North Island has proportionately more land under low relief, but it also has an area of central highlands along whose lower slopes lie the pastures of New Zealand's chief dairying district. Hence, while Australia's land lies relatively low in elevation and exhibits much low relief, New Zealand's is on the average high and is dominated by rugged relief.

Human Spatial Organization

Thus the most promising areas for habitation are the lower-lying slopes and lowland fringes on both islands. On the North Island, the largest urban area, Auckland (1.2 million), occupies a comparatively low-lying peninsula. On the South Island, the largest lowland is the agricultural Canterbury Plain, centered on Christchurch. What makes these lower areas so attractive, apart from their availability as cropland, is their magnificent pastures. The range of soils and pasture plants allows both summer and winter

FROM THE FIELD NOTES

"The drive from Christchurch to Arthur's Pass on the South Island of New Zealand was a lesson in physiography and biogeography. Here, on the east side of the Southern Alps, you leave the Canterbury Plain and its agriculture and climb into the rugged topography of the glacier-cut, snowcapped mountains. A last pasture lies on a patch of flatland in the foreground; in the background is the unmistakable wall of a U-shaped valley sculpted by ice. Natural vegetation ranges from pines to ferns, becoming even more luxuriant as you approach the moister western side of the island."

grazing. Moreover, the Canterbury Plain, the chief farming region, also produces a wide variety of vegetables, cereals, and fruits. About half of all New Zealand is pasture land, and much of the farming provides fodder for the pastoral industry. Sixty million sheep and eight million cattle dominate these livestock-raising activities, with wool, meat, and dairy products providing nearly two-thirds of the islands' export revenues.

Despite their contrasts in size, shape, physiography, and history, New Zealand and Australia have much in common. Apart from their joint British heritage, they share a sizeable pastoral economy, a small local market, the problem of great distances to world markets, and a desire to stimulate (through protection) domestic manufacturing. The high degree of urbanization in New Zealand (77 percent of the total population) again resembles Australia: substantial employment in city-based industries, mostly the processing and packing of livestock and farm products, and government jobs.

More remote even than Australia, New Zealand also has been affected by Pacific Rim development, though

FIGURE 11-7

less so than its giant neighbor. Still, by the mid-1990s, Japan and other Pacific Rim economies were buying more of New Zealand's exports (mainly agricultural products) than either Australia or the United States. Britain today takes only about 6 percent of New Zealand's exports. Australia and the United States still send New Zealand most of its imports (about 40 percent combined), but the Pacific Rim's contribution has been rising. As Figure G-11 and Table G-1 show, New Zealand is a high-income economy, but it is not in the upper tier of the world's richer economies.

Spatially, New Zealand shares with Australia its pattern of **peripheral development** (Fig. G-9), imposed not by desert but by high rugged mountains. The country's major cities—Auckland and the capital of Wellington (together with its satellite of Hutt) on the North Island, and Christchurch and Dunedin on the South Island—are all located on the coast, and the entire railway and road system is peripheral in its configuration (Fig. 11-7). This is more pronounced on the South Island than in the north because the Southern Alps are New Zealand's most formidable barrier to surface communications.

The Maori Factor and New Zealand's Future

One of New Zealand's historic sites is the wooden Treaty House at Waitangi, in the northern peninsula of the North Island. There, following a series of brutal conflicts, the British and the Maori signed the Treaty of Waitangi in 1840. The treaty granted the British sovereignty over New Zealand, but it also guaranteed Maori rights over tribal lands. February 6, the date of the signing of the treaty, is a public holiday in New Zealand called Waitangi Day. A ceremony at the Treaty House, attended by the prime minister and by Maori leaders, is a highlight of the day's events. Visitors to the ceremony in 1995 were in for a shock, however. Maoris disrupted the proceedings; the prime minister was jostled by members of the crowd; and an attempt was made to burn the Treaty House. For the first time in more than half a century, the ceremony had to be abandoned.

What precipitated this show of anger? In fact, it was only the latest in a series of events signaling a rise in Maori ethnic consciousness. Maori leaders insist that the terms of the Waitangi Treaty (partly abrogated in 1862) be enforced, that large tracts of land in urban as well as rural areas, amounting to more than half of the national territory, be awarded to their rightful Maori owners, and that Maori fishing rights be paramount over offshore waters. Judicial rulings during the 1990s supported the Maori position, and Maori successes in one arena have led to new claims in others. Today, the Maori question has become the leading national issue in New Zealand, and Maori activism has risen sharply. In this context, the events of February 6, 1995 were no surprise.

The Maori appear to have reached New Zealand during the tenth century A.D., and by the time the European colonists arrived they had had an enormous impact on the islands' ecosystems, especially on the North Island where most Maori lived. One of the Maori complaints is the persistently slow pace of integration of the Maori minority of under 400,000 into modern New Zealand society. In fact, the Maori have been joined by other Pacific islanders, so that Auckland today not only is New Zealand's largest city; it also may be the largest urban concentration of Polynesians anywhere, with over 100,000 Maori as well as Samoans, Cook Islanders, Tongans, and others making up one-sixth of the metropolitan population. Their neighborhoods give Auckland an ethnic patchwork that evinces the Maoris' complaint about nonintegration.

The Maori issue has other geographic overtones. Nearly all Maori still live on the North Island (where three-quarters of all New Zealanders reside), but Maori leaders are making huge claims, based on historic primacy, over much of the South Island. Remote and com-

FROM THE FIELD NOTES

"It was Sunday morning in Christchurch on New Zealand's South Island, and the city center was quiet. As I walked along Linwood Street I heard a familiar sound, but on an unfamiliar instrument: the Bach sonata for unaccompanied violin in G minor—played magnificently on a guitar. I followed the sound to the artist, a Maori musician of such technical and interpretive capacity that there was something new in every phrase, every line, every tempo. I was his only listener; there were a few coins in his open guitar case. Shouldn't he be playing before thousands, in schools, maybe abroad? No, he said, he was happy here, he did alright. A world-class talent, a street musician playing Bach on a Christchurch side street, where tourists from around the world were his main source of income. Talk about globalization."

paratively conservative, South Islanders have reacted to these claims much as their counterparts in Western Australia have to the Mabo and Wik decisions on Aboriginal rights.

The Waitangi Treaty was intended as the framework for a partnership between the British and the Maoris. If the current disputes lead to a reconsideration and fulfillment of the treaty's terms in the context of modern New Zealand's human geography, then the country's future could be shaped by a harmonious, mainly bicultural society. If this effort fails, the specter of serious ethnic polarization threatens against a demographic background that projects the rapidly growing Polynesian population of New Zealand will double to 25 percent of the national total by 2012.

Dominant cultural heritage and prevailing cultural landscape form two criteria on which the delimitation of the Austral Realm is based. But in both Australia and New Zealand, the cultural mosaic is changing, and the convergence with neighboring realms is proceeding.

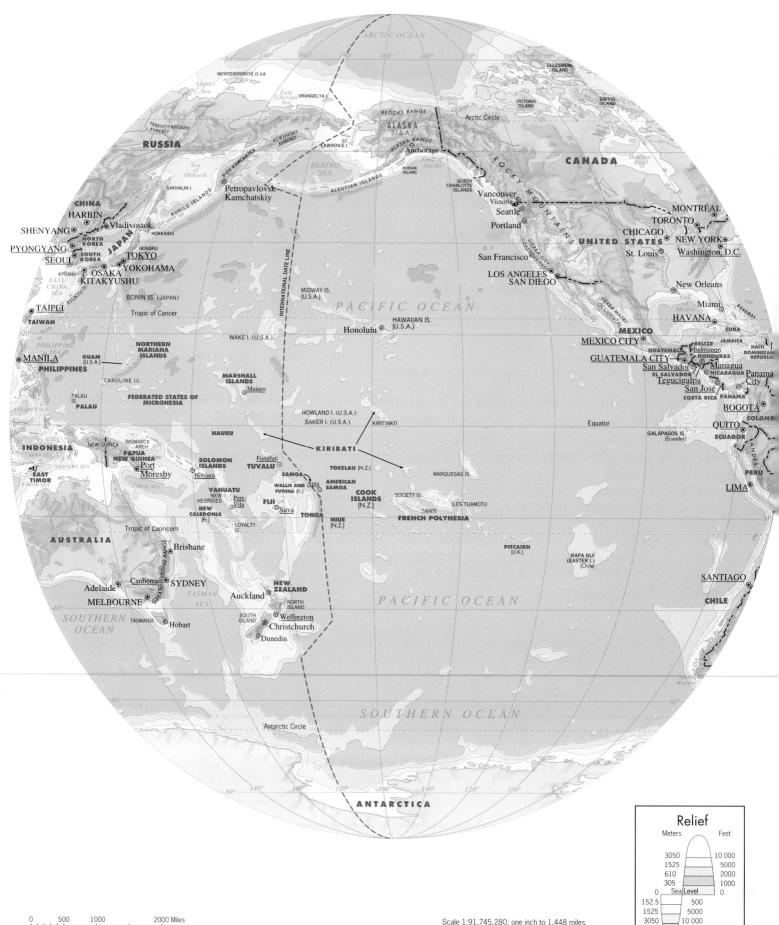

FIGURE 12-1

Scale 1:91,745,280; one inch to 1,448 miles
Apinus II projection
Elevations and depressions are given in feet

Relief

Meters		Feet
3050		10 000
1525		5000
610		2000
305		1000
0	Sea Level	0
152.5		500
1525		5000
3050		10 000
6100		20 000

0   500   1000   2000 Miles

0   1000   2000 Kilometers

The Pacific Realm

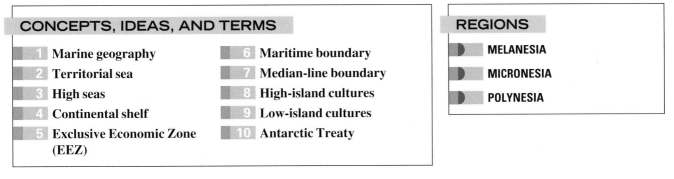

CONCEPTS, IDEAS, AND TERMS

1 Marine geography
2 Territorial sea
3 High seas
4 Continental shelf
5 Exclusive Economic Zone (EEZ)
6 Maritime boundary
7 Median-line boundary
8 High-island cultures
9 Low-island cultures
10 Antarctic Treaty

REGIONS

- MELANESIA
- MICRONESIA
- POLYNESIA

DEFINING THE REALM

BETWEEN THE AMERICAS to the east and the western Pacific Rim to the west lies the vast Pacific Ocean, larger than all the world's land areas combined. In this greatest of all oceans lie tens of thousands of islands, some large (New Guinea is by far the largest), most small (many are uninhabited). Together, the land area of these islands is a mere 376,000 square miles (974,000 sq km), about the size of Texas plus New Mexico, and over 90 percent of this lies in New Guinea.*

The Pacific geographic realm—land and water—covers nearly an entire hemisphere of this world, the one commonly called the Sea Hemisphere (Fig. 12-1). This Sea Hemisphere meets the Russian and North American realms in the far north and merges into the Southern Ocean in the south. Despite the preponderance of water, this fragmented, culturally complex realm does possess

regional identities. It includes the Hawaiian Islands, Tahiti, Tonga, and Samoa—fabled names in a world apart.

In terms of modern cultural and political geography, Indonesia and the Philippines are not part of the Pacific Realm, although Indonesia's political system reaches into it; nor are Australia and New Zealand part of it. Before the European invasion and colonization, Australia would have been included because of its Aboriginal population and New Zealand because of its Maori population's Polynesian affinities. But the Europeanization of their countries has engulfed black Australians and Maori New Zealanders, and the regional geography of Australia and New Zealand today is decidedly not Pacific. In New Guinea, on the other hand, Pacific peoples remain numerically and culturally the dominant element.

The Pacific islands were colonized by the French, British, and Americans; an indigenous Polynesian kingdom in the Hawaiian Islands was annexed by the United States and is now the fiftieth State. Still, today, the political map is an assemblage of independent and colonial

*The figures in Table G-1 do not match these totals because only the political entity of Papua New Guinea is listed, not the Indonesian part of the island (West Papua). Here, as Figure 10-2 showed, the political and the realm boundaries do not coincide.

The Pacific Realm

1. The Pacific Realm's total area is the largest of all geographic realms. Its land area, however, is the smallest, as is its population.

2. The island of New Guinea, with 7.7 million people, alone contains over 80 percent of the Pacific Realm's population.

3. The Pacific Realm, with its wide expanses of water and numerous islands, has been strongly affected by United Nations Law of the Sea provisions regarding states' rights over economic assets in their adjacent waters.

4. The highly fragmented Pacific Realm consists of three regions: Melanesia (including New Guinea), Micronesia, and Polynesia.

5. Melanesia forms the link between Papuan and Melanesian cultures in the Pacific.

6. The Pacific Realm's islands and cultures may be divided into volcanic *high-island* cultures and coral-based *low-island* cultures.

7. In Micronesia, U.S. influence has been particularly strong and continues to affect local societies.

8. In Polynesia, local cultures are nearly everywhere severely strained by external influences. In Hawai'i, as in New Zealand, indigenous culture has been largely submerged by Westernization.

9. Indigenous Polynesian culture continues to exhibit a remarkable consistency and uniformity throughout the Polynesian region, its enormous dimensions and dispersal notwithstanding.

territories. Paris controls New Caledonia and French Polynesia. The United States administers Guam and American Samoa, the Line Islands, Wake Island, Midway Islands, and several smaller islands; the United States also has special relationships with other territories, former dependencies that are now nominally independent. The British, through New Zealand, have responsibility for the Pitcairn group of islands, and New Zealand administers and supports the Cook, Tokelau, and Niue Islands. Easter Island, that storied speck of land in the southeastern Pacific, is part of Chile. Indonesia rules West Papua.

Other island groups have become independent states. The largest are Fiji, once a British dependency, the Solomon Islands (also formerly British), and Vanuatu (until 1980 ruled jointly by France and Britain). Also on the current map, however, are such microstates as Tuvalu, Kiribati, Nauru, and Palau. Foreign aid is crucial to the survival of most of these countries. Tuvalu, for example, has a total area

of about 10 square miles, a population of some 10,000, and a per-capita GNP of about $600, derived from fishing, copra sales, and some tourism. But what really keeps Tuvalu going is an international trust fund set up by Australia, New Zealand, the United Kingdom, Japan, and South Korea. Annual grants from that fund, as well as money sent back to families by workers who have left for New Zealand and elsewhere, allow Tuvalu to survive.

THE PACIFIC REALM AND ITS MARINE GEOGRAPHY

Certain land areas may not be part of the Pacific Realm (we cited the Philippines and New Zealand), but the Pacific Ocean extends from the shores of North and South America to mainland East and Southeast Asia and from the Bering Sea to the Subtropical Convergence. This means that several seas, including the Sea of Japan (East Sea), the East China Sea, and the South China Sea, are part of the Pacific Ocean. As we will see later, this relationship matters. Pacific coastal countries, large and small, mainland and island, compete for jurisdiction over the waters that bound them.

The Pacific Realm and its ocean, therefore, form an ideal place to focus on **marine geography**. This field encompasses a variety of approaches to the study of oceans and seas; some marine geographers focus on the biogeography of coral reefs, others on the geomorphology of beaches, and still others on the movement of currents and drifts. A particularly interesting branch of marine geography has to do with the definition and delimitation of political boundaries at sea. Here geography meets political science and maritime law.

Littoral (coastal) states do not end where atlas maps suggest they do. States have claimed various forms of jurisdiction over coastal waters for centuries, closing off bays and estuaries and ordering foreign fishing fleets to stay away from nearby fishing grounds. Thus arose the notion of the **territorial sea**, where all the rights of a coastal state would prevail. Beyond lay the **high seas**, free, open, and unfettered by national interests.

It was in the interest of colonizing, mercantile states to keep territorial seas narrow and high seas wide, thus interfering as little as possible with their commercial fleets. In the seventeenth and eighteenth centuries, the territorial sea was 3, 4, or at most 6 nautical miles wide, and the colonizing powers claimed the same widths for their colonies (1 nautical mile = 1.15 statute miles [1.85 km]).

In the twentieth century these constraints weakened. States without trading fleets saw no reason to limit their territorial seas. States with nearby fishing grounds traditionally exploited by their own fleets wanted to keep the increasing number of foreign trawlers away. States with shallow **continental shelves**, offshore continuations of coastal plains, wished to control the resources on and below the seafloor, which were made more accessible by improving technology. States disagreed on the methods by which offshore boundaries, whatever their width, should be defined. Early efforts by the League of Nations in the 1920s to resolve these issues had only partial success, mainly in the technical area of boundary delimitation.

In 1945, the United States helped precipitate what has become known as the "scramble for the oceans." President Harry S Truman issued a proclamation that claimed U.S. jurisdiction and control over all the resources "in and on" the continental shelf down to its margin, around 100 fathoms (600 ft) deep. In some areas, the shallow continental shelf of the United States extends more than 200 miles offshore, and Washington did not want foreign countries drilling for oil just beyond the 3-mile territorial sea.

Few observers foresaw the impact the Truman Proclamation would have, not only on U.S. waters but on the oceans everywhere, including the Pacific. It set off a rush of other claims. In 1952, a group of South American countries, some with little continental shelf to claim, issued the Declaration of Santiago, claiming exclusive fishing rights up to a distance of 200 nautical miles (230 statute miles) off their coasts. Meanwhile, as part of the Cold War competition, the Soviet Union urged its allies to claim a 12-mile territorial sea.

The UNCLOS Convention

Now the United Nations intervened, and a series of UNCLOS (United Nations Conference on the Law of the Sea) meetings began. These meetings addressed issues ranging from the closure of bays to the width and delimitation of the territorial sea, and after three decades of negotiations they achieved a convention that changed the political and economic geography of the oceans forever. Among its key provisions were the authorization of a 12-mile territorial sea for all countries and the establishment of a 200-mile (230 statute-mile) **Exclusive Economic Zone** (**EEZ**) over which a coastal state has certain economic rights. It has total control over the resources on and beneath the seabed, such as oil, natural gas, or manganese nodules. It also has substantial rights to the fish in the EEZ, but the UNCLOS Convention does stipulate that coastal states must cooperate through international organizations to ensure the conservation of highly migratory species such as tuna, marlin, and swordfish.

These provision had a far-reaching impact on the world's oceans and seas (Fig. 12-2), especially on the Pacific. Unlike the Atlantic Ocean, the Pacific is studded with islands large and small, and a microstate consisting of one small island suddenly acquired an EEZ covering 166,000 square nautical miles. European colonial powers still holding minor Pacific possessions (notably France) saw their maritime jurisdictions vastly expanded. Small low-income archipelagos could now bargain with large fishing nations over fishing rights in their EEZs. Under

FROM THE FIELD NOTES

"Having arrived late at night, I did not have a perspective yet of the setting of Pago Pago—but even before I got up I knew what the leading industry here had to be. An intense, overpowering smell of fish filled the air. Sure enough, daylight revealed a huge tuna processing plant right across the water, with fishing boats arriving and departing continuously. American Samoa has a huge Pacific Ocean EEZ, one of the richest fishing grounds in the whole realm, and seafood exports rank ahead of tourism as the leading source of external income for the territory, an industry that is mostly American-owned. As the photograph shows, American Samoa is a group of 'high-islands' of volcanic origin, with considerable relief and environmental variety. Tuvalu, the country I had just left, was the complete opposite—a 'low-island' territory whose EEZ has been leased to Japanese fishing fleets, which do their processing on board and employ no locals."

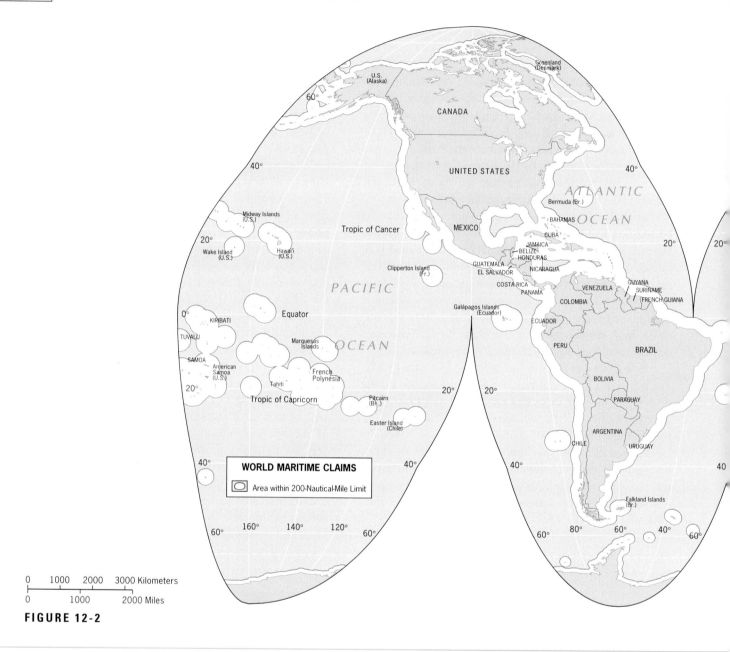

WORLD MARITIME CLAIMS

▢ Area within 200-Nautical-Mile Limit

0   1000   2000   3000 Kilometers
0   1000   2000 Miles

FIGURE 12-2

UNCLOS rules, such bargaining should involve (in the case of the Pacific) such organizations as the South Pacific Forum Fisheries Agency, and the United States insists on this procedure, which exists to protect highly migratory species. But some other countries, notably Japan and Taiwan, flout that rule and negotiate directly with the island-states over fishing rights in their EEZs. In May 2003, the journal *Nature* published the results of a study that reported a 90-percent decline in large species over the past 50 years as fishing nations with high-technology equipment are scouring the oceans and destroying marine ecosystems even in remote reaches of the globe.

The extension of the territorial sea to 12 nautical miles and the EEZ to an additional 188 nautical miles (both are measured from the coastline) created new **maritime boundary** problems. Waters less than 24 miles [6] wide separate many countries all over the world, so that **median lines**, equidistant from opposite shores, have been [7] delimited to establish their territorial seas. And even more countries lie closer than 400 nautical miles apart, requiring further maritime-boundary delimitation to determine their EEZs. In such maritime regions as the North Sea, the Caribbean Sea, and the Japan, East China, and South China Seas, a maze of maritime boundaries emerged, some of them subject to dispute. Political changes on land can lead to significant modifications at sea and sometimes result in serious disputes. As we noted in Chapter 10, when East Timor became an independent state in 2002, this seemed to void the maritime boundary between Indonesia and Australia negotiated bilaterally and delimited in 1972 (Fig. 10-13).

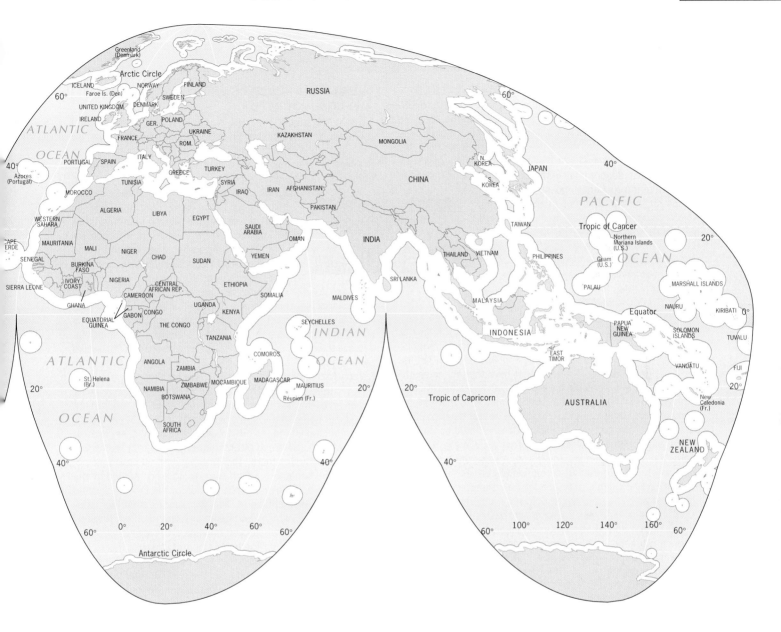

According to UNCLOS principles, the new boundary between East Timor and Australia would be the median line between the two countries. But Australia was unwilling to give up the wealth it gained from the 1972 boundary. It not only withdrew from UNCLOS, but also announced that it would not be subject to any maritime-boundary decisions made by the International Court of Justice. As far as Australia was concerned, the 1972 boundary would stay in place, and most of the reserves in the Timor Gap and elsewhere in the Timor Sea would continue to be exploited by Australia. A weak, newly independent but poor state would have little hope of defying a powerful wealthy one, exposing the weakness of the UNCLOS Convention.

The UNCLOS provisions also created opportunities for some states to expand their spheres of influence. Wider territorial-sea and EEZ allocations raised the stakes: claiming an island now entailed potential control over a huge maritime area. In Chapters 9 and 10, we referred to the large number of island disputes off mainland East Asia: between Japan and Russia, Japan and South Korea, Japan and China, China and Vietnam, and China and the Philippines. Ownership of many islands there is uncertain, and small specks of island territory have become large stakes in the scramble for the oceans. In the case of the Spratly Islands (Fig. 10-2), six countries claim ownership, including both Taiwan and China. China's island claims in the South China Sea support Beijing's contention that this body of water is part and parcel of the Chinese state—a position that worries other states with coasts facing it.

Who Should Own the Oceans?

WE WHO LIVE HERE SHOULD OWN THE WATERS

"WITHOUT EXCLUSIVE FISHING rights in our waters up to 200 nautical miles offshore, our economy would be dead in the water, no pun intended, sir. As it is, it's bad enough. I'm an official in the Fisheries Department of the Ministry of Natural Resources Development here in Tuvalu, and this modest building you're visiting is the most modern in all of the capital, Funafuti. That's because we're the only ones making some money for the state other than our Internet domain, which has been the big story around here since we commercialized it. We don't have much in the way of natural resources and we export some products from our coconut and breadfruit trees, but otherwise we need donors to help us and major fishing nations to pay us for the use of our 750,000 square kilometers of ocean.

"Which brings me to one of our big concerns. Three-quarters of a million square kilometers sound like a lot, but look at the map. It's a mere speck in the vast Pacific Ocean. Enormous stretches of our ocean seem to belong to nobody. That is, they're open to fishing by nations that have the technology to exploit them, but we don't have those fast boats and factory ships they use nowadays. Still, the fish they take there might have come to our exclusive grounds, but we don't get paid for them. It seems to us that the Pacific Ocean should be divided among the countries that exist here, not just on the basis of nearby Exclusive Economic Zones but on some other grounds. For example, islands that are surrounded by others, like we are, ought to get additional maritime territory someplace else. Islands that aren't surrounded should get much more than just 200 nautical miles. To us the waters of the Pacific mean much more than they do to other countries, not only economically but historically too. I don't like Japanese and American and Norwegian fishing fleets on these nearby 'high seas.' They don't obey international regulations on fishing, and they overfish wherever they go. When there's nothing left, they'll simply end their agreement with us, and we'll be left without income from a natural resource over which we never had control.

"I don't expect anything to change, but we feel that the era of the 'high seas' should end and all waters and subsoil should be assigned to the nearest states. The world doesn't pay much attention to our small countries, I realize. But we've put up with colonialism, world wars, nuclear weapons testing, pollution. It's time we got a break."

THE OCEAN SHOULD BE OPEN AND UNRESTRICTED

"Ours is the Blue Planet, the planet of oceans and seas, and the world has fared best when its waters were free and open for the use of all. Encroachment from land in the form of 'territorial' seas (what a misnomer!) and other jurisdictions only interfered with international trade and commercial exploitation and caused far more problems than they solved. Already much of the open ocean is assigned to coastal countries that have no maritime tradition and don't know what to do with it other than to lease it. And when coastal countries lie closer than 400 nautical miles to each other, they divide the waters between them for EEZs and don't even leave a channel for international passage. Yes, I realize that there are guarantees of uninhibited passage through EEZs. But mark my words: the time will come when small coastal states will start charging vessels, for example cruise ships, for passage through 'their' waters. That's why, as the first officer of a passenger ship, I am against any further allocation of maritime territory to coastal states. We've gone too far already.

"In the 1960s, when those UNCLOS meetings were going on, a representative from Malta named Arvid Pardo came up with the notion that the whole ocean and seafloor beyond existing jurisdictions should be declared a 'common heritage of mankind' to be governed by an international agency that would control all activities there. The idea was that all economic activity on and beneath the high seas would be controlled by the UN, which would ensure that a portion of the proceeds would be given to 'Geographically Disadvantaged States' like landlocked countries, which can't claim any seas at all. Fortunately nothing came of this—can you imagine a UN agency administering half the world for the benefit of humanity?

"So you can expect us mariners to oppose any further extensions of land-based jurisdiction over the seas. The maritime world is complicated enough already, what with choke points, straits, narrows, and various kinds of prohibitions. To assign additional maritime territory to states on historic or economic grounds is to complicate the future unnecessarily, and to invite further claims. From time to time I see references to something called the 'World Lake Concept,' the preliminary assignment of all the waters of the world to the coastal states through the delimitation of median lines. Such an idea may be the natural progression from the EEZ concept, but it gives me nightmares."

Figure 12-2 reveals what EEZ regulations have meant to Pacific Realm countries such as Tuvalu, Kiribati, and Fiji, with nearly circular EEZs now surrounding the clusters of islands in this vast ocean space. Japan, Taiwan, and other fishing nations have purchased fishing rights in these EEZs from the island governments, although violations of EEZ rights do occur. Recently, Vanuatu and the Philippines were at odds over unauthorized Filipino fishing in Vanuatu's EEZ.

The process of boundary delimitation continues. In an earlier edition of this book, we included a map of the South China Sea and nearby waters off East and Southeast Asia, showing the median-line boundaries delimited according to UNCLOS specifications. But that map changed when coastal states engaged in bilateral and multilateral negotiations—and argued over island ownership with major boundary implications. The Pacific—and world—map of maritime boundaries remains a work in progress (see the Issue Box entitled "Who Should Own the Oceans?").

This point is illustrated by a development that occurred late in 1999, when the implications of provisions in Article 76 of the 1982 UNCLOS Convention were reexamined. Where a continental shelf extends beyond the 200-mile limit of the EEZ, the Convention apparently allows a coastal state to claim that extension as a "natural prolongation" of its landmass. Although there is as yet no regulation that would also allow the state to extend its EEZ to the edge of the continental shelf, it is not difficult to foresee such an amendment to the Convention. As it stands, states are now delimiting their proposed "natural prolongations" under a 2009 deadline, which puts poorer states at a disadvantage since the required marine surveys are very expensive. In any case, an article in *Science* in December 2002 reports that the big winners are likely to include the United States, Canada, Australia, New Zealand, Russia, and India—although a total of 60 coastal countries could ultimately benefit to some extent from the provision. Thus the "scramble for the oceans" continues, and with it the constriction of the world's high seas and open waters.

REGIONS OF THE REALM

In this realm as a whole, tourism may well be the largest overall revenue earner. Sail across the Pacific Ocean, and one spectacular vista follows another. Dormant and extinct volcanoes, sculpted by erosion into basalt spires draped by luxuriant tropical vegetation and encircled by reefs and lagoons, tower over azure waters. Low atolls with nearly snow-white beaches, crowned by stands of palm trees, seem to float in the water. Pacific islanders, where foreign influences have not overtaken them, appear to take life with enviable ease.

More serious investigations reveal that such Pacific sameness is more apparent than real. Even the Pacific Realm, with its long sailing traditions, its still-diffusing populations, and its historic migrations, has a persistent regional framework. Figure 12-3 outlines the three regions that constitute the Pacific Realm:

Melanesia West Papua (Indonesia), Papua New Guinea, Solomon Islands, Vanuatu, New Caledonia (France), Fiji

Micronesia Palau, Federated States of Micronesia, Northern Mariana Islands, Republic of the Marshall Islands, Nauru, western Kiribati, Guam (U.S.)

Polynesia Hawaiian Islands (U.S.), Samoa, American Samoa, Tuvalu, Tonga, eastern Kiribati, Cook and other New Zealand-administered islands, French Polynesia

Ethnic, linguistic, and physiographic criteria are among the bases for this regionalization of the Pacific Realm, but we should not lose sight of the dimensions. Not only is the land area small, but also in 2004 the population of this entire realm (including Indonesia's West Papua) was only about 11 million—8.5 million without West Papua—about the same as one very large city. Fewer people live in this realm than in another vast area of far-flung settlements, the oases of North Africa's Sahara.

MAJOR CITIES OF THE REALM

City	Population* (in millions)
Honolulu, Hawai'i (U.S.)	0.9
Nouméa, New Caledonia	0.2
Port Moresby, Papua New Guinea	0.4
Suva, Fiji	0.3

*Based on 2004 estimates.

FIGURE 12-3

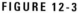

MELANESIA

The large island of New Guinea lies at the western end of a Pacific region that extends eastward to Fiji and includes the Solomon Islands, Vanuatu, and New Caledonia (Fig. 12-3). The human mosaic here is complex, both ethnically and culturally. Most of the 7.7 million people of New Guinea (including the Indonesian part, West Papua, and the independent state of Papua New Guinea) are Papuans, and a large minority is Melanesian. Altogether there are as many as 700 communities speaking different languages; the Papuans are most numerous in the densely forested highland interior and in the south, while the Melanesians inhabit the north and east. The region as a whole has more than 7 million inhabitants, making this the most populous Pacific region by far.

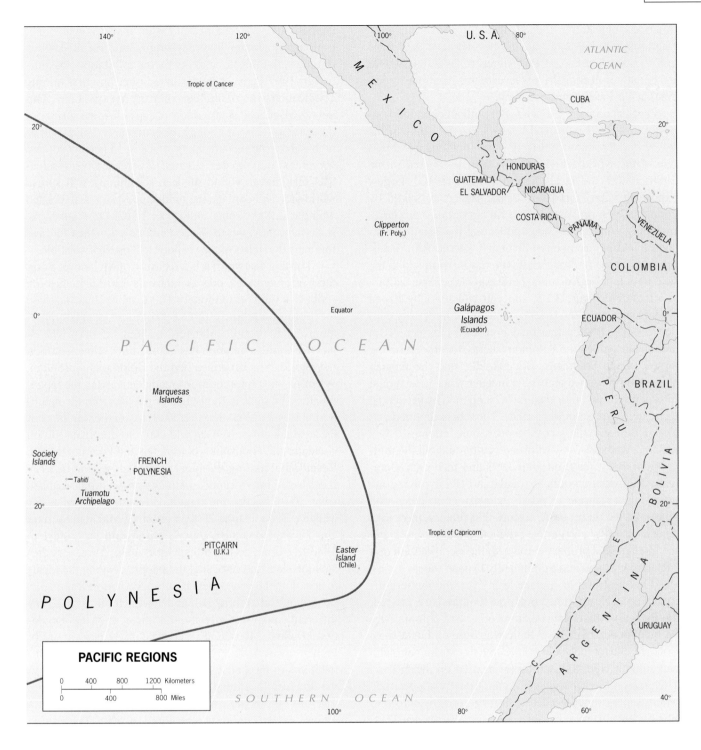

PACIFIC REGIONS

0   400   800   1200 Kilometers

0   400   800 Miles

With 5.2 million people today, **Papua New Guinea (PNG)** became a sovereign state in 1975 after nearly a century of British and Australian administration. Almost all of PNG's limited development is taking place along the coasts, while most of the interior remains hardly touched by the changes that transformed neighboring Australia. Perhaps four-fifths of the population lives in what we may describe as a self-sufficient subsistence economy, growing root crops and hunting wildlife, raising pigs, and gathering forest products. Old traditions of the kind lost in Australia persist here, protected by remoteness and the rugged terrain.

Welding this disparate population into a nation is a task hardly begun, and PNG faces numerous obstacles in addition to its cultural complexity. Not only are hundreds of languages in use, but more than half the population is illiterate. English, the official language, is used by the educated minority but is of little use beyond the coastal zone and its towns. The capital, Port Moresby, has about 380,000 inhabitants, reflecting the low level of urbanization (15 percent) in this low-income economy.

Yet Papua New Guinea is not without economic opportunities. As Figure G-11 shows, it now ranks in the lower-middle-income group of states (not long ago it was one of the world's poorest). Oil was discovered in the 1980s, and by the late 1990s crude oil was PNG's largest export by value. Gold now ranks second, followed by copper, silver, timber, and several agricultural products, including coffee and cocoa, reflecting the country's resource and environmental diversity. Pacific Rim developments have affected even PNG: most exports go to the nearest neighbor, Australia, but Japan ranks close behind.

Examine Figure 12-3, and you will see why Papua New Guinea has been beset by centrifugal forces since independence. PNG's maritime boundary cuts across an archipelago called the Solomon Islands, incorporating the largest of the Solomons, Bougainville, into the Papuan state. Almost immediately after independence, a separatist insurgency arose on Bougainville, supported by irredentist elements in the Solomon Islands. The rebels captured and closed Bougainville's large copper mine, cutting PNG's export revenues. A costly insurgency throughout the mid-1990s was not resolved until 1997 and then only tentatively following a political scandal in PNG involving the government's hiring of white mercenaries to defeat the rebels. At the outset of the twenty-first century, there was tenuous stability and copper production had resumed, but the fundamental problem remained: the incorporation of a Melanesian island into a dominantly Papuan state.

This destructive episode highlights the vulnerability of Papua New Guinea in the postcolonial era. Indeed, the Bougainville issue is a direct legacy of colonialism. At independence in 1975 there was little enthusiasm in Bougainville for inclusion in PNG, but the Australians, who had administered both territories, insisted on the merger. Their reason was economic: revenues from the copper mine would supply much-needed cash from foreign sources to PNG's new government. But it was a forced marriage, leading to decades of strife costing more than the mine could ever repay.

As Figures 12-2 and 12-3 show, Papua New Guinea has a large EEZ, especially to the northeast where the large islands of New Britain and New Ireland encircle the Bismarck Sea. This is an area of tectonic-plate contact (see Fig. G-4), and such contact is associated with seafloor volcanic activity. Under certain circumstances, volcanic vents and upwelling, superheated waters combine to deposit rich lodes of dissolved minerals. When volcanic activity decreases, it leaves behind mounds of copper, silver, gold, zinc, and other minerals in concentrations far greater than anything found on land. Now mining technology is making their exploitation possible, and in 1997 Papua New Guinea issued the first mining certification to an Australian company to exploit the *chimneys* (vent-related stacks of ore) deep below the surface of the sea (Fig. 12-3). Although the economic rewards will be high—the eastern claim has as much as 15 percent copper, 19 percent iron, 26 percent zinc, 7 percent silver—scientists fear that this kind of deep-sea mining will spread worldwide and damage the unique ecosystems associated with the still-hot vents. But under UNCLOS regulations, states have the rights to all economic assets in the EEZ, and for many the riches of the seafloor will prove irresistible.

Turning eastward, it is a measure of Melanesia's cultural fragmentation that as many as 120 languages are spoken in the approximately 1000 islands that make up the **Solomon Islands** (about 80 of these islands support almost all the people, numbering about 475,000). The islands' natural separation has helped limit ethnic conflict, but World War II bequeathed the Solomons a problem. When American forces took Guadalcanal from the Japanese, they brought in thousands of people from the neighboring island of Malaita (Fig. 12-3). After the war, most of these Malaitans stayed on and came to dominate life on Guadalcanal, controlling commerce and buying up land. Resentment among the local people grew, and in 1999 it exploded into violence. Attacks on Malaitans caused some 10,000 to flee the capital, Honiara; many sailed for Malaita. Then in June 2000, a group of Malaitans still on Guadalcanal took the country's prime minister hostage in a bid to extract concessions. Once again, foreigners created the state and relocated the people, leaving the locals to deal with the consequences after decolonization.

New Caledonia, still under French rule, is in a very different situation. Only about 45 percent of the population of about 240,000 are Melanesian; 37 percent are of French ancestry, many of them descended from the inhabitants of the penal colony France established here in the nineteenth century. Nickel mines, based on reserves that rank among the world's largest, dominate New Caledonia's export economy (see photo, p. 561). The mining industry attracted additional French settlers, and social problems arose. Most of the French population lives in or near the capital city of Nouméa, steeped in French cultural landscapes, in the southeastern quadrant of the island. The miners, mainly Melanesians, live in dormitory-style buildings near the huge refinery in the capital; those not directly involved farm for subsistence in the island's middle and northern areas. Melanesian demands for an end to colonial rule have led to violence, and the two communities are still in the process of coming to terms.

FROM THE FIELD NOTES

"Arriving in the capital of New Caledonia, Nouméa, in 1996, was an experience reminiscent of French Africa 40 years earlier. The French tricolor was much in evidence, as were uniformed French soldiers. European French residents occupied hillside villas overlooking palm-lined beaches, giving the place a Mediterranean cultural landscape. And New Caledonia, like Africa, is a source of valuable minerals. It is one of the world's largest nickel producers, and from this vantage point you could see the huge treatment plants, complete with concentrate ready to be shipped (left, under conveyor). What you cannot see here is how southern New Caledonia has been ravaged by the mining operations, which have denuded whole mountainsides. Working in the mines and in this facility are the local Kanaks, Melanesians who make up about 45 percent of the population of about 240,000. Violent clashes between Kanaks and French have obstructed government efforts to change New Caledonia's political status in such a way as to accommodate pressures for independence as well as continued French administration."

On its eastern margins, Melanesia includes one of the Pacific Realm's most interesting countries, **Fiji**. On two larger and over 100 smaller islands live nearly one million Fijians, of whom 51 percent are Melanesians and 44 percent South Asians. The South Asians were brought to Fiji from India during the British colonial occupation to work on the sugar plantations. When Fiji achieved independence in 1970, the native Fijians owned most of the land and held political control, while the Indians were concentrated in the towns (chiefly Suva, the capital) and dominated commercial life. It was a recipe for trouble, and it was not long in coming when, in a later election, the politically active Indians outvoted the Fijians for seats in the parliament. A coup by the Fijian military was followed by a revision of the constitution, which awarded a majority of seats to ethnic Fijians. Before long, however, the Fijian majority splintered, but a coalition government for some time proved the constitution workable—until 2000. In May 1999, Fiji's first prime minister of Indian ancestry had taken office, angering some ethnic Fijians to the point of staging another coup. The prime minister and members of the government were taken hostage at the parliament building in Suva, and the perpetrators demanded that Fijians would henceforth govern the country.

This action, and Fiji's inability to counter it, had a devastating impact on the country. Foreign trading partners stopped buying Fijian products. The tourist industry suffered severely. And in the end, although the coup leaders were ousted and arrested, they secured a deal that gave ethnic Fijians the control they had sought. Fiji's future thus remains clouded.

Melanesia, the most populous region in the Pacific Realm, also is bedeviled by centrifugal forces of many kinds. No two countries present the same form of multiculturalism; each has its own challenges to confront, and some of these challenges spill over into neighboring (or more distant foreign) islands.

"Back in Suva, Fiji more than 20 years after I had done a study of its CBD, I noted that comparatively little had changed, although somehow the city seemed more orderly and prosperous than it was in 1978. At the Central Market I watched a Fijian woman bargain with the seller over a batch of taro, the staple starch-provider in local diets. I asked her how she would serve it. 'Well, it's like your potato,' she said. 'I can make a porridge, I can cut it up and put it in a stew, and I can even fry pieces of it and make them look like the French fries they give you at the McDonald's down the street.' Next I asked the seller where his taro came from. 'It grows

over all the islands,' he said. 'Sometimes, when there's too much rain, it may rot—but this year the harvest is very good.' . . . Next I walked into the crowded Indian part of the CBD. On a side street I got a reminder of the geographic concept of agglomeration. Here was a colonial-period building, once a hotel, that had been converted into a business center. I counted 15 enterprises, ranging from a shoestore to a photographer and including a shop where rubber stamps were made and another selling diverse tobacco products. Of course (this being the Indian sector of downtown Suva) tailors outnumbered all other establishments."

▶ MICRONESIA

North of Melanesia and east of the Philippines lie the islands that constitute the region known as Micronesia (Fig. 12-3). The name (*micro* means small) refers to the size of the islands: the 2000-plus islands of Micronesia are not only tiny (many of them no larger than 1 square mile), but they are also much lower-lying, on an average, than those of Melanesia. Some are volcanic islands (**high islands**, as the people call them), but they are outnumbered by islands composed of coral, the **low islands** that barely lie above sea level. Guam, with 210 square miles (550 sq km), is Micronesia's largest island, and no island elevation anywhere in Micronesia reaches 3300 feet (1000 m).

The high-island/low-island dichotomy is useful not only in Micronesia, but also throughout the realm. Both the physiographies of these islands and the economies they support differ in crucial ways. High islands wrest substantial moisture from the ocean air; they tend to be well watered and have good volcanic soils. As a result, agricultural products show some diversity, and life is reasonably secure. Populations tend to be larger on these high islands than on the low islands, where drought is the rule and fishing and the coconut palm are the mainstays of life. Small communities cluster on the low islands, and over time, many of these have died out. The major migrations, which sent fleets to populate islands from Hawai'i to New Zealand, tended to originate in the high islands.

Until the mid-1980s, Micronesia was largely a United States Trust Territory (the last of the post–World War II trusteeships supervised by the United Nations), but that status has now changed. As Figure 12-3 shows, today Micronesia is divided into countries bearing the names of independent states. The **Marshall Islands**, where the United States tested nuclear weapons (giving prominence to the name *Bikini*), now is a republic in "free association" with the United States, having the same status as the Federated States of Micronesia and (since 1994) Palau. The **Northern Mariana Islands** are a commonwealth "in political union" with the United States. In effect, the United States provides billions of dollars in assistance to these countries, in return for which they commit themselves to avoid foreign policy actions that are contrary to U.S. interests. There are other conditions: **Palau**, for example, granted the United States rights to existing military bases for 50 years following independence.

Also part of Micronesia are the U.S. territory of **Guam**, where independence is not in sight and where U.S. military installations and tourism provide the bulk of income, and the remarkable Republic of **Nauru**. With a population of barely 12,000 and 8 square miles of land, this microstate got rich by selling its phosphate deposits to Australia and New Zealand, where they are used as fertilizer. Per-capita incomes rose to U.S. $10,000, making Nauru one of the Pacific's high-income societies. But the phosphate deposits ran out, and an island scraped bare faced an economic crisis.

In this region of tiny islands, most people subsist on farming or fishing, and virtually all the countries need infusions of foreign aid to survive. The natural economic complementarity between the high-island farming cultures and the low-island fishing communities all too often is negated by distance, spatial as well as cultural. Life here may seem idyllic to the casual visitor, but for the Micronesians it often is a daily challenge.

▶ POLYNESIA

To the east of Micronesia and Melanesia lies the heart of the Pacific, enclosed by a great triangle stretching from the Hawaiian Islands to Chile's Easter Island to New Zealand. This is Polynesia (Fig. 12-3), a region of numerous islands (*poly* means many), ranging from volcanic mountains rising above the Pacific's waters (Mauna Kea on Hawai'i reaches nearly 13,800 feet [over 4200 m]), clothed by luxuriant tropical forests and drenched by well over 100 inches of rainfall each year, to low coral atolls where a few palm trees form the only vegetation and where drought is a persistent problem. The Polynesians have somewhat lighter-colored skin and wavier hair than do the other peoples of the Pacific Realm; they are often also described as having an excellent physique. Anthropologists differentiate between these original Polynesians and a second group, the Neo-Hawaiians, who are a blend of Poly-

nesian, European, and Asian ancestries. In the U.S. State of Hawai'i—actually an archipelago of more than 130 islands—Polynesian culture has been both Europeanized and Orientalized.

Its vastness and the diversity of its natural environments notwithstanding, Polynesia clearly constitutes a geographic region within the Pacific Realm. Polynesian culture, though spatially fragmented, exhibits a remarkable consistency and uniformity from one island to the next, from one end of this widely dispersed region to the other. This consistency is particularly expressed in vocabularies, technologies, housing, and art forms. The Polynesians are uniquely adapted to their maritime environment, and long before European sailing ships began to arrive in their waters, Polynesian seafarers had learned to navigate their wide expanses of ocean in huge double canoes as long as 150 feet (45 m). They traveled hundreds of miles to favorite fishing zones and engaged in inter-island barter trade, using maps constructed from bamboo sticks and cowrie shells and navigating by the stars. However, modern descriptions of a Pacific Polynesian paradise of emerald seas, lush landscapes, and gentle people distort harsh realities. Polynesian society was forced to get used to much loss of life at sea when storms claimed their boats; families were ripped apart by accident as well as migration; hunger and starvation afflicted the inhabitants of smaller islands; and the island communities were often embroiled in violent conflicts and cruel retributions.

The political geography of Polynesia is complex. In 1959, the Hawaiian Islands became the fiftieth State to join the United States. The State's population is now 1.3 million, with over 80 percent living on the island of Oahu. There, the superimposition of cultures is symbolized by the panorama of Honolulu's skyscrapers against the famous extinct volcano at nearby Diamond Head. The Kingdom of **Tonga** became an independent country in 1970 after seven decades as a British protectorate; the British-administered Ellice Islands were renamed **Tuvalu**, and along with the Gilbert Islands to the north (now renamed **Kiribati**), they received independence from Britain in 1978. Other islands continued under French control (including the Marquesas Islands and Tahiti), under New Zealand's administration (Rarotonga), and under British, U.S., and Chilean flags.

In the process of politico-geographical fragmentation, Polynesian culture has suffered severe blows. Land developers, hotel builders, and tourist dollars have set **Tahiti** on a course along which Hawai'i has already traveled far. The Americanization of eastern **Samoa** has created a new society different from the old. Polynesia has lost much of its ancient cultural consistency; today, the region is a patchwork of new and old—the new often bleak and barren, with the old under intensifying pressure.

The countries and cultures of the Pacific Realm lie in an ocean on whose rim a great drama of economic and political transformation will play itself out during the twenty-first century. Already, the realm's own former margins—in Hawai'i in the north and in New Zealand in the south—have been so recast by foreign intervention that little remains of the kingdoms and cultures that once prevailed. Now the Pacific world faces changes far greater even than those brought here by European colonizers. Once upon a time the shores and waters of the Mediterranean Sea formed an arena of regional transformation that changed the world. Then it was the Atlantic, avenue of the Industrial Revolution and stage of fateful war. Now it seems to be the turn of the Pacific as the world's largest country and next superpower (China) faces the richest and most powerful (the United States). Giants will jostle for advantage in the Pacific; how will the weak societies of the Pacific Realm fare?

A FINAL CAVEAT: PACIFIC AND ANTARCTIC

South of the Pacific Realm lies Antarctica and its encircling Southern Ocean. The combined area of these two geographic expanses constitutes 40 percent of the entire planet—two-fifths of the Earth's surface containing one one-thousandth of the world's population.

Is Antarctica a geographic realm? In physiographic terms, yes, but not on the basis of the criteria we use in this book. Antarctica is a continent, nearly twice as large as Australia, but virtually all of it is covered by a dome-

shaped ice sheet nearly 2 miles (3.2 km) thick near its center. The continent often is referred to as the "white desert" because, despite all of its ice and snow, annual precipitation is low (less than 6 inches [15 cm] per year). Temperatures are frigid, with winds so strong that Antarctica also is called the "home of the blizzard." For all its size, no functional regions have developed here, no towns, no transport networks except the supply lines of research stations. And Antarctica still is a frontier, even a scientific frontier still slowly giving up its secrets. Underneath all that ice lie some 70 lakes of which Lake Vostok is the largest with over 5400 square miles (14,000 sq km). It may be as much as 2000 feet (600 m) deep. No one has yet seen a sample of its water.

Like virtually all frontiers, Antarctica has always attracted pioneers and explorers. Whale and seal hunters destroyed huge populations of Southern Ocean fauna during the eighteenth and nineteenth centuries, and explorers planted the flags of their countries on Antarctic shores. Between 1895 and 1914, the quest for the South Pole became an international obsession; Roald Amundsen, the Norwegian, reached it first in 1911. All this led to national claims in Antarctica during the interwar period (1918–1939).

The geographic effect was the partitioning of Antarctica into pie-shaped sectors centered on the South Pole (Fig. 12-4). In the least frigid area of the continent, the Antarctic Peninsula (highlighted in the inset map), British, Argentinean, and Chilean claims overlapped—and still do. One sector, Marie Byrd Land, was never formally claimed by any country.

Why should states be interested in territorial claims in so remote and difficult an area? Both land and sea contain raw materials that may some day become crucial: proteins in the waters, and fuels and minerals beneath the

FROM THE FIELD NOTES

"The Antarctic Peninsula is geologically an extension of South America's Andes Mountains, and this vantage point in the Gerlach Strait leaves you in no doubt: the mountains rise straight out of the frigid waters of the Southern Ocean. We have been passing large, flat-topped icebergs, but these do not come from the peninsula's shores; rather, they form on the leading edges of ice shelves or on the margins of the mainland where continental glaciers slide into the sea. Here along the peninsula, the high-relief topography tends to produce jagged, irregular icebergs. The mountain range that forms the Antarctic Peninsula continues across Antarctica under thousands of feet of ice and is known as the Transantarctic Mountains. Looking at this place you are reminded that beneath all this ice lies an entire continent, still little-known, with fossils, minerals, even lakes yet to be discovered and studied."

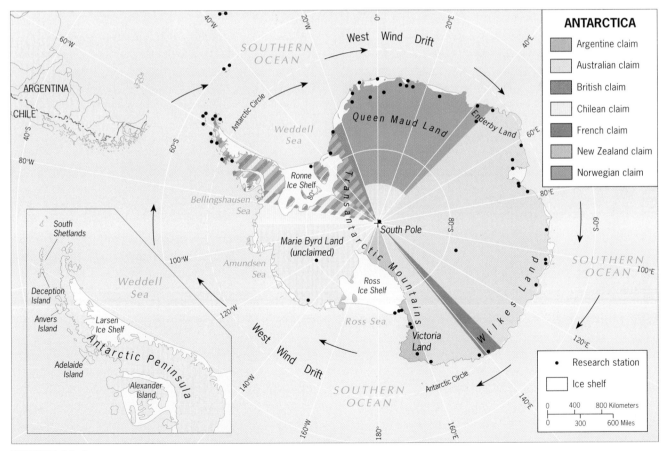

FIGURE 12-4

ice. Antarctica (5.5 million sq mi/14.2 million sq km) is almost twice as large as Australia, and the Southern Ocean (see box, p. 533) is nearly as large as the North and South Atlantic. However distant actual exploitation may be, countries want to keep their stakes here.

But the claimant states (those with territorial claims) recognize the need for cooperation. During the late 1950s, they joined in the International Geophysical Year (IGY) that launched major research programs and established a number of permanent research stations throughout the continent. This spirit of cooperation led to the 1961 signing of the **Antarctic Treaty**, which ensures continued scientific collaboration, prohibits military activities, protects the environment, and holds national claims in abeyance. In 1991, when the treaty was extended under the terms of the Wellington Agreement, concerns were raised that it does not do enough to control future resource exploitation.

In an age of growing national self-interest and increasing consumption of raw materials, the possibility exists that Antarctica and its offshore waters may yet become an arena for international rivalry. Until now, its remoteness and its forbidding environments have saved it from that fate. The entire world benefits from this because evidence is mounting that Antarctica plays a critical role in the global environmental system, so that human modifications may have global (and unpredictable) consequences.

Using the Maps

*T*F, AS IS OFTEN SAID, a picture is worth a thousand words, then a map is worth a million. This book contains more than 200 maps, locational and thematic, to help you understand the text. Please make it a point to study each map when it is referred to, and do so as carefully as you would a paragraph of text.

Maps display locations and patterns in geographic space. They are a kind of shorthand through which encoded spatial messages are transmitted to the reader. Such shorthand is necessary because the real world is so complex that a great deal of geographic information must be compressed into the small confines of maps that can fit onto the pages of this book. Cartographers (mapmakers) must carefully choose which information to include; these decisions force them to omit many other things in order to prevent cluttering a map with less relevant information. For instance, Figure A shows several city blocks in central London but avoids mapping individual buildings because they would interfere with the key information being presented—the spatial distribution of cholera deaths in that part of London during an epidemic.

FIGURE A

MAP READING

Deciphering the coded messages contained in our maps—*map reading*—is not difficult, but a few rules of the road do help. The need to miniaturize portions of the world on small maps is discussed in the section on map scale (pp. 7–8), and Figures A and B offer two additional and contrasting examples.

Orientation, or direction, on maps can usually be discerned by reference to the geographic grid of latitude and longitude. *Latitude* is measured from 0° to 90° north and south of the equator (parallels of latitude are always drawn in an east-west direction), with the equator being 0° and the North and South Poles being 90°N and 90°S, respectively. Meridians of *longitude* (always drawn north-south) are measured 180° east and west of the *prime meridian* (0°),

which passes through the Greenwich Observatory next to London; the 180th meridian, for the most part, serves as the *international date line* that lies in the middle of the Pacific Ocean (see Fig. 12-1). Inspection of Figure B shows that north is not automatically at the top of a map; instead, the direction of north curves along every meridian, with all such lines of longitude converging at the North Pole. The many minor directional distortions in this map are unavoidable: it is geometrically impossible to transfer the grid of a three-dimensional sphere (globe) onto a two-dimensional flat map. Therefore, compromises in the form of *map projections* must be devised in which, for example, properties such as areal size and distance are preserved but directional constancy is sacrificed.

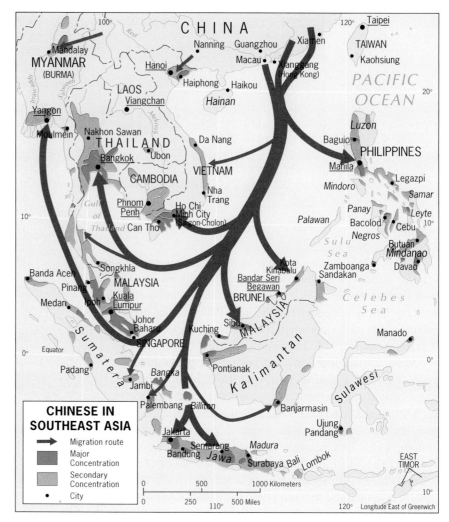

FIGURE B

Map legend:

CHINESE IN SOUTHEAST ASIA

→ Migration route

■ Major Concentration

■ Secondary Concentration

• City

MAP SYMBOLS

Once the background mechanics of scale and orientation are understood, the main task of decoding the map's content can proceed. The content of most maps in this book is organized within the framework of point, line, and area symbols, which are made especially clear through the use of color. These symbols are usually identified in the map's *legend*, but sometimes accompanying text must tell the map reader more. In Figure A, for example, we are informed that the dots reflect cholera deaths, but we are not informed that each ℗ symbol represents a municipal water pump at the time of this epidemic in 1854. *Point symbols*, shown as dots on maps, can tell us two things: the location of each phenomenon, and, sometimes, its quantity. Look, for example, at Figure 1-14 (p. 64). Not only does this map show the locations of Western European cities, but also their relative population size.

Line symbols connect places or areas between which some sort of movement or flow has occurred or is occurring. Figure B, taken from the chapter on Southeast Asia,

shows the migration of Chinese into this realm, a mass movement that originated mainly in southeastern China and produced coastal Chinese communities on the islands and peninsulas of this multicultural part of the world. Note that we make no attempt to rank the sizes of Southeast Asian cities, so as not to obscure the key theme. *Area symbols* are used to classify two-dimensional spaces. Primary and secondary concentrations of Chinese residents in Southeast Asia are differentiated by color in Figure B. Area symbols can also be used to communicate quantitative information: for example, the light-tan-colored zones in Figure G-7 (p. 12) delimit semiarid areas within which annual precipitation averages from 12 to 20 inches (30 to 50 cm).

MAP INTERPRETATION

The explanation of cartographic patterns is one of the geographer's most important tasks. Although that task is performed for you throughout this book, readers should be aware that today's practitioners use many sophisticated techniques and cutting-edge computer technology to analyze vast quantities of areal data. These modern methods notwithstanding, geographic inquiry still focuses on the search for meaningful *spatial relationships*. This longstanding concern of the discipline is classically demonstrated in Figure A. By showing on his map that cholera fatalities clustered around municipal water pumps, Dr. John Snow was able to persuade city authorities to shut them off; almost immediately, the number of new disease victims dwindled to zero, thereby confirming Snow's theory that contaminated drinking water was crucial in the spread of cholera.

Sometimes maps can reveal information or insights beyond those anticipated or intended by the cartographer. Take a look at the ethnolinguistic map on page 366, Fig. 7-20. That map was drawn before the events of September 11, 2001 and subsequent developments in and around Afghanistan. Its relevance and meaning now exceed its original, illustrative intent. So study maps such as this in detail, and you will vastly increase what you get out of this book and this course.

Opportunities in Geography

THE CHAPTERS OF THIS BOOK give an idea of the wide range of topics and interests that geographers pursue, particularly in the many discussions of concepts (regional and otherwise) and frequent references to the content of geography's systematic fields. There are specializations within each of those topical fields as well as in regional studies—but an introductory book such as this lacks enough space to discuss all of them. This appendix, therefore, is designed to help you, should you decide to major or minor in geography and/or to consider it as a career option.

AREAS OF SPECIALIZATION

As in all disciplines, areas of concentration or specialization change over time. In North American geography early in the twentieth century, there was a period when most geographers were physical geographers, and the natural landscape was the main object of geographic analysis. Then the pendulum swung toward human (cultural) geography, and students everywhere focused on the imprints of human activity on the surface of the Earth. Still later the analysis of spatial organization became a major concern. In the meantime, geography's attraction for some students lay in technical areas: in cartography, in aerial and satellite remote sensing, in computer-assisted spatial data analysis, and, most recently, in geographic information systems (GIS) and GIScience.

All this meant that geography posed (and continues to pose) a challenge to its professionals. New developments require that we keep up to date, but we must also continue to build on established foundations.

Regional Geography

One of these established foundations, of course, is regional geography, which encompasses a large group of specializations. Some geographers specialize in the theory of regions: how they should be defined, how they are structured, and how their internal components work. This leads in the direction of *regional science*, and some geographers have preferred to call themselves regional scientists. But make no mistake: regional science is regional geography.

Another, older approach to regional geography involves specialization in an area of the world ranging in size from a geographic realm to a single region or even a State or part of a State. There was a time when regional geographers, because of the interdisciplinary nature of their knowledge, were sought after by government agencies. Courses in regional geography abounded in universities' geography departments; regional geographers played key roles in international studies and research programs. But then the drive to make geography a more rigorous science and to search for universal (rather than regional) truths contributed to a decline in regional geography. The results were not long in coming, and in recent years you have probably seen the issue of "geographic illiteracy" discussed in newspapers and magazines. Now the pendulum is swinging back again, and regional geography and regional specialization are reviving. This is a propitious time to consider regional geography as a professional field.

Your Personal Interests

Geography, as we have pointed out, is united by several bonds, of which regional geography is but one. Regional geography exemplifies the spatial view that all geographers hold; the spatial approach to study and research binds physical and human geographers, regionalists, and topical specialists. Another unifying theme is an abiding interest in the relationships between human societies and natural environments. We have referred to that topic frequently in this book; as an area of specialization it has gone through difficult times. Perhaps more than anything, geography remains a field of *synthesis*, of understanding interrelationships.

Geography also is a field science, using "field" in another context. In the past, almost all major geography departments required a student's participation in a "field camp" as part of a master's degree program; thus many undergraduate programs included field experience. It was one of those bonding practices in which students and faculty with diverse interests met, worked together, and learned from one another. Today, few field camps of this sort are offered, but that does not change what geography is all about. If you see an opportunity for field experience with professional geographers—even just a one-day reconnaissance—take it. But realize this: a few days in the field with geographic instruction may hook you for life.

Some geographers, in fact, are far better field-data gatherers than analysts or writers. In this respect they are not alone: this also happens in archeology, geology, and biology (among other field disciplines). This does not mean that these field workers do not contribute significantly to knowledge. Often on a research team some of the members are better in the field, and others excel in subsequent analysis. From all points of view, however, fieldwork is important.

Geography, then, is practiced in the field and in the office, in physical and human contexts, in generality and detail. Small wonder that so many areas of specialization have developed! If you check the undergraduate catalogue of your college or university, you will see some of these specializations listed as semester-length courses. But no geography department, no matter how large, could offer them all.

How does an area of specialization develop, and how can one become a part of it? The way in which geographic specializations have developed tells us much about the entire discipline. Some major areas, now old and established, began as research and theory building by one scholar and his or her students. These graduate students dispersed to the faculties of other universities and began teaching what they had learned. Thus, for example, did Carl Sauer's cultural geography spread from the University of California, Berkeley during the middle decades of the twentieth century.

It is one of the joys of geography that the basics and methods, once learned, are applicable to so many features of the human and physical world. Geographers have specialized in areas as disparate as shopping-center location and glacier movement, tourism and coastal erosion, real estate and wildlife, retirement communities and sports. Many of these specializations began with the interests and energies of a single scholar. Some thrived and grew into major geographic pursuits; others remained one-person shows, but with potential. When you discuss your own interests with a faculty advisor, you may refer to a university where you would like to do graduate work. "Oh yes," the answer

may be, "Professor X is in their geography department, working on just that." Or perhaps your advisor will suggest another university where a member of the faculty is known to be working on the topic in which you are interested. That is the time to write a letter or e-mail of inquiry. What is the professor working on now? Are graduate students involved? Are research funds available? What are the career prospects after graduation?

The Association of American Geographers, or **AAG** (1710 16th Street, N.W., Washington, D.C. 20009-3198 [**www.aag.org**]), recognizes no less than 52 so-called Specialty Groups. In academic year 2003–2004, the Specialty Group roster included the following branches of geography:

Africa

Aging and the Aged

Applied Geography

Asian Geography

Bible

Biogeography

Canadian Studies

Cartography

China

Climate

Coastal and Marine Geography

Cryosphere (Earth's Ice System)

Cultural Ecology

Cultural Geography

Disability

Economic Geography

Energy and Environment

Environmental Perception and Behavioral Geography

Ethics, Justice, and Human Rights

Ethnic Geography

European Geography

Geographic Information Systems (GIS)

Geographic Perspectives on Women

Geography Education

Geography of Religions and Belief Systems

Geomorphology (Landform Analysis)

Hazards

Historical Geography

History of Geography

Human Dimensions of Global Change

Indigenous Peoples

Latin American Geography

Medical Geography

Microcomputers

Middle East Geography

Military Geography

Mountain Geography

Political Geography

Population

Qualitative Research

Recreation, Tourism, and Sport

Regional Development and Planning

Remote Sensing

Rural Geography

Russian, Central Eurasian, and East European
  Geography

Sexuality and Space

Socialist Geography

Spatial Analysis and Modeling

Transportation Geography

Urban Geography

Water Resources

Worldwide Web

If you contact the AAG at the Internet address given above, click on the "Specialty Groups" icon on the home page and then scroll down the alphabetical listing to obtain the e-mail address of the current chairperson of any group. We encourage you to contact the professional geographer who chairs the group(s) you are interested in. These elected leaders are ready, enthusiastic, and willing to provide you with the information you are looking for.

All this may seem far in the future. Still, the time to start planning for graduate school is now. Applications for admission and financial assistance must be made shortly after the *beginning* of your senior year. That makes your junior year a year of decision.

AN UNDERGRADUATE PROGRAM

The most important concern for any geography major or minor is basic education and training in the field. An undergraduate curriculum contains all or several of the following courses (titles may vary):

1. Physical Geography of the Global Environment (weather and climate, natural landscapes, landforms, soils, moisture and surface water flows, elementary biogeography)
2. Introduction to Human Geography (basic principles of cultural and economic geography)
3. World Regional Geography (world geographic realms)

These beginning courses are followed by more specialized courses, including both methodological and substantive ones:

4. Geographic Information Technology
5. Cartographic Theory and Techniques
6. Introduction to Quantitative Methods of Analysis
7. Introduction to Geographic Information Systems (GIS)
8. Analysis of Remotely Sensed Data
9. Cultural Geography
10. Political Geography
11. Urban Geography
12. Economic Geography
13. Historical Geography
14. Geomorphology
15. Geography of United States-Canada, Europe, and/or other world geographic realms

You can see how Physical Geography of the Global Environment would be followed by the more specialized analysis of landforms covered in Geomorphology, and how Human Geography now divides into such areas as intermediate and/or advanced cultural and economic geography. As you progress, the focus becomes even more specialized. Thus Economic Geography may be followed by:

16. Industrial Geography
17. Transportation Geography
18. Geography of Development

At the same time, regional concentrations may come into sharper focus:

19. Geography of Western Europe (or other major regions)

In these more advanced courses, you will use the technical knowledge acquired from courses numbered **4** through **8** (and perhaps others). Now you can avail yourself of the opportunity to develop these skills further. Many departments offer such courses as these:

20. Advanced Quantitative Methods
21. Computer Cartography
22. Intermediate GIS
23. Advanced GIS
24. Advanced Satellite Imagery Interpretation

From this list (which represents only part of a comprehensive curriculum), it is evident that you cannot, even in four years of undergraduate study, register for all courses. The geography major in many universities requires a minimum of only 30 (semester-hour) credits—just 10 courses, fewer than half those listed here. This is another reason to begin thinking about specialization at an early stage.

Because of the number and variety of possible geography courses, most departments require that their majors complete a core program that includes courses in substantive areas as well as theory and methods. That core program is important, and you should not be tempted to put off these courses until your last semesters. What you learn in the core program will make what follows (or should follow) much more meaningful.

You should also be aware of the flexibility of many undergraduate programs, something that can be especially important to geographers. Imagine that you are majoring in geography and develop an interest in Southeast Asia. But the regional specialization of the geographers in your department may be focused somewhere else—say, Middle and South America. However, courses on Southeast Asia are indeed offered by other departments, such as anthropology, history, or political science. If you are going to be a regional specialist, those courses will be very useful and should be part of your curriculum—but you are not able to receive geography credit for them. After successfully completing those courses, however, you may be able to register for an independent study or reading course in the Geography of Southeast Asia if a faculty member is willing to guide you. Always discuss such matters with the undergraduate advisor or chairperson of your geography department.

LOOKING AHEAD

By now, as a geography major, you will be thinking of the future—either in terms of graduate school or a salaried job. In this connection, if there is one important lesson to keep in mind, it is to *plan ahead* (a redundancy for emphasis!). The choice of graduate school is one of the most important you will make in your life. The professional preparation you acquire as a graduate student will affect your competitiveness in the job market for years to come.

Choosing a Graduate School

Your choice of a graduate school hinges on several factors, and the geography program it offers is one of them. Possibly, you are constrained by residency factors, and your choice involves the schools of only one State. Your undergraduate record and grade point average affect the options. As a geographer, you may have strong feelings in favor of or against particular parts of the country. And although some schools may offer you financial support, others may not.

Certainly the programs and specializations of the prospective graduate department are extremely impor-

tant. If you have settled on your own area of interest, it is wise to find a department that offers opportunities in that direction. If you have yet to decide, it is best to select a large department with several options. Some students are so impressed by the work and writings of a particular geographer that they go to his or her university solely to learn from and work with that scholar. In every case, information and preparation are crucial. Many a prospective graduate student has arrived on campus eager to begin work with a favorite professor, only to find that the professor is away on a sabbatical leave!

Fortunately, information can be acquired with little difficulty. One of the most useful publications of the AAG is its *Guide to Geography Programs in North America*, published yearly. A copy of this annual directory should be available in the office of your geography department, but if you plan to enter graduate school, a personal copy would be an asset. Not only does the *Guide* describe the programs, requirements, financial aid, and other aspects of geography departments in the United States, Canada, and Mexico, but it also lists all faculty members and their current research and teaching specializations. Moreover, it also contains a complete listing of the Association's approximately 7,000 members and their specializations. And Internet addresses are now provided as well for accessing the Web sites of geography departments, which they increasingly utilize to supplement their entries in the *Guide* and provide even more detailed information about their programs and faculty. Although you will find the *Guide* to be very useful in your decision-making, we also strongly recommend that you explore the Web sites of departments you wish to consider.

You may discover one particular department that stands out as the most interesting and most appropriate for you. But do not limit yourself to one school. Consult the *Guide*'s 25-page table (entitled *Program Specialties*) that links 35 geographic specializations to every department listed, and then explore each appropriate match-up. After careful investigation, it is best to rank a half-dozen schools (or more), write or e-mail all of them for admission application forms, and apply to several. Multiple applications are a bit costly, but the investment is worth it.

Assistantships and Scholarships

One reason to apply to several universities has to do with financial (and other) support for which you may be eligible. If you have a reasonably well-rounded undergraduate program behind you and a good record of achievement, you are eligible to become a teaching assistant (TA) in a graduate department. Such assistantships usually offer full or partial tuition plus a monthly stipend (during the nine-month academic year). Conditions vary, but this position may make it possible for you to attend a university that would otherwise be out of reach. A tuition waiver alone can be well over $10,000 annually. Application for an assistantship is made directly to the department; you should write or e-mail the department contact listed in the latest AAG *Guide* (or on the department's Web site), who will either respond directly to you or forward your inquiry to the departmental committee that evaluates applications.

What does a TA do? The responsibilities vary, but often teaching assistants are expected to lead discussion sections (of a larger class taught by a professor), laboratories, or other classes. They prepare and grade examinations and help undergraduates deal with problems arising from their courses. This is an excellent way to determine your own ability and interest in a teaching career.

In some instances, especially in larger geography departments, research assistantships are available. When a member of the faculty is awarded a large grant (e.g., by the National Science Foundation) for a research project, that grant may make possible the appointment of one or more research assistants (RAs). These individuals perform tasks generated by the project and are rewarded by a modest salary (usually comparable to that of a TA). Normally, RAs do not receive tuition waivers, but sometimes the department and the graduate school can arrange a waiver to make the research assistantship more attractive to better students. Usually, RAs are chosen from among graduate students already on campus who have proven their interest and ability. Sometimes, however, an incoming student is appointed. Always ask about opportunities.

Geography students also are among those eligible for many scholarships and fellowships offered by universities and off-campus organizations. When you write your introductory letter, be sure to inquire about all forms of financial aid.

JOBS FOR GEOGRAPHERS

Upon completing your bachelor's degree, you may decide to take a job rather than going on to graduate school. Again, this is a decision best made early in your junior year, for two main reasons: (1) so that you can tailor your curriculum for a vocational objective, and (2) so that you can start searching for a job well before graduation.

Internships

A very good way to enter the job market—and to become familiar with the working environment—is by taking an internship in an agency, office, or firm. Many organizations find it useful to have interns. Internships help organizations train beginning professionals and give companies

an opportunity to observe the performance of trainees. Many an intern has ultimately been employed by his or her organization. Some employers have even suggested what courses the intern should take in the next academic year to improve future performance. For example, an urban or regional planning agency that employs an intern might suggest that the intern add a relevant GIS course or urban-planning course to his or her program of study.

Some organizations will appoint interns on a continuing basis, say, two afternoons a week around the year; others make full-time, summer-only appointments available. Occasionally, an internship can be linked to a departmental curriculum, yielding academic credit as well as vocational experience. Your department undergraduate advisor or the chairperson will be the best source of assistance.

One of the most interesting internship programs is offered by the National Geographic Society in Washington, D.C. Every year, the Society invites three groups of about eight interns each to work with the permanent staff of its various departments. Application forms are available in every geography department in the United States, and competition is strong. The application itself is a useful exercise, as it tells you what the Society (and other organizations) look for in your qualifications.

Professional Opportunities

A term you will sometimes see in connection with jobs is *applied geography*. This, one supposes, distinguishes geography teaching from practical geography. In fact, however, all professional geography in education, business, government, and elsewhere is "applied." In the past a large majority of geography graduates became teachers—in elementary and high schools, colleges, and universities. More recently, geographers have entered other arenas in increasing numbers. In part, this is related to the decline in geographic education in schools, but it also reflects the growing recognition of geographic skills by employers in business and government.

Nevertheless, what you as a geographer can contribute to a business is not yet as clear to the managers of many companies as it should be. The anybody-can-do-geography attitude is a form of ignorance you will undoubtedly confront. (This is so even in precollegiate education, where geography was submerged in social studies—and often taught by teachers who had never taken a course in the discipline!)

So, once employed, you may have to prove not only yourself but also the usefulness of the skills and capacities you bring to your job. There is a positive side to this. Many employers, once dubious about hiring a geographer, learn how geography can contribute—and become enthusiastic users of geographic talent. Where one geographer is employed, whether in a travel firm, publishing house, or planning office, you will soon find more.

Geographers are employed in business, government, and education. Planning is a profession that employs many geographers. Employment in business has grown in recent years (see the following section). Governments at national, State, and local levels have always been major employers of geographers. And in education, where geography was long in decline, the demand for geography teachers will grow again.

For more detailed information about employment, you should contact the AAG (at the address given earlier) for the inexpensive booklet entitled *Careers in Geography*; or check with your geography department, whose office is very likely to have a copy.

Business and Geography

With their education in global and international affairs, their knowledge of specialized areas of interest to business, and their training in cartography, GIS, methods of quantitative analysis, and writing, geographers with a bachelor's degree are attractive business-employment prospects. Some undergraduate students have already chosen the type of business they will enter; for them, there are departments that offer concentrations in their areas of interest. Several geography departments, for instance, offer a curriculum that concentrates on tourism and travel—a business in which geographic skills are especially useful.

These days, many companies want graduates with a strong knowledge of international affairs and fluency in at least one foreign language in addition to their skills in other areas. The business world is quite different from academia, and the transition is not always easy. Your employer will want to use your abilities to enhance profits. You may at first be placed in a job where your geographic skills are not immediately applicable, and it will be up to you to look for opportunities to do so.

One of our students was in such a situation some years ago. She was one of more than a dozen new employees doing what was essentially clerical work. (Some companies will use this type of work to determine a new employee's punctuality, work habits, adaptability, and productivity.) One day she heard that the company was considering establishing a branch to sell its product in East Africa. The student had a regional interest in East Africa as an undergraduate and had even taken a year of Swahili-language training. On her own time she wrote a carefully documented memorandum to the company's president and vice presidents, describing factors that should be taken into consideration in the projected expansion. That report showcased her regional skills, locational insights, and knowledge of the local market and transport problems, along with the probable cultural reaction to the company's product, the country's political circumstances, and (last but significantly) this employee's ability to present such issues effectively. She supported her report with good maps and several illustrations. Soon she received a special assignment to participate in the planning process, and her rise in the company's ranks had begun. She had seized the opportunity and demonstrated the utility of her geographic skills.

Geography graduates have established themselves in businesses of all kinds: banking, international trade, manufacturing, retailing, and many more. Should you join a large firm, you may be pleasantly surprised by the number of other geographers who hold jobs there—not under the title of geographer but under countless other titles ranging from analyst and cartographer to market researcher and program manager. These are positions for which the appointees have competed with other graduates, including business graduates. As we noted earlier, once an employer sees the assets a geographer can bring, the role of geographers in the company is assured.

Government and Geography

Government has long been a major employer of geographers at the national and State as well as local level. *Careers in Geography* (the AAG booklet) estimates that at least 2,500 geographers are working for governments, about half of them for the federal government. In the U.S. State Department, for instance, there is an Office of the Geographer staffed by professional geographers. Other agencies where geographers are employed include the Defense Mapping Agency, the Bureau of the Census, the U.S. Geological Survey, the Central Intelligence Agency, and the Army Corps of Engineers. Still other employers are the Library of Congress, the National Science Foundation, and the Smithsonian Institution. Many other branches of government also have positions for which geographers are eligible.

Opportunities also exist at State and local levels. Several States now have their own Office of the State Geographer; all States have agencies engaged in planning, resource analysis, environmental protection, and transportation policy-making. All these agencies need geographers who have skills in cartography, remote sensing, spatial database analysis, and the operation of geographic information systems.

Securing a position in government requires early action. If you want a job with the federal government, start at the beginning of your senior year. Every State capital and many other large cities have a Federal Job Information Center (FJIC). (The Washington office is at 1900 E Street, N.W., Washington, D.C. 20415.) You may request information about a particular agency and its job opportunities; it is appropriate to write or e-mail directly to the personnel office of the agency or agencies in which you are interested. You may also write to the Office of Personnel Management (OPM), Washington, D.C. 20415, which has offices in most large cities.

Planning and Geography

Planning has become one of geography's allied professions. The planning process is a complex one comprising people trained in many fields. Geographers, with their cartographic, locational, regional, and analytical skills, are sought after by planning agencies. Many an undergraduate student gets that first professional opportunity as an intern in a planning office.

Planning is done by many agencies and offices at levels ranging from the federal to the municipal. Cities have planning offices, as do regional authorities. Working in a planning office can be a very rewarding experience because it involves the solving of social and economic problems, the conservation and protection of the environment, the weighing of diverse and often conflicting arguments and viewpoints, and much interaction with workers trained in other fields. Planning is a superb learning experience.

A career in planning can be much enhanced by a background in geography, but you will have to adjust your undergraduate curriculum to include courses in such areas as public administration, public finance, and other related fields. Thus a career in planning itself requires early planning on your part. At many universities, the geography department is closely associated with the planning department, and your faculty advisor can inform you about course requirements. But if you have your eye on a particular office or agency, you should also request information from its director about desired and required skills.

Planning is by no means a monopoly of government. Government-related organizations such as the Agency for International Development (AID), the World Bank, and the International Monetary Fund (IMF) have planning offices, as do nongovernmental organizations (NGOs) such as banks, airline companies, industrial firms, multinational corporations, and research institutes. Private-sector opportunities for planners have been expanding, and you may wish to explore them. An important organization is the planners' equivalent of the geographers' AAG: the American Planning Association, 1776 Massachusetts Avenue, N.W., Washington, D.C. 20006. The AAG's *Careers in Geography* provides additional information on this expanding field and where you might pursue its study.

Teaching Geography

If you are presently a freshman or sophomore, your graduation may coincide with the end of a long decline in the geography-teaching profession. Just a few decades ago, teaching geography in elementary or high school was the goal of thousands of undergraduate geography majors. But then began the merging of geography into the hybrid field called "social studies," and prospective teachers no longer needed to have any training in geography in many States. In Florida, for instance, teachers were formerly required to take courses in regional geography and conservation (as taught in geography departments), but in the 1970s those requirements were dropped. Education planners fell victim to the myth that the teaching of geography does not require any training. What was left of geography often was taught by teachers whose own fields were history, civics, or even basketball coaching!

If you read the daily newspapers, you have seen reports of the predictable results. As mentioned earlier, "geographic illiteracy" has become a common complaint (often made by the same education planners who pushed geography into the social studies program and eliminated teacher-education requirements). Now States are returning to the education requirement so that teachers will learn some geography. And geography is returning to elementary and high school curricula-assisted in particular by the National Geographic Society, which supports State-level Geographic Alliances of educators and academic geographers. The need for geography teachers will soon be on the upswing again.

So this may be a good time to consider teaching geography as a career. You should do some research, however, because States vary in their progressiveness in this arena. You should also visit the School of Education in your college or university and ask questions about instructional opportunities in geography. In addition, contact not only the AAG but also the *National Council for Geographic Education (NCGE)* at 206-A Martin Hall, Jacksonville State University, Jacksonville, AL 36265-1602 (**www.ncge.org**). Your State geographic society or Geographic Alliance also may be helpful; ask your department advisor or chairperson for details.

SOME FINAL THOUGHTS

The opportunities in geography are many, but often they are not as obvious as those in other fields. You will find that good and timely preparation produces results and, frequently, unexpected rewards.

The discussion in this appendix has provided a comprehensive answer to the oft-asked question, "What can one do with geography?" If you wish to explore this question further, along with the AAG's *Careers in Geography* we recommend *On Becoming a Professional Geographer*, a book edited by Martin S. Kenzer (Merrill/Macmillan, 1989).

We wish you every success in all your future endeavors. And, speaking for the entire community of professional geographers, we would be delighted to have you join our ranks should you choose a geography-related career.

Pronunciation Guide

Abacha (ah-BAH-chah)
Abadan (ahba-DAHN)
Abakan (ahb-uh-KAHN)
Abdullah (ahb-DOO-lah)
Abidjan (abb-ih-JAHN)
Abkhazia (ahb-KAHZ-zee-uh)
Aboriginal (abb-uh-RIDGE-uh-null)
Abu Dhabi (ah-boo-DAH-bee)
Abu Sayyaf (AH-boo sye-YAHF)
Abuja (uh-BOO-juh)
Abyssinia (abb-ih-SINNY-uh)
Acacia (uh-KAY-shuh)
Acadia (uh-KAY-dee-uh)
Acapulco (ah-kah-POOL-koh)
Accra (uh-KRAH)
Aceh (ah-CHEH)
Acehnese (ah-chay-NEEZ)
Acre (AH-kray)
Acropolis (uh-CROP-uh-liss)
Adamawa (add-uh-MAH-wuh)
Adan (AH-dahn/AYD'N)
Adis Abeba (adda-SAB-uh-buh)
Adelaide (ADDLE-ade)
Adirondack (add-ih-RONN-dak)
Adriatic (ay-dree-ATTIK)
Adygeya (ah-duh-GAY-uh)
Aegean (uh-JEE-un)
Afars (uh-FAHRZ)
Afghanistan (aff-GHANN-uh-stan)
Afrikaans (uff-rih-KAHNZ)
Afrikaner (uff-rih-KAHN-nuh)
Agios Dimitrios (AH-jee-ohss deh-
    MEE-tree-ohss)
Agra (AHG-ruh)
Agri (AH-gree)
Agurchand (ah-ghoor-SHAHND)
Ahad (ah-HAHD)
Ahaggar (uh-HAH-gahr)
Ahmadabad (AH-muh-duh-bahd)
Ahvaz (ah-VAHZ)
Ainu (EYE-noo)
Ajaria (uh-JAR-ree-uh)
Akan (AH-kahn)
Akashi Kaikyo (AH-kah-shee KYE-
    kyoh)

Akhbari (ahk-BAH-ree)
Akihito (ah-kee-HEE-toh)
Al Andalus (ahl ahnda-LOOSE)
Al Hudaydah (ahl-hoh-DAY-duh)
Al Madinah (ahl-mah-DEENA)
Alagoas (ah-LAH-goh-ahss)
Alawite (AH-lah-wyte)
Alcazar (AHL-kah-zahr)
Aleijadinho (ah-lay-jah-DEEN-yoh)
Aleppo (uh-LEP-poh)
Alevis (ah-LEE-veez)
Alexandria (ah-leg-ZAN-dree-uh)
Algae (AL-jee)
Algeria (al-JEERY-uh)
Algiers (al-JEERZ)
Ali (ah-LEE)
Allah (AHL-ah)
Allahabad (ALLA-huh-bahd)
Allemanni (alla-MAH-nee)
Alluvial (uh-LOO-vee-ull)
Alma-Ata (ahl-muh-uh-TAH)
Almaty (ahl-mah-TUH)
Alpaca (al-PAK-uh)
Al-Qaeda (ahl-KYE-duh)
Al-Quds (ahl-KOOTSS)
Al-Salam (ahl-sah-LAHM)
Altaic (al-TAY-ik)
Altay (AL-tye)
Altaya (al-TYE-uh)
Altiplano (ahl-tee-PLAH-noh)
Altun (ahl-TOON)
Al-Wahhab (ahl-wah-HAHB)
Amador (AMMA-dor)
Amazonas (ahma-ZOAN-ahss)
Ambarawa (am-bah-RAH-wah)
Ambeno (AHM-bay-noh)
Ambon (ahm-BON)
Amerind (AMMER-rind)
Amhara (am-HAH-ruh)
Amharic (am-HAH-rik)
Amin, Idi (uh-MEEN, iddy)
Amman (uh-MAHN)
Amoy (ah-MOY)
Amritsar (um-RIT-sahr)
Amu-Darya (uh-moo-DAHR-yuh)

Amundsen, Roald (AH-moon-sun,
    ROH-ahl)
Amur (uh-MOOR)
An Najaf (un-nuh-JAHF)
Anatolia (anna-TOH-lee-uh)
Ancona (ahng-KOH-nuh)
Andalusia (ahn-duh-loo-SEE-uh)
Andaman (ANN-duh-mun)
Andes (ANN-deez)
Andhra Pradesh (ahn-druh pruh-DESH)
Angara (ahng-guh-RAH)
Angkor [Wat] (ANG-kor [WOT])
Angola (ang-GOH-luh)
Anhui (ahn-HWAY)
Ankara (ANG-kuh-ruh)
Annam (uh-NAHM)
Annam[ese]/[ite] (anna-
    [MEEZE]/[MITE])
Anshan (ahn-SHAHN)
Antalya (ahn-tahl-YAH)
Antananarivo (ahn-TAH-nah-nah-
    REEV)
Antecedent (an-tee-SEE-dunt)
Antigua (an-TEE-gwuh)
Antilles (an-TILL-eeze)
Antimony (ANN-tih-moh-nee)
Antioquia (ahn-tee-OH-kee-ah)
Antofagasta (untoh-fah-GAHSS-tah)
Antwerp (ANN-twerp)
Anuradhapura (UH-NOO-RAH-thah-
    poor-uh)
Anyang (ahn-YAHNG)
Aozou (OO-zoo)
Apapa (uh-PAHP-uh)
Apartheid (APART-hate)
Appennines (APP-uh-neinz)
Apure (ah-POOR-ray)
Apurimac (ah-poo-REE-mahk)
Aqaba (AH-kuh-buh)
Aqmola (ahk-moh-LAH)
Aquifer (ACK-kwuh-fer)
Arakan (ah-ruh-KAHN)
Aral (ARREL)
Aramco (uh-RAM-koh)
Arauca (ah-RAU-kah)

Arawak   (ARRA-wak)

Arc de Triomphe   (ark duh tree-AWMFF)

Archangel   (ARK-ain-jull)

Archipelago   (ark-uh-PELL-uh-goh)

Ardennes   (ar-DEN)

Arequipa   (ah-ray-KEE-pah)

Argentine   (AR-jen-tyne)

Arguedas, Alcides   (ahr-GWAY-dahss, ahl-SEE-duss)

Arica   (ah-REE-kah)

Aridisol   (uh-RIDDY-sol)

Aristide, Jean-Betrand   (ah-ree-STEED, zhawn-bair-TRAWN)

Arkhangelsk   (ahr-KAHN-gyil-sk)

Armenia   (ar-MEENY-uh)

Armonk   (AHR-munk)

Armuelles   (ahr-MWAY-yace)

Artesian   (ahr-TEE-zhun)

Aruba   (uh-ROO-buh)

Arunachal Pradesh   (AHRA-NAHTCH-ull pruh-DESH)

Aryan   (AHR-yun)

Aryeetey-Attoh   (ahr-YEE-tee ATTOH)

As Sulaymaniyah   (ahss soo-lee-mah-NEE-yah)

Asante   (ah-SAHN-tay)

Ashanti   (ah-SHAHN-tee)

Ashgabat   (AHSH-gah-baht)

Ashkelon   (AHSH-kih-lonn)

Aśoka   (uh-SHOH-kuh)

Assad, Bashar   (uh-SAHD, buh-SHAR)

Assad, Hafez al-   (uh-SAHD, huh-FEZ ahl)

Assam   (uh-SAHM)

Assyrian   (uh-SEERY-un)

Astana   (ah-STAH-nah)

Astrakhan   (ASTRA-kahn)

Asturias   (uh-STOOR-ree-uss)

Asunción   (ah-soohn-see-OAN)

Aswan   (as-SWAHN)

Atacama   (ah-tah-KAH-mah)

Atatürk   (ATTA-tyoork)

Atheism   (AYTHEE-izm)

Athena   (uh-THEENA)

Atoll   (AY-toal)

Attenuated   (uh-TEN-yoo-ay-ted)

Atyrau   (ah-tah-RAU)

Auckland   (AWK-lund)

Augelli   (aw-JELLY)

Austral   (AW-strull)

Australopithecene   (aw-strallo-PITH-uh-seen)

Autocracy   (aw-TOCK-ruh-see)

Avenida de 9 Julio   (ah-veh-NEEDA day noo-AY-vay HOO-lee-oh)

Avenida Nicolas de Pierola   (ah-veh-NEEDA NEEKO-lahss day pyair-ROH-lah)

Avignon   (ah-veen-YAW)

Axum   (AHKSS-oom)

Ayatollah   (eye-uh-TOH-luh)

Aymara   (ai-MAH-rah)

Ayodhya   (uh-YODE-yuh)

Azerbaijan   (ah-zer-bye-JAHN)

Azeri   (ah-ZAREY)

Azov   (uh-ZAWF)

Ba'ath   (BATH)

Bab el Mandeb   (bab ull MAN-dub)

Babylon   (BABBA-lon)

Bac Bo   (BAHK-BOH)

Baden-Württemberg   (BAHDEN VEERT-um-bairk)

Baganda   (bah-GAHN-duh)

Bahamas   (buh-HAH-muzz)

Bahasa   (bah-HAH-sah)

Bahia   (bah-EE-yah)

Bahr el Ghazal   (bar-ahl-gah-ZAHL)

Bahrain   (bah-RAIN)

Baht   (BAHT)

Baja   (BAH-hah)

Bajau   (bah-YAU)

Bakassi   (bah-KAH-see)

Baki   (bah-KEE)

Baku   (bah-KOO)

Bali [nese]   (BAH-lee [NEEZ])

Balikpapan   (bah-LIK-pah-pahn)

Balkan   (BAWL-kun)

Balkhan   (bahl-KAHN)

Balsas   (BAHL-suss)

Baluchistan   (buh-loo-chih-STAHN)

Bamako   (BAH-mah-koh)

Banda   (BAHN-duh)

Bandar Lampung   (BAHN-dahr lahm-PUNG)

Bandar Seri Begawan   (BUN-dahr SERRY buh-GAH-wun)

Bandung   (BAHN-doong)

Bangalore   (BANG-guh-loar)

Bangladesh [i]   (bang-gluh-DESH [ee])

Banja Luka   (BUN-yuh LOO-kuh)

Bantu   (ban-TOO)

Banyamulenge   (bahn-yah-moo-LENG-geh)

Baotou   (bao-TOH)

Barbados   (bar-BAY-dohss)

Barcelona   (bar-suh-LOH-nuh)

Barents   (BARRENS)

Barim   (buh-REEM)

Barquisimeto   (bar-key-suh-MAY-toh)

Barranquilla   (bah-rahn-KEE-yah)

Barrio [s]   (BAHR-ree-oh[ss])

Basalt   (buh-SAWLT)

Bashkortostan   (bahsh-KORT-uh-stahn)

Basilan   (bah-see-LAHN)

Basilicata   (bah-ZEE-lee-kah-tah)

Basque   (BASK)

Basra   (BAHZ-ruh)

Batak   (buh-TAHK)

Batavia   (buh-TAY-vee-uh)

Batista, Fulgencio   (bah-TEESTA, fool-HEN-see-oh)

Batticaloa   (buh-tih-kuh-LOH-uh)

Batumi   (bah-TOO-mee)

Bauxite   (BAWKS-site)

Bavaria   (buh-VAIRY-uh)

Baykal [iya]   (bye-KAHL [ee-ah]))

Bayou   (BYE-yoo)

Bayóvar   (bye-YOH-vahr)

Bayu-Undan   (bye-YOO oon-DAHN)

Beaumont   (BOH-mont)

Bedouin   (BEH-doo-in)

Beihai   (bay-HYE)

Beijing   (bay-ZHING)

Beira   (BAY-ruh)

Beirut   (bay-ROOT)

Bekaa   (buh-KAH)

Bekasi   (beh-KAH-see)

Belarus   (bella-ROOSE)

Belarussia   (bella-RUSH-uh)

Belém   (bay-LEM)

Belgian   (BEL-jun)

Belize   (beh-LEEZE)

Belmopan   (bell-moh-PAN)

Belo Horizonte   (BAY-loh haw-ruh-ZONN-tee)

Bengal [i]   (beng-GAHL [ee])

Benghazi   (ben-GAH-zee)

Benguela   (beng-GWELLA)

Beni   (BAY-nee)

Benin   (beh-NEEN)

Benue   (BANE-way)

Berber   (BERR-berr)

Berimbau   (beh-RIM-bau)

Bering   (BERRING)

Bern   (BAIRN)

Betsimisaraka   (bye-tsee-mee-SAH-rah-kah)

Beyoglu   (bay-uh-GLOO)

Bharatiya Janata   (bah-rah-TEE-uh jah-NAH-tah)

Bhutan   (boo-TAHN)

Bhutia   (BOOH-tee-uh)

Biafra   (bee-AH-fruh)

Bihar   (bih-HAHR)

Bihe   (bee-HAY)

Bilbao   (bil-BAU)

Bilharzia   (bill-HARZEE-uh)

Bin Laden, Usama   (bin-LAHD-nn, oo-SAH-mah)

Bioko   (bee-OH-koh)

Biota   (bye-OH-tuh)

Bishkek   (BISH-kek)

Bitumen   (bye-TOO-men)

Boeing   (BOH-ing)

Boer   (BOOR)

Bogor   (BOH-goar)

Bogotá   (boh-goh-TAH)

Bohai   (bwoh-HYE)

Bohemia   (boh-HEE-mee-uh)

Bolívar, Simón   (boh-LEE-vahr, see-MOAN)

Bolshevik   (BOAL-shuh-vick)

Bom Bahia (bom-bah-EE-yah)
Bombay (bom-BAY)
Bonaire (bun-AIR)
Bora Bora (boar-ruh-BOAR-ruh)
Boran (boar-RAN)
Bordeaux (boar-DOH)
Borneo (BOAR-nee-oh)
Bosnia-Herzegovina (BOZ-nee-uh
  hert-suh-goh-VEE-nuh)
Bosporus (BAHSS-puh-russ)
Botha (BAW-tuh)
Botswana (bah-TSWAHN-uh)
Bouchard, Lucien (boo-SHAR, looss-
  YAN)
Bougainville (BOO-gun-vil)
Boulangerie (boo-LAHN-zheh-ree)
Boulevard St. Laurent (boo-leh-VAHR
  san-law-RAH)
Bourgogne (boor-GOAN-yuh)
Bové, José (boh-VAY, zhoh-ZAY)
Brahmaputra (brahm-uh-POOH-truh)
Brandenburg (BRAHN-den-boorg)
Brasília (bruh-ZEAL-yuh)
Bratislava (BRUDDIS-lahva)
Bratsk (BRAHTSK)
Brazzaville (BRAHZ-uh-veel)
Brcko (BIRCH-koh)
Bremen (BRAY-mun)
Brisbane (BRIZZ-bun)
Brno (BURR-noh)
Brunei (BROO-nye)
Brunswick (BROON-svik)
Bubonic (byoo-BON-nik)
Bucharest (BOO-kuh-rest)
Budapest (BOODA-pest)
Buddha (BOOD-uh)
Buddhism (BOOD-izm)
Buddhist (BOOD-ist)
Buenos Aires (BWAY-nohss EYE-
  race)
Buganda (boo-GAHN-duh)
Bulawayo (boo-luh-WAY-oh)
Bulgaria (bahl-GHAIR-ree-uh)
Bund (BUND)
Bunia (BOO-nee-uh)
Bureya (buh-RAY-yuh)
Burkina Faso (ber-keena FAHSSO)
Burkinabe (ber-kee-NAH-bay)
Burmans (BURR-munz)
Burundi (buh-ROON-dee)
Buryat [iya] (boor-YAHT [ee-uh])
Busan (BOO-SAHN)
Buthelezi, Mangosuthu (boo-teh-LAY-
  zee, mungo-SOO-too)
Butte (BYOOT)
Byzantine (BIZZ-un-teen)
Cabimas (kah-BEE-mahss)
Cabinda (kuh-BIN-duh)
Cabora Bassa (kuh-boar-rah
  BAHSSA)
Cabramatta (kabbra-MADDA)

Cacao (kuh-KAY-oh/kuh-KAU)
Cadre (KAH-dray)
Caicos (KAY-kohss)
Cairo (KYE-roh)
Calabria (kuh-LAH-bree-uh)
Calais (kah-LAY)
Calcutta (kal-KUTT-uh)
Caldas (KAHL-dahss)
Calgary (KAL-guh-ree)
Cali (KAH-lee)
Caliph (KAY-liff)
Caliphate (KALLA-fate)
Callao (kah-YAH-oh)
Cambodia (kam-BOH-dee-uh)
Campeche (kahm-PAY-chee)
Canberra (KAN-burruh)
Candomblé (kan-DOM-blay)
Caño, Limón (KAHN-yoh lee-MONE)
Cantabria (kahn-TAH-bree-uh)
Canterbury (KAN-ter-berry)
Canton (kan-TONN)
Capoeira (kah-poh-AY-ruh)
Caprivi (kuh-PREE-vee)
Caquetá (kah-kway-TAH)
Caracas (kah-RAH-kuss)
Carajás (kuh-ruh-ZHUSS)
Carara (kuh-RAHR-ruh)
Caribbean (kuh-RIB-ee-un/karra-BEE-
  un)
CARICOM (CARRY-komm)
Carioca (kah-ree-OH-kah)
Caritas (kuh-REE-tuss)
Carpathians (kar-PAY-thee-unz)
Cartagena (karta-HAY-nuh)
Casamance (KAH-zah-mahnss)
Casanare (kah-sah-NAH-ray)
Caspian (KASS-spee-un)
Caste (CAST)
Castile (kuh-STEEL)
Castile-La Mancha (kuh-STEEL luh-
  MAHN-chuh)
Castile-Leon (kuh-STEEL lay-OAN)
Catalan (katta-LAHN)
Catalonia (katta-LOH-nee-uh)
Cauca (KOW-kah)
Caucasian (kaw-KAY-zhun)
Caucasus (KAW-kuh-zuss)
Cay (KEE)
Cayambe (kah-YAHM-bee)
Ceará (see-ah-RAH)
Ceausescu, Nicolae (chow-SHESS-
  koo, NICK-oh-lye)
Cebu (seh-BOO)
Celebes (SELL-uh-beeze)
Celt [ic] (KELT [ick])
Cenozoic (senno-ZOH-ik)
Centavo (sen-TAH-voh)
Central (sen-TRAHL)
Centrifugal (sen-TRIFFA-gull)
Centripetal (sen-TRIPPA-tull)
Centro-Oeste (SENTRO oh-ESS-tee)

Cerrado (seh-RAH-doh)
Cerro de Pasco (serro day PAH-skoh)
Ceuta (see-YOO-tah)
Ceyhan (jay-HAHN)
Ceylon (seh-LONN)
Chaco (CHAH-koh)
Chaebol (JAY-BOAL)
Champagne (shahm-PAHN-yuh)
Champs Elysées (SHAWZ elly-ZAY)
Chang Jiang (chang jee-AHNG)
Ch'angan (CHAHNG-GAHN)
Changchun (CHAHNG-CHOON)
Chao Phraya (CHOW pruh-yah)
Chaparé (chah-pah-RAY)
Charisma (kuh-RIZZ-muh)
Charleroi (SHARL-rwah)
Chateau Frontenac (shat-TOH
  FRAWN-teh-nahk)
Chavez (SHAH-vez)
Chechen (CHEH-chen)
Chechenya-Ingushetiya (cheh-CHEN-
  yuh in-goo-SHETTY-uh)
Chechnya (CHETCH-nee-uh)
Chelyabinsk (chel-YAH-bunsk)
Chengdu (chung-DOO)
Chennai (cheh-NYE)
Cherkessk (cher-KESK)
Chernobyl (see Chornobyl)
Chhattisgarh (CHADDISS-gahr)
Chiang Kai-shek (jee-AHNG kye-
  SHECK)
Chiang Mai (chee-AHNG mye)
Chiapas (chee-AHP-uss)
Chihuahua (chuh-WAH-wah)
Chile (CHILLI/CHEE-lay)
Chiloé (chee-luh-WAY)
Chilung [Jilong] (JEE-LOONG)
Chisinau (kee-shih-NAU)
Chittagong (CHITT-uh-gahng)
Cholla (JOH-luh)
Cholon (choh-LONN)
Cholula (choh-LOO-lah)
Choluteca (choh-loo-TAY-kah)
Chongqing (chong-CHING)
Chordata (kor-DATTA)
Chornobyl (CHAIR-noh-beel)
Chota Nagpur (choat-uh-NAHG-poor)
Chubu (CHOO-BOO)
Chukotka (chuh-KAHT-kuh)
Chukotskiy (chuh-KAHT-skee)
Chungyang (joong-YAHNG)
Chuquicamata (choo-kee-kah-MAH-
  tah)
Chuvash [iya] (choo-VAHSH [ee-uh])
Ciliwung (SILL-uh-wong)
Ciudad Alta (see-yoo-DAHD AHL-
  tuh)
Ciudad Baixa (see-yoo-DAHD bah-
  ZHEE-uh)
Ciudad del Este (see-yoo-DAHD del-
  ESS-stay)

Ciudad Guayana (see-yoo-DAHD gwuh-YAHNA)
Ciudad Juárez (see-you-DAHD WAH-rez)
Ciudades perdidas (see-you-DAH-dayss pair-DEE-duss)
Cliburn (KLYE-bern)
Clonshaugh (KLONZ-haw)
Coahuila (koh-uh-WEE-luh)
Coatzacoalcos (koh-aht-sah-koh-AHL-kohss)
Cochabamba (koh-chah-BUM-bah)
Cochin (KOH-chin)
Cocos (KOH-kuss)
Collor de Mello, Fernanado (KAW-lohr deh MELLOH, fair-NAHN-doh)
Colombo (kuh-LUM-boh)
Colón (kuh-LOAN)
Comodoro Rivadavia (comma-DORE-oh ree-vah-DAH-vee-ah)
CONAIE (koh-NYE)
Conakry (koh-NAK-ree)
Confucius (kun-FEW-shuss)
Coniferous (kuh-NIFF-uh-russ)
Conquistador (koan-KEE-stuh-doar)
Constantinople (kon-stant-uh-NOPLE)
Conurbation (konner-BAY-shun)
Copenhagen (koh-pen-HAHGEN)
Copra (KOH-pruh)
Corcovado (koar-koh-VAH-doh)
Cordillera (kor-dee-YERRA)
Cordillera del Litoral (kor-dee-YERRA dale lee-toh-RAHL)
Córdoba (KORR-doh-buh)
Coromandel (kor-uh-MANDLE)
Corsica (KORR-sih-kuh)
Cortés, Hernán (kor-TAYSS, air-NAHN)
Costa del Sol (koh-stuh-del-SOAL)
Costa Rica (koh-stuh-REE-kuh)
Côte D'Ivoire (KOAT deev-WAH)
Coup d'état (koo-day-TAH)
Coveñas (koh-VAIN-yahss)
Creole [s] (KREE-oal[z])
Crete (KREET)
Crimea (cry-MEE-uh)
Crimean (cry-MEE-un)
Croat (KROH-aht/KROH-at)
Croatia (kroh-AY-shuh)
Cruz (KROOZ)
Cuiabá (koo-yuh-BAH)
Curaçao (koor-uh-SAU)
Cusiana-Cupiagua (koo-see-AHNA, koopee-YAH-gwah)
Cuyo (KOO-yoh)
Cuzco (KOO-skoh)
Cyclades (SIK-luh-deeze)
Cyprus (SYE-pruss)
Cyrenaica (sear-uh-NAY-ih-kuh)
Cyrillic (suh-RILL-ick)
Czar (ZAHR)

Czarina (zah-REE-nuh)
Czech (CHECK)
Czechoslovakia (check-uh-sloh-VAH-kee-uh)
Dagestan (dag-uh-STAHN)
Dahomey (dah-HOH-mee)
Dakar (duh-KAHR)
Dalai Lama (dah-lye LAHMA)
Dalian (dah-lee-ENN)
Dalit (DAH-lit)
Dalmatia (dal-MAY-shuh)
Damascus (duh-MASK-uss)
Damietta (dam-ee-ETTA)
Danube (DAN-yoob)
Daoism (DAU-ism)
Daqing (dah-CHING)
Dar es Salaam (dahr ess suh-LAHM)
Dardanelles (dahr-duh-NELZ)
Darien (dar-YEN)
Daw Aung San Suu Kyi (daw-ong-SAHN soo-CHEE)
Dayak (DYE-ack)
De Gaulle, Charles (duh-GAWL, SHARL)
Deccan (DECKEN)
Delhi (DELLY)
Delphi (DELL-fye)
Deltaic (dell-TAY-ick)
Deng Xiaoping (DUNG shau-PING)
Département (day-part-MAW)
Devolution (dee-voh-LOOH-shun)
Dhahran (dah-RAHN)
Dhaka (DAHK-uh)
Dhow (DAU)
Diaspora (dee-ASP-uh-ruh)
Dichotomy (dye-KOTT-uh-mee)
Dien Bien Phu (d'yen-b'yen-FOOH)
Dijon (dee-ZHAW)
Dili (DIH-lee)
Dinaric (dih-NAHR-rick)
Diyarbakir (dih-yahr-buh-KEER)
Djezira-al-Maghreb (juh-ZEER-uh ahl-mahg-GRAHB)
Djibouti (juh-BOODY)
Djouf (JOOF)
Dnieper (duh-NYEPPER)
Dniester (duh-NYESS-truh)
Dnipropetrovsk (duh-nep-roh-puh-TRAWFSSK)
Dodoma (DOH-duh-mah)
Dominica (duh-MIN-ih-kuh)
Donbas (DAHN-bass)
Donets (duh-NETTS)
Donetsk (duh-NETTSK)
Dostoyevsky (doss-stoy-YEFF-skee)
Douala (doo-AHLA)
Drakensberg (DRAHK-unz-berg)
Dravidian (druh-VIDDY-un)
Dresden (DREZ-den)
Druzba (DROOZE-bah)
Druze (DROOZE)

Dubai (doo-BYE)
Duchy (DUTCH-ee)
Dunedin (duh-NEED-nn)
Durban (DER-bun)
Dushanbe (doo-SHAHM-buh)
Dvina (duh-vee-NAH)
Dzibilchaltun (zeeb-eel-chahl-TOON)
Dzungaria (joong-GAH-ree-uh)
Ebola (ee-BOH-luh)
Ecuador (ECK-wah-dor)
Ecumene (ECK-yoo-meen)
Edgecumbe (EDGE-kum)
Edinburgh (EDDIN-burruh)
Edo (EDD-oh)
Eelam (EE-lum)
Efik (EFFIK)
Egoli (ee-GOH-lee)
Eilat (AY-laht)
Eire (AIR)
Ejido[s] (eh-HEE-doh [ss])
El Aqsa (el AHK-suh)
El Cojo (ell-KOH-hoh)
El Niño (ell-NEEN-yoh)
El Paso (ell-PASSO)
El Puente (el-PWEN-tay)
El Qahira (el-KAH-hee-ruh)
Elam (EE-lum)
Elbe (ELB)
Elburz (el-BOORZ)
Ellesmere (ELZ-mear)
Ellinikon (eh-LINNY-konn)
Emirate (EMMA-rate)
Endemism (en-DEM-izm)
Endoreic (en-doh-RAY-ick)
England (ING-glund)
Enkare (en-KAH-ray)
Entebbe (en-TEBBA)
Entre Rios (en-truh-REE-ohss)
Entrepôt (AHNTRA-poh)
Erdogan (er-DAU-un)
Eritrea (erra-TRAY-uh)
Erlitou (air-lee-TOH)
Esmeraldas (ezz-may-RAHL-dahss)
Estonia (eh-STOH-nee-uh)
Ethiopia (eeth-ee-OH-pea-uh)
Etorofu (etta-ROH-foo)
Eucalyptus (yoo-kuh-LIP-tuss)
Eunuch (YOO-nuck)
Euphrates (yoo-FRATE-eeze)
Eurasia (yoo-RAY-zhuh)
Evens (eh-VENSS)
Extremadura (ess-truh-muh-DOORA)
Façade (fuh-SAHD)
Faeroe (FAR-roh)
Faisalabad (fye-SAHL-ah-bahd)
Falkland [s] (FAWK-lund[z])
Farakka (fah-RAH-kuh)
Fare, Wahiba (FAH-ray, wah-HEE-buh)
Farghona (fahr-GOH-nuh)
Faro del Comercio (FAH-roh del koh-MAIR-see-oh)

Fauna (FAW-nuh)
Favela [s] (fah-VAY-lah[ss])
Fazenda (fah-ZENN-duh)
Fellaheen (fella-HEEN)
Fergana (fahr-guh-NAH)
Ferronorte (feh-roh-NOR-tay)
Feyzabad (FAY-zuh-bahd)
Fezzan (fuh-ZANN)
Ficus (FYE-kuss)
Fiji (FEE-jee)
Fijian (fuh-JEE-un)
Filipinos (filla-PEA-noze)
Finno-Ugric (finno-YOO-grick)
Fjord (FYORD)
Flores (FLAW-rihss)
Florianópolis (flaw-ree-ah-NOH-poh-leese)
Foederis (feh-DARE-iss)
Fonseca (fahn-SAY-kuh)
Formosa (for-MOH-suh)
Fortaleza (for-tuh-LAY-zuh)
Fox, Vicente (FOX, vee-SEN-tay)
Foz do Iguaçu (fahz-doo-ig-wah-SOO)
Francophone (FRANK-uh-foan)
Frankfurt (FRUNK-foort)
Fría (FREE-uh)
Fuji (FOODGY)
Fujian (foo-jee-ENN)
Fujimori, Alberto (foo-jee-MAW-ree, ahl-BAIR-toh)
Fukuoka (foo-kuh-WOH-kuh)
Fulani (foo-LAH-nee)
Funafuti Atoll (foo-nah-FOO-tee AY-toal)
Fungi (FUN-jye)
Fushun (foo-SHUN)
Futa Jallon (food-uh juh-LOAN)
Fuzhou (foo-ZHOH)
Gabon (gah-BAW)
Gadhafi, Muammar (guh-DAHFI, MOO-uh-mar)
Gaelic (GALE-ick)
Gaillard (gil-YARD)
Galicia (guh-LEE-see-uh)
Galla (GAH-lah)
Gambia (GAM-bee-uh)
Gandhi (GONDY)
Ganga (GUNG-guh)
Ganges/Gangetic (GAN-jeez/gan-JETTICK)
Gansu (gahn-SOO)
Gao (GAU)
Gaoliang (gow-lee-AHNG)
Gaoxiong (see Kaohsiung)
Garagum (gah-rah-GOOM)
Garonne (guh-RON)
Gasohol (GAS-uh-hoal)
Gatún (guh-TOON)
Gauteng (GAU-teng)
Gavan (guh-VAHN)
Gaza (GAH-zuh)

Gdansk (guh-DAHNSK)
Gdynia (guh-DINNY-uh)
Gedo (GEDDO)
Geiger (GHYE-gherr)
Geneva (jeh-NEE-vuh)
Genoa (JENNO-uh)
Georges (ZHOR-zh)
Georgia (GEORGE-uh)
Gerardi, Juan José (heh-RAHR-dee, HWAHN hoh-ZAY)
Gerlach (GHERR-lahk)
Gezira (juh-ZEER-uh)
Ghana (GAH-nuh)
Ghanaian (gah-NAY-un)
Ghats (GAHTSS)
Gibraltar (jih-BRAWL-tuh)
Gilgit (GILL-gutt)
Ginza (GHIN-zuh)
Giralda (hih-RAHL-duh)
Giza (GHEE-zuh)
Glasgow (GLASS-skau)
Glasnost (GLUZZ-nost)
Goa (GOH-uh)
Gobi (GOH-bee)
Godthab (GAWT-hawb)
Goiás (goy-AHSS)
Golan (goh-LAHN)
Gondwana (gond-WON-uh)
Gorbachev, Mikhail (GOR-buh-choff, meek-HYLE)
Gorkiy (GORE-kee)
Gorno-Altay (gore-noh-AL-tye)
Gorno-Badakhshan (gore-noh-bah-dahk-SHAHN)
Gorod (guh-RAHD)
Göteborg (GOAT-uh-borg)
Gouda (GOO-dah)
Grande Arche (GRAWND-ARSH)
Grande Carajás (GRUNN-dee kuh-ruh-ZHUSS)
Greenwich (GREN-itch)
Grenada (gruh-NAY-duh)
Groznyy (GRAWZ-nee)
Guadalajara (gwah-duh-luh-HAHR-uh)
Guadalcanal (gwaddle-kuh-NAL)
Guadalquivir (gwahddle-kee-VEER)
Guadeloupe (GWAH-duh-loop)
Guajira (gwah-HEAR-ah)
Guam (GWAHM)
Guanabara (gwah-nah-BAR-uh)
Guando (GWAHN-doh)
Guangdong (gwahng-DUNG)
Guangxi Zhuang (gwahng-shee JWAHNG)
Guangzhou (gwahng-JOH)
Guaraní (gwah-rah-NEE)
Guatemala (gwut-uh-MAH-lah)
Guayaquil (gwye-ah-KEEL)
Guayas (GWYE-ahss)
Guelph (GWELF)
Guerrero (geh-RARE-roh)

Guiana [s] (ghee-AH-nah[z])
Guilin (gway-LIN)
Guinea (GHINNY)
Guinea-Bissau (ghinny-bih-SAU)
Guizhou (gway-JOH)
Gujarat (goo-juh-RAHT)
Gulag (GHOO-lahg)
Guri (GOOR-ree)
Gurkha (GHOOR-kuh)
Guryev (GHOOR-yeff)
Guyana (guy-AHNA)
Habibie (huh-BEE-bee)
Habomai (HAH-boh-mye)
Hacienda (ah-see-EN-duh)
Hägerstrand, Torsten (HAYGER-strand, TOR-stun)
Hague (HAIG)
Haifa (HYE-fah)
Haikou (HYE-KOH)
Hainan (HYE-NAHN)
Haiphong (hye-FONG)
Haiti (HATE-ee)
Haitian (HAY-shun)
Halaib (huh-LABE)
Hamad (hah-MAHD)
Hamas (hah-MAHSS)
Hamburg (HAHM-boorg)
Hamoud (hah-MOOD)
Han (HAHN)
Hang Bai (HAHNG-bye)
Hangzhou (hahng-JOH)
Hankou (hahn-KOH)
Hanoi (han-NOY)
Hanseatic (han-see-ATTIK)
Hanyang (hahn-YAHNG)
Harappa (huh-RAP-uh)
Harare (huh-RAH-ray)
Harbin (HAR-bin)
Harer (HAH-rahr)
Hargeysa (hahr-GAY-suh)
Harijan (hah-ree-JAHN)
Hartshorne (HARTSS-horn)
Haryana (hah-ree-AHNA)
Hashish (hah-SHEESH)
Hausa (HOW-sah)
Hawai'i (huh-WAH-ee)
Hazara (huh-ZAHR-ah)
Hebei (huh-BAY)
Hegemony (heh-JEH-muh-nee)
Heian (HAY-ahn)
Heilongjiang (hay-long-jee-AHNG)
Hejira (heh-JEER-ruh)
Helsinki (hel-SINKEE)
Hema (HAY-muh)
Henan (heh-NAHN)
Herat (heh-RAHT)
Herodotus (heh-RODDA-tuss)
Hesse (HESS)
Hexi (huh-SHEE)
Hidrovia (ee-DROH-vee-uh)
Hierarchical (hire-ARK-uh-kull)

Himachal Pradesh (huh-MAHTCH-ull pruh-DESH)
Himalayas (him-AHL-yuzz/himma-LAY-uzz)
Hindu (HIN-doo)
Hindu Kush (hin-doo KOOSH)
Hindustan (hin-doo-STAHN)
Hindustani (hin-doo-STAHN-nee)
Hindutva (hin-DOOT-vah)
Hippocrates (hih-POCK-ruh-teez)
Hiroshima (hirra-SHEE-muh/huh-ROH-shuh-muh)
Hispaniola (iss-pahn-YOH-luh)
Hizbollah (HIZZ-boh-luh)
Hohhot (huh-HOO-tuh)
Hokkaido (hoh-KYE-doh)
Holistic (hoh-LISS-tick)
Holocene (HOLLO-seen)
Hominid (HOM-ih-nid)
Homo sapiens (hoh-moh SAY-pea-enz)
Homogeneity (hoh-moh-juh-NAY-uh-tee)
Honduras (hon-DURE-russ)
Hongkou (hong-KOH)
Honiara (hoh-nee-AHR-ah)
Honolulu (honn-uh-LOO-loo)
Honshu (HONN-shoo)
Hooghly (HOO-glee)
Hormuz (hoar-MOOZE)
Hotan (HOH-TAN)
Houghton (HOH-t'n)
Houphouët-Boigny, Félix (oo-FWAY bwah-NYEE, fay-LEEKS)
Hsinchu [Xinzhu] (shin-JOO)
Hu Jintao (hoo-jin-DAU)
Huai (HWYE)
Huallaga (wah-YAH-gah)
Huancayo (wahn-KYE-oh)
Huang [He] (HWAHNG [huh])
Huangpu (hwahng-POO)
Hubei (hoo-BAY)
Hué (HWAY)
Humboldt (HUMM-bolt)
Humus (HYOO-muss)
Hunan (hoo-NAHN)
Husayn (hoo-SINE)
Hussein, Saddam (hoo-SAIN, suh-DAHM)
Hutu (HOO-too)
Hwange (WAHNG-ghee)
Hyderabad (HIDE-uh-ruh-bahd)
Hyundai (HUN-dye)
Ibadan (ee-BAHD'N)
Iberia (eye-BEERY-uh)
Ibn Saud (ib'n sah-OOD)
Ibo [land] (EE-boh [land])
Ife (EE-fay)
Iguaçu (EE-gwah-soo)
Iki (EE-kee)
Ikoyi (ee-KOY-yee)
Île de France (EEL duh-FRAWSS)

Île de la Cité (EEL duh-la-see-TAY)
Illyrian (ih-LEER-ree-un)
Ilmen (yill-MEN)
Ilyich (ILL-yitch)
Imam (ih-MAHM)
Inchon (int-CHON)
Indonesia (indo-NEE-zhuh)
Ingush (in-GOOSH)
Ingushetiya (in-goo-SHETTY-uh)
Inkatha (in-KAH-tah)
Inle (IN-lay)
Interahamwe (in-terra-HAHM-way)
Interdigitate (intuh-DID-juh-tate)
Intramuros (in-trah-MOOR-rohss)
Inuit (IN-yoo-it)
Ipanema (ee-pah-NAY-muh)
Iquitos (ih-KEE-tohss)
Iran (ih-RAN/ih-RAHN)
Iranian (ih-RAIN-ee-un)
Iraq (ih-RAK/ih-RAHK)
Iraqi (ih-RAKKY)
Irbil (EAR-beel)
Irian Jaya (IH-ree-ahn JYE-uh)
Irkutsk (ear-KOOTSK)
Irrawaddy (ih-ruh-WODDY)
Irredentism (irruh-DEN-tism)
Irtysh (ear-TISH)
Isfahan (iz-fuh-HAHN)
Iskenderun (iz-ken-duh-ROON)
Islam (iss-LAHM)
Islamabad (iss-LAHM-uh-bahd)
Islington (IZZ-ling-tunn)
Ismaili (izz-MYE-lee)
Isohyet (EYE-so-hyatt)
Israel (IZ-rail)
Issas (ih-SAHZ)
Istanbul (iss-tum-BOOL)
Isthmus/isthmian (ISS-muss/ ISS-mee-un)
Itaipu (ee-TYE-pooh)
Ituri (ih-TOORY)
Ivanovo (ee-VAH-nuh-voh)
Ivoirian (ih-VWAH-ree-un)
Izhevsk (EE-zheffsk)
Jabal (JAB-ull)
Jabotabek (juh-BOH-tah-bek)
Jaffna (JAHF-nuh)
Jains (JYE-nz)
Jakarta (juh-KAHR-tuh)
Jakota (juh-KOH-tuh)
Jalisco (huh-LISS-koh)
Jamaica (juh-MAKE-uh)
Jammu (JUH-mooh)
Jamshedpur (JAHM-shed-poor)
Java (JAH-vuh)
Jawa (JAH-vuh)
Jawanese (jah-vuh-NEEZE)
Jayapura (jye-uh-POOR-ruh)
Jazirah (juh-ZEER-uh)
Jebel Ali (JEH-bel ah-LEE)
Jerusalem (juh-ROO-suh-lum)

Jesuit (JEH-zoo-it)
Jharkhand (JAHR-kahnd)
Jiang Zemin (jee-AHNG zuh-MIN)
Jiangsu (jee-ahng-SOO)
Jiangxi (jee-ahng-SHEE)
Jihad (jee-HAHD)
Jilin (jee-LIN)
Jilong (jee-LUNG)
Jinmen Dao (JIN-MEN-dau)
Joao, Dom (ZHWOW, dom)
Johannesburg (joh-HANNIS-berg)
Johor (juh-HOAR)
Jomon (JOH-mon)
Jubail (joo-BILE)
Judaic (joo-DAY-ick)
Judaism (JOODY-ism)
Junggar (JOONG-gahr)
Junta (HOON-tah)
Jurong (juh-RONG)
Jutland (JUT-lund)
Kabardino-Balkar [iya] (kabber-DEE-noh bawl-KAR [ree-uh])
Kabila, Laurent (kuh-BEE-luh law-RAH)
Kabol (KAH-bull)
Kabylia (kuh-BEE-lee-uh)
Kachins (kuh-CHINZ)
Kaduna (kah-DOO-nah)
Kai Tak (KYE TAK)
Kaifeng (kye-FUNG)
Kakadu (KAH-kuh-doo)
Kalaallit Nunaat (kuh-LAHT-lit noo-NAT)
Kalahari (kalla-HAH-ree)
Kalenjin (kuh-LEN-jin)
Kalgoorlie (kal-GHOOR-lee)
Kalimantan (kalla-MAN-tan)
Kaliningrad (kuh-LEEN-in-grahd)
Kalmyk (KAL-mik)
Kalmykiya (kal-MIK-ee-uh)
Kamba (KAHM-bah)
Kambalda (kahm-BAHL-duh)
Kamchatka (kum-CHAHT-kuh)
Kamehameha (kah-MAY-hah-may-hah)
Kampala (kahm-PAH-luh)
Kampuchea (kahm-pooh-CHEE-uh)
Kanaks (KAH-nahkss)
Kanarese (KAHN uh reece)
Kandahar (KAHN-duh-hahr)
Kannada (KAHN-uh-duh)
Kano (KAH-noh)
Kansai (KAHN-SYE)
Kanto (KAN-toh)
Kaohsiung [Gaoxiong] (GAU-see-OOHNG)
Kara (KAHR-ruh)
Kara Kum (kahr-ruh KOOM)
Karachay (kah-ruh-CHYE)
Karachayevo-Cherkessiya (kahra-CHAH-yeh-vuh cheer-KESS-ee-uh)

Karachi (kuh-RAH-chee)
Karafuto (kahra-FOO-toh)
Karaganda (karra-gun-DAH)
Karakalpak (karra-kal-PAK)
Karakoram (kahra-KOR-rum)
Karamay (kah-RAH-may)
Kareliya (kuh-REE-lee-uh)
Karen (kuh-REN)
Karens (kuh-RENZ)
Kariba (kuh-REE-buh)
Karnataka (kahr-NAHT-uh-kuh)
Karpas (KAHR-pahss)
Karzai, Hamid (KAR-zye, HAH-mid)
Kashagan (KAH-shah-gahn)
Kashgar (KAHSH-gahr)
Kashi (KAH-shee)
Kashmir (KASH-meer)
Katanga (kuh-TAHNG-guh)
Kathmandu (kat-man-DOOH)
Katowice (kah-toh-VEE-tsuh)
Kattegat (KAT-ih-gat)
Kauai (KAU-eye)
Kawasaki (kah-wah-SAH-kee)
Kazakh (KUZZ-uck)
Kazakhstan (KUZZ-uck-STAHN/KUZZ-uck-stahn)
Kazan (kuh-ZAHN)
Kefamenanu (kuh-fahm-uh-NAH-noo)
Kelang (kuh-LAHNG)
Kelantan (keh-LAHN-tahn)
Kemal, Mustafa (keh-MAHL, moo-stah-FAH)
Kenya (KEN-yuh)
Kerala (KEH-ruh-luh)
Keratsinion (keh-rut-SINNY-onn)
Kerinci (kuh-REEN-chee)
Khabarovsk (kuh-BAHR-uffsk)
Khakassiya (kuh-KAHSS-ee-uh)
Khalistan (kahl-ee-STAHN)
Khan, Genghis (KAHN, JING-guss)
Khanate (KAHN-ate)
Khartoum (kar-TOOM)
Khatami (kah-TAH-mee)
Khiva (KEE-vuh)
Khmer [Rouge] (kuh-MAIR [ROOZH])
Khoi (KHOY)
Khoikhoi (KHOY-khoy)
Khoisan (khoy-SAHN)
Kholmsk (KAWLMSK)
Khomeini (hoh-MAY-nee)
Khorat (koh-RAHT)
Khulna (KOOL-nah)
Khurasan (koor-uh-SAHN)
Khuzestan (KOOH-ih-stahn)
Khyber (KYE-burr)
Kiev (Kyyiv) (KEE-yeff)
Kievan (kee-EVAN)
Kigali (kih-GAH-lee)
Kikuyu (kee-KOO-yoo)
Kikwit (KIK-wit)

Kilimanjaro (kil-uh-mun-JAH-roh)
Kilinochchi (kih-luh-NOCK-chee)
Kilwa (KEEL-wah)
Kinabalu (kin-ah-BAH-loo)
Kindu (KIN-doo)
Kinki (kin-KEE)
Kinmen (kin-MEN)
Kinneret (kin-uh-RETH)
Kinshasa (kin-SHAH-suh)
Kiosk (KEE-osk)
Kirghiz (keer-GEEZE)
Kiribati (KIH-ruh-bahss)
Kirkuk (keer-KOOK)
Kiruna (kih-ROONA)
Kisangani (kee-sahn-GAH-nee)
Kismaayo (keese-MYE-oh)
Kitakyushu (kee-TAH-KYOO-shoo)
Kivu (KEE-voo)
Kiwi (KEE-wee)
Klaipeda (KLYE-puh-duh)
Klyuchevskaya (klee-ooh-CHEFF-skuh-yuh)
Koala (kuh-WAH-luh)
Kobe (KOH-bay)
Kolkata (kol-KUTTA)
Kolkhoz (KOLL-koze)
Kolok (KOH-lok)
Kolyma (koh-LEE-mah)
Komi (KOH-mee)
Kompong Som (kahm-pong SAWM)
Kompong Thom (kahm-pong TAWM)
Komsomolsk (komm-suh-MAWLSK)
Kongfuzi (kung-FOODZEE)
Kongzi (KUNG-dzee)
Königsberg (KAY-nix-bairk)
Konkan (KAHNG-kun)
Köppen (KER-pun)
Koran (kaw-RAHN)
Kordofan (kor-duh-FAN)
Korea (kuh-REE-uh)
Korla (KOOR-LAH)
Koryakiya (kor-YAH-kee-uh)
Koryakskaya (kor-YAHK-skuh-yuh)
Kosciusko (kuh-SHOO-skoh)
Kosovar (KAW-suh-vahr)
Kosovo (KAW-suh-voh)
Kostunica, Vojislav (koss-too-NEE-chah, VOY-slahv))
Kota Kinabalu (KOH-tuh kin-ah-BAH-loo)
Kourou (koo-ROO)
Koutoubia (koo-TOO-bee-yuh)
Kowloon (kau-LOON)
Kra Isthmus (KRAH ISS-muss)
Krakow (KRAH-koov)
Krasnodar (KRASS-nuh-dahr)
Krasnovodsk (kruzz-noh-VAUGHTSK)
Krasnoyarsk (krass-nuh-YARSK)
Kribi (KREE-bee)
Krivoy Rog (krih-voy-ROAG)

Kronstadt (KROAN-shtaht)
Kruger (KROO-guh)
Krugersdorp (KROO-guzz-dorp)
Kryvyy Rih (kree-VEE-REE)
Kuala Lumpur (KWAHL-uh LOOM-poor)
Kublai Khan (koob-lye KAHN)
Kufra (KOO-fruh)
Kumartuli (koo-MAR-too-lee)
Kunashiri (koo-NAH-shuh-ree)
Kunlun (KOON-LOON)
Kunming (koon-MING)
Kura (KOOR-uh)
Kurd [istan] (KERD [uh-stahn])
Kure (KOOH-ray)
Kurile (KYOOR-reel)
Kuroshio (koo-roh-SHEE-oh)
Kush (KOOSH)
Kushites (KOO-sheits)
Kuta (KOO-tuh)
Kuwait (koo-WAIT)
Kuybyshev (KWEE-buh-sheff)
Kuzbas (kooz-BASS)
Kuznetsk (kooz-NETSK)
Kwangju (GWONG-JOO)
Kwazulu (kwah-ZOO-loo)
Kyongju (GYOONG-JOO)
Kyongsang (GYOONG-SAHN)
Kyoto (kee-YOH-toh)
Kyrgyz (KEER-geeze)
Kyrgyzstan (KEER-geeze-stahn)
Kyushu (kee-YOO-shoo)
Kyyiv [See Kiev]
La Coruña (lah-kor-ROON-yah)
La Défense (lah-day-FAWSS)
La Paz (lah-PAHZ)
La Violencia (lah vee-oh-LENN-see-uh)
Ladakh (luh-DAHK)
Ladino (luh-DEE-noh)
Laem Chabang (lay-EMM chuh-BAHNG)
Lafaiete (lah-fuh-YAY-tuh)
Lagos (LAY-gohss)
Lahore (luh-HOAR)
Lamanai (LAH-mah-nay)
Lancang (LAHN-ZAHNG)
Land (LAHNT)
Länder (LEN-derr)
Lantau (LAHN-DAU)
Lanzhou (lahn-JOH)
Lao (LAU)
Laos (LAUSS)
Laotian (lay-OH-shun)
Las Colinas (lahss-koh-LEE-nahss)
Latvia (LATT-vee-uh)
Lautoka (lau-TOH-kuh)
Laval (lah-VAHL)
Le Havre (luh-HAHV)
Le Chateau Frontenac (luh shat-TOH frawn-teh-NUCK)

Legume (LEG-gyoom)
Leipzig (LYPE-sik)
Leith (LEETH)
Leizhou (lay-JOH)
Lemur (LEE-mer)
Lena (LAY-nuh)
Lendu (LEN-doo)
Lenin (LENNIN)
León (lay-OAN)
Lesotho (leh-SOO-too)
Levant (luh-VAHNT)
Lhasa (LAH-suh)
Li (LEE)
Lianyungang (lee-en-yoong-GAHNG)
Liao (lee-AU)
Liaodong (lee-au-DUNG)
Liaoning (lee-au-NING)
Liberia (lye-BEERY-uh)
Libreville (LEE-bruh-veel)
Lichens (LYE-kenz)
Liechtenstein (LIK-ten-shtine)
Liège (lee-EZH)
Lima (LEE-muh)
Limpopo (lim-POH-poh)
Lingala (ling-GAH-lah)
Lingua franca (LEENG-gwuh FRUNK-uh)
Linguae francae (LEENG-gwee FRUNK-kee)
Lisas (LEE-suss)
Litani (lih-TAH-nee)
Lithuania (lith-oo-AINY-uh)
Littoral (LIT-oh-rull)
Livingstonia (lih-ving-STOH-nee-uh)
Ljubljana (lee-oo-blee-AHNA)
Llama (LAH-muh)
Llano [s] (YAH-noh [ss])
Loess (LERSS)
Loihi (loh-EE-hee)
Loire (luh-WAHR)
Lombardy (LOM-bar-dee)
Lombok (LAHM-bahk)
Lourenço Marques (loh-REN-soh mahr-KESS)
Louvre (LOOV)
Lozi (LOH-zee)
Lualaba (loo-uh-LAH-buh)
Luanda (loo-AN-duh)
Lubumbashi (loo-boom-BAH-shee)
Lucknow (LUCK-nau)
Luhya (LOO-yuh)
Lujiazui (loo-jee-ah-ZWAY)
Lula da Silva, Luiz Inácio (LOO-lah dah SIL-vah, loo-EESE ee-NAH-see-oh)
Luleå (LOO-lee-oh)
Lund (LOOND)
Luo (LOO-oh)
Lusitania (loo-see-TANE-yuh)
Luxembourg (LUX-em-borg)
Luxor (LUK-soar)

Luzon (loo-ZAHN)
Lyon (lee-AW)
Maale (MAH-lay)
Maas (MAHSS)
Maasai (muh-SYE)
Maastricht (mah-STRICT)
Mabo (MAY-boh)
Macau (muh-KAU)
Macedonia (massa-DOH-nee-uh)
Machu Picchu (MAH-choo PEEK-choo)
Mackinder, Halford (muh-KIN-der, HAL-ferd)
Mactan (mahk-TAHN)
Macumba (mah-KOOM-bah)
Madagascar (madda-GAS-kuh)
Madhya Pradesh (mahd-yuh-pruh-DESH)
Madras (muh-DRAHSS)
Madrid (muh-DRID)
Madura (muh-DOORA)
Madurai (mahd-uh-RYE)
Madurese (muh-dooh-REECE)
Magaz (muh-GAHZ)
Magdalena (mahg-dah-LAY-nah)
Magellan (muh-JELL-un)
Maghreb (mahg-GRAHB)
Magyar (MAG-yahr)
Mahakam (MAH-hah-kahm)
Maharajah (mah-hah-RAH-juh)
Maharashtra (mah-huh-RAH-shtra)
Mahathir Mohamad (MAH-hah-theer moh-HAH-mud)
Maize (MAYZ)
Makhachkala (muh-kahtch-kuh-LAH)
Makkah (MEK-ah)
Makung (mah-GOONG)
Mala Strana (mah-lah-STRAH-nah)
Malabar (MAL-uh-bahr)
Malabo (muh-LAH-boh)
Malacca (muh-LAH-kuh)
Málaga (MAHL-uh-guh)
Malagasy (malla-GASSY)
Malaita(ns) (mah-LAY-tah [nz])
Malawi (muh-LAH-wee)
Malay (muh-LAY)
Malaya (muh-LAY-uh)
Malayalam (mal-uh-YAH-lum)
Malaysia (muh-LAY-zhuh)
Maldives (MAWL-deevz)
Mali (MAH-lee)
Malmö (MAHL-meh)
Maluku(s) (mah-LOO-koo[z])
Malvinas (mahl-VEE-nahss)
Manado (muh-NAH-doh)
Managua (mah-NAH-gwuh)
Manaus (muh-NAUSS)
Manchu (man-CHOO)
Manchukuo (mahn-JOH-kwoh)
Manchuria (man-CHOORY-uh)
Mandalay (man-duh-LAY)

Mandarin (MAN-duh-rin)
Mandela (man-DELLA)
Manipur (man-uh-POOR)
Manitoba (manna-TOH-buh)
Manzanillo (mun-zuh-NEE-yoh)
Mao Zedong (MAU zee-DUNG)
Maori (MAH-aw-ree/MAU-ree)
Maputo (mah-POOH-toh)
Maquiladora (mah-kee-luh-DORR-uh)
Mara (MAHRA)
Maracaibo (mah-rah-KYE-boh)
Maracay (mah-rah-KYE)
Marajó (mah-rah-ZHOH)
Mariana (marry-ANNA)
Mariinsk (muh-ryee-YEENSK)
Maritsa (muh-REET-suh)
Mariy [-el] (MAH-ree [el])
Marmagao (marma-GAU)
Marmara (MAH-muh-ruh)
Marquesas (mahr-KAY-suzz)
Marrakech (mahr-uh-KESH)
Marseille (mar-SAY)
Marsupial (mar-SOOPY-ull)
Martinique (mahr-tih-NEEK)
Massif (mass-SEEF)
Matadi (muh-TAH-dee)
Matão (mah-TAU)
Mato Grosso (mutt-uh-GROII-soh)
Mato Grosso do Sul (mutt-uh-GROH-soh duh-SOOL)
Matsu (mah-TSOO)
Matsu Tao (mah-TSOO DAU)
Mauna Kea (mau-nuh-KAY-uh)
Mauritania (maw-ruh-TAY-nee-uh)
Mauritius (maw-REE-shuss)
Mauryan (MAW-ree-un)
Maya (MYE-uh)
Mayan (MYE-un)
Mazar-e-Sharif (mah-ZAHR-ee-shah-REEF)
Mbeki, Thabo (mm-BEH-kee, TAH-boh)
Mecca (MEK-ah)
Mecsek (MAH-chek)
Medan (MAY-dahn)
Medellín (meh-deh-YEEN)
Medina (muh-DEENA)
Megalopolis (meh-guh-LOPP-uh-liss)
Meghalaya (may-guh-LAY-uh)
Meghna (MAIG-nuh)
Meiji (may-EE-jee)
Mejía, Hipólito (meh-HEE-yah, ee-POH-lee-toh)
Mekong (MAY-kong)
Melaka (see Malacca)
Melanesia (mella-NEE-zhuh)
Melbourne (MEL-bun)
Melilla (meh-LEE-yuh)
Mendoza (men-DOH-zah)
Menem, Carlos (MENNEM, KAHR-lohss)

Mengkabong (MENG-kah-bong)
Menshevik (MEN-shuh-vick)
Merauke (muh-RAU-kuh)
Mercosul (mair-koh-SOOL)
Mercosur (mair-koh-SOOR)
Mérida (MAY-ree-dah)
Meridian (meh-RIDDY-un)
Merina (meh-REE-nuh)
Meroe (MEH-roh-ay)
Meru (MAY-roo)
Mesa (MAY-suh)
Mesabi (meh-SAH-bee)
Meseta (meh-SAY-tuh)
Meshketian (mesh-KETTY-un)
Mesoamerica (MEZZOH-america)
Mesopotamia (messo-puh-TAY-mee-uh)
Mesquita (meh-SKEE-tuh)
Mestizo (meh-STEE-zoh)
Meuse (MERZZ)
Mezzogiorno (met-soh-JORR-noh)
Miao (m'YOW)
Michoacán (mee-chuh-wah-KAHN)
Micronesia (mye-kroh-NEE-zhuh)
Milan (mih-LAHN)
Millau (mee-LOH)
Milosevic, Slobodan (mih-LAW-suh-vitch, SLOH-boh-dahn)
Minahasa (MEE-nah-hah-sah)
Minaret (MINNA-ret)
Minas Gerais (MEE-nuss zhuh-RICE)
Mindanao (min-duh-NAU)
Minh, Ho Chi (MINN, hoh chee)
Minifundia (minny-FOON-dee-uh)
Mirador(es) (meera-DOAR [ayss])
Miraflores (meera-FLAW-rayss)
Miskito (mih-SKEE-toh)
Mitú (mee-TOO)
Mius (mee-OOS)
Mizoram (mih-ZOR-rum)
Mobutu (moh-BOO-too)
Moçambique (moh-sum-BEEK)
Mogadishu (moo-gah-dee-SHOH)
Mohenjo Daro (moh-hen-joh-DAHRO)
Moi (MOY)
Mojave (moh-HAH-vee)
Moldavia (moal-DAY-vee-uh)
Moldova (moal-DOH-vuh)
Moluccans (muh-LUCK-unz)
Molybdenum (muh-LIB-dun-um)
Mombasa (mahm-BAHSSA)
Mongol (MUNG-goal)
Mongolia (mung-GOH-lee-uh)
Monogamy (muh-NOG-ah-mee)
Montagnards (MON-tun-yardz)
Montaña (mon-TAHN-yah)
Montego (mon-TEE-goh)
Montenegro (mon-teh-NAY-groh)
Monterrey (mon-teh-RAY)
Montevideo (moan-tay-vee-DAY-oh)
Montparnasse (mawn-pahr-NAHSS)
Montpellier (maw-pell-YAY)

Montreal (mun-tree-AWL)
Montserrat (mont-seh-RAHTT)
Moped (MOH-ped)
Moravia (more-RAY-vee-uh)
Mordoviya (mor-DOH-vee-uh)
Moscow (MAW-skau)
Moselle (moh-ZELL)
Mosque (MOSK)
Mount Isa (mount EYE-suh)
Mpumalanga (mm-pooma-LAHNG-guh)
Muara Muntai (moo-WAH-rah MOON-tye)
Mugabe (moo-GAH-bay)
Muglad (moo-GLAHD)
Muhajir (MOO-hah-jeer)
Muhammad (moo-HAH-mid)
Mujahideen (moo-jah-heh-DEEN)
Mulatto (moo-LAH-toh)
Mullah (MOOL-ah)
Multan (mool-TAHN)
Mulukas (MAH-loo-kooz)
Mumbai (MOOM-bye)
Munich (MYOO-nik)
Murmansk (moor-MAHNTSK)
Murray (MUH-ree)
Musandam (muh-SAND-um)
Muscat (MUH-skaht)
Muscovy (muh-SKOH-vee)
Museveni (moo-SAY-veh-nee)
Musharraf, Pervez (moo-SHAR-rahf, PER-vez)
Muslim (MUZZ-lim)
Mutrah (MUTT-truh)
Muuga (MOO-guh)
Muwahhidun (moo-wah-hih-DOON)
Myanmar (mee-ahn-MAH)
Mymensingh (mye-men-SING)
Naga [land] (NAHGA [-land])
Nagasaki (nah-guh-SAHKEE)
Nagorno-Karabakh (nuh-GORE-noh KAH-ruh-bahk)
Nagoya (nuh-GOYA)
Nagpur (NAHG-poor)
Nairobi (nye-ROH-bee)
Najaf (nuh-JAHF)
Nakhodka (nuh-KAUGHT-kuh)
Nakhon Sawan (NUH-KAWN suh-WOON)
Nam Bo (nahm-BOH)
Namib (nah-MEEB)
Namibia (nuh-MIBBY-uh)
Nanjing (nahn-ZHING)
Naoroji (nau-ROH-jee)
Nara (NAHRA)
Nassau (NASS-saw)
Nasser, Gamal Abdel (NASS-er, guh-MAHL AB-dul)
Natal (nuh-TAHL)
Nauru (nah-OO-roo)
Nautical (NAW-dih-kull)

Navi (NAH-vee)
Naxcivan (nah-kee-chuh-VAHN)
Nazi (NAH-tsee)
Nazran (nahz-RAHN)
Ndebele (en-duh-BEH-leh)
N'Djamena (en-jah-MAY-nah)
Ndola (en-DOH-luh)
Negev (NEH-ghev)
Negro (NAY-groh)
Nehru, Jawaharlal (NAY-roo, juh-WAH-hur-lahl)
Nei Mongol (nay-MUNG-goal)
Nepal (nuh-PAHL)
Neva (NAY-vuh)
Nevis (NEE-vuss)
Nevsky Prospekt (NEFF-skee PROSS-spekt)
New Caledonia (noo-kalla-DOAN-yuh)
New Guinea (noo-GHINNY-uh)
Newfoundland (NYOO-fun-lund)
Ngodi (eng-GOH-dee)
Ngorongoro (eng-gore-ong-GORE-roh)
Nha Trang (nah-TRAHNG)
Niamey (NEE-ah-may)
Nicaragua (nick-uh-RAH-gwuh)
Niger [Country] (nee-ZHAIR)
Niger [River] (NYE-jer)
Nigeria (nye-JEERY-uh)
Niger-Kordofanian (NYE-jer kor-doh-FAN-ee-un)
Nilotic (nye-LODDIK)
Ningbo (ning-BWOH)
Ningxia Hui (NING-shee-AH HWAY)
Nistru (NEE-stroo)
Niue (nee-OOH-ay)
Nizhniy Novgorod (NIZH-nee NAHV-guh-rahd)
Nkrumah, Kwame (en-KROO-muh, KWAH-mee)
Nobi (NOH-bee)
Nord-Pas de Calais (NORD pah-duh-kah-LAY)
Noriega (noar-ree-AY-gah)
Norilsk (nuh-REELSK)
Norte (NOR-tay)
Notre Dame (NOH-truh DAHM)
Nouakchott (noo-AHK-shaht)
Nouméa (noo-MAY-uh)
Nouveau Quebec (noo-VOH kwih-BECK)
Nova Scotia (nova-SKOH-shuh)
Novaya Zemlya (NOH-vuh-yuh zem-lee-AH)
Novgorod (NAHV-guh-rahd)
Novi Arbat (NOH-vee AHR-baht)
Novokuznetsk (noh-voh-kooz-NETSK)
Novorossiysk (noh-voh-ruh-SEESK)
Novosibirsk (noh-voh-suh-BEERSK)
Nubia (NOO-bee-uh)
Nueva Mendez (noo-AY-vuh MEN-dez)

Nuevo León   (noo-AY-voh lay-OAN)
Nunavut   (NOON-uh-voot)
Nungambakkam   (noon-GAHM-bah-KAHM)
Nuuk   (NEWK)
Nyasaland   (nye-ASSA-land)
Nyika   (nye-YEE-kuh)
Oahu   (uh-WAH-hoo)
Oaxaca   (wuh-HAH-kuh)
Obasanjo   (oh-BAH-sahn-joh)
Oblast   (OB-blast)
Obsidian   (ob-SIDDY-un)
Occidental   (oak-see-den-TAHL)
Ocho Rios   (oh-choh REE-ohss)
Ochre   (OH-ker)
Ocussi-Ambeno   (oh-KOOH-see AHM-bay-noh)
Oder-Neisse   (OH-der NEISS)
Odesa   (oh-DESSA)
Ogaden   (oh-gah-DEN)
Ogoni   (oh-GOH-nee)
Oguz   (uh-GOOZ)
Ohmae, Kenichi   (OH-may, keh-NEE-chee)
Oirot   (AW-ih-rut)
Oise   (WAHZ)
Okavango   (oh-kuh-VAHNG-goh)
Okhotsk   (oh-KAHTSK)
Okrug   (AW-krook)
Olduvai   (OLE-duh-way)
Oman   (oh-MAHN)
Omdurman   (om-duhr-MAHN)
Omote Nippon   (oh-MOH-tay NIP-on)
Oran   (aw-RAHN)
Orangutan   (aw-RANG-gyoo-tan)
Ordos   (ORD-uss)
Ordzhonikidze   (or-johnny-KIDD-zuh)
Øresund   (ERR-uh-sun)
Organisasi Papua Merdeka   (or-kah-nee-SAH-see pahp-OO-uh mair-DAY-kah)
Oriental   (orry-en-TAHL)
Oriente   (orry-EN-tay)
Orinoco   (orry-NOH-koh)
Orissa   (aw-RISSA)
Oromo   (AW-ruh-moh)
Orontes   (aw-RAHN-teeze)
Osaka   (oh-SAH-kuh)
Öskemen   (ERSKEE-min)
Oslo   (OZ-loh)
Oslofjord   (OZ-loh-fyord)
Osman   (oz-MAHN)
Ossetia   (oh-SEE-shuh)
Ossetians   (oh-SEE-shunz)
Ossies   (OH-seez)
Ostrava   (AW-struh-vuh)
Otavalo   (oh-tah-VAH-loh)
Ottawa   (OTTA-wuh)
Oweinat   (oh-WEE-naht)
Oxisol   (OXY-soll)
Oyo   (OH-joh)

Padania   (pah-DAHN-yah)
Pago Pago   (PAH-goh PAH-goh)
Pagoda   (puh-GOH-duh)
Pahlavi, Shah Muhammad Reza   (puh-LAH-vee, shah mooh-HAH-mid RAY-zuh)
Pakhtuns   (puck-TOONZ)
Pakistan   (PAH-kih-stahn)
Paktia   (pahk-TEE-uh)
Palau   (puh-LAU)
Palawan   (pah-LAH-wahn)
Pale   (PAH-lay)
Palembang   (pah-LEM-bahng)
Palenque   (puh-LENG-kay)
Palestine   (PAL-uh-stine)
Palestinian   (pal-uh-STINNY-un)
Palk   (PAWK)
Palo Alto   (PALLOH AL-toh)
Palu   (pah-LOO)
Pamir   (pah-MEER)
Pamirs   (pah-MEERZ)
Pampa   (PAHM-pah)
Panacea   (pan-uh-SEE-uh)
Panache   (pah-NAHSH)
Panaji   (pah-NAH-jee)
Pangaea   (pan-GAY-uh)
Papeete   (pahp-ee-ATE-tee)
Papua New Guinea   (pahp-OO-uh noo-GHINNY)
Papua [ns]   (pahp-OO-uh [-unz])
Papuan   (pahp-OO-un)
Pará   (puh-RAH)
Paraguay   (PAHRA-gwye)
Paraíba   (pah-rah-EE-buh)
Paramillos   (pah-rah-MEE-yohss)
Paraná   (pah-rah-NAH)
Paria   (PAHR-yah)
Pariah   (puh-RYE-uh)
Parliament   (PAR-luh-ment)
Paseo del Norte   (pah-SAY-oh del NOR-tay)
Pashtuns [see Pushtuns]
Pasig   (PAH-sig)
Patagonia   (patta-GOH-nee-uh)
Pathan   (puh-TAHN)
Pattani   (pah-TAH-nee)
Pattaya   (puh-TYE-uh)
Paulista [s]   (pow-LEASH-tah [ss])
Pechora   (peh-CHORE-ruh)
Pedro Miguel   (PAY-droh mee-GWELL)
Peking   (pea-KING)
Peloponnesus   (pelloh-puh-NEEZE-uss)
Pelourinho   (peh-loo-REEN-yoh)
Penghu   (pung-HOO)
Peón   (pay-OAN)
Peones   (pay-OH-nayss)
Perejil   (peh-reh-HEEL)
Perestroika   (perra-STROY-kuh)
Perm   (PAIRM)
Permyakiya   (pairm-YAH-kee-uh)

Pernambuco   (pair-nahm-BOO-koh)
Perón, Juan   (puh-ROAN, WAHN)
Peronista   (peh-roh-NEE-stah)
Persepolis   (per-SEPP-uh-luss)
Persia   (PER-zhuh)
Peshawar   (puh-SHAH-wahr)
Peso   (PAY-soh)
Petén   (peh-TEN)
Petrograd   (PETTRO-grahd)
Petronas   (peh-TROH-nuss)
Petropavlovsk-Kamchatskiy   (pit-ruh-PAHV-lufsk kahm-CHAHT-skee)
Pettah   (PET-uh)
Phaleron   (fuh-LEER-un)
Pharaonic   (fair-ray-ONNICK)
Philippines   (FILL-uh-peenz)
Phnom Penh   (puh-NOM PEN)
Phoenicians   (fuh-NEE-shunz)
Phu Bia   (POO-bee-uh)
Phuket   (POO-KETT)
Physiography   (fizzy-OGG-ruh-fee)
Phytogeography   (FYE-toh-jee-OG-gruh-fee)
Piazza   (pea-AH-tsuh)
Piedmont   (PEED-mont)
Pietermaritzburg   (peeh-tuh-MAHR-utss-berg)
Pilipino   (pill-uh-PEA-noh)
Pinang   (puh-NANG)
Pinatubo   (pin-uh-TOO-boh)
Pinheiros   (PEEN-yay-ross)
Pinyin   (pin-YIN)
Piraeus   (puh-RAY-uss)
Pisac   (pea-SAHK)
Pitcairn   (PITT-kairn)
Pizarro, Francisco   (pea-SAHRO, frahn-SEECE-koh)
Place de la Concorde   (PLAHSS duh lah kon-KORD)
Placer [ing]   (PLASS-uh [ring])
Plata, Rio de la   (PLAH-tah, REE-oh day-lah)
Platypus   (PLATT-uh-pus)
Plaza Azoguejo   (PLAH-sah ah-soh-GAY-hoh)
Plaza de Armas   (PLAH-sah day AR-mahss)
Plaza Bolívar   (PLAH-sah boh-LEE-vahr)
Plaza de Mayo   (PLAH-sah day MYE-oh)
Plaza de San Martin   (PLAH-sah day sahn mahr-TEEN)
Plaza Mayor   (PLAH-sah mye-YOAR)
Pleistocene   (PLY-stoh-seen)
Plymouth   (PLIH-muth)
Pointe-Noire   (pwahnt-nuh-WAHR)
Politburo   (POLLIT-byoor-roh)
Polonoroeste   (POLLOH-nuh-roh-ESS-tee)
Polygamy   (puh-LIG-uh-mee)

Polynesia (polla-NEE-zhuh)
Polytheism (polly-THEE-izm)
Pontianak (pahn-tee-AH-nahk)
Popocatépetl (poh-poh-kah-TEH-peh-til)
Port Moresby (port MORZ-bee)
Port Said (port-sah-EED)
Port-au-Prince (por-toh-PRANSS)
Porteños (por-TAIN-yohss)
Pôrto Alegre (POR-too uh-LEG-ruh)
Pôrto Velho (POR-too VELL-yoo)
Posavina (poh-suh-VEENA)
Potgietersrus (POT-gheeters-roost)
Potosí (poh-toh-SEE)
Povolzhye (puh-VOLL-zhuh)
Prague (PRAHG)
Prayagraj (PRAY-ag-rahdge)
Pretoria (prih-TOR-ree-uh)
Primorskiy (pree-MOHR-skee)
Príncipe (PREEN-see-pea)
Progeny (PRAH-juh-nee)
Prokofiev (pruh-KOFFEY-ev)
Proselytism (PRAH-sell-eh-tizm)
Provence (pro-VAHNSS)
Prudhoe (PROO-doh)
Prut (PRROOT)
Pudong (poo-DUNG)
Puebla (poo-EBB-luh)
Puerto Barrios (pwair-toh bah-REE-uss)
Puerto Caldera (pwair-toh kahl-DERRA)
Puerto Rico (pwair-toh REE-koh)
Puerto Montt (pwair-toh-MAWNT)
Puget (PYOO-jet)
Punjab (pun-JAHB)
Punjabi (poon-JAH-bee)
Punta Arenas (POON-tah ah-RAY-nahss)
Punta del Este (POON-tah del ESS-tay)
Pushtun (PAH-shtoon)
Putin, Vladimir (POO-t'n, VLAH-duh-meer))
Putrajaya (poo-truh-JYE-UH)
Pyatigorsk (pyih-tyih-GORSK)
Pygmy (PIG-mee)
Pyongyang (pea-AWHNG-yahng)
Pyrenees (PEER-uh-neez)
Qaidam (CHYE-DAHM)
Qanat (KAH-naht)
Qaraghandy (kah-rah-GONDY)
Qasim (kah-SEEM)
Qatar (KOTTER)
Qattarah (kuh-TAR-ruh)
Qiang (chee-AHNG)
Qin (CHIN)
Qing (CHING)
Qingdao (ching-DAU)
Qinghai (ching-HYE)
Qingyi (ching-YEE)
Qinhuangdao (chin-hwahng-DAU)

Qiqihar (chee-CHEE-har)
Qoraqalpog (kora-kal-PAHG)
Quaternary (kwuh-TER-nuh-ree)
Quebec (kwih-BECK)
Quebecer (kwih-BECK-er)
Quebecois (kay-beh-KWAH)
Quebracho (kay-BROTCH-oh)
Quechua (KAYTCH-wah)
Quemoy (keh-MOY)
Quetzal (kay-DZAHL)
Quezon (KAY-soan)
Quinary (KWYE-nuh-ree)
Quito (KEE-toh)
Quran (kor-RAHN)
Raisina (rye-SEENA)
Raj (RAHDGE)
Rajang (rah-JAHNG)
Rajasthan (RAH-juh-stahn)
Rajin-Sonbong (RAH-jin SAWM-bong)
Rajiv (ruh-ZHEEV)
Rajshahi (rahdge-SHAH-hee)
Ramadan (rahma-DAHN)
Randstad (RUND-stud)
Rangoon (rang-GOON)
Rangpur (RHANG-poor)
Rarotonga (rarra-TAHNG-guh)
Rashid (ruh-SHEED)
Ratzel, Friedrich (RAHT-sull, FREED-rish)
Ravenstein (RAVVEN-steen)
Rawalpindi (rah-wull-PIN-dee)
Real (ray-AHL)
Recife (ruh-SEE-fuh)
Reg (REGG)
Réunion (ray-yoon-YAW)
Reykjavik (RAKE-yah-veek)
Rhaeto-Romansch (RAY-shoh roh-MAHN-ssh)
Rhine (RYNE)
Rhodesia (roh-DEE-zhuh)
Rhône-Alpes (ROAN AHLP)
Rhône-Saône (ROAN say-OAN)
Riau (REE-au)
Riga (REEGA)
Rio Branco (REE-oh BRUNG-koh)
Rio de Janeiro (REE-oh day zhah-NAIR-roh)
Rio Grande do Norte (REE-oh GRUN-dee duh NORTAH)
Rio Grande do Sul (REE-oh GRUN-dee duh SOOL)
Rioja (ree-OH-hah)
Riverine (RIVER-reen)
Riyadh (ree-AHD)
Rocinho (haw-SEEN-yah)
Roma (ROH-muh)
Romania (roh-MAIN-yuh)
Rondônia (roh-DOAN-yuh)
Roraima (raw-RYE-muh)
Rostow (ROSS-stoff)

Rouen (roo-AW)
Rub al Khali (roob ahl KAH-lee)
Rue Notre Dame (ROO noh-truh DAHM)
Rue Saint-Louis (ROO sah-loo-WEE)
Rue Saint Louis (roo-saw-loo-EE)
Ruhr (ROOR)
Ruijin (rway-JEEN)
Rumailah (roo-MYE-luh)
Rus (ROOSE)
Ruwenzori (roo-when-ZOHR-ree)
Rwanda (roo-AHN-duh)
Ryukyu (ree-YOO-kyoo)
Saarland (ZAHR-lunt)
Saba (SAY-buh/SAH-buh)
Sabah (SAHB-ah)
Sabinas (sah-BEE-nuss)
Saddhu (SAH-dooh)
Sahara (suh-HARRA)
Sahel (suh-HELL)
Saigon (sye-GAHN)
Saint Dominique (san-dommi-NEEK)
Saint-Denis (san-deh-NEE)
Sakartvelos (sah-KART-vuh-lohss)
Sakha (SAH-kuh)
Sakhalin (SOCK-uh-leen)
Salalah (sah-LAH-luh)
Salina Cruz (sah-LEE-nuh CROOZ)
Salinas (sah-LEE-nuss)
Salisbury (SAWLZ-buh-ree)
Salta (SAHL-tah)
Salvador (SULL-vuh-dor)
Salween (SAHL-wain)
Samara (suh-MAH-ruh)
Samarinda (sah-mah-RIN-dah)
Samarqand (sah-mahr-KAHND)
Samarra (sah-MAH-rah)
Samoa (suh-MOH-uh)
Samoan (suh-MOH-un)
Samsung (SAHM-SOONG)
San (SAHN)
San Cristóbal (sahn kree-STOH-bahl)
San Joaquin (san-wah-KEEN)
San José (sahn hoh-ZAY)
San Juan (sahn HWAHN)
San Pedro Sula (sahn pay-droh-SOO-luh)
San Rafael (sahn rah-fye-ELL)
San Martin, José de (sahn-mahr-TEEN, hoh-ZAY day)
San'a (suh-NAH)
Sandinista (sahn-dee-NEE-stuh)
Santa Catarina (SUN-tuh kuh-tuh-REE-nuh)
Santa Cruz (SAHN-tah CROOZ)
Santa Tecla (SAHN-tah TEK-lah)
Santiago (sahn-tee-AH-goh)
Santos (SUNT-uss)
Sanxia (sahn-SHAH)
São Francisco (sau frahn-SEECE-koh)
São Luis (sau loo-EECE)

São Paulo (sau PAU-loh)
São Tomé (sau TOH-may)
Sarajevo (sahra-YAY-voh)
Saratov (suh-RAHT-uff)
Sarawak (suh-RAH-wahk)
Sardinia (sahr-DINNY-uh)
Sarmatian (sahr-MAY-shee-un)
Saskatchewan (suss-KATCH-uh-wunn)
Saudi Arabia (SAU-dee uh-RAY-bee-uh)
Sauer (SAU-er)
Savimbi (suh-VIM-bee)
Savoy (suh-VOY)
Saxony (SAX-uh-nee)
Scania (SKAIN-yuh)
Schengen (SHENG-gen)
Schistosomiasis (shistoh-soh-MYE-uh-siss)
Scythian (SITH-ee-un)
Sedentary (SEDDEN-terry)
Seikan (say-KAHN)
Seine (SENN)
Semarang (seh-MAHR-rahng)
Semeru (suh-MAY-roo)
Semitic (seh-MITTICK)
Sendero Luminoso (sen-DARE-oh loo-mee-NOH-soh)
Senegal (sen-ih-GAWL)
Sen-en (sen-NENN)
Seoul (SOAL)
Sepoy (SEE-poy)
Sepulchre (SEP-ul-kurr)
Serbia (SER-bee-uh)
Serer (seh-RAIR)
Sergipe (SAIR-zhee-pay)
Serov (SAIR-roff)
Serra dos Carajás (SERRA doo kuh-ruh-ZHUSS)
Sertão (sair-TOWNG)
Sesame (SESS-uh-mee)
Seto (SET-oh)
Sevastopol (seh-VASS-toh-pawl)
Seville (suh-VILL)
Seychelles (say-SHELLZ)
Sha Mian (shah mee-AHN)
Shaanxi (shahn-SHEE)
Shaba (SHAH-bah)
Shan (SHAHN)
Shandong (shahn-DUNG)
Shanghai (shang-HYE)
Shang-Yin (SHANG-YIN)
Shantou (SHAHN-TOH)
Shanxi (shahn-SHEE)
Shari (SHAH-ree)
Sharia (SHAH-ree-uh)
Shatt al Arab (shot ahl uh-RAHB)
Sheikdom (SHAKE-dum)
Shenyang (shun-YAHNG)
Shenzhen (shun-ZHEN)
Shevardnadze, Eduard (sheh-vart-NAHD-zeh, ED-wahrd)

Shi (SHERR)
Shia (SHEE-uh)
Shi'a (SHEE-uh)
Shihezi (shee-HAY-zee)
Shi'ism (SHEE-izm)
Shi'ite (SHEE-ite)
Shikoku (shick-KOH-koo)
Shikotan (shee-koh-TAHN)
Shimonoseki (shim-uh-noh-SECKEE)
Shiraz (shih-RAHZ)
Shirioshi (shih-ree-OH-shee)
Shogun (SHOH-goon)
Shona (SHOH-nuh)
Shostakovich (shosta-KOH-vitch)
Shwedogon (SHWAY-duh-gonn)
Siam (sye-AMM)
Siberia (sye-BEERY-uh)
Sichuan (zeh-CHWAHN)
Sicily (SISS-uh-lee)
Siddhartha (sid-DAHR-tuh)
Sidra (SIDD-ruh)
Siem Reap (SERM-REEP)
Sierra Madre Occidental (see-ERRA MAH-dray oak-see-den-TAHL)
Sierra Madre Oriental (see-ERRA MAH-dray aw-ree-en-TAHL)
Sierra Maestra (see-ERRA mah-AY-strah)
Sikh (SEEK)
Sikhism (SEEK-izm)
Siking (see-KING)
Sikkim (SICK-um)
Silesia (sye-LEE-zhuh)
Sinai (SYE-nye)
Singapore (SING-uh-poar)
Sinhala (sin-HAHLA)
Sinhalese (sin-hah-LEEZE)
Sinicization (sine-ih-sye-ZAY-shun)
Sinicized (SYE-nuh-sized)
Sinuiji (seen-WEE-joo)
Sioux (SOO)
Sisal (SYE-sull)
Sitka (SIT-kuh)
Sivas (SEE-vahss)
Sjaelland (ZEE-lund)
Skagerrak (SKAG-uh-rak)
Skopje (SKAWP-yay)
Slamet (SLAH-met)
Slav (SLAHV)
Slavic (SLAH-vick)
Slovakia (sloh-VAH-kee-uh)
Slovenia (sloh-VEE-nee-uh)
Sofala (soh-FAH-luh)
Sofia (SOH-fee-uh)
Somali (suh-MAH-lee)
Somalia (suh-MAHL-yuh)
Somoza (suh-MOH-zah)
Song Qiang (SAWNG chee-AHNG)
Songhai (SAWNG-hye)
Songhua (SAWNG-hwah)
Sonora (suh-NORA)

Sotho (SOO-too)
Soufrière (soo-free-AIR)
Sovkhoz (SOV-koze)
Soweto (suh-WETTO)
Spatial (SPAY-shull)
Spratly (SPRAT-lee)
Spykman (SPIKE-mun)
Sri Lanka (sree-LAHNG-kuh)
Srinagar (srih-NUG-arr)
Srpska (SERP-skuh)
St. Eustatius (saint yoo-STAY-shuss)
St. Louis (sah-loo-EE)
St. Lucia (saint LOO-shuh)
St. Maarten (sint MAHRT-un)
Stalin (STAH-lin)
Stamboul (STAHM-bool)
Stanovoy (stuh-nah-VOY)
Steppe (STEP)
Stockholm (STOCK-hoam)
Strasbourg (STRAHSS-boorg)
Stupa (STOO-puh)
Stuttgart (SHTOOT-gart)
Sudan (soo-DAN)
Sudanese (soo-duh-NEEZE)
Sudd (SOOD)
Sudeten (soo-DAYTEN)
Suez (SOO-ez)
Suharto (soo-HAHR-toh)
Sui (SWAY)
Sukarno (soo-KAR-noh)
Sukarnoputri, Megawati (soo-kar-noh-POO-tree, mega-WATTY)
Sulawesi (soo-luh-WAY-see)
Suleyman (SOO-lay-mahn)
Sultan (SULL-tun)
Sultanate (SULL-tuh-nut)
Sulu (SOO-loo)
Sumatera (suh-MAH-tuh-ruh)
Sumatra (suh-MAH-truh)
Sumbawa (soom-BAH-wuh)
Sumer (SOO-mer)
Sun Yat-sen (SOON yaht-SENN)
Sunda [nese] (SOON-duh [NEEZ])
Sung (SOONG)
Sunni (SOO-nee)
Supsa (SOOP-suh)
Suriname (soor-uh-NAHM-uh)
Suryavarman (soory-AHVA-mun)
Susten (SOOS-ten)
Sutlej (SUTT-ledge)
Suva (SOO-vuh)
Suzhou (SOO-JOH)
Swahili (swah-HEE-lee)
Swazi [land] (SWAH-zee [land])
Syktyvkar (sik-tiff-KAR)
Syr-Darya (seer-DAHR-yuh)
Syria (SEARY-uh)
Szczecin (SHCHE-tseen)
Tabasco (tuh-BAH-skoh)
Tagalog (tuh-GAH-log)
Tahiti (tuh-HEET-tee)

Taiga   (TYE-guh)
Taipa   (TYE-pah)
Taipei [Taibei]   (tye-BAY)
Taiwan   (tye-WAHN)
Tajik   (TUDGE-ick)
Tajikistan   (tah-JEEK-ih-stahn)
Takla Makan   (tahk-luh-muh-KAHN)
Taliban   (TALLA-ban)
Tallinn   (TALLEN)
Tamaulipas   (tah-mau-LEEPUS)
Tamil [Nadu]   (TAMMLE [NAH-doo])
Tampere   (TAHM-puh-ray)
Tampico   (tam-PEEK-oh)
Tana   (TAHNA)
Tang   (TAHNG)
Tanganyika   (tan-gun-YEEKA)
Tangerang   (tahn-guh-RAHNG)
Tangshan   (tahng-SHAHN)
Tannu Tuva   (tan-ooh-TOO-vuh)
Tanzania   (tan-zuh-NEE-uh)
Tao   (DAU)
Tarim   (TAH-REEM)
Tashkent (Toshkent)   (tahsh-KENT)
Tasman   (TAZZ-mun)
Tasmania   (tazz-MAY-nee-uh)
Tatar   (TAHT-uh)
Tatarstan   (TAHT-uh-STAHN)
Tatra   (TAHT-truh)
Taxco   (TAHSS-koh)
Tayshet   (tye-SHET)
Tbilisi   (tuh-BILL-uh-see)
Tchaikovsky   (chye-KOFF-skee)
Teatro Colón   (tay-AH-troh koh-LOAN)
Tecnópolis   (tek-NOH-poh-leece)
Teda   (TAY-duh)
Tegucigalpa   (tuh-goose-ih-GAHL-puh)
Tehran   (tay-uh-RAHN)
Tehuantepec   (tuh-WHAHN-tuh-pek)
Tel Aviv-Jaffa   (tella-VEEVE-JOFF-uh)
Telugu   (TELLOO-goo)
Tema   (TAY-muh)
Templada   (tem-PLAH-dah)
Tengah   (TENG-gah)
Tengiz   (TEN-ghiz)
Tenochtitlán   (tay-noh-chit-LAHN)
Teotihuacán   (tay-uh-tee-wah-KAHN)
Terai   (teh-RYE)
Terek   (TEH-rek)
Terengganu   (teh-reng-GAH-noo)
Tertiary   (TER-shuh-ree)
Tete   (TATE-uh)
Thailand   (TYE-land)
Thames   (TEMZ)
Thamesmead   (TEMZ-meed)
Thebes   (THEEBZ)
Theocracy   (thee-OCK-ruh-see)
Thessaloniki   (thess-uh-luh-NEE-kee)
Thimphu   (thim-POOH)
Thünian   (TOO-nee-un)

Tian Shan   (TYAHN SHAHN)
Tiananmen   (TYAHN-un-men)
Tianjin   (tyahn-JEEN)
Tiber   (TYE-ber)
Tibet   (tuh-BETT)
Tierra caliente   (tee-ERRA kahl-YEN-tay)
Tierra del Fuego   (tee-ERRA dale FWAY-goh)
Tierra fría   (tee-ERRA FREE-uh)
Tierra helada   (tee-ERRA ay-LAH-dah)
Tierra nevada   (tee-ERRA neh-VAH-dah)
Tierra templada   (tee-ERRA tem-PLAH-dah)
Tigrean   (tih-GRAY-un)
Tigris   (TYE-gruss)
Tijuana   (tee-WHAHN-uh)
Tikhvin   (TIK-vun)
Tikrit   (tih-KREET)
Tilbury   (TILL-buh-ree)
Timbuktu   (tim-buck-TOO)
Timor   (TEE-moar)
Timur   (TEE-moor)
Tirane   (tih-RAHN-uh)
Titicaca   (tiddy-KAH-kuh)
Tlingit   (TLING-git)
Tobago   (tuh-BAY-goh)
Tocantins   (toke-un-TEENS)
Tokaido   (toh-KYE-doh)
Tokelau   (TOH-kuh-lau)
Tokyo   (TOH-kee-oh)
Toledo   (toh-LAY-doh)
Tolédo, Alejandro   (toh-LAY-doh, ah-lay-HAHN-droh)
Tolstoy   (TAWL-stoy)
Tolyatti   (tawl-YAH-tee)
Tomolon   (TOM-oh-loan)
Tondano   (tonn-DAH-noh)
Tonga   (TAHNG-guh)
Tongan   (TONG-gun)
Tonkin   (TAHN-KIN)
Tonle Sap   (tahn-lay SAP)
Toponym   (TOH-poh-nim)
Toponymy   (toh-PONN-uh-mee)
Tordesillas   (tor-day-SEE-yahss)
Torii   (taw-REE)
Torres   (TOAR-russ)
Torrijos   (tor-REE-hohss)
Toshka   (TOSH-kah)
Toshkent   (tahsh-KENT)
Toyama   (toh-YAH-muh)
Train à grande vitesse   (TRAN ah-grawnd-vee-TESS)
Trajan   (TRAY-junn)
Transcaucasia   (tranz-kaw-KAY-zhuh)
Transdniestra   (tranz-duh-NYESS-truh)
Transhumance   (tranz-hyoo-MANSS)
Transvaal   (TRUNZ-vahl)
Trias Monge, José   (TREE-ahss MON-hay, hoh-ZAY)

Trichardt   (TREE-shart)
Tripoli   (TRIPPA-lee)
Tripolitania   (trip-olla-TANEY-yuh)
Tripura   (TRIP-uh-ruh)
Trondheim   (TRAHN-hame)
Trung Bo   (TROONG BOH)
Trypanosomiasis   (try-pan-noh-soh-MYE-uh-sis)
Tselinograd   (seh-LEENO-grahd)
Tsetse   (TSETT-see)
Tsugaru   (tsoo-GAH-roo)
Tsukuba   (tsoo-KOOB-uh)
Tsumeb   (SOO-meb)
Tsunami   (tsoo-NAH-mee)
Tsushima   (tsoo-SHEE-muh)
Tswana   (TSWAHN-uh)
Tswana   (SWAH-nuh)
Tuareg   (TWAH-regg)
Tubarão   (too-buh-RAWNG)
Tuchueh   (too-CHOO-way)
Tucson   (TOO-sonn)
Tucumán   (too-koo-MAHN)
Tucuruí   (too-koo-roo-EE)
Tufo   (TOO-foh)
Tukiu   (too-KYOO)
Tula   (TOO-luh)
Tun Abdul Razak   (toon ahb-DOOL rah-ZAHK)
Tunis   (TOO-niss)
Tunisia   (too-NEE-zhuh)
Turfan   (TER-fan)
Turin   (TOOR-rin)
Turkana   (ter-KANNA)
Turkestan   (TER-kuh-stahn)
Türkmenbashi   (tyoork-men-BAH-shee)
Turkmenistan   (terk-MEN-uh-stahn)
Turku   (TOOR-koo)
Tuscany   (TUSS-kuh-nee)
Tutsi   (TOOTSIE)
Tuva   (TOO-vuh)
Tuvalu   (too-VAHL-oo)
Tuzla   (TOOZ-lah)
Twa   (TOO-wah)
Tyrol   (tih-ROLL)
Tyumen   (tyoo-MEN)
Tyva   (TOO-vuh)
Tyvinian   (too-VINNY-un)
Ubangi   (oo-BANG-ghee)
Ubundu   (oo-BOON-doo)
Udaipur   (uh-DYE-poor)
Udmurt [iya]   (ood-MOORT [ee-uh])
Ufa   (oo-FAH)
Uganda   (yoo-GAHN-duh/yoo-GANDA)
Uhuru   (oo-HOO-roo)
Ujungpandang   (oo-JUNG PAHN-dahng)
Ukraine   (yoo-CRANE)
Ulaanbaatar   (oo-lahn-BAH-tor)
Ulsan   (OOL-SAHN)
Uluru   (ooh-LOO-roo)

Ulyanov (ool-YAH-noff)
UNCLOS (UNN-klohss)
Ungava (ung-GAH-vuh)
Ura Nippon (OOH-ruh NIP-on)
Urabá (oo-rah-BAH)
Ural (YOOR-ull)
Ural-Altaic (YOOR-ull-al-TAY-ick)
Urdu (OOR-doo)
Uribe, Alvaro (oo-REE-bay, AHL-vah-roh)
Uruguay (OO-rah-gwye)
Ürümqi (oo-ROOM-chee)
Ushuaia (oo-SWAH-yah)
Usinsk (oo-SINSK)
Ussuri (ooh-SOOR-ree)
Ust-Kamenogorsk
Usuli (oo-SOO-lee)
Utrecht (YOO-trekt)
Uttar Pradesh (ootar-pruh-DESH)
Uttaranchal (OO-tahr-rahn-CHARL)
Uyghur (WEE-ghoor)
Uzbek (OOZE-beck)
Uzbekistan (ooze-BECK-ih-stahn)
Uzen (oo-ZEN)
Vaal (VAHL)
Vakhan (wah-KAHN)
Valencia (vuh-LENN-see-uh)
Valle Central (VAH-yay sen-TRAHL)
Valletta (vuh-LEH-tuh)
Valparaíso (vahl-pah-rah-EE-so)
Vanino (VAH-nih-noh)
Vanni (VAH-nee)
Vanuatu (vahn-uh-WAH-too)
Varanasi (vuh-RAHN-uh-see)
Varangian (vuh-RANGE-ee-un)
Veld (VELT)
Veneto (VENN-uh-toh)
Venezuela (veh-neh-SWAY-lah)
Venizelos, Eleftherios (venni-SAY-lohss, ellef-TEER-ree-ohss)
Ventspils (VENT-spilz)
Veracruz (verra-CROOZE)
Verde (VER-dee)
Vereeniging (fuh-REEN-ih-king)
Versailles (vair-SYE)
Viangchan (vyung-CHAHN)
Vienna (vee-ENNA)
Vietnam (vee-et-NAHM)
Vietnamese (vee-et-nuh-MEEZE)
Villahermosa (vee-yuh-air-MOH-suh)
Villa-Lobos, Heitor (vill-ah-LOH-bohss, AY-tohr)
Vilnius (VILL-nee-uss)
Vindhya (VIN-dyuh)
Visayan (vuh-SYE-un)
Vistula (VIST-yulluh)
Vladikavkaz (vlad-uh-kuff-KAHZ)
Vladivostok (vlad-uh-vuh-STAHK)
Vltava (VULL-tuh-vuh)
Vojvodina (VOY-vuh-deena)
Volgograd (VOLL-guh-grahd)

Vologda (VAW-lug-duh)
Volta Redonda (vahl-tuh rih-DONN-duh)
Von Thünen, Johann Heinrich (fon-TOO-nun, YOH-hahn HINE-rish)
Voortrekker (FOR-trecker)
Wabenzi (wah-BENZ-zee)
Wahhabism (wah-HAH-b'izm)
Wahid, Abdurrahman (wah-HEED, ahb-doo-RAH-mun)
Waikiki (wye-kuh-KEE)
Waitangi (WYE-tonggy)
Wakisihu (wah-kee-SEE-hoo)
Walachia (wuh-LAH-kee-yuh)
Wallaby (WALLA-bee)
Wallonia (wah-LOANY-uh)
Walloon (wah-LOON)
Walvis (WAHL-vuss)
Wasatch (WAW-satch)
Weber (VAY-buh)
Wegener (VAY-ghenner)
Wei (WAY)
Wenzhou (whunn-JOH)
Weser (VAY-zuh)
Westermann (VESS-tair-munn)
Westphalia (west-FAIL-yuh)
Wildebeest (WIL-duh-beest)
Willamette (wuh-LAMM-ut)
Windhoek (VINT-hook)
Winnipeg (WIN-uh-peg)
Witbank (WHIT-bank)
Witwatersrand (WITT-waw-terz-rand)
Wolof (WOH-loff)
Wroclaw (VROH-tswahf)
Wuchang (woo-CHAHNG)
Wuhan (woo-HAHN)
Wuzhou (woo-JOH)
Xenophobia (zee-nuh-FOH-bee-uh)
Xhosa (SHAW-suh)
Xi (SHEE)
Xi Jiang (SHEE jee-AHNG)
Xia (SHAH)
Xiamen (shah-MEN)
Xian (shee-AHN)
Xianggang (see-AHNG-gahng)
Xiangkhoang (SHEE-AHNG-kwahng)
Xiaolangdi (SH-YAU-lahng-dee)
Xinjiang (shin-jee-AHNG)
Xizang (sheedz-AHNG)
Yacyretá (yah-see-ray-TAH)
Yakut [sk] (yuh-KOOT [sk])
Yakutiya (yuh-KOOTY-uh)
Yamalo-Nenetskiy (yuh-MAH-luh-nuh-NET-skee)
Yamoussoukro (yahm-uh-SOO-kroh)
Yamuna (YAH-muh-nuh)
Yanan (yen-AHN)
Yanbu (YAN-boo)
Yangon (yahn-KOH)
Yangpu (YAHNG-poo)
Yangzi [Yangtze] (YANG-dzee)

Yanomami (yah-noh-MAH-mee)
Yantai (yahn-TYE)
Yaohan (yau-HAHN)
Yaoundé (yown-DAY)
Yaroslavl (yar-uh-SLAHV-ull)
Yawata (yuh-WAH-tuh)
Yayoi (yah-YOH-ee)
Yeates (YATES)
Yekaterinburg (yeh-KAHTA-rin-berg)
Yeltsin, Boris (YELT-sinn, BAW-reese)
Yemen (YEMMEN)
Yemeni (YEH-meh-nee)
Yenisey (yen-uh-SAY)
Yerba maté (YAIR-bah mah-TAY)
Yerevan (yair-uh-VAHN)
Yerushalayim (yeh-roo-shuh-LYE-im)
Yevreyskaya (yev-RAY-sky-yah)
Yibin (EE-BIN)
Yichang (yee-CHAHNG)
Yi-Luo (YEE-loo-oh)
Yogyakarta (yah-gyuh-KAR-tuh)
Yokohama (yoh-kuh-HAH-muh)
Yongbyon (yong-B'YON)
Yonne (YAHN)
Yoruba (YAH-rooba)
Yuan (YOO-ahn)
Yucatán (yoo-kuh-TAIIN)
Yugoslavia (yoo-goh-SLAH-vee-uh)
Yulara (yoo-LAH-rah)
Yumen (YOO-mun)
Yungas (YOONG-gahss)
Yunnan (yoon-NAHN)
Zagreb (ZAH-grebb)
Zagros (ZAH-gruss)
Zaïre (zah-EAR)
Zambezi (zam-BEEZY)
Zambia (ZAM-bee-uh)
Zamboanga (zahm-boh-AHNG-guh)
Zamfara (zahm-FAHR-rah)
Zanzibar (ZANN-zih-bar)
Zapata, Emiliano (zah-PAH-tah, eh-mee-lee-AH-noh)
Zapatista (zah-pah-TEESTA)
Zedillo, Ernesto (zeh-DEE-yoh, air-NES-toh)
Zeeland (ZAY-lund)
Zhanjiang (JAHN-jee-AHNG)
Zhejiang (JEJ-ee-AHNG)
Zhou (JOH)
Zhou Enlai (JOH en-lye)
Zhuhai (joo-HYE)
Zimbabwe (zim-BAHB-way)
Zionism (ZYE-un-izm)
Zoogeography (ZOH-oh-jee-OG-gruh-fee)
Zoroaster (zorro-AST-tuh)
Zoroastrian (zorro-ASTREE-un)
Zuider Zee (ZYDER ZEE)
Zulu (ZOO-loo)
Zürich (ZOOR-ick)

References and Further Readings

(Bullets [●] denote basic introductory works)

INTRODUCTION

Brunn, Stanley D., Williams, Jack F., & Zeigler, Donald J. *Cities of the World: World Regional Urban Development* (Lanham, Md.: Rowman & Littlefield, 3rd rev. ed., 2003).

Clark, Gordon L., et al., eds. *The Oxford Handbook of Economic Geography* (New York: Oxford University Press, 2000).

Claval, Paul, ed. *Introduction to Regional Geography* (Malden, Mass.: Blackwell, 1998).

● de Blij, H. J. & Murphy, Alexander B. *Human Geography: Culture, Society, and Space* (New York: John Wiley & Sons, 7th rev. ed., 2002).

● de Blij, H. J., Muller, Peter O., & Williams, Richard S., Jr. *Physical Geography: The Global Environment* (New York: Oxford University Press, 3rd rev. ed., 2004).

Douglas, Ian, Huggett, Richard, & Robinson, Mike, eds. *Companion Encyclopedia of Geography: The Environment and Mankind* (London & New York: Routledge, 1996).

Fenneman, Nevin M. "The Circumference of Geography," *Annals of the Association of American Geographers*, 9 (1919): 3–11.

Gaile, Gary L. & Willmott, Cort J., eds. *Geography in America at the Dawn of the Twenty-First Century* (New York: Oxford University Press, 2nd rev. ed., 2004).

● *Goode's World Atlas* (Skokie, Ill.: Rand McNally, 20th rev. ed., 2000).

Grove, Jean M. *The Little Ice Age* (London & New York: Methuen, 1988). Quotation taken from pp. 1–2.

Haggett, Peter. *Geography: A Global Synthesis* (Harlow, UK: Pearson Education, 2001).

Hargittai, Eszter & Centeno, Miguel A., guest eds. "Mapping Globalization," *American Behavioral Scientist*, 44 (June 2001): 1541–1767.

Holt-Jensen, Arild. *Geography: History and Concepts–A Student's Guide* (Thousand Oaks, Cal.: Sage, 3rd rev. ed., trans. Brian Fullerton, 1999).

Jackson, John Brinckerhoff. *Discovering the Vernacular Landscape* (New Haven, Conn.: Yale University Press, 1984).

Johnston, Ron J., et al., eds. *The Dictionary of Human Geography* (Malden, Mass.: Blackwell, 4th rev. ed., 2000).

Jordan, Terry G. "The Concept and Method," in Lich, Glen E., ed., *Regional Studies: The Interplay of Land and People* (College Station, TX: Texas A & M Press, 1992), pp. 8–24.

Livingstone, David. *The Geographical Tradition* (Cambridge, Mass.: Blackwell, 1992).

● Muehrcke, Phillip C. & Muehrcke, Juliana C. *Map Use: Reading-Analysis-Interpretation* (Madison, Wisc.: JP Publications, 3rd rev. ed., 1992).

National Research Council. *Rediscovering Geography: New Relevance for Science and Society* (Washington, D.C.: National Academy Press, 1997).

● Pattison, William D. "The Four Traditions of Geography," *Journal of Geography*, 63 (1964): 211–216.

Rogers, Alisdair & Viles, Heather A., eds. *The Student's Companion to Geography* (Malden, Mass.: Blackwell, 2nd rev. ed., 2003).

Sauer, Carl Ortwin. "Cultural Geography," *Encyclopedia of the Social Sciences*, Vol. 6 (New York: Macmillan, 1931), pp. 621–623.

Sheppard, Eric & Barnes, Trevor, eds. *A Companion to Economic Geography* (Malden, Mass.: Blackwell, 2000).

Wegener, Alfred. *The Origin of Continents and Oceans* (New York: Dover, reprint of the 1915 original, trans. John Biram, 1966).

● Wheeler, James O., Muller, Peter O., Thrall, Grant I., & Fik, Timothy J. *Economic Geography* (New York: John Wiley & Sons, 3rd rev. ed., 1998).

Whittlesey, Derwent S., et al. "The Regional Concept and the Regional Method," in James, Preston E. & Jones, Clarence F., eds., *American Geography: Inventory and Prospect* (Syracuse, N.Y.: Syracuse University Press, 1954), pp. 19–68.

CHAPTER 1

Aslund, Anders. *Building Capitalism: The Transformation of the Former Soviet Bloc* (New York: Cambridge University Press, 2001).

Baldersheim, Harald & Stahlberg, Krister, eds. *Nordic Region-Building in a European Perspective* (Brookfield, Vt.: Ashgate, 1999).

- Berentsen, William H., ed. *Contemporary Europe: A Geographic Analysis* (New York: John Wiley & Sons [7th rev. ed. of George W. Hoffman, *Europe in the 1990s: A Geographic Analysis*], 1997).

Bernstein, Richard. "An Aging Europe May Find Itself on the Sidelines," *New York Times,* June 29, 2003, 3.

Blacksell, Mark & Williams, Allan M., eds. *The European Challenge: Geography and Development in the European Community* (New York: Oxford University Press, 1994).

Bruni, Frank. "Persistent Drop in Fertility Reshapes Europe's Future," *New York Times*, December 26, 2002, A1, A10.

Burtenshaw, David, et al. *The European City: A Western Perspective* (New York: John Wiley & Sons, 1991).

Butlin, Robin A. & Dodgshon, R. A. *An Historical Geography of Europe* (New York: Oxford University Press, 1999).

Carter, Frank W. & Turnock, David. *Environmental Problems in Eastern Europe* (London & New York: Routledge, 2nd rev. ed., 1997).

Champion, Tony, et al. *The New Regional Map of Europe* (Amsterdam & New York: Elsevier, 1996).

Chisholm, Michael. *Rural Settlement and Land Use: An Essay in Location* (London: Hutchinson University Library, 3rd rev. ed., 1979).

- Clout, Hugh D., et al. *Western Europe: Geographical Perspectives* (New York: Wiley/Longman, 3rd rev. ed., 1994).

- Cole, John P. & Cole, Francis. *A Geography of the European Union* (London & New York: Routledge, 2nd rev. ed., 1997).

Delamaide, Darrell. *The New Superregions of Europe* (New York: Dutton, 1994).

- Diem, Aubrey. *Western Europe: A Geographical Analysis* (New York: John Wiley & Sons, 1979).

Embleton, Clifford, ed. *Geomorphology of Europe* (New York: Wiley-Interscience, 1984).

Emerson, Michael. *Redrawing the Map of Europe* (New York: St. Martin's Press, 1998).

Fernandez-Armesto, Felipe, ed. *The Times Guide to the Peoples of Europe* (Boulder, Colo.: Westview Press, 1995).

Glebe, Günther & O'Loughlin, John, eds. *Foreign Minorities in Continental European Cities* (Wiesbaden, West Germany: Franz Steiner Verlag, 1987).

- Gottmann, Jean. *A Geography of Europe* (New York: Holt, Rinehart & Winston, 4th rev. ed., 1969).

Graham, Brian, ed. *Modern Europe: Place, Culture, and Identity* (New York: Oxford University Press, 1998).

Grove, Alfred T. & Rackham, Oliver. *The Nature of Mediterranean Europe: An Ecological History* (New Haven, Conn.: Yale University Press, 2001).

Guttman, Robert J., ed. *Europe in the New Century: Visions of an Emerging Superpower* (Boulder, Colo.: Lynne Rienner, 2001).

Haggett, Peter. *Locational Analysis in Human Geography* (London: Edward Arnold, 1965). Definition from p. 19.

Hall, Derek & Danta, Darrick, eds. *Reconstructing the Balkans: A Geography of the New Southeast Europe* (New York: John Wiley & Sons, 1996).

Hall, Ray & White, Paul, eds. *Europe's Population: Towards the Next Century* (London: UCL Press/Taylor & Francis, 1995).

"Happy Family? A Survey of the Nordic Countries," *The Economist*, January 23, 1999, special insert, 16 pp.

Heffernan, Michael. *The Meaning of Europe: Geography and Geopolitics* (New York: Oxford University Press, 1998).

Heffernan, Michael. *Twentieth-Century Europe: A Political Geography* (New York: John Wiley & Sons, 1997).

Hoffman, Eva. *Exit into History: A Journey Through the New Eastern Europe* (New York: Viking Press, 1993).

Hoggart, Keith, et al. *Rural Europe: Identity and Change* (New York: John Wiley & Sons, 1995).

Hudson, Ray & Williams, Allan M., eds. *Divided Europe: Society and Territory* (Thousand Oaks, Cal.: Sage, 1998).

Hunter, Shireen T., ed. *Islam, Europe's Second Religion: The New Social, Cultural, and Political Landscape* (Westport, Conn.: Praeger, 2002).

Hupchick, Dennis P. & Cox, Harold E., eds. *A Concise Historical Atlas of Eastern Europe* (New York: St. Martin's Press, 1996).

Jefferson, Mark. "The Law of the Primate City," *Geographical Review*, 29 (1939): 226–232.

Johnson, Lonnie R. *Central Europe: Enemies, Neighbors, Friends* (New York: Oxford University Press, 1996).

Jones, Philip N. & Wild, Trevor. "Regional and Local Variations in the Emerging Economic Landscape of the New German *Länder*," *Applied Geography*, 17 (1997): 283–299.

Jönsson, Christer, Tägil, Sven, & Törnqvist, Gunnar. *Organizing European Space* (Thousand Oaks, Calif.: Sage, 2000).

- Jordan-Bychkov, Terry G. & Bychkova Jordan, Bella. *The European Culture Area: A Systematic Geography* (Lanham, Md.: Rowman & Littlefield, 4th rev. ed., 2001).

Keating, Michael. *Nations Against the State: The New Politics of Nationalism in Quebec, Catalonia, and Scotland* (New York: Macmillan, 1996).

Keating, Michael. *The New Regionalism in Western Europe: Territorial Restructuring and Political Change* (Northampton, Mass.: Edward Elgar, 1998).

King, Russel. *Mass Migration in Europe: The Legacy and the Future* (New York: John Wiley & Sons, 1995).

Kucera, Tomas, et al., eds. *New Demographic Faces of Europe* (New York: Springer Verlag, 2000)

Kürti, Laszlo & Langman, Juliet, eds. *Beyond Borders: Remaking Cultural Identities in the New East and Central Europe* (Boulder, Colo.: Westview Press, 1997).

Livi-Bacci, Massimo. *The Population of Europe* (Malden, Mass.: Blackwell, trans. Carl Ipsen, 2000).

Matvejevic, Predag. *The Mediterranean: A Cultural Landscape* (Berkeley, Cal.: University of California Press, trans. Michael H. Heim, 1999).

McCarthy, Linda & Danta, Darrick. "Cities of Europe," in Brunn, Stanley D., Williams, Jack F., & Zeigler, Donald J. *Cities of the World: World Regional Urban Development* (Lanham, Md.: Rowman & Littlefield, 3rd rev. ed., 2003), pp. 176–221.

● McDonald, James R. *The European Scene: A Geographic Perspective* (Upper Saddle River, N.J.: Prentice-Hall, 1997).

Mortimer, Edward & Fine, Robert, eds. *People, Nation and State: The Meaning of Ethnicity and Nationalism* (London: I.B. Tauris, 1999).

Murphy, Alexander B. "Emerging Regional Linkages Within the European Community: Challenging the Dominance of the State," *Tijdschrift voor Economische en Sociale Geografie*, 84 (1993): 103–118.

Murphy, Alexander B. "Rethinking the Concept of European Identity," in Herb, Guntram H. & Kaplan, David H., eds., *Nested Identities: Nationalism, Territory, and Scale* (Lanham, Md.: Rowman & Littlefield, 1999), pp. 53–73.

O'Dowd, Liam & Wilson, Thomas M., eds. *Borders, Nations and States: Frontiers of Sovereignty in the New Europe* (Brookfield, Vt.: Ashgate, 1996).

Ohmae, Kenichi. *The End of the Nation-State: The Rise of Regional Economies* (New York: Free Press, 1996).

Ohmae, Kenichi. "The Rise of the Region State," *Foreign Affairs*, Spring 1993, 78–87.

Petrakos, George, ed. *Integration and Transition in Europe: Economic Geography of Interaction* (London & New York: Routledge, 2000).

Pinder, David, ed. *The New Europe: Economy, Society and Environment* (Chichester, UK & New York: John Wiley & Sons, 1998).

Pond, Elizabeth. *The Rebirth of Europe* (Washington, DC: Brookings Institution Press, 1999).

"Poverty in Eastern Europe: The Land That Time Forgot," *The Economist*, September 23, 2000, 27–30.

Rhodes, Martin, ed. *Regions and the New Europe: Patterns in Core and Periphery Development* (Manchester, UK: Manchester University Press, 1996).

● Shaw, Denis J. B., ed. *The Post-Soviet Republics: A Systematic Geography* (New York: John Wiley & Sons/ Longman, 1995).

Townsend, Alan R. *Making a Living in Europe: Human Geographies of Economic Change* (London & New York: Routledge, 1997).

Turnock, David, ed. *East Central Europe and the Former Soviet Union: Environment and Society* (London: Arnold, 2001).

Turnock, David. *The Human Geography of East-Central Europe* (London & New York: Routledge, 2003).

Unwin, Tim, ed. *A European Geography* (Harlow, UK & New York: Longman, 1998).

Wheeler, James O., Muller, Peter O., Thrall, Grant I., & Fik, Timothy J. *Economic Geography* (New York: John Wiley & Sons, 3rd rev. ed., 1998), Chapter 13.

Williams, Allan M. *The West European Economy: A Geography of Post-War Development* (Savage, Md.: Rowman & Littlefield, 1988).

Wintle, Michael, ed. *Culture and Identity in Europe: Perceptions of Divergence and Unity in Past and Present* (Brookfield, Vt.: Ashgate, 1996).

CHAPTER 2

"A Caspian Gamble: A Survey of Central Asia," *The Economist*, February 7, 1998, special insert, 18 pp.

Aslund, Anders. *Building Capitalism: The Transformation of the Former Soviet Bloc* (New York: Cambridge University Press, 2001).

● Bater, James H. *Russia and the Post-Soviet Scene: A Geographical Perspective* (New York: John Wiley & Sons, 1996).

Brawer, Moshe. *Atlas of Russia and the Independent Republics* (New York: Simon & Schuster, 1994).

● Brown, Archie, et al., eds. *The Cambridge Encyclopedia of Russia and the Former Soviet Union* (New York: Cambridge University Press, 1994).

Brooke, James. "Russia's Latest Oil and Gas Oasis: Sakhalin Fields Have Japan, China, and Korea Lining Up," *New York Times*, May 13, 2003, W1, W7.

Bychkova Jordan, Bella & Jordan-Bychkov, Terry G. *Siberian Village: Land and Life in the Sakha Republic* (Minneapolis: University of Minnesota Press, 2001).

Chew, Allen F. *Atlas of Russian History: Eleven Centuries of Changing Borders* (New Haven, Conn.: Yale University Press, 1967).

Chinn, Jeff & Kaiser, Robert. *Russians as the New Minority: Ethnicity and Nationalism in the Soviet Successor States* (Boulder, Colo.: Westview Press, 1996).

Davis, Sue. *The Russian Far East: The Last Frontier* (London & New York: Routledge, 2002).

Demko, George J., et al., eds. *Population Under Duress: The Geodemography of Post-Soviet Russia* (Boulder, Colo.: Westview Press, 1998).

Dmitrieva, O. *Regional Development: The U.S.S.R. and After* (London: University College London Press, 1996).

Duncan, W. Raymond & Holman, G. Paul, Jr., eds. *Ethnic Nationalism and Regional Conflict: The Former Soviet Union and Yugoslavia* (Boulder, Colo.: Westview Press, 1994).

Dunlop, John B. *The Rise of Russia and the Fall of the Soviet Empire* (Princeton, N.J.: Princeton University Press, 1993).

Engelmann, Kurt E. & Pavlakovic, Vjeran, eds. *Rural Development in Eurasia and the Middle East: Land Reform, Demographic Change, and Environmental Constraints* (Seattle: University of Washington Press, 2001).

Forsyth, James. *A History of the Peoples of Siberia: Russia's North Asian Colony, 1581–1990* (New York: Cambridge University Press, 1992).

Gachechiladze, Revaz. *The New Georgia: Space, Society, Politics* (College Station, Tex.: Texas A&M University Press, 1995).

Hanson, Philip & Bradshaw, Michael, eds. *Regional Change in Russia* (Northampton, Mass.: Edward Elgar, 2000).

Harris, Chauncy D. "A Geographic Analysis of Non-Russian Minorities in Russia and Its Ethnic Homelands," *Post-Soviet Geography*, 34 (1993): 543–597.

Horensma, Pier. *The Soviet Arctic* (London & New York: Routledge, 1991).

Hosking, Geoffrey. *Russia and the Russians: A History* (Cambridge, Mass.: Belknap/Harvard University Press, 2001).

Howe, G. Melvyn. *The Soviet Union: A Geographical Study* (London & New York: Longman, 2nd rev. ed., 1986).

Hunter, Shireen T. *Transcaucasia in Transition: Nation-Building and Conflict* (Boulder, Colo.: Westview Press, 1994).

Huttenbach, Henry. *The Caucasus: A Region in Crisis* (Boulder, Colo.: Westview Press, 1996).

Ioffe, Gregory & Nefedova, Tatyana. *Continuity and Change in Rural Russia* (Boulder, Colo.: Westview Press, 1997).

Kaiser, Robert J. *The Geography of Nationalism in Russia and the U.S.S.R.* (Princeton, N.J.: Princeton University Press, 1994).

Kramer, Mark. *Travels With a Hungry Bear: A Journey to the Russian Heartland* (Boston: Houghton Mifflin, 1996).

Kraus, Michael & Liebowitz, Ronald D., eds. *Russia and Eastern Europe After Communism: The Search for New Political, Economic, and Security Systems* (Boulder, Colo.: Westview Press, 1995).

Lincoln, W. Bruce. *The Conquest of a Continent: Siberia and the Russians* (New York: Random House, 1994).

Lloyd, John. "The Russian Devolution," *The New York Times Magazine*, August 15, 1999, 34–41, 52, 61, 64.

● Lydolph, Paul E. *Geography of the U.S.S.R.* (Elkhart Lake, Wisc.: Misty Valley Publishing, 1990).

Lydolph, Paul E. *Climates of the Soviet Union* (Amsterdam: Elsevier, 1977).

Mackinder, Halford J. *Democratic Ideals and Reality* (New York: Holt, 1919).

Mastyugina, Tatiana & Perepelkin, Lev. *An Ethnic History of Russia: Pre-Revolutionary Times to the Present* (Westport, Conn.: Greenwood Press, 1996).

Mote, Victor L. *Siberia: Worlds Apart* (Boulder, Colo.: Westview Press, 1998).

Nijman, Jan. *The Geopolitics of Power and Conflict: Superpowers in the International System, 1945–1992* (New York: John Wiley & Sons/Belhaven, 1993).

Peterson, D. J. *Troubled Lands: The Legacy of Soviet Environmental Destruction* (Boulder, Colo.: Westview Press, 1993).

Pryde, Philip R., ed. *Environmental Resources and Constraints in the Former Soviet Republics* (Boulder, Colo.: Westview Press, 1995).

Remnick, David. *Lenin's Tomb: The Last Days of the Soviet Empire* (New York: Random House, 1993).

Remnick, David. *Resurrection: The Struggle to Build a New Russia* (New York: Random House, 1997).

● Shaw, Denis J. B. *Russia in the Modern World: A New Geography* (Malden, Mass.: Blackwell, 1999).

● Shaw, Denis J. B., ed. *The Post-Soviet Republics: A Systematic Geography* (New York: John Wiley & Sons/Longman, 1995).

Shlapentokh, Vladimir, et al. *From Submission to Rebellion: The Provinces Versus the Center in Russia* (Boulder, Colo.: Westview Press, 1997).

Smith, Graham. *The Post Soviet States: Mapping the Politics of Transition* (London: Arnold, 1999).

Stewart, John Massey, ed. *The Soviet Environment: Problems, Policies and Politics* (New York: Cambridge University Press, 1992).

● Symons, Leslie, ed. *The Soviet Union: A Systematic Geography* (London & New York: Routledge, 2nd rev. ed., 1990).

"The Caucasus: Where Worlds Collide," *The Economist*, August 19, 2000, 17–19.

Thompson, John M. *Russia and the Soviet Union: An Historical Introduction from the Kievan State to the Present* (Boulder, Colo.: Westview Press, 3rd rev. ed., 1994).

Thornton, Judith & Ziegler, Charles E., eds. *Russia's Far East: A Region at Risk* (Seattle: University of Washington Press, 2002).

Thubron, Colin. *Among the Russians* (New York: HarperCollins, 2000).

Thubron, Colin. *In Siberia* (New York: HarperCollins, 2000).

Tomikel, John & Henderson, Bonnie. *Russia and the Near Abroad* (Elgin, Penn.: Allegheny Press, 2 rev. ed., 2003).

Trenin, Dmitri. *The End of Eurasia: Russia on the Border Between Geopolitics and Globalization* (Washington: Carnegie Endowment for International Peace, 2002).

Turnock, David, ed. *East Central Europe and the Former Soviet Union: Environment and Society* (London: Arnold, 2001).

Wixman, Ronald. *The Peoples of the U.S.S.R.: An Ethnographic Handbook* (Armonk, N.Y.: M. E. Sharpe, 1984).

CHAPTER 3

"A Survey of Canada," *The Economist*, July 24, 1999, special insert, 18 pp.

Adams, John S. "Residential Structure of Midwestern Cities," *Annals of the Association of American Geographers*, 60 (1970): 37–62. Model diagram adapted from p. 56.

Agnew, John A. & Smith, Jonathan M., eds. *American Space/American Place: Geographies of the Contemporary United States* (London & New York: Routledge, 2002).

Allen, James P. & Turner, Eugene J. *We the People: An Atlas of America's Ethnic Diversity* (New York: Macmillan, 1987).

Atlas of North America: Space Age Portrait of a Continent (Washington, D.C.: National Geographic Society, 1985).

● Atwood, Wallace W. *The Physiographic Provinces of North America* (New York: Ginn, 1940).

Berry, Brian J. L. "The Decline of the Aging Metropolis: Cultural Bases and Social Process," in Sternlieb, George & Hughes, James W., eds., *Post-Industrial America: Metropolitan Decline and Inter-Regional Job Shifts* (New Brunswick, N.J.: Center for Urban Policy Research, Rutgers University, 1975), pp. 175–185.

● Birdsall, Stephen S., Florin, John W., & Price, Margo L. *Regional Landscapes of the United States and Canada* (New York: John Wiley & Sons, 5th rev. ed., 1999).

● Boal, Frederick W. & Royle, Stephen A., eds. *North America: A Geographical Mosaic* (London: Arnold, 1999).

● Bone, Robert M. *The Regional Geography of Canada* (New York: Oxford University Press, 2 rev. ed., 2002).

Bone, Robert M. *The Geography of the Canadian North: Issues and Challenges* (Toronto: Oxford University Press, 1992). Quotation taken from p. 12.

Borchert, John R. "American Metropolitan Evolution," *Geographical Review*, 57 (1967): 301–332.

Borchert, John R. "Futures of American Cities," in Hart, John Fraser, ed., *Our Changing Cities* (Baltimore, Md.: Johns Hopkins University Press, 1991), pp. 218–250.

Brewer, Cynthia A. & Suchan, Trudy A. *Mapping Census 2000: The Geography of U.S. Diversity* (Redlands, Cal.: ESRI Press, 2001).

Britton, John N. H., ed. *Canada and the Global Economy: The Geography of Structural and Technological Change* (Montreal: McGill-Queen's University Press, 1996).

Castells, Manuel & Hall, Peter. *Technopoles of the World: The Making of Twenty-First-Century Industrial Complexes* (London & New York: Routledge, 1994).

Conrad, Margaret & Hiller, James. *Atlantic Canada: A Region in the Making* (New York: Oxford University Press, 2001).

DePalma, Anthony. *Here: A Biography of the North American Continent* (New York: Public Affairs, 2001).

Garreau, Joel. *Edge City: Life on the New Frontier* (New York: Doubleday, 1991).

Garreau, Joel. *The Nine Nations of North America* (Boston: Houghton Mifflin, 1981).

Gastil, Raymond D. *Cultural Regions of the United States* (Seattle: University of Washington Press, 1975).

Gaustad, Edwin Scott & Barlow, Philip L. *New Historical Atlas of Religion in America* (New York: Oxford University Press, 2001).

● Getis, Arthur & Getis, Judith, eds. *The United States and Canada: The Land and the People* (Dubuque, Iowa: Wm. C. Brown, 1995).

Gober, Patricia. "Americans on the Move," *Population Bulletin*, 48 (November 1993): 1–40.

Gottmann, Jean. *Megalopolis: The Urbanized Northeastern Seaboard of the United States* (New York: Twentieth Century Fund, 1961).

Harris, R. Cole. "Regionalism and the Canadian Archipelago," in McCann, Lawrence D., ed., *Heartland and Hinterland: A Geography of Canada* (Scarborough, Ont.: Prentice-Hall Canada, 1982), pp. 458–484.

Hartshorn, Truman A. *Interpreting the City: An Urban Geography* (New York: John Wiley & Sons, 2nd rev. ed., 1992).

Hart, John Fraser. *The Land That Feeds Us* (New York: W.W. Norton, 1991).

Hartshorn, Truman A. "The Changed South, 1947–1997," *Southeastern Geographer*, 37 (November 1997): 122–139.

Harvey, Thomas. "The Changing Face of the Pacific Northwest," *Journal of the West*, 37 (July 1998): 22–32.

Historical Atlas of the United States (Washington, D.C.: National Geographic Society, centennial ed., 1988).

Hodge, Gerald & Robinson, Ira M. *Planning Canadian Regions* (Vancouver: University of British Columbia Press, 2001).

● Hudson, John C. *Across This Land: A Regional Geography of the United States and Canada* (Baltimore: Johns Hopkins University Press, 2002).

● Hunt, Charles B. *Natural Regions of the United States and Canada* (San Francisco: W. H. Freeman, 2nd rev. ed., 1974).

Janelle, Donald G., ed. *Geographical Snapshots of North America* (New York: Guilford Press, 1992).

Kotkin, Joel. *The New Geography: How the Digital Revolution is Reshaping the American Landscape* (New York: Random House, 2000).

Krauss, Clifford. "Quebec Seeking to End Its Old Cultural Divide," *New York Times*, April 13, 2003, A6.

Lemco, Jonathan. *Turmoil in the Peaceable Kingdom: The Quebec Sovereignty Movement and Its Implications for Canada and the United States* (Toronto: University of Toronto Press, 1994).

Magocsi, Paul Robert, ed. *Encyclopedia of Canada's Peoples* (Toronto: University of Toronto Press, 1998).

Martin, Philip & Midgley, Elizabeth. "Immigration: Shaping and Reshaping America," *Population Bulletin,* 58 (June 2003): 1–44.

Martin, Philip & Widgren, Jonas. "International Migration: Facing the Challenge," *Population Bulletin,* 57 (March 2002): 1–40.

● McCann, Lawrence D. & Gunn, Angus, eds. *Heartland and Hinterland: A Regional Geography of Canada* (Scarborough, Ont.: Prentice Hall Canada, 3 rev. ed., 1998).

McIlwraith, Thomas F. & Muller, Edward K., eds. *North America: The Historical Geography of a Changing Continent* (Lanham, Md.: Rowman & Littlefield, 2nd rev. ed., 2001).

McKee, Jesse O., ed. *Ethnicity in Contemporary America: A Geographical Appraisal* (Lanham, Md.: Rowman & Littlefield, 2nd rev. ed., 2000).

● McKnight, Tom L. *Regional Geography of the United States and Canada* (Upper Saddle River, N.J.: Prentice-Hall, 4th rev. ed., 2004).

Meinig, Donald W. *The Shaping of America: A Geographical Perspective on 500 Years of History; Vol. 1: Atlantic America, 1492–1800* (New Haven, Conn.: Yale University Press, 1986).

Meinig, Donald W. *The Shaping of America: A Geographical Perspective on 500 Years of History; Vol. 2: Continental America, 1800–1867* (New Haven, Conn.: Yale University Press, 1993).

Meinig, Donald W. *The Shaping of America: A Geographical Perspective on 500 Years of History. Vol. 3: Transcontinental America, 1850–1915* (New Haven, Conn.: Yale University Press, 1998).

Muller, Peter O. *Contemporary Suburban America* (Englewood Cliffs, N.J.: Prentice-Hall, 1981).

Ohmae, Kenichi. *The End of the Nation-State: The Rise of Regional Economies* (New York: Free Press, 1996).

Orme, Antony, ed. *The Physical Geography of North America* (New York: Oxford University Press, 2002).

Pacione, Michael. *Urban Geography: A Global Perspective* (London & New York: Routledge, 2001).

Pollard, Kelvin M. & O'Hare, William P. "America's Racial and Ethnic Minorities," *Population Bulletin,* 54 (September 1999): 1–48.

● Rooney, John F., Jr., et al., eds. *This Remarkable Continent: An Atlas of United States and Canadian Society and Cultures* (College Station, Tex.: Texas A&M University Press, 1982).

Shelley, Fred M., et al. *Political Geography of the United States* (New York: Guilford, 1996).

"The New Geography of the Information Technology Industry," *The Economist,* July 19, 2003, 47–49.

"The New Map of High-Tech: From Billville to Silicon Alley, The 13 Hottest Regions in America," *Wall Street Journal,* November 23, 1999, B1, B12.

Vance, James E., Jr. *This Scene of Man: The Role and Structure of the City in the Geography of Western Civilization* (New York: Harper's College Press, 1977). Urban realms model discussed on pp. 411–416.

Wallace, Iain. *A Geography of the Canadian Economy* (New York: Oxford University Press, 2002).

Ward, David. *Cities and Immigrants: A Geography of Change in Nineteenth-Century America* (New York: Oxford University Press, 1971).

● Warkentin, John. *Canada: A Regional Geography* (Scarborough, Ont.: Prentice Hall Canada, 1997).

"Welcome to Amexia: The Border is Vanishing Before Our Eyes, Creating a New World for All of Us," *Time,* June 11, 2001, 36–79.

Wheeler, James O., Muller, Peter O., Thrall, Grant I., & Fik, Timothy J. *Economic Geography* (New York: John Wiley & Sons, 3rd rev. ed., 1998).

Yeates, Maurice H. *The North American City* (New York: Harper & Row, 4th rev. ed., 1990). Canadian urban evolution model discussed on pp. 60–67.

Yeates, Maurice H. *Main Street: Windsor to Quebec City* (Toronto: Macmillan of Canada, 1975).

Zelinsky, Wilbur. *Exploring the Beloved Country: Geographic Forays into American Society and Culture* (Iowa City, Iowa: University of Iowa Press, 1995).

● Zelinsky, Wilbur. *The Cultural Geography of the United States: A Revised Edition* (Englewood Cliffs, N.J.: Prentice-Hall, 2nd rev. ed, 1992).

CHAPTER 4

"A Silicon Republic . . . Lessons From [Costa Rica's] High-Tech Frontier," *Newsweek,* August 28, 2000, 42–44.

"After the Revolution: A Survey of Mexico," *The Economist,* October 28, 2000, special supplement, 16 pp.

Alleyne, Mervyn C. *The Construction and Representation of Race and Ethnicity in the Caribbean* (Kingston, Jamaica: University of the West Indies Press. 2002).

Arreola, Daniel D. & Curtis, James R. *The Mexican Border Cities: Landscape Anatomy and Place Personality* (Tucson: University of Arizona Press, 1993).

Augelli, John P. "The Rimland-Mainland Concept of Culture Areas in Middle America," *Annals of the Association of American Geographers,* 52 (1962): 119–129.

Barker, David, et al., eds. *A Reader in Caribbean Geography* (Kingston, Jamaica: Ian Randle, 1998).

Barton, Jonathan R. *A Political Geography of Latin America* (London & New York: Routledge, 1997).

● Blakemore, Harold & Smith, Clifford T., eds. *Latin America: Geographical Perspectives* (London & New York: Methuen, 2nd rev. ed., 1983).

● Blouet, Brian W. & Blouet, Olwyn M. *Latin America and the Caribbean: A Systematic and Regional Survey* (New York: John Wiley & Sons, 4th rev. ed., 2002).

Boswell, Thomas D. & Conway, Dennis. *The Caribbean Islands: Endless Geographical Diversity* (New Brunswick, N.J.: Rutgers University Press, 1992).

Brea, Jorge A. "Population Dynamics in Latin America," *Population Bulletin*, 58 (March 2003): 1–36.

Butler, Edgar W., et al. *Mexico and Mexico City in the World Economy* (Boulder, Colo.: Westview, 2001).

● Clawson, David L. *Latin America and the Caribbean: Lands and Peoples* (Dubuque, Iowa: WCB/McGraw- Hill, 2nd rev. ed., 2000).

Collier, Simon, et al., eds. *The Cambridge Encyclopedia of Latin America and the Caribbean* (New York: Cambridge University Press, 2nd rev. ed., 1992).

Davidson, William V. & Parsons, James J., eds. *Historical Geography of Latin America* (Baton Rouge, La.: Louisiana State University Press, 1980).

Elbow, Gary S. "Regional Cooperation in the Caribbean: The Association of Caribbean States," *Journal of Geography*, 96 (January/February 1997): 13–22.

Forero, Juan. "As China Gallops, Mexico Sees Factory Jobs Slip Away," *New York Times,* September 3, 2003, A 3.

Galloway, Jock H. "The Lesser Antilles," in Donald G. Janelle, ed., *Geographical Snapshots of North America* (New York: Guilford, 1992), pp. 16–19.

Gilbert, Alan. *Latin America* (London & New York: Routledge, 1990).

Gilbert, Alan, ed. *The Mega-City in Latin America* (New York & Tokyo: United Nations University Press, 1996).

Goodbody, Ivan & Hope, Elizabeth T., eds. *Natural Resource Management for Sustainable Development in the Caribbean* (Kingston, Jamaica: University of the West Indies Press. 2001).

Greenfield, Gerald M., ed. *Latin American Urbanization: Historical Profiles of Major Cities* (Westport, Conn.: Greenwood Press, 1994).

Griffin, Ernst C. & Ford, Larry R. "Cities of Latin America," in Brunn, Stanley D. & Williams, Jack F., eds., *Cities of the World: World Regional Urban Development* (New York: HarperCollins, 2nd rev. ed., 1993), pp. 224–265.

Gwynne, Robert N. & Kay, Cristobal. *Latin America Transformed: Globalization and Modernity* (New York: Oxford University Press, 1999).

Hall, Carolyn & Brignoli, Hector Perez. *Historical Atlas of Central America* (Norman, Okla.: University of Oklahoma Press, 2003).

Herzog, Lawrence A. *From Aztec to High-Tech: Architecture and Landscape Across the Mexico-United States Border* (Baltimore: Johns Hopkins University Press, 1999).

Hillman, Richard S. & D'Agostino, Thomas J., eds. *Understanding the Contemporary Caribbean* (Boulder, Colo.: Lynne Rienner, 2003).

● James, Preston E. & Minkel, Clarence W. *Latin America* (New York: John Wiley & Sons, 5th rev. ed., 1986), Chaps. 2–16.

Knapp, Gregory, ed. *Latin America in the Twenty-First Century: Challenges and Solutions* (Austin, Tex.: University of Texas Press, 2002).

Kopinak, Kathryn. *Desert Capitalism: Maquiladoras in North America's Western Industrial Corridor* (Tucson: University of Arizona Press, 1996).

Lewis, Patsy. *Surviving Small Size: Regional Integration in Caribbean Ministates* (Kingston, Jamaica: University of the West Indies Press, 2002).

Lowenthal, David. *West Indian Societies* (New York: Oxford University Press, 1972).

MacLachlan, Ian & Aguilar, Adrian G. "Maquiladora Myths: Locational and Structural Change in Mexico's Export Manufacturing Industry," *The Professional Geographer*, 50 (August 1998): 315–323.

Malkin, Elisabeth. "Manufacturing Jobs are Exiting Mexico: Business Leaders Try to Stop the Exodus of Factories to China," *New York Times*, November 5, 2002, W1, W7.

McCullough, David G. *The Path Between the Seas: The Creation of the Panama Canal, 1870–1914* (New York: Simon & Schuster, 1977).

"Mexico's Indians: One Nation, or Many?," *The Economist*, January 20, 2001, 33–34.

Myers, Norman. *The Primary Source: Tropical Forests and Our Future* (New York: W. W. Norton, 1984).

Payne, Anthony & Sutton, Paul. *Charting Caribbean Development* (Gainesville, FL: University Press of Florida, 2001).

Pick, James B. & Butler, Edgar W. *Mexico Megacity* (Boulder, Colo.: Westview Press, 1997).

Portes, Alejandro, et al., eds. *The Urban Caribbean: Transition to a New Global Economy* (Baltimore: Johns Hopkins University Press, 1997).

Potter, Robert B. *The Urban Caribbean in an Era of Global Change* (Brookfield, Vt.: Ashgate, 2000).

● Preston, David A., ed. *Latin American Development: Geographical Perspectives* (London & New York: Longman, 2nd rev. ed., 1995).

● Richardson, Bonham C. *The Caribbean in the Wider World, 1492–1992: A Regional Geography* (New York: Cambridge University Press, 1992).

Rohter, Larry. "A Tiger in a Sea of Pussy Cats: Trinidad and Tobago Bid Goodbye to Oil, Hello to Gas," *New York Times*, September 4, 1998, C1–C2.

Rohter, Larry. "Asia Moves in on the Big Ditch," *New York Times*, December 19, 1999, WK 3.

Sargent, Charles S., Jr. "The Latin American City," in Blouet, Brian W. & Blouet, Olwyn M., eds., *Latin America and the Caribbean: A Systematic and Regional Survey* (New York: John Wiley & Sons, 3rd rev. ed., 1997), pp. 139–180. Diagram adapted from p. 173.

Sealey, Neil. *Caribbean World: A Complete Geography* (New York: Cambridge University Press, 1992).

"The Caribbean: Trouble in Paradise," *The Economist*, November 23, 2002, 36–37.

"The Mexicans Treated as Aliens in Their Own Country: The Right Not to be Hispanic," *The Economist*, March 7, 1998, 88–89.

Thompson, Ginger. "Chasing Mexico's Dream Into Squalor: Misery on the Border," *New York Times*, February 11, 2001, 1, 6.

Trias Monge, José. *Puerto Rico: The Trials of the Oldest Colony in the World* (New Haven, Conn.: Yale University Press, 1997).

Ward, Peter M. *Mexico City* (Boston: G. K. Hall, 1990).

Warren, Kay B. & Jackson, Jean E. *Indigenous Movements, Self-Representation, and the State in Latin America* (Austin, Tex.: University of Texas Press, 2003).

Watts, David. *The West Indies: Patterns of Development, Culture and Environmental Change Since 1492* (New York: Cambridge University Press, 1987).

West, Robert C. *Sonora: Its Geographical Personality* (Austin, Tex.: University of Texas Press, 1993).

● West, Robert C., Augelli, John P., et al. *Middle America: Its Lands and Peoples* (Englewood Cliffs, N.J.: Prentice-Hall, 3rd rev. ed., 1989).

Winn, Peter. *Americas: The Changing Face of Latin America and the Caribbean* (Berkeley, Cal.: University of California Press, updated ed., 1999).

CHAPTER 5

Augelli, John P. "The Controversial Image of Latin America: A Geographer's View," *Journal of Geography*, 62 (1963): 103–112. Quotation from p. 111.

Barton, Jonathan R. *A Political Geography of Latin America* (London & New York: Routledge, 1997).

Becker, Bertha K. & Egler, Claudio A. G. *Brazil: A New Regional Power in the World-Economy: A Regional Geography* (New York: Cambridge University Press, 1992).

● Blouet, Brian W. & Blouet, Olwyn M. *Latin America and the Caribbean: A Systematic and Regional Survey* (New York: John Wiley & Sons, 4th rev. ed., 2002).

Box, Ben. "Latin America's New Transportation Links," *1999 Britannica Book of the Year* (Chicago: Encyclopedia Britannica, 1999), pp. 412–413.

Brawer, Moshe. *Atlas of South America* (New York: Simon & Schuster, 1991).

"Brazil: The Northeast—Politics, Water and Poverty," *The Economist*, August 29, 1998, 36–38.

Brea, Jorge A. "Population Dynamics in Latin America," *Population Bulletin*, 58 (March 2003): 1–36.

Brockerhoff, Martin P. "An Urbanizing World," *Population Bulletin*, 55 (September 2000): 1–44.

● Bromley, Rosemary D. F. & Bromley, Ray. *South American Development: A Geographical Introduction* (New York: Cambridge University Press, 2nd rev. ed., 1988).

Brooke, James. "The New South Americans: Friends and Partners," *New York Times*, April 8, 1994, A 3.

Brookfield, Harold. *Exploring Agrodiversity* (New York: Columbia University Press, 2001).

Browder, John D. & Godfrey, Brian J. *Rainforest Cities: Urbanization, Globalization, and Development of the Amazon Basin* (New York: Columbia University Press, 1997).

Brunn, Stanley D., Williams, Jack F., & Zeigler, Donald J. *Cities of the World: World Regional Urban Development* (Lanham, Md.: Rowman & Littlefield, 3 rev. ed., 2003).

Caviedes, César N. *The Southern Cone: Realities of the Authoritarian State* (Totowa, N.J.: Rowman & Allanheld, 1984).

● Caviedes, César N. & Knapp, Gregory. *South America* (Englewood Cliffs, N.J.: Prentice Hall, 1995).

● Clawson, David L. *Latin America and the Caribbean: Lands and Peoples* (Dubuque, Iowa: WCB/McGraw-Hill, 2nd rev. ed., 2000).

Clawson, Patrick L. & Lee, Rensselaer W. *The Andean Cocaine Industry* (New York: St. Martin's Press, 1996).

Collier, Simon, et al., eds. *The Cambridge Encyclopedia of Latin America and the Caribbean* (New York: Cambridge University Press, 2nd rev. ed., 1992).

Denevan, William M. *Cultivated Landscapes of Native Amazonia and the Andes: Triumph Over the Soil* (New York: Oxford University Press, 2001).

Eakin, Marshall C. *Brazil: The Once and Future Country* (New York: St. Martin's Press, 1997).

Ford, Larry R. "A New and Improved Model of Latin American City Structure," *Geographical Review*, 86 (July 1996): 437–440.

Forero, Juan. "Colombia: The Next Oil Patch," *New York Times*, November 1, 2001, W 1, W 8.

Gade, Daniel W. *Nature and Culture in the Andes* (Madison, Wisc.: University of Wisconsin Press, 1999).

● Gilbert, Alan. *Latin America* (London & New York: Routledge, 1990).

Gilbert, Alan, ed. *The Mega-City in Latin America* (New York & Tokyo: United Nations University Press, 1996).

Godfrey, Brian J. "Brazil," *Focus*, Summer 1999, 28 pp.

Godfrey, Brian J. "Revisiting Rio de Janeiro and Sao Paulo," *Geographical Review*, 89 (January 1999): 94–121.

Greenfield, Gerald M., ed. *Latin American Urbanization: Historical Profiles of Major Cities* (Westport, Conn.: Greenwood Press, 1994).

Griffin, Ernst. "Testing the von Thünen Theory in Uruguay," *Geographical Review*, 63 (October 1973): 500–516.

Griffin, Ernst C. & Ford, Larry R. "Cities of Latin America," in Brunn, Stanley D. & Williams, Jack F., eds. *Cities of the World: World Regional Urban Development* (New York: HarperCollins, 2nd rev. ed., 1993), pp. 224–265.

Griffin, Ernst C. & Ford, Larry R. "A Model of Latin American City Structure," *Geographical Review*, 70 (1980): 397–422. Model diagram adapted from p. 406.

Gugler, Josef, ed., *The Urban Transformation of the Developing World* (New York: Oxford University Press, 1996).

Gwynne, Robert N. & Kay, Cristobal. *Latin America Transformed: Globalization and Modernity* (New York: Oxford University Press, 1999).

Hansis, Randall. *The Latin Americans: Understanding Their Legacy* (New York: McGraw-Hill, 1997).

● James, Preston E. & Minkel, Clarence W. *Latin America* (New York: John Wiley & Sons, 5th rev. ed., 1986), Chaps. 17–37.

Kelly, Philip. *Checkerboards and Shatterbelts: The Geopolitics of South America* (Austin: University of Texas Press, 1997).

Knapp, Gregory, ed. *Latin America in the Twenty-First Century: Challenges and Solutions* (Austin, Tex.: University of Texas Press, 2002).

McColl, Robert W. 'The Insurgent State: Territorial Bases of Revolution," *Annals of the Association of American Geographers*, 59 (1969): 613–631.

● Morris, Arthur S. *South America* (Totowa, N.J.: Barnes & Noble, 3rd rev. ed., 1987).

Myers, Norman. *The Primary Source: Tropical Forests and Our Future* (New York: W. W. Norton, 1984).

Pacione, Michael. *Urban Geography: A Global Perspective* (London & New York: Routledge, 2001).

Potter, Robert B. *Third World Urbanization: Contemporary Issues in Geography* (New York: Oxford University Press, 1991).

● Preston, David A., ed. *Latin American Development: Geographical Perspectives* (London & New York: Longman, 2nd rev. ed., 1995).

Ramos, Alcida Rita. "South America's Indigenous Peoples," *2000 Britannica Book of the Year* (Chicago: Encyclopedia Britannica, 2000), pp. 520–521.

Ribeiro, Darcy. *The Brazilian People: The Formation and Meaning of Brazil* (Gainesville, Fla.: University Press of Florida, trans. Gregory Rabassa, 2000).

Rohter, Larry. "Weave of Drugs and Strife in Colombia," *New York Times*, April 21, 2000, A1, A10, A 11.

Schiffer, Sueli Ramos. "São Paulo: Articulating a Cross-Border Region," in Saskia Sassen, ed., *Global Networks, Linked Cities* (London & New York: Routledge, 2002), pp. 209–236.

Smith, Nigel J. H. *Amazon Sweet Sea: Land, Life, and Water at the River's Mouth* (Austin, Tex.: University of Texas Press, 2002).

Smith, Nigel J. H. *The Amazon River Forest: A Natural History of Plants, Animals, and People* (New York: Oxford University Press, 1999).

Smith, Nigel J. H., et al. *Amazonia: Resiliency and Dynamism of the Land and Its People* (New York & Tokyo: United Nations University Press, 1995).

Sponsel, Leslie E., ed. *Indigenous Peoples and the Future of Amazonia* (Tucson, Ariz.: University of Arizona Press, 1995).

Sternberg, Rolf. "Brazilian Amazonia: A Metamorphosis in Progress," *Revista Geografica*, 125 (Enero-Junio 1999): 5–47.

Tenenbaum, Barbara, ed. *Encyclopedia of Latin American History and Culture* (New York: Scribners, 5 vols., 1996).

"The Andean Coca Wars: A Crop That Refuses to Die," *The Economist*, March 4, 2000, 23–25.

Warren, Kay B. & Jackson, Jean E. *Indigenous Movements, Self-Representation, and the State in Latin America* (Austin, Tex.: University of Texas Press, 2003).

Wilkie, Richard W. *Latin American Population and Urbanization Analysis: Maps and Statistics, 1950–1982* (Westwood, Calif.: UCLA Latin American Center, 1984).

Winn, Peter. *Americas: The Changing Face of Latin America and The Caribbean* (Berkeley, Cal.: University of California Press, updated ed., 1999).

Wood, Charles H. & Porro, Roberto, eds. *Deforestation and Land Use in the Amazon* (Gainesville, Fla.: University Press of Florida, 2002).

World Bank. *World Resources, 1996–1997: The Urban Environment* (New York: Oxford University Press, 1996).

CHAPTER 6

Adams, William M. & Mortimore, Michael J. *Working the Sahel: Environment and Society in Northern Nigeria* (London & New York: Routledge, 1999).

Adams, William M., Goudie, Andrew S., & Orme, Antony R., eds. *The Physical Geography of Africa* (New York: Oxford University Press, 1996).

"Africa for Africans: A Survey of Sub-Saharan Africa," *The Economist*, September 7, 1996, special insert, 18 pp.

Agyei-Mensah, Samuel & Casterline, John B., eds. *Reproduction and Social Context in Sub-Saharan Africa* (Westport: Conn.: Praeger, 2003).

● Aryeetey-Attoh, Samuel, ed. *Geography of Sub-Saharan Africa* (Upper Saddle River, N.J.: Prentice-Hall, 2 rev. ed., 2003).

Azarya, Victor. *Nomads and the State in Africa: The Political Roots of Marginality* (Brookfield, Vt.: Ashgate, 1996).

Bassett, Thomas J. & Crummey, Donald, eds. *African Savannas: Global Narratives and Local Knowledge of Environmental Change* (Westport, Conn.: Praeger, 2003).

Berkeley, Bill. *The Graves Are Not Yet Full: Race, Tribe and Power in the Heart of Africa* (New York: Basic Books, 2001).

● Best, Alan C. G. & de Blij, H. J. *African Survey* (New York: John Wiley & Sons, 1977).

Binns, Tony. *Tropical Africa* (London & New York: Routledge, 1994).

Binns, Tony, ed. *People and Environment in Africa* (New York: John Wiley & Sons, 1995).

● Bohannan, Paul & Curtin, Philip D. *Africa and Africans* (Prospect Heights, Ill.: Waveland Press, 4th rev. ed., 1996).

Brookfield, Harold. *Exploring Agrodiversity* (New York: Columbia University Press, 2001).

Chapman, Graham P. & Baker, Kathleen M., eds. *The Changing Geography of Africa and the Middle East* (London & New York: Routledge, 1992).

Christopher, Anthony J. *Colonial Africa: An Historical Geography* (Totowa, N.J.: Barnes & Noble, 1984).

Christopher, Anthony J. *The Atlas of Changing South Africa* (London & New York: Routledge, 2 rev. ed., 2001).

Cloudsley-Thompson, J. L., ed. *Sahara Desert* (Oxford, UK: Pergamon Press, 1984).

Curtin, Philip D. *The Atlantic Slave Trade* (Madison, Wisc.: University of Wisconsin Press, 1969).

Curtin, Philip D., et al. *African History: From Earliest Times to Independence* (London & New York: Longman, 2nd rev. ed., 1995).

Curtis, Sarah & Taket, Ann. *Health and Societies: Changing Perspectives* (London & New York: Edward Arnold, 1996).

● de Blij, H. J. "Africa's Geomosaic Under Stress," *Journal of Geography*, 90 (January/February, 1991): 2–9.

Ezzell, Carol. "Care for a Dying Continent," *Scientific American*, May 2000, 96–105.

Fardon, Richard & Furniss, Graham, eds. *African Languages, Development, and the State* (London & New York: Routledge, 1994).

Foster, Harold D. *Health, Disease and the Environment* (New York: John Wiley & Sons, 1992).

Fox, Roddy & Rowntree, Kate, eds. *The Geography of South Africa in a Changing World* (New York: Oxford University Press, 2000).

Freeman-Grenville, G. S. P. *The New Atlas of African History* (New York: Simon & Schuster, 1991).

Gates, Henry Louis, Jr. & Appiah, Kwame A., eds. *Africana: Encyclopedia of the African and African-American Experience* (New York: Basic Books, 1999).

Gesler, Wilbert M. *The Cultural Geography of Health Care* (Pittsburgh: University of Pittsburgh Press, 1991).

Gooneratne, Wilbert & Obudho, Robert A., eds. *Contemporary Issues in Regional Development Policy: Perspectives from Eastern and Southern Africa* (Brookfield, Vt.: Ashgate, 1997).

● Gould, Peter R. *The Slow Plague: A Geography of the AIDS Pandemic* (Cambridge, Mass.: Blackwell, 1993).

Griffith, Daniel A. & Newman, James L., eds. *Eliminating Hunger in Africa: Technical and Human Perspectives* (Syracuse, N.Y.: Maxwell School, Syracuse University, Foreign and Comparative Studies/African Series 45, 1994).

Griffiths, Ieuan L. L. *The Atlas of African Affairs* (London & New York: Routledge, 2nd rev. ed., 1994).

● Grove, Alfred T. *The Changing Geography of Africa* (New York: Oxford University Press, 2nd rev. ed., 1994).

Gurdon, Charles G., ed. *The Horn of Africa* (New York: St. Martin's Press, 1994).

Haggett, Peter. *The Geographical Structure of Epidemics* (New York: Oxford University Press, 2000).

Harrison Church, Ronald J. *West Africa: A Study of the Environment and Man's Use of It* (London: Longman, 8th rev. ed., 1980).

Huke, Robert E. "The Green Revolution," *Journal of Geography*, 84 (1985): 248–254.

Kalipeni, Ezekiel, et al., eds. *HIV and AIDS in Africa: Beyond Epidemiology* (Malden, Mass.: Blackwell, 2003).

Kapuscinski, Ryszard. *The Shadow of the Sun* (New York: Knopf, trans. Klara Glowczewska, 2001).

Kearns, Robin A. & Gesler, Wilbert M., eds. *Putting Health Into Place: Landscape, Identity, and Well-Being* (Syracuse, N.Y.: Syracuse University Press, 1998).

Knight, C. Gregory & Newman, James L., eds. *Contemporary Africa: Geography and Change* (Englewood Cliffs, N.J.: Prentice-Hall, 1976).

Lamb, David. *The Africans* (New York: Random House, 1983).

Learmonth, Andrew T. A. *Disease Ecology: An Introduction* (New York: Blackwell, 1988).

Lemon, Anthony, ed. *Geography of Change in South Africa* (New York: John Wiley & Sons/Longman, 1995).

Lemon, Anthony & Rogerson, C.M., eds. *Geography and Economy in South Africa and Its Neighbours* (Brookfield, Vt.: Ashgate, 2002).

Lewis, Laurence A. & Berry, Leonard. *African Environments and Resources* (Winchester, Mass.: Unwin Hyman, 1988).

Martin, Esmond Bradley & de Blij, H. J., eds. *African Perspectives: An Exchange of Essays on the Economic Geography of Nine African States* (London & New York: Methuen, 1981).

McEvedy, Colin. *The Penguin Atlas of African History* (New York: Penguin Putnam, rev. ed., 2000).

Meade, Melinda S. & Earickson, Robert J. *Medical Geography* (New York, Guilford, 2nd rev. ed., 2000).

Mehretu, Assefa. *Regional Disparity in Sub-Saharan Africa: Structural Readjustment of Uneven Development* (Boulder, Colo.: Westview Press, 1989).

Middleton, John, ed. *Encyclopedia of Sub-Saharan Africa* (New York: Simon & Schuster, 4 vols., 1994).

Moon, Graham & Jones, Kelvyn. *Health, Disease and Society: An Introduction to Medical Geography* (New York: Routledge & Kegan Paul, 1988).

- Mountjoy, Alan & Hilling, David. *Africa: Geography and Development* (Totowa, N.J.: Barnes & Noble, 1987).

Murdock, George P. *Africa: Its Peoples and Their Culture History* (New York: McGraw-Hill, 1959).

Newman, James L. *The Peopling of Africa: A Geographic Interpretation* (New Haven, Conn.: Yale University Press, 1995).

O'Connor, Anthony M. *Poverty in Africa: A Geographical Approach* (New York: Columbia University Press, 1991).

Olson, James S. *The Peoples of Africa: An Ethnohistorical Dictionary* (Westport, Conn.: Greenwood Press, 1996).

Onishi, Norimitsu. "Rising Muslim Power in Africa Causes Unrest in Nigeria and Elsewhere," *New York Times*, November 1, 2001, A 12.

Pakenham, Thomas. *The Scramble for Africa: The White Man's Conquest of the Dark Continent from 1876 to 1912* (New York: Random House, 1991).

Press, Robert M. *The New Africa: Dispatches From a Changing Continent* (Gainesville, Fla.: University Press of Florida, 1999).

Pritchard, J. M. *Landform and Landscape in Africa* (London: Edward Arnold, 1979).

Rakodi, Carole, ed. *The Urban Challenge in Africa: Growth and Management of Its Large Cities* (New York & Tokyo: United Nations University Press, 1996).

Reader, John. *Africa: A Biography of a Continent* (New York: Knopf, 1998).

Saff, Grant R. *Changing Cape Town: Urban Dynamics, Policy and Planning During the Political Transition in South Africa* (Lanham, Md.: University Press of America, 1998).

Senior, Michael & Okunrotifa, P. *A Regional Geography of Africa* (London & New York: Longman, 1983).

Siddle, David & Swindell, Ken. *Rural Change in Tropical Africa* (Cambridge, Mass.: Blackwell, 1994).

- Stock, Robert. *Africa South of the Sahara: A Geographical Interpretation* (New York: Guilford Press, 1995).

Stren, Richard & White, Rodney, eds. *African Cities in Crisis: Managing Rapid Urban Growth* (Boulder, Colo.: Westview Press, 1989).

Theroux, Paul. *Dark Star Safari: Overland from Cairo to Cape Town* (Boston: Houghton Mifflin, 2003).

Turner, Bill L. II, Hyden, Goran, & Kates, Robert W., eds. *Population Growth and Agricultural Change in Africa* (Gainesville, Fla.: University Press of Florida, 1993).

Waldmeir, Patti. *Anatomy of a Miracle: The End of Apartheid and the Birth of the New South Africa* (New York: W. W. Norton, 1997).

CHAPTER 7

Amery, Hussein A. & Wolf, Aaron T. *Water in the Middle East: A Geography of Peace* (Austin, Tex.: University of Texas Press, 2000).

Ahmed, Akbar S. *Islam Today: A Short Introduction to the Muslim World* (London & New York: I.B. Tauris, 2 rev. ed., 1999).

- Anderson, Ewan W. *The Middle East: Geography and Geopolitics* (London & New York: Routledge, 2000).

"At the Crossroads: A Survey of Central Asia," *The Economist*, July 26, 2003, special insert, 16 pp.

Atlas of the Middle East (Washington, D.C.: National Geographic Society, 2003).

- Beaumont, Peter, et al. *The Middle East: A Geographical Study* (New York: John Wiley & Sons/Halsted, 2nd rev. ed., 1988).

Blake, Gerald H., et al. *The Cambridge Atlas of the Middle East and North Africa* (New York: Cambridge University Press, 1988).

Bonine, Michael E., ed. *Population, Poverty, and Politics in Middle East Cities* (Gainesville, Fla.: University Press of Florida, 1997).

Chapman, Graham P. & Baker, Kathleen M., eds. *The Changing Geography of Africa and the Middle East* (London & New York: Routledge, 1992).

Chinn, Jeff & Kaiser, Robert J. *Russians as the New Minority: Ethnicity and Nationalism in the Soviet Successor States* (Boulder, Colo.: Westview Press, 1996).

Cloudsley-Thompson, J. L., ed. *Sahara Desert* (Oxford, UK: Pergamon Press, 1984).

Cohen, Saul B. "Middle East Geopolitical Transformation: The Disappearance of a Shatter Belt," *Journal of Geography*, 91 (January/February 1992): 2–10.

Collins, Robert O. *The Nile* (New Haven, Conn.: Yale University Press, 2002).

- Cressey, George B. *Crossroads: Land and Life in Southwest Asia* (Philadelphia: J. B. Lippincott, 1960).

Dekmejian, R. Hrair & Simonian, Hovann H. *Troubled Waters: The Geopolitics of the Caspian Region* (London & New York: I.B. Tauris, 2001).

Dowry, Alan. *The Jewish State: A Century Later* (Berkeley, Cal.: University of California Press, 1998).

- Drysdale, Alasdair & Blake, Gerald H. *The Middle East and North Africa: A Political Geography* (New York: Oxford University Press, 1985).

Ebel, Robert & Menon, Rajan, eds., *Energy and Conflict in Central Asia and the Caucasus* (Lanham, Md.: Rowman & Littlefield, 2000).

Engelmann, Kurt E. & Pavlakovic, Vjeran, eds. *Rural Development in Eurasia and the Middle East: Land Reform, Demographic Change, and Environmental Constraints* (Seattle: University of Washington Press, 2001).

Esposito, John L. *What Everyone Needs to Know About Islam* (New York: Oxford University Press, 2002).

- Fisher, William B. *The Middle East: A Physical, Social and Regional Geography* (London & New York: Methuen, 7th rev. ed., 1978).

Freeman-Grenville, G. S. P. *The Historical Atlas of the Middle East* (New York: Simon & Schuster, 1993).

Fuller, Graham E. & Francke, Rend Rahim. *The Arab Shi'a: The Forgotten Muslims* (New York: St. Martin's Press, 2000).

Gilbert, Martin. *Israel* (New York: William Morrow, 1998).

Gleason, Gregory. *The Central Asian States: Discovering Independence* (Boulder, Colo.: Westview Press, 1997).

Goldscheider, Calvin. *Israel's Changing Society: Population, Ethnicity, and Development* (Boulder, Colo.: Westview Press, 2 rev. ed., 2002).

Goldschmidt, Arthur Jr. *A Concise History of the Middle East* (Boulder, Colo.: Westview, 7 rev. ed., 2001).

● Gould, Peter R. *Spatial Diffusion* (Washington, D.C.: Association of American Geographers, Commission on College Geography, Resource Paper No. 4, 1969).

Gradus, Yehuda & Lipshitz, Gabi, eds. *The Mosaic of Israeli Geography* (Beersheva, Israel: Ben Gurion University Press, 1996).

Heathcote, Ronald L. *The Arid Lands: Their Use and Abuse* (London & New York: Longman, 1983).

● Held, Colbert C. *Middle East Patterns: Places, Peoples, and Politics* (Boulder, Colo.: Westview Press, 3rd ed., 2000).

Hourani, Albert H. *A History of the Arab Peoples* (Cambridge, Mass.: Belknap/Harvard University Press, 1991).

Joffé, George, ed. *North Africa: Nation, State and Region* (London & New York: Routledge, 1993).

Khalidi, Rashid I. "The Middle East as an Area in an Era of Globalization," in Ali Mirsepassi, et al., eds., *Localizing Knowledge in a Globalizing World: Recasting the Area Studies Debate* (Syracuse, N.Y.: Syracuse University Press, 2003), pp. 171–190.

Kliot, Nurit. *Water Resources and Conflict in the Middle East* (London & New York: Routledge, 1994).

Lamb, David. *The Arabs: Journeys Beyond the Mirage* (New York: Random House, 1987).

Lemarchand, Philippe, ed. *The Arab World, The Gulf, and the Middle East: An Atlas* (Boulder, Colo.: Westview Press, 1992).

Lewis, Bernard. *What Went Wrong? Western Impact and Middle East Response* (New York: Oxford University Press, 2002).

Lewis, Robert A., ed. *Geographic Perspectives on Soviet Central Asia* (London & New York: Routledge, 1992).

● Longrigg, Stephen H. *The Middle East: A Social Geography* (Chicago: Aldine, 2nd rev. ed., 1970).

Matvejevic, Predag. *The Mediterranean: A Cultural Landscape* (Berkeley, Cal.: University of California Press, trans. Michael H. Heim, 1999).

Meiselas, Susan. *Kurdistan: In the Shadow of History* (New York: Random House, 1998).

Mostyn, Trevor, ed. *The Cambridge Encyclopedia of the Middle East and North Africa* (New York: Cambridge University Press, 1988).

Newman, David. *The Dynamics of Territorial Change: A Political Geography of the Arab-Israeli Conflict* (Boulder, Colo.: Westview Press, 1998).

Polat, Necati. *Boundary Issues in Central Asia* (Ardsley, N.Y.: Transnational Publishers, 2002).

Prescott, J. R. V. *Political Frontiers and Boundaries* (Winchester, Mass.: Allen & Unwin, 1987).

Pryde, Philip R., ed. *Environmental Resources and Constraints in the Former Soviet Republics* (Boulder, Colo.: Westview Press, 1995).

Rashid, Ahmed. *Jihad: The Rise of Militant Islam in Central Asia* (New Haven, Conn.: Yale University Press, 2002).

Rashid, Ahmed. *Taliban: Militant Islam, Oil, and Fundamentalism in Central Asia* (New Haven: Yale University Press, 2000).

Robinson, Francis, ed. *Atlas of the Islamic World Since 1500* (New York: Facts on File, 1982).

Rubin, Barnett R. *The Fragmentation of Afghanistan* (New Haven, Conn.: Yale University Press, 2 rev. ed., 2002).

Sicker, Martin. *The Middle East in the Twentieth Century* (Westport, Conn.: Praeger, 2001).

Smart, Ninian. *The World's Religions* (New York: Cambridge University Press, 2nd rev. ed., 1998).

Soffer, Arnon. *Rivers of Fire: The Conflict Over Water in the Middle East* (Lanham, Md.: Rowman & Littlefield, trans. Murray Rosovsky & Nina Copaken, 1999).

Swearingen, Will D. & Bencherifa, Abdellatif, eds. *The North African Environment at Risk* (Boulder, Colo.: Westview Press, 1996).

Waines, David. *An Introduction to Islam* (New York: Cambridge University Press, 1995).

Wines, Michael. "Grand Soviet Scheme for Sharing Water in Central Asia is Foundering," *New York Times,* December 9, 2002, A14.

Wright, Robin. *Sacred Rage: The Wrath of Militant Islam* (New York: Simon & Schuster, 2 rev. ed., 2002).

Wright, Robin. *The Last Great Revolution: Turmoil and Transformation in Iran* (New York: Knopf, 2000).

Zoubir, Yahia H., ed. *North Africa in Transition: State, Society, and Economic Transformation in the 1990s* (Gainesville, Fla.: University Press of Florida, 1999).

CHAPTER 8

"A Survey of India's Economy: The Plot Thickens," *The Economist,* June 2, 2001, special insert, 22 pp.

Baxter, Craig. *Bangladesh: From a Nation to a State* (Boulder, Colo.: Westview Press, 1997).

Bennett Jones, Owen. *Pakistan: The Eye of the Storm* (New Haven, Conn.: Yale University Press, 2002).

Chapman, Graham P. *The Geopolitics of South Asia: From Early Empires to India, Pakistan and Bangladesh* (Brookfield, Vt.: Ashgate, 2 rev. ed., 2003).

Chapman, Graham P., Dutt, Ashok K., & Bradnock, Robert W., eds. *Urban Growth and Development in Asia. Volume I: Making the Cities; Volume II: Living in the Cities* (Brookfield, Vt.: Ashgate, 1999).

Corbridge, Stuart E., guest ed. "India: 1947–1997," *Environment and Planning A*, Vol. 27, No. 12, 1997.

Corbridge, Stuart E. & Harriss, John. *Reinventing India: Liberalization, Hindu Nationalism and Popular Democracy* (Malden, Mass.: Blackwell/Polity, 2000).

Crossette, Barbara. *So Close to Heaven: The Vanishing Buddhist Kingdoms of the Himalayas* (New York: Knopf, 1995).

Crossette, Barbara. *India: Facing the Twenty-First Century* (Bloomington, Ind.: Indiana University Press, 1993).

Deshpande, C. D. *India: A Regional Interpretation* (New Delhi: Northern Book Centre/Indian Council of Social Science Research, 1992).

Dumont, Louis. *Homo Hierarchus: The Caste System and Its Implications* (Chicago: University of Chicago Press, 1970).

Dutt, Ashok K. & Pomeroy, George M. "Cities of South Asia," in Brunn, Stanley D., Williams, Jack F., & Ziegler, Donald J., eds., *Cities of the World: World Regional Urban Development* (Lanham, Md.: Rowman & Littlefield, 3 rev. ed., 2003), pp. 330–371.

Dutt, Ashok K. & Geib, Margaret. *An Atlas of South Asia* (Boulder, Colo.: Westview Press, 1987).

Er-Rashid, Haroun. *Geography of Bangladesh* (Boulder, Colo.: Westview Press, 1977).

● Farmer, B. H. *An Introduction to South Asia* (London & New York: Routledge, 2nd rev. ed, 1993).

Gupta, Akhil. *Postcolonial Developments: Agriculture in the Making of Modern India* (Durham, N.C.: Duke University Press, 1998).

Haque, C. Emdad. *Hazards in a Fickle Environment: Bangladesh* (Hingham, Mass.: Kluwer, 1998).

Hasan, Mushirul. *Legacy of a Divided Nation: India's Muslims from Independence to Ayodhya* (Boulder, Colo.: Westview Press, 1997).

Huke, Robert E. "The Green Revolution," *Journal of Geography*, 84 (1985): 248–254.

Ives, Jack D. & Messerli, Bruno. *Himalayan Dilemma: Reconciling Development and Conservation* (London & New York: Routledge, 1989).

Jaffrelot, Christopher. *The Hindu Nationalist Movement in India* (New York: Columbia University Press, 1997).

Javed, Shaid. *Pakistan: Fifty Years of Nationhood* (Boulder, Colo.: Westview, 3 rev. ed., 1999).

Johnson, Basil L. C. *Bangladesh* (Totowa, N.J.: Barnes & Noble, 2nd rev. ed., 1982).

Johnson, Basil L. C. *India: Resources and Development* (Totowa, N.J.: Barnes & Noble, 1979).

Johnson, Gordon, ed. *Cultural Atlas of India* (New York: Facts on File, 1996).

Karan, Pradyumna P. & Ishii, Hiroshi. *Nepal: Development and Change in a Landlocked Himalayan Kingdom* (Tokyo: Institute for the Studies of Languages and Cultures of Asia and Africa, Tokyo University of Foreign Studies, 1994).

Karan, Pradyumna P. & Ishii, Hiroshi, eds. *Nepal: A Himalayan Kingdom in Transition* (New York & Tokyo: United Nations University Press, 1996).

Khilnani, Sunil. *The Idea of India* (New York: Farrar, Straus and Giroux, 1997).

Kosinski, Leszek A. & Elahi, K. Maudood, eds. *Population Redistribution and Development in South Asia* (Hingham, Mass.: D. Reidel, 1985).

Lall, Arthur S. *The Emergence of Modern India* (New York: Columbia University Press, 1981).

Lipner, Julius J. *Hinduism* (London & New York: Routledge, 1994).

Lukacs, John R., ed. *The People of South Asia: The Biological Anthropology of India, Pakistan, and Nepal* (New York: Plenum Press, 1984).

● McFalls, Joseph A., Jr. "Population: A Lively Introduction," *Population Bulletin*, 53 (September 1998): 1–48.

McGowan, William. *Only Man Is Vile: The Tragedy of Sri Lanka* (New York: Farrar, Straus & Giroux, 1992).

Mitra, Subrata K., ed. *Subnational Movements in South Asia* (Boulder, Colo.: Westview Press, 1994).

Mumtaz, Khavar, et al. *Pakistan* (Sterling: Va.: Stylus, 2 rev. ed., 2003).

Murphey, Rhoads. *A History of Asia* (New York: Harper-Collins, 1992).

● Muthiah, S., ed. *An Atlas of India* (New York: Oxford University Press, 1992).

Newman, James L. & Matzke, Gordon E. *Population: Patterns, Dynamics, and Prospects* (Englewood Cliffs, N.J.: Prentice-Hall, 1984).

Noble, Allen G. & Dutt, Ashok K., eds. *India: Cultural Patterns and Processes* (Boulder, Colo.: Westview Press, 1982).

Paz, Octavio. *In Light of India* (Orlando, Fla.: Harcourt Brace, trans. Eliot Weinberger, 1997).

Roberts, Paul W. *Empire of the Soul: Some Journeys in India* (New York: Riverhead, 1999).

Robinson, Francis, ed. *The Cambridge Encyclopedia of India, Pakistan, Bangladesh, Sri Lanka, Nepal, Bhutan, and the Maldives* (New York: Cambridge University Press, 1989).

Samad, Yunas. *A Nation in Turmoil: Nationalism and Ethnicity in Pakistan* (Thousand Oaks, Cal.: Sage, 1995).

Sanderson, Warren C. & Tan, Jee-Peng. *Population in Asia* (Brookfield, Vt.: Ashgate, 1996).

Schwartzberg, Joseph E. *A Historical Atlas of South Asia: 2nd Impression, With Additional Material* (New York: Oxford University Press, 1992).

Shurmer-Smith, Pamela. *India: Globalization and Change* (London: Arnold, 2000).

Sopher, David E., ed. *An Exploration of India: Geographical Perspectives on Society and Culture* (Ithaca, N.Y.: Cornell University Press, 1980).

● Spate, Oskar H. K. & Learmonth, Andrew T. A. *India and Pakistan: A General and Regional Geography* (London: Methuen, 2 vols., 1971).

● Spencer, Joseph E. & Thomas, William L., Jr. *Asia, East by South: A Cultural Geography* (New York: John Wiley & Sons, 2nd rev. ed., 1971).

Srinivas, Smritri. *Landscapes of Urban Memory: The Sacred and the Civic in India's High-Tech City* (Minneapolis: University of Minnesota Press, 2001).

Thapar, Valmik. *Land of the Tiger: A Natural History of the Indian Subcontinent* (Berkeley, Cal.: University of California Press, 1998).

"Time to Let Go: A Survey of India," *The Economist*, February 22, 1997, special insert, 26 pp.

● Weightman, Barbara A. *Dragons and Tigers: Geography of South, East, and Southeast Asia* (New York: John Wiley & Sons, 2002).

Wirsing, Robert G. *India, Pakistan, and the Kashmir Dispute: On Regional Conflict and Its Resolution* (New York: St. Martin's Press, 1994).

Wirsing, Robert G. *War or Peace on the Line of Control? The India-Pakistan Dispute Over Kashmir Turns Fifty* (Durham, UK: International Boundaries Research Unit, University of Durham, Vol. 2, Boundary and Territory Briefing No. 5, 1998).

● Wolpert, Stanley. *India* (Berkeley, Cal.: University of California Press, updated ed., 1999).

Zachariah, K. C. & Rajan, S. Irudaya, eds. *Kerala's Demographic Transition* (Thousand Oaks, Cal.: Sage, 1998).

Zurick, David & Karan, Pradyumna P. *Himalaya: Life on the Edge of the World* (Baltimore: Johns Hopkins University Press, 1999).

CHAPTER 9

Atlas of Population, Environment, and Sustainable Development of China (Beijing & New York: Science Press, 2000).

Barnett, A. Doak. *China's Far West: Four Decades of Change* (Boulder, Colo.: Westview Press, 1993).

Benewick, Robert & Donald, Stephanie. *The State of China Atlas* (New York: Penguin Putnam, 1999).

● Burks, Ardath W. *Japan: A Postindustrial Power* (Boulder, Colo.: Westview Press, 3rd rev. ed., 1991).

Cartier, Carolyn. *Globalizing South China* (Malden, Mass.: Blackwell, 2001).

Chapman, Graham P., Dutt, Ashok K., & Bradnock, Robert W., eds. *Urban Growth and Development in Asia. Volume I: Making the Cities; Volume II: Living in the Cities* (Brookfield, Vt.: Ashgate, 1999).

Chinoy, Mike. *China Live: Two Decades in the Heart of the Dragon* (Atlanta: Turner Publishing, 1997).

Cook, Ian G., et al., eds. *Fragmented Asia: Regional Integration and National Disintegration in Pacific Asia* (Brookfield, Vt.: Ashgate, 1996).

Cressey, George B. *Asia's Lands and Peoples: A Geography of One-Third of the Earth and Two-Thirds of Its People* (New York: McGraw-Hill, 3rd rev. ed., 1963).

Cumings, Bruce. *Korea's Place in the Sun: A Modern History* (New York: W. W. Norton, 1997).

Cybriwsky, Roman A. *Tokyo: The Changing Profile of an Urban Giant* (Boston: G. K. Hall, 1991).

Dietrich, Craig. *People's China: A Brief History* (New York: Oxford University Press, 3 rev. ed., 1998).

Dirlik, Arif, ed. *What Is In a Rim? Critical Perspectives on the Pacific Region Idea* (Lanham, Md.: Rowman & Littlefield, 2nd rev. ed., 1998).

Dwyer, Denis J., ed. *China: The Next Decades* (New York: John Wiley & Sons/Longman, 1993).

Eccleston, Bernard, Dawson, Michael, & McNamara, Deborah, eds. *The Asia-Pacific Profile* (London & New York: Routledge, 1998).

Ginsburg, Norton S., ed. *The Pattern of Asia* (Englewood Cliffs, N.J.: Prentice-Hall, 1958).

"Go West, Young Han: Plans to Develop China's Western Provinces are About More than Economics," *The Economist*, December 23, 2000, 45–46.

Harrell, Stevan, ed. *Cultural Encounters on China's Ethnic Frontiers* (Seattle: University of Washington Press, 1995).

Harris, Chauncy D. "The Urban and Industrial Transformation of Japan," *Geographical Review*, 72 (January 1982): 50–89.

Hillel, Daniel. "Lash of the Dragon: China's Yellow River Remains Untamed," *Natural History*, August 1991, 28–37.

Hoare, James & Pares, Susan. *Korea: An Introduction* (London & New York: Routledge, 1988).

● Hodder, Rupert. *The West Pacific Rim: An Introduction* (New York: John Wiley & Sons, 1992).

Hoh, Erling. "The Long River's Journey Ends: China's Three Gorges Dam Will Soon Transform the Yangtze," *Natural History*, July 1996, 28–38.

"Hong Kong and Shanghai: Rivals More Than Ever," *The Economist*, March 30, 2002, 19–21.

Hook, Brian, ed. *The Cambridge Encyclopedia of China* (New York: Cambridge University Press, 2nd rev. ed., 1994).

Hsieh, Chiao-min & Hsieh, Jean Kan. *China: A Provincial Atlas* (New York: Macmillan, 1992).

Hsieh, Chiao-Min & Lu, Max. *Changing China: A Geographical Appraisal* (Boulder, Colo.: Westview Press, 1998).

Hsu, Francis L. K. & Serrie, Hendrick. *The Overseas Chinese: Ethnicity in National Context* (Lanham, Md.: University Press of America, 1998).

Kam Wing Chan. *Cities With Invisible Walls: Reinterpreting Urbanization in Post-1949 China* (New York: Oxford University Press, 1994).

Karan, Pradyumna P. & Stapleton, Kristin, eds. *The Japanese City* (Lexington, Ky.: University Press of Kentucky, 1997).

Knapp, Ronald G., ed. *Chinese Landscapes: The Village as Place* (Honolulu: University of Hawai'i Press, 1992).

● Kornhauser, David H. *Japan: Geographical Background to Urban-Industrial Development* (London & New York: Longman, 2nd rev. ed., 1982).

Kristof, Nicholas D. & WuDunn, Sheryl. *Thunder From the East: Portrait of a Rising Asia* (New York: Knopf, 2000).

● Leeming, Frank. *The Changing Geography of China* (Cambridge, Mass.: Blackwell, 1993).

Le Heron, Richard & Park, Sam Ock, eds. *The Asian Pacific Rim and Globalization: Enterprise, Governance, and Territoriality* (Brookfield, Vt.: Ashgate, 1995).

Lin, George C. S. *Red Capitalism in South China: Growth and Development of the Pearl River Delta* (Vancouver: University of British Columbia Press, 1997).

Linge, Godfrey J.R., ed. *China's New Spatial Economy: Heading Towards 2020* (New York: Oxford University Press, 1997).

Linge, Godfrey J. R. & Forbes, D. K., eds. *China's Spatial Economy* (New York: Oxford University Press, 1990).

Lo, Chor-Pang. *Hong Kong* (New York: John Wiley & Sons/Belhaven, 1992).

Lo, Fu-chen & Yeung, Yue-man, eds. *Emerging World Cities in Pacific Asia* (New York & Tokyo: United Nations University Press, 1996).

MacDonald, Donald. *A Geography of Modern Japan* (Ashford, UK: Paul Norbury, 1985).

Mackerras, Colin. *China's Minority Cultures: Identities and Integration Since 1912* (New York: St. Martin's Press, 1995).

Mackerras, Colin. *The New Cambridge Handbook of Contemporary China* (New York: Cambridge University Press, 2001).

Marton, Andrew M. *China's Spatial Economic Development: Restless Landscapes in the Lower Yangzi Delta* (London & New York: Routledge, 2000).

McColl, Robert W., guest ed. "Special Theme Issue: Mongolia," *Focus,* Fall 2002, 1–37.

Murphey, Rhoads. *A History of Asia* (New York: HarperCollins, 1992).

● Noh, Toshio & Kimura, John C., eds. *Japan: A Regional Geography of an Island Nation* (Tokyo: Teikoku-Shoin, 1985).

Nolan, Peter. *Transforming China: Globalization, Transition and Development* (Sterling, Va.: Stylus, 2003).

Ohmae, Kenichi. *The End of the Nation-State: The Rise of Regional Economies* (New York: Free Press, 1996).

Olds, Kris, et al., eds. *Globalization and the Asia Pacific: Contested Territories* (London & New York: Routledge, 2000).

Pannell, Clifton W. "China's Urban Transition," *Journal of Geography,* 94 (May/June 1995): 394–403.

● Pannell, Clifton W. & Ma, Laurence J. C. *China: The Geography of Development and Modernization* (New York: Halsted Press/V. H. Winston, 1983).

● Pannell, Clifton W. & Veeck, Gregory. *East Asia* (Lanham, Md.: Rowman & Littlefield, 1999).

Porter, Jonathan. *Macau: The Imaginary City* (Boulder, Colo.: Westview Press, 1996).

Potter, Robert B., et al. *Geographies of Development* (Harlow, UK & New York: Longman, 1999).

Preston, Peter W. *Pacific Asia in the Global System: An Introduction* (Malden, Mass.: Blackwell, 1998).

Rozman, Gilbert, ed. *Japan and Russia: The Tortuous Path to Normalization, 1949–1999* (New York: St. Martin's Press, 2000).

Rumley, Dennis, et al., eds. *Global Geopolitical Change and the Asia-Pacific: A Regional Perspective* (Brookfield, Vt.: Ashgate, 1996).

Salisbury, Harrison E. *The Long March: The Untold Story* (New York: Harper & Row, 1985).

"Shanghai: FT Survey." *Financial Times,* special supplement, October 16, 2001, 10 pp.

Shen, Jianfa. *Internal Migration and Regional Population Dynamics in China* (Amsterdam & New York: Elsevier, 1996).

Si-ming Li & Wing-shing Tang, eds. *China's Regions, Polity, and Economy: A Study of Spatial Transformation in the Post-Reform Era* (Hong Kong: The Chinese University Press, 2000).

Sit, Victor F. S. *Beijing: The Nature and Planning of a Chinese Capital City* (New York: John Wiley & Sons, 1995).

● Smith, Christopher J. *China in the Post-Utopian Age* (Boulder, Colo.: Westview Press, 2000).

Smith, Patrick. *Japan: A Reinterpretation* (New York: Pantheon, 1997).

Spence, Jonathan D. *The Chan's Great Continent: China in Western Minds* (New York: W. W. Norton, 1998).

● Spencer, Joseph E. & Thomas, William L., Jr. *Asia, East by South: A Cultural Geography* (New York: John Wiley & Sons, 2nd rev. ed., 1971).

Starr, John Bryan. *Understanding China: A Guide to China's Economy, History, and Political Structure* (New York: Hill & Wang, 1997).

Terrill, Ross. *The New Chinese Empire: Beijing's Political Dilemma and What it Means for the United States* (Boulder, Colo.: Westview Press, 2003).

Theroux, Paul. "Going to See the Dragon: A Journey Through South China, Land of Capitalist Miracles, Where Yesterday's Rice Paddy Becomes Tomorrow's Metropolis, and a Thousand Factories Bloom," *Harper's Magazine*, October 1993, 33–56.

Trewartha, Glenn T. *Japan: A Geography* (Madison, Wisc.: University of Wisconsin Press, 2nd rev. ed., 1965).

van Kemenade, Willem. *China, Hong Kong, Taiwan Inc.: The Dynamics of a New Empire* (New York: Knopf, 1997). Quotation taken from p. 184.

Veeck, Gregory, ed. *The Uneven Landscape: Geographic Studies in Post-Reform China* (Baton Rouge, La.: Geoscience Publications, 1991).

Watters, R.F. & McGee, T.G. *Asia-Pacific: New Geographies of the Pacific Rim* (London: Hurst, 1997).

Wei, Yehua Dennis. *Regional Development in China: States, Globalization, and Inequality* (London & New York: Routledge, 2000).

● Weightman, Barbara A. *Dragons and Tigers: Geography of South, East, and Southeast Asia* (New York: John Wiley & Sons, 2002).

Weiping Wu. *Pioneering Economic Reform in China's Special Economic Zones: The Promotion of Foreign Investment and Technology Transfer in Shenzhen* (Brookfield, Vt.: Ashgate, 1999).

Witherick, M. & Carr, M. *The Changing Face of Japan* (London: Hodder & Stoughton, 1993).

Wittwer, Sylvan, et al. *Feeding a Billion: Frontiers of Chinese Agriculture* (East Lansing, Mich.: Michigan State University Press, 1987).

Yatsko, Pamela. *New Shanghai: The Rocky Rebirth of China's Legendary City* (Singapore: John Wiley & Sons [Asia], 2001).

Yeung, Yue-man & Chu, David K.Y., eds. *Guangdong: Survey of a Province Undergoing Rapid Change* (Hong Kong: Chinese University Press, 1994).

Yeung, Yue-man & Hu, Xu-Wei, eds. *China's Coastal Cities: Catalysts for Modernization* (Honolulu: University of Hawai'i Press, 1992).

Yeung, Yue-man & Yun-wing, Sung, eds. *Shanghai: Transformation and Modernization Under China's Open Policy* (Hong Kong: Chinese University Press, 1996).

● Zhao Songqiao. *Geography of China: Environment, Resources, Population, and Development* (New York: John Wiley & Sons, 1994).

Zhao Songqiao. *Physical Geography of China* (New York & Beijing: John Wiley & Sons/Science Press, 1986).

CHAPTER 10

Broek, Jan O. M. "Diversity and Unity in Southeast Asia," *Geographical Review*, 34 (1944): 175–195.

Brookfield, Harold C., et al. *In Place of the Forest: Environmental and Socio-Economic Transformation in Borneo and the Eastern Malay Peninsula* (New York: United Nations University Press, 1995).

Bunnell, Tim. *Malaysia, Modernity and the Multimedia Super Corridor: The Construction of a High-Tech City-Region* (London & New York: Routledge, 2002).

Cleary, Mark & Eaton, Peter. *Tradition and Reform: Land Tenure and Rural Development in South-East Asia* (New York: Oxford University Press, 1997).

Cohen, Saul B. *Geopolitics of the World System* (Lanham, Md.: Rowman & Littlefield, 2002).

Cox, Christopher R. *Chasing the Dragon: Into the Heart of the Golden Triangle* (New York: Holt/Marion Wood, 1996).

Daws, Gavan & Fujita, Marty. *Archipelago: The Islands of Indonesia* (Berkeley, Calif.: University of California Press, 1999).

● Dixon, Chris. *South East Asia in the World Economy* (New York: Cambridge University Press, 1991).

Dixon, Chris & Drakakis-Smith, David, eds. *Uneven Development in Southeast Asia* (Brookfield, Vt.: Ashgate, 1998).

Dwyer, Denis J., ed. *South East Asian Development* (New York: John Wiley & Sons/Longman, 1990).

● Fisher, Charles A. *Southeast Asia: A Social, Economic and Political Geography* (New York: E. P. Dutton, 2nd rev. ed., 1966).

Friend, Theodore. *Indonesian Destinies* (Cambridge, Mass.: Belknap/Harvard University Press, 2003).

● Fryer, Donald W. *Emerging South-East Asia: A Study in Growth and Stagnation* (New York: John Wiley & Sons, 2nd rev. ed., 1979).

Gargan, Edward A. *The River's Tale: A Year on the Mekong* (New York, Knopf, 2002).

Ginsburg, Norton S., et al., eds. *The Extended Metropolis: Settlement Transition in Asia* (Honolulu: University of Hawai'i Press, 1991).

● Glassner, Martin I. & de Blij, H. J. *Systematic Political Geography* (New York: John Wiley & Sons, 4th rev. ed., 1989).

"Good Fences: Borders are Arbitrary Abstractions, Economic Impediments and Surprisingly Ineradicable," *The Economist*, December 19, 1998, 19–22.

Hartshorne, Richard. "Suggestions on the Terminology of Political Boundaries," *Annals of the Association of American Geographers*, 26 (1936): 56–57.

Hill, Ronald D., ed. *South-East Asia: A Systematic Geography* (New York: Oxford University Press, 1979).

Hodder, Rupert. *The West Pacific Rim: An Introduction* (New York: John Wiley & Sons, 1992).

Hsu, Francis L. K. & Serrie, Hendrick. *The Overseas Chinese: Ethnicity in National Context* (Lanham, Md.: University Press of America, 1998).

Karnow, Stanley. *In Our Image: America's Empire in the Philippines* (New York: Random House, 1989).

Leinbach, Thomas R. & Sien, Chia Lin, eds. *South-East Asian Transport* (New York: Oxford University Press, 1989).

- Leinbach, Thomas R. & Ulack, Richard, eds. *Southeast Asia: Diversity and Development* (Upper Saddle River, N.J.: Prentice Hall, 2000).

Logan, William S., ed. *The Disappearing 'Asian' City: Protecting Asia's Urban Heritage in a Globalizing World* (New York: Oxford University Press, 2002).

McGee, Terence G. *The Mega-Urban Regions of Southeast Asia* (Vancouver, Canada: University of British Columbia Press, 1995).

McGee, Terence G. *The Southeast Asian City: A Social Geography* (New York: Praeger, 1967).

Mulder, Neils. *Inside Southeast Asia: Religion, Everyday Life, and Cultural Change* (Seattle: University of Washington Press, 2001).

Murray, Geoffrey & Perera, Audrey. *Singapore: The Global City-State* (New York: St. Martin's Press, 1996).

Parnwell, Michael J. G. & Bryant, Raymond L., eds. *Environmental Change in South-East Asia: People, Politics and Sustainable Development* (London & New York: Routledge, 1996).

- Prescott, J. R. V. *Political Frontiers and Boundaries* (Winchester, Mass.: Allen & Unwin, 1987).

Preston, Peter W. *Pacific Asia in the Global System: An Introduction* (Malden, Mass.: Blackwell, 1998).

Reynolds, Craig J. "A New Look at Old Southeast Asia," *Journal of Asian Studies*, 54 (1995): 419–446.

Rigg, Jonathan. *Southeast Asia—A Region in Transition: A Thematic Human Geography of the ASEAN Region* (New York: HarperCollins Academic, 1991).

- Rigg, Jonathan. *Southeast Asia: The Human Landscape of Modernization and Development* (London & New York: Routledge, 2 rev. ed., 2003).

Rodan, Gary, ed. *Singapore* (Brookfield, Vt.: Ashgate, 2001).

SarDesai, D. R. *Southeast Asia: Past and Present* (Boulder, Colo.: Westview Press, 5th rev. ed., 2003).

- Spencer, Joseph E. & Thomas, William L., Jr. *Asia, East by South: A Cultural Geography* (New York: John Wiley & Sons, 2nd rev. ed., 1971).

"Southeast Asia: The Geography of Terror," *The Economist*, August 23, 2003, 31–32.

Taylor, John G. *East Timor: The Price of Freedom* (London: Zed Books, 2000).

"The Faltering Firefighter: A Survey of Indonesia," *The Economist*, July 8, 2000, special insert, 16 pp.

Tyner, James. "Cities of Southeast Asia," in Brunn, Stanley D., Williams, Jack F., & Zeigler, Donald J., eds., *Cities of the World: World Regional Urban Development* (Lanham, Md.: Rowman & Littlefield, 3rd rev. ed., 2003), pp. 370–411.

Ulack, Richard & Pauer, Gyula. *Atlas of Southeast Asia* (New York: Macmillan, 1988).

Watters, R.F. & McGee, Terence G. *Asia-Pacific: New Geographies of the Pacific Rim* (London: Hurst, 1997).

- Weightman, Barbara A. *Dragons and Tigers: Geography of South, East, and Southeast Asia* (New York: John Wiley & Sons, 2002).

Wing-Kai Chiu, Stephen, et al. *City States in the Global Economy: Industrial Restructuring in Hong Kong and Singapore* (Boulder, Colo.: Westview Press, 1996).

Wurfel, David & Burton, Bruce, eds. *Southeast Asia in the New World Order* (New York: St. Martin's Press, 1996).

CHAPTER 11

"A Survey of Australia," *The Economist*, September 9, 2000, special insert, 16 pp.

Bambrick, Susan, ed. *The Cambridge Encyclopedia of Australia* (New York: Cambridge University Press, 1994).

- Barrett, Rees D. & Ford, Roslyn A. *Patterns in the Human Geography of Australia* (South Melbourne: Macmillan of Australia, 1987).

Beadle, N. C. W. *The Vegetation of Australia* (New York: Cambridge University Press, 1981).

Britton, Steve, et al., eds. *Changing Places in New Zealand: A Geography of Restructuring* (Dunedin, N.Z.: Allied Press, 1992).

Burnley, Ian. *The Impact of Immigration in Australia: A Demographic Approach* (New York: Oxford University Press, 2001).

Camm, J. C. R. & McQuilton, J., eds. *Australia: An Historical Atlas* (Broadway, N.S.W.: Fairfax, Syme & Weldon Associates, 1987).

Connell, John, ed. *Sydney: The Emergence of a World City* (New York: Oxford University Press, 2000).

Courtenay, Percy P. *Northern Australia: Patterns and Problems of Tropical Development in an Advanced Country* (London & New York: Longman, 1983).

Cumberland, Kenneth B. & Whitelaw, James S. *New Zealand* (Chicago: Aldine, 1970).

"Environment and Development in Australia," *Australian Geographer*, 19 (May 1988): 3–220.

Fitzpatrick, Judith M., ed. *Endangered Peoples of Oceania: Struggles to Survive and Thrive* (Westport, Conn.: Greenwood Press, 2000).

Forster, Clive. *Australian Cities: Continuity and Change* (Melbourne: Oxford University Press, 2 rev. ed., 1999).

Greiner, Alyson L. & Jordan-Bychkov, Terry G. *Anglo-Celtic Australia: Colonial Immigration and Cultural Regionalism* (Chicago: University of Chicago Press, 2003).

Head, Lesley. *Second Nature: The History and Implications of Australia as Aboriginal Landscape* (Syracuse, N.Y.: Syracuse University Press, 2000).

- Heathcote, Ronald L. *Australia* (New York: John Wiley & Sons/Longman, 2nd rev. ed., 1994).

Heathcote, Ronald L., ed. *The Australian Experience: Essays in Australian Land Settlement and Resource Management* (Melbourne: Longman Cheshire, 1988).

Heathcote, Ronald L. & Mabbutt, J. A. *Land, Water and People: Essays in Australian Resource Management* (Winchester, Mass.: Allen & Unwin, 1988).

Hofmeister, Burkhard. *Australia and Its Urban Centers* (Berlin: Gebrüder Borntraeger, 1988).

Holland, P. G. & Johnston, W. B., eds. *Southern Approaches: Geography in New Zealand* (Christchurch, N.Z.: New Zealand Geographical Society, 1987).

Hughes, Robert. *The Fatal Shore* (New York: Knopf, 1987).

Hughes, Robert. "The Real Australia," *Time*, September 11, 2000, 98–111.

Inglis, C., et al., eds. *Asians in Australia: The Dynamics of Migration and Settlement* (Sydney: Allen & Unwin, 1992).

● Jeans, Dennis N., ed. *Australia: A Geography. Volume 1: The Natural Environment* (Sydney: Sydney University Press, 1986); *Volume 2: Space and Society* (Sydney: Sydney University Press, 1987).

● MacDonald, Glen M. *Biogeography: Introduction to Space, Time and Life* (New York: Wiley, 2003).

McKnight, Tom L. *Australia's Corner of the World* (Englewood Cliffs, N.J.: Prentice-Hall, 1970).

● McKnight, Tom L. *Oceania: The Geography of Australia, New Zealand, and the Pacific Islands* (Englewood Cliffs, N.J.: Prentice Hall, 1995).

Meinig, Donald W. *On the Margins of the Good Earth: The South Australian Wheat Frontier, 1869–1884* (Chicago: Rand McNally, 1962).

● Morton, Harry & Johnston, Carol M. *The Farthest Corner: New Zealand, A Twice Discovered Land* (Honolulu: University of Hawai'i Press, 1989).

● O'Connor, Kevin, Stimson, Robert, & Daly, Maurice. *Australia's Changing Economic Geography: A Society Dividing* (South Melbourne: Oxford University Press, 2001).

Powell, Joseph M. *An Historical Geography of Modern Australia: The Restive Fringe* (New York: Cambridge University Press, 1988).

Powell, Joseph M. "Revisiting the Australian Experience: Transmillennial Conjurings," *Geographical Review*, 90 (January 2000): 1–17.

Rich, David C. *The Industrial Geography of Australia* (Sydney: Methuen, 1987).

● Robinson, Guy M., Loughran, Robert J., & Tranter, Paul J. *Australia and New Zealand: Economy, Society and Environment* (New York: Oxford University Press, 2000).

Serventy, J. *Landforms of Australia* (New York: American Elsevier, 1968).

● Spate, Oskar H. K. *Australia* (New York: Praeger, 1968).

Stratford, Elaine, ed. *Australian Cultural Geographies* (New York: Oxford University Press, 1999).

Walmsley, D. J. & Sorenson, A. D. *Contemporary Australia: Explorations in Economy, Society and Geography* (Melbourne: Longman Cheshire, 2nd rev. ed., 1993).

Young, Ann. *Environment and Change in Australia Since 1788* (New York: Oxford University Press, 2 rev. ed., 2000).

CHAPTER 12

Arnberger, Hertha & Arnberger, Erik. *The Tropical Islands of the Indian and Pacific Oceans* (Seattle: University of Washington Press, 2001).

● Bier, James A., cartographer. *Reference Map of Oceania: The Pacific Islands of Micronesia, Polynesia, Melanesia* (Honolulu: University of Hawai'i Press, 1995).

Blake, Gerald H., ed. *Maritime Boundaries: World Boundaries, Vol. 5* (London & New York: Routledge, 1994).

Blake, Gerald H., ed. *Maritime Boundaries and Ocean Resources* (Totowa, N.J.: Rowman & Littlefield, 1988).

Blank, Paul W. & Spier, Fred, eds. *Defining the Pacific: Opportunities and Constraints* (Brookfield, Vt.: Ashgate, 2002).

Brookfield, Harold C. *Exploring Agrodiversity* (New York: Columbia University Press, 2001).

Brookfield, Harold C., ed. *The Pacific in Transition: Geographical Perspectives on Adaptation and Change* (New York: St. Martin's Press, 1973).

Brookfield, Harold C. & Hart, Doreen. *Melanesia: A Geographical Interpretation of an Island World* (New York: Barnes & Noble, 1971).

Bunge, Frederica M. & Cooke, Melinda W., eds. *Oceania: A Regional Study* (Washington, D.C.: U.S. Government Printing Office, 1984).

Carter, John, ed. *Pacific Islands Yearbook* (Sydney: Pacific Publications, annual).

● Chaturvedi, Sanjay. *The Polar Regions: A Political Geography* (New York: John Wiley & Sons, 1997).

Cole, George M. *Water Boundaries* (New York: John Wiley & Sons, 1997).

Connell, John. *Papua New Guinea: The Struggle for Development* (London & New York: Routledge, 1997).

Couper, Alastair D., ed. *Development and Social Change in the Pacific Islands* (London & New York: Routledge, 1989).

Crossley, L. *Explore Antarctica* (New York: Cambridge University Press, 1995).

Damas, David. *Bountiful Island: A Study of Land Tenure on a Micronesian Atoll* (Waterloo, Ont., Canada: Wilfrid Laurier University Press, 1996).

de Blij, H. J. "A Regional Geography of Antarctica and the Southern Ocean," *University of Miami Law Review*, 33 (1978): 299–314.

Dodds, Klaus J. *Geopolitics of Antarctica: Views From the Southern Oceanic Rim* (New York: John Wiley & Sons, 1997).

Fitzpatrick, Judith M., ed. *Endangered Peoples of Oceania: Struggles to Survive and Thrive* (Westport, Conn.: Greenwood Press, 2000).

Freeman, Otis W., ed. *Geography of the Pacific* (New York: John Wiley & Sons, 1951).

Friis, Herman R., ed. *The Pacific Basin: A History of Its Geographical Exploration* (New York: American Geographical Society, Special Publication No. 38, 1967).

• Glassner, Martin I. *Neptune's Domain: A Political Geography of the Sea* (Winchester, Mass.: Unwin Hyman, 1990).

Grossman, Lawrence S. *Peasants, Subsistence Ecology, and Development in the Highlands of Papua New Guinea* (Princeton, N.J.: Princeton University Press, 1984).

Hanlon, David & White, Geoffrey M., eds. *Voyaging Through the Contemporary Pacific* (Lanham, Md.: Rowman & Littlefield, 2000).

Howard, A., ed. *Polynesia: Readings on a Culture Area* (Scranton, Penna.: Chandler, 1971).

Howlett, Diana. *Papua New Guinea: Geography and Change* (Melbourne, Australia: Thomas Nelson, 1973).

Johnston, Douglas M. & Saunders, Philip M. *Ocean Boundary-Making: Regional Issues and Developments* (London & New York: Routledge, 1988).

Karolle, Bruce G. *Atlas of Micronesia* (Honolulu: Bess Press, 2nd rev. ed., 1995).

Kirch, Patrick V. *On the Road of Winds: An Archaeological History of the Pacific Islands Before European Contact* (Berkeley, Cal.: University of California Press, 2000).

Kirch, Patrick V. *The Wet and the Dry: Irrigation and Agricultural Intensification in Polynesia* (Chicago: University of Chicago Press, 1994).

Kissling, Christopher C., ed. *Transport and Communications for Pacific Microstates: Issues in Organization and Management* (Suva, Fiji: University of the South Pacific, Institute of Pacific Studies, 1984).

Kluge, P. F. *The Edge of Paradise: America in Micronesia* (New York: Random House, 1991).

Lea, John & Connell, John. *Urbanization in the Pacific* (London & New York: Routledge, 2001).

Leibowitz, Arnold H. *Embattled Island: Palau's Struggle for Independence* (Westport, Conn.: Praeger, 1996).

Lockhart, Douglas G., et al., eds. *The Development Process in Small Island States* (London & New York: Routledge, 1993).

• McEvedy, Colin. *The Penguin Historical Atlas of the Pacific* (New York: Penguin Putnam, 1999).

• McKnight, Tom L. *Oceania: The Geography of Australia, New Zealand, and the Pacific Islands* (Englewood Cliffs, N.J.: Prentice Hall, 1995).

Mitchell, Andrew. *The Fragile South Pacific: An Ecological Odyssey* (Austin: University of Texas Press, 1991).

"Mobility and Identity in the Island Pacific," special issue, *Pacific Viewpoint*, 26, No. 1 (1985).

Morgan, Joseph R. *Hawai'i: A Unique Geography* (Honolulu: Bess Press, 1996).

Nunn, Patrick. *Oceanic Islands* (Cambridge, Mass.: Blackwell, 1994).

Oliver, Douglas L. *Native Cultures of the Pacific Islands* (Honolulu: University of Hawai'i Press, 1989).

• Peake, Martin. *Pacific People and Society* (New York: Cambridge University Press, 1992).

Prescott, J. R. V. *The Maritime Political Boundaries of the World* (London & New York: Methuen, 1986).

Robillard, Albert B., ed. *Social Change in the Pacific Islands* (London & New York: Kegan Paul International, 1991).

Sager, Robert J. "The Pacific Islands: A New Geography," *Focus*, Summer 1988, pp. 10–14.

Spate, Oskar H. K. *The Pacific Since Magellan, Vol. II: Monopolists and Freebooters* (Minneapolis: University of Minnesota Press, 1983).

Spate, Oskar H. K. *The Pacific Since Magellan, Vol. III: Paradise Found and Lost* (Minneapolis: University of Minnesota Press, 1989).

Spate, Oskar H. K. *The Spanish Lake: A History of the Pacific Since Magellan, Vol. I* (Beckenham, UK: Croom Helm, 1979).

Sugden, David E. *Arctic and Antarctic: A Modern Geographical Synthesis* (Totowa, N.J.: Barnes & Noble, 1982).

Theroux, Paul. *The Happy Isles of Oceania: Paddling the Pacific* (New York: G. P. Putnam's Sons, 1992).

Ward, R. Gerard, ed. *Man in the Pacific Islands: Essays on Geographical Change in the Pacific* (New York: Oxford University Press, 1972).

Woodcock, Deborah W., ed. *Hawai'i: New Geographies* (Manoa, Hawai'i: University of Hawai'i, Department of Geography, 1999).

Zurick, David N. "Preserving Paradise," *Geographical Review*, 85 (April 1995): 157–172.

Glossary

Aboriginal population   See **indigenous peoples**.*

Absolute location   The position or place of a certain item on the surface of the Earth as expressed in degrees, minutes, and seconds of **latitude**, 0° to 90° north or south of the equator, and **longitude**, 0° to 180° east or west of the *prime meridian* passing through Greenwich, England (a suburb of London).

Accessibility   The degree of ease with which it is possible to reach a certain location from other locations. *Inaccessibility* is the opposite of this concept.

Acculturation   Cultural modification resulting from intercultural borrowing. In **cultural geography**, the term refers to the change that occurs in the culture of **indigenous peoples** when contact is made with a society that is technologically superior.

Agglomeration   Process involving the clustering or concentrating of people or activities.

Agrarian   Relating to the use of land in rural communities, or to **agricultural** societies in general.

Agriculture   The purposeful tending of crops and livestock in order to produce food and fiber.

Alluvial   Refers to the mud, silt, and sand (collectively *alluvium*) deposited by rivers and streams. *Alluvial plains* adjoin many larger rivers; they consist of these renewable deposits that are laid down during floods, creating fertile and productive soils. Alluvial **deltas** mark the mouths of rivers such as the Nile and the Ganges.

Altiplano   High-elevation plateau, basin, or valley between even higher mountain ranges, especially in the Andes of South America.

Altitudinal zonation   Vertical regions defined by physical-environmental zones at various elevations (see Fig. 4-12), particularly in the highlands of South and Middle America. See *tierra caliente*, *tierra templada*, *tierra fría*, *tierra helada*, and *tierra nevada*.

American Manufacturing Belt   North America's near-rectangular Core Region, whose corners are Boston, Milwaukee, St. Louis, and Baltimore. Dominated the industrial geography of the U.S. and Canada during the industrial age; still a formidable economic powerhouse that remains the realm's geographic heart.

Antecedent boundary   A political boundary that existed before the **cultural landscape** emerged and stayed in place while people moved in to occupy the surrounding area.

Anthracite coal   Hardest and highest carbon-content coal, and therefore of the highest quality.

Apartheid   Literally, *apartness*. The Afrikaans term for South Africa's pre-1994 policies of racial separation, a system that produced highly segregated socio-geographical patterns.

Aquaculture   The use of a river segment or an artificial pond for the raising and harvesting of food products, including fish, shellfish, and even seaweed (particularly in Japan).

Aquifer   An underground reservoir of water contained within a porous, water-bearing rock layer.

Arable   Land fit for cultivation by one farming method or another. See **physiologic density**.

Archipelago   A set of islands grouped closely together, usually elongated into a *chain*.

Area   A term that refers to a part of the Earth's surface with less specificity than **region**. For example, *urban area* alludes generally to a place where urban development has occurred, whereas *urban region* requires certain specific criteria upon which such a designation is based (e.g., the spatial extent of commuting or the built townscape).

Areal functional organization   A geographic principle for understanding the evolution of regional organization, whose five interrelated tenets are applied to the spatial development of Japan on p. 471.

Areal interdependence   A term related to **functional specialization**. When one area produces certain goods or has certain raw materials and another area has a different set of raw materials and produces different goods, their needs may be *complementary*; by exchanging raw materials and products, they can satisfy each other's requirements.

Arithmetic density   A country's population, expressed as an average per unit area, without regard for its **distribution** or the limits of **arable** land—see also **physiologic density**.

Aryan   From the Sanskrit *Arya* ("noble"), a name applied to an ancient people who spoke an Indo-European language and who moved into northern India from the northwest.

Atmosphere   The Earth's envelope of gases that rests on the oceans and land surface and penetrates open spaces within soils. This layer of nitrogen (78 percent), oxygen (21 percent), and traces of other gases is densest at the Earth's surface and thins with altitude.

Autocratic   A government that holds absolute power, often ruled by one person or a small group of persons who control the country by despotic means.

Balkanization   The fragmentation of a **region** into smaller, often hostile political units.

Barrio   Term meaning "neighborhood" in Spanish. Usually refers to an urban community in a Middle or South American city; also applied to low-income, inner-city concentrations of Hispanics in such western U.S. cities as Los Angeles.

Bauxite   Aluminum ore; usually deposited at shallow depths in the wet tropics.

Biogeography   The study of *flora* (plant life) and *fauna* (animal life) in spatial perspective.

Birth rate   The *crude birth rate* is expressed as the annual number of births per 1000 individuals within a given population.

Bituminous coal   Softer coal of lesser quality than **anthracite**, but of higher grade than **lignite**. When heated and converted to coking coal or *coke*, it is used to make steel.

*Words in boldface type are defined elsewhere in this Glossary.

G1

Break-of-bulk point A location along a transport route where goods must be transferred from one carrier to another. In a port, the cargoes of oceangoing ships are unloaded and put on trains, trucks, or perhaps smaller river boats for inland distribution.

Buffer state See **buffer zone**.

Buffer zone A country or set of countries separating ideological or political adversaries. In southern Asia, Afghanistan, Nepal, and Bhutan were parts of a buffer zone between British and Russian-Chinese imperial spheres. Thailand was a *buffer state* between British and French colonial domains in mainland Southeast Asia.

Caliente See **tierra caliente**.

Cartogram A specially transformed map not based on traditional representations of **scale** or area.

Cartography The art and science of making maps, including data compilation, layout, and design. Also concerned with the interpretation of mapped patterns.

Caste system The strict **social stratification** and segregation of people—specifically in India's Hindu society—on the basis of ancestry and occupation.

Cay A low-lying small island usually composed of coral and sand. Pronounced *kee*, and often spelled "key."

Central business district (CBD) The downtown heart of a central city, the CBD is marked by high land values, a concentration of business and commerce, and the clustering of the tallest buildings.

Centrality The strength of an urban center in its capacity to attract producers and consumers to its facilities; a city's "reach" into the surrounding region.

Centrifugal forces A term employed to designate forces that tend to divide a country—such as internal religious, linguistic, ethnic, or ideological differences.

Centripetal forces Forces that unite and bind a country together—such as a strong national culture, shared ideological objectives, and a common faith.

Chaebol Giant corporation controlling numerous companies and benefitting from government connections and favors, dominant in South Korea's **economic geography**; key to the country's **development** into an **economic tiger**, but more recently a barrier to free-market growth.

Charismatic Personal qualities of certain leaders that enable them to capture and hold the popular imagination, to secure the allegiance and even the devotion of the masses. Gandhi, Mao Zedong, and Franklin D. Roosevelt were good examples in the 20th century.

China Proper The eastern and northeastern portions of China that contain most of the country's huge population; mapped in Figure 9-4.

Choke point A narrowing of an international waterway causing marine traffic congestion, requiring reduced speeds and/or sharp turns, and increasing the risk of collision as well as vulnerability to attack. When the waterway narrows to a distance of less than 24 miles (38 km), this necessitates the drawing of a **median line (maritime) boundary**. Examples are the Hormuz Strait between Oman and Iran at the entrance to the Persian Gulf, and the Strait of Malacca between Malaysia and Indonesia.

City-state An independent political entity consisting of a single city with (and sometimes without) an immediate **hinterland**. The ancient city-states of Greece have their modern equivalent in Singapore.

Climate The long-term conditions (over at least 30 years) of aggregate **weather** over a region, summarized by averages and measures of variability; a synthesis of the succession of weather events we have learned to expect at any given location.

Climate change theory An alternative to the **hydraulic civilization theory**; holds that changing **climate** (rather than a monopoly over **irrigation** methods) could have provided certain cities in the ancient **Fertile Crescent** with advantages over others.

Climatology The geographic study of **climates**. Includes not only the classification of climates and the analysis of their regional distribution, but also broader environmental questions that concern climate change, interrelationships with soil and vegetation, and human-climate interaction.

Coal See **anthracite coal**, **bituminous coal**, and **lignite**.

Collectivization The reorganization of a country's **agriculture** under communism that involves the expropriation of private holdings and their incorporation into relatively large-scale units, which are farmed and administered cooperatively by those who live there.

Colonialism See **imperialism**.

Commercial agriculture For-profit **agriculture**.

Common market A **free-trade area** that not only has created a **customs union** (a set of common tariffs on all imports from outside the area) but also has eliminated restrictions on the movement of capital, labor, and enterprise among its member countries.

Compact state A politico-geographical term to describe a **state** that possesses a roughly circular, oval, or rectangular territory in which the distance from the geometric center to any point on the boundary exhibits little variance. Poland and Cambodia are examples of this shape category.

Complementarity Exists when two regions, through an exchange of raw materials and/or finished products, can specifically satisfy each other's demands.

Confucianism A philosophy of ethics, education, and public service based on the writings of Confucius (*Kongfuzi*); traditionally regarded as one of the cornerstones of Chinese **culture**.

Congo See the footnote on p. 271, which makes the distinction between the two Equatorial African countries, Congo and The Congo.

Coniferous forest A forest of cone-bearing, needleleaf evergreen trees with straight trunks and short branches, including spruce, fir, and pine.

Contagious diffusion The distance-controlled spreading of an idea, innovation, or some other item through a local population by contact from person to person—analogous to the communication of a contagious illness.

Conterminous United States The 48 **contiguous** or adjacent States that occupy the southern half of the North American realm. Alaska is not contiguous to these States because western Canada lies in between; neither is Hawai'i, separated from the mainland by over 2000 miles of ocean.

Contiguous Adjoining; adjacent.

Continental drift The slow movement of continents controlled by the **processes** associated with **plate tectonics**.

Continental shelf Beyond the coastlines of many landmasses, the ocean floor declines very gently until the depth of about 660 feet (200 m). Beyond the 660-foot line the sea bottom usually drops off sharply, along the *continental slope*, toward the much deeper mid-oceanic basin. The submerged continental margin is called the continental shelf, and it extends from the shoreline to the upper edge of the continental slope.

Continentality The variation of the continental effect on air temperatures in the interior portions of the world's landmasses.

The greater the distance from the moderating influence of an ocean, the greater the extreme in summer and winter temperatures. Continental interiors also tend to be dry when the distance from oceanic moisture sources becomes considerable.

Conurbation General term used to identify a large multi-metropolitan complex formed by the coalescence of two or more major **urban areas**. The Atlantic Seaboard **Megalopolis**, extending along the northeastern U.S. coast from southern Maine to Virginia, is a classic example.

Copra The dried-out, fleshy interior of a coconut that is used to produce coconut oil.

Cordillera Mountain chain consisting of sets of parallel ranges, especially the Andes in northwestern South America.

Core See **core area**; **core-periphery relationships**.

Core area In geography, a term with several connotations. *Core* refers to the center, heart, or focus. The core area of a **nation-state** is constituted by the national heartland, the largest population cluster, the most productive region, and the part of the country with the greatest **centrality** and **accessibility**—probably containing the capital city as well.

Core-periphery relationships The contrasting spatial characteristics of, and linkages between, the *have* (core) and *have-not* (periphery) components of a national or regional **system**.

Corridor In general, refers to a spatial entity in which human activity is organized in a linear manner, as along a major transport route or in a valley confined by highlands. More specifically, the politico-geographical term for a land extension that connects an otherwise **landlocked state** to the sea.

Cultural diffusion The **process** of spreading and adoption of a cultural element, from its place of origin across a wider area.

Cultural ecology The multiple interactions and relationships between a **culture** and its **natural environment**.

Cultural environment See **cultural ecology**.

Cultural geography The wide-ranging and comprehensive field of geography that studies spatial aspects of human **cultures**.

Cultural landscape The forms and artifacts sequentially placed on the **natural landscape** by the activities of various human occupants. By this progressive imprinting of the human presence, the physical (natural) landscape is modified into the cultural landscape, forming an interacting unity between the two.

Cultural pluralism See **plural(istic) society**.

Cultural revival The regeneration of a long-dormant **culture** through internal renewal and external infusion.

Culture The sum total of the knowledge, attitudes, and habitual behavior patterns shared and transmitted by the members of a society. This is anthropologist Ralph Linton's definition; hundreds of others exist.

Culture area See **culture region**.

Culture hearth Heartland, source area, innovation center; place of origin of a major **culture**.

Culture region A distinct, culturally discrete spatial unit; a **region** within which certain cultural norms prevail.

Customs union A **free-trade area** in which member countries set common tariff rates on imports from outside the area.

Death rate The *crude death rate* is expressed as the annual number of deaths per 1000 individuals within a given population.

Deciduous A deciduous tree loses its leaves at the beginning of winter or the onset of the dry season.

Definition In political geography, the written legal description (in a treaty-like document) of a boundary between two countries or territories—see **delimitation**.

Deforestation The clearing and destruction of forests (especially tropical rainforests) to make way for expanding settlement frontiers and the exploitation of new economic opportunities.

Deglomeration Deconcentration.

Delimitation In political geography, the translation of the written terms of a boundary treaty (the **definition**) into an official cartographic representation (map).

Delta **Alluvial** lowland at the mouth of a river, formed when the river deposits its alluvial load on reaching the sea. Often triangular in shape, hence the use of the Greek letter whose symbol is Δ.

Demarcation In political geography, the actual placing of a political boundary on the **cultural landscape** by means of barriers, fences, walls, or other markers.

Demographic transition model Multi-stage model, based on Western Europe's experience, of changes in population growth exhibited by countries undergoing industrialization. High **birth rates** and **death rates** are followed by plunging death rates, producing a huge net population gain; this is followed by the convergence of birth and death rates at a low overall level. See Figure 8-7.

Demography The interdisciplinary study of population—especially **birth rates** and **death rates**, growth patterns, longevity, **migration**, and related characteristics.

Desert An arid area supporting sparse vegetation, receiving less than 10 inches (25 cm) of precipitation per year. Usually exhibits extremes of heat and cold because the moderating influence of moisture is absent.

Desertification The **process** of **desert** expansion into neighboring **steppelands** as a result of human degradation of fragile semiarid environments.

Development The economic, social, and institutional growth of national **states**.

Devolution The **process** whereby regions within a **state** demand and gain political strength and growing autonomy at the expense of the central government.

Dhows Wooden boats with characteristic triangular sails, plying the seas between Arabian and East African coasts.

Dialect Regional or local variation in the use of a major language, such as the distinctive accents of many residents of the U.S. South or New England.

Diffusion The spatial spreading or dissemination of a **culture** element (such as a technological innovation) or some other phenomenon (e.g., a disease outbreak). For the various channels of outward geographic spread from a source area, see **contagious**, **expansion**, **hierarchical**, and **relocation diffusion**.

Distance decay The various degenerative effects of distance on human spatial structures and interactions.

Diurnal Daily.

Divided capital In political geography, a country whose central administrative functions are carried on in more than one city is said to have divided capitals. The Netherlands and South Africa are examples.

Domestication The transformation of a wild animal or wild plant into a domesticated animal or a cultivated crop to gain control over food production. A necessary evolutionary step in the development of humankind: the invention of **agriculture**.

Domino theory The belief that political destabilization in one **state** can result in the collapse of order in a neighboring state, triggering a chain of events that, in turn, can affect a series of **contiguous** states.

Double cropping The planting, cultivation, and harvesting of two crops successively within a single year on the same plot of farmland.

Doubling time The time required for a population to double in size.

Dry canal An overland rail and/or road **corridor** across an **isthmus** dedicated to performing the transit functions of a canalized waterway. Best adapted to the movement of containerized cargo, there must be a port at each end to handle the necessary **break-of-bulk** unloading and reloading.

Ecology The study of the many interrelationships between all forms of life and the natural environments in which they have evolved and continue to develop. The study of *ecosystems* focuses on the interactions between specific organisms and their environments. See also **cultural ecology**.

Economic geography The field of geography that focuses on the diverse ways in which people earn a living, and how the goods and services they produce are expressed and organized spatially.

Economic tiger One of the burgeoning beehive countries of the western **Pacific Rim**. Following Japan's route since 1945, these countries have experienced significant modernization, industrialization, and Western-style economic growth since 1980. Three leading economic tigers today are South Korea, Taiwan, and Singapore. Term is increasingly used more generally to describe any fast-developing economy.

Economies of scale The savings that accrue from large-scale production wherein the unit cost of manufacturing decreases as the level of operation enlarges. Supermarkets operate on this principle and are able to charge lower prices than small grocery stores.

Ecosystem See **ecology**.

Ecumene The habitable portions of the Earth's surface where permanent human settlements have arisen.

Elite A small but influential upper-echelon social class whose power and privilege give it control over a country's political, economic, and cultural life.

El Niño-Southern Oscillation (ENSO) A periodic, large-scale, abnormal warming of the sea surface in the low latitudes of the eastern Pacific Ocean that has global implications, disturbing normal **weather** patterns in many parts of the world, especially South America.

Elongated state A **state** whose territory is decidedly long and narrow in that its length is at least six times greater than its average width. Chile and Vietnam are two classic examples.

Emigrant A person **migrating** away from a country or area; an out-migrant.

Empirical Relating to the real world, as opposed to theoretical abstraction.

Enclave A piece of territory that is surrounded by another political unit of which it is not a part.

Endemism Refers to a disease in a host population that affects many people in a kind of equilibrium without causing rapid and widespread deaths.

Entrepôt A place, usually a port city, where goods are imported, stored, and transshipped; a **break-of-bulk point**.

Environmental degradation The accumulated human abuse of a region's **natural landscape** that, among other things, can involve air and water pollution, threats to plant and animal ecosystems, misuse of **natural resources**, and generally upsetting the balance between people and their habitat.

Epidemic A local or regional outbreak of a disease.

Escarpment A cliff or very steep slope; frequently marks the edge of a plateau.

Estuary The widening mouth of a river as it reaches the sea; land subsidence or a rise in sea level has overcome the tendency to form a **delta**.

Ethnic cleansing The slaughter and/or forced removal of one **ethnic** group from its homes and lands by another, more powerful ethnic group bent on taking that territory.

Ethnicity The combination of a people's **culture** (traditions, customs, language, and religion) and racial ancestry.

European Union (EU) **Supranational** organization constituted by 15 European countries to further their common economic interests. These countries are: Germany, France, Italy, the United Kingdom, Belgium, the Netherlands, Luxembourg, Denmark, Finland, Sweden, Austria, Ireland, Portugal, Spain, and Greece. In 2004, 10 additional countries were admitted, raising the EU total to 25. These countries include Estonia, Latvia, Lithuania, the Czech Republic, Slovakia, Poland, Hungary, Slovenia, Malta, and Cyprus.

European state model A **state** consisting of a legally defined territory inhabited by a population governed from a capital city by a representative government.

Exclave A bounded (non-island) piece of territory that is part of a particular **state** but lies separated from it by the territory of another state. Alaska is an exclave of the United Sates.

Exclusive Economic Zone (EEZ) An oceanic zone extending up to 200 **nautical miles** from a shoreline, within which the coastal **state** can control fishing, mineral exploration, and additional activities by all other countries.

Expansion diffusion The spreading of an innovation or an idea through a fixed population in such a way that the number of those adopting grows continuously larger, resulting in an expanding area of dissemination.

Extraterritoriality Politico-geographical concept suggesting that the property of one **state** lying within the boundaries of another actually forms an extension of the first state.

Fatwa Literally, a legal opinion or proclamation issued by an Islamic cleric, based on the holy texts of Islam, long applicable only in the *Umma*, the realm ruled by the laws of Islam. In 1989 the Iranian ayatollah Khomeini extended the reach of the *fatwa* by condemning to death a British citizen and author living in the United Kingdom.

Favela Shantytown on the outskirts or even well within an urban area in Brazil.

Fazenda Coffee plantation in Brazil.

Federal state A political framework wherein a central government represents the various subnational entities within a **nation-state** where they have common interests—defense, foreign affairs, and the like—yet allows these various entities to retain their own identities and to have their own laws, policies, and customs in certain spheres.

Federation See **federal state**.

Fertile Crescent Crescent-shaped zone of productive lands extending from near the southeastern Mediterranean coast through Lebanon and Syria to the **alluvial** lowlands of Meso-potamia (in Iraq). Once more fertile than today, this is one of the world's great source areas of **agricultural** and other innovations.

First Nations Canada's indigenous peoples of American descent, whose U.S. counterparts are called Native Americans.

Fjord Narrow, steep-sided, elongated, and inundated coastal valley deepened by glacier ice that has since melted away, leaving the sea to penetrate.

Floodplain Low-lying area adjacent to a mature river, often covered by **alluvial** deposits and subject to the river's floods.

Forced migration Human **migration** flows in which the movers have no choice but to relocate.

Formal region A type of **region** marked by a certain degree of homogeneity in one or more phenomena; also called *uniform region* or *homogeneous region*.

Forward capital Capital city positioned in actually or potentially contested territory, usually near an international border; it confirms the **state's** determination to maintain its presence in the region in contention.

Fossil fuels The energy resources of **coal**, natural gas, and petroleum (oil), so named collectively because they were formed by the geologic compression and transformation of tiny plant and animal organisms.

Four Motors of Europe Rhône-Alpes (France), Baden-Württemberg (Germany), Catalonia (Spain), and Lombardy (Italy). Each is a high-technology-driven region marked by exceptional industrial vitality and economic success not only within Europe but on the global scene as well.

Fragmented state A **state** whose territory consists of several separated parts, not a **contiguous** whole. The individual parts may be isolated from each other by the land area of other states or by international waters. The United States and Indonesia are examples.

Francophone French-speaking. Quebec constitutes the heart of Francophone Canada.

Free-trade area A form of economic integration, usually consisting of two or more **states**, in which members agree to remove tariffs on trade among themselves. Usually accompanied by a **customs union** that establishes common tariffs on imports from outside the trade area, and sometimes by a **common market** that also removes internal restrictions on the movement of capital, labor, and enterprise.

Fría See **tierra fría**.

Frontier Zone of advance penetration, usually of contention; an area not yet fully integrated into a national **state**.

FTAA (Free Trade Area of the Americas) The ultimate goal of **supranational** economic integration in North, Middle, and South America: the creation of a single-market trading bloc (perhaps as soon as 2005) that would involve every country in the Western Hemisphere between the Arctic shore of Canada and Cape Horn at the southern tip of Chile.

Functional region A **region** marked less by its sameness than its dynamic internal structure; because it usually focuses on a central node, also called *nodal region* or *focal region*.

Functional specialization The production of particular goods or services as a dominant activity in a particular location.

Fundamentalism See **revivalism (religious)**.

Gentrification The upgrading of an older residential area through private reinvestment, usually in the downtown area of a central city. Frequently this involves the displacement of established lower-income residents, who cannot afford the heightened costs of living, and conflicts are not uncommon as such neighborhood change takes place.

Geographic realm The basic spatial unit in our world regionalization scheme. Each realm is defined in terms of a synthesis of its total human geography—a composite of its leading cultural, economic, historical, political, and appropriate environmental features.

Geography of development The field of geography concerned with spatial aspects and regional expressions of **development**.

Geometric boundaries Political boundaries **defined** and **delimited** (and occasionally **demarcated**) as straight lines or arcs.

Geomorphology The geographic study of the configuration of the Earth's solid surface—the world's landscapes and their constituent landforms.

Ghetto An intraurban region marked by a particular **ethnic** character. Often an inner-city poverty zone, such as the black ghetto in U.S. central cities. Ghetto residents are involuntarily segregated from other income and racial groups.

Glaciation See **Pleistocene Epoch**.

Globalization The gradual reduction of regional contrasts at the world scale, resulting from increasing international cultural, economic, and political exchanges.

Green Revolution The successful recent development of higher-yield, fast-growing varieties of rice and other cereals in certain developing countries.

Gross domestic product (GDP) The total value of all goods and services produced in a country by that state's economy during a given year.

Gross national product (GNP) The total value of all goods and services produced in a country by that state's economy during a given year, plus all citizens' income from foreign investment and other external sources.

Growth pole An urban center with certain attributes that, if augmented by a measure of investment support, will stimulate regional economic development in its **hinterland**.

Hacienda Literally, a large estate in a Spanish-speaking country. Sometimes equated with the **plantation**, but there are important differences between these two types of agricultural enterprise (see pp. 193–194).

Heartland theory The hypothesis, proposed by British geographer Halford Mackinder during the early twentieth century, that any political power based in the heart of Eurasia could gain sufficient strength to eventually dominate the world. Further, since Eastern Europe controlled access to the Eurasian interior, its ruler would command the vast "heartland" to the east.

Hegemony The political dominance of a country (or even a region) by another country. The former Soviet Union's postwar grip on Eastern Europe, which lasted from 1945 to 1990, was a classic example.

Helada See **tierra helada**.

Hierarchical diffusion A form of **diffusion** in which an idea or innovation spreads by trickling down from larger to smaller adoption units. An urban **hierarchy** is usually involved, encouraging the leapfrogging of innovations over wide areas, with geographic distance a less important influence.

Hierarchy An order or gradation of phenomena, with each level or rank subordinate to the one above it and superior to the one below. The levels in a national urban hierarchy are constituted by hamlets, villages, towns, cities, and (frequently) the **primate city**.

High island Volcanic islands of the Pacific Realm that are high enough in elevation to wrest substantial moisture from the tropical ocean air (see **orographic precipitation**). They tend to be well watered, their volcanic soils enable productive agriculture, and they support larger populations than **low islands**—which possess none of these advantages and must rely on fishing and the coconut palm for survival.

High sea Areas of the oceans away from land, beyond national jurisdiction, open and free for all to use.

Highveld A term used in Southern Africa to identify the high, grass-covered plateau that dominates much of the region. The lowest-lying areas in South Africa are called *lowveld*; areas that lie at intermediate elevations are the *middleveld*.

Hinterland Literally, "country behind," a term that applies to a surrounding area served by an urban center. That center is the focus of goods and services produced for its hinterland and is its dominant urban influence as well. In the case of a port city, the hinterland also includes the inland area whose trade flows through that port.

Historical inertia A term from manufacturing geography that refers to the need to continue using the factories, machinery, and equipment of heavy industries for their full, multiple-decade lifetimes to cover major initial investments—even though these facilities may be increasingly obsolete.

Holocene The current *interglaciation* epoch (the warm period of glacial contraction between the glacial expansions of an **ice age**); extends from 10,000 years ago to the present. Also known as the *Recent Epoch*.

Humus Dark-colored upper layer of a soil that consists of decomposed and decaying organic matter such as leaves and branches, nutrient-rich and giving the soil a high fertility.

Hydraulic civilization theory The theory that cities able to control **irrigated** farming over large **hinterlands** held political power over other cities. Particularly applies to early Asian civilizations based in such river valleys as the Chang (Yangzi), the Indus, and those of Mesopotamia.

Hydrologic cycle The **system** of exchange involving water in its various forms as it continually circulates among the **atmosphere**, the oceans, and above and below the land surface.

Ice age A stretch of geologic time during which the Earth's average atmospheric temperature is lowered; causes the equatorward expansion of continental ice sheets in the higher latitudes and the growth of mountain glaciers in and around the highlands of the lower latitudes.

Iconography The identity of a region as expressed through its cherished symbols; its particular **cultural landscape** and personality.

Immigrant A person **migrating** into a particular country or area; an in-migrant.

Imperialism The drive toward the creation and expansion of a colonial empire and, once established, its perpetuation.

Import-substitution industries The industries local entrepreneurs establish to serve populations of remote areas when transport costs from distant sources make these goods too expensive to import.

Inaccessibility See **accessibility**.

Indentured workers Contract laborers who sell their services for a stipulated period of time.

Indigenous peoples Native or *aboriginal* peoples; often used to designate the inhabitants of areas that were conquered and subsequently colonized by the **imperial** powers of Europe.

Industrial Revolution The term applied to the social and economic changes in agriculture, commerce, and especially manufacturing and urbanization that resulted from technological innovations and specialization in late-eighteenth-century Europe.

Informal sector Dominated by unlicensed sellers of homemade goods and services, the primitive form of capitalism found in many developing countries that takes place beyond the control of government.

Infrastructure The foundations of a society: urban centers, transport networks, communications, energy distribution systems, farms, factories, mines, and such facilities as schools, hospitals, postal services, and police and armed forces.

Insular Having the qualities and properties of an island. Real islands are not alone in possessing such properties of **isolation**: an **oasis** in the middle of a **desert** also has qualities of insularity.

Insurgent state Territorial embodiment of a successful guerrilla movement. The establishment by anti-government insurgents of a territorial base in which they exercise full control; thus a state within a **state**.

Intercropping The planting of several types of crops in the same field; commonly used by **shifting cultivators**.

Interglaciation See **Pleistocene Epoch**.

Intermontane Literally, between mountains. The location can bestow certain qualities of natural protection or **isolation** to a community.

Internal migration **Migration** flow within a country, such as ongoing westward and southward movements toward the **Sunbelt** in the United States.

International migration **Migration** flow involving movement across an international boundary.

Intervening opportunity In trade or **migration** flows, the presence of a nearer opportunity that greatly diminishes the attractiveness of sites farther away.

Inuit **Indigenous** peoples of North America's Arctic zone, formerly known as Eskimos.

Irredentism A policy of cultural extension and potential political expansion by a **state** aimed at a community of its nationals living in a neighboring state.

Irrigation The artificial watering of croplands.

Isohyet A line connecting points of equal rainfall total.

Isolation The condition of being geographically cut off or far removed from mainstreams of thought and action. It also denotes a lack of receptivity to outside influences, caused at least partially by poor **accessibility**.

Isotherm A line connecting points of equal temperature.

Isthmus A **land bridge**; a comparatively narrow link between larger bodies of land. Central America forms such a link between Mexico and South America.

Jakota Triangle The easternmost region of the East Asian realm, consisting of *Ja*pan, (South) *Ko*rea, and *Tai*wan.

Juxtaposition Contrasting places in close proximity to one another.

Karst The distinctive natural landscape associated with the chemical erosion of soluble limestone rock.

Land alienation One society or culture group taking land from another. In Subsaharan Africa, for example, European **colonialists** took land from **indigenous** Africans and put it to new uses.

Land bridge A narrow **isthmian** link between two large landmasses. They are temporary features—at least in terms of geologic time—subject to appearance and disappearance as the land or sea level rises and falls.

Land reform The spatial reorganization of **agriculture** through the allocation of farmland (often expropriated from landlords) to **peasants** and tenants who never owned land.

Land tenure The way people own, occupy, and use land.

Landlocked An interior **state** surrounded by land. Without coasts, such a country is disadvantaged in terms of **accessibility** to international trade routes, and in the scramble for possession of areas of the **continental shelf** and control of the **exclusive economic zone** beyond.

Latitude Lines of latitude are **parallels** that are aligned east-west across the globe, from 0° latitude at the equator to 90° North and South latitude at the poles.

Leached soil Infertile, reddish-appearing, tropical soil whose surface consists of oxides of iron and aluminum; all other soil nutrients have been dissolved and transported downward into the subsoil by percolating water associated with heavy rainfall.

Leeward The protected or downwind side of a **topographic** barrier with respect to the winds that flow across it.

Lignite Low-grade, brown-colored variety of coal.

Lingua franca A "common language" prevalent in a given area; a second language that can be spoken and understood by many peoples, although they speak other languages at home.

Littoral Coastal or coastland.

Llanos The interspersed **savanna** grasslands and scrub woodlands of the Orinoco River's wide basin that covers much of interior Colombia and Venezuela.

Location theory A logical attempt to explain the locational pattern of an economic activity and the manner in which its producing areas are interrelated. The agricultural location theory that underlies the **von Thünen model** is a leading example.

Loess Deposit of very fine silt or dust that is laid down after having been windborne for a considerable distance. Notable for its fertility under **irrigation** and its ability to stand in steep vertical walls.

Longitude Angular distance (0° to 180°) east or west as measured from the *prime meridian* (0°) that passes through the Greenwich Observatory in suburban London, England. For much of its length across the mid-Pacific Ocean, the 180th meridian functions as the *international date line*.

Low island Low-lying coral islands of the Pacific Realm that—unlike **high islands**—cannot wrest sufficient moisture from the tropical ocean air to avoid chronic drought. Thus productive agriculture is impossible and their modest populations must rely on fishing and the coconut palm for survival.

Madrassa Revivalist (**fundamentalist**) religious school where the curriculum focuses on Islamic religion and law and requires rote memorization of the Qu'ran (Koran), Islam's holy book. Founded in former British India, these schools were most numerous in present-day Pakistan but have diffused to Turkey in the west and Indonesia in the east.

Maghreb The region occupying the northwestern corner of Africa, consisting of Morocco, Algeria, and Tunisia.

Main Street Canada's dominant **conurbation** that is home to nearly two-thirds of the country's inhabitants; extends southwestward from Quebec City in the mid-St. Lawrence Valley to Windsor on the Detroit River.

Mainland-Rimland framework Twofold regionalization of the Middle American realm based on its modern cultural history. The Euro-Amerindian *Mainland*, stretching from Mexico to Panama (minus the Caribbean coastal strip), was a self-sufficient zone dominated by **hacienda land tenure**. The Euro-African *Rimland*, consisting of that Caribbean coastal zone plus all of the Caribbean islands to the east, was the zone of the **plantation** that heavily relied on trade with Europe.

Maquiladora The term given to modern industrial plants in Mexico's U.S. border zone. These foreign-owned factories assemble imported components and/or raw materials, and then export finished manufactures, mainly to the United States. Import duties are disappearing under **NAFTA**, bringing jobs to Mexico and the advantages of low wage rates to the foreign entrepreneurs.

Marchland An area or **frontier** of uncertain boundaries that is subject to various national claims and an unstable political history. Refers specifically to the movement of various armies across such zones.

Marine geography The geographic study of oceans and seas. Its practitioners investigate both the physical (e.g., coral-reef **biogeography**, ocean-**atmosphere** interactions, coastal **geomorphology**) and human (e.g., **maritime boundary-making**, fisheries, beachside development) aspects of oceanic environments.

Maritime boundary An international boundary that lies in the ocean. Like all boundaries, it is a vertical plane, extending from the seafloor to the upper limit of the air space in the atmosphere above the water.

Median line boundary An international **maritime boundary** drawn where the width of a sea is less than 400 **nautical miles**. Because the **states** on either side of that sea claim **exclusive economic zones** of 200 nautical miles, it is necessary to reduce those claims to a (median) distance equidistant from each shoreline. **Delimitation** on the map almost always appears as a set of straight-line segments that reflect the configurations of the coastlines involved.

Medical geography The study of health and disease within a geographic context and from a spatial perspective. Among other things, this field of geography examines the sources, **diffusion** routes, and distributions of diseases.

Megacity Informal term referring to the world's most heavily populated cities; in this book, the term refers to a **metropolis** containing a population of greater than 10 million.

Megalopolis When spelled with a lower-case *m*, a synonym for **conurbation**, one of the large coalescing supercities forming in diverse parts of the world. When capitalized, refers specifically to the multi-metropolitan corridor that extends along the northeastern U.S. seaboard from north of Boston to south of Washington, D.C. (Fig. 3-9).

Mercantilism Protectionist policy of European **states** during the sixteenth to the eighteenth centuries that promoted a state's economic position in the contest with rival powers. The acquisition of gold and silver and maintaining a favorable trade balance (more exports than imports) were central to the policy.

Meridian Line of **longitude**, aligned north-south across the globe, that together with **parallels** of **latitude** forms the global grid system. All meridians converge at both poles and are at their maximum distances from each other at the equator.

Mestizo Derived from the Latin word for *mixed*, refers to a person of mixed white and Amerindian ancestry.

Metropolis Urban **agglomeration** consisting of a (central) city and its suburban ring. See **urban (metropolitan) area**.

Metropolitan area See **urban (metropolitan) area**.

Migration A change in residence intended to be permanent. See also **forced**, **internal**, **international**, and **voluntary migration**.

Migratory movement Human relocation movement from a source to a destination without a return journey, as opposed to cyclical movement (see **nomadism**).

Model An idealized representation of reality built to demonstrate its most important properties. A **spatial** model focuses on a geographical dimension of the real world, such as the **von Thünen model** that explains agricultural location patterns in a commercial economy.

Monsoon Refers to the seasonal reversal of wind and moisture flows in certain parts of the subtropics and lower-middle latitudes. The *dry monsoon* occurs during the cool season when dry offshore winds prevail. The *wet monsoon* occurs in the hot summer months, which produce onshore winds that bring large amounts of rainfall. The air-pressure differential over land and sea is the triggering mechanism, with windflows always moving from areas of relatively higher pressure toward areas of relatively lower pressure. Monsoons make their greatest regional impact in the coastal and near-coastal zones of South Asia, Southeast Asia, and East Asia.

Mosaic culture The emerging cultural-geographic framework of the United States, dominated by the fragmentation of specialized social groups into homogeneous communities of interest marked not only by income, race, and ethnicity but also by age, occupational status, and lifestyle. The result is an increasingly heterogeneous socio-spatial complex, which resembles an intricate mosaic composed of myriad uniform—but separate—tiles.

Mulatto A person of mixed African (black) and European (white) ancestry.

Multinationals Internationally active corporations that can strongly influence the economic and political affairs of many countries they operate in.

Muslim An adherent of the Islamic faith.

Muslim Front A term used by certain scholars for the African Transition Zone of northern Africa, which is primarily regarded as a still-expanding frontier of Islam that affects countries from Guinea in the west to the African Horn in the east (see Fig. 6-19).

NAFTA (North American Free Trade Agreement) The **free-trade area** launched in 1994 involving the United States, Canada, and Mexico.

Nation Legally a term encompassing all the citizens of a **state**, it also has other connotations. Most definitions now tend to refer to a group of tightly-knit people possessing bonds of language, **ethnicity**, religion, and other shared **cultural** attributes. Such homogeneity actually prevails within very few states.

Nation-state A country whose population possesses a substantial degree of **cultural** homogeneity and unity. The ideal form to which most **nations** and **states** aspire—a political unit wherein the territorial state coincides with the area settled by a certain national group or people.

NATO (North Atlantic Treaty Organization) Established in 1950 at the height of the Cold War as a U.S.-led **supranational** defense pact to shield postwar Europe against the Soviet military threat. NATO is now in transition, expanding its membership while modifying its objectives in the post-Soviet era.

Natural hazard A natural event that endangers human life and/or the contents of a **cultural landscape**.

Natural increase rate Population growth measured as the excess of live births over deaths per 1000 individuals per year. Natural increase of a population does not reflect either **emigrant** or **immigrant** movements.

Natural landscape The array of landforms that constitutes the Earth's surface (mountains, hills, plains, and plateaus) and the physical features that mark them (such as water bodies, soils, and vegetation). Each **geographic realm** has its distinctive combination of natural landscapes.

Natural resource Any valued element of (or means to an end using) the environment; includes minerals, water, vegetation, and soil.

Nautical mile By international agreement, the nautical mile—the standard measure at sea—is 6076.12 feet in length, equivalent to approximately 1.15 statute miles (1.85 km).

Neocolonialism The term used by developing countries to underscore that the entrenched colonial system of international exchange and capital flow has not changed in the postcolonial era—thereby perpetuating the huge economic advantages of the developed world.

Network (transport) The entire regional **system** of transportation connections and nodes through which movement can occur.

Nevada See **tierra nevada**.

Nomadism Cyclical movement among a definite set of places. Nomadic peoples mostly are **pastoralists**.

Nucleation Cluster; **agglomeration**.

Oasis An area, small or large, where the supply of water (from an **aquifer** or a major river such as the Nile) permits the transformation of the adjacent **desert** into productive cropland.

Occidental Western. Also see *Oriental*.

Offshore banking Term referring to financial havens for foreign companies and individuals, who channel their earnings to accounts in such a country (usually an "offshore" island-state) to avoid paying taxes in their home countries.

OPEC (Organization of Petroleum Exporting Countries) The international oil *cartel* or syndicate formed by a number of producing countries to promote their common economic interests through the formulation of joint pricing policies and the limitation of market options for consumers. The 11 member-states (as of mid-2003) are: Algeria, Indonesia, Iran, Iraq, Kuwait, Libya, Nigeria, Qatar, Saudi Arabia, United Arab Emirates (UAE), and Venezuela.

Organic theory Friedrich Ratzel's theory of **state** development that conceptualized the state as a biological organism whose life—from birth through maturation to eventual senility and collapse—mirrors that of any living thing.

Oriental The root of the word *oriental* is from the Latin for *rise*. Thus it has to do with the direction in which one sees the sun "rise"—the east; *oriental* therefore means Eastern. *Occidental* originates from the Latin for *fall*, or the "setting" of the sun in the west; *occidental* therefore means Western.

Orographic precipitation Mountain-induced precipitation, especially where air masses are forced to cross **topographic** barriers. Downwind areas beyond such a mountain range experience the relative dryness known as the **rain shadow effect**.

Outback The name given by Australians to the vast, peripheral, sparsely-settled interior of their country.

Outer city The non-central-city portion of the American **metropolis**; no longer "sub" to the "urb," this outer ring was transformed into a full-fledged city during the late twentieth century.

Pacific Rim A far-flung group of countries and parts of countries (extending clockwise on the map from New Zealand to Chile) sharing the following criteria: they face the Pacific Ocean; they evince relatively high levels of economic development, industrialization, and urbanization; their imports and exports mainly move across Pacific waters.

Pacific Ring of Fire Zone of crustal instability along tectonic **plate** boundaries, marked by earthquakes and volcanic activity, that ring the Pacific Ocean basin.

Paddies (paddyfields) Ricefields.

Pandemic An outbreak of a disease that spreads worldwide.

Pangaea A vast, singular landmass consisting of most of the areas of the present-day continents. This supercontinent began to break up more than 200 million years ago when still-ongoing **plate** divergence and **continental drift** became dominant processes (see Fig. 6-3).

Parallel An east-west line of **latitude** that is intersected at right angles by **meridians** of **longitude**.

Pastoralism A form of **agricultural** activity that involves the raising of livestock.

Peasants In a **stratified** society, peasants are the lowest class of people who depend on **agriculture** for a living. But they often own no land at all and must survive as tenants or day workers.

Peninsula   A comparatively narrow, finger-like stretch of land extending from the main landmass into the sea. Florida and Korea are examples.

Peon (*peone*)   Term used in Middle and South America to identify people who often live in serfdom to a wealthy landowner; landless **peasants** in continuous indebtedness.

Per capita   Capita means *individual*. Income, production, or some other measure is often given per individual.

Perforated state   A **state** whose territory completely surrounds that of another state. South Africa, which encloses Lesotho and is perforated by it, is a classic example.

Periodic market   Village market that opens every third day or at some other regular interval. Part of a regional network of similar markets in a preindustrial, rural setting where goods are brought to market on foot and barter remains a major mode of exchange.

Periphery   See **core-periphery relationships**.

Permafrost   Permanently frozen water in the near-surface soil and bedrock of cold environments, producing the effect of completely frozen ground. Surface can thaw during brief warm season.

Physical geography   The study of the geography of the physical (natural) world. Its subfields include **climatology**, **geomorphology**, **biogeography**, soil geography, **marine geography**, and water **resources**.

Physical landscape   Synonym for **natural landscape**.

Physiographic political boundaries   Political boundaries that coincide with prominent physical features in the **natural landscape**—such as rivers or the crest ridges of mountain ranges.

Physiographic region (province)   A **region** within which there prevails substantial **natural-landscape** homogeneity, expressed by a certain degree of uniformity in surface **relief**, **climate**, vegetation, and soils.

Physiography   Literally means *landscape description*, but commonly refers to the total **physical geography** of a place; includes all of the natural features on the Earth's surface, including landforms, **climate**, soils, vegetation, and water bodies.

Physiologic density   The number of people per unit area of **arable** land.

Pilgrimage   A journey to a place of great religious significance by an individual or by a group of people (such as a pilgrimage to Mecca for **Muslims**).

Plantation   A large estate owned by an individual, family, or corporation and organized to produce a cash crop. Almost all plantations were established within the tropics; in recent decades, many have been divided into smaller holdings or reorganized as cooperatives.

Plate tectonics   Plates are bonded portions of the Earth's mantle and crust, averaging 60 miles (100 km) in thickness. More than a dozen such plates exist (see Fig. G-4), most of continental proportions, and they are in motion. Where they meet one slides under the other, crumpling the surface crust and producing significant volcanic and earthquake activity; a major mountain-building force.

Pleistocene Epoch   Recent period of geologic time that spans the rise of humankind, beginning about 2 million years ago. Marked by *glaciations* (repeated advances of continental ice-sheets) and milder *interglaciations* (ice sheet contractions). Although the last 10,000 years are known as the **Holocene** Epoch, Pleistocene-like conditions seem to be continuing and we are most probably now living through another Pleistocene interglaciation; thus the glaciers likely will return.

Plural(istic) society   A society in which two or more population groups, each practicing its own **culture**, live adjacent to one another without mixing inside a single **state**.

Polder   Land reclaimed from the sea adjacent to the shore of the Netherlands by constructing dikes and then pumping out the water trapped behind them.

Political geography   The study of the interaction of geographical area and political **process**; the spatial analysis of political phenomena and processes.

Pollution   The release of a substance, through human activity, which chemically, physically, or biologically alters the air or water it is discharged into. Such a discharge negatively impacts the environment, with possible harmful effects on living organisms—including humans.

Population density   The number of people per unit area. Also see **arithmetic density** and **physiologic density** measures.

Population distribution   The way people have arranged themselves in geographic space. One of human geography's most essential expressions because it represents the sum total of the adjustments that a population has made to its natural, cultural, and economic environments.

Population explosion   The rapid growth of the world's human population during the past century, attended by ever-shorter **doubling times** and accelerating *rates* of increase.

Population geography   The field of geography that focuses on the spatial aspects of **demography** and the influences of demographic change on particular places.

Population implosion   The opposite of the **population explosion**, refers to the declining populations of many European countries and Russia in which the **death rate** exceeds the **birth rate** and **immigration** rate.

Population movement   See **migration**; **migratory movement**.

Population projection   The future population total that demographers forecast for a particular country. For example, in Table G-1 such projections are given for all the world's countries for 2025.

Population (age-sex) structure   Graphic representation (*profile*) of a population according to age and gender.

Postindustrial economy   Emerging economy, in the United States and a handful of other highly advanced countries, as traditional industry is increasingly eclipsed by a higher-technology productive complex dominated by services, information-related, and managerial activities.

Primary economic activity   Activities engaged in the direct extraction of **natural resources** from the environment such as mining, fishing, lumbering, and especially **agriculture**.

Primate city   A country's largest city—ranking atop the urban **hierarchy**—most expressive of the national culture and usually (but not always) the capital city as well.

Process   Causal force that shapes a spatial pattern as it unfolds over time.

Protruded state   Territorial shape of a **state** that exhibits a narrow, elongated land extension (or *protrusion*) leading away from the main body of territory. Thailand is a leading example.

Push-pull concept   The idea that **migration** flows are simultaneously stimulated by conditions in the source area, which tend to drive people away, and by the perceived attractiveness of the destination.

Qanat   In **desert** zones, particularly in Iran and western China, an underground tunnel built to carry **irrigation** water by gravity flow from nearby mountains (where **orographic precipitation** occurs) to the arid flatlands below.

Quaternary economic activity   Activities engaged in the collection, processing, and manipulation of *information*.

Quinary economic activity Managerial or control-function activity associated with decision-making in large organizations.

Rain shadow effect The relative dryness in areas downwind of mountain ranges caused by **orographic precipitation**, wherein moist air masses are forced to deposit most of their water content as they cross the highlands.

Realm See **geographic realm**.

Region A commonly used term and a geographic concept of central importance. An **area** on the Earth's surface marked by specific criteria.

Regional boundary In theory, the line that circumscribes a **region**. But razor-sharp lines are seldom encountered, even in nature (e.g., a coastline constantly changes depending upon the tide). In the **cultural landscape**, not only are regional boundaries rarely self-evident, but when they are ascertained by geographers they most often turn out to be **transitional** borderlands.

Regional complementarity See **complementarity**.

Regional disparity The spatial unevenness in standard of living that occurs within a country, whose "average," overall income statistics invariably mask the differences that exist between the extremes of the wealthy **core** and the poorer **periphery**.

Regional state A "natural economic zone" that defies political boundaries, and is shaped by the global economy of which it is a part; its leaders deal directly with foreign partners and negotiate the best terms they can with the national governments under which they operate.

Relative location The regional position or **situation** of a place relative to the position of other places. Distance, **accessibility**, and connectivity affect relative location.

Relict boundary A political boundary that has ceased to function, but the imprint of which can still be detected on the **cultural landscape**.

Relief Vertical difference between the highest and lowest elevations within a particular area.

Relocation diffusion Sequential **diffusion process** in which the items being diffused are transmitted by their carrier agents as they relocate to new areas. The most common form of relocation diffusion involves the spreading of innovations by a **migrating** population.

Revivalism (religious) Religious movement whose objectives are to return to the foundations of that faith and to influence **state** policy. Often called **religious fundamentalism**; but in the case of Islam, **Muslims** prefer the term revivalism.

Rift valley The trough or trench that forms when a strip of the Earth's crust sinks between two parallel faults (surface fractures).

Sahel Semiarid **steppeland** zone extending across most of Africa between the southern margins of the arid Sahara and the moister tropical **savanna** and forest zone to the south. Chronic drought, **desertification**, and overgrazing have contributed to severe famines in this area since 1970.

Savanna Tropical grassland containing widely spaced trees; also the name given to the tropical wet-and-dry climate (*Aw*).

Scale Representation of a real-world phenomenon at a certain level of reduction or generalization. In **cartography**, the ratio of map distance to ground distance; indicated on a map as a bar graph, representative fraction, and/or verbal statement. *Macroscale* refers to a large area of national proportions; *microscale* refers to a local area no bigger than a county.

Scale economies See **economies of scale**.

Secondary economic activity Activities that process raw materials and transform them into finished industrial products. The *manufacturing* sector.

Sedentary Permanently attached to a particular area; a population fixed in its location. The opposite of **nomadic**.

Separate development The spatial expression of South Africa's "grand" **apartheid** scheme, whereby nonwhite groups were required to settle in segregated "homelands." The policy was dismantled when white-minority rule collapsed in the early 1990s.

Sequent occupance The notion that successive societies leave their cultural imprints on a place, each contributing to the cumulative **cultural landscape**.

Shantytown Unplanned slum development on the margins of cities in developing countries, dominated by crude dwellings and shelters mostly made of scrap wood, iron, and even pieces of cardboard.

Sharecropping Relationship between a large landowner and farmers on the land whereby the farmers pay rent for the land they farm by giving the landlord a share of the annual harvest.

Sharia The criminal code based in Islamic law that prescribes corporal punishment, amputations, stonings, and lashing for both major and minor offenses. Its occurrence today is associated with the spread of **religious revivalism** in **Muslim** societies.

Shatter belt Region caught between stronger, colliding external cultural-political forces, under persistent stress, and often fragmented by aggressive rivals. Eastern Europe and Southeast Asia are classic examples.

Shifting agriculture Cultivation of crops in recently cut and burned tropical-forest clearings, soon to be abandoned in favor of newly cleared nearby forest land. Also known as *slash-and-burn agriculture*.

Sinicization Giving a Chinese cultural imprint; Chinese **acculturation**.

Site The internal locational attributes of an urban center, including its local spatial organization and physical setting.

Situation The external locational attributes of an urban center; its **relative location** or regional position with reference to other non-local places.

Social stratification See **stratification (social)**.

Spatial Pertaining to space on the Earth's surface. Synonym for *geographic(al)*.

Spatial diffusion See **diffusion**.

Spatial interaction See **complementarity**, **transferability**, and **intervening opportunity**.

Spatial model See **model**.

Spatial process See **process**.

Spatial system The components and interactions of a **functional region**, which is defined by the areal extent of those interactions. Also see **system**.

Special Economic Zone (SEZ) Manufacturing and export center within China, created in the 1980s to attract foreign investment and technology transfers. Six SEZs—all located on southern China's Pacific coast—currently operate: Shenzhen, adjacent to Hong Kong; Zhuhai; Shantou; Xiamen; Hainan Island, in the far south; and still-building Pudong, across the river from Shanghai.

Squatter settlement See **shantytown**.

State A politically organized territory that is administered by a sovereign government and is recognized by a significant portion of the international community. A state must also contain a permanent resident population, an organized economy, and a functioning internal circulation system.

State capitalism Government-controlled corporations competing under free-market conditions, usually in a tightly regimented society. South Korea is a leading example. Also see *chaebol*.

State planning   Involves highly centralized control of the national planning process, a hallmark of communist economic systems. Soviet central planners mainly pursued a grand political design in assigning production to particular places; their frequent disregard of **economic geography** contributed to the eventual collapse of the U.S.S.R.

State territorial morphology   A **state's** geographical shape, which can have a decisive impact on its spatial cohesion and political viability. A **compact** shape is most desirable; among the less efficient shapes are those exhibited by **elongated**, **fragmented**, **perforated**, and **protruded** states.

Stateless nation   A **national** group that aspires to become a **nation-state** but lacks the territorial means to do so; the Palestinians and Kurds of Southwest Asia are classic examples.

Steppe   Semiarid grassland; short-grass prairie. Also the name given to the semiarid climate type (*BS*).

Stratification (social)   In a layered or stratified society, the population is divided into a **hierarchy** of social classes. In an industrialized society, the working class is at the lower end; **elites** that possess capital and control the means of production are at the upper level. In the traditional **caste system** of Hindu India, the "untouchables" form the lowest class or caste, whereas the still-wealthy remnants of the princely class are at the top.

Subduction   In **plate tectonics**, the **process** that occurs when an oceanic plate converges head-on with a plate carrying a continental landmass at its leading edge. The lighter continental plate overrides the denser oceanic plate and pushes it downward.

Subsequent boundary   A political boundary that developed contemporaneously with the evolution of the major elements of the **cultural landscape** through which it passes.

Subsistence   Existing on the minimum necessities to sustain life; spending most of one's time in pursuit of survival.

Subtropical Convergence   A narrow marine **transition zone**, girdling the globe at approximately latitude 40°S, that marks the equatorward limit of the frigid Southern Ocean and the poleward limits of the warmer Atlantic, Pacific, and Indian Oceans to the north.

Suburban downtown   In the United States (and increasingly in other advantaged countries), a significant concentration of major urban activities around a highly **accessible** suburban location, including retailing, light industry, and a variety of leading corporate and commercial operations. The largest are now coequal to the American central city's **central business district** (**CBD**).

Sunbelt   The popular name given to the southern tier of the United States, which is anchored by the mega-States of California, Texas, and Florida. Its warmer climate, superior recreational opportunities, and other amenities have been attracting large numbers of relocating people and activities since the 1960s; broader definitions of the Sunbelt also include much of the western U.S., particularly Colorado and the coastal Pacific Northwest.

Superimposed boundary   A political boundary emplaced by powerful outsiders on a developed human landscape. Usually ignores pre-existing cultural-spatial patterns, such as the border that still divides North and South Korea.

Supranational   A venture involving three or more **states**—political, economic, and/or cultural cooperation to promote shared objectives. The **European Union** is one such organization.

System   Any group of objects or institutions and their mutual interactions. Geography treats systems that are expressed **spatially**, such as in **functional regions**.

Systematic geography   Topical geography: **cultural, political, economic geography**, and the like.

Taiga   The subarctic, mostly **coniferous** snowforest that blankets northern Russia and Canada south of the **tundra** that lines the Arctic shore.

Takeoff   Economic concept to identify a stage in a country's **development** when conditions are set for a domestic Industrial Revolution.

Taxonomy   A **system** of scientific classification.

Technopole   A planned techno-industrial complex (such as California's Silicon Valley) that innovates, promotes, and manufactures the products of the **postindustrial** informational economy.

Tectonics   See **plate tectonics**.

Templada   See **tierra templada**.

Terracing   The transformation of a hillside or mountain slope into a step-like sequence of horizontal fields for intensive cultivation.

Territoriality   A country's or more local community's sense of property and attachment toward its territory, as expressed by its determination to keep it inviolable and strongly defended.

Territorial sea   Zone of seawater adjacent to a country's coast, held to be part of the national territory and treated as a segment of the sovereign **state**.

Tertiary economic activity   Activities that engage in *services*— such as transportation, banking, retailing, education, and routine office-based jobs.

Tierra caliente   The lowest of the **altitudinal zones** into which the human settlement of Middle and South America is classified according to elevation. The *caliente* is the hot humid coastal plain and adjacent slopes up to 2,500 feet (750 m) above sea level. The natural vegetation is the dense and luxuriant tropical rainforest; the crops include sugar and bananas in the lower areas, and coffee, tobacco, and corn along the higher slopes.

Tierra fría   The cold, high-lying **altitudinal zone** of settlement in Andean South America, extending from about 6,000 feet (1,800 m) in elevation up to nearly 12,000 feet (3,600 m). **Coniferous** trees stand here; upward they change into scrub and grassland. There are also important pastures within the *fría*, and wheat can be cultivated.

Tierra helada   In Andean South America, the highest-lying habitable **altitudinal zone**—ca. 12,000 to 15,000 feet (3,600 to 4,500 m)—between the tree line (upper limit of the *tierra fría*) and the snow line (lower limit of the *tierra nevada*). Too cold and barren to support anything but the grazing of sheep and other hardy livestock.

Tierra nevada   The highest and coldest **altitudinal zone** in Andean South America (lying above 15,000 feet [4,500 m]), an uninhabitable environment of permanent snow and ice that extends upward to the Andes' highest peaks of more than 20,000 feet (6,000 m).

Tierra templada   The intermediate **altitudinal zone** of settlement in Middle and South America, lying between 2,500 feet (750 m) and 6,000 feet (1,800 m) in elevation. This is the "temperate" zone, with moderate temperatures compared to the *tierra caliente* below. Crops include coffee, tobacco, corn, and some wheat.

Topography   The surface configuration of any segment of **natural landscape**.

Toponym   Place name.

Transculturation   Cultural borrowing and two-way exchanges that occur when different **cultures** of approximately equal complexity and technological level come into close contact.

Transferability   The capacity to move a good from one place to another at a bearable cost; the ease with which a commodity may be transported.

Transhumance Seasonal movement of people and their livestock in search of pastures. Movement may be vertical (into highlands during the summer and back to lower elevations in winter) or horizontal, in pursuit of seasonal rainfall.

Transition zone An area of **spatial** change where the peripheries of two adjacent **realms** or **regions** join; marked by a gradual shift (rather than a sharp break) in the characteristics that distinguish these neighboring geographic entities from one another.

Transmigration The policy of the Indonesian government to induce residents of the overcrowded, **core-area** island of Jawa to move to the country's other islands.

Treaty ports **Extraterritorial enclaves** in China's coastal cities, established by European colonial invaders under unequal treaties enforced by gunboat diplomacy.

Tropical deforestation See **deforestation**.

Tropical savanna See **savanna**.

Tsunami A seismic (earthquake-generated) sea wave that can attain gigantic proportions and cause coastal devastation.

Tundra The treeless plain that lies along the Arctic shore in northernmost Russia and Canada, whose vegetation consists of mosses, lichens, and certain hardy grasses.

Turkestan Northeasternmost region of the North Africa/Southwest Asia realm. Known as Soviet Central Asia before 1992, its five (dominantly Islamic) former S.S.R.'s have become the independent countries of Kazakhstan, Uzbekistan, Turkmenistan, Kyrgyzstan, and Tajikistan. Today Turkestan has expanded to include a sixth state, Afghanistan.

Unitary state A **nation-state** that has a centralized government and administration that exercises power equally over all parts of the state.

Urbanization A term with several connotations. The proportion of a country's population living in urban places is its level of urbanization. The **process** of urbanization involves the movement to, and the clustering of, people in towns and cities—a major force in every geographic realm today. Another kind of urbanization occurs when an expanding city absorbs rural countryside and transforms it into suburbs; in the case of cities in disadvantaged countries, this also generates peripheral **shantytowns**.

Urban (metropolitan) area The entire built-up, non-rural area and its population, including the most recently constructed suburban appendages. Provides a better picture of the dimensions and population of such an area than the delimited municipality (central city) that forms its heart.

Urban realms model A spatial generalization of the contemporary large American city. It is shown to be a widely dispersed, multi-centered **metropolis** consisting of increasingly independent zones or *realms*, each focused on its own **suburban downtown**; the only exception is the shrunken central realm, which is focused on the **central business district** (see Figs. 3-11 and 3-12).

Veld See **highveld**.

Voluntary migration Population movement in which people relocate in response to perceived opportunity, not because they are forced to migrate.

Von Thünen's Isolated State model Explains the location of agricultural activities in a commercial economy. A **process** of spatial competition allocates various farming activities into concentric rings around a central market city, with profit-earning capability the determining force in how far a crop locates from the market. The original (1826) Isolated State model now applies to the continental scale (see Fig. 1-6).

Weather The immediate and short-term conditions of the **atmosphere** that impinge on daily human activities.

Wet monsoon See **monsoon**.

Windward The exposed, upwind side of a **topographic** barrier that faces the winds that flow across it.

World geographic realm See **geographic realm**.

List of Maps and Figures

List of Photo Credits

Index